Mastering
AutoCAD® Civil 3D® 2014

Mastering
AutoCAD® Civil 3D® 2014

Louisa Holland

Cyndy Davenport

Eric Chappell

Senior Acquisitions Editor: Willem Knibbe
Development Editor: Sara Barry
Technical Editor: Tommie Richardson
Production Editor: Rebecca Anderson
Copy Editors: Judy Flynn and Kathy Grider-Carlyle
Editorial Manager: Pete Gaughan
Production Manager: Tim Tate
Vice President and Executive Group Publisher: Richard Swadley
Vice President and Publisher: Neil Edde
Book Designers: Maureen Forys, Happenstance Type-O-Rama; Judy Fung
Compositor: Cody Gates, Happenstance Type-O-Rama
Proofreaders: Dan Aull and Louise Watson, Word One New York
Indexer: Ted Laux
Project Coordinator, Cover: Katherine Crocker
Cover Designer: Ryan Sneed
Cover Image: ©istockphoto.com/asterix0597

Copyright © 2013 by John Wiley & Sons, Inc., Indianapolis, Indiana
Published simultaneously in Canada

ISBN: 978-1-118-60381-9
ISBN: 978-1-118-78681-9 (ebk.)
ISBN: 978-1-118-79127-1 (ebk.)

No part of this publication may be reproduced, stored in a retrieval system or transmitted in any form or by any means, electronic, mechanical, photocopying, recording, scanning or otherwise, except as permitted under Sections 107 or 108 of the 1976 United States Copyright Act, without either the prior written permission of the Publisher, or authorization through payment of the appropriate per-copy fee to the Copyright Clearance Center, 222 Rosewood Drive, Danvers, MA 01923, (978) 750-8400, fax (978) 646-8600. Requests to the Publisher for permission should be addressed to the Permissions Department, John Wiley & Sons, Inc., 111 River Street, Hoboken, NJ 07030, (201) 748-6011, fax (201) 748-6008, or online at http://www.wiley.com/go/permissions.

Limit of Liability/Disclaimer of Warranty: The publisher and the author make no representations or warranties with respect to the accuracy or completeness of the contents of this work and specifically disclaim all warranties, including without limitation warranties of fitness for a particular purpose. No warranty may be created or extended by sales or promotional materials. The advice and strategies contained herein may not be suitable for every situation. This work is sold with the understanding that the publisher is not engaged in rendering legal, accounting, or other professional services. If professional assistance is required, the services of a competent professional person should be sought. Neither the publisher nor the author shall be liable for damages arising herefrom. The fact that an organization or Web site is referred to in this work as a citation and/or a potential source of further information does not mean that the author or the publisher endorses the information the organization or Web site may provide or recommendations it may make. Further, readers should be aware that Internet Web sites listed in this work may have changed or disappeared between when this work was written and when it is read.

For general information on our other products and services or to obtain technical support, please contact our Customer Care Department within the U.S. at (877) 762-2974, outside the U.S. at (317) 572-3993 or fax (317) 572-4002.

Wiley publishes in a variety of print and electronic formats and by print-on-demand. Some material included with standard print versions of this book may not be included in e-books or in print-on-demand. If this book refers to media such as a CD or DVD that is not included in the version you purchased, you may download this material at http://booksupport.wiley.com. For more information about Wiley products, visit www.wiley.com.

Library of Congress Control Number: 2013941608

TRADEMARKS: Wiley, Sybex, and the Sybex logo are trademarks or registered trademarks of John Wiley & Sons, Inc. and/or its affiliates, in the United States and other countries, and may not be used without written permission. AutoCAD and Civil 3D are registered trademarks of Autodesk, Inc. All other trademarks are the property of their respective owners. John Wiley & Sons, Inc. is not associated with any product or vendor mentioned in this book.

10 9 8 7 6 5 4 3 2

Dear Reader,

Thank you for choosing *Mastering AutoCAD Civil 3D 2014*. This book is part of a family of premium-quality Sybex books, all of which are written by outstanding authors who combine practical experience with a gift for teaching.

Sybex was founded in 1976. More than 30 years later, we're still committed to producing consistently exceptional books. With each of our titles, we're working hard to set a new standard for the industry. From the paper we print on, to the authors we work with, our goal is to bring you the best books available.

I hope you see all that reflected in these pages. I'd be very interested to hear your comments and get your feedback on how we're doing. Feel free to let me know what you think about this or any other Sybex book by sending me an email at nedde@wiley.com. If you think you've found a technical error in this book, please visit http://sybex.custhelp.com. Customer feedback is critical to our efforts at Sybex.

    Best regards,

    Neil Edde
    Vice President and Publisher
    Sybex, an Imprint of Wiley

Acknowledgments

Writing *Mastering AutoCAD® Civil 3D® 2014* was not just an exercise in creating a useful manual for Civil 3D users, it was a labor of love. Many people (with day jobs) put in late nights and long hours to bring readers the content contained in this book. We'd like to thank all of our readers for helping this book evolve over the years. And of course we'd like to thank the Wiley crew: Willem Knibbe, Paul Gaughan, and Becca Anderson. Thanks to our editors Sara Barry, Tommie Richardson, Judy Flynn, Kathy Grider-Carlyle, Dan Aull, and Louise Watson for dotting the i's and crossing the t's (and there was a *lot* of that!).

—*Louisa Holland, Cyndy Davenport,* and *Eric Chappell*

Oh my gosh—where to start? I'd like to offer special thanks this year to Willem Knibbe and Paul Gaughan for their patience while I made an unexpected cross-country move. Don Quinn of Eagle Point—thanks for getting me into this side of the industry; I both thank you and blame you. Thanks to coworkers past and present who have encouraged me, put up with me, and taught me oodles of AutoCAD and Civil 3D tricks over the years. Thanks to Cyndy Davenport and Eric Chappell for jumping in with late notice to get this book out. Hugs and kisses to my sisters (my own little chapter of SWE), whom I love dearly. To my parents who still don't quite know what Civil 3D is, thank you for everything. Most important, thanks to my husband, Mark, for his home network administration and love. Smooches!

—*Louisa Holland*

I'd like to thank my parents, Grey & Dood, for continuously supporting and believing in me, through the good and the bad, decade after decade: Sorry about those gray hairs, but if it makes you feel any better, I have them too. Without the dedication, ambition, candidness, and sense of humor you passed on to me, I would have never been able to take on this labor of love. I'd like to thank Hellboy for giving me the space, the time, and the peace needed to work on this project, thereby putting up with an empty fridge and finding creative things to do with ramen noodles many a night; Son, you are the center of my life and my inspiration.

Thanks to my boss, Michael Stys, "the Director," for making it possible to make my deadlines and cracking me up daily with his crazy sense of humor. I'm even grateful for his merry band of lunatics, Doctor Pfaff, the BIM-ster, and the GIS-ster, for keeping it even more crazy and making every day at work seem like a *Saturday Night Live* skit. It's not too often that you find yourself in a workplace surrounded by family. Their genuineness and superb guruship make me feel like I'm among greatness, which boosted me to meet the challenge of this book. Many folks at Bowman Consulting inspired me to come up with creative solutions and pulled me down into the weeds with the software: James, Danny, Kerri, Dorris, Victor, Robin, Karen, Tracy, Jeremy, Cody, Greg, the countless Michaels, Brian, David, and Tombstone Shawn, to name a few. Without their willingness to explore strange new worlds, I wouldn't be the übergeek I am today.

Last, I'd like to thank Rumpus Kat and Yeller Dawg for giving me a reason to get up in the morning and get on with my day; even though meals and walks are the most important parts of their day, the joy they bring me is priceless.

—*Cyndy Davenport*

I'd like to thank Autodesk for creating great software to write about; the Civil 3D community for all its great ideas, enthusiasm, and help throughout the years; and most of all my wife and children for being my inspiration and the reason for all that I do.

—*Eric Chappell*

About the Authors

Louisa "Lou" Holland is a LEED-accredited civil engineer currently living in San Francisco, California. She has trained users on Eagle Point Software and AutoCAD® since 2001 and on AutoCAD Civil 3D since 2006. She has worked extensively with the Wisconsin Department of Transportation, CalTrans, and various consultants on AutoCAD Civil 3D implementations. Louisa is an Autodesk Approved Instructor (AAI), an AutoCAD Civil 3D Certified Professional, and a regular speaker at Autodesk University, Autodesk User Group International, and other industry events. In her next life, Louisa would like to be reincarnated as an Orange County housewife.

Cyndy Davenport has been working in the land development industry for 27 years in the capacity of design, technology management, implementation, training, and support. She's earned repeated Civil 3D certifications in implementation, support, and training over the years. Cyndy is a regular speaker at Autodesk University and occasionally contributes workflow solutions on her blog to the Civil 3D community (c3dcougar.typepad.com).

Eric Chappell, a recognized expert in the world of AutoCAD® Civil 3D® software, has been working, teaching, writing, and consulting in the world of civil engineering software for over 20 years. He has written training materials and performed training for end users, trainers, and Autodesk employees around the globe and worked with Autodesk in authoring and developing two Autodesk certification exams. He is the design systems manager for Timmons Group, a civil engineering and surveying firm. Eric is also a highly rated instructor at Autodesk University. Eric lives in the Richmond, Virginia, area.

Contents at a Glance

Introduction . . . . *xxiii*

Chapter 1 • The Basics 1

Chapter 2 • Survey 57

Chapter 3 • Points 97

Chapter 4 • Surfaces 125

Chapter 5 • Parcels 197

Chapter 6 • Alignments 247

Chapter 7 • Profiles and Profile Views 299

Chapter 8 • Assemblies and Subassemblies 365

Chapter 9 • Basic Corridors 409

Chapter 10 • Advanced Corridors, Intersections, and Roundabouts 453

Chapter 11 • Superelevation 525

Chapter 12 • Cross Sections and Mass Haul 555

Chapter 13 • Pipe Networks 587

Chapter 14 • Grading 669

Chapter 15 • Plan Production 713

Chapter 16 • Advanced Workflows 751

Chapter 17 • Quantity Takeoff 777

Chapter 18 • Label Styles 803

Chapter 19 • Object Styles 861

Appendix A • The Bottom Line 919

Appendix B • Autodesk Civil 3D 2014 Certification 963

Index *967*

Contents

Introduction . . . . *xxiii*

Chapter 1 • The Basics **1**
The Interface 1
   Toolspace 2
   Panorama 25
   Ribbon 25
Civil 3D Templates 26
   Starting New Projects 27
   Importing Styles 30
Creating Basic Lines and Curves 32
   COGO Line Commands 32
   Direction-Based Line Commands 34
   Re-creating a Deed Using Line Tools 39
Creating Curves 40
   Standard Curves 42
   Best Fit Entities 47
   Attach Multiple Entities 49
   Adding Line and Curve Labels 50
Using Transparent Commands 52
   Standard Transparent Commands 52
   Matching Transparent Commands 53
The Bottom Line 55

Chapter 2 • Survey **57**
Setting Up the Databases 57
   Survey Database Defaults 58
   The Equipment Database 60
   The Figure Prefix Database 61
   The Linework Code Set Database 63
Description Keys: Field to Civil 3D 64
   Creating a Description Key Set 67
   The Main Event: Your Project's Survey Database 71
   Under the Hood in Your Survey Network 77
   Other Survey Features 86
   The Coordinate Geometry Editor 89
Using Inquiry Commands 91
The Bottom Line 94

Chapter 3 • Points 97

Anatomy of a Point 97
   COGO Points vs. Survey Points 98
Creating Basic Points 98
   Point Settings 98
   Importing Points from a Text File 101
   Converting Points from Non-Civil 3D Sources 104
   A Closer Look at the Create Points Toolbar 106
Basic Point Editing 112
   Graphic Point Edits 112
   Panorama and Prospector Point Edits 113
   Point Groups: Don't Skip This Section! 114
   Changing Point Elevations 118
Point Tables 119
User-Defined Properties 121
The Bottom Line 123

Chapter 4 • Surfaces 125

Understanding Surface Basics 125
Creating Surfaces 127
   Additional Surface Creation Methods 130
   Surface Approximations 138
Refining and Editing Surfaces 145
   Surface Properties 145
Surface Additions 149
   Breaklines 152
   Surface Boundaries 155
Surface Analysis 171
   Elevation Banding 171
   Slopes and Slope Arrows 176
   Visibility Checker 178
Comparing Surfaces 179
   TIN Volume Surface 180
Labeling the Surface 184
   Contour Labeling 184
   Additional Surface Label Types 186
Point Cloud Surfaces 189
   Importing a Point Cloud 189
   Working with Point Clouds 192
   Creating a Point Cloud Surface 193
The Bottom Line 195

Chapter 5 • Parcels 197

Introduction to Sites 197
   Think Outside of the Lot 197
   Creating a New Site 202

Creating a Boundary Parcel 204
    Using Parcel Creation Tools 206
    Creating a Right-of-Way Parcel. 208
    Adding a Cul-de-Sac Parcel. 211
Creating Subdivision Lot Parcels Using Precise Sizing Tools. 214
    Attached Parcel Segments. 214
    Parcel Sizing Settings. 215
    Parcel Sizing Tools 216
Editing Parcels by Deleting Parcel Segments 223
Best Practices for Parcel Creation 227
    Forming Parcels from Segments 227
    Parcels Reacting to Site Objects 228
    Constructing Parcel Segments with the Appropriate Vertices 233
Labeling Parcel Areas 234
Labeling Parcel Segments. 238
    Labeling Multiple-Parcel Segments. 238
    Labeling Spanning Segments 241
    Adding Curve Tags to Prepare for Table Creation 242
    Creating a Table for Parcel Segments 244
The Bottom Line. 246

Chapter 6 • Alignments. 247

Alignment Concepts 247
    Alignments and Sites 247
    Alignment Entities 248
Creating an Alignment 249
    Creating from a Line, Arc, or Polyline. 250
    Creating by Layout. 255
    Best Fit Alignments 261
    Reverse Curve Creation 265
    Creating with Design Constraints and Check Sets. 267
Editing Alignment Geometry 271
    Grip Editing. 271
    Tabular Design 273
    Component-Level Editing 274
    Understanding Alignment Constraints 275
    Changing Alignment Components 279
Alignments As Objects 280
    Alignment Properties. 281
    The Right Station. 284
    Assigning Design Speeds 286
    Labeling Alignments 288
    Alignment Tables 292
The Bottom Line. 296

Chapter 7 • Profiles and Profile Views 299

The Elevation Element. 299
   Surface Sampling 300
   Layout Profiles. 308
   The Best Fit Profile 320
   Creating a Profile from a File 321
Editing Profiles. 322
   Grip-Editing Profiles 322
   Editing Profiles using Profile Layout Parameters 323
   Editing Profiles using Profile Grid View. 325
   Component-Level Editing 326
   Other Profile Edits 328
Profile Views. 329
   Creating Profile Views during Sampling 329
   Creating Profile Views Manually. 330
   Splitting Views 331
Editing Profile Views. 340
   Profile View Properties 340
   Profile View Labeling Styles 353
Profile Labels. 355
   Applying Labels 355
   Using Profile Label Sets 358
Profile Utilities 359
   Superimposing Profiles 359
   Projecting Objects in Profile View 360
   Creating a Quick Profile 363
The Bottom Line. 363

Chapter 8 • Assemblies and Subassemblies 365

Subassemblies. 365
   The Tool Palettes. 366
   The Corridor Modeling Catalogs 367
   Adding Subassemblies to a Tool Palette 368
Building Assemblies 368
   Creating a Typical Road Assembly 369
   Subassembly Components. 377
   Jumping into Help 379
   Commonly Used Subassemblies 381
   Editing an Assembly 384
   Creating Assemblies for Non-road Uses. 388
Specialized Subassemblies 393
   Using Generic Links. 394
   Daylighting with Generic Links. 396
   Working with Daylight Subassemblies 397

Advanced Assemblies 402
   Offset Assemblies 402
   Marked Points and Friends 402
Organizing Your Assemblies 404
   Storing a Customized Subassembly on a Tool Palette 405
   Storing a Completed Assembly on a Tool Palette 405
The Bottom Line 407

Chapter 9 • Basic Corridors 409

Understanding Corridors 409
Recognizing Corridor Components 410
   Baseline 411
   Regions 411
   Assemblies 411
   Frequency 411
   Targets 412
   Corridor Feature Lines 412
   Rebuilding Your Corridor 420
   Tweaking Corridors 421
Working with Corridor Feature Lines 424
Understanding Targets 429
   Using Target Alignments and Profiles 430
Editing Sections 435
Creating a Corridor Surface 438
   The Corridor Surface 438
   Corridor Surface Creation Fundamentals 439
   Adding a Surface Boundary 442
Performing a Volume Calculation 447
Building Non-Road Corridors 448
The Bottom Line 452

Chapter 10 • Advanced Corridors, Intersections, and Roundabouts . . . 453

Using Multiregion Baselines 453
Modeling a Cul-de-Sac 456
   Using Multiple Baselines 456
   Establishing EOP Design Profiles 457
   Putting the Pieces Together 458
   Troubleshooting Your Cul-de-Sac 462
Moving Up to Intersections 466
   Using the Intersection Wizard 468
   Manually Modeling an Intersection 476
   Creating an Assembly for the Intersection 478
   Adding Baselines, Regions, and Targets for the Intersections 480
   Troubleshooting Your Intersection 485
   Checking and Fine-Tuning the Corridor Model 486

Using an Assembly Offset 493
Understanding Corridor Utilities 500
    Using Corridor Utilities in Practice 502
Using a Feature Line as a Width and Elevation Target 504
Tackling Roundabouts: The Mount Everest of Corridors 510
    Drainage First 510
    Roundabout Alignments 511
    Center Design 518
    Profiles for All 520
    Tie It All Together 520
    Finishing Touches 522
The Bottom Line 524

Chapter 11 • Superelevation 525

Preparing for Superelevation 525
    Design Criteria Files 527
    Ready Your Alignment 531
    Super Assemblies 531
Applying Superelevation to the Design 537
    Start with the Alignment 538
    Transition Station Overlap 541
Oh Yes, You Cant 545
    Workin' on the Railroad 545
    Creating a Rail Assembly 546
    Applying Cant to the Alignment 548
Superelevation and Cant Views 548
    Using a Superelevation View to Edit Data 550
The Bottom Line 552

Chapter 12 • Cross Sections and Mass Haul 555

Section Workflow 555
    Sample Lines vs. Frequency Lines 555
    Creating Sample Lines 556
    Editing the Swath Width of a Sample Line Group 561
Creating Section Views 563
    Creating a Single-Section View 564
    Creating Multiple Section Views 567
    Section Views and Annotation Scale 570
It's a Material World 574
    Creating a Materials List 575
    Creating a Volume Table in the Drawing 576
    Adding Soil Factors to a Materials List 577
    Generating a Volume Report 578
Section View Final Touches 579
    Sample More Sources 579
    Cross-Section Labels 580

Mass Haul 581
    Taking a Closer Look at the Mass Haul Diagram 581
    Create a Mass Haul Diagram 582
    Editing a Mass Haul Diagram 583
The Bottom Line 585

Chapter 13 • Pipe Networks 587

Pipe Network Setup 587
    Parts List — Sewer Systems 588
    Planning a Typical Pipe Network 588
    Part Rules 590
    Putting Your Parts List Together 597
Creating a Sanitary Sewer Network 600
    Establishing Pipe Network Parameters 601
    Using the Network Layout Creation Tools 602
    Creating a Storm Drainage Pipe Network from a Feature Line 609
Editing a Pipe Network 612
    Changing Flow Direction 612
    Editing Your Network in Plan View 613
    Using the Pipe Network Vista Effectively 615
    Pipe Network Contextual Tab Edits 616
    Editing with the Network Layout Tools Toolbar 621
Creating an Alignment from Network Parts 624
Drawing Parts in Profile View 626
    Vertical Movement Edits Using Grips in Profile 628
    Removing a Part from Profile View 630
    Showing Pipes That Cross the Profile View 631
Adding Pipe Network Labels 633
    Creating a Labeled Pipe Network Profile with Crossings 633
    Pipe and Structure Labels 635
Creating an Interference Check 636
Creating Pipe Tables 639
    Exploring the Table Creation Dialog 639
    The Table Panel Tools 642
Under Pressure 644
    Pressure Network Parts List 644
    Creating a Pressure Network 648
    Design Checks 658
Part Builder 660
Part Builder Orientation 661
    Understanding the Organization of Part Builder 661
    Exploring Part Families 664
    Adding a Part Size Using Part Builder 664
    Sharing a Custom Part 666
    Adding an Arch Pipe to Your Part Catalog 667
The Bottom Line 668

Chapter 14 • Grading ... 669
Working with Grading Feature Lines ... 669
   Accessing Grading Feature Line Tools ... 669
   Creating Grading Feature Lines ... 671
   Editing Feature Line Information ... 679
   Labeling Feature Lines ... 699
Grading Objects ... 701
   Creating Gradings ... 701
   Editing Gradings ... 705
   Creating Surfaces from Grading Groups ... 707
The Bottom Line ... 711

Chapter 15 • Plan Production ... 713
Preparing for Plan Sets ... 713
   Prerequisite Components ... 713
Using View Frames and Match Lines ... 714
   The Create View Frames Wizard ... 715
   Creating View Frames ... 723
   Editing View Frames and Match Lines ... 725
Creating Plan and Profile Sheets ... 729
   The Create Sheets Wizard ... 729
   Managing Sheets ... 735
Creating Section Sheets ... 740
   Creating Multiple Section Views ... 740
   Creating Section Sheets ... 744
Drawing Templates ... 746
The Bottom Line ... 749

Chapter 16 • Advanced Workflows ... 751
Data Shortcuts ... 751
   Getting Started ... 753
   Setting a Working Folder and Data Shortcut Folder ... 753
   Creating Data Shortcuts ... 755
   Creating a Data Reference ... 757
   Updating References ... 763
Using LandXML ... 772
The Bottom Line ... 775

Chapter 17 • Quantity Takeoff ... 777
Employing Pay Item Files ... 777
   Pay Item Favorites ... 778
   Searching for Pay Items ... 781
Keeping Tabs on the Model ... 785
   AutoCAD Objects as Pay Items ... 785
   Pricing Your Corridor ... 787

Pipes and Structures as Pay Items 791
Highlighting Pay Items 797
Inventorying Your Pay Items 799
The Bottom Line 801

Chapter 18 • Label Styles 803

Label Styles 803
General Labels 803
Frequently Seen Tabs 804
General Note Labels 818
Point Label Styles 821
Line and Curve Labels 824
Single Segment Labels 824
Spanning Segment Labels 826
Curve Labels 827
Pipe and Structure Labels 829
Pipe Labels 829
Structure Labels 831
Profile and Alignment Labels 834
Label Sets 834
Alignment Labels 835
Advanced Style Types 851
Table Styles 852
Code Set Styles 854
The Bottom Line 860

Chapter 19 • Object Styles 861

Getting Started with Object Styles 861
Frequently Seen Tabs 863
General Settings 867
Point and Marker Object Styles 868
Linear Object Styles 872
Alignment Style 875
Parcel Styles 876
Feature Line Styles 877
Surface Styles 877
Contour Style 878
Triangles and Points Surface Style 882
Analysis Styles 885
Pipe and Structure Styles 889
Pipe Styles 889
Structure Styles 897
Profile View Styles 900
Profile View Bands 908
Section View Styles 911
Group Plot Styles 912
The Bottom Line 917

Appendix A • The Bottom Line **919**

Appendix B • Autodesk Civil 3D 2014 Certification **963**

Index *967*

Introduction

The AutoCAD® Civil 3D® program was introduced in 2004 as a trial product. Over the past few years, the AutoCAD Civil 3D series have evolved from the wobbly baby introduced on those first trial discs to a mature platform used worldwide to handle the most complex dynamic engineering designs. With this change, many engineers still struggle with how to make the transition. The civil engineering industry as a whole is an old dog learning new tricks.

We hope this book will help you in this journey. As the user base grows and users get beyond the absolute basics, more materials are needed, offering a multitude of learning opportunities. While this book is starting to move away from the basics and truly become a Mastering book, we hope that we are headed in that direction with the general readership. We know we cannot please everyone, but we do listen to your comments—all toward the betterment of this book.

Designed to help you get past the steepest part of the learning curve and teach you some guru-level tricks along the way, *Mastering AutoCAD Civil 3D 2014* is the ideal addition to any AutoCAD Civil 3D user's bookshelf.

Who Should Read This Book

The Mastering book series is designed with specific users in mind. In the case of *Mastering AutoCAD Civil 3D 2014*, we expect you'll have a solid knowledge of AutoCAD in general and some basic engineering knowledge as well. A basic understanding of AutoCAD Civil 3D will be helpful, although there are explanations and examples to cover many needs and experience levels. We expect this book will appeal to a large number of AutoCAD Civil 3D users, but we envision a few primary users:

Beginning Users Looking to Make the Move to Using AutoCAD Civil 3D   These people understand AutoCAD and some basics of engineering, but they are looking to learn AutoCAD Civil 3D on their own, broadening their skill set to make themselves more valuable in their firms and in the market.

AutoCAD Civil 3D Users Looking for a Desktop Reference   With the digitization of the official help files, many users still long for a book they can flip open and keep beside them as they work. These people should be able to jump to the information they need for the task at hand, such as further information about a confusing dialog or troublesome design issue.

Users Looking to Prepare for the Autodesk Certification Exams   This book focuses on the elements you need to pass the Associate and Professional exams with flying colors and includes margin icons to note topics of interest. Just look for the icon.

Classroom Instructors Looking for Better Materials   This book was written with real data from real design firms. We've worked hard to make many of the examples match the real-world problems we have run into as engineers. This book also goes into greater depth than any other available text, allowing short classes to review the basics (and leave the in-depth material for self-discovery) and longer classes can cover the full material presented.

This book can be used front to back as a self-teaching or instructor-based instruction manual. Each chapter has a number of exercises and most (but not all) build on the previous exercise. You can also skip to almost any exercise in any chapter and jump right in. We've created a large number of drawing files that you can download from www.sybex.com/go/masteringcivil3d2014 to make choosing your exercises a simple task.

What You Will Learn

This book isn't a replacement for training. There are too many design options and parameters to make any book a good replacement for training from a professional. This book teaches you to use the tools, explores a large number of the options, and leaves you with an idea of how to use each tool. At the end of the book, you should be able to look at any design task you run across, consider a number of ways to approach it, and have some idea of how to accomplish the task. To use one of our common analogies, reading this book is like walking around your local home-improvement warehouse. You see a lot of tools and use some of them, but that doesn't mean you're ready to build a house.

What You Need

Before you begin learning AutoCAD Civil 3D, you should make sure your hardware is up to snuff. Visit the Autodesk website, www.autodesk.com, and review graphic requirements, memory requirements, and so on. One of the most frustrating things that can happen is to be ready to learn only to be stymied by hardware-related crashes. AutoCAD Civil 3D is a hardware-intensive program, testing the limits of every computer on which it runs. You'll also want to download any service packs available.

We also strongly recommend using either a wide format or dual-monitor setup. The number of dialogs, palettes, and so on make AutoCAD Civil 3D a real estate hog. By having the extra space to spread out, you'll be able to see more of your design along with the feedback provided by the program itself.

You need to visit www.sybex.com/go/masteringcivil3d2014 to download all of the data and sample files. We recommend that you save these files locally on your computer in `C:/Mastering` unless told otherwise.

> **FREE AUTODESK SOFTWARE FOR STUDENTS AND EDUCATORS**
>
> The Autodesk Education Community is an online resource with more than five million members that enables educators and students to download—for free (see website for terms and conditions)—the same software used by professionals worldwide. You can also access additional tools and materials to help you design, visualize, and simulate ideas. Connect with other learners to stay current with the latest industry trends and get the most out of your designs. Get started today at www.autodesk.com/joinedu.

The Mastering Series

The Mastering series from Sybex provides outstanding instruction for readers with intermediate and advanced skills in the form of top-notch training and development for those already working in their field and clear, serious education for those aspiring to become pros. Every *Mastering* book includes the following features:

- Real-world scenarios ranging from case studies to interviews that show how the tool, technique, or knowledge presented is applied in actual practice
- Skill-based instruction, with chapters organized around real tasks rather than abstract concepts or subjects
- A self-review section called The Bottom Line, so you can be certain you're equipped to do the job right

What Is Covered in This Book

This book contains 19 chapters and two appendices:

- Chapter 1, "The Basics," introduces you to the interface and many of the common dialogs in AutoCAD Civil 3D. This chapter discusses navigating the interface and customizing your drawing's settings. You will also explore various tools for creating linework.

- Chapter 2, "Survey," examines the Survey tab of Toolspace and the unique toolset it contains for handling field surveying and for field book data handling. You will also look at various surface and surveying relationships.

- Chapter 3, "Points," introduces AutoCAD Civil 3D points and the various methods of creating them. You will also spend some time exploring the control of AutoCAD Civil 3D points with description keys and groups.

- Chapter 4, "Surfaces," introduces the various methods of creating surfaces, using free and low-cost data to perform preliminary surface creation. Then you will investigate the various surface editing and analysis methods. The chapter also discusses point clouds and their use.

- Chapter 5, "Parcels," examines the best practices for keeping your parcel topology tight and your labeling neat. It examines the various editing methods for achieving the desired results for the most complicated plats.

- Chapter 6, "Alignments," introduces the basic horizontal layout element. This chapter also examines using layout tools that maintain the relationships between the tangents, curves, and spiral elements that create alignments.

- Chapter 7, "Profiles and Profile Views," examines the vertical aspect of road design from the establishment of the existing profile to the design and editing of the proposed profile. In addition, you will explore how profile views can be customized to meet the required format for your design and plans.

- Chapter 8, "Assemblies and Subassemblies," introduces the building blocks of AutoCAD Civil 3D cross-sectional design. You will look at the many subassemblies available in the tool palettes and look at how to build full design sections for use in any design environment.

- Chapter 9, "Basic Corridors," introduces the basics of corridors—building full designs from horizontal, vertical, and cross-sectional design elements. You will look at the various components to understand how corridors work before moving to a more complex design set.

- Chapter 10, "Advanced Corridors, Intersections, and Roundabouts," further examines using corridors in more complex situations. You will learn about building surfaces, intersections, and other areas of corridors that make them powerful in any design situation.

- Chapter 11, "Superelevation," takes a close look at the tools used to add superelevation to roadways and railways. This functionality has changed greatly in the last few years, and you will have a chance to use the axis of Rotation (AOR) subassemblies that can pivot from several design points.

- Chapter 12, "Cross Sections and Mass Haul," looks at slicing sections from surfaces, corridors, and pipe networks using alignments and the mysterious sample line group. Working with the wizards and tools, you will see how to make your sections to order. You will explore mass haul functionality to demonstrate the power of AutoCAD Civil 3D for creation of the mass haul diagrams.

- Chapter 13, "Pipe Networks," gets into the building blocks of the pipe network tools. You will look at modifying an existing part to add new sizes and then building parts lists for various design situations. You will then work with the creation tools for creating pipe networks and plan and profile views to get your plans looking like they should.

- Chapter 14, "Grading," examines both feature lines and grading objects. You will look at creating feature lines to describe critical areas and then using grading objects to describe mass grading.

- Chapter 15, "Plan Production," walks you through the basics of creating view frame groups, sheets, and templates used to automate the plan and profile drawing sheet process. In addition, you will look at creating section views and section sheets.

- Chapter 16, "Advanced Workflows," looks at the various ways of sharing and receiving data. We describe the data-shortcut mechanism for sharing data between AutoCAD Civil 3D users. We also consider other methods of importing and exporting, such as XML.

- Chapter 17, "Quantity Takeoff," shows you the ins and outs of assigning pay items to corridor codes, blocks, areas, and pipes. You learn how to set up new pay items and generate quantity takeoff reports.

- Chapter 18, "Label Styles," is devoted to editing and creating label styles. You learn to navigate the Text Component Editor and how to master label style conundrums you may come across.

- Chapter 19, "Object Styles," examines editing and creating object styles. You will learn how to create styles for surfaces, profile views, and other objects to match your company standards.

- Appendix A, "The Bottom Line," gathers together all the Master It problems from the chapters and provides a solution for each.
- Appendix B, "AutoCAD® Civil 3D® Certification," points you to the chapters in this book that will help you master the objectives for the Certified Professional Exam.

How to Contact the Authors

We welcome feedback from you about this book and/or about books you'd like to see from us in the future. Feel free to connect with us on LinkedIn:

- www.linkedin.com/in/louisaholland
- http://www.linkedin.com/pub/cyndy-davenport/13/61b/1a9

You can also keep up with Cyndy Davenport on Twitter (C3DCougar) and email Eric Chappell at civilessentials@gmail.com.

Sybex strives to keep you supplied with the latest tools and information you need for your work. Please check their website at www.sybex.com/go/masteringcivil3d2014, where we'll post additional content and updates that supplement this book if the need arises.

Thanks for purchasing *Mastering AutoCAD Civil 3D 2014*. We appreciate it and look forward to exploring AutoCAD Civil 3D with you!

Chapter 1

The Basics

It takes patience and time to truly become a "master" of the AutoCAD® Civil 3D® program, and your first step will be to understand the basics. There are numerous dialogs, ribbons, menus, and icons to pore over. They might seem daunting at first glance, but as you use them, you will gain familiarity with their location and use. In this chapter, you will explore the interface and learn terminology that will be used throughout this book.

In addition, we will introduce the Lines and Curves commands, which offer loads of options for drawing lines and curves accurately.

In this chapter, you will learn to:

- Find any Civil 3D object with just a few clicks
- Modify the drawing scale and default object layers
- Navigate the ribbon's contextual tabs
- Create a curve tangent to the end of a line
- Label lines and curves

The Interface

If you are new to Civil 3D or are coming from Civil 3D 2009 or prior, this part of the chapter is especially for you. If you have used newer versions of Civil 3D, this section will help you understand the terminology used throughout this book. Civil 3D uses a ribbon-based interface, which is where you will access many of the tools. The ribbon consists of tabs and panels that organize tools into logical groups. When working in Civil 3D 2014, you will spend the majority of your time on the Home tab, shown in Figure 1.1.

FIGURE 1.1
The Home tab of the ribbon runs horizontally across the top of your screen and is your first stop for creating new objects.

When you click on a Civil 3D object, you will see a context-specific *contextual tab* appear in the ribbon. Figure 1.2 shows the Civil 3D palette sets along with the AutoCAD tool palettes and ribbon displayed in a typical environment.

FIGURE 1.2
Overview of the Civil 3D environment. Toolspace is docked to the left, and tool palettes float over the drawing window. The ribbon is at the top of the workspace.

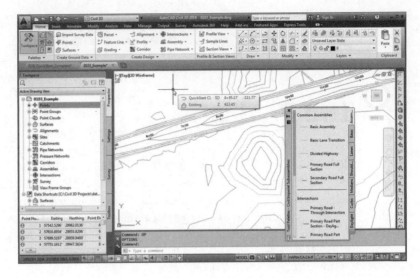

Panels are subgroups within each tab of the ribbon that further organize your tools. For example, the Palettes panel on the Home tab (shown in Figure 1.3) is where you can toggle on or off the elements you are about to examine. These icons will become highlighted in blue when the palette is visible.

FIGURE 1.3
Palettes panel of the Home tab. Icons will be blue when the palette is displayed.

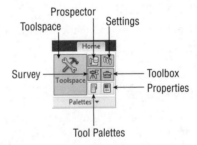

Toolspace

Toolspace is a set of palettes that is specific to Civil 3D. You will want to have the palette visible anytime you are working in Civil 3D. If you do not see it, click the Toolspace button on the Palettes panel of the Home tab.

Toolspace has four tabs to manage user data, as follows:

- Prospector
- Settings
- Survey
- Toolbox

The tabs can be turned on or off by toggling the display on the Palettes panel, but it is perfectly fine to have them all up all the time.

THE INTERFACE

Each tab has a unique role to play in working with Civil 3D. Prospector and Settings will be your most frequently visited tabs. Survey and Toolbox are used for special tasks that you will examine in the following sections.

Prospector

Prospector's job is to show you information about specific Civil 3D objects. In the top portion of Prospector, you will find drawing-specific information. Civil 3D objects are listed in workflow order, starting at the top of the listing. From the Data Shortcuts listing down, the information you see is a listing of data available to you regardless of the drawing it is in (you will learn how to work with data shortcuts in Chapter 16, "Advanced Workflows"). Each main grouping under the drawing name is referred to as a *collection*. If you expand a collection by clicking the plus sign next to the name, you will see the contents of that group.

Because all Civil 3D data is dynamically linked, you will see object dependencies as well. You can learn details about an individual object by expanding the tree and selecting an object (Figure 1.4).

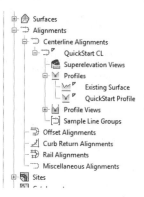

FIGURE 1.4
A look at the Alignment branch of the Prospector tab. Profiles are linked to alignments; therefore, they appear under alignments.

Right-clicking the collection name allows you to select various commands that apply to all the members of that collection. For example, right-clicking the Point Groups collection brings up the menu shown in Figure 1.5 (left).

In addition, right-clicking the individual object in the list view offers many commands unique to Civil 3D, such as Zoom To and Pan To, shown in Figure 1.5 (right). By using these commands, you can find any parcel, point, cross section, or other Civil 3D object in your drawing almost instantly.

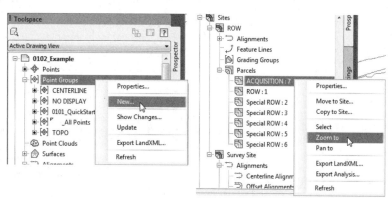

FIGURE 1.5
Context-sensitive menus in Prospector for creating new elements (left) and zooming to a specific object (right)

For example, if you are interested in locating a parcel named ACQUISITION 7 using the Zoom To command, locate the Sites collection on the Prospector tab of Toolspace. Expand Proposed Site and highlight Parcels. At the bottom of Prospector, you will see the parcel listing. To locate ACQUISITION 7 graphically, right-click it and select Zoom To.

Near the top of the Toolspace you will see a pull-down giving you the options Active Drawing view and Master view.

Active Drawing view will show you the following items:

- The current drawing
- Data shortcuts

Master view will show you these items:

- Open drawings
- Data shortcuts
- Drawing templates
- Refresh icon

Master view will list every drawing you have open as well as its contents and templates. If you use Master view, the name of the drawing you are working with appears at the top of the list in bold. To make a drawing current, right-click its name in Prospector and select Switch To.

Many users prefer to use the *Active Drawing view*. You can have more than one drawing open, but Prospector displays only one set of Civil 3D data at a time. Active Drawing view will change to reflect whichever drawing is current.

In addition to the branches, Prospector has a series of icons across the top that toggle various settings on and off. Let's take a closer look at those icons:

Item Preview Toggle   Turn this on if you wish to see a graphic preview of an item at the bottom of Prospector when certain items are selected.

Preview Area Display Toggle   This icon will be active only when Toolspace is undocked. This button moves the preview area from the right of the tree view to beneath the tree view area.

Panorama Display Toggle   This button provides one of several ways to turn on and off the display of the Panorama window. This button will be grayed out if there are no active warnings or if you have not yet viewed data in the Panorama window.

You can always return to the Panorama regardless of your warning status, by clicking the Event Viewer button from the Home tab ➤ Palettes panel.

Help   Don't underestimate how helpful Help can be!

> ### Help Using Help
>
> At any time during your use of Civil 3D, you can use the F1 key to bring up the help file relevant to the dialog you are working in.
>
> Even for seasoned users, Help provides a comprehensive reference to objects and options. The most difficult part of using Help is knowing what terminology is used to describe the task you are trying to perform. Luckily, you have this book to assist you with that!

As you navigate the tabs of Toolspace, you will encounter many symbols to help you along the way. Table 1.1 shows you a few that you should familiarize yourself with.

TABLE 1.1: Common Toolspace symbols and meanings

SYMBOL	MEANING
▽	The object or style is in use. Also appears when there is a dependency to the object or if the style has child styles. For example, you will see this icon on a surface when a profile has been created from it.
⊞	Clicking this will expand the branch of Toolspace.
⊟	Clicking this will collapse the branch of Toolspace.
◉	Data resides in this branch and more information can be found at the bottom of Toolspace.
⚠	Object needs to be rebuilt or updated. Can also indicate broken data reference.
▨	Civil 3D may still be processing the object or the branch of Prospector needs to be refreshed.
↗	This symbol represents a data reference in a drawing, and it's shown next to the data shortcuts section of the Prospector tab.

> ### HIT THE ROAD RUNNING: QUICK START PROJECT
>
> Most new users are eager to get started on their first project before reading the entirety of this book. Author Louisa Holland says that if she had her druthers, she'd sequester every new Civil 3D user until they've had a chance to work through every exercise in this book, cover to cover. Alas, time, money, and several abduction laws prevent this dream from becoming a reality.
>
> This exercise will give you a chance to work through a basic project. Unless otherwise specified, don't change the dialog box default options. After each relevant step, you will see where to go for in-depth explanation.
>
> 1. Open the drawing `0101_QuickStart.dwg` (`0101_QuickStart_METRIC.dwg`). You can download this and all other files related to this book from this book's web page, www.sybex.com/go/masteringcivil3d2014.
>
>   See the section "Civil 3D Templates" in this chapter to read about the importance of styles, settings, and starting with a Civil 3D drawing template.
>
>   This drawing contains an assembly, which you will learn to create in Chapter 8, "Assemblies and Subassemblies."

2. From the Home tab of the ribbon, open the Create Ground Data panel and click Import Survey Data.

See Chapter 2, "Survey," to learn more about importing survey data.

3. Click Create New Survey Database.
4. In the New Local Survey Database dialog, name the new database **QuickStart** and click OK.

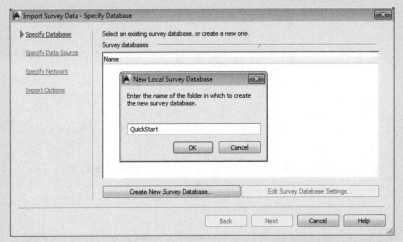

5. Click Next.

    Where is this survey data stored? Find out in Chapter 2 in the section "The Main Event: Your Project's Survey Database."

6. In the Import Survey Data – Specify Data Source dialog, follow these steps:

    a. Set Data Source Type to Point File.

    b. Click the plus sign to the right of the selected files box.

    c. Set your Files of Type option to Text/Template/Extract File (*.txt) and browse for 0101_QuickStart.txt (0101_QuickStart_METRIC.txt) and click OK.

    d. Set Specify Point File Format to PNEZD (Comma Delimited).

    e. Click Next.

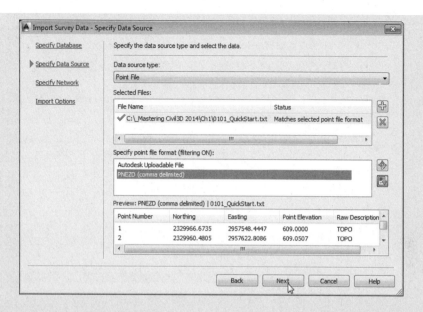

7. Click Create New Network.
8. Name the new network **QuickStart Network** and click OK.
9. Highlight QuickStart Network and click Next.

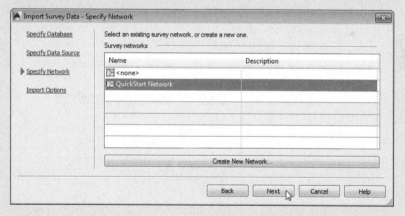

10. In the Import Survey Data – Import Options dialog, follow these steps:
    a. Place a check mark across from Process Linework During Import.
    b. Place a check mark across from Insert Figure Objects.

c. Place a check mark across from Insert Survey Points.
   d. Leave all other options at the default settings and click Finish.

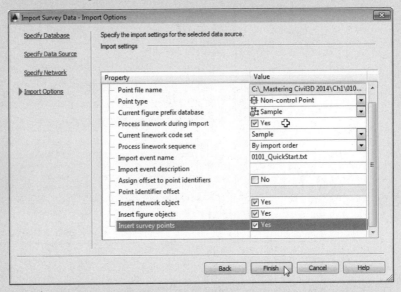

   See the sections "The Figure Prefix Database" and "The Linework Code Set Database" in Chapter 2 for more information on creating figures on importing survey data.

   Did you notice that shots with the description TOPO look different from other survey points in the drawing? Find out why in Chapter 2 in the section "Description Keys: Field to Civil 3D."

11. From the Home tab of the ribbon, open the Create Ground Data panel and click Surfaces ➢ Create Surface.

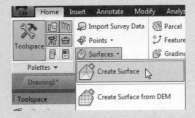

12. In the Create Surface dialog, change the name to **Existing**.
13. Click OK.
14. In the Prospector tab of Toolspace, expand Surfaces ➢ Existing ➢ Definition.

   Chapter 4, "Surfaces," contains the section "Creating Surfaces," which describes the different types of data that can be used to define elevation in a surface model.

15. Right-click Point Groups and select Add.

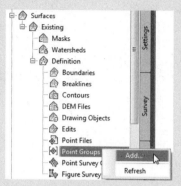

For an in-depth look at the importance of Point Groups, be sure to read the section "Point Groups: Don't Skip this Section!" in Chapter 3, "Points."

16. Select _All Points and click OK.

    At this point you should see contours and the surface border. See Chapter 4 for more information on creating, editing and displaying surfaces.

17. On the Survey tab of Toolspace, right-click Figures and select Create Breaklines.

18. In the Create Breaklines dialog, note that you are adding breaklines to the surface you created earlier. Click the Check-All option and click OK.
19. In the Add Breaklines dialog, click OK.

    See the section in Chapter 4 called "Adding Breakline Information" for a detailed explanation.
20. Click Save.

21. In the Prospector tab of Toolspace, right-click Point Groups and select Update.

A list of Toolspace symbols and their meanings can be found in Table 1.1.

22. From the Home tab of the ribbon, open the Create Design panel and click Alignment ➢ Create Best Fit Alignment.

The many methods for creating and editing alignments are found in Chapter 6, "Alignments."

23. In the Create Best Fit Alignment dialog, do the following:

  a. Change the input type to COGO Points.

  b. Change Path 1 Point Group to CENTERLINE.

  c. Change the alignment name to **QuickStart CL**.

  d. Clear the check box for Show Report.

  e. Click OK.

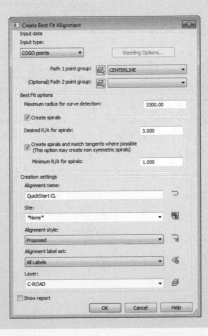

24. Select the new alignment (the green line).

   Want to know why the alignment appears green? See the section "Linear Object Styles" in Chapter 19, "Object Styles," for more information.

25. From the Alignment contextual tab ➢ Launch Pad panel, click Surface Profile.

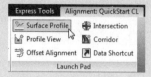

26. In the Create Profile From Surface dialog, click Add.
27. Click Draw In Profile View.

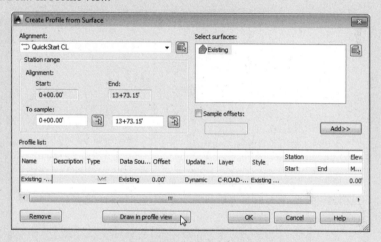

28. In the Create Profile View – General dialog, click Create Profile View.
29. Click anywhere to the north of the site, outside of the surface area.
30. Save the drawing.

   You should now see the profile in the profile view. Chapter 7, "Profiles and Profile Views," will take you through the details of these objects.

31. From the Home tab of the ribbon, open the Create Design panel, click Profile ➢ Create Best Fit Profile.

32. When prompted to select a profile view, click the grid of the profile view you created in the previous steps.
33. In the Create Best Fit Profile dialog, follow these steps:
    a. Set the input type to Surface Profile (the surface profile will automatically go to Existing – Surface (4)).
    b. Change the profile name to **QuickStart Profile**.
    c. Change the profile style to Design Profile.
    d. Clear the check box for Show Report.
    e. Click OK.

34. From the Home tab of the ribbon, open the Create Design panel and click Corridor.
35. In the Create Corridor dialog, do the following:
    a. Set the name of the corridor to **QuickStart Corridor**.
    b. Verify that the alignment is set to QuickStart CL.
    c. Set Profile to QuickStart Profile.
    d. Set Assembly to Shoulder Widening.
    e. Set Target Surface to Existing.
    f. Clear the check box for Set Baseline And Region Parameters.
    g. Click OK.

36. If you receive any Event Viewer warnings, dismiss the Panorama window by clicking the green check mark.

    You will learn all about Corridor creation and the meanings of various warnings in Chapter 9, "Basic Corridors."

Now that you've had your first taste of the power behind Civil 3D, you are ready to buckle down and get more in depth with the details. This is just a sampling of the functionality of Civil 3D. You may want to learn about pipe networks (Chapter 13), plan production (Chapter 15), or grading (Chapter 14).

Settings

The Settings tab of Toolspace controls all things aesthetic and the default behavior of the commands. Text placed by Civil 3D is controlled by *label styles*. *Object styles* control the look of design elements such as surface contours or pipes. These settings and styles should be set in your template drawing. Every time you start a project with your company's Civil 3D–specific template, items such as an alignment's color and linetype will already be set. Chapter 18, "Label Styles," and Chapter 19, "Object Styles," are dedicated to building these styles. Later on in this chapter you will learn more about templates.

Drawing Settings

At the top of the Settings tab you will see the name of the drawing. There are some important settings you should verify before proceeding with a project. Right-click on the name of the drawing and click Edit Drawing Settings, as shown in Figure 1.6, to access the Drawing Settings dialog.

FIGURE 1.6
Accessing the Drawing Settings dialog

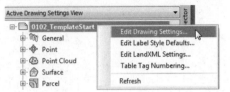

Each tab in this dialog controls a different aspect of the drawing. Most of the time, you'll pick up the settings on the Object Layers, Abbreviations, and Ambient Settings tabs from a company-wide template. However, the drawing scale and coordinate information change for every job, so you'll visit the Units And Zone and Transformation tabs frequently.

The Units And Zone Tab

On the Units And Zone tab, you specify metric or Imperial units for your drawing as well as set an appropriate coordinate system for the file. You'll notice that when a coordinate zone is selected from the Zone portion of the dialog, the Imperial To Metric Conversion option becomes grayed out. This is because the drawing coordinate system will take care of any conversion for you. Note that by default, this setting is international feet rather than survey feet.

This tab also includes the options Scale Objects Inserted From Other Drawings and Set AutoCAD Variables To Match. The Set AutoCAD Variables To Match option sets the base AutoCAD angular units, linear units, block insertion units, hatch pattern, and linetype units to match the values placed in this dialog. As shown in Figure 1.7, you do want these options selected.

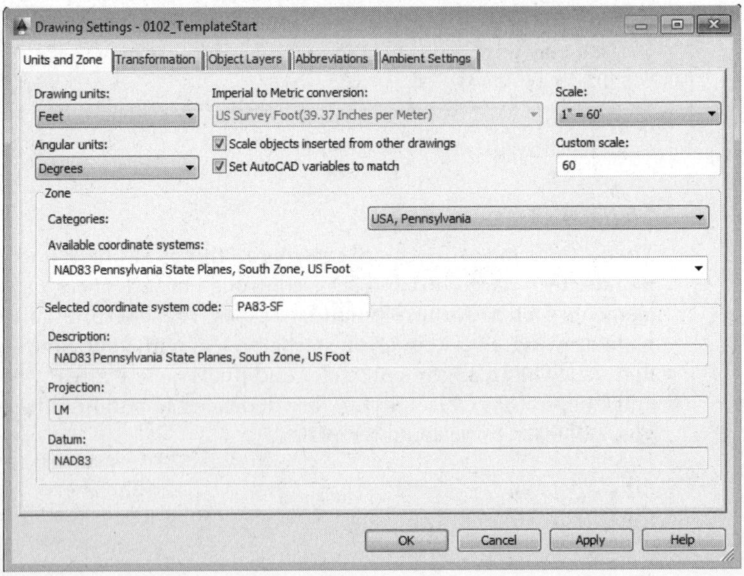

FIGURE 1.7
Before placing any project-specific information in a drawing, set the coordinate system in the Units And Zone tab of the Drawing Settings dialog.

The scale that you see on the right side of the Units And Zone tab is the same as your *annotation scale*. You can change it here, but it is much easier to select your annotation scale from the bottom of the drawing window.

If you choose to work in assumed coordinates, you can leave Zone set to No Datum, No Projection. To set the coordinate system for your locale, first set the category from the long list of possibilities. Civil 3D is used worldwide; therefore, most recognized surveying coordinate systems (including obsolete ones) can be found in the Units And Zone tab of the Drawing Settings dialog.

Try the following quick exercise to practice setting a drawing coordinate system:

1. Open the drawing `0102_TemplateStart.dwg` (`0102_TemplateStart_METRIC.dwg`). You can download this and all other files related to this book from this book's web page, www.sybex.com/go/masteringcivil3d2014.

2. Switch to the Settings tab of Toolspace.

3. Right-click the filename and select Edit Drawing Settings.

4. Switch to the Units And Zone tab to display the options shown previously in Figure 1.7.

5. Select USA, Pennsylvania from the Categories drop-down menu on the Units And Zone tab.

6. Select NAD83 Pennsylvania State Planes, South Zone, US Foot (NAD83 Pennsylvania State Planes, South Zone, Meter) from the Available Coordinate Systems drop-down menu.

7. Place a check mark next to both Scale Objects Inserted From Other Drawings and Set AutoCAD Variables To Match. Click OK when complete.

    You could have also typed **PA83-SF** (**PA83-S**) in the Selected Coordinate System Code box.

8. Save the drawing for use in an upcoming exercise

Notice that once you have set the coordinate system, the geographic marker symbol becomes visible (if you don't see it, zoom to the extents of the drawing). This is a graphic indication that a coordinate system is set. It will not plot, and its size is always a fixed percentage of your screen size.

If you wish to hide the geometric marker, you can click the red pin icon at the bottom of the screen. Clicking this icon toggles the GEOMARKERVISIBILTY variable on or off.

You will also see the active coordinate system displayed at the bottom of the screen. The pin and coordinate system display are new features in Civil 3D 2014.

The Transformation Tab

Most survey-grade GPS equipment takes care of the transformation to local grid coordinates for you. In the United States, state plane coordinate systems already have regional projections taken into account. In the rare case that surveyors need to manually transform local observations from geoid to ellipsoid and ellipsoid to grid, the Transformation tab enables access to enter transformation factors.

With a base coordinate system selected, you can do any further refinement you'd like using the Transformation tab, shown in Figure 1.8. The coordinate systems on the Units And Zone tab can be refined to meet local ordinances, tie in with historical data, complete a grid-to-ground transformation, or account for minor changes in coordinate system methodology. These changes can be made with the following options:

Apply Sea Level Scale Factor This value is known in some circles as *elevation factor* or *orthometric height scale*. The sea level scale factor takes into account the mean elevation of the site and the spheroid radius that is currently being applied as a function of the selected zone ellipsoid.

Grid Scale Factor At any given point on a projected map, there is a distortion between the "flat" measurement and the measurement on the ellipsoid. Grid Scale Factor is based on a 1:1 value, a user-defined uniform scale factor, a reference point scaling, or a prismoidal formula transformation in which every point in the grid is adjusted by a unique amount.

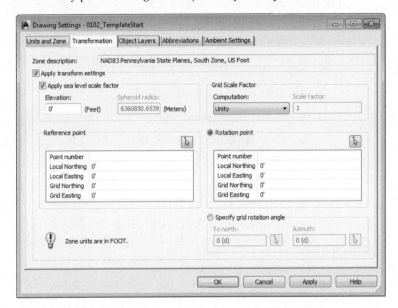

FIGURE 1.8
The Transformation tab

Reference Point To apply the grid scale factor and the sea level factor correctly, you need to tell Civil 3D where you are on Earth. Reference Point can be used to set a singular point in the drawing field via pick or Point number, Local Northing and Easting, or Grid Northing and Easting Values.

Rotation Point Rotation Point can be used to set the reference point for rotation via the same methods as the reference point.

Specify Grid Rotation Angle Some people may know this as the *convergence angle*. This is the angle between Grid North and True North. Enter an amount or set a line to north by picking an angle or deflection in the drawing. You can use this same method to set the azimuth if desired.

It should be noted that this is not the place to transform assumed coordinates to a predefined coordinate system. See Chapter 2 to learn how to translate a survey.

The Object Layers Tab

Civil 3D and AutoCAD layers have a love-hate relationship with each other. Civil 3D is built on top of AutoCAD; therefore, all the objects do reside on layers. However, Civil 3D is not traditional CAD. Your surfaces, corridors, points, profiles, and everything else generated by Civil 3D are dynamic *objects* rather than simple lines, arcs, or circles.

When you create an alignment in Chapter 6, for example, you will not have to think about the current layer. This is because Civil 3D styles "push" objects and labels to the correct layer as part of their intelligence.

Layers are found in several areas of the Civil 3D template. The first location you will examine is the Drawing Settings area. The layers listed here represent overall layers where the objects will be created. For those of you who are familiar with AutoCAD blocks, it is useful to think of these layers in the same way as a block's insertion layer.

In the Object Layers tab, every Civil 3D object must have a layer set, as shown in Figure 1.9. Do not leave any object layers set to 0. An optional modifier can be added to the beginning (*prefix*) or end (*suffix*) of the layer name to further separate items of the same type.

FIGURE 1.9
Every object is on a layer; the corridor layer contains a modifier.

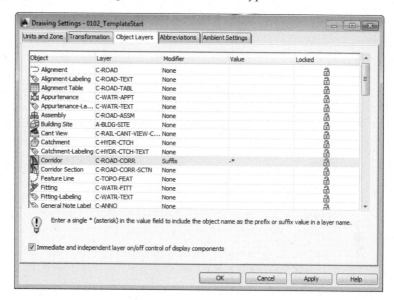

A common practice is to add wildcard suffixes to corridor, surface, pipe, and structure layers to make it easier to manipulate them separately. For example, if the layer for a corridor is specified to be C-ROAD-CORR and a suffix of -* (dash asterisk, as shown in Figure 1.9) is added as the modifier value, a new layer will automatically be created when a new corridor is created. The resulting layer will take on the name of the corridor in place of the asterisk. If the corridor is called 13th Street, the new layer name will be C-ROAD-CORR-13th Street. This new layer is created once and is not dynamic to the object name. In other words, if you decide to change the name of 13th Street to Holland Avenue, the layer remains C-ROAD-CORR-13th Street.

If the main layer name you are after does not exist in the drawing, you can create it as you work through the Object Layers dialog. Click the New button, and set up the layer as needed, including color, lineweight, linetype, and so forth, as shown in Figure 1.10.

FIGURE 1.10
Click New to add a new layer.

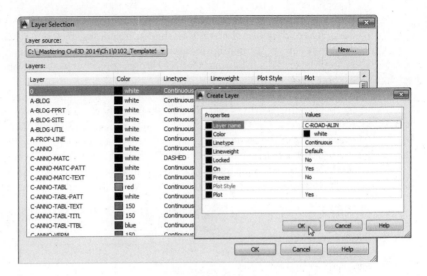

"Immediate And Independent Layer On/Off Control Of Display Components" is a setting you will want to have selected. As mentioned earlier, the layers listed here are like an object's insertion layer. However, you will encounter more layers within the object's display style (these are the object styles you will learn about in Chapter 19). Having this option selected allows you to turn off components within an object's style without turning off the entire object. For example, consider a surface whose object layer is set as C-TOPO. When that surface has contours displayed, the major contours might be on C-TOPO-MAJR and the minor contours may be on C-TOPO-MINR. With this option selected, you could turn off the C-TOPO-MINR independently from the overall object.

In the following exercise, you set object layers in a template:

1. Continue working in the drawing 0102_TemplateStart.dwg (0102_TemplateStart_METRIC.dwg). It is not necessary to have completed the previous exercise.

2. From the Settings tab of Toolspace, right-click on the name of the drawing and select Edit Drawing Settings.

3. Switch to the Object Layers tab.

4. Click in the Layer field next to Alignment, and click New to create a new layer.

5. Create a new layer called C-ROAD-ALIN. Leave other layer settings at the defaults.

6. Set the newly created layer as the layer for the Alignment object.

7. Set the layer for Building Site to A-BLDG-SITE.

8. Set the layer for Catchment-Labeling to C-HYDR-CTCH-TEXT.

9. For the corridor layer, keep the main layer as C-ROAD-CORR.

    ♦   Set the modifier to Suffix.

    ♦   Set the modifier value to -*.

The asterisk acts as a wildcard that will add the corridor name as part of a unique layer for each corridor, as previously described.

10. Scroll down to locate the Pipe object listing.

11. Create several new layers and add suffix information:
    - For Pipe, create a layer called C-NTWK-PIPE with a modifier of Suffix and a value of -*.
    - For Pipe-Labeling, create a new layer called C-NTWK-PIPE-TEXT.
    - For Pipe And Structure Table, set the layer to C-NTWK-PIPE-TEXT.
    - For Pipe Network Section, create a new layer called C-NTWK-SECT.
    - For Pipe or Structure Profile, create a new layer called C-NTWK-PROF.

12. Scroll down a bit further and create a new layer for Structure called C-NTWK-STRC.
    - Add a modifier of Suffix and a value of -*.

13. For Structure-Labeling, create a new layer called C-NTWK-STRC-TEXT.
    - Add a modifier of Suffix to the Tin Surface object layer and a value of -*.

    Your layers and suffixes should now resemble Figure 1.11.

FIGURE 1.11
Examples of the completed layer names in the Object Layers tab

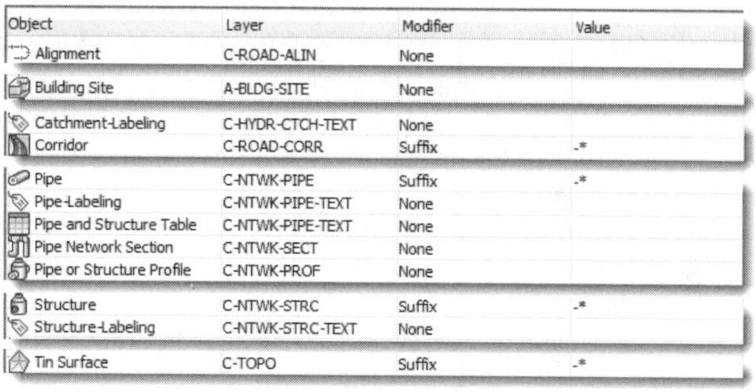

Object	Layer	Modifier	Value
Alignment	C-ROAD-ALIN	None	
Building Site	A-BLDG-SITE	None	
Catchment-Labeling	C-HYDR-CTCH-TEXT	None	
Corridor	C-ROAD-CORR	Suffix	-*
Pipe	C-NTWK-PIPE	Suffix	-*
Pipe-Labeling	C-NTWK-PIPE-TEXT	None	
Pipe and Structure Table	C-NTWK-PIPE-TEXT	None	
Pipe Network Section	C-NTWK-SECT	None	
Pipe or Structure Profile	C-NTWK-PROF	None	
Structure	C-NTWK-STRC	Suffix	-*
Structure-Labeling	C-NTWK-STRC-TEXT	None	
Tin Surface	C-TOPO	Suffix	-*

14. Place a check mark next to "Immediate And Independent Layer On/Off Control Of Display Components."

    As described previously, this setting will allow you to use the On/Off toggle in Layer Manager to work with Civil 3D objects.

15. Click Apply and then OK.

16. Save the drawing for use in the next exercise.

The Abbreviations Tab

When you add labels to certain objects, Civil 3D automatically uses the abbreviations assigned in this tab to indicate geometry features. For example, left is *L* and right is *R*. Figure 1.12 shows a sampling of customizable abbreviations.

FIGURE 1.12
Features are customizable down to the letter on the Abbreviations tab.

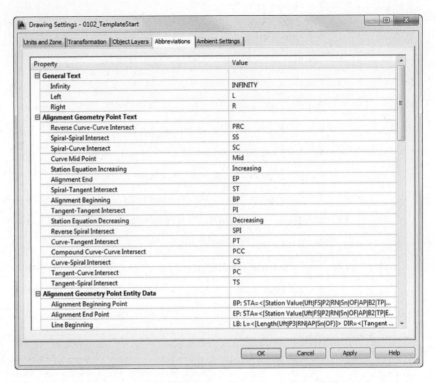

Civil 3D uses industry-standard abbreviations wherever they are found. If necessary, you can easily change VPI to PVI for Point of Vertical Intersection. In most cases, changing an abbreviation is as simple as clicking in the Value field and typing a new one.

The Ambient Settings Tab

Examine the settings in the Ambient Settings tab to see what can be set here. The main options you'll want to adjust are in the General category, and the display precision settings are in the subsequent categories. You will also want to visit the angle and direction categories to verify the format of the angles.

The level of precision that you see in this dialog does not change the precision in labels. What you see here is the number of decimal places reported to you in various dialog boxes.

Being familiar with the way this tab works will help you further down the line because almost every other settings dialog box in the program works like the one shown in Figure 1.13.

You can set the following options in the General category:

Plotted Unit Display Type Civil 3D knows you want to plot at the end of the day. In this case, it's asking how you would like your plotted units measured. For example, would you like that bit of text to be 0.25g tall or ¼g high? Most engineers are comfortable with the Leroy method of text heights (L80, L100, L140, and so on), so the decimal option is the default.

FIGURE 1.13
Ambient Settings at the main drawing level

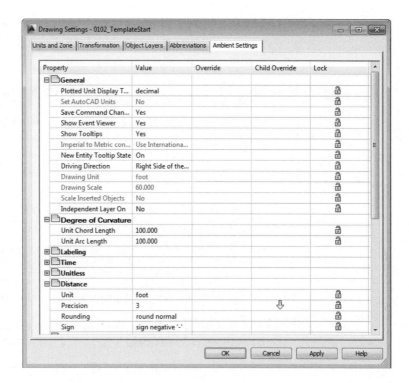

Set AutoCAD Units   This option specifies whether Civil 3D should attempt to match AutoCAD drawing units, as specified on the Units And Zone tab. This setting is specified on the Units And Zone tab but is displayed here for reference and so you can lock it if desired.

Save Command Changes To Settings   Set this to Yes. This setting is incredibly powerful but a secret to almost everyone. By setting it to Yes, you ensure that your changes to commands will be remembered from use to use. This means if you make changes to a command during use, the next time you call that Civil 3D command, you won't have to make the same changes. It's frustrating to do work over because you forgot to change one of the five things that needed changing, so this setting is invaluable.

Show Event Viewer   Event Viewer is the main Civil 3D feedback mechanism, especially when things go wrong. Event Viewer uses the Panorama interface to display warnings such as when a surface contains crossing breaklines. Event Viewer will pop up with informational messages as well. If you have multiple monitors, it is a good idea to leave Panorama on but set aside for review.

Show Tooltips   One of the cool features that people remark on when they first use Civil 3D is the small pop-up that displays relevant design information when the cursor is paused on the screen. This includes things such as station-offset information, surface elevation, section information, and so on. Once a drawing contains numerous bits of information, this display can be overwhelming; therefore, Civil 3D offers the option to turn off these tooltips universally with this setting. A better approach is to control the tooltips at the object type by editing the individual feature settings. You can also control the tooltips by pulling up the properties for any individual object and looking at the Information tab.

Imperial To Metric Conversion   This setting displays the conversion method specified on the Units And Zone tab. The two options are US Survey Foot and International Foot.

New Entity Tooltip State   This setting controls whether the tooltip is turned on at the object level for new Civil 3D objects. If you change this setting to Off partway through a project, the tooltip will not be displayed for any Civil 3D objects created after the change.

Driving Direction   This specifies the side of the road that forward-moving vehicles use for travel. This setting is important in terms of curb returns and intersection design.

Drawing Unit, Drawing Scale, and Scale Inserted Objects   These settings were specified on the Units And Zone tab but are displayed here for reference and so that you can lock them if desired.

Independent Layer On   This is the same control that was set on the Object Layers tab. Yes is the recommended setting, as described previously.

The ambient settings in the Direction category offer the following choices:

- Unit: Degree, Radian, and Grad.
- Precision: 0 through 8 decimal places.
- Rounding: Round Normal, Round Up, and Truncate.
- Format: Decimal, two types of DDMMSS, and Decimal DMS. In most cases, people want to display DD°MM'SS.SS". Whether you want spaces between the subdivisions is up to you.
- Direction: Short Name (spaced or unspaced) and Long Name (spaced or unspaced).
- Capitalization: You can display as typed or force uppercase, lowercase, or title caps.
- Sign: Gives you your choice of how negative numbers are displayed. You can use a negative sign to denote negative numbers only, use a parenthesis to denote a negative, or use a sign regardless of value. The latter option will show a plus for positive values and a minus for negative values.
- Measurement Type: Bearings, North Azimuth, and South Azimuth.
- Bearing Quadrant: This should be left at the industry standard. 1-NE, 2-SE, 3-SW, 4-NW.

Certification Objective

When you're using the Bearing Distance transparent command, for example, these settings control how you input your quadrant, your bearing, and the number of decimal places in your distance.

Explore the other categories, such as Angle, Lat Long, and Coordinate, and customize the settings to how you work.

At the bottom of the Ambient Settings tab is a Transparent Commands category. These settings control how (or if) you're prompted for the following information:

Prompt For 3D Points   Controls whether you're asked to provide a z elevation after x and y have been located.

Prompt For Y Before X   For transparent commands that require x and y values, this setting controls whether you're prompted for the y-coordinate before the x-coordinate. Most users prefer this value set to False so they're prompted for an x-coordinate and then a y-coordinate.

Prompt For Easting Then Northing   For transparent commands that require Northing and Easting values, this setting controls whether you're prompted for Easting first and Northing second. Most users prefer this value set to False so they're prompted for Northing first and then Easting.

Prompt For Longitude Then Latitude   For transparent commands that require longitude and latitude values, this setting controls whether you're prompted for longitude first and latitude second. Most users prefer this set to False so they're prompted for latitude and then longitude.

The settings that are applied here can also be changed at the object level. For example, you may typically want elevation to be shown to two decimal places, but when looking at surface elevations, you might want just one. The Override and Child Override columns give you feedback about these types of changes. See Figure 1.14.

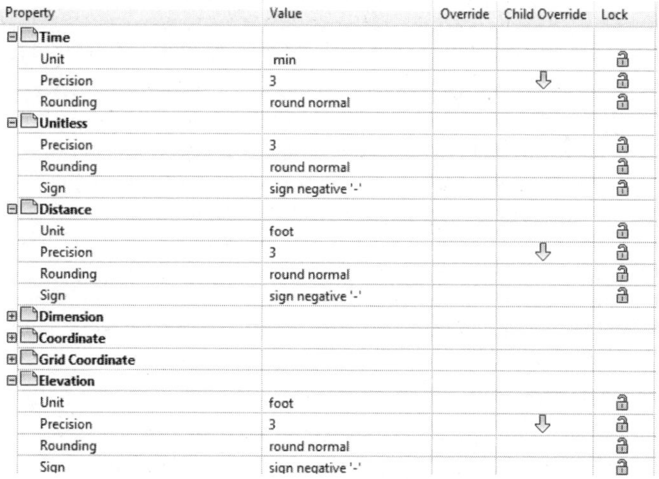

FIGURE 1.14
The Child Override indicator in the Time, Distance, and Elevation values

Property	Value	Override	Child Override	Lock
⊟ Time				
Unit	min			🔒
Precision	3		⬇	🔒
Rounding	round normal			🔒
⊟ Unitless				
Precision	3			🔒
Rounding	round normal			🔒
Sign	sign negative '-'			🔒
⊟ Distance				
Unit	foot			🔒
Precision	3		⬇	🔒
Rounding	round normal			🔒
Sign	sign negative '-'			🔒
⊞ Dimension				
⊞ Coordinate				
⊞ Grid Coordinate				
⊟ Elevation				
Unit	foot			🔒
Precision	3		⬇	🔒
Rounding	round normal			🔒
Sign	sign negative '-'			🔒

The Override column shows whether the current setting is overriding something higher up. Because you're at the Drawing Settings level, these are clear. However, the Child Override column displays a down arrow, indicating that one of the objects in the drawing has overridden this setting. After a little investigation of the objects, you'll find the override in the Edit Feature Settings dialog of the Profile view, as shown in Figure 1.15.

FIGURE 1.15
The profile elevation settings and the Override indicator

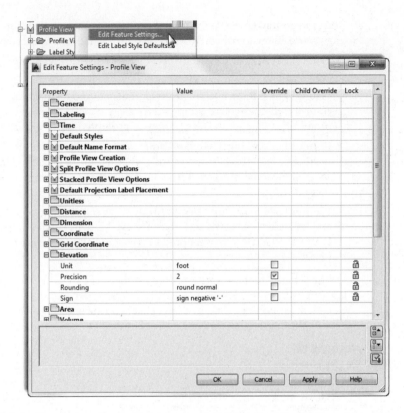

Notice that in this dialog, the box for the Precision setting is checked in the Override column. This indicates that you're overriding the settings mentioned earlier, and it's a good alert that things have changed from the general drawing settings to this object level setting.

But what if you don't want to allow those changes? Each settings dialog includes one more column: Lock. At any level, you can lock a setting, graying it out for lower levels. This can be handy for keeping users from changing settings at the lower level that perhaps should be changed at a drawing level, such as sign or rounding methods.

Survey

The Survey tab of Toolspace is displayed optionally and controls the use of the survey, equipment, and figure prefix databases. Surveying is an essential part of land-development projects. Because of the complex nature of this tab, all of Chapter 2 is devoted to it.

Toolbox

The Toolbox tab of Toolspace is a launching point for add-ons and reporting functions. To display the Toolbox, from the Home tab in the ribbon, select Toolspace ➢ Palettes and click the Toolbox icon (as shown previously in Figure 1.3). Out of the box, the Toolbox contains reports created by Autodesk, but you can expand its functionality to include your own macros or reports. The buttons on the top of the Toolbox, shown in Figure 1.16, allow you to customize the report settings and add new content. If you are an Autodesk® Subscription customer, new goodies released throughout the year are frequently accessed from this area.

FIGURE 1.16
The Toolbox with the Edit Toolbox Content icon highlighted

Panorama

The Panorama window is the Civil 3D feedback and tabular editing mechanism. It's designed to be a common interface for a number of different Civil 3D–related tasks, and you can use it to provide information about the creation of profile views, to edit pipe or structure information, or to run basic volume analysis between two surfaces. For an example of Panorama in action, open it up by going to the Home tab ➢ Palettes panel flyout and clicking the Event Viewer icon. You'll explore and use Panorama more during this book's discussion of specific objects and tasks.

Ribbon

As with AutoCAD, the ribbon is the primary interface for accessing Civil 3D commands and features. When you select an AutoCAD Civil 3D object, the ribbon displays commands and features related to that object in a *contextual tab*. If several object types are selected, the Multiple contextual tab is displayed. Use the following procedure to familiarize yourself with the ribbon:

1. Open 0103_Example.dwg (0103_Example_METRIC.dwg), which you will find at www.sybex.com/go/masteringcivil3d2014.

2. Select one of the parcel labels (the labels in the middle of the lot areas).

    Notice in the Parcel contextual tab that the Labels & Tables, General Tools, Modify, and Launch Pad panels are displayed, as shown in Figure 1.17.

FIGURE 1.17
The contextual tab in the ribbon

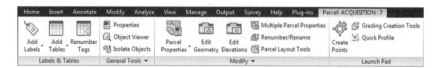

3. Select a parcel line and notice the display of the Multiple contextual tab (Figure 1.18).

FIGURE 1.18
When more than one object is selected, the Multiple contextual tab appears.

4. Use the Esc key to cancel all selections.
5. Reselect a parcel by clicking one of the numeric labels.
6. Select the down arrow next to the Modify panel name.
7. Click the pin at the bottom-left corner of the panel to keep it open.
8. Select the Properties command in the General Tools panel to open the AutoCAD Properties palette.

   Notice that the Modify panel remains open and pinned until the current selection set changes.

> **STYLES AND MORE STYLES**
>
> Civil 3D uses *styles* to change the look of objects and labels. Styles control everything from which layer your surface contours will be created on to the number of decimal places displayed in a label.
>
> Civil 3D has an unbelievable number of options when it comes to how you want your design elements to look. It is easy to get bogged down in the intricacies of object and label style creation. The authors have decided to separate styles into separate chapters so you can focus on learning functionality first. Once you have an understanding of how the tools operate, you can then adjust how your designs are represented graphically.
>
> In this chapter and throughout the book, you will be using styles that have already been created for you. For an in-depth look at styles, refer to Chapter 18 and Chapter 19.

Civil 3D Templates

Styles and settings should come from your template. Ideally, that template will have all the styles you need for the type of project you are working on. If you find you are constantly changing style settings, reexamine your workflow.

When starting a project, or continuing a project from an outside source, it is important to start with a Civil 3D template file. Right after installing the software you will see two usable Civil 3D–specific templates (Figure 1.19).

FIGURE 1.19
Selecting a Civil 3D template by going through the Application menu

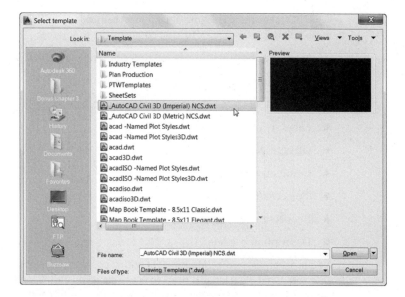

> **TEMPLATES USED IN THIS BOOK**
>
> Throughout this book, when you start a file from scratch, you will use one of the templates that come with Civil 3D when you install it. The templates that come with Civil 3D may not be exactly what you want initially, but they are a great starting point when you are customizing your projects.
>
> The following Civil 3D–specific templates install with Civil 3D:
>
> ♦ _AutoCAD Civil 3D (Imperial) NCS.dwt
>
> ♦ _AutoCAD Civil 3D (Metric) NCS.dwt

Starting New Projects

When you start a project with the correct template (DWT file), the repetitive task of defining the basic framework for your drawing is already completed. Base AutoCAD DWT files contain the following:

- Unit type (architectural or decimal) and insertion scale (meters or feet)
- Layers and their respective linetypes, colors, and other properties
- Text, dimension, and multileader styles
- Layouts and plot setups
- Block definitions

Civil 3D takes the base AutoCAD template and kicks it up several notches. In addition to the items just listed, a Civil 3D template contains the following:

- More specific unit information (international feet, survey feet, or meters)
- Civil object layers
- Ambient settings
- Label styles and formulas (expressions)
- Object styles
- Command settings
- Object naming templates
- Report settings
- Description key sets

You must always start new projects with a proper Civil 3D template. If you receive a drawing from a non–Civil 3D user and need to continue it in Civil 3D, you must import the styles and settings. Without suitable styles and settings, all of your object and label styles will show up with the name Standard, as shown in Figure 1.20. You do not want objects and labels to use the Standard style, as it is the Civil 3D equivalent of drawing on layer 0. Items that use the Standard style appear on layer 0, and will contain the most basic display settings.

FIGURE 1.20
A non–Civil 3D DWG will list all styles as Standard, which is the Civil 3D equivalent to drawing on layer 0.

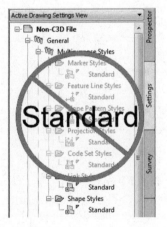

 Real World Scenario

BEST PRACTICES FOR RECEIVING A NON–CIVIL 3D DRAWING

Say someone sends you a drawing that was not done in Civil 3D. Perhaps it was exported from a non-AutoCAD-based product or created in an outdated civil drafting program. You now have the task of creating Civil 3D objects, but making this task even more difficult is that there are no Civil 3D styles present. Perhaps the drawing was created in Civil 3D but your organization's styles look completely different.

The best course of action to take when receiving an outside drawing is to insert it into a blank file that you started with your Civil 3D template. When you insert a drawing, Civil 3D is doing several things to help you:

- The Insert command will detect the units of the incoming drawing and scale it to match your drawing.
- If both drawings have a coordinate system defined, Civil 3D will place the incoming drawing by geographic data.
- All of your styles and settings will stay intact, including a few items that do not get imported using the Import Styles command.

The following exercise walks you through exactly what you need to do in this situation:

1. Start by choosing Application ➢ New ➢ Drawing. Select either `_AutoCAD Civil 3D (Imperial) NCS.dwt` or `_AutoCAD Civil 3D (Metric) NCS.dwt` and click Open.
2. Click the Save icon from the Quick Access toolbar.
3. Save the drawing with the rest of your Mastering Civil 3D files as **Project_0104.dwg**.
4. Go to the Insert tab on the ribbon. From the Block panel, click Insert.
5. Click Browse and locate the file `0104_MysteryFile.dwg` that is part of the dataset for this chapter. Click Open.
6. Be sure that the Insertion Point, Scale, and Rotation check boxes are clear.
7. Select the Explode check box. Your Insert dialog should look like this.

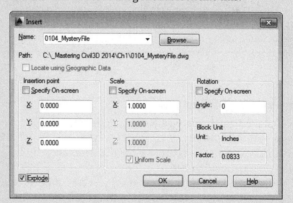

Notice that AutoCAD has picked up the units of the `0104_MysteryFile.dwg` file and is automatically scaling it as needed. If you used the English units template, you will see the conversion scale factor as 0.0833. If you use the metric template drawing, you will see the scale factor as 0.025.

8. Click OK. Double-click your middle mouse wheel to zoom extents.

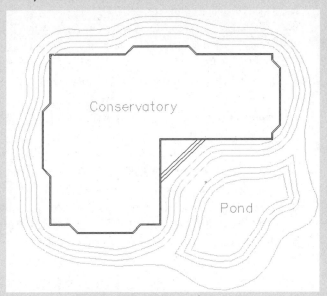

A quick measurement of the north wall of the conservatory building should reveal that it is 200′ (61 m) in length.

At this point you can now work on the drawing in Civil 3D without re-creating any established standards. Everything from the outside source has come in, including blocks, layers, and dimension styles, but they will not override any of your blocks, layers or styles if they happen to have the same name.

Importing Styles

If you have a batch of styles you would like to move between drawings, you can find the Import Styles button on the Styles panel of the Manage tab in the ribbon. When you click the button, you will be prompted to browse for the template (DWT) or drawing (DWG) that contains the styles you are looking for.

You must save the drawing before you can import styles. If you forget, you will be prompted to save the drawing before you can proceed with the import.

Once you select the file whose styles you will import, a dialog box similar to Figure 1.21 appears.

The dialog has the following options:

Import Settings Notice the Import Settings option at the bottom of the dialog. This option is turned on by default. As you look through the list of styles to be imported, you will notice that some items are grayed out and can't be modified.

FIGURE 1.21
Import Civil 3D Styles dialog

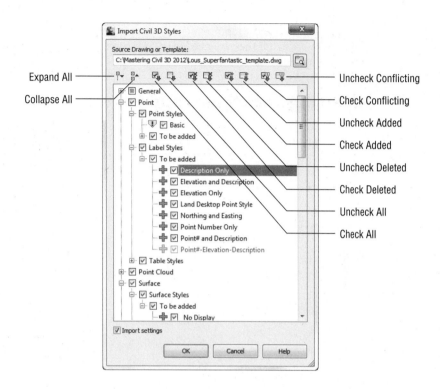

The grayed-out styles represent items in command settings. Styles referred to in command settings must be imported if the Import Settings option is turned on. You will read about command settings in the upcoming section.

Uncheck Conflicting   You will also notice items with a warning symbol (see the Basic point style in Figure 1.21). The warning symbol indicates that there is a style in the current drawing with the same name as a style in the batch to be imported. Use the Uncheck Conflicting button if you do not want styles in the destination drawing to be overwritten. If you leave these items selected, the incoming styles "win." If you are not sure if there is a difference between the styles, pause your cursor over the style name and a tooltip will tell you what (if any) difference exists.

Uncheck Added   Use the Uncheck Added button if you only want styles with the same name to come in. Wherever possible, Civil 3D will release items in the To Be Added categories. In cases where a style is used by a setting, you will not be able to uncheck it unless you do not import settings.

Uncheck Deleted   A style in the current drawing will be deleted if the source drawing does not contain a style with the same name. Use the Uncheck Deleted button to prevent the style from being deleted.

The Import Styles command does not replace the best practice of starting with a proper template. Note that many critical items will not get transferred with this tool. Description key sets, expressions and predefined point groups will not import. For description keys and point groups, dragging and dropping items between drawings will work. Unfortunately, expressions must be re-created if they are needed in additional drawings. Last, pipe network parts lists do not transfer using this tool either.

Creating Basic Lines and Curves

You can draw lines many ways in an AutoCAD-based environment. The tools found on the Draw panel of the Home tab in the ribbon create lines that are the same as those created by the standard AutoCAD Line command. How the Civil 3D lines differ from those created by the regular Line command isn't in the resulting entity but in the process of creating them. Figure 1.22 shows the available line commands.

FIGURE 1.22
Line creation tools

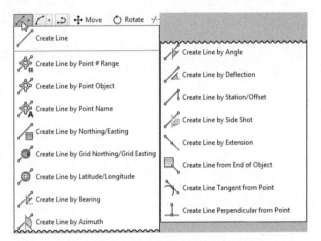

Later on in this chapter you will take a look at *transparent commands*. Transparent commands allow you to draw any object using similar "civil-friendly" techniques.

COGO Line Commands

The next few commands, discussed in the following sections, help you create a line using Civil 3D points and/or coordinate inputs. Each command requires you to specify a Civil 3D point, a location in space, or a typed coordinate input. These line tools are useful when your drawing includes Civil 3D points that will serve as a foundation for linework, such as the edge of pavement shots, wetlands lines, or any other points you'd like to connect with a line.

Create Line Command

The Create Line command on the Draw panel of the Home tab in the ribbon issues the standard AutoCAD Line command. It's equivalent to typing **line** on the command line or clicking the Line tool on the Draw toolbar.

CREATING BASIC LINES AND CURVES | 33

CREATE LINE BY POINT # RANGE COMMAND

The Create Line By Point # Range command prompts you for a point number. You can type in an individual point number, press ↵, and then type in another point number. A line is drawn connecting those two points. You can also type in a range of points, such as **601-607**. Civil 3D draws a line that connects those points in numerical order—from 601 to 607, and so on (see Figure 1.23). The line that is created by this method connects point to point regardless of the description or type of point. Each endpoint will inherit the elevation of the point to which it is created.

FIGURE 1.23
Lines created using 601-607 as input

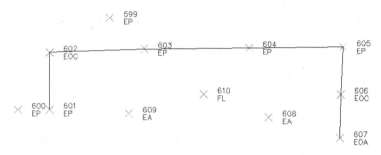

Alternatively, you can enter a list of points such as **601, 603, 610, 605** (Figure 1.24). Civil 3D draws a line that connects the point numbers in the order of input. This approach is useful when your points were taken in a zigzag pattern (as is commonly the case when cross-sectioning pavement) or when your points appear so far apart in the AutoCAD display that they can't be readily identified.

FIGURE 1.24
Lines created using 601, 603, 610, 605 as input

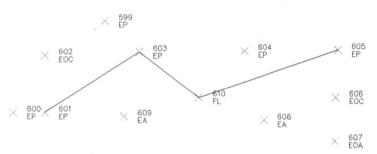

CREATE LINE BY POINT OBJECT COMMAND

The Create Line By Point Object command prompts you to select a point object. To select a point object, locate the desired start point and click any part of the point. This tool is similar to using the regular Line command and a Node object snap (also known as an Osnap); however, it will work on only Civil 3D points. Each endpoint will inherit the elevation of the point to which it is created.

Create Line By Point Name Command

The Create Line By Point Name command prompts you for a point name. A *point name* is a field in point properties, not unlike the point number or description. The difference between a point name and a *point description* is that a point name must be unique. It is important to note that some survey instruments name points rather than number points as is the norm.

To use this command, enter the names of the points you want to connect with linework. Each endpoint will inherit the elevation of the point to which it is created.

Create Line By Northing/Easting and Create Line By Grid Northing/Grid Easting Commands

The Create Line By Northing/Easting and Create Line By Grid Northing/Grid Easting commands let you input northing (y) and easting (x) coordinates as endpoints for your linework. The Create Line By Grid Northing/Grid Easting command requires that the drawing have an assigned coordinate system.

Create Line By Latitude/Longitude Command

The Create Line By Latitude/Longitude command prompts you for geographic coordinates to use as endpoints for your linework. This command also requires that the drawing have an assigned coordinate system. Enter the latitude and longitude as separate entries at the command line using degrees, minutes, and seconds.

> ### Important Notes on Entering Data into Civil 3D
>
> In this chapter, you will get your first taste of keying in data to the command line. Always keep an eye out for what is on your command line (or your tooltip, if dynamic input is on). Sometimes you are asked to confirm an option; other times you are asked for input.
>
> As stated previously, most people prefer to enter angles by degrees, minutes, and seconds rather than decimal degrees. The ambient setting format will affect exactly how Civil 3D displays this information but is more flexible when it comes to your data entry. To enter angles into Civil 3D, you can use a *DD.MMSS* format. To input 15°21'35", you can use **15.2135**. Any numbers beyond four decimal places will be considered decimal seconds. To input 6°5'2", enter **6.0502**. You could also use **6d5′2″** at the command line, but most people find it faster to use the former method.
>
> When entering station values into Civil 3D, it is not necessary to use station notation. In English units, if you are asked to enter station 3+25, you can simply enter **325**. In metric, a station of 0+110 can be entered as **110**. Similarly, a metric station of 2+450 can be entered as **2450**.

Direction-Based Line Commands

The next few commands help you specify the direction of a line. Each of these commands requires you to choose a start point for your line before you can specify the line direction. You can specify your start point by physically choosing a location, using an Osnap, or using one of the point-related line commands discussed earlier.

Many of these line commands require a line or arc and will not work with a polyline. Those commands include Create Line By Sideshot, Create Line By Extension, Create Line From End Of Object, Create Line Tangent From Point, and Create Line Perpendicular From Point.

CREATE LINE BY BEARING COMMAND

The Create Line By Bearing command will likely be one of your most frequently used line commands.

This command prompts you for a start point, followed by prompts to input the Quadrant, Bearing, and Distance values. You can enter values on the command line for each input, or you can graphically choose inputs by picking them onscreen. The glyphs at each stage of input guide you in any graphical selections. After creating one line, you can continue drawing lines by bearing, or you can switch to any other method by clicking one of the other Line By commands on the Draw panel (see Figure 1.25).

FIGURE 1.25
The tooltips for a quadrant (top), a bearing (middle), and a distance (bottom)

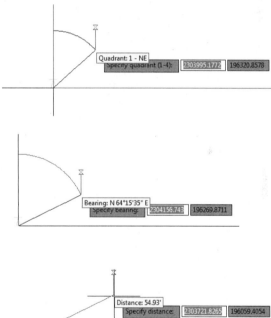

CREATE LINE BY AZIMUTH COMMAND

The Create Line By Azimuth command prompts you for a start point, followed by a north azimuth, and then a distance (Figure 1.26).

FIGURE 1.26
The tooltip for the Create Line By Azimuth command

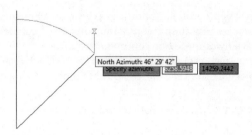

CREATE LINE BY ANGLE COMMAND

The Create Line By Angle command prompts you for a start point, an ending point to establish a backsight direction, a turned angle, and then a distance (Figure 1.27). By default this command assumes the angle-right surveying convention (clockwise from starting direction). However, an option to turn counterclockwise is offered at the command line if needed.

FIGURE 1.27
The tooltip for the Create Line By Angle command

CREATE LINE BY DEFLECTION COMMAND

By definition, a *deflection angle* is the amount of angular deviation (usually measured clockwise) from a backsight direction. In other words, 180 added to a given deflection angle would be an equivalent angle-right. When you use the Create Line By Deflection command, the command line and tooltips prompt you for a deflection angle followed by a distance (Figure 1.28).

FIGURE 1.28
The tooltip for the Create Line By Deflection command

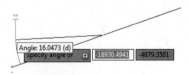

CREATE LINE BY STATION/OFFSET COMMAND

To use the Create Line By Station/Offset command, you must have a Civil 3D Alignment object in your drawing. The line created from this command allows you to start and end a line on the basis of a station and offset from an alignment.

You're prompted to choose the alignment and then input a station and offset value. The line *begins* at the station and offset value. You will not see a line form until at least two points are specified.

When prompted for the station, you're given a tooltip that tracks your position along the alignment, as shown in Figure 1.29. You can graphically choose a station location by clicking it in the drawing. Alternatively, you can enter a station value on the command line.

FIGURE 1.29
The Create Line By Station/Offset command provides a tooltip for you to track stationing along the alignment.

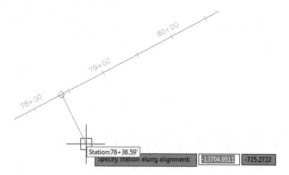

Once you've selected the station, you're given a tooltip that is locked on that particular station and tracks your offset from the alignment (see Figure 1.30). You can graphically choose an offset by clicking the station in the drawing, or you can type an offset value on the command line. A negative value for offset indicates an offset left of the alignment.

FIGURE 1.30
The Create Line By Station/Offset command provides a tooltip that helps you track the offset from the alignment.

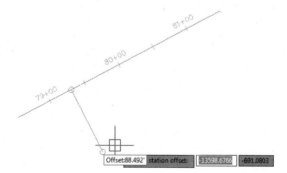

CREATE LINE BY SIDE SHOT COMMAND

The Create Line By Side Shot command starts by asking you to select a line (or two points) that will be the backsight creating a new line. After you select the first line, you will see a yellow glyph indicating your occupied point (see Figure 1.31). By default, Civil 3D is looking for an angle-right and distance to establish the first point of the new line. However, you can follow the command prompts to change the angle entry to bearing, deflection, or azimuth, if needed. You can also use the command line options to change to counterclockwise.

FIGURE 1.31
The tooltip for the Create Line By Side Shot command tracks the angle, bearing, deflection, or azimuth of the side shot.

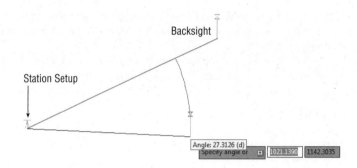

Once you have entered data for the first point, you will be asked a second time for an angle and a distance to place the second point of the new line.

CREATE LINE BY EXTENSION COMMAND

The Create Line By Extension command is similar to the AutoCAD Lengthen command. This command allows you to add length to a line or specify a desired total length of the line. Figure 1.32 shows the summary report that will pop up indicating the changes made to the line.

FIGURE 1.32
The Create Line By Extension command provides a summary of the changes to the line.

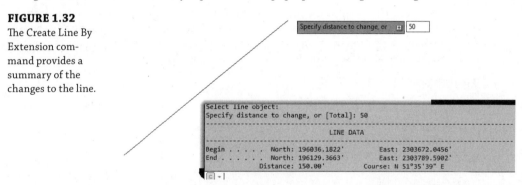

The advantage of the Create Line By Extension command over simply using the Lengthen command is the summary report that appears at the command line. The summary report shown in Figure 1.33 shows the same beginning coordinate as in Figure 1.32 but a different end coordinate, resulting in a total length of 100′ (30.5 m).

FIGURE 1.33
The summary report on a line where the command specified a total distance

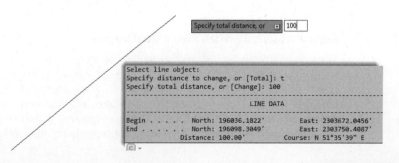

Create Line From End Of Object Command

The Create Line From End Of Object command lets you draw a line tangent to the end of a line or arc (but not polyline) of your choosing as shown in Figure 1.34.

FIGURE 1.34
The Create Line From End Of Object command lets you add a tangent line to the end of an arc.

Create Line Tangent From Point Command

The Create Line Tangent From Point command is similar to the Create Line From End Of Object command, but Create Line Tangent From Point allows you to choose a point of tangency that isn't the endpoint of the line or arc. Use the Nearest or Midpoint Osnap to pick a point on the line. If you pick an endpoint, the command behaves the same as Create Line From End Of Object.

Create Line Perpendicular From Point Command

Using the Create Line Perpendicular From Point command, you can specify that you'd like a line drawn perpendicular to any point of your choosing along a line or arc. In the example shown in Figure 1.35, a line is drawn perpendicular to the endpoint of the arc.

FIGURE 1.35
A perpendicular line is drawn from the endpoint of an arc using the Create Line Perpendicular From Point command.

Re-creating a Deed Using Line Tools

The upcoming exercise will help you apply some of the tools you've learned so far to reconstruct the overall parcel.

When you open up the file for the exercise, you will see a legal description with the following information (the metric file will have different lengths, of course):

```
From the POINT OF BEGINNING at a location of N 186156.65', E 2305474.07'
Thence, S 11° 45' 41.4" E for a distance of 693.77 feet to a point on a line.
Thence, S 73° 10' 54.4" W for a distance of 265.45 feet to a point on a line.
Thence, S 05° 59' 04.4" E for a distance of 185.89 feet to a point on a line.
Thence, S 45° 55' 02.4" W for a distance of 68.73 feet to a point on a line.
Thence, N 06° 04' 37.0" W for a distance of 217.80 feet to a point on a line.
```

```
Thence, N 73° 21' 22.5" E for a distance of 4.22 feet to a point on a line.
Thence, N 06° 04' 51.6" W for a distance of 200.14 feet to a point on a line.
Thence, S 87° 32' 10.4" W for a distance of 121.22 feet to a point on a line.
Thence, N 02° 25' 32.2" W for a distance of 168.91 feet to a point on a line.
Thence, N 15° 38' 57.5" E for a distance of 283.16 feet to a point on a line.
Thence, N 06° 19' 22.4" W for a distance of 79.64 feet to a point  on a line.
N 76° 55' 49.8" E a distance of 250.00 feet Returning to the POINT OF BEGINNING;
Containing 5.58 acres (more or less)
```

Follow these steps (Note that the legal description for metric users is located in the DWG file as text):

1. Open the 0105_Legal.dwg (0105_Legal_METRIC.dwg) file, which you can download from this book's web page at www.sybex.com/go/masteringcivil3d2014.

2. Turn off dynamic input by pressing F12 or by toggling the icon off at the status bar.

3. From the Draw panel on the Home tab in the ribbon, select the Line drop-down and choose the Create Line By Bearing command.

4. At the Select first point: prompt, use endpoint object snap to snap to the arrow head of the note indicating POB.

5. At the >>Specify quadrant (1-4): prompt, enter **2** to specify the SE quadrant, and then press ↵.

6. At the >>Specify bearing: prompt, enter **11.45414**, and press ↵.

7. At the >>Specify distance: prompt, enter **693.77'** (**211.4615** m), and press ↵.

8. Repeat steps 5 through 7 for the rest of the courses and distances.

9. Press Esc to exit the Create Line By Bearing command.

   The finished linework should look like Figure 1.36.

10. Save your drawing. You'll need it for the next exercise.

Creating Curves

Curves are an important part of surveying and engineering geometry. The curves you create in this chapter are no different from AutoCAD arcs. What make the curve commands unique from the basic AutoCAD commands isn't the resulting arc entity but the inputs used to draw the arc. Civil 3D wants you to provide directions to the arc commands using land surveying terminology rather than with generic Cartesian parameters. Figure 1.37 shows the Create Curves menu options.

FIGURE 1.36
The finished linework

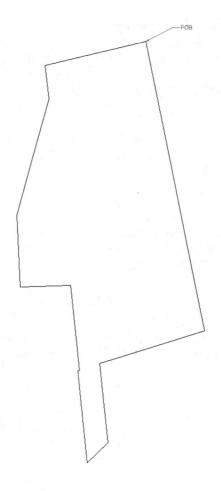

FIGURE 1.37
Create Curves commands

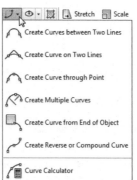

Standard Curves

When re-creating legal descriptions for roads, easements, and properties, users such as engineers, surveyors, and mappers often encounter a variety of curves. Although standard AutoCAD arc commands could draw these arcs, the AutoCAD arc inputs are designed to be generic to all industries. The following curve commands have been designed to provide an interface that more closely matches land surveying, mapping, and engineering language.

CREATE CURVE BETWEEN TWO LINES COMMAND

The Create Curve Between Two Lines command is much like the standard AutoCAD Fillet command, except that you aren't limited to a radius parameter. The command draws a curve that is tangent to two lines of your choosing. This command also trims or extends the original tangents so their endpoints coincide with the curve endpoints. The lines are trimmed or extended to the resulting PC (point of curve, which is the beginning of a curve) and PT (point of tangency, or the end of a curve). You may find this command most useful when you're creating foundation geometry for road alignments, parcel boundary curves, and similar situations.

The command prompts you to choose the first tangent and then the second tangent. The command line gives the following prompt:

```
Select entry [Tangent/External/Degree/Chord/Length/Mid-Ordinate/
miN-dist/Radius]<Radius>:
```

Pressing ↵ at this prompt lets you input your desired radius. As with standard AutoCAD commands, pressing T changes the input parameter to Tangent, pressing C changes the input parameter to Chord, and so on.

As with the Fillet command, your inputs must be geometrically possible. For example, your two lines must allow for a curve of your specifications to be drawn while remaining tangent to both. Figure 1.38 shows two lines with a 25' (7.6 m) radius curve drawn between them. Note that the tangents have been trimmed so their endpoints coincide with the endpoints of the curve. If either line had been too short to meet the endpoint of the curve, that line would have been extended.

FIGURE 1.38
Two lines using the Create Curve Between Two Lines command

CREATE CURVE ON TWO LINES COMMAND

The Create Curve On Two Lines command is identical to the Create Curve Between Two Lines command, except that the Create Curve On Two Lines command leaves the chosen tangents intact. The lines aren't trimmed or extended to the resulting PC and PT of the curve.

Figure 1.39, for example, shows two lines with a 25′ (7.6 m) radius curve drawn on them. The tangents haven't been trimmed and instead remain exactly as they were drawn before the Create Curve On Two Lines command was executed.

FIGURE 1.39
The original lines stay the same after you execute the Create Curve On Two Lines command.

Create Curve Through Point Command

The Create Curve Through Point command lets you choose two tangents for your curve followed by a pass-through point. This tool is most useful when you don't know the radius, length, or other curve parameters but you have two tangents and a target location. It isn't necessary that the pass-through location be a true point object; it can be any location of your choosing.

This command also trims or extends the original tangents so their endpoints coincide with the curve endpoints. The lines are trimmed or extended to the resulting PC and PT of the curve.

Figure 1.40, for example, shows two lines and a desired pass-through point. Using the Create Curve Through Point command allows you to draw a curve that is tangent to both lines and that passes through the desired point. In this case, the tangents have been trimmed to the PC and PT of the curve.

FIGURE 1.40
The first image shows two lines with a desired pass-through point. In the second image, the Create Curve Through Point command draws a curve that is tangent to both lines and passes through the chosen point.

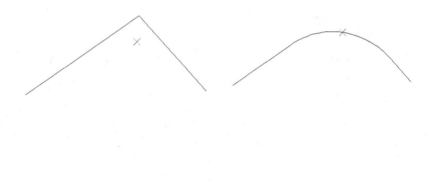

Create Multiple Curves Command

The Create Multiple Curves command lets you create several curves that are tangentially connected. The resulting curves have an effect similar to an alignment spiral section. This command can be useful when you are re-creating railway track geometry based on field-survey data.

The command prompts you for the two tangents. Then, the command line prompts you as follows:

```
Enter Number of Curves:
```

The command allows for up to 10 curves between tangents.

One of your curves must have a flexible length that's determined on the basis of the lengths, radii, and geometric constraints of the other curves. Curves are counted clockwise, so enter the number of your flexible curve:

```
Enter Floating Curve #:
Enter the length and radii for all your curves:
Enter curve 1 Radius:
Enter curve 1 Length:
```

The floating curve number will prompt you for a radius but not a length.

As with all other curve commands, the specified geometry must be possible. If the command can't find a solution on the basis of your length and radius inputs, it returns no solution (see Figure 1.41).

FIGURE 1.41
Two curves were specified with the #2 curve designated as the floating curve.

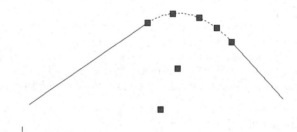

CREATE CURVE FROM END OF OBJECT COMMAND

The Create Curve From End Of Object command enables you to draw a curve tangent to the end of your chosen line or arc.

The command prompts you to choose an object to serve as the beginning of your curve. You can then specify a radius and an additional parameter (such as Delta or Length) for the curve or the endpoint of the resulting curve chord (see Figure 1.42).

FIGURE 1.42
A curve, with a 25′ (7.6 m) radius and a 30′ (9.1 m) length, drawn from the end of a line

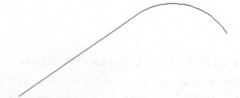

CREATE REVERSE OR COMPOUND CURVE COMMAND

The Create Reverse Or Compound Curve command allows you to add additional curves to the end of an existing curve. Reverse curves are drawn in the opposite direction (i.e., a curve to the right tangent to a curve to the left) from the original curve to form an S shape. In contrast, compound curves are drawn in the same direction as the original curve (see Figure 1.43). This tool can be useful when you are re-creating a legal description of a road alignment that contains reverse and/or compound curves.

FIGURE 1.43
A tangent and curve before adding a reverse or compound curve (left); a compound curve drawn from the end of the original curve (right).

THE CURVE CALCULATOR

Sometimes you may not have enough information to draw a curve properly. Although many of the curve-creation tools assist you in calculating the curve parameters, you may find an occasion where the deed you're working with is incomplete.

The Curve Calculator found in the Curves drop-down on the Draw panel helps you calculate a full collection of curve parameters on the basis of your known values and constraints.

The Curve Calculator can remain open on your screen while you're working through commands. You can send any value in the Calculator to the command line by clicking the button next to that value (see Figure 1.44).

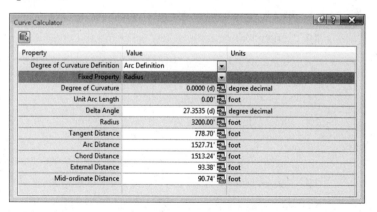

FIGURE 1.44
The Curve Calculator

The button at the upper left of the Curve Calculator allows you to pick an existing arc in the drawing, and the drop-down menu in the Degree Of Curve Definition selection field allows you to choose whether to calculate parameters for an arc or a chord definition.

The drop-down menu in the Fixed Property selection field also gives you the choice of fixing your radius or delta angle when calculating the values for an arc or a chord, respectively. The parameter chosen as the fixed value is held constant as additional parameters are calculated.

To send any value in the Curve Calculator to the command line, use the button next to that value. This ability is most useful while you're active in a curve command and would like to use a certain parameter value to complete the command.

To test your new line and curve knowledge, follow these steps. (You do not need to have completed the previous exercises to complete this one.)

1. Continue working in the file `0105_Legal.dwg` (`0105_Legal_METRIC.dwg`).
2. Start the Line By Northing/Easting command.
3. For Northing, enter **184898.42** (**56357.038** for metric users) and press ↵.
4. For Easting, enter **2305136.46** (**702605.593** for metric users) and press ↵.

   You will not see any graphic indication that the line has been started. You will not see the segment until after you complete step 5.

5. You will again be prompted for Northing and Easting. Using the same procedure you used in step 2, type in the endpoint of the line at Northing = **185059.94** (**56406.270** for metric users), Easting = **2305413.52** (**702690.041** for metric users).
6. Press Esc once to end the Northing and Easting entry. Press Esc again to exit the line command.
7. Start the Create Curve From End Of Object command.
8. You are then prompted to select the Line or Arc object. Pan to the left, and click on the east end of the line you created in steps 2–5.

   You will then be prompted to select the entry.

9. Press ↵ to select the default Radius option.
10. Enter a Radius value of **550'** (**167.6** m), and press ↵.

    Next you will see the `Select entry [Tangent/Chord/Delta/Length/External/Mid-ordinate] <Length>:` prompt.

11. Enter **D** (for Delta) and press ↵.
12. Enter a delta angle of **40** and press ↵. (Note: The degree symbol is not needed in angular entry.)

    You should now have an arc tangent to the first segment.

13. Return to the Curves menu and select Create Reverse Or Compound Curve.
14. Select the arc you created in the previous steps.
15. When prompted, `Select [Compound/Reverse] <Compound>:`.

16. Press **R** then ↵ to specify a reverse curve.
17. Enter a radius of **630′** (**192.0** m) and press ↵.
18. When you see the prompt `Select entry [Tangent/Chord/Delta/Length/External/Mid-ordinate] <Length>:`, press ↵ to accept the default.
19. Enter a length of curve value of **400′** (**121.9** m), and press ↵.
20. For the last tangent segment, return to the Lines menu and select Create Line From End Of Object.
21. Select the east end of the second arc you created in this exercise.
22. At the `Specify Distance:` prompt, enter **150′** (**45.7** m) and press ↵.

Your completed lines and arcs will look like Figure 1.45, and they will be located just south of the property you entered in the previous exercise.

FIGURE 1.45
Lines and arcs of the completed exercise

Best Fit Entities

AutoCAD Civil 3D provides many tools for relating surveyed information to a graphic. Best fit tools like the ones in the following sections perform analysis on different types of irregular geometry to create uniform shapes (see Figure 1.46). Similar tools are available for alignment creation, as discussed in chapter 6, "Alignments."

FIGURE 1.46
The Create Best Fit Entities menu options

Create Best Fit Line Command

The Create Best Fit Line command under the Best Fit drop-down on the Draw panel takes a series of Civil 3D points, AutoCAD points, entities, or drawing locations and draws a single best-fit line segment from this information. In Figure 1.47, the Create Best Fit Line command draws a best-fit line through a series of points that aren't quite collinear. The best-fit line will adjust as more points are selected.

FIGURE 1.47
A preview line drawn through points that are not collinear

After you press Enter to complete the selection, a Panorama window appears with a regression data chart showing information about each point in the selection, as shown in Figure 1.48.

FIGURE 1.48
The Panorama window lets you optimize your best fit.

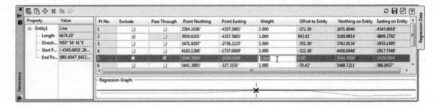

This interface allows you to optimize your best fit by adding more points, excluding points, selecting the check box in the Pass Through column to force one of your points on the line, or adjusting the value under the Weight column.

CREATE BEST FIT ARC COMMAND

The Create Best Fit Arc command under the Best Fit drop-down works the same as the Create Best Fit Line command, except that the resulting entity is a single arc segment as opposed to a single line segment (see Figure 1.49).

FIGURE 1.49
Preview of the curve created by best fit

CREATE BEST FIT PARABOLA COMMAND

The Create Best Fit Parabola command under the Create Best Fit Entities option is similar to the line and arc commands in how it works. After you select this command, the Parabola By Best Fit dialog appears (see Figure 1.50).

FIGURE 1.50
The Parabola By Best Fit dialog

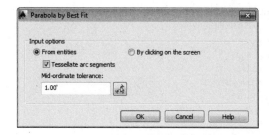

You can select inputs from entities (such as lines, arcs, polylines, or profile objects) or by clicking onscreen. The command then draws a best-fit parabola on the basis of this information. In Figure 1.51, the shots were represented by AutoCAD points; more points were added by selecting the By Clicking On The Screen option and using the Node Osnap to pick each point.

FIGURE 1.51
The best-fit preview line changes as more points are picked.

Once you've selected your points, a Panorama window (shown in Figure 1.52) appears, showing information about each point you chose. Also note the information in the right pane regarding K-value, curve length, grades, and so forth. You can optimize your K-value, length, and other values by adding more points, selecting the check box in the Pass Through column to force one of your points on the line, or adjusting the value under the Weight column.

FIGURE 1.52
The Panorama window lets you make adjustments to your best-fit parabola.

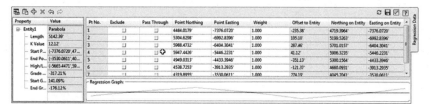

Attach Multiple Entities

The Attach Multiple Entities command (found on the Home tab in the ribbon and the extended Draw panel pull-down) is a combination of the Line From End Of Object command and the Curve From End Of Object command. Each entity created with this tool is tangent to the previous segment. Using this command saves you time because you don't have to switch between the Line From End Of Object command and the Curve From End Of Object command (see Figure 1.53).

FIGURE 1.53
The Attach Multiple Entities command draws a series of lines and arcs so that each segment is tangent to the previous one.

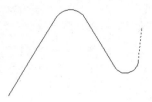

Adding Line and Curve Labels

Although most robust labeling of site geometry is handled using parcel or alignment segment labels, limited line- and curve-annotation tools are available in Civil 3D. The line and curve labels are composed much the same way as other Civil 3D labels, with marked similarities to parcel and alignment segment labels.

Our next exercise leads you through labeling the deed you entered earlier in this chapter:

1. Continue working in the file `0105_Legal.dwg` (`0105_Legal_METRIC.dwg`). If you did not complete the previous exercises, use `0105_Legal_Complete.dwg` (`0105_Legal_METRIC_Complete.dwg`) from the book's web page.

2. On the Annotate tab of the ribbon, select Labels And Tables ➢ Add Labels.

   The Add Labels dialog appears.

3. Choose Line And Curve from the Feature drop-down menu.

   Your Add Labels dialog will look similar to Figure 1.54.

FIGURE 1.54
The Add Labels dialog, with Label Type set to Multiple Segment

4. Choose Multiple Segment from the Label Type drop-down menu.

   The Multiple Segment option places the label at the midpoint of each selected line or arc.

5. Confirm that Line Label Style is set to Bearing Over Distance and that Curve Label Style is set to Distance-Radius And Delta (as shown in Figure 1.54).

6. Click the Add button.

7. At the Select Entity: prompt, select each line from the deed and the lines and arcs that you drew in the previous exercises.

   A label appears on each entity at its midpoint, as shown in Figure 1.55.

FIGURE 1.55
The labeled linework

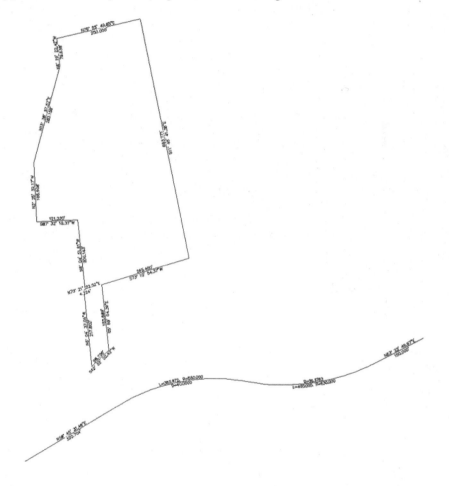

8. Save the drawing.

Using Transparent Commands

You might be surprised to know that you have already been using a form of the transparent commands if you used the Line By Point # Range command or the Line By Bearing command.

Transparent commands are "helper" commands that make base AutoCAD input more surveyor- and civil designer-friendly. Transparent commands can only be used when another command is in progress. They are "commands within commands" that allow you to choose your input based on what information you have.

For example, you may have a plat whose information you are trying to input into AutoCAD. You may need to insert a block at a specific Northing and Easting. While the Line command has a built-in option for Northing Easting, the Insert command does not, and a transparent command will help you. In this case, you would start the Insert command with the Specify On Screen option checked. Before you click to place the block in the graphic, click the Northing Easting transparent command. The command line will then walk you through placing the block at your desired location.

While a transparent command is active, you can press Esc once to leave the transparent mode but stay active in your current command. You can then choose another transparent command if you'd like. For example, you can start a line using the Endpoint Osnap, activate the Angle Distance transparent command, draw a line-by-angle distance, and then press Esc, which takes you out of angle-distance mode but keeps you in the line command. You can then draw a few more segments using the Point Object transparent command, press Esc, and finish your line with a Perpendicular Osnap.

You can activate the transparent commands using keyboard shortcuts or the Transparent Commands toolbar. Be sure you include the Transparent Commands toolbar (shown in Figure 1.56) in all your Civil 3D and survey-oriented workspaces.

FIGURE 1.56
The Transparent Commands toolbar

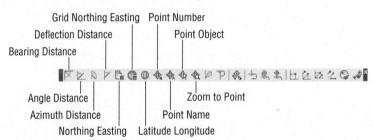

The six profile-related transparent commands will be covered in Chapter 7.

Standard Transparent Commands

Certification Objective

The transparent commands shown in Table 1.2 behave the same as their like-named counterparts from the Draw panel (discussed earlier in this chapter). The difference is that you can call up these transparent commands in any AutoCAD or Civil 3D draw command, such as a line, polyline, alignment, parcel segment, feature line, or pipe-creation command.

TABLE 1.2: The transparent commands used in plan view

TOOL ICON	MENU COMMAND	KEY-IN	TOOL ICON	MENU COMMAND	KEY-IN
	Angle Distance	'AD		Latitude Longitude	'LL
	Bearing Distance	'BD		Point Number	'PN
	Azimuth Distance	'ZD		Point Name	'PA
	Deflection Distance	'DD		Point Object	'PO
	Northing Easting	'NE		Zoom To Point	'ZTP
	Grid Northing Grid Easting	'GN		Side Shot	'SS
				Station Offset	'SO

Matching Transparent Commands

You may have construction or other geometry in your drawing that you'd like to match with new lines, arcs, circles, alignments, parcel segments, or other entities.

While actively drawing an object that has a radius parameter, such as a circle, an arc, an alignment curve, or a similar object, you can choose the Match Radius transparent command and then select an object in your drawing that has your desired radius. Civil 3D draws the resulting entity with a radius identical to that of the object you chose during the command.

The Match Length transparent command works the same as the Match Radius transparent command except that it matches the length parameter of your chosen object.

Some people find working with transparent commands awkward at first. It feels strange to move your mouse away from the drawing area and click an icon while another command is in progress. If you prefer, each transparent command has a corresponding command line entry.

Try a few transparent commands to get the feel for how they operate:

1. Open the drawing 0106_Transparent_Commands.dwg (0106_Transparent_Commands_METRIC.dwg).

2. On the Home tab of the ribbon, click the Draw flyout ➢ 3DPOLY.

3. From the Transparent Commands toolbar, click the Point Number Transparent command.

4. At the Enter Point number prompt, type **100-112** and press ↵.

   The 3D polyline will connect the points in the drawing.

5. Press Esc once.

   Notice that you are now back in the 3D polyline command with no transparent commands active.

6. Press Esc a second time.

   This will exit the 3D polyline command.

7. Choose Insert ➢ Block ➢ Insert.

8. From the Name pull-down, locate the block Fire Hydrant 01.

   You do not need to browse because this block is already defined in this example drawing.

9. Check the Specify On-Screen option for Insertion Point.

10. Uncheck the Explode option.

11. Verify that your settings match what is shown in Figure 1.57, and click OK.

FIGURE 1.57
Using the Insert command to place a block

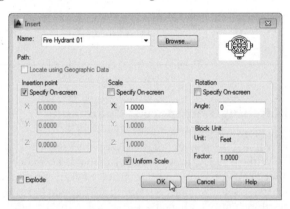

12. At the Specify insertion point: prompt, select the Station Offset transparent command.

    You are then prompted to select an alignment.

13. Click the alignment to the south of the site. (Hint: Be sure to click the green line rather than the labels.)

14. At the Specify station along alignment: prompt, enter **450′ (175 m)** and press ↵.

15. At the Specify station offset: prompt, enter **-30′ (-10 m)** and press ↵.

You should now see the fire hydrant symbol at the specified station, on the left of the alignment.

The Bottom Line

Find any Civil 3D object with just a few clicks. By using Prospector to view object data collections, you can minimize the panning and zooming that are part of working in a CAD program. When common subdivisions can have hundreds of parcels or a complex corridor can have dozens of alignments, jumping to the desired one nearly instantly shaves time off everyday tasks.

Master It Open `0103_Example.dwg` (`0103_Example_METRIC.dwg`) from www.sybex.com/go/masteringcivil3d2014, and find parcel number 6 without using any AutoCAD commands or scrolling around on the drawing screen. (Hint: Take a look at Figure 1.5.)

Modify the drawing scale and default object layers. Civil 3D understands that the end goal of most drawings is to create hard-copy construction documents. When you set a drawing scale, Civil 3D removes much of the mental gymnastics that other programs require when you're sizing text and symbols. When you set object layers for the entire drawing, Civil 3D makes uniformity of drawing files easier than ever to accomplish.

Master It Change the Annotation scale in the model tab of `0103_Example.dwg` from the 50-scale drawing to a 100-scale drawing. (For metric users: Use `0103_Example_METRIC.dwg` and change the scale from 1:250 to 1:1000.)

Navigate the ribbon's contextual tabs. As with AutoCAD, the ribbon is the primary interface for accessing Civil 3D commands and features. When you select an AutoCAD Civil 3D object, the ribbon displays commands and features related to that object. If several object types are selected, the Multiple contextual tab is displayed.

Master It Continue working in the file `0103_Example.dwg` (`0103_Example_METRIC.dwg`). It is not necessary to have completed the previous exercise to continue. Using the ribbon interface, access the Alignment properties for QuickStart Alignment and rename it Existing CL.

Create a curve tangent to the end of a line. It's rare that a property stands alone. Often, you must create adjacent properties, easements, or alignments from their legal descriptions.

Master It Open the drawing `MasterIt0101.dwg` (`MasterIt0101_METRIC.dwg`). Create a curve tangent to the east end of the line labeled in the drawing. The curve should meet the following specifications:

- Radius: 200.00′ (60 m)
- Arc Length: 66.580′ (20 m)

Label lines and curves. Although converting linework to parcels or alignments offers you the most robust labeling and analysis options, basic line- and curve-labeling tools are available when conversion isn't appropriate.

Master It Add line and curve labels to each entity created in `MasterIt0101.dwg` or `MasterIt0101_METRIC.dwg`. It is recommended that you complete the previous exercise so you will have a curve to work with. Choose a label that specifies the bearing and distance for your lines and length, radius, and delta of your curve.

Chapter 2

Survey

The AutoCAD® Civil 3D® software supports a collaborative workflow in many aspects of the design process, but especially in the survey realm. Accurate data starts outdoors. A survey that has been consistently and correctly coded in the field can save hours of drafting time. Surveyors can collect line information such as swales, curbs, or even pavement markings and communicate this digitally to data collectors.

Civil 3D can often eliminate the need for third-party survey software because it can download and process survey data directly from a data collector. To enter data in a manner that is easily digested by Civil 3D, your survey process should incorporate the information from this chapter.

In this chapter, you will learn to:

- Properly collect field data and import it into Civil 3D
- Set up description key and figure databases
- Translate surveys from assumed coordinates to known coordinates
- Perform traverse analysis

Setting Up the Databases

Before any project-specific data is imported, there is a bit of initial setup that will improve the translation between the field and the office. For this chapter you will need to see your Survey tab in Toolspace. If you do not see this tab, click the Survey button on the Home tab ➢ Palettes panel.

Your survey database defaults, equipment database, linework code set, and the figure prefix database should be in place before you import your first survey. You can find the location of these files by going to Toolspace ➢ Survey tab and clicking the Survey User Settings button in the upper-left corner. The dialog shown in Figure 2.1 opens.

FIGURE 2.1
Survey User Settings dialog

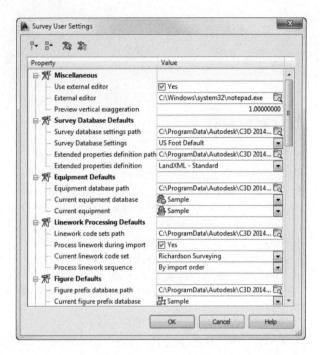

The Survey Database Settings Path points to the locations of the Equipment Database, Figure Prefix Database, and Linework Code Sets. Note that these files are separate from the Civil 3D template. These databases are separate files that reside in `C:\ProgramData\Autodesk\C3D 2014\enu\Survey\` by default. It is common practice to place these files on a network server so your organization can share them. These paths can be set during the software installation process. Otherwise, change the paths in the Survey User Settings dialog on each individual computer and they will "stick" regardless of which drawing you have open.

Survey Database Defaults

Every survey database that is created for a project has settings that you will need to examine and verify. To create a new survey database for a project, click the Import Survey Data button from the Home tab of the ribbon or right-click Survey Databases from Toolspace ➤ Survey tab and select New Local Survey Database.

Once the new database is created, you will be able to edit its properties, as outlined in the upcoming exercise. The Survey Database Settings dialog contains these settings:

Units   Most likely, the Units setting is the only one you will ever need to modify in the Survey Database Settings dialog. This section is where you set your master coordinate zone for the database. Potentially, your drawing and your incoming survey data may have different coordinate systems. If you insert any information in the database into a drawing with a different coordinate zone, the program will automatically translate that data to the drawing coordinate zone (upon initial import only). Your coordinate zone units will lock the distance units in the Units section. Although usually not necessary, you can also set the angle, direction, temperature, and pressure specific to the survey database here.

FIGURE 2.2
Survey Database Settings dialog

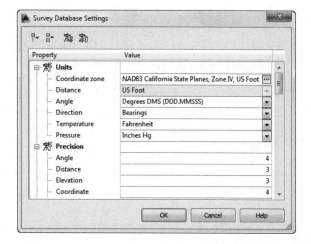

Precision This section is where you define and store the precision information of angles, distance, elevation, coordinates, and latitude and longitude specific to the database. Note that this affects display precision for the survey interface and is independent from label precision and drawing precision set in the Drawing Settings dialog discussed in Chapter 1, "The Basics."

Measurement Type Defaults This section lets you tell Civil 3D what type of information to expect when importing survey data from a file. The information can be measurement types, such as angle type, distance type, vertical type, and target type.

Measurement Corrections This section is used to define the methods (if any) for correcting measurements. You will probably not need to change anything in this section because most data collectors will have processed this for you.

Traverse Analysis Defaults This section is where you choose what type(s) of traverse analysis you want to perform. You can control the method you wish to use and required precision and tolerances for each. There are four types of 2D traverse analysis methods: Compass Rule, Transit Rule, Crandall Rule, and Least Squares Analysis.

There are three potential types of 3D traverse analyses: Length Weighted Distribution, Equal Distribution, and Least Squares Vertical. Vertical options for Least Squares will be available only if it is first set as Horizontal Adjustment Method. Of course, you can always choose None to omit that calculation from the analysis.

Least Squares Analysis Defaults If you are performing a Least Squares analysis, you must specify 2-Dimensional or 3-Dimensional adjustment type. Use 3-Dimensional if you are performing both horizontal and vertical Least Squares adjustment.

Survey Command Window In the rare event that you'll need it, the Survey Command window is the interface for manual survey tasks and for running survey batch files. This section lets you define the default settings for this window.

Error Tolerance Set tolerances for the survey database in this section. If you perform an observation more than one time and the tolerances established here are not met, an error will appear in the Survey Command window and you will be asked what action you want to take.

Extended Properties   You may work with LandXML files that contain information beyond traditional "P,N,E,Z,D" data. If this is the case, you will want to turn the Extended Properties options to Yes. Create New Definitions Automatically will add extended properties to your survey database and populate the fields from the imported file. Display Warnings For Missing Required Fields will display the Panorama if there is missing information in the imported file.

Change Reporting   It is a great idea to turn this option to Yes by setting Logging Enabled to Yes. This will create an audit trail of changes to the database that occur after import. The changes to the database are stored in a LOG file (*.log) located in the same directory as your survey database. At any time, you can access the contents of the log by right-clicking the name of the survey database and selecting Display Change Report.

When you first configure your survey database settings, it is a good idea to create a test database for setting the defaults. Because survey database settings are independent of which drawing you are in, you can perform these steps with any file open. To create the test database, follow these steps:

1. In the Toolspace ➢ Survey tab, right-click Survey Databases.
2. Select Set Working Folder.

   Civil 3D will create a working folder to contain your survey database. Ideally, this will be stored in a network location for your organization's projects. For examples in this book, this will be set to your local C drive.

3. Verify that the `C:\Civil 3D Projects` folder is highlighted and click OK.
4. In the Toolspace ➢ Survey tab, right-click Survey Databases and choose New Local Survey Database.
5. Name the new database **Test**, and click OK to continue.
6. Right-click the new Test database and select Edit Survey Database Settings.
7. Set your desired defaults for units, precision, and other options.

8. Click the Export Settings To A File button.
9. Save the settings to the folder specified in the Survey User Settings dialog (see Figure 2.2) as `MySettings.sdb_set` and then click OK.

To delete your test database, you will need to close the database from the Survey tab. To do so, locate the database in Toolspace ➢ Survey tab. Right-click it and select Close Survey Database. Using Windows Explorer, you can then browse to the working folder containing the database and delete it. There is no way to delete a survey database from within the software.

The Equipment Database

The equipment database is where you set up the various types of survey equipment that you are using in the field. Doing so allows you to apply the proper correction factors to your traverse analyses when it is time to balance your traverse. Civil 3D comes with a sample piece of equipment for you to inspect to see what information you will need when it comes time to create your equipment. The Equipment Database Manager dialog provides all the default settings for the sample equipment in the equipment database. On the Survey tab of Toolspace ➢ Equipment

Databases, right-click Sample and click Manage Equipment Database to access this dialog, shown in Figure 2.3.

Figure 2.3 shows the settings for a specific model of total station—the Trimble S8. When you input this data to an equipment database, consult your instrument's datasheet for specifications. The specifics of total station equipment will vary by manufacturer and model.

FIGURE 2.3
Use Equipment Database Manager.

You will want to create your own equipment entries and enter the specifications for your particular total station. Add a new piece of equipment to the database by clicking the plus sign at the top of the Equipment Database Manager window. If you are unsure of the settings to enter, refer to the user documentation that you received when you purchased your total station.

The Figure Prefix Database

The figure prefix database is used to translate descriptions in the field to lines in CAD. These survey-generated lines are called *figures*. If a description matches a listing in the figure prefix database, the figure is assigned the properties and style dictated by the database (see Figure 2.4).

FIGURE 2.4
The Figure Prefix Database Manager

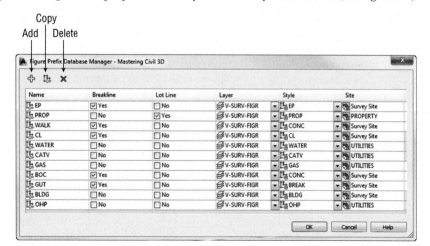

The Figure Prefix Database Manager contains these columns:

Name   The figure name is important because it is used to match lines coded within the imported survey file with an entry in the figure prefix database. For example, if a survey contains linework for a FLOWLINE and the figure prefix database contains a FLOWLINE entry, the match will cause the figure to take on the properties from the table.

Breakline   Placing a check mark in this column will allow you to flag a figure as a breakline. The most powerful use of this setting is in survey queries (discussed later in this chapter). If your survey query pulls in figures that have this setting on, you can readily add them to the definition of a surface model. Even if you don't flag a figure as a breakline, you can still add it to a surface model manually.

Lot Line   This column specifies whether the figure should behave as a parcel segment. If a closed area is formed with figures of this kind, a parcel is formed automatically.

Layer   This column specifies the insertion layer of the figure. If the layer already exists in the drawing, the figure will be placed on that layer. If the layer does not exist in the drawing, the layer will be created and the figure placed on the newly created layer.

Style   This column specifies the style to be used for each figure. Figure styles will override color and linetype if they are set to something other than ByLayer. See the section on linear object styles in Chapter 19, "Object Styles," for more information on this type of style.

Site   This column specifies which site the figures should reside on when inserted into the drawing. As with previous settings, if the site exists in the drawing, the figure will be inserted into that site. If the site does not exist in the drawing, a site will be created with that name and the figure will be inserted into the newly created site.

Remember that figure prefix databases are not drawing specific. The only reason a drawing is needed is to access the styles that the figures will use.

You'll explore these settings in a practical exercise:

1. Open the drawing `0201_FigurePrefix.dwg` or `0201_FigurePrefix_METRIC.dwg`, which you can download from this book's web page, www.sybex.com/go/masteringcivil3d2014.

   This file contains the survey figure styles needed to complete this exercise.

2. In Toolspace ➢ Survey tab, right-click Figure Prefix Databases and select New.

   The New Figure Prefix Database dialog opens.

3. Enter **Mastering Civil 3D** in the Name text box, and click OK to dismiss the dialog.

   If you expand the Figure Prefix Database listing, you will see the Mastering Civil 3D entry.

4. Right-click the newly created Mastering Civil 3D figure prefix database and select Manage Figure Prefix Database.

   The Figure Prefix Database Manager will appear.

5. Select the white + symbol in the upper-left corner of the Figure Prefix Database Manager to create a new figure prefix.

6. Click the Sample name and change the name of the figure prefix to **EP** (for Edge of Pavement).

7. Click the check box next to the word *No* in the Breakline column. This will turn into a Yes, indicating that it will contain the breakline property.

   Leave the box in the Lot Line column unchecked so that the figure will not be treated as a parcel segment.

8. Under the Layer column, select V-SURV-FIGR.

9. Under the Style column, select EP.

10. Under the Site column, leave the name of the site set to Survey Site.

11. Complete the figure prefixes table with the values shown in Table 2.1.

TABLE 2.1: Figure settings

NAME	BREAKLINE	LOT LINE	LAYER	STYLE	SITE
PROP	No	Yes	V-SURV-FIGR	PROP	PROPERTY
CL	Yes	No	V-SURV-FIGR	CL	SURVEY SITE
WATER	No	No	V-SURV-FIGR	WATER	UTILITIES
BOC	Yes	No	V-SURV-FIGR	CONC	SURVEY SITE
GUT	Yes	No	V-SURV-FIGR	BREAK	SURVEY SITE
BLDG	No	No	V-SURV-FIGR	BLDG	SURVEY SITE

Hint: For figures with similar properties, select a figure definition already in the database and then use the Copy Figure Prefix button.

12. Click OK to dismiss the Figure Prefix Database Manager.

You can choose whether lines are automatically formed in the linework code set when they match one of these figure prefixes. When a figure is generated, the site will also be created in the drawing.

The Linework Code Set Database

The linework code set (Figure 2.5) lists what designators are used to start, stop, continue, or add additional geometry to lines. For example, the B code that is typically used to begin a line can be replaced by a code of your choosing, a decimal (.) can be used for a right-turn value, and a minus sign (–) can be used for a left-turn value. Linework code sets allow a survey crew to customize their data collection techniques based on methods used by various types of software not related to Civil 3D.

FIGURE 2.5
The Edit Linework Code Set dialog

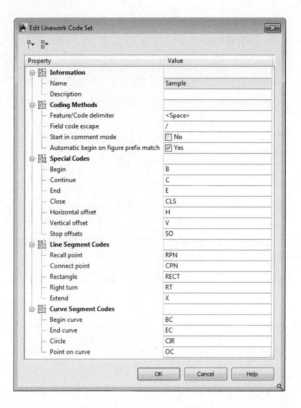

> **LINEWORK OPTION: BEGIN ON FIGURE PREFIX MATCH**
>
> There is a setting in the linework code set called Automatic Begin On Figure Prefix Match (as you can see in Figure 2.5). If you select this option, lines will start when a shot description matches one of your figure prefixes (such as EP or CL in Figure 2.4), with no additional coding needed.
>
> Depending on your survey department's method for picking up linework, Automatic Begin On Figure Prefix Match can be a good thing or a bad thing. If the survey crew has fastidiously collected each line with its own name, such as CL1, CL2, EP1, EP2, and so on with no repeats, it can work well. If they use start and stop codes but reuse line names throughout the survey, it can produce figures that connect unintentionally.

Description Keys: Field to Civil 3D

Certification Objective

Description keys bridge the gap between the field and the office. Unlike the linework code set and figure prefix database, which are each external files, description keys should be created and saved in your Civil 3D template. *The description key set* is a listing of field descriptions and how they should look and behave once they are imported into Civil 3D.

For example, a surveyor may code in **BM** to indicate a benchmark. When the file is imported into Civil 3D either through the survey database or by point import from a text file (as you will do in Chapter 3, "Points"), it will be checked against the description key set. If BM exists in the list as it is in Figure 2.6, then the styles, format, layer, and several other parameters are applied to the point as it is placed in the drawing.

FIGURE 2.6
Description key set

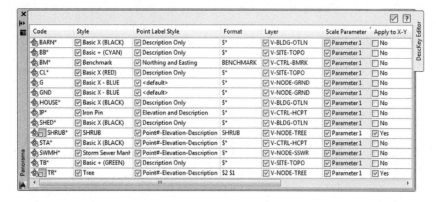

To access your description key set, you will need to expand Toolspace ➢ Settings ➢ Point ➢ Description Key Sets.

The Description Key Editor will open (using the Panorama interface) with the following columns for you to edit:

Code   The raw description or field code entered by the person collecting or creating the points. The code works as an identifier for matching the point with the correct description key. Click inside this field to activate it, and then type your desired code. Wildcards are useful when more information is added to the shot in addition to the field code. Right-click an existing code to copy or create a new description key or delete an existing one.

Codes are case sensitive! The code bm is read differently from BM. A match to the Description Key Set will not be made if the capitalization does not match perfectly.

Style   Style refers to the point style that will be applied to points that meet the code criteria. Check the box, and then click inside the field to activate a point style selection dialog. By default, styles set here will take precedence over styles set elsewhere (unless overridden in a point group). For more information on creating or modifying point styles, see Chapter 19.

Point Label Style   The point label style that will be applied to points that meet the code criteria. Check the box, and then click inside the field to activate a style selection dialog. By default, styles set here will take precedence over styles set elsewhere (unless overridden in a point group). For more information on creating or modifying point label styles, see Chapter 18, "Label Styles."

Format   The Format column can convert a surveyor's shorthand into something that is more drafter friendly. In Civil 3D terms, the Format column converts the *raw description* to the *full description*. The default of $* means the raw description and full description will have the same value. You can also use $+, which means that information after the main description will appear as the full description. In Figure 2.6, Format will convert all codes starting with BM to a full description of BENCHMARK.

If a survey crew is consistent in coding, even fancier formats can be used. The code should always come first, but the crew can use a space to indicate a *parameter*.

Consider this example raw description: TR 30 PINE ELIM. TR is the code, or $0. Parameter 1 is 30, or $1 in the Format field. PINE is the second bit of information after the code referred to as parameter 2, or $2. ELIM is the third item after the code, so it is $3. Based on the example description key set in Figure 2.6, this would translate to a full description of PINE 30. You can have up to nine parameters after the code if your survey crew is feeling verbose. Table 2.2 shows some example formats and the corresponding full description.

TABLE 2.2: Format examples

RAW DESCRIPTION	FORMAT	FULL DESCRIPTION
TR 30 PINE ELIM	Tree	Tree
TR 30 PINE ELIM	$*	TR 30 PINE ELIM
TR 30 PINE ELIM	$0	TR
TR 30 PINE ELIM	$1"	30"
TR 30 PINE ELIM	$2	PINE
TR 30 PINE ELIM	$3	ELIM
TR 30 PINE ELIM	$2 $1	PINE 30
TR 30 PINE ELIM	$+	30 PINE ELIM

Layer Points that match a description key will be inserted on the layer specified here. Click inside this field to activate a layer selection dialog. The layer set here will take precedence over layer defaults set in the point command settings or the point creation tools.

Scale Parameter The Scale parameter is used to tell Civil 3D which bit of information after the code will be used to scale the symbol. By default it is checked, but it won't do anything unless Apply To X-Y is also selected. Once you enable Apply To X-Y (or Apply To Z, which is less frequently used), you can change which parameter contains scale information.

In our example, TREE 30 PINE ELIM, 30 is the Scale parameter.

Fixed Scale Factor Fixed Scale Factor is an additional scale multiplier that can be applied to the symbol size. A common use of Fixed Scale Factor is to convert a field measurement of inches to feet. If the 30 in our example represents a dripline measurement and is meant to be feet, no Fixed Scale Factor is needed. However, if the 30 represents inches (i.e., a trunk diameter), you would need to turn on Fixed Scale Factor and set the value to 0.0833.

Use Drawing Scale In most cases, you will leave this option unchecked. By default, marker styles dictate that they will grow or shrink based on the annotative scale of the drawing. Generally, this setting is not needed unless you want to scale your point symbol based on a parameter in addition to the scale factor.

Apply To X-Y   If you wish to scale symbols based on information in the field code, you need to turn this option on by placing a check mark in the box. This option works with the marker style and the Scale parameter to increase the size of an item to a scale indicated by the surveyor in the raw description.

Apply To Z   In most cases, you will leave this option unchecked. Most marker symbols are 2D blocks, so selecting this option will have no effect on the point. If your marker symbol consists of a 3D block, it will be stretched by the parameter value, which is rarely needed.

Marker Rotate Parameter, Marker Fixed Rotation, Label Rotate Parameter, Label Fixed Rotation, and Rotation Direction   These options are similar to the scale factor parameter except they dictate the rotation of a symbol or label. They are not widely used, however, because it is often more time effective to have the drafter rotate the points in CAD than to have the surveyor key in a rotation. If you would like to rotate the points for readability, the better method to rotate the text is in the point label style (discussed in Chapter 18).

Creating a Description Key Set

Description key sets appear on the Toolspace ➢ Settings tab, under the Point branch. You can create a new description key set by right-clicking the Description Key Sets collection and choosing New, as shown in Figure 2.7.

FIGURE 2.7
Creating a new description key set on the Settings tab of Toolspace

In the resulting Description Key Set dialog, give your description key set a meaningful name, such as your company name, and click OK. You'll create the actual description keys in another dialog.

Creating Description Keys

To enter the individual description key codes and parameters, go to Toolspace ➢ Settings tab ➢ Point, right-click the description key set, as illustrated in Figure 2.8, and select Edit Keys. The DescKey Editor in Panorama appears.

FIGURE 2.8
Editing a description key set

To enter new codes, right-click a row with an existing key in the DescKey Editor, and choose New or Copy from the context menu, as shown in Figure 2.9.

FIGURE 2.9
Creating or copying a description key

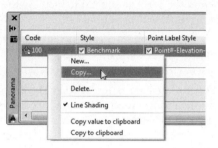

Activating a Description Key Set

Once you've created a description key set, you should verify the settings for your commands so that Civil 3D knows to match your newly created points with the appropriate key.

In Toolspace ➢ Settings tab, expand Point ➢ Commands and right-click CreatePoints. Select Edit Command Settings, as shown in Figure 2.10.

Using Wildcards in Description Key Sets

Civil 3D can use several special characters to do more complex matching or exclusions in your Description Key Set Code field. You can use combinations of these wildcards to build intelligence into the translation between the field and Civil 3D.

- Most commonly, the asterisk (*) acts as a general wildcard and is recommended after most codes. An asterisk before or after your code will allow a surveyor to add more information to a shot without compromising the matching on the Civil 3D end.

- A hash or pound sign (#) signifies a single-digit numeric value 0 through 9.

- An "at" symbol (@) signifies any letter of the alphabet A through Z.

- A question mark (?) signifies any letter of the alphabet or a number.

- A period in a code signifies any character that is not alphanumeric, such as an ampersand (&) or a plus sign (+).

- A tilde (~) is an operator you can use to exclude characters or groups of characters. A tilde inside square brackets tells Civil 3D to not match characters inside the brackets.

- The use of square brackets is supported and allows for additional logic. Multiple characters inside brackets is like telling the description keys, "Any of these will work." You can even use a range such as [1-9] or [A-N]. Add the tilde to the brackets to signify, "None of these will work."

For the @ symbol, hash mark, question mark, and period, a character must be in place; blanks will not form the match.

Code	Example Matches	Will not match
G*	GR G GAS	g X 7
CL#	CL1 CL2 CL9	CL cl1 CLZ
1@	1A 1z	12 1
A.	A+ A! A<	A1 AB A
F?	F7 FZ	F F+
~*ASB*	XINL GR FL	INL-ASB ASB ASBCL
T[+-]	T+ T-	T TR
[1-5]FL	1FL 4FL	FL 6FL
[~XE]SMH	ASMH 1SMH +SMH	XSMH ESMH SMH

Don't forget that description keys are case sensitive. Additionally, you can have multiple description key sets as discussed in the section "Activating a Description Key Set." With the knowledge you have gained so far in this chapter, you can create a powerful tool for managing survey points.

FIGURE 2.10
Right-click Create-Points and choose Edit Command Settings.

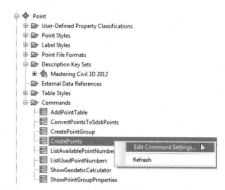

In the Edit Command Settings – CreatePoints dialog, expand the Points Creation category and ensure that Match On Description Parameters is set to True and that Disable Description Keys is set to False, as shown in Figure 2.11.

FIGURE 2.11
Verify that Match On Description Parameter is set to True and Disable Description Keys is set to False.

It is common to have multiple description key sets in your template. You can leverage description key sets for multiple clients or external survey firms that you work with. If you have multiple description key sets, they are all active, but if a set has a duplicate key, the first one Civil 3D runs across will take precedence. For example, if one set uses FL for flowline but a second set uses FL for fence line, the second occurrence of the FL key gets ignored.

You can control the search order from the Toolspace ➢ Settings tab by right-clicking Description Keys Sets and selecting Properties. Figure 2.12 shows the Description Key Sets Search Order dialog. Choose a description key set in the list and then use the arrows on the right side of the dialog to set the order. The set listed first takes first priority, then the second, and so on. Note that the listing in the Settings tab may not reflect the true listing in the properties.

FIGURE 2.12
The Description Key Sets Search Order dialog

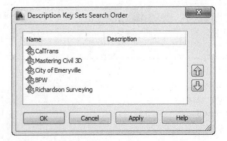

The Main Event: Your Project's Survey Database

Now that you know how to get everything set up, you are probably eager to get some real, live data into your drawing.

First, set your survey working folder to your desired survey storage location. The Civil 3D survey database is a set of external database files that reside in your survey working folder. By default, this folder is located in C:\Civil 3d Projects\. This is another item you will want to send to a network location in your own projects.

On the Toolspace ➢ Survey tab, right-click Survey Databases and select Set Working Folder, as shown in Figure 2.13.

FIGURE 2.13
Set the survey working folder.

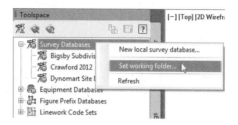

In the current version of Civil 3D, this is a different, independent setting from the working folder for data shortcuts (as discussed in Chapter 16, "Advanced Workflows"). Most folks want this on a network location, tucked neatly into the survey folder for the project.

 To create a survey database, you can either right-click and select New Local Survey Database from Toolspace ➢ Survey tab as mentioned earlier, or you can select Import Survey Data from the Home tab's Create Ground Data panel.

The contents of a survey database are organized into the following categories:

 Import Events Import events provide a framework for viewing and editing specific survey data, and they are created each time you import data into a survey database. The default name for the import event is the same as the imported filename. The Import Event collection contains the networks, figures, and survey points that are referenced from a specific import event and provides an easy way to remove, reimport, and reprocess survey data in the current drawing.

 Survey Queries Survey queries allow users to search for specific information within a survey database. You can use a query to locate and group related survey points and figures. For example, you may want all figures representing power lines in a query with survey points representing power poles. If you create a query containing your utilities, you can isolate specific figures and points to insert into another drawing.

The most powerful use of a survey query is when they are used in conjunction with a surface. When a query containing points and figures is added to a surface model, the figures are automatically added as breaklines. Figure 2.14 shows an example query that might be used for generating surface data.

FIGURE 2.14
Survey database query

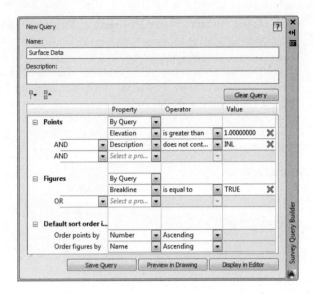

If the Survey Query you create is worthy of using in other databases, you can export them for use in other databases. Right-click on Survey Queries to Export or Import the query.

Networks A survey network is a collection of related data that is collected in the field. The network consists of setups, control points, non-control points, directions, and traverses. A survey database can have multiple networks. For example, you can use different networks for different phases of a project.

Network Groups Network groups are collections of various survey networks within a survey database. These groups can be created to facilitate inserting multiple networks into a drawing at once simply by dragging and dropping.

Figures Figures are the linework created by codes and commands entered into the raw data file during data collection. The figure names typically come from the descriptor or description of a point.

Figure Groups Similar to network groups, figure groups are collections of individual figures. These groups can be created to facilitate quick insertion of multiple figures into a drawing.

Survey Points One of the most basic components of a survey database, points form the basis for each and every survey. Survey points look just like regular Civil 3D point objects, and their visibility can be controlled just as easily. However, one major difference is that a survey point cannot be edited within a drawing. Survey points are locked by the survey database, and the only way of editing them is to edit the observation that collected the data for the points. This provides the surveyor with the confidence that points will not be accidentally erased or edited. Like figures, survey points can be inserted into a drawing by either dragging and dropping from the Toolspace ➢ Survey tab or by right-clicking Surveying Points and selecting the Points ➢ Insert Into Drawing option.

Survey Point Groups Just like network groups and figure groups, survey point groups are collections of points that can be easily inserted into a drawing. When these survey point groups are inserted into the drawing, a Civil 3D point group is created with the same name as the survey point group. This point group can be used to control the visibility or display properties of each point in the group.

In the following exercise, you'll create an import event and import an ASCII file with survey data. The survey data includes linework.

1. Create a new drawing from the _AutoCAD Civil 3D (Imperial) NCS.dwt or _AutoCAD Civil 3D (Metric) NCS.dwt template file. If you did not complete the previous exercise, copy the file Mastering Civil 3D.fdb_xdef from the dataset to C:\ProgramData\Autodesk\C3D 2014\enu\Survey. Note that this directory is hidden in Windows by default, so you will need to change the folder view options or type this path into the address bar of Windows Explorer.

2. From the Home tab of the ribbon ➤ Create Ground Data panel, click Import Survey Data to open the Import Survey Data Wizard.

3. Click Create New Survey Database.

4. Enter **Roadway** as the name of the folder in which your new database will be stored. Click OK.

    Roadway is now added to the list of survey databases.

5. With Roadway highlighted, click Edit Survey Database Settings and verify that the units are set to US Foot (or meter). Click OK to dismiss the dialog.

6. Click Next.

7. Set the Data Source Type pull-down to Point File.

8. Click the white plus sign on the right side of the Selected Files list to browse to the 0202_import points.txt or 0202_import points_METRIC.txt file, which you can download from this book's web page.

9. Select the file and then click Open, and the name will be listed in the dialog, as shown in Figure 2.15.

10. Under Specify Point File Format, scroll through the list until you find PNEZD (Comma Delimited).

11. Highlight the format.

    The preview will show that the correct data type is selected, as shown at the bottom of Figure 2.15.

12. Click Next to continue.

13. Click Create New Network.

    Note that if you forget to create a network to place your points, you will not be able to manipulate this group apart from other points.

FIGURE 2.15
Select the correct source type, file, and format in the Import Survey Data Wizard.

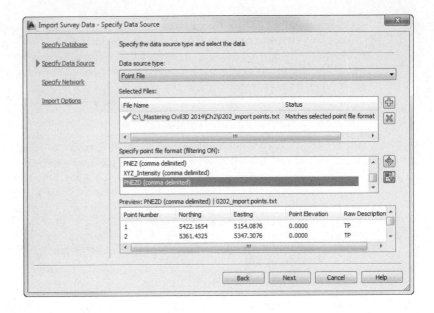

14. Enter **Roadway 4-22-2013** as the name of the network. Highlight the new network and click OK, and click Next to continue.

15. Set Current Figure Prefix Database to Mastering Civil 3D. Set the Process Linework During Import option to Yes. Be sure Insert Survey Points is set to Yes. When your import options match what is shown in Figure 2.16, click Finish.

FIGURE 2.16
The Import Options page in the Import Survey Data wizard

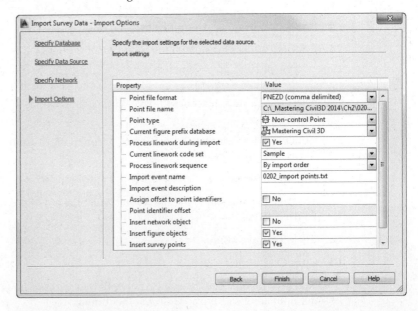

16. Save the file as **0202_RoadwayFigures.dwg** to the same location as the rest of your downloaded example files.

    You will need it in the next exercise.

The data is imported and the linework is drawn; however, the building is missing the left side. The following steps will resolve this issue:

1. Continue working in the drawing from the previous exercise. In Toolspace ➢ Survey tab, expand Survey Databases, and then select Roadway ➢ Networks ➢ Roadway 4-22-2013 ➢ Non-Control Points.

2. Right-click Non-Control Points and select Edit to bring up the Non-Control Points Editor in Panorama.

3. Scroll to the bottom of the point list and notice the last line in the file describing point Number 34. You will be editing the description of this point.

4. Double-click your cursor on the Description field. Use the arrow keys on your keyboard to move your cursor to the end of the description. Add a space and type **CLS** ↵, as shown in Figure 2.17. This is the default Close Figure command.

FIGURE 2.17
Editing the import event to add the CLS command to close the building geometry

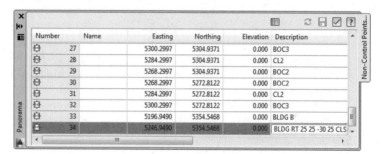

5. Click the Save icon in the upper-right corner of Panorama, and then click the green check box in the upper right of the palette to apply changes and save your edits.

    At this point you should see a yellow exclamation mark symbol next to the network name. This is indicating that a change has been made to the survey database that has not yet been processed in the drawing. In the next steps, you will update the network to reflect the change graphically.

6. Click Yes if you are asked to apply your changes.

7. While still working in Survey Databases ➢ Roadway, expand Import Events, and select **0202_import points.txt** (**0202_import points_METRIC.txt**).

8. Right-click the import event, and select Process Linework to bring up the Process Linework dialog.

9. Click OK to reprocess the linework with your updated point description.

10. Right-click the network name (Roadway 4-22-2013) and select Update Network.

11. Right-click the Figures branch and select Update Figures.

All warnings indicating that the database needs to be updated should now be cleared.

The building figure line and your drawing should look something like Figure 2.18.

FIGURE 2.18
After editing and reprocessing the linework

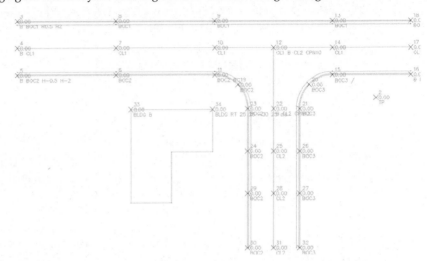

When you make modifications to survey data, only the Civil 3D version is changed. The original file remains untouched. The survey database doesn't use the file unless you reimport the data.

If you edit raw data in the Survey tab, Civil 3D will recalculate all affected information. For example, if you modify an instrument height, all elevations that need to be updated will automatically adjust.

Keep in mind that if you edit the source file and reimport, the Civil 3D survey (and any edits you made) will be overwritten.

FINE-TUNING FIGURES

Survey figures are usually 3D lines (unless you have flattened them in the figure style). Figures can be edited with the same type of tools you use on feature lines, which you will learn about in Chapter 14, "Grading." An editing tool unique to survey figures can be found in the Figure contextual tab of the ribbon.

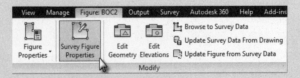

Select the figure you wish to edit and select Survey Figure Properties.

You will then see the details of the figure in tabular form.

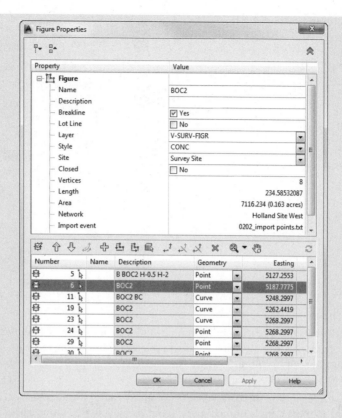

Once you are in the Figure Properties dialog, you can fix errors created by improper data collection or change fundamental qualities about the line, such as its name or breakline status. Explore the tools in this dialog to locate ways to modify vertices, add or delete vertices, reverse figure direction, and break, trim, and join figures.

When you have completed editing your figures with this method, clicking Update Survey Data From Drawing will commit the change to the survey database.

Under the Hood in Your Survey Network

Once you import survey data into a network, expand the branch to see how Civil 3D helps you make sense of it. In each network, data is organized by type, as shown in Figure 2.19:

Control Points Control points are created when data from an FBK file is imported. Inside the FBK, control points are prefaced by NE, NEZ, or LAT LONG. You can force any point to be listed as a control point by right-clicking this branch and selecting New.

Non-Control Points Keyed-in points, GPS-collected points, and any point brought in through an ASCII file will appear as non-control points. This is the default type if no other information is known about the point.

FIGURE 2.19
A typical survey database network with data

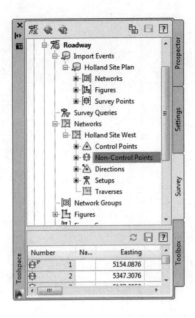

 Directions The direction from one point to another must be manually entered into the data collector for the direction to show up later in the survey network for editing. The direction can be as simple as a compass shot between two initial traverse points that serves as a rough basis of bearings for a survey job.

Setups If you imported data that contains setups and observations, Setups is where the meat of the data is found. Every setup, as well as the points (side shots) located from that setup, can be found listed here. Setups will contain two components: the station (or occupied point) and the backsight. To see or edit the observations located from the setup, right-click Setups and select Edit Setups That Observe. The interface for editing setups is shown in Figure 2.20. Angles and instrument heights can also be changed in this dialog.

FIGURE 2.20
Setups can be changed in the Setups Editor.

Station Point	Backsight Point	Backsight Dire...	Backsight Orie...	Backsight Face1	Backsight Face2	Instrument Hei...	Instrument Ele...
2	1	79.2900	0.0003	0.0000	180.0005	5.075	785.815
3	2	336.3913	0.0003	0.0000	180.0005	4.950	778.976
4	3	278.4248	0.0000	0.0000	180.0010	4.825	774.627
1	4	199.1753	0.0005	0.0000	180.0010	5.020	779.655

Traverses

The Traverses section is where new traverses are created or existing ones are edited. These traverses can come from your data collector, or they can be manually entered from field notes via the Traverse Editor, as shown in Figure 2.21. You can view or edit each setup in the Traverse Editor, as well as the traverse stations located from that setup.

FIGURE 2.21
The Traverse Editor

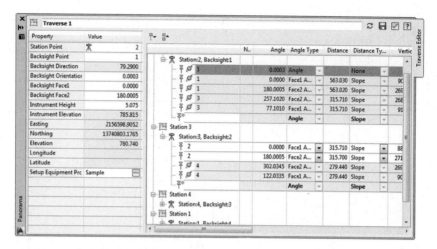

Once you have defined a traverse, you can adjust it by right-clicking its name and selecting Traverse Analysis. You can adjust the traverse either horizontally or vertically, using a variety of methods. The traverse analysis can be written to text files to be stored, and the entire network can be adjusted on the basis of the new values of the traverse, as you'll do in the following exercise:

1. Create a new drawing from the _AutoCAD Civil 3D (Imperial) NCS.dwt (_AutoCAD Civil 3D (Metric) NCS.dwt) template file.
2. Navigate to the Toolspace ➢ Survey tab.
3. Right-click Survey Databases and select New Local Survey Database.

   The New Local Survey Database dialog opens.
4. Enter **Traverse** as the name of the folder in which your new database will be stored.
5. Click OK to dismiss the dialog.

   The Traverse survey database is created as a branch under the Survey Databases branch.
6. Expand the Traverse branch, right-click Networks, and select New.

   The New Network dialog opens.
7. Expand the Network branch in the dialog if needed.
8. Name your new network **Traverse Practice**, and click OK.

   The Traverse Practice network is now listed as a branch under Toolspace ➢ Prospector ➢ Networks.
9. Right-click the Traverse Practice network and select Import ➢ Import Field Book.
10. Select the file 0203_traverse.fbk (0203_traverse_METRIC.fbk), which you can download from this book's web page, and click Open.

    The Import Field Book dialog opens.

11. Make sure you have checked the boxes shown in Figure 2.22.

FIGURE 2.22
The Import Field Book dialog

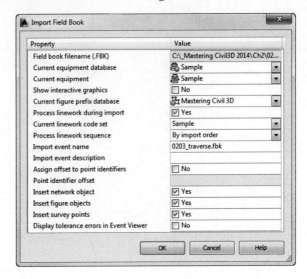

There is no linework in this file because it is just traverse shots.

12. Click OK. Double-click your middle mouse wheel to zoom extents to get a look at the imported traverse.

13. Save the drawing as **0203_Traverse.dwg** for the next exercise.

Still looking at the Toolspace ➢ Survey tab, expand the network you created earlier and inspect the data. You have one control point in the northwest corner that was manually entered into the data collector. There is one direction, and there are four setups. Each setup combines to form a closed polygonal shape that defines the traverse. Notice that there is no traverse definition. In the following exercise, you'll create that traverse definition for analysis.

Continue working in the drawing from the previous exercise:

1. Go to Toolspace ➢ Survey ➢ Survey Databases ➢ Traverse ➢ Networks ➢ Traverse Practice; then right-click Traverses and select New to open the New Traverse dialog.

2. Name the new traverse **Traverse 1**.

3. Type **2** as the Initial Station point number and press Enter.

The traverse will now pick up the rest of the stations in the traverse and enter them in the next box.

4. Verify that the points on the traverse match what is shown in Figure 2.23.

5. Click OK.

6. Right-click Traverse 1 in the bottom portion of Toolspace. Select Traverse Analysis.

FIGURE 2.23
Defining a new traverse

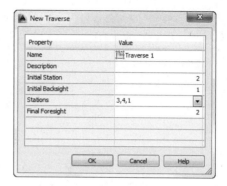

7. In the Traverse Analysis dialog, ensure that Yes is selected for Do Traverse Analysis and Do Angle Balance.

8. Select Least Squares for both Horizontal Adjustment Method and Vertical Adjustment Method.

9. Set both Horizontal Closure Limit 1:X and Vertical Closure Limit 1:X to **20,000**.

10. Leave Angle Error Per Set at the default.

11. Make sure the option Update Survey Database is set to Yes.

    The Traverse Analysis dialog will look like Figure 2.24.

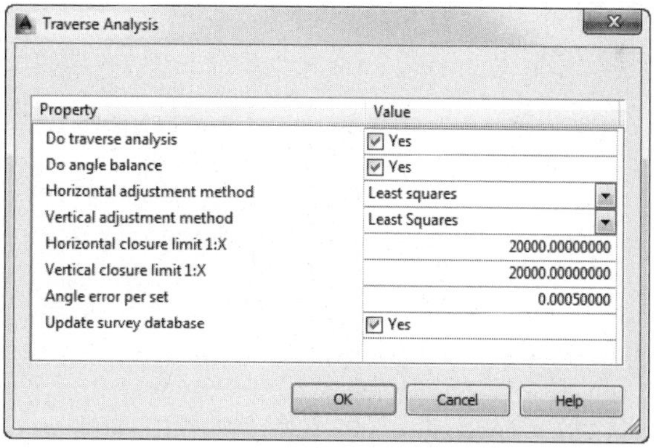

FIGURE 2.24
Specify the adjustment method and closure limits in the Traverse Analysis dialog.

12. Click OK.

The analysis is performed, and four text files are displayed that show the results of the adjustment. These files are automatically saved in the survey working folder under the same directory as the survey database (in this example, it should be C:\Civil 3DProjects\Traverse\Traverse 1\). Note that if you look back at your survey network, all points are now control points because the analysis has upgraded all the points to control point status.

Figure 2.25 shows the `Traverse1 Raw Closure.trv` and `Traverse 1 Vertical Adjustment.trv` files that are generated from the analysis. The raw closure file shows that your new precision is well within the tolerances set in step 8. The vertical adjustment file describes how the elevations have been affected by the procedure.

FIGURE 2.25
Horizontal and vertical traverse analysis results

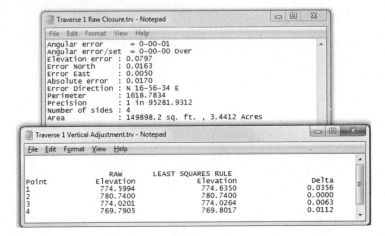

The third text file is shown in three separate portions. The first portion is shown in Figure 2.26. This portion of the file displays the various observations along with their initial measurements, standard deviations, adjusted values, and residuals. You can view other statistical data at the beginning of the file.

FIGURE 2.26
Statistical and observation data portion of text file

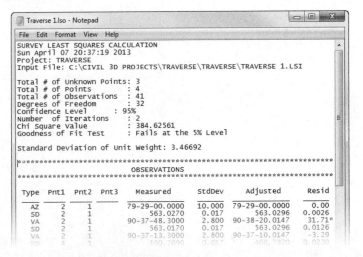

Figure 2.27 shows the second portion of this text file (you will need to scroll down to see it) and displays the adjusted coordinates, the standard deviation of the adjusted coordinates, and information related to error ellipses displayed in the drawing. If the deviations are too high for your acceptable tolerances, first check the instrument settings and tolerances in the equipment database. If everything is set correctly, you may need to redo the work or edit the field book.

FIGURE 2.27
Adjusted coordinate information portion of text file

```
                       ADJUSTED   COORDINATES

Point        Northing              Easting             Elevation
  1       13740905.9355         2157152.4444            774.6350
  3       13740513.3864         2156723.9875            774.0264
  4       13740471.0583         2157000.1693            769.8017

               Standard Deviations - Adjusted Coordinates
                          STANDARD DEVIATIONS
Point          North                 East                Elevation
  1          0.026865              0.008341              0.003265
  3          0.008517              0.015068              0.003629
  4          0.020368              0.017793              0.003828

                    Least Squares Error Ellipses
                       at 95% Confidence Level

Point     Semi-Major Axis       Semi-Minor Axis        NE-Axis Azimuth
  1        0.2438418757          0.0607871778            169-29-00
  3        0.1432969460          0.0580918243             67-58-16
```

Figure 2.28 displays the final portion of this text file—Blunder Detection/Analysis. Civil 3D will look for and analyze data in the network that is obviously wrong and choose to keep it or throw it out of the analysis if it doesn't meet your criteria. If a blunder (or bad shot) is detected, the program will not fix it. You will have to edit the data manually, whether by going out in the field and collecting the correct data or by editing the FBK file.

FIGURE 2.28
Blunder analysis portion of text file

```
                       Blunder Detection/Analysis

                                                    Reliability   Tests
Type Pnt1 Pnt2 Pnt3     Adjusted     Resid   Redun  Estimate  Marg  Ext
 AZ   2    1         79-29-00.0000   0.000   0.000   <None>    F    F
 SD   2    1              563.030    0.003   0.840   -0.003    P    P
 VA   2    1         90-38-20.0147  31.715   0.818  -38.789    F    F
 SD   2    1              563.030    0.013   0.840   -0.015    P    P
 VA   2    1         90-37-10.0147  -3.285   0.818    4.018    P    P
 SD   4    1              460.792    0.023   0.844   -0.027    P    P
 VA   4    1         89-24-13.8005   8.701   0.805  -10.802    P    F
```

OTHER METHODS OF MANIPULATING SURVEY DATA

Often, it is necessary to edit the entire survey network at one time. For example, rotating a network to a known bearing or azimuth from an assumed one happens quite frequently. To find this hidden gem of functionality, go to Toolspace ➢ Survey tab ➢ Survey Databases, right-click the name of the database you wish to modify, and select Translate Survey Database as shown in Figure 2.29.

FIGURE 2.29
The elusive yet indispensable Translate Survey Database command

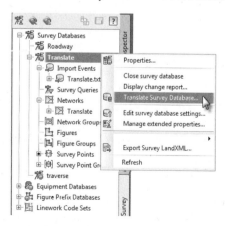

Real World Scenario

MANIPULATING THE NETWORK

Frequently, surveys performed in assumed coordinate systems need to be adjusted to match a known coordinate system. Along with changing coordinate systems, you will probably need to rotate the survey to the correct orientation. Additionally, networks may need to be adjusted from assumed elevations to a known datum. In this example, the DWG you will be working with contains two known points from the desired coordinate system. The file you will import is initially in an assumed coordinate system. You will use the Translate command to move and rotate the network accordingly.

1. Open the file 0204_Translate.dwg (0204_Translate_METRIC.dwg).
2. In the Toolspace ➢ Survey tab, right-click Survey Databases and select New Local Survey Database.

   The New Local Survey Database dialog opens.
3. Enter **Translate** in the text box.

   This is the name of the folder for the new database.
4. Click OK, and the Translate database will now be listed under the Survey Databases branch.
5. Select Networks under the new Translate branch.
6. Right-click and select New to open the New Network dialog.
7. Enter **Translate** as the name of this new network.
8. Click OK to dismiss the dialog.
9. Right-click the Translate network, and select Import ➢ Import Point File.
10. Navigate to the file 0204_Translate.txt (0204_Translate_METRIC.txt), which you can download from this book's web page, and click Open.

    The Import Point File dialog opens.
11. Verify that the point file format is PNEZD (Comma Delimited).
12. Verify that the Insert Survey Points option is set to Yes and click OK.
13. Update the Point Groups collection in Prospector so that the points show up in the drawing if needed. Zoom extents to view the points.

    Points 1 and 47 from the newly imported network correspond to 10000 and 10001 from the actual coordinate system. This is all the information needed to perform a translation.
14. Make sure you are able to locate these points for the next steps by panning and zooming with your mouse.

    It will also be helpful to turn on your Node object snap.
15. In the Toolspace ➢ Survey tab, right-click the name of the survey database and select Translate Survey Database, as shown earlier in Figure 2.29.

16. For Base Point Number, enter **1** and press ↵.

    Civil 3D will pick up the assumed coordinate of this point.

17. Click Next.

18. For the rotation angle, click the Pick In Drawing button in the bottom-left corner of the dialog.

    You will be prompted for an initial direction.

19. For Initial Direction, use the Node Osnap to pick point 1 and then point 47.

    You will then be prompted to specify a new direction.

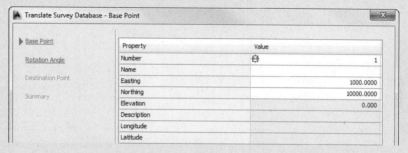

20. Use the Node Osnap to pick points 10000 and 10001.

    The resulting rotation angle should be 45°.

21. Click Next.
22. In the Destination Point screen, click Pick In Drawing.
23. Use the Node Osnap to select point 10000.
24. Click Next.
25. Verify your translation setup on the Summary screen and then click Finish.

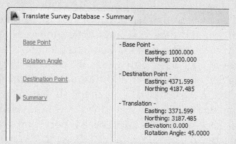

26. Save and close the drawing.

The changes were made directly to the database and the network can be imported into any drawing.

Other Survey Features

Other components of the survey functionality included with Civil 3D 2014 are the Astronomic Direction Calculator, the Geodetic Calculator, Mapcheck reports, and Coordinate Geometry Editor. All of these features are accessed from the ribbon under Analyze tab ➢ Ground Data ➢ Survey.

The Astronomic Direction Calculator, shown in Figure 2.30, is used to calculate sun shots or star shots. For the obscure art of these types of observations, Civil 3D has all the ephemeris data built in, making this a very convenient tool.

FIGURE 2.30
The Astronomic Direction Calculator

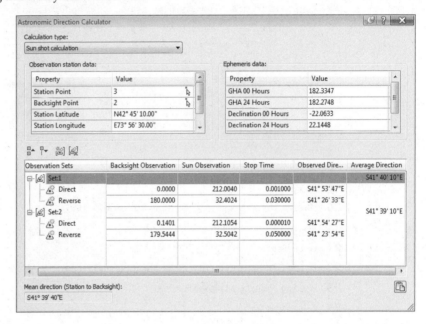

The Geodetic Calculator is used to calculate and display the latitude and longitude of a selected point as well as its local and grid coordinates. It can also be used to calculate unknown points. If you know the grid coordinates, the local coordinates, or the latitude and longitude of a point, you can enter it in the Geodetic Calculator and create a point at that location. Note that the Geodetic Calculator only works if a coordinate system is assigned to the drawing in the Drawing Settings dialog. In addition, any transformation settings specified in this dialog will be reflected in the Geodetic Calculator, shown in Figure 2.31.

Remember those labels you added to your lines in Chapter 1, "The Basics"? Put them to work in a Mapcheck report. The Mapcheck report computes closure based on line, curve, or parcel segment labels, as you'll see in the following exercise:

1. Open the 0205_Mapcheck.dwg (0205_Mapcheck-_METRIC.dwg) file, which you can download from this book's web page.

FIGURE 2.31
The Geodetic Calculator

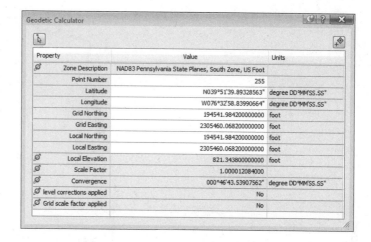

2. Change to the Analyze tab of the ribbon, and select Survey ➢ Mapcheck from the Ground Data panel to display the Mapcheck Analysis palette.

3. Click the New Mapcheck button at the top of the menu bar.

4. At the Enter name of mapcheck: prompt, type **Record Deed** and press Enter.

5. At the Specify point of beginning (POB): prompt, choose the north endpoint of the line representing the east line of the parcel (the longest line in the file).

    A red glyph will appear to represent the POB.

6. Working clockwise, select each parcel label one at a time.

    Be sure not to skip the small segment in the southwest portion of the site.

    In the northwest portion of the site, you will encounter a label whose bearing is flipped. You will use the Mapcheck Reverse command to fix this in your Mapcheck report.

7. At the Select a label or [Clear/Flip/New/Reverse]: prompt, type **R** and then press ↵ to reverse direction.

    The Mapcheck glyph will now appear in the correct location along the segment, as shown in Figure 2.32.

8. Select the last line label along the north section of the parcel.

9. Press ↵ to complete the Mapcheck entry.

    The completed parcel should have 12 sides.

10. Select the output view as shown in Figure 2.33 to verify closure.

FIGURE 2.32
The Mapcheck glyph verifies that your input is correct.

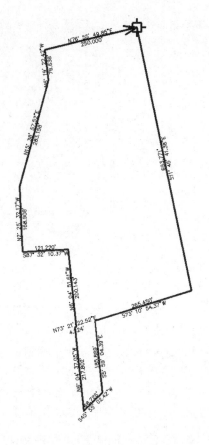

FIGURE 2.33
The completed deed in the Mapcheck Analysis palette

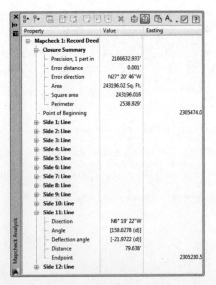

The Coordinate Geometry Editor

Certification Objective

The Coordinate Geometry Editor (Figure 2.34) is a powerhouse tool that makes creating and evaluating 2D boundaries easier than before. The functionality introduced with this feature supplants entering parcel data one segment at a time using the Line By Bearing And Distance command. Traverse analysis can be performed on manually entered segments, polylines, or COGO point objects without needing to define them in a survey database.

FIGURE 2.34
Your new best friend, the Coordinate Geometry Editor

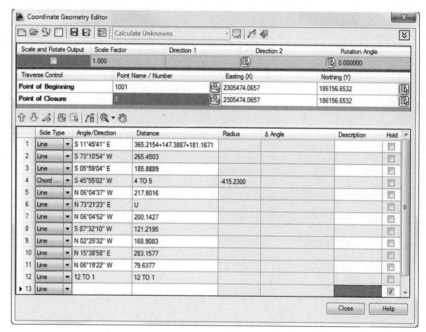

Boundary data can be entered in the Coordinate Geometry Editor using a mix of methods. As shown in the first line of the traverse seen in Figure 2.34, you can use formulas to enter data. In the example shown, multiple segments with the same bearing have been consolidated into a single entry.

If a value is unknown, such as the distance in line 6 of Figure 2.34, you can enter a **U**. Civil 3D will calculate the unknown value when you generate the traverse report.

To enter data using points, use the Pick COGO Points In Drawing button to select the points in the direction of the traverse side. Note that the direction and distance are entered independently of each other, so you will need to repeat the selection for each column of the table. You can also copy and paste between columns.

If a line has been entered in error, the Coordinate Geometry Editor offers a variety of tools for fixing problems. To remove a line of the table, highlight the row, right-click, and select Delete Row, as shown in Figure 2.35.

FIGURE 2.35
Removing unwanted traverse data

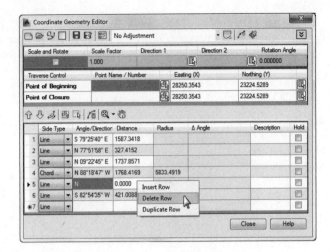

Similar to the glyphs you saw in the Mapcheck command, the Coordinate Geometry glyph will appear in the graphic showing the side directions and point of closure, as seen in Figure 2.36.

FIGURE 2.36
Temporary graphics, or "glyphs," to help you identify your boundary

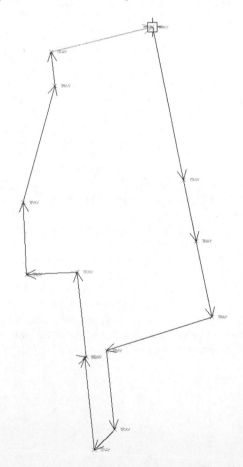

When you want to run a traverse report, set the report type you wish to run from the top of the Coordinate Geometry Editor. If you have unknowns in your traverse, your only option will be to calculate the unknown values. Click the Display Report button to view the results of your entries. Depending on the type of adjustment you chose, your results should resemble Figure 2.37.

FIGURE 2.37
Traverse report created by the Coordinate Geometry Editor

Traverse Report

Closure

Total Traverse Length	2538.928
Error in Closure	0.001
Closure is one part in	1866445.9895
Error in North(Y)	0.0013
Error in East(X)	0.0004
Direction of Error	N 19°00'01" E

Traverse Control

	Point Name	Northing	Easting
Point of Beginning	1001	186156.6532	2305474.0657
Point of Closure	1	186156.6532	2305474.0657

Input Data

Side	Angle/Direction	Distance	Radius	#Delta Angle	Description
1	S 11°45'41" E	365.2154			
2	13 TO 14	13 TO 14			
3	S 11°45'41" E	181.1671			
4	S 73°10'54" W	265.4503			
5	S 05°59'04" E	185.8889			
6	S 45°55'02" W	68.7261	-393.6557		
7	N 06°04'37" W	217.8016			

Using Inquiry Commands

A large part of a surveyor's work involves querying lines and curves for their length, direction, and other parameters.

The Inquiry commands panel (Figure 2.38) is on the Analyze tab of the ribbon, and it makes a valuable addition to your Civil 3D and survey-related workspaces. Remember, panels can be dragged away from the ribbon and set in the graphics environment much like a toolbar.

FIGURE 2.38
The Inquiry commands panel

The Inquiry Tool (shown in Figure 2.39) provides a diverse collection of commands that assist you in studying Civil 3D objects. You can access the Inquiry Tool by going to the Analyze tab of the ribbon ➢ Inquiry panel and clicking the Inquiry Tool button.

FIGURE 2.39
Choosing an Inquiry type from the Inquiry Tool palette

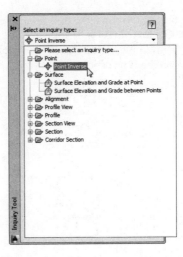

To use the Point Inverse option in the Inquiry commands, first set the Select An Inquiry Type pull-down to Point ➢ Point Inverse as shown in Figure 2.40.

FIGURE 2.40
Point Inverse results

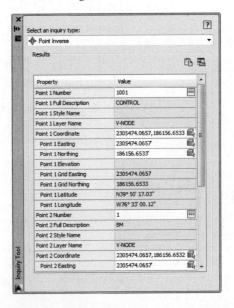

You can enter the point number or use the Pick In CAD icon to select the points you wish to examine. If no point exists, the Pick In CAD option will pick up the northing and easting of the location. You could also type in a northing and easting if desired.

The other Inquiry commands that are specific to Civil 3D are also handy to the survey process.

The List Slope tool provides a short command-line report that lists the elevations and slope of an entity (or two points) that you choose, such as a line or feature line.

The Line And Arc Information tool provides a short report about the line or arc of your choosing (see Figure 2.41). This tool also works on parcel segments and alignment segments. Alternatively, you can type **P** and press Enter for points at the command line to get information about the apparent line that would connect two points on screen.

FIGURE 2.41
Command-line results of a line inquiry and arc inquiry

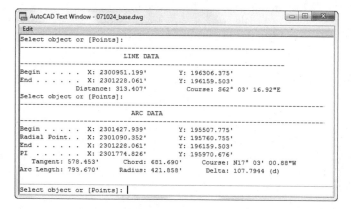

Don't Get Burned by AutoCAD Angles

When looking for information about a line, use the Line And Arc tool rather than the base AutoCAD LIST command. The Civil 3D Line And Arc Information tool (CGLIST) works on more object types and is not affected by rotated coordinate systems.

When entering angles in base AutoCAD, the full N50d10′10″E is needed to denote a bearing. In Civil 3D commands, N50.1010E will denote the same thing and is much faster to type. However, as every surveyor knows, there is a huge difference between 50.1010° and 50°10′10″.

A setting you will want to change in your Civil 3D template is the angular entry method for general angles, as shown here. This setting mainly affects the Angle Distance and Bearing Distance Transparent commands. Check back in Chapter 1, "The Basics," for more information on template settings and transparent commands.

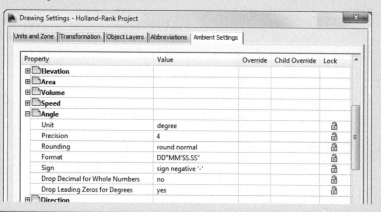

The Angle Information tool lets you pick two lines (or a series of points on the screen). It provides information about the acute and obtuse angles between those two lines. Again, this also works for alignment segments and parcel segments.

The Continuous Distance tool provides a sum of distances between several points on your screen or one base point and several points.

The Add Distances tool is similar to the Continuous Distance command, except the points on your screen do not have to be continuous.

The Bottom Line

Properly collect field data and import it into Civil 3D. Once survey data has been collected, you will want to pull it into Civil 3D via the survey database. This will enable you to create lines and points that correctly reflect your field measurements.

Master It Create a new drawing based on the template of your choice and a new survey database and import the `MasterIt_0201.txt` (or `MasterIt_0201_METRIC.txt`) file into the drawing. The format of this specific file is PNEZD (Comma Delimited).

Set up description key and figure databases. Proper setup is key to working successfully with the Civil 3D survey functionality.

Master It Create a new description key set and the following description keys using the default styles. Make sure all description keys are going to layer V-Node:

- CL*
- EOP*
- TREE*
- BM*

Change the description key search order so that the new description key set takes precedence over the default.

Create a figure prefix database called MasterIt containing the following codes:

- CL
- EOP
- BC

Test the new description key set and figure prefix database by importing the file `MasterIt_CodeTest_0202.txt` (use the same file for both US and metric units). Note that this file is a comma-delimited PNEZD file.

Translate surveys from assumed coordinates to known coordinates. Understanding how to manipulate data once it is brought into Civil 3D is important to making your field measurements match your project's coordinate system.

Master It Create a new drawing based on the template of your choice and start a new survey database. Import `0203_traverse.fbk` (or `0203_traverse_METRIC.fbk`). When you import the file, turn on the Insert Network Object option. Translate the database based on the following settings:

- Base Point 1
- Rotation Angle of 10.3053°

Perform traverse analysis. Traverse analysis is needed for boundary surveys to check for angular accuracy and closure. Civil 3D will generate the reports that you need to capture these results.

Master It Use the survey database and network from the previous Master It exercise. Analyze and adjust the traverse using the following criteria:

- Use an Initial Station value of 2 and an Initial Backsight value of 1.
- Use the Compass Rule option for Horizontal Adjustment.
- Use Length Weighted Distribution Method for Vertical Adjustment.
- Use a Horizontal Closure Limit value of 1:25,000.
- Use a Vertical Closure Limit value of 1:25,000.

Chapter 3

Points

In the previous chapter, "Survey," you looked at a specific method for bringing in points and figures. In this chapter, you will take a closer look at creating and organizing points.

Most commonly, points are used to identify the location of existing features, such as trees and property corners; topography, such as ground shots; or stakeout information, such as road geometry points. However, points can be used for much more. This chapter will both focus on traditional point uses and introduce ideas to apply the dynamic power of point editing, labeling, and grouping to other applications.

In this chapter, you will learn to:

- Import points from a text file using description key matching
- Create a point group
- Export points to LandXML and ASCII format
- Create a point table

Anatomy of a Point

AutoCAD® Civil 3D® *points* (see Figure 3.1) are intelligent objects that represent x, y, and z locations in space. Each point has a unique number and, optionally, a unique name that can be used for additional identification and labeling.

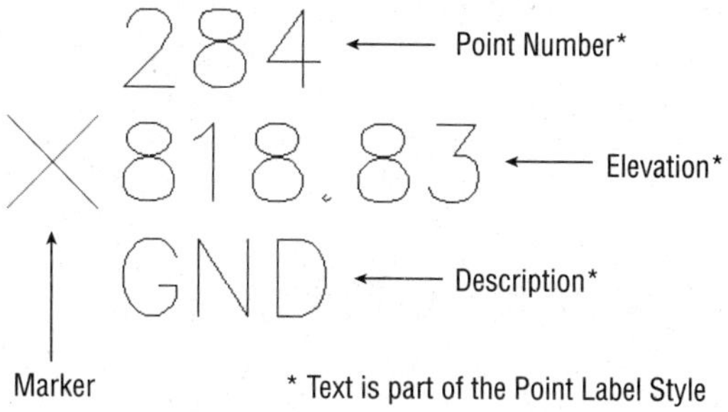

FIGURE 3.1
A typical point object showing a marker, a point number, an elevation, and a description

284 ← Point Number*
✕ 818.83 ← Elevation*
↑ GND ← Description*
Marker   * Text is part of the Point Label Style

> **A Quick Word on Styles**
>
> Separating the point functionality discussed in this chapter from the styles that make them look the way they do is difficult. Chapter 18, "Label Styles," and Chapter 19, "Object Styles," will get into the nitty-gritty of creating and manipulating label styles and point styles. In this chapter, you will work with styles that are already part of a drawing. This is true for points, labels, and tables.

COGO Points vs. Survey Points

In Chapter 2, you imported survey data that contained points. Points brought in through the methods described in Chapter 2 are referred to as *survey points*. In this chapter, you will import points from a delimited text file and place them in CAD using the point creation tools. Points created in this manner are referred to as *COGO points*. Figure 3.2 shows the contextual tab differences between points brought in as COGO points (top) and points brought in through a database (bottom).

FIGURE 3.2
The context-sensitive ribbon reflects similarities and differences between COGO points (top) and survey points (bottom).

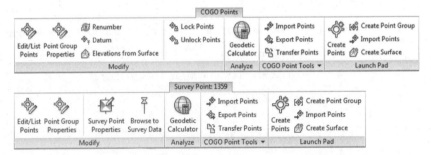

The differences between COGO points and survey points are subtle but important to note. A COGO point is unlocked by default—meaning it can readily be edited both graphically and through a tabular view in Panorama. A survey point, on the other hand, must be unlocked if a user wishes to edit the point graphically. To edit a survey point in Panorama, you can manipulate it through the Survey tab of Toolspace. A survey point stays tied to the database from which it came, whereas a COGO point maintains no tie to the originating text file or object it was created against. Regardless of their origin, both COGO points and survey points obey the principles outlined in this chapter.

Creating Basic Points

Certification Objective

You can create points many ways using the Points menu in the Create Ground Data panel on the Home tab of the ribbon. Points can also be imported from text files or external databases or converted from AutoCAD points or other legacy point types.

Point Settings

Point settings are our first glimpse at what is known as *command settings*. All Civil 3D objects have command settings tucked away in the Settings tab of Toolspace. In the case of points, it is handy to have these settings readily available for on-the-fly modifications. Whether or not the changes you make on the fly are remembered the next time you create points depends on your template settings.

To make sure that the settings you change will hold every time you create points, follow these steps:

1. On the Settings tab of Toolspace, locate the Point collection and click the plus sign to expand the branch.

2. Click the plus sign to expand Commands.

3. As shown in Figure 3.3, right-click the command CreatePoints and select Edit Command Settings.

FIGURE 3.3
In your Civil 3D template, make sure Save Command Changes To Settings is set to Yes for Points.

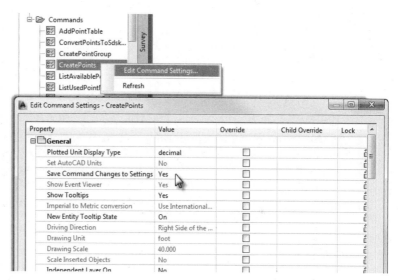

4. Expand the General section and verify that Save Command Changes To Settings is set to Yes.

If you explore the command settings further, you will see the options Default Layer and Points Creation. By setting the Save Command Changes To Settings option to Yes, you are ensuring that changes you make on the CreatePoints toolbar will be saved for the next time you use the command.

To access the Create Points toolbar:

1. Go to the Home tab of the ribbon.

2. In the Create Ground Data panel, select Points ➢ Point Creation Tools.

3. Expand the toolbar by clicking the chevron button on the far-right side.

Default Layer

For most Civil 3D objects, the object layer is established in the drawing settings. In the case of points, the layer depends on whether or not it matches a description key. The layer set by the description key set takes precedence over other settings. If there is no description key match, the default layer is set in the command settings for point creation and can be changed in the Create Points dialog (see Figure 3.4). Review Chapter 2 for more information about Description Key Sets.

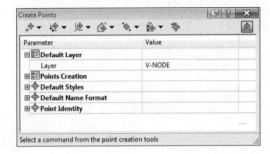

FIGURE 3.4
Verify the point object layer before creating points.

Prompt for Elevations, Names, and Descriptions

When creating points in your drawing, you have the option of being prompted for elevations, names, and descriptions (see Figure 3.5). In the case of creating points from alignment geometry, you can choose to have points pick up descriptions using the Automatic – Object option. Initially, the Default Description option is blank. In many cases, you'll want to leave these options set to Manual. The command line will ask you to assign an elevation and description for every point you create. It is best to avoid using the point names in general (as described later in this section), therefore leave the Prompt For Point Names option to None.

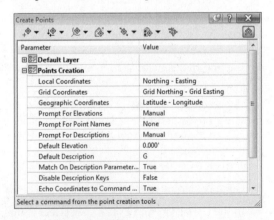

FIGURE 3.5
You can change the elevation, point name, and description settings from Manual to Automatic. You can also use the None option to omit this information.

If you're creating a batch of points that have the same description or elevation, you can change the Prompt toggle from Manual to Automatic and then provide the description and elevation in the default cells. For example, if you're setting a series of trees at an elevation of 10', you can establish settings as shown in Figure 3.6.

FIGURE 3.6
Default settings for placing tree points at an elevation of 10′

Points Creation	
Local Coordinates	Northing - Easting
Grid Coordinates	Grid Northing - Grid Easting
Geographic Coordinates	Latitude - Longitude
Prompt For Elevations	Automatic
Prompt For Point Names	None
Prompt For Descriptions	Automatic
Default Elevation	10.000′
Default Description	TREE
Match On Description Parameters (...	True

Note that these settings apply only to points created from this toolbar. The settings do not affect the elevation or description of points imported from a file.

> **WHAT'S IN A NAME?**
>
> Users often confuse a point's *name* with its *description*. The name is a unique, alphanumeric sequence that some software programs use in lieu of a point number. The description refers to the all-important code given to a point out in the field. The Civil 3D workflow favors the use of point numbers, so the Prompt for Point Names option is usually set to None. If you do have a point file that uses a name instead of a point number, you will need to create a custom point format, as described later in this chapter.

Importing Points from a Text File

Certification Objective

One of the most common means of creating points in your drawing is to import an external text file (see Figure 3.7).

FIGURE 3.7
The Import Points and Point File Formats dialogs

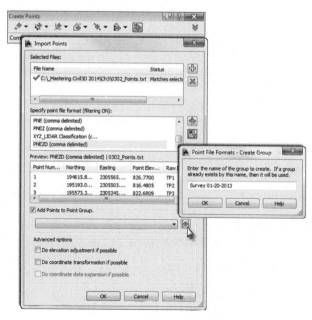

To add a file to your Import Points dialog, click the white plus sign to browse. You can add multiple files at once if they are in the same point format—such as a comma delimited point number, northing, easting, elevation, description (i.e. PNEZD). The import process supports most text formats as well as Microsoft Access database (MDB) files. Later in this chapter, you will experience adding your own text format.

When your file is listed in the top of the dialog, a green check mark will indicate that Civil 3D can parse the information. Be careful, though, because Civil 3D does not know the difference between a northing and an easting or a point number and an elevation. You still need to select the correct file format.

The file format filter is there to help you. Civil 3D recognizes how the file is delimited (i.e., tab, comma, space) and only shows you the formats that apply. If you don't want the help, you can turn the filtering off by clicking the Filter icon.

Custom Point File Formats and Advanced Options

If the file format you need is not available, or you wish to use the adjustment and transformation capabilities, you can do so by clicking the Manage Formats button. Many types of data are available for adding to a point format, including LIDAR classification and various latitude-longitude types.

As you can see in Figure 3.7 (shown previously), advanced options are available at the bottom of the Import Points dialog. These options will not affect your points unless the point file format has adjustment, transformation, or expansion built into it.

These three advanced options apply only if the point file format you choose has built-in adjustment or expansion:

- Do Elevation Adjustment If Possible
- Do Coordinate Transformation If Possible
- Do Coordinate Expansion If Possible

You can make an elevation adjustment if the point file contains additional columns for thickness, Z+, or Z-. To get Civil 3D to recognize these extra data columns, a new point file format is needed. You can perform a coordinate system transformation if a coordinate system has been assigned both to your drawing (under the drawing settings) and as part of a custom point format. In this case, the program can also do a coordinate data expansion, which calculates the latitude and longitude for each point.

 Real World Scenario

The Name Game

A common use of the point file formats is to import a name-based point file. Point file formats are stored with the drawing, so this is another item that would be wise to have set up in your Civil 3D template. In the following example, you will create a new file format to accommodate names (instead of point numbers):

1. Open the drawing 0301_PointFileFormat.dwg (0301_PointFileFormat_METRIC .dwg), which you can download from this book's web page at www.sybex.com/go/ masteringcivil3d2014.

CREATING BASIC POINTS | 103

2. Select the Settings tab of Toolspace and choose Point ➢ Point File Formats.

You can also access this functionality on the fly from the Import Points dialog.

3. Right-click Point File Formats and select New.
4. Select User Point File and click OK.
5. Name the format **Name-NEZD (Comma Delimited)**.
6. Toggle on the Delimited By option and place a comma in the field.
7. Click the first <unused> column heading, select Name from the Column Name pull-down, and click OK.
8. Click the next <unused> column and select Northing from the Column Name pull-down.
9. Leave the Invalid Indicator and Precision fields at their defaults and click OK.
10. Repeat the process for Easting, Point Elevation, and Raw Description.

    Be sure to add the columns in the correct order or the format will not work with the example file. You may notice that as you use an item from the Column Name list, it is no longer available for use in another column.

11. To test the format, click Load, select the file 0301_TestFormat.txt (0301_TestFormat_METRIC.txt) and click Open.
12. Click Parse.

If the format has been created successfully, you will see the file preview, as shown here.

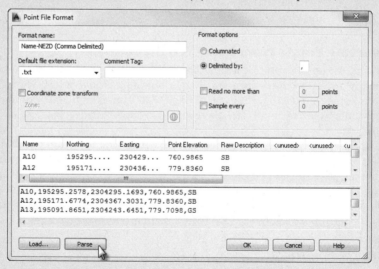

13. Click OK to complete the format.

After you learn how, revisit this drawing and import points. You will see that Civil 3D assigns point numbers in addition to using the names.

To compare your work with a completed example, see 0301_PointFileFormat_FINISHED.dwg (0301_PointFileFormat_METRIC_FINISHED.dwg).

Importing a Text File of Points

In this exercise, you'll learn how to import a TXT file of points into Civil 3D:

1. Open the `0302_PointImport.dwg` (`0302_PointImport_METRIC.dwg`) file, which you can download from this book's web page.

2. On the Home tab of the ribbon ➢ Create Ground Data panel ➢ Points, select Point Creation Tools.

3. On the Create Points toolbar, click the Import Points button.

4. Click the white + button to the right of the Selected Files field, and navigate out to locate the `0302_Points.txt` (`0302_Points_METRIC.txt`) file.

5. Select the file and click Open.

6. In the Specify Point File Format field, set the format to PNEZD (Comma Delimited).

7. Place a check mark in the box next to Add Points To Point Group and click the Create Point Group icon. Name the point group **Survey 01-20-2013** and click OK.

8. Leave the boxes under Advanced Options unchecked.

9. Click OK.

You may have to use Zoom Extents to see the imported points. (Hint: Double-click your middle mouse wheel for zooming extents.)

10. Save the drawing.

Converting Points from Non-Civil 3D Sources

Base AutoCAD points, Land Desktop points, and Softdesk points are types of data that Civil 3D can readily convert from within a drawing. Whenever possible, your best course of action would be to request the survey in text format or LandXML. However, if you need to migrate a drawing that contains legacy point data, there are many tools to convert old points to Civil 3D point objects.

A Land Desktop point database (the `Points.mdb` file found in the `COGO` folder in a Land Desktop project) can be directly imported into Civil 3D in the same interface in which you'd import a text file.

Land Desktop point objects, which appear as AECC_POINTs in the AutoCAD Properties palette, can also be converted to Civil 3D points (see Figure 3.8). Upon conversion, this tool gives you the opportunity to assign styles, create a point group, and more.

Converting Points

In this exercise, you'll convert Land Desktop point objects and AutoCAD point entities into Civil 3D points:

1. Open the `0303_ConvertPoints.dwg` (`0303_ConvertPoints_METRIC.dwg`) file, which you can download from this book's web page.

2. Use the List command or the AutoCAD Properties palette to confirm that most of the objects in this drawing are AECC_POINTs, which are points from Land Desktop.

FIGURE 3.8
The Convert Land Desktop Points option (left) opens the Convert Autodesk Land Desktop Points dialog (right).

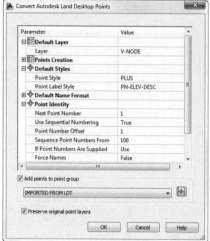

3. On the Home tab of the ribbon ➢ Create Ground Data panel ➢ Points menu, select Convert Land Desktop Points.

   Note that the Convert Autodesk Land Desktop Points dialog allows you to choose point creation settings and styles.

4. Place a check mark next to Add Points To Point Group.

5. Click the Create A New Point Group button.

6. Name the group **Converted From LDT** and click OK.

7. Clear the Preserve Original Point Layers check box.

   This option will move the resulting points to the layer specified in the description key set if there is a match. If there is no match, the point will go to the default layer.

8. Click OK to complete the conversion process.

   Civil 3D scans the drawing looking for Land Desktop point objects.

9. Once Civil 3D has finished the conversion, zoom in on any of the former Land Desktop points.

   The points should now show as COGO points in the AutoCAD Properties palette, confirming that the conversion has taken place. The Land Desktop points have been replaced with Civil 3D points, and the original Land Desktop points are no longer in the drawing.

10. In Toolspace ➢ Prospector tab, expand the Point Groups category.

    Notice that there is a yellow exclamation shield symbol indicating that the Converted From LDT point group needs to be updated.

11. Right-click Point Groups and select Update.

106 CHAPTER 3 POINTS

12. Zoom in on the cyan AutoCAD point objects.
13. On the Home tab of the ribbon ➢ Create Ground Data panel ➢ Points menu, select Convert AutoCAD Points.

    The command line reads `Select AutoCAD Points`.
14. Use a crossing window to select all the cyan-colored AutoCAD points, and then press ↵.
15. At the command-line prompt, enter a description of **GS** (for Ground Shot) and press ↵ for each point.

    Hint: Use the up arrow on your keyboard to recall the last typed entry and avoid unnecessary typing. When prompted to select AutoCAD points, this is a sign that all the points have been converted. Press Enter again to end the command.
16. Zoom in on one of the converted points, and select it. The contextual tab confirms that it has been converted to a Civil 3D point.

    Also, note that the original AutoCAD points have been erased from the drawing.

A Closer Look at the Create Points Toolbar

Certification Objective

In Civil 3D 2013, you can find point creation tools directly under the Home tab of the ribbon ➢ Create Ground Data Panel ➢ Points drop-down as well as in the Create Points toolbar. The toolbar is *modeless*, which means it stays on your screen even when you switch between tasks. Figure 3.9 shows the toolbar with the point creation methods labeled.

FIGURE 3.9
The Create Points toolbar

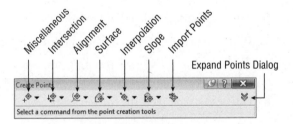

As you place points using these tools, a few general rules apply to all of them. If you place a point on an object with elevation, the point will automatically inherit the elevation of the object. If you use the surface options, the point will automatically inherit the elevation of the surface you choose.

Miscellaneous Point Creation Options   The options in the Miscellaneous category are based on manually selecting a location or on an AutoCAD entity, such as a line, pline, and so on. Some common examples include placing points at intervals along a line or polyline as well as converting Softdesk points or AutoCAD points (see Figure 3.10).

CREATING BASIC POINTS | **107**

FIGURE 3.10
Miscellaneous point creation options

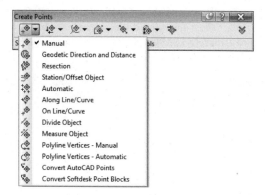

Intersection Point Creation Options The options in the Intersection category allow you to place points at a certain location without having to draw construction linework. For example, if you needed a point at the intersection of two bearings, you could draw two construction lines using the Bearing Distance transparent command, manually place a point where they intersect, and then erase the construction lines. Alternatively, you could use the Direction/Direction tool in the Intersection category (see Figure 3.11).

FIGURE 3.11
Intersection point creation options

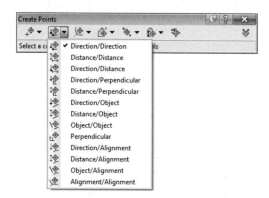

Alignment Point Creation Options The options in the Alignment category are designed for creating stakeout points based on a road centerline or other alignments. You can also set profile geometry points along the alignment using a tool from this menu. See Figure 3.12.

FIGURE 3.12
Alignment point creation options

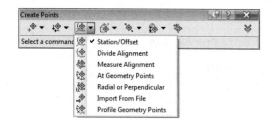

Surface Point Creation Options The options in the Surface category let you set points that harvest their elevation data from a surface. Note that these are points, not labels, and therefore aren't dynamic to the surface. You can set points manually, along a contour or a polyline, or in a grid (see Figure 3.13).

FIGURE 3.13
Surface point creation options

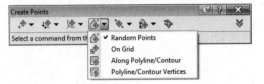

Interpolation Point Creation Options The Interpolation category lets you fill in missing information from survey data or establish intermediate points for your design tasks. For example, suppose your survey crew picked up centerline road shots every 100′ (30 m) and you'd like to interpolate intermediate points every 25′ (8 m). Instead of doing a manual slope calculation, you could use the Incremental Distance tool to create additional points (see Figure 3.14).

FIGURE 3.14
The Interpolation point creation options

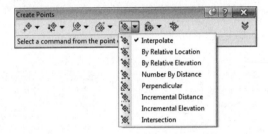

Another use would be to set intermediate points along a pipe stakeout. You could set a point for the starting and ending invert and then set intermediate points along the pipe to assist the field crew.

Slope Point Creation Options The Slope category allows you to set points between two known elevations by setting a slope or grade. Similar to the options in the Interpolation and Intersection categories, these tools save you time by eliminating construction geometry and hand calculations (see Figure 3.15).

FIGURE 3.15
Slope point creation options

Using the Automatic – Object Option

The description option in the point settings, Automatic – Object, can be used only when placing points along an alignment. For example, when you're placing points using the At Geometry Points option, the point will inherit the alignment's name, the station value of the point, and the type of geometry as its description.

For all other point placement options, Automatic – Object will behave exactly the same as Automatic.

Creating Points

In this exercise, you'll learn how to create points along a parcel segment and along a surface contour:

1. Open 0304_PointCreation.dwg (0304_PointCreation_METRIC.dwg), which you can download from this book's web page.

   Note that the drawing includes an alignment, a series of parcels, and an existing ground surface.

2. On the Home tab of the ribbon ➢ Create Ground Data panel ➢ Points menu, click Point Creation Tools.

3. If it is not already expanded, click the chevron icon on the right to expand the dialog.

4. Expand the Points Creation category.

5. Change the Prompt For Elevations value to None and the Prompt For Descriptions value to Automatic by clicking in the respective cell in the Value column, clicking the down arrow, and selecting the appropriate option.

6. Enter **LOT** for Default Description (see Figure 3.16).

FIGURE 3.16
Point creation settings in the Create Points dialog

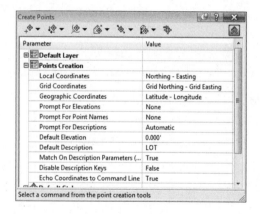

This will save you from having to enter a description and elevation each time. Because you're setting stakeout points for rear lot corners, you will disregard elevation for now.

7. Select the Automatic tool from the Miscellaneous flyout (the first button flyout on the top left of the Create Points toolbar).

8. Select all of the blue property lines in the drawing. Press ↵.

9. Press Esc to exit the command.

   A point is placed at each property corner and at the endpoints of each curve.

10. Select the Measure Object tool from the Miscellaneous flyout.

11. Click anywhere on the parcel boundary for Property 10.

    After you select the parcel boundary for Property 10, this tool prompts you for starting and ending stations.

12. Press ↵ twice to accept the default starting and ending stationing and again to accept the default of 0 for the offset.
13. At the Interval prompt, enter **25** if you are working in feet and **10** if you are working in metric units.
14. Press Esc to exit the command.

    A point is placed at 25′ (10 m) intervals along the property boundary.

    Next you'll experiment with the Direction/Direction option from the Intersection Point flyout. Be sure your Endpoint Osnap is on for the next steps.

15. Click the Direction/Direction icon and click the southeast endpoint of the "floating" parcel line.
16. Click the opposite endpoint to establish the direction of the line.

    The yellow arrow that appears indicates the direction.

17. Press ↵ to specify the default of 0 for the offset.
18. Click the south corner of Property 6 and then the west corner of Property 6.
19. Press ↵ to specify a zero offset.

    A point is generated where the two lines would intersect if they were to be extended. Press Esc to exit the command.

20. In the point creation settings, change Prompt For Descriptions to Automatic – Object.
21. From the Alignments flyout, choose At Geometry Points.
22. When prompted for an alignment, select the green centerline alignment.

    You are now prompted for a profile.

23. Select Layout (1) from the drop-down and click OK.
24. Press ↵ to confirm the starting station value along the alignment. For the ending station, key in **4+60.90** (metric users use **0+140.40**).
25. Press Esc to exit the command.

    You should now see points whose names are based on alignment information.

26. Return to the Point Settings category and change Prompt For Elevations to Manual and Default Description to **EG**.

    The next round of points you'll set will be based on the existing ground elevation.

27. Select the Along Polyline/Contour tool in the Surface flyout.
28. When prompted `Distance between points<10.0000>`, key in **25′ (10 m)** and press Enter to continue.
29. Click the yellow driveway lines near HOUSE2.

30. Save the drawing and keep it open for use in the next exercise.

    Experiment with the plethora of point placement tools available to you!

DOUBLE TROUBLES

Civil 3D does not allow two points to share the same point number or name. If a duplicate point number is detected, Civil 3D will warn you and ask you how you would like to handle it.

The Duplicate Point number dialog will pop up and allow you to take action. You have the choice of several options in the Resolution drop-down:

Add An Offset   This option will allow you to add a value to all incoming points. Specifying an offset of 1000 would turn 1, 2, and 5 into 1001, 1002, and 1005.

Merge   If the existing point has a description but no elevation, and the incoming point has an elevation and no description, Civil 3D will fill in the gaps with the incoming information. If there is no missing data and the coordinates are identical, the incoming point is ignored. Be careful using Merge; it will behave similar to the Overwrite option if the coordinates don't match.

Overwrite   This option deletes the existing point and replaces it with the incoming point.

Sequence From   This option will restart the numbering at a higher value. Unlike with adding an offset, the original point number is ignored. Setting a sequence from 1000 would turn 1, 2, and 5 into 1001, 1002, and 1003.

Use Next Point Number   The default option in the Resolution drop-down, Use Next Point Number, finds the next available point number and imports the point.

Point numbers are assigned using the Point Identity settings in the Create Points dialog or the point file from which they originated. To list available point numbers, enter **ListAvailablePointNumbers** on the command line, or select any point to open the COGO Point contextual tab and choose COGO Point Tools ➢ List Available Point Numbers.

> In the case of duplicate point names, options exist to ensure that you keep these unique as well.
>
> The Duplicate Point Name dialog also gives you options to resolve duplicates. In the Resolution drop-down you will find the following options:
>
> **Counter** This option will take the name and add a numeric counter. For example, if you are attempting to import a duplicate number A100, the new name will become A100(1).
>
> **Specify** This option will give you a chance to type in the name you want for the new point. Even though names are optional for points in general, this option will force you to key in a name before allowing you to proceed.
>
> **Use Name Template** This is the default option which will look back at the point-creation command settings. The existing point name is ignored and overridden by the setting from the point name template.

Basic Point Editing

Despite your best efforts, points will often be placed in the wrong location or need additional editing after their initial creation. Points may need to be rotated as a group to match a different horizontal datum. Points may need to be raised or lowered to match a different benchmark.

Graphic Point Edits

Points can be moved, copied, rotated, deleted, and more using standard AutoCAD commands and grip edits. When you pause your cursor over a grip, a special grip menu will appear with different options. Figure 3.17 (left) shows the grip menu options for the label. Figure 3.17 (right) shows the options available directly on the point. Using the options shown here, you can move the point and rotate it independently of the text and rotate the text label independently of the marker.

FIGURE 3.17
The top grip allows label modifications (left); the center grip allows marker modifications (right).

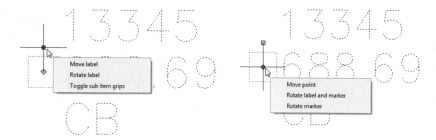

Panorama and Prospector Point Edits

You can access many point properties through the Point Editor in Panorama:

1. Continue working in the drawing from the previous exercise, or open `0304_PointCreation_Finished.dwg` (`0304_PointCreation_METRIC.dwg`) and choose a point (or several points).

Edit/List Points

2. From the COGO Point contextual tab ➢ Modify panel, select Edit/List Points.

   Panorama brings up information for the selected point(s) (see Figure 3.18).

FIGURE 3.18
Edit points in Panorama

Point Nu...	Northing	Easting	Point Elev...	Name	Raw Description	Full Description	Description For
4	196006.3211'	2305168.6365'	821.711'		TP4	TP4	
11	196006.4760'	2305168.5780'	821.615'		OPUS 3109-11	OPUS 3109-11	
708	196040.0894'	2305170.0278'	821.314'		CL DYL	CL DYL	$*
709	196009.9786'	2305179.6236'	821.945'		CL DYL	CL DYL	$*
710	195992.5522'	2305186.0707'	822.370'		CL DYL	CL DYL	$*
711	195964.3460'	2305197.2031'	822.867'		CL	CL	$*
742	195979.0835'	2305179.1747'	822.310'		EM	EM	
743	195999.1276'	2305171.5996'	821.936'		EM	EM	
744	196012.7283'	2305166.3933'	821.588'		EM DW	EM DW	
745	196015.4935'	2305161.0280'	821.718'		EM DW	EM DW	
746	196036.3229'	2305158.2661'	821.153'		EM DW	EM DW	
747	196036.1508'	2305157.8396'	821.407'		EM DW	EM DW	
748	196036.8479'	2305174.6820'	821.425'		EM	EM	
749	196021.1647'	2305188.1071'	821.291'		EM	EM	
750	196020.2548'	2305188.0515'	821.290'		EM	EM	
751	196010.7712'	2305195.5571'	821.282'		EM	EM	
752	196007.1122'	2305202.3786'	821.337'		FC GUTTER	FC GUTTER	

You can access a similar interface in the Prospector tab of Toolspace by following these steps:

1. Highlight the Points collection (see Figure 3.19).

2. In either location, right-click the point or points you wish to examine and select Zoom To.

FIGURE 3.19
Prospector lets you view your entire Points collection at once.

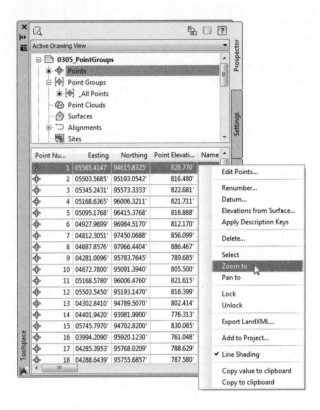

Prospector works like a spreadsheet in many ways. You can sort by any column by clicking the heading of the column. Clicking the column a second time will sort the data in the opposite order. For example, if you'd like to list your points alphabetically by description, click the Description column heading.

3. Click the Raw Description column heading to resort the points.

Point Groups: Don't Skip This Section!

Certification Objective

Working with point groups is one of the most powerful techniques you will learn from this chapter. Want to turn all your points off without touching layers? Make a point group! Want to move last week's survey up by the blown instrument height difference? Make a point group! Want to show all your Topo shots as dots rather than Xs? Want to prevent invert shots from throwing off your surface model? Point group! Point group!

A *point group* is a collection of points that has been filtered for a certain criterion. You can use any point property or combination of properties such as description, elevation, and point number, or you can select specific points in the drawing.

Civil 3D creates the _All Points group for you, which contains every point in the drawing. It cannot be renamed or deleted, nor can it have its properties modified to exclude any points. Create point groups for collections of points you might wish to separate from others, as shown in Figure 3.20.

FIGURE 3.20
An example of useful point groups in Prospector

Point groups can (and should!) be created upon import of a text file, as shown previously in Figure 3.7. That way, if a problem comes to light about that group of points (such as incorrect instrument height), they can be isolated and dealt with apart from other points. Create a new point group by right-clicking the main point group category and selecting New.

In this exercise, you'll learn how to use point groups to separate points into usable categories:

1. Open the drawing `0305_PointGroups.dwg` (`0305_PointGroups_METRIC.dwg`), which you can download from this book's web page.

2. In Prospector, right-click Point Groups and select New.

3. On the Information tab, name the point group **Vegetation**.

4. Set Point Style to Tree.

5. Set Point Label Style to Description Only.

6. Switch to the Include tab and place a check mark next to With Raw Descriptions Matching.

7. In the Raw Descriptions Matching field, type **TREE***, **SHRUB***, **TL***.

   The asterisk acts as a wildcard to include points that may have additional information after the description. You are adding multiple descriptions by separating them with a comma, as shown in Figure 3.21.

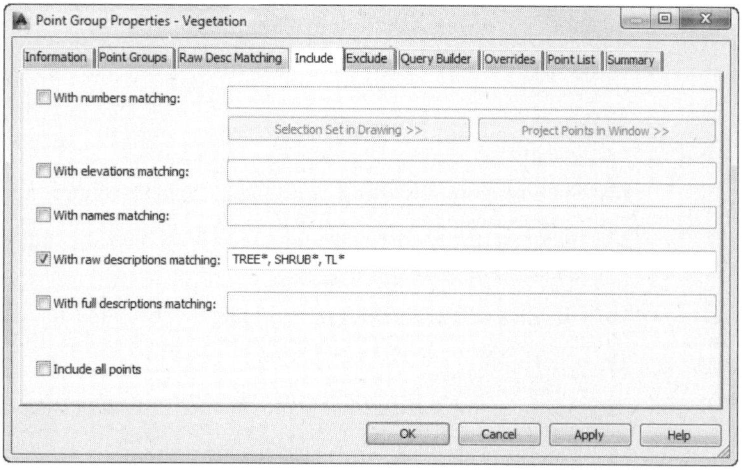

FIGURE 3.21
The Include tab of the Vegetation point group properties

8. Switch to the Overrides tab.
9. Place a check mark next to Style and Point Label Style, as shown in Figure 3.22.

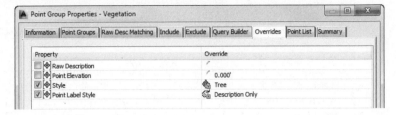

FIGURE 3.22
Overrides force the styles to conform by point group rather than the description key.

Doing so ensures that the point group, instead of the description keys, will control the styles.

10. Switch to the Point List tab and examine the points that have been picked up by the group.

   Only points beginning with SHRUB, TREE, and TL should appear in the list.

11. Click OK.
12. Again, right-click Point Groups and select New.
13. On the Information tab, name the group **NO DISPLAY**, as shown in Figure 3.23.

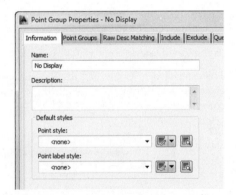

FIGURE 3.23
Most drawings should contain a NO DISPLAY point group with styles set to <none>.

14. Set both Point Style and Point Label Style to <none>.
15. Switch to the Include tab, and at the bottom of the dialog, put a check mark next to Include All Points.
16. Switch to the Overrides tab and place a check mark next to Style and Point Label Style.
17. Click OK.

   All the points are hidden from view as a result of the point group.

18. Create another point group called **Topo**.
19. Set Point Style to Basic X (BLACK) and Point Label Style to Elevation And Description.

BASIC POINT EDITING   117

20. Switch to the Exclude tab and place a check mark next to With Elevations Matching.
21. Type **<=0** in the accompanying field.
22. Place a check mark next to With Raw Descriptions Matching.
23. Type **INV*, HYD*** in the accompanying field, as shown in Figure 3.24, and click OK. You may need to use the AutoCAD REGEN command to refresh the graphic and show your points again.

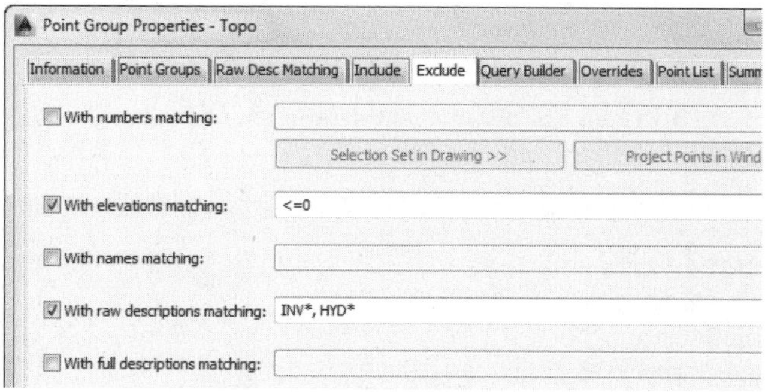

FIGURE 3.24
Use Exclude to create a Topo point group.

24. Save the drawing.

Point Groups will dictate point styles only if the overrides are set or if no matches exist in the description key set. Otherwise, the description key set sets the style. For example, if your point placement options have a layer set but you place a point that matches a description key with a layer set to something different, the description key set "wins." In the point group creation examples, you set the Overrides tab to have Style and Point Label Style selected (shown previously in Figure 3.22). Those settings wrestle control of the styles away from the description key and into the hands of the point group.

BEST PRACTICE: CONTROL POINT DISPLAY USING POINT GROUPS RATHER THAN LAYERS

Civil 3D drawings will have many layers in them. It is much easier to switch the display of the point groups rather than create layer states for each point visibility scenario.

A point can belong to more than one group at once. For instance, a water valve cover with elevation may be in a Topo group, a Utilities group, and the _All Points group. In these cases, the order in which the point groups are displayed in Prospector determines which point group a point is "listening to" for its properties.

In this exercise, we will walk you through an example of how point group display order works:

1. Open the drawing `0306_PointDisplay.dwg` (`0306_PointDisplay_METRIC.dwg`), which you can download from this book's web page.
2. In Prospector, right-click Point Groups and select Properties (see Figure 3.25).

FIGURE 3.25 Select Point Groups ➢ Properties to change point group display precedence.

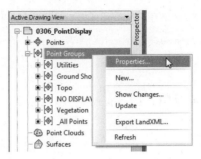

3. Using the arrows on the far right of the Point Groups listing, select Vegetation and move it to the top of the list.

4. Move NO DISPLAY so that it is listed directly below Vegetation, as shown in Figure 3.26.

FIGURE 3.26 The order in which the point groups appear in this list controls precedence.

5. Click OK.

Notice that only the Vegetation group is visible.

6. Experiment with changing the order of the point groups using the properties.

Changing Point Elevations

Points placed on or along an object that has elevation will automatically inherit the object's elevation. Points placed with tools in the Surface flyout will automatically inherit the elevation of a surface model. If you have chosen to place points with manual elevation entry and press ↵ when prompted to specify an elevation, the elevation will be null (no elevation).

You are never stuck with a COGO point's elevation. They can be changed individually or as a group using the Panorama window. Additional tools are available for manipulating points (see Figure 3.27) in the COGO Point contextual tab that opens when you select a point object.

FIGURE 3.27 Point-editing commands in the ribbon

Elevations From Surface is an extremely handy tool for forcing points to a surface elevation (see Figure 3.28).

FIGURE 3.28
Shrub points as placed (a); shrub points moved up to surface elevation (b)

(a) 13346 0.00 SHRUB 3 | 13347 0.00 SHRUB 5 | 13348 0.00 SHRUB 6 | 13349 0.00 SHRUB 2 | 13350 0.00 SHRUB 3 | 13351 0.00 SHRUB 4 | 13352 0.00 SHRUB 2

(b) 13346 823.03 SHRUB 3 | 13347 822.95 SHRUB 5 | 13348 822.83 SHRUB 6 | 13349 822.81 SHRUB 2 | 13350 822.83 SHRUB 3 | 13351 822.84 SHRUB 4 | 13352 822.98 SHRUB 2

In the event that you change the datum, you are most likely going to move a group of points' elevations. Right-click on the name of the point group in Prospector and select Edit Points. Panorama will appear for your point-editing delight.

Use Windows keyboard tricks to control which points are selected for modification. Pressing Ctrl+A will select all points in the Panorama listing, as shown in Figure 3.29. When you are done selecting points, right-click and choose Datum. The command line will prompt you to specify the change in elevation you require.

FIGURE 3.29
Right-click to access point modification tools from Panorama.

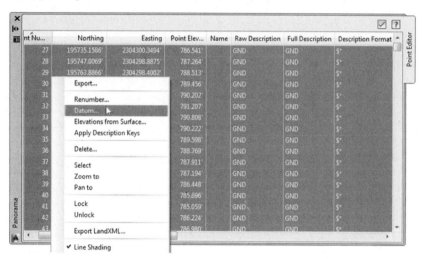

Point Tables

You've seen some of the power of dynamic point editing; now let's look at how those dynamic edits can be used to your advantage in point tables.

Most commonly, you may need to create a point table for survey or stakeout data; it could be as simple as a list of point numbers, northing, easting, and elevation. These types of tables are easy to create using the standard point-table styles and the tools located in the Points menu under the Add Tables option. Follow these steps to create a point table:

1. Open the `0307_PointTable.dwg` (`0307_PointTable_METRIC.dwg`) file, which you can download from this book's web page.

This file will appear empty, but it isn't. First, you will review reordering point group properties to change the display of points. You don't need to see points to make a point table from a group, but this will help you see that the table reflects the specific group.

2. In Prospector, expand Point Groups.

   You will see that NO DISPLAY is listed on the top, which means the styles set in its properties are taking over the other point groups.

3. Right-click Point Groups and select Properties.

4. In the listing, move the group Trees To Be Removed to the top by using the arrows on the right. Click OK.

5. Use Zoom Extents to see the cluster of trees you are working with.

6. Switch to the Annotate ribbon tab and click Add Tables ➤ Add Point Table.

7. Verify that Table Style is set to Tree Removal, and click the Point Group icon.

8. Select Trees To Be Removed, and click OK.

9. Verify that the check box next to Split Table is clear. When all your settings match those in Figure 3.30, click OK.

FIGURE 3.30
Point Table Creation options

10. Click anywhere in the graphic to place the table.

User-Defined Properties

Standard point properties include items such as number, easting, northing, elevation, name, description, and the other entries you see when examining points in Prospector or Panorama. But what if you'd like a point to know more about itself?

It's common to receive points from a soil scientist that list additional information, such as groundwater elevation or subsurface elevation. Surveyed manhole points often include invert elevations or flow data. Tree points may also contain information about species or caliber measurements. All this additional information can be added as user-defined properties to your point objects. You can then use user-defined properties in point labeling, analysis, point tables, and more.

How Can Civil 3D Work with Soil Boring Data?

There are several options for users that want to work with sub-surface data in Civil 3D. One option is the Autodesk Geotechnical Module. This subscription-only tool allows users to import borehole data in several formats including AGS31, AGS4, CSV and from a Keynetix HoleBase database. The boreholes created with this tool are not point objects. For more information about the Autodesk Geotechnical Module, log into subscription.autodesk.com.

The option discussed in this inset involves using user-defined properties on points to enter subsurface data. This is an option that does not require any special downloads.

In the following example, you will add user-defined properties to some soil boring points and leverage point groups to work with the data. The skills you learn in this example can be applied to multiple soil boring values for the purposes of creating subsurface data.

1. Open the file 0308_SoilBorings.dwg (0308_SoilBorings_METRIC.dwg).

    This file contains an existing ground surface along with several point groups, including one containing soil boring points.

2. Take a moment to examine the soil boring points and the current elevation listing.

3. From the Settings tab of Toolspace, expand the Point collection, right-click User-Defined Property Classifications and select New.

4. Name the new classification **Soil Borings** and click OK.

5. Expand the User-Defined Property Classifications category (if it is not already), and right-click Soil Borings. Select New.

6. Name the new property **Watertable Elevation**.

7. Set Property Field Type to Elevation, uncheck Default Value, and then click OK.

8. Jump back to the Toolspace ➢ Prospector tab, and highlight the main Point Groups listing. At the very bottom of Toolspace you will see a listing of all the point groups, as seen here.

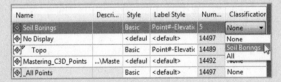

9. Set the classification as shown in the screen shot.
10. Right-click the Soil Borings point group and select Edit Points.
11. Scroll over in Panorama until you locate the new classification column named Watertable Elevation.

    This is the information you added in the previous steps.
12. Add the Watertable Elevation entries in Panorama, as shown here. Dismiss Panorama when you're done.

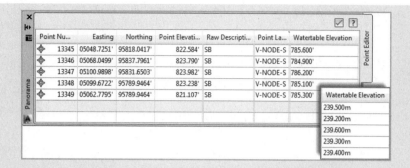

13. Right-click the Soil Borings group and select Properties. Switch to the Overrides tab, as shown here.

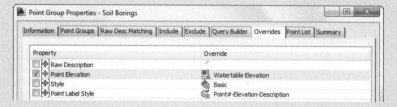

14. Place a check mark next to Point Elevation on the Overrides tab.

15. Click the tiny pencil icon twice, or until it turns into the user-defined property icon. Initially the value will be <none>.

16. Click the field next to the icon to set the value to Watertable Elevation.

17. Click OK to dismiss the Point Group Properties dialog.

Notice that the elevation labels for the five points are listed as the water table elevations.

The Bottom Line

Import points from a text file using description key matching. Most engineering offices receive text files containing point data at some time during a project. Description keys provide a way to automatically assign the appropriate styles, layers, and labels to newly imported points.

Master It Create a new drawing from _AutoCAD Civil 3D (Imperial) NCS.dwt or _AutoCAD Civil 3D (Metric) NCS.dwt. Revise the Civil 3D description key set to contain only the parameters listed here:

Code	Point style	Point label style	Format	Layer
GS*	Basic	Elevation Only	Ground Shot	V-NODE
GUY*	Guy Pole	Elevation and Description	Guy Pole	V-NODE
HYD*	Hydrant (existing)	Elevation and Description	Existing Hydrant	V-NODE-WATR
TOP*	Basic	Point#-Elevation-Description	Top of Curb	V-NODE
TREE*	Tree	Elevation and Description	Existing Tree	V-NODE-TREE

Import the PNEZD (space delimited) file `MasterIt0301.txt` (`MasterIt0301_METRIC.txt`). Confirm that the description keys made the appropriate matches by looking at a handful of points of each type. Do the trees look like trees? Do the hydrants look like hydrants?

Save the resulting file for use in the remaining exercises.

Create a point group. Building a surface using a point group is a common task. Among other criteria, you may want to filter out any points with zero or negative elevations from your Topo point group.

Master It Create a new point group called Topo that includes all points except those with elevations of zero or less. Use the DWG created in the previous Master It exercise or start with `MasterIt0301_FINISHED.dwg` (`MasterIt0301_METRIC_FINISHED.dwg`).

Export points to LandXML and ASCII format. It's often necessary to export a LandXML or ASCII file of points for stakeout or data-sharing purposes. Unless you want to export every point from your drawing, it's best to create a point group that isolates the desired point collection.

Master It Create a new point group that includes all the points with a raw description of TOP. Export this point group via LandXML to a PNEZD comma-delimited text file. Use the DWG created in the previous Master It exercise or start with `MasterIt0302_FINISHED.dwg` (`MasterIt0302_METRIC_FINISHED.dwg`).

Create a point table. Point tables provide an opportunity to list and study point properties. In addition to basic point tables that list number, elevation, description, and similar options, you can customize point table formats to include user-defined property fields.

Master It Use the DWG created in the previous Master It exercise or start with `MasterIt0303_FINISHED.dwg` (`MasterIt0303_METRIC_FINISHED.dwg`). Create a point table for the Topo point group using the PNEZD format table style.

Chapter 4

Surfaces

One of the most fundamental elements in a three-dimensional model of any design is the surface. As you learned in the previous chapter, once survey information is gathered and points are set with elevations, you can proceed to turn some of that information into an intelligent surface. This chapter examines various methods of surface creation and editing. Then it moves into discussing ways to view, analyze, and label surfaces and explores how they interact with other parts of your project.

In this chapter, you will learn to:

- Create an existing ground surface using points
- Modify and update a TIN surface
- Prepare a slope analysis
- Label surface contours and spot elevations
- Import a point cloud into a drawing and create a surface model

Understanding Surface Basics

A surface in the AutoCAD® Civil 3D® program is generated using the principle of geometric triangulation. At the very simplest, a surface consists of points. The computer generates a triangular plane using a group of three points. Each triangular plane shares an edge with another, and a continuous surface is made. This methodology is referred to as a *triangulated irregular network (TIN)*, as shown in Figure 4.1.

FIGURE 4.1
A triangulated irregular network, or TIN

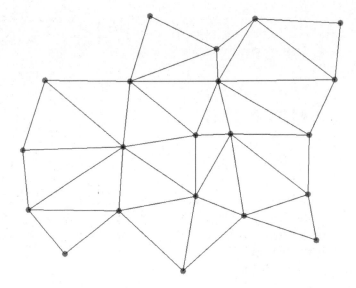

For any given (*x*,*y*) point, there can be only one unique *z* value within the surface (since slope is equal to rise over run, when the run is equal to 0, the result is "undefined"). What does this mean to you? It means surfaces created by Civil 3D have several rules:

No Thickness Modeled surfaces can be thought of as a sheet draped over a surface; they have no thickness in the vertical direction associated with them.

No Vertical Faces Vertical faces cannot exist in a TIN because two points on the surface cannot have the same (*x*,*y*) coordinate pair. Vertical walls or curb structures must have a slight offset of at least 0.001′ (0.0001 m); otherwise they will appear flattened.

No Caves or Tunnels The triangular planes that make up a TIN cannot overlap. If a design requires any tunnel-like structures, you will need to accomplish this with several surfaces instead.

> **PAGING DR. DELAUNAY**
>
> The math behind many software surface generation tools can be attributed to Russian mathematician and avid mountaineer Dr. Boris Delaunay. Civil 3D surface calculations are based on the Delaunay triangulation algorithm. You don't need to understand the algebra behind this technique, but it is the reason behind both the accuracy and quirks of surface modeling.
>
> Delaunay triangulation is essentially a set of geometry rules for optimizing the relationship between points and works by finding groups of three points that are used to form triangles. Once you have three points, you can also define a circle that passes through those points.
>
> These special circles, called *circumcircles*, don't appear graphically in Civil 3D, but they determine the initial relationship between points (before we intervene with the surface editing tools, of

course). There can't be any points inside of the circles, and all points must be accounted for with a circumcircle. This is why some triangle legs near the edges of a site will connect two points that, from a surveying perspective, don't relate to each other, such as back of curb and centerline of a road.

The Delaunay triangulation is just the starting point. Many of the surface tools you will learn about in this chapter force the Delaunay triangulation rules out the window. Surface properties, breaklines, boundaries, and other edits all affect how those triangles are formed.

There are three main categories of surfaces in Civil 3D: standard surfaces, volume surfaces, and corridor surfaces. A standard surface is based on a single set of points, whereas a volume surface is built by measuring vertical distances between two standard surfaces. Each of these two categories of surfaces can also be a grid or TIN surface.

The grid version is still a TIN upon calculation of planar faces, but the data points are arranged in a regularly spaced grid of information. The TIN version is made from randomly located points that may or may not follow any pattern to their location. A corridor surface is generated from a corridor and will be discussed further in Chapter 9, "Basic Corridors."

Creating Surfaces

Certification Objective

When you first create a surface in Civil 3D, you will give it a name and set its style. Initially, your surface will contain no data; your next step is to add data to the surface definition.

In Prospector, you can view the contents of the surface by expanding its branch and then further expanding the Definition area, as shown in Figure 4.2.

FIGURE 4.2
Create a new surface (left). Expand a surface's definition to add or modify elevation data (right).

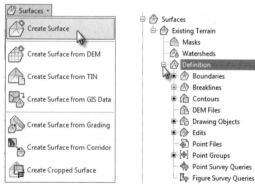

The following components can be used as part of a surface definition:

Boundaries *Boundaries* are closed polylines that determine the visibility of the TIN inside the polyline. A boundary can be a 2D polyline, a 3D polyline, or even a feature line, but only the horizontal information will be used to generate the boundary—the elevation of the polyline is ignored. If the polyline that created the boundary is modified, the surface will become out of date, thus requiring a rebuild. More detail about the types of boundaries that can be created is found in the section later in this chapter called "Refining and Editing Surfaces."

> **BOUNDARY VS. BORDER**
>
> In Civil 3D, the terms *boundary* and *border* mean two different things. A *boundary* is a user-created restriction on triangulation and is optional in your surface definition. A *border* refers to the outermost triangle legs of a surface. Borders always exist, even if they are not displayed graphically. In the first exercise of this chapter, the surface you create will not have a boundary, but you will see the border displayed in yellow.

Breaklines *Breaklines* are used for creating hard-coded triangulation paths, even when those paths violate the Delaunay algorithms for normal TIN creation. They can describe anything from the top of a ridge to the flowline of a curb section. A TIN line may not cross the path of a breakline. A breakline cannot be added to a grid surface. Breaklines can be defined using 2D polylines, 3D polylines, survey figures, or feature lines. Similar to boundaries, if a breakline is modified, the surface will become out of date, thus requiring a rebuild. Breaklines will be discussed in more detail later in this chapter.

Contours Sometimes a specific contour is desired, and it can be inserted into the surface as a 2D polyline at an elevation. Points will be placed along the contour to be used in the triangulation process. This process will be discussed in more detail later in this chapter. Similar to a breakline, a contour cannot be added to a grid surface. Similar to boundaries, if a breakline is modified, the surface will become out of date, thus requiring a rebuild. Adding contour data will be discussed further later in this chapter.

DEM Files *Digital Elevation Model (DEM)* files are the standard format files from governmental agencies and GIS systems. These files are typically very large in scale but can be great for planning purposes.

Drawing Objects AutoCAD objects that have an insertion point at an elevation (e.g., text, blocks, lines, AutoCAD points, 3D faces, or polyfaces) can be used to populate a surface with elevation data. For text and blocks, the text insertion point z position is used as the elevation. Changes to these objects will not cause the out of date flag to appear on the surface.

Edits Any manipulation after the surface is completed, such as adding or removing triangles or changing the datum, will be part of the edit history. These changes can be viewed in the surface properties, where edits can be toggled on and off individually to make reviewing changes simple. They can also be reordered, since edits are implemented in the order that they are added.

Point Files Point files work well when you're working with large data sets where the points themselves don't necessarily contain extra information. Examples include laser scanning or aerial surveys. A drawing will stay referenced to a point file. If the point file is moved or deleted, the reference in the drawing will be broken.

Point Groups Civil 3D point groups or survey point groups can be used to build a surface from their respective members and maintain the link between the membership in the point group and being part of the surface. In other words, if a point is removed from a group used in the creation of a surface, it is also removed from the surface.

Point Survey Queries and Figure Survey Queries Adding a Point Survey Query and a Figure Survey Query perform similar tasks. A saved survey query created on the Survey Tab of Toolspace can be used similarly to a point group for populating surface elevation data. If the query contains both points and figures, you can choose to use both types of data, only points, or only figures. Unlike a point group, data from the query can be added without the points or figures being inserted in the drawing.

Working with all these elements, you can model and render almost any surface you'd find in the world. In the exercise that follows, you'll build your first basic surface:

1. Open the file 0401_SurfaceFromPointGroup.dwg (0401_SurfaceFromPointGroup_METRIC.dwg), which you can download from this book's web page, www.sybex.com/go/masteringcivil3d2014.

2. On the Home tab of the ribbon ➢ Create Ground Data panel, select Surfaces ➢ Create Surface.

3. In the Create Surface dialog, click in the field for name.

4. Remove the default text by highlighting it and replace with the name **Existing Surface**.

5. Leave all other options as shown in Figure 4.3 and click OK.

FIGURE 4.3
Creating your first new TIN surface

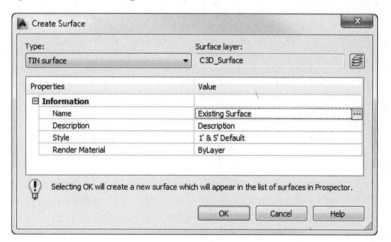

6. In Prospector, expand the Surfaces branch. You should now see your new surface in the listing.

7. Expand Existing Surface and then expand Definition by clicking the tiny plus sign to the left of the listing in Prospector.

8. Right-click Point Groups and select Add.

9. In the Point Groups dialog, highlight TOPO Shots and click OK. You should now see contours for your surface.

10. Save your drawing at the conclusion of the exercise. Compare your work with 0401_SurfaceFromPointGroup_FINISHED.dwg or 0401_SurfaceFromPointGroup_METRIC_FINISHED.dwg if desired.

> **THE YELLOW EXCLAMATION POINT FLAG**
>
> At some point you are bound to see a yellow exclamation point status icon in Prospector. This is a flag showing you that some elements are out of date and require rebuilding. In the image shown here, the EG surface needs to be rebuilt because the Point Files branch is out of date.
>
>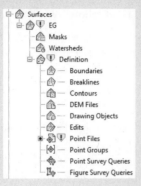
>
> No matter what type of definition in a surface is out of date, to rebuild the surface, right-click the surface's name (in this example that would be EG) and select Rebuild. You could also select Rebuild Automatic, which would result in the surface always rebuilding when required instead of you always having to manually select Rebuild.

Additional Surface Creation Methods

Many forms of data can be digested by Civil 3D to create a surface. The data types discussed in the following sections are common when working with organizations that are not using Civil 3D.

File types like TIN and LandXML are often best for moving data from outdated civil design software programs into Civil 3D. DEM files are the data type of choice when working with many government organizations.

TIN Files

Typically, a TIN file comes from a land development project on which you or a peer has worked. These files contain the baseline TIN information from the original surface and can be used to replicate it easily.

LandXML Files

These typically come from an outside source or are exported from another project. LandXML has become a common means of communicating data in the land development industry. These files include information about points and triangulation, making replication of the original surface as easy as a few mouse clicks.

Digital Elevation Models

Digital Elevation Model (DEM) files are used by the US Department of the Interior's United States Geological Survey (USGS) and are commonly produced by government organizations for their GIS systems. The DEM format can be read directly by Civil 3D, but the USGS typically distributes the data in a complex format called Spatial Data Transfer Standard (SDTS). The files can be converted using a freely available program named sdts2dem. This DOS-based program converts the files from the SDTS format to the DEM format you need. Once you are in possession of a DEM file, creating a surface from it is relatively simple, as you'll see in this exercise:

1. Start a new drawing from the _AutoCAD Civil 3D (Imperial) NCS template that ships with Civil 3D. Metric users should use the _AutoCAD Civil 3D (Metric) NCS template.

2. Switch to the Settings tab of Toolspace, right-click the drawing name, and select Edit Drawing Settings.

   The Drawing Settings dialog appears.

3. Imperial users should set the zone category to USA, Pennsylvania, and set the coordinate system to NAD83 Pennsylvania State Planes, South Zone, US Foot (PA83-SF), as shown in Figure 4.4, via the Units And Zone tab of the Drawing Settings dialog. Metric users should set the zone category to USA, Pennsylvania, and set the coordinate system to NAD83 Pennsylvania State Planes, South Zone, Meter (PA83-S).

FIGURE 4.4
Imperial coordinate settings for DEM import

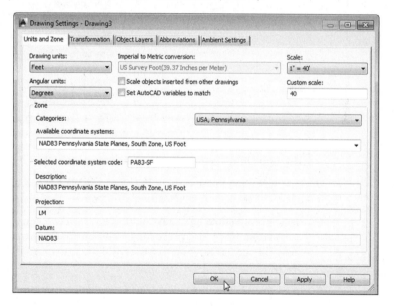

4. All users, accept all other defaults. Click OK.

   The coordinate system of the DEM file that you will import will be set to adjust to the coordinate system of the drawing.

5. From the Home tab ➢ Create Ground Data panel, choose Surfaces ➢ Create Surface.

    The Create Surface dialog appears. You may notice that there is also a Create DEM Surface option under the Surfaces drop-down. While this may seem like a prudent option since you are using DEM data, the drawback to that method is that no coordinate transformation is possible—and you need it for this example.

6. Accept the options in the dialog, and click OK to create the surface.

    This surface is added as Surface1 to the Surfaces collection.

7. In Prospector, expand the Surfaces ➢ Surface1 ➢ Definition branch.

8. Right-click DEM Files and select the Add option (see Figure 4.5).

FIGURE 4.5
Adding DEM data to a surface

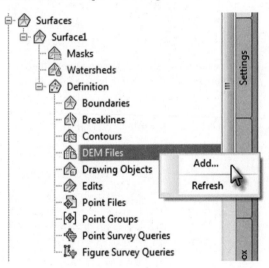

The Add DEM File dialog appears.

9. Use the button to the right of the DEM File Name area to navigate to the 0402_Stewartstown_PA.DEM file and click Open.

    Remember, all data and drawing files for this book can be downloaded from www.sybex.com/go/masteringcivil3d2014. The DEM file information will populate in the Add DEM File dialog showing that the DEM file you are using is UTM Zone 18, NAD27 Datum, Meters.

10. In the Add DEM File dialog, click in the Value column next to CS Code to display the ellipsis button; click that button to display the Select Coordinate Zone dialog.

11. Both Imperial and metric users, set the coordinate system code (CS code) to match the DEM file by selecting UTM With NAD27 Datum, Zone 18, Meter; Central Meridian 75d W (UTM27-18), as shown in Figure 4.6, and click OK.

FIGURE 4.6
Setting the 0402_Stewartstown_PA.dem coordinate zone

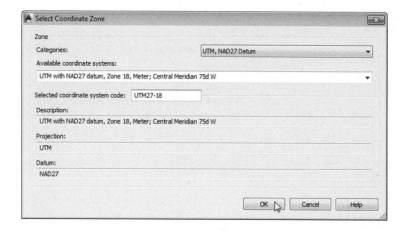

This information is necessary to properly translate the DEM's coordinate system to the drawing's coordinate system.

The Add DEM File dialog should now match the one shown in Figure 4.7.

FIGURE 4.7
Setting the 0402_Stewartstown_PA.DEM file properties

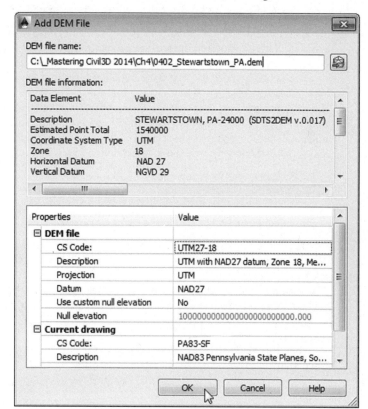

12. Click OK in the Add DEM File dialog. Importing this DEM file might take a few minutes. You can check the status of the import in the lower-left corner of the application window. The message will be "Reading points from file" and then "Adding points to surface" with a progress gauge.

    You may receive a message regarding the level of detail, which is discussed later in this chapter. If a Level Of Detail dialog appears, click the Do Not Enter Level Of Detail Display Mode option.

13. In Prospector, right-click Surface1 and select Zoom To to bring the surface into view.

14. Select the surface. From the Surface contextual tab of the ribbon ➤ Modify panel, click Surface Properties.

    The Surface Properties dialog appears. Earlier, in the Create Surface dialog, you allowed the default surface name to be used, which created the name of Surface1. Since the default name does not provide much information, you will now revise the default name to something that offers more information to the user.

15. On the Information tab, change the Name field entry to **Stewartstown PA**.

16. Change the Surface Style drop-down list to Contours And Triangles, and then click OK to accept the settings in the Surface Properties dialog.

Once you have the DEM data imported, you can pause over any portion of the surface and see that feedback showing the surface elevation is provided through a tooltip. This surface can be used for preliminary planning purposes but isn't accurate enough for construction purposes.

The main drawback to DEM data is the sheer bulk of the surface size and point count. The `0402_Stewartstown_PA.DEM` file you just imported contains 1.4 million points and covers more than 55 square miles. This much data can be overwhelming, and it covers an area much larger than the typical site. If you try zooming in and out on the surface, you will notice that the computer will be slow as it tries to regenerate the surface with each change. To ease the processing and activate the Level Of Detail display, do the following:

1. Switch to the View tab.

2. Expand the bottom of the Views panel.

3. Select Level Of Detail.

After these steps are complete you will notice that a new icon appears in the upper-left corner of your model space, showing you that Level Of Detail is activated. To turn off Level Of Detail, follow the same steps.

Turning on the Level Of Detail display does not change the data in the surface but simply changes what is viewable at the different zoom levels. Figure 4.8 shows the same area of the surface before turning on Level Of Detail (left) and after (right). You'll look at some data reduction methods later in this chapter.

When this exercise is complete, you may close the drawing. Due to the large file size, a finished state of this drawing is not available for download on the book's web page.

FIGURE 4.8
DEM surface: shown without Level Of Detail (left), zoomed out with Level Of Detail (right)

 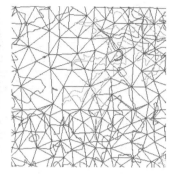

SURFACE FROM GIS DATA

Whether the source is internal or from an outside firm, elevation data from *Geographic Information Systems (GIS)* is becoming increasingly common. Civil 3D understands GIS and can work with the data given. In this section, we'll show you how to import GIS data pertaining to surfaces:

1. Start a new drawing by using the `_AutoCAD Civil 3D (Imperial) NCS` template and set the coordinate system to NAD83 Georgia State Planes, West Zone, US Foot (GA83-WF). Metric users should use the `_AutoCAD Civil 3D (Metric) NCS` template and set the coordinate system to NAD83, Georgia State Planes, West Zone, Meter (GA83-W).

2. From the Home tab ➢ Create Ground Data panel, choose Surfaces ➢ Create Surface From GIS Data.

   The Create Surface From GIS Data – Object Options page appears.

3. Set Name to **GIS Data**, change Description to **Import from GIS Data**, and set the style to Contours 5′ and 25′ (Background), or Contours 2 m and 10 m (Background) for metric users as shown in Figure 4.9, and click the Next button.

FIGURE 4.9
The Create Surface From GIS Data – Object Options page

The Create Surface From GIS Data – Connect To Data page appears.

4. You are importing a SHP file, so change Data Source Type to SHP.

5. Click the ellipsis next to SHP Path. Locate the 0403_contours2008.shp file (which you'll find at www.sybex.com/go/masteringcivil3d2014).

The path is now populated with the location of the SHP file, as shown in Figure 4.10.

FIGURE 4.10
The Create Surface From GIS Data – Connect To Data page

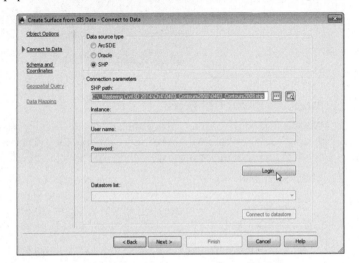

6. Click the Login button.

Don't worry; you won't actually need a username or password to log in.

The Create Surface From GIS Data – Schema And Coordinates page now appears (Figure 4.11).

FIGURE 4.11
The Create Surface From GIS Data – Schema And Coordinates page

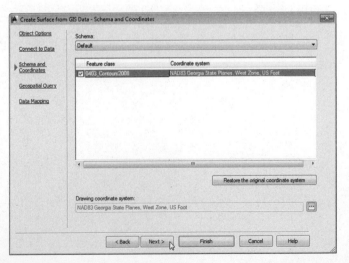

You will notice that the name of the file appears as well as the coordinate system in which the SHP was created. If you are an Imperial user, the NAD83 Georgia State Planes, West Zone, US Foot matches what you set the drawing up with, as shown in Figure 4.11. For metric users, the coordinate system of the SHP file is different than the drawing coordinate system.

7. Verify that the 0403_Contours2008 check box is checked under Feature Class and click Next.

8. On the Create Surface From GIS Data – Geospatial Query page, look at the settings for future reference but do not make any changes (Figure 4.12). Click Next.

FIGURE 4.12
The Create Surface From GIS Data – Geospatial Query page

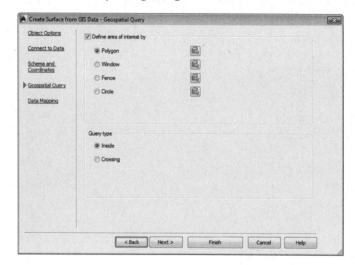

9. On the Create Surface From GIS Data – Data Mapping page, click the drop-down list next to Elevation in the GIS Field column and select Elevation, as shown in Figure 4.13.

FIGURE 4.13
The Create Surface From GIS Data – Data Mapping page

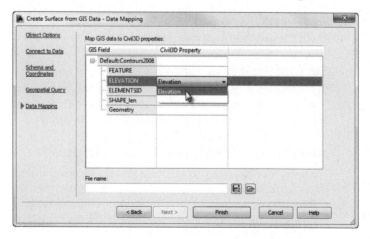

Many dialogs throughout the software use tables such as those shown on this page for you to input data. If at any time the column is not wide enough for you to view all of the contents, you may modify the column width by clicking between the column headings.

At the bottom of the Create Surface From GIS Data – Data Mapping page you will notice a File Name field as well as save and open icons. That means you can save the current data mapping information that you've set for future use.

10. Click the Finish button.
11. Dismiss Panorama and zoom extents to see the surface based on the SHP file (Figure 4.14).

FIGURE 4.14
The finished imported GIS contours

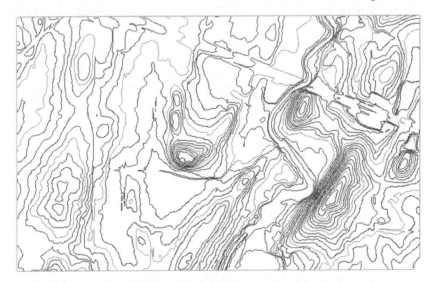

When this exercise is complete, you may close the drawing. A saved finished copy of this drawing is available from the book's web page with the filename 0403_SurfaceGIS_FINISHED.dwg or 0403_SurfaceGIS_METRIC_FINISHED.dwg.

This is just another avenue for getting data from other sources into Civil 3D.

Surface Approximations

In the following sections, you'll work with polylines at elevation. You may recognize these polylines as representing surface elevations, but Civil 3D will not recognize these lines as a surface object until you tell the software what those lines mean.

Surfaces from Polyline Information

The quality of the data you receive if you are working with contours varies greatly. The contour interval and smoothing factors will affect the result when imported to Civil 3D. The resulting surface may not accurately reflect the original survey data. This is because elevation information is provided along the contour lines but not in between the contour lines, causing the interpolation between the contours to lack accuracy. Civil 3D includes a series of surface algorithms that work very well at matching the resulting surface to the original contour data by providing additional derived data points. You'll look at those surface edits in this series of exercises:

1. Open the `0404_SurfaceFromPolylines.dwg` file (or the `0404_SurfaceFromPolylines_METRIC.dwg` file).

   Note that the contours in this file are composed of polylines with elevation values.

2. In Prospector, right-click the Surfaces branch and select the Create Surface option.

   The Create Surface dialog appears.

3. Leave the Type field set to TIN Surface but change the Name value to **EG-Polylines**.

4. Change Description to **Surface From Polylines**.

5. Click in the Value column next to Style to display the ellipsis button; once it's visible, click the ellipsis button to display the Select Surface Style dialog.

6. From the drop-down list, select Contours 1' And 5' (Background), or Contours 0.2 m And 1 m (Background) for metric users, and click OK to close the Select Surface Style dialog.

7. Click OK to close the Create Surface dialog.

8. In Prospector, expand the Surfaces ➢ EG-Polylines ➢ Definition branch.

9. Right-click Contours and select the Add option.

   The Add Contour Data dialog appears.

10. Set Description to **Polylines**. Under Weeding Factors, set Distance to **15** (or **5** for metric users) and Angle to **4** degrees. Under Supplementing Factors, set Distance to **100** (or **30** for metric users) and Mid-Ordinate Distance to **1** (or **0.3** for metric users).

11. Verify that none of the check boxes are checked, as shown in Figure 4.15, and click OK.

FIGURE 4.15
The Add Contour Data dialog

In this example it is okay to leave all the Minimize Flat Areas By options unchecked. You will return to the Minimize Flat Areas By options in a bit.

12. At the `Select contours:` prompt, enter **ALL** ↵ to select all the entities in the drawing and press ↵ again to end the command.

You can dismiss Panorama if it appears by clicking the green check mark. Save the drawing and keep it open for the next portion of the exercise.

The contour data has some tight curves and flat spots where the basic contouring algorithms simply fail. Zoom into any portion of the site, and you can see these areas by looking for the blue and cyan original contours not matching the new Civil 3D–generated contour, as shown in Figure 4.16.

FIGURE 4.16
Contour surface without minimizing flat areas

You'll fix that now.

13. In Prospector, expand the Definition branch of the EG-Polylines surface if it's not already open from the previous exercise and right-click Edits.

14. Select the Minimize Flat Areas option to open the Minimize Flat Areas dialog.

    Note that the dialog has the same options found in that portion of your original Add Contour Data dialog.

15. Click OK to accept the defaults.

    Save the drawing and keep it open for the next portion of the exercise.

    Now the contours displayed more closely match the original contour information, as shown in Figure 4.17. There might be a few instances where gaps exist between old and new contour lines, but in a cursory analysis, none was off by more than 0.4′ in the horizontal direction—not bad when you're dealing with almost a square mile of contour information.

FIGURE 4.17
Contour surface with minimizing flat areas

16. Zoom into an area with a dense contour spacing and select the surface to make the contextual tab associated with the TIN Surface: EG-Polylines appear.

17. From the TIN Surface contextual tab ➢ Modify panel, choose Surface Properties to display the Surface Properties dialog.

18. On the Information tab, set Surface Style to Contours And Points and click OK to see a drawing similar to Figure 4.18.

FIGURE 4.18
Surface data points and derived data points

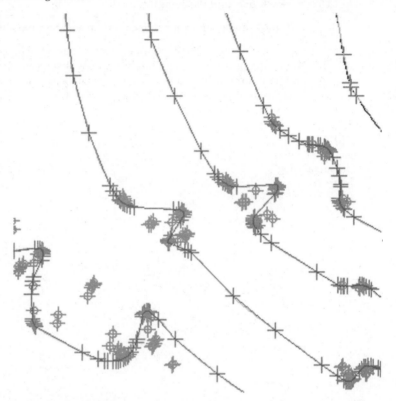

In Figure 4.18, you're seeing the points the TIN is derived from, with some styling applied to help you understand the creation source of the points. Although you can't see the colors in the printed figure, on the screen the points are shown in color to help you differentiate them. Each point shown as a red + symbol is a point picked up from the contour data itself. The magenta points shown with a circle symbol circumscribed over a + symbol are all added data on the basis of the Minimize Flat Areas edits. These points make it possible for the Civil 3D surface to match almost exactly the input contour data.

At this point, the original polylines are still present in the drawing, and you may find yourself tempted to delete them. Before doing so, you will want to change the surface property Copy Deleted Dependent Objects as discussed in the section "Surface Build Options" later in this chapter.

When this exercise is complete, you may close the drawing. A completed version of this drawing is available from the book's web page with the filename `0404_SurfaceFromPolylines_FINISHED.dwg` or `0404_SurfaceFromPolylines_METRIC_FINISHED.dwg`.

SURFACES FROM EXTERNAL TEXT FILES

In the first exercise of this chapter, you created a surface from points that were displayed in the drawing and part of a point group. The advantage of this method is that point groups can be used to exclude shots that are not valid for elevation information. However, some point files are so large that including the point data as part of the DWG would cause the resulting file to become very large.

In the exercise that follows, you will leverage the ability to work with point files while keeping the original text file external to the DWG. The advantage to this method is that the resulting surface model is created without needing to import the points first. There are several disadvantages to this method, however. First, because you are not using point groups, it is more difficult to exclude non-topo data (e.g., utilities and points with no elevation) from the surface model. The second thing to watch out for is the surface's connection to the original text file. If the path to the text file changes, or if the text file is no longer available, the surface will also be missing. Be sure to use *surface snapshots* to avoid this situation.

1. Create a folder on your computer called `C:\Mastering`.

2. Place the `0405_ConcordCommons.txt` file (or `0405_ConcordCommons_METRIC.txt` file for metric users) in your newly created folder.

   Doing so ensures that future exercises will function properly.

3. Create a new drawing using the `_AutoCAD Civil 3D (Imperial) NCS` template. Metric users should use the `_AutoCAD Civil 3D (Metric) NCS` template.

4. Imperial users, change the coordinate system to NAD83 Pennsylvania, South Zone, US Foot (PA83-SF). Metric users, change the coordinate system to NAD83 Pennsylvania, South Zone, Meters (PA83-S).

5. From the Home tab ➤ Create Ground Data panel, choose Surfaces ➤ Create Surface.

   The Create Surface dialog appears.

6. Change the Name value to **EG**, and click OK to close the dialog.

7. In Prospector, expand the Surfaces ➤ EG ➤ Definition branch.

8. Right-click Point Files and select the Add option.

   The Add Point File dialog, shown in Figure 4.19, appears.

FIGURE 4.19
Adding a point file to the surface definition

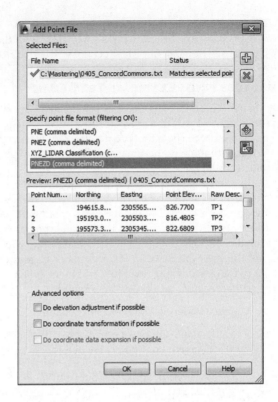

9. Set the Specify Point File Format option to PNEZD (Comma Delimited).

10. Click the Browse button.

11. Navigate to the previously created C:\Mastering folder, and select the 0405_ConcordCommons.txt file (or 0405_ConcordCommons_METRIC.txt file). Click Open.

12. In the Advanced Options area of the dialog, clear all check boxes.

    For more information on the advanced options, see Chapter 3, "Points."

13. Click OK to accept the settings in the Add Point File dialog and build the surface.

    Panorama will appear, but you can dismiss it by clicking the green check mark in the upper-right corner.

14. Right-click EG Surface in Prospector and select the Zoom To option to view the newly created surface.

When this exercise is complete, you may save the drawing and keep it open to continue on to the next exercise. A completed version of this drawing is available from the book's web page (0405_SurfaceFromFile_FINISHED.dwg or 0405_SurfaceFromFile_METRIC_FINISHED.dwg).

Now that you have learned how to add several types of data to a surface definition, you will need to fine-tune your surface models for better accuracy.

Surface Snapshots

A *surface snapshot* captures the state of the surface at the time it was created. If there is a snapshot in a surface, a surface's rebuild will start at the snapshot since the snapshot summarizes all of the previous build operations.

As mentioned previously, a drawing will stay referenced to any external data used to create it, most commonly a point file, LandXML file, or breakline file. If the external data is moved or deleted, the reference in the drawing will be broken. To prevent this from affecting the surface, you can create a snapshot of the surface while the surface is still working as intended. Then if anything happens to the external data, the surface will remain intact if a snapshot exists.

If you right-click a surface's name in Prospector, you will see three options related to snapshots:

Create Snapshot By creating a snapshot, you add a build operation that captures the surface information in the current state. Once a snapshot is created, the icon next to Definition in the Prospector tree will change to a camera icon.

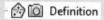

Remove Snapshot This option will remove the snapshot from the build operation. After a snapshot is removed, the surface will behave as if the snapshot was never there.

Rebuild Snapshot If the operations prior to the snapshot become outdated, you will see a yellow status icon next to its node in the Prospector tree. If you rebuild the snapshot, a new snapshot is created incorporating all changes to the surface model from the point of its initial creation. In many cases, you will want to leave the snapshot in place if you do not want the changes to affect the surface.

Refining and Editing Surfaces

Certification Objective

Once a basic surface is built, and in some cases even before it is built, you can do some cleanup and modification to the TIN construction that make it much more usable and realistic. Some of these edits include limiting the input data, tweaking the triangulation, adding breakline information, and hiding areas from view. In the following sections, you'll explore a number of ways of refining surfaces to end up with the best possible model from which to build.

Surface Properties

The most basic steps you can perform in making a better model are right in the Surface Properties dialog. The surface object contains information about the build and edit operations, along with some values used in surface calculations.

Surface Build Options

There are two main places you can set surface build options. The first location is on the Settings tab of Toolspace under Surfaces ➢ Commands ➢ Create Surface. If you right-click Create Surface, you will find an option to edit command settings. You can set these at the beginning of a project to automatically set the options for all surfaces as they are created. It is a great idea to have the command settings stored in your template so your desired triangulation defaults will be set on all projects.

The second place you will see the build options for a surface is in Surface Properties on the Definition tab. Changing the settings under Surface Properties will affect only the individual surface you are working with.

The basic options are listed here:

Copy Deleted Dependent Objects   When you select Yes and an object (such as a surface boundary or breakline) that is part of the surface definition is deleted, the information derived from that object is copied into the surface definition. Setting this option to Yes in the EG Surface Properties dialog will let you erase the polylines from the drawing file while still maintaining the surface information. If this option is set to No for the EG surface, when elevation source objects (such as polylines added as contours) are deleted, they will be removed from the surface definition when the surface is rebuilt.

Exclude Elevations Less Than   Setting Exclude Elevations Less Than to Yes allows you to set a surface's lowest valid elevation in the Elevation < field. In the example surface, EG, there is data at elevation zero, causing problems that can be solved by verifying that this option is set to Yes. Note that Elevation < is a "less than" operation, not "less than or equal to," so to rectify the problem caused by items at 0, you will need to set this to 1, for example.

Exclude Elevations Greater Than   Setting this to Yes will allow you to set the Elevation > field. This option allows you to set a highest valid elevation value.

Use Maximum Angle   Setting this option to Yes will allow you to set a value for Maximum Angle Between Adjacent TIN Lines. This setting will limit the number of narrow "sliver" triangles with one large obtuse angle and two acute angles that can indicate dubious surface data.

Use Maximum Triangle Length   Setting this option to Yes will allow you to set a value for Maximum Triangle Length. If a triangle leg is excessively long, it is often an indicator that valid elevation data for triangulation is sparse, as often occurs near the edges of a site.

Convert Proximity Breaklines To Standard   Toggling this option to Yes will create breaklines out of the lines and entities used as proximity breaklines.

Allow Crossing Breaklines   Setting this option to Yes allows you to set the Elevation To Use option. These options specify what Civil 3D should do if two breaklines in a surface definition intersect in the X,Y plane. An (x,y) coordinate pair cannot have two z values, so some decision must be made about the elevation at the crossing. Leaving this option set to No will cause the surface to totally throw out both offending breaklines.

In this exercise, you'll go through a couple of the basic surface-building controls that are available. You'll use them one at a time in order to observe their effects on the final surface display.

1. Continue working in the drawing from the previous exercise. You need to have completed the previous exercise to proceed.

2. Locate the portion of the surface that contains poor elevation data. Use your mouse to pan and zoom in the drawing.

You will know you have found the correct location when you zoom into the portion of the surface that resembles Figure 4.20. This blob is a batch of densely packed contours where the surface is incorrectly picking up elevations at zero.

FIGURE 4.20
EG surface showing a batch of densely packed contours, indicating bad elevations

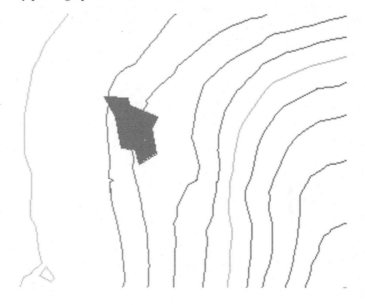

3. Select the surface by clicking any part of it in the graphic.
4. From the TIN Surface EG contextual tab, click Surface Properties.

   The Surface Properties dialog appears.
5. Select the Definition tab.
6. Under Definition Options at the top of the dialog (Figure 4.21), expand the Build category by clicking the + symbol.

FIGURE 4.21
Surface Properties Definition options

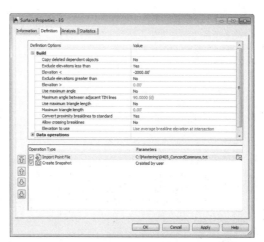

7. Verify that the Exclude Elevations Less Than value is Yes.
8. Set the value to **200** (**60** for metric users) and click OK to accept the settings in the dialog.

   A warning message will appear. Civil 3D is warning you that your surface definition has changed.
9. Click Rebuild The Surface to rebuild the surface.
10. You may see a message appear in Panorama indicating that three points were ignored because they are below the given limit. If this is the case, close the Panorama window by clicking the green check box.
11. Zoom extents to view the full surface by double-clicking your middle mouse wheel.

   When this step is complete, the surface will look similar to Figure 4.22. Save the drawing and keep it open for the next portion of the exercise.

FIGURE 4.22
EG surface after ignoring low elevations

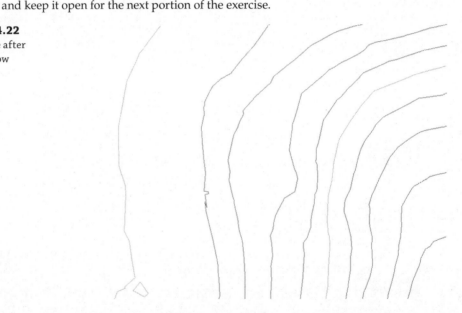

Although this surface is improving, there are still large areas being contoured that shouldn't be. By changing the style to review the surface, you can see where you still have some issues.

12. Open the Surface Properties dialog again, and switch to the Information tab.
13. Change the Surface Style field to Contours And Triangles.
14. Click Apply. Doing so makes the changes without exiting the dialog.
15. Drag the dialog to the side so you can see the site.

   On the outer edges of the site, you can see some long triangles formed in areas where there was no survey taken but the surface decided to connect the triangles anyway (Figure 4.23, left).

FIGURE 4.23
EG surface before Use Maximum Triangle Length was applied (left) and after (right). Note that the Level of Detail display option is on in these figures.

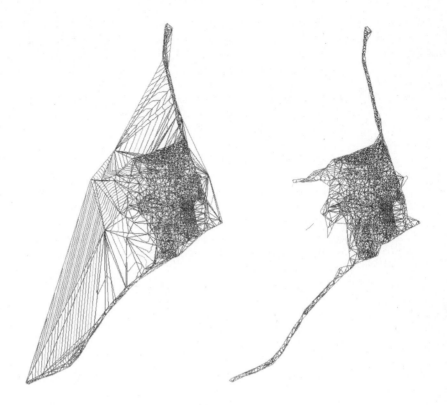

16. In the Surface Properties dialog, switch to the Definition tab.
17. Expand the Build category by clicking the + symbol.
18. Set the Use Maximum Triangle Length value to Yes.
19. In the Maximum Triangle Length value field, enter **300** (or for metric, **70**).
20. Click OK to apply the settings and close the dialog.
21. Click Rebuild The Surface to update and dismiss the warning message to see the revised surface (Figure 4.23, right).

When this exercise is complete, you may close the drawing. A completed version of this drawing is available from the book's web page with the filename `0406_SurfaceProperties_FINISHED.dwg` or `0406_SurfaceProperties_METRIC_FINISHED.dwg`.

Surface Additions

Beyond the simple changes to the way the surface is built, you can look at editing the pieces that make up the surface. With your drawing so far, you have merely been building from points. Although this is fine for small surfaces, you need to go further with this surface. In the following sections, you'll add a few breaklines and a border and then perform some manual edits to your site.

Real World Scenario

REORDERING BUILD OPERATIONS

The order in which you add data to a surface can make a significant difference in how elevation data is processed. You want the surface to process the most reliable data first (such as surveyed points and breaklines) and then work its way down to the least reliable data (contours from polylines, for example). Additionally, you want the surface to process boundaries last so that the triangulation gets reigned in as expected. Luckily, if you forget to add items in the correct order, you can easily change how Civil 3D processes elevation information.

In the following example, you will fix surface problems simply by changing the order in which elevation data is processed. The Operation Type listing is displayed at the bottom of the Definition tab in the Surface Properties dialog.

1. Open the file `0407_SurfaceBuildOrder.dwg` or `0407_SurfaceBuildOrder_METRIC.dwg`.

    This surface comprises a mélange of data types. It contains contours from polylines, points, a breakline, 3D faces, and a boundary.

    Examining the surface, you will see that there are some strange contour formations due to the order in which data was added to the surface definition. In the southwest part of the surface, contours shoot across a gap in the data where the boundary is not being respected.

2. Select the surface by clicking any part of it in the graphic.

3. From the TIN Surface: Existing contextual tab, click Surface Properties.

4. Switch to the Definition tab.

    The Operation Type listing at the bottom of the dialog shows us the order in which Civil 3D is using surface elevation information. Items near the top of the list are processed first if conflicting elevation information is added. In this example, the elevations from contours are conflicting with the point data.

5. Highlight the listing for Point Group and use the topmost arrow on the left side of the dialog to move it to the top of the Operation Type list.

    You'll see the warning symbol appear to the left of all the data types whose place in Operation Type order is affected by the change.

6. Click OK and then click Rebuild The Surface when prompted to do so.

    Examine the surface again. The contours are now forming as expected in the areas where surveyed point data is taking precedence over the rest of the elevation data.

7. Return to Surface Properties by clicking the surface and clicking Surface Properties from the TIN Surface: Existing contextual tab.

8. Use the arrow buttons to reorder the Operation Type listing to match what is shown in the following image.

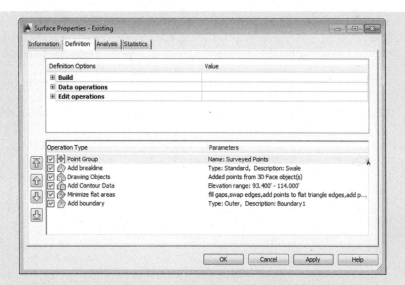

9. Click OK and then click Rebuild The Surface when prompted to do so.

The surface with completed edits will look like the following image.

You will need to use your professional judgment and a little common sense to determine the best order of operations for different surface data scenarios. If you would like to see what the surface would look like without a particular operation, you can clear the check box next to its listing in the Definition tab. You can completely remove operations from the listing by right-clicking the item in the listing and selecting Remove From Definition.

Breaklines

Breaklines change the triangulation of a surface by forcing triangle edges to follow along the segments of the breakline. Breaklines represent changes in grade due to constructed elements such as ditches, curbs, and retaining walls, just to name a few.

There are several methods for adding breaklines to a surface. On the Prospector tab of Toolspace, you can locate the Breakline listing of the surface definition, right-click it, and select Add. Additionally, you can easily add survey figures by going to the Survey tab of Toolspace, right-clicking the Figures branch, and selecting Create Breaklines. When either figures or feature lines are selected, each have an Add To Surface As Breakline option in their contextual tabs.

No matter which method you use to add breaklines to a surface's definition, you will need to choose the type in the Add Breaklines dialog (as seen in Figure 4.24). The breakline types are as follows:

Standard When the breakline you wish to add contains valid elevation data, use the Standard option. The vertex elevations of a standard breakline are used as data points and no triangle legs will cross the breakline.

Proximity Use the proximity breakline type when you wish to affect triangulation but not add elevation data. When a line is added as a proximity breakline, the original vertex elevations are ignored and replaced by the nearest—in the X,Y plane—surface triangle vertex elevations. For example, consider a survey firm whose drafter connects road centerline survey shots using a polyline at elevation zero. The survey shots are already part of the Civil 3D surface, so the vertex elevation for the polyline are already accounted for. When the polyline is added as a proximity breakline, the road centerline will be just as clean as it would have been if it had been a survey figure added as a standard breakline. (The survey drafter would have saved herself quite a bit of time by simply using automated linework features of survey figures, but for the purposes of explaining proximity breaklines, we assumed a less efficient workflow.)

Wall Wall breaklines are perfect for situations where a nearly vertical wall must be included in the surface, such as, for instance, a retaining wall where only the toe of the wall has been collected in the survey data. Lines added as wall breaklines must have valid elevation data at each vertex. When the line is added to the surface as a wall breakline, Civil 3D will ask for the height of the wall. In a situation in which the toe of the wall is the surveyed data, the height will be a positive number (i.e., the wall is going up). If the top of wall had been surveyed instead, a negative value for the height of the wall would be used. The wall height can be uniform for the entire length of the breakline or can be different from segment to segment. The end result of adding a wall breakline is the same as adding two standard breaklines separated by the elevation difference that you specify and offset by 0.001' (0.001 m).

From File The From File option is only needed if you have an FLT text file containing breakline data. This file type is usually the result of output from another program and contains XYZ data on each vertex for the lines described in the file. Once the FLT file has been imported, the effect on the surface is identical to a standard breakline.

Non-destructive Non-destructive breaklines neither change elevation nor change grades of the surface model. Non-destructive breaklines simply force additional triangulation along the line you add. You may find yourself needing a non-destructive breakline in anticipation of cleaning up triangle data, as discussed later in this chapter, in the section "Manual Surface Edits."

By far, standard is the most frequently used type of breakline, followed by proximity and wall breaklines. In this example, you'll add in some breaklines that describe road and surface features:

1. Open the `0408_SurfaceBreaklines.dwg` file or the `0408_SurfaceBreaklines_METRIC.dwg` file.

2. Thaw the _Polylines-Road layer.

   The roads are the color red.

3. Select the surface, and then pan and zoom to see that the triangulation lines do not appear along all the breaklines.

4. Unselect the surface by pressing Esc.

5. Select and then right-click one of the red polylines and choose the Select Similar option.

   All the red polylines are now highlighted.

6. In Prospector, expand the Surfaces ➢ EG ➢ Definition branches.

7. Right-click Breaklines and select the Add option.

   The Add Breaklines dialog appears.

8. Enter a description if you wish and the settings shown in Figure 4.24.

FIGURE 4.24
The Add Breaklines dialog

9. Click OK to accept the settings and close the dialog.

   You can dismiss Panorama if it appears.

10. Thaw the _Polylines-Surface layer.

    The surface polylines are the color green.

11. Repeat steps 5 through 9, but instead of the red polylines, select the green lines on the _Polylines-Surface layer. Change the description to **Surface Polylines**.

As shown in Figure 4.25, by adding the breaklines, you force the TIN lines to align with them, thus cleaning up the contours and making them follow the ridgelines of the road centerline, gutterlines, and shoulders as well as the changes in grade around the small detention area.

FIGURE 4.25
EG surface with only contours displayed before breaklines added (left) and after (right)

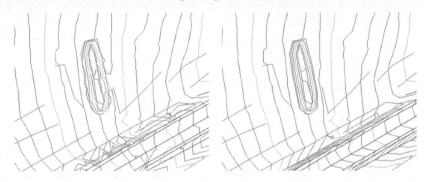

When this exercise is complete, you may close the drawing. A completed version of this drawing is available from the book's web page with the filename `0408_SurfaceBreaklines_FINISHED.dwg` or `0408_SurfaceBreaklines_METRIC_FINISHED.dwg`.

Crossing Breaklines

Invariably, you will see Panorama pop up with a message about crossing breaklines. In general, Civil 3D does not like breaklines that cross themselves. The Resolve Crossing Breaklines tool will let you examine those situations:

1. Click on a surface with breaklines.
2. From the TIN Surface contextual tab ➢ Analyze panel, choose Resolve Crossing Breaklines.
3. At the `Please specify the types of breakline you want to find or [surveyDatabase Figure Surface]:` prompt, enter **S** ↵ to select the surface option.

   The Crossing Breaklines tab on Panorama lists the crossing breaklines. You can decide how you want to resolve them using Use Higher Elevation, Use Lower Elevation, Use Average Elevation, or Specify Elevation, as shown here.

   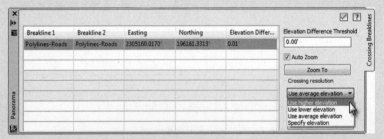

4. Click on each breakline and click the Resolve button, which is located underneath the drop-down list shown in the preceding screen shot; the conflict disappears from the Crossing Breaklines conflict list.

Surface Boundaries

Earlier in this chapter, you restricted the number of surface triangles formed in a surface model by changing the setting for maximum triangle length. You can also restrict triangle formation by changing the maximum angle between triangle legs.

When general triangle restrictions do not give the result that is needed, the next step to reigning in triangles is creating a surface boundary. Many object types can be used as surface boundaries, such as survey figures, feature lines, 2D polylines, and 3D polylines. Whatever type of line you choose as your boundary, it must be closed and cannot cross itself (i.e., no "loop-the-loops"). Elevations of the boundaries are ignored.

There are four types of surface boundaries:

Outer Use this type to define the outer edge of the surface model.

Show Use this type to show the surface inside a hide boundary, essentially creating a reverse donut effect in the surface display. This boundary type is also useful when you have a surface that includes portions that don't touch. Because there can only be one outer boundary per surface, one portion of the surface can be reined in by the outer boundary, and any nonadjacent portions can be displayed using a show boundary.

Hide Use this type to punch a hole in the surface display for tasks like building footprints or wetlands areas that are not to be touched by design. Hidden surface areas are *not* deleted but merely not displayed; therefore, the surface inside a hide boundary is still used for calculations such as area or cut/fill.

Data Clip Data clip boundaries place limits on data that will be considered part of the surface from that point going *forward*. This type is different from an outer boundary in that the data clip boundary will keep the data from ever being built into the surface as opposed to limiting it after the build.

With the exception of data clip boundaries, you'll want to have your boundaries among the last operations in your surface-building process. Therefore, as future edits are made, you may want to move the Add Boundary build operation back to the bottom of the operations list on the Definition tab in the Surface Properties dialog, as discussed earlier in this chapter.

The addition of every boundary is considered a separate part of the building operations. This means that the order in which the boundaries are applied controls their final appearance. For example, a show boundary selected before a hide boundary will be overridden by that hide operation.

The Extract Objects From Surface utility allows you to re-create any displayed surface element (contours, border, etc.) as an independent AutoCAD entity. It is important to note that only the objects that are currently visible in the surface style are extractable. In this exercise, you'll extract the existing surface boundary as a starting point for creating a more refined boundary that will limit triangulation:

1. Open the `0409_SurfaceBoundary.dwg` or `0409_SurfaceBoundary_METRIC.dwg` file.

2. Select the surface.

Destructive vs. Non-destructive Boundaries

The boundary types outer, show, and hide all have the option to be destructive (i.e., the Non-destructive Breakline check box is cleared) or non-destructive (i.e., the Non-destructive Breakline option is checked). What Civil 3D is "destroying" are the triangles that underlie all TIN surfaces.

The following image shows a schematic of a surface model before a boundary is added. For illustration purposes, triangle vertices have been highlighted with different shapes. The stars represent surface data points inside the boundary, and the circles represent points outside of the boundary.

Use the Non-destructive Breakline option when the data you are working with is valid right up to the boundary edge. A boundary added as a non-destructive breakline will retriangulate the surface and create triangles up to the boundary, as shown in the following image. The squares show locations where Civil 3D has interpolated a surface data point at the boundary. The portion of the surface that would be hidden by adding a boundary is being shown in light gray for illustration purposes.

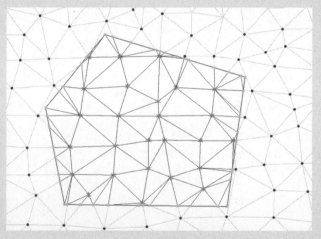

Use a destructive boundary if the line you are adding as a boundary is an approximation of the area you want to reign in. A boundary added as destructive will remove any triangle lines it crosses. The following image shows the illustration surface with the same boundary as before, but added as destructive. The portion of the surface that would be hidden by adding a boundary is being shown in gray for illustration purposes. Notice how the outermost triangle legs of the surface are all inside the boundary.

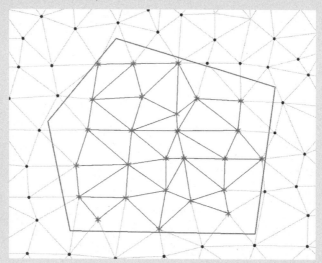

Consider a surface that needs an outer boundary, for example. If you create a rough polyline that encloses the points you wish to include as your surface model, add the polyline to the surface as an outer boundary with the non-destructive option cleared.

If you deliberately created the polyline by snapping each vertex of the boundary to a surveyed point, for example, you could add it as a non-destructive breakline.

You will need to use your professional judgment to know when to use the non-destructive option and when to use the Destructive Breakline option, but a good rule of thumb is that outer boundaries are usually destructive. Hide and show boundaries are usually non-destructive.

Extract Objects

3. From the TIN Surface contextual tab ➢ Surface Tools panel, choose Extract Objects to open the Extract Objects From Surface dialog.

4. Leave the Border object selected and deselect the Major Contour and Minor Contour options, as shown in Figure 4.26.

FIGURE 4.26
Extracting the border from the surface object

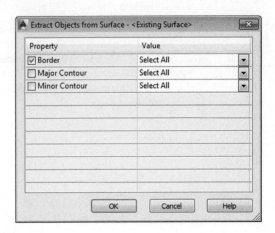

5. Click OK to finish the process. Press Esc to deselect the surface.

   A 3D polyline has been created from the surface border.

6. Pick the border 3D polyline.

   This polyline will form the basis for your outer surface boundary. By extracting the border polyline from the existing surface, you save a lot of time playing connect the dots along the points that are valid. Next, you'll refine this polyline and add it back into the surface as a boundary.

7. In Prospector, right-click Point Groups and select the Properties option. The Point Groups dialog appears.

8. Select _All Points from the list of point groups and then move _All Points to the top of the list using the Move The Selected Item To The Top Of The Order arrow on the right side of the dialog.

9. Click OK to display all your points on the screen.

10. Working your way around the site, grip-edit the polyline you created in step 6 to exclude some of the area at the southwest of the site where there are no points. Use the Add Vertex option from the grip menu as shown in Figure 4.27. Note that this process does not require you to use object snaps. Because we will be using the destructive breakline option, this boundary only needs to be approximate.

    On a large site, you can see that this is a time-consuming process but worth the effort to clean up the site nicely (Figure 4.27). Thankfully, there are other methods you can use to clean up the surface border; we will discuss these methods in later exercises.

11. In Prospector, right-click Point Groups and select the Properties option.

    The Point Groups dialog appears.

FIGURE 4.27
Using the grips to adjust the border

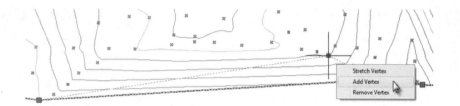

12. Select the No Display point group from the list and then move No Display to the top of the list using the Move The Selected Item To The Top Of The Order arrow on the right side of the dialog.
13. Click OK to turn off the display of all your points on the screen.

    When completed, your polyline should look like Figure 4.28.

FIGURE 4.28
Revised surface border polyline

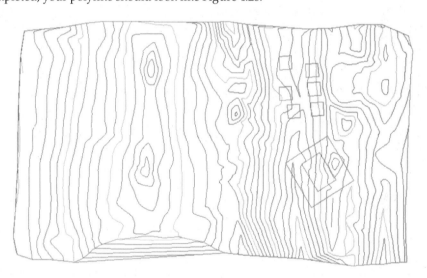

14. In Prospector, expand the Surfaces branch.
15. Right-click Existing Surface and select the Surface Properties option.

    The Surface Properties dialog appears.
16. On the Information tab, change the Surface Style to 1' and 5' TIN Editing (Ex.), or 1 m and 5 m TIN Editing (Ex.) for the metric users; then click OK.
17. In Prospector, expand the Surfaces ➢ Existing Surface ➢ Definition branches.
18. Right-click Boundaries and select the Add option.

    The Add Boundaries dialog opens (Figure 4.29).

FIGURE 4.29
Add Boundaries dialog

19. Enter the name **LOD** and keep the check mark next to the Non-destructive Breakline option; then click OK.
20. Pick the 3D polyline border that you previously extracted and edited. Notice the immediate change.
21. Zoom in on the southwest portion of your site, as shown in Figure 4.30. Note that the exact number and position of the non-destructive points will vary depending on the position of your 3D polyline boundary.

FIGURE 4.30
A non-destructive outer boundary in action

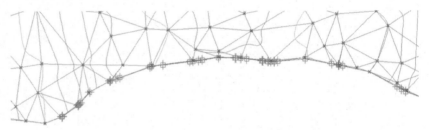

Save the drawing and keep it open for the next portion of the exercise.

Notice how the edge of the triangulation includes points shown as a square with a + symbol; these are the additional points created along the boundary line where it intersects with the triangles it crosses. The points you attempted to exclude from the surface are still being included in the calculation of elevations for this point; they are just excluded from the display and calculations. This isn't the result you were after, so let's fix it now.

22. In Prospector, expand the Surfaces ➢ Existing Surface ➢ Definition branches and select Boundaries.

    A listing of the boundaries appears in the preview area of Toolspace.

23. Right-click the boundary you just created and select the Delete option.
24. Click OK in the warning dialog that tells you the selected definition items will be permanently removed from the surface.

    In Prospector, you will now see the yellow exclamation point status flag next to the Existing Surface branch as well as the Definition branch. This is because the surface needs to be rebuilt.

25. Right-click the Existing Surface branch and select the Rebuild option to rebuild the surface without the boundary definition.

    You can dismiss Panorama if it appears.

26. In Prospector, right-click Boundaries and select the Add option again.

    The Add Boundaries dialog appears.

27. This time, leave the Non-destructive Breakline option unchecked, and click OK.

28. Pick the border 3D polyline again.

    Notice that no triangles intersect your boundary now where it does not connect points, as shown in Figure 4.31.

FIGURE 4.31
A destructive outer boundary in action

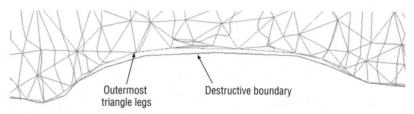

Next you will add some hide boundaries to the surface.

29. In Prospector, right-click Boundaries and select the Add option again.

30. Name the boundary Buildings and change Type to Hide. Make sure that Non-destructive Breakline is checked as shown in Figure 4.32, and click OK.

FIGURE 4.32
Adding buildings as non-destructive, hide boundaries

31. Select the six small squares in the graphic and the large outer square toward the south side of the site. Press Enter when complete.

    You should now have voids in your surface where the building pads in the drawing exist.

32. Right-click Boundaries and select the Add option one last time.

33. Change the name of the boundary to Courtyard, set the type to Show and verify that Non-destructive Breakline is checked.

34. Click OK and then click the inner square at the southeast portion of the site.

You should now see an island of surface data inside the previously hidden location.

When this exercise is complete, you may close the drawing. A finished copy of this drawing is available from the book's web page with the filename 0409_SurfaceBoundary_FINISHED.dwg or 0409_SurfaceBoundary_METRIC_FINISHED.dwg.

> **BIG ONES**
>
> Surfaces are data-rich objects that can be very large—geographically and data-wise. Large surfaces can be unwieldy in a drawing, testing your patience when you want to perform memory-intensive procedures on them. There are several options for working with large surfaces depending on the situation:
>
> **Data clip boundaries** As discussed earlier, data clip boundaries are a type of boundary that can be added before any elevation data is added. This is the best option to use if you have information covering a large geographic area but are only working in a smaller area.
>
> **Cropped surface** The Create Cropped Surface command can be found in the contextual tab of your surface, hidden in the Surface Tools panel flyout. This command will break a piece of a surface model off and allow you to send the smaller piece to a new drawing. This is a fast way to create a new drawing containing the desired surface data. The major disadvantage to the cropped surface tool is that there is no connection between the original surface and its spawn. In other words, if the original surface is changed, the new surface will not be affected.
>
> **Data shortcuts** The best of both worlds, data shortcuts allow you to work in a new file but remain connected to the surface's source data. Another major advantage of using data shortcuts is that multiple users can access the surface data without duplicating it. Data shortcuts are discussed at length in Chapter 16, "Advanced Workflows."
>
> Here are some other hints that will be helpful in working with large surfaces:
>
> ♦ Turn on the Level Of Detail option in the View tab.
>
> ♦ Do not use Rebuild-Automatic option.
>
> ♦ Turn off the selection preview on the Selection tab of the Options dialog. (Search AutoCAD Help for "visual effect settings" for more information.)
>
> ♦ Clear the Show Tooltips option in the Surface Properties dialog.

MANUAL SURFACE EDITS

If boundaries and breaklines are not available, or don't achieve the result you desire, manual edits can help. Manual edits allow further surface refinement by changing the surface at its core—the triangles. A number of manual edits can be performed on a surface. These edit options become part of the definition of the surface and include the following:

Add Line Connects two points where a triangle did not exist before. This option essentially adds a breakline to the surface, so adding a breakline would generally be a better solution. This option is not available on grid surfaces.

Delete Line   Removes the connection between two points. In addition to using outer boundaries, this option is used frequently to clean up the edge of a surface or to remove internal data where a surface should have no triangulation at all. This can be an area such as a building pad or water surface.

Swap Edge   Changes the direction of the triangulation methodology. For any four points, there are two solutions to the internal triangulation, and the Swap Edge option alternates from one solution to the other. The necessity of numerous swap edge operations can be limited by the use of appropriate breaklines. This option is not available on grid surfaces.

Add Point   Allows for the manual addition of surface data. This function is often used to add a peak to a digitized set of contours that might have a flat spot at the top of a hill.

Delete Point   Allows for the manual removal of a data point from the surface definition. Generally, it's better to fix the source of the bad data, but this option can be a fix if the original data is not editable (in the case of a LandXML file, for example).

Modify Point   Modify Point allows for changing the elevation of a surface point. Only the TIN point is modified, not the original data input.

Move Point   Move Point is limited to horizontal movement. Like Modify Point, only the TIN point is modified, not the original data input. This option is not available on grid surfaces.

Minimize Flat Areas   Performs the edits you saw earlier in this chapter to add supplemental information to the TIN and to create a more accurate surface, forcing triangulation to work in the z direction instead of creating flat planes. This option is not available on grid surfaces.

Raise/Lower Surface   A simple arithmetic operation that moves the entire surface in the positive or negative z direction. This option is useful for testing rough grading schemes for balancing dirt or for adjusting entire surfaces after a new benchmark has been observed.

Smooth Surface   Presents a pair of methods for supplementing the surface TIN data (note that this option is not available on grid surfaces). Both smoothing methods work by extrapolating more information from the current TIN data, but they are distinctly different in their methodology:

> **Natural Neighbor Interpolation (NNI)**   Adds points to a surface on the basis of the weighted average of nearby points. This data generally works well to refine contouring that is sharply angular because of limited information or long TIN connections. NNI works only within the bounds of a surface; it cannot extend beyond the original data.
>
> **Kriging**   Adds points to a surface based on one of five distinct algorithms to predict the elevations at additional surface points. These algorithms create a trending for the surface beyond the known information and can therefore be used to extend a surface beyond even the available data. Kriging is very volatile, and you should understand the full methodology before applying this information to your surface. Kriging is frequently used in subsurface exploration industries such as mining, where surface (or strata) information is difficult to come by and the distance between points can be higher than desired.

Paste Surface   Pulls in the TIN information from the selected surface and replaces the TIN information in the host surface with this new information while keeping the dynamic relationship to the original surface. This option is helpful in creating composite surfaces that reflect both the original ground and the design intent. This option is not available on grid surfaces. We'll look at pasting in Chapter 14, "Grading."

Simplify Surface   Allows you to reduce the amount of TIN data being processed while maintaining the accuracy of the surface. This is done using one of two methods: Edge Contraction, wherein Civil 3D tries to collapse two points connected by a line to one point, or Point Removal, which removes selected surface points based on algorithms designed to reduce data points that are similar. This option is not available on grid surfaces.

Manual editing should always be the last step in updating a surface. Fixing the surface is a poor substitution for fixing the underlying data the TIN is built from, but in some cases, it is the quickest and easiest way to make a more accurate surface.

Point and Triangle Editing

In this section, you'll remove triangles manually, and then finish your surface by correcting what appears to be a blown survey shot.

1. Open the `0410_SurfaceEdits.dwg` file (or the `0410_SurfaceEdits_METRIC.dwg` file). Confirm that the EG surface style is set to Contours And Triangles.

    Note that if you have Level Of Detail turned on, a red circle warning will appear if you are not zoomed in enough to view the true triangulation. You will want to turn Level Of Detail off by going to the View tab ➢ Views panel flyout. If the Level Of Detail button appears blue, that means it is on. Click the Level Of Detail button to turn it off.

2. Select the surface model by clicking on it.

3. From the TIN Surface: EG contextual tab ➢ Modify panel click Edit Surface ➢ Delete Line.

4. At the `Select edges:` prompt, enter **F** ↵ to use the Fence selection mode.

    In Fence selection mode you will draw a multisegment selection line and any objects that cross the line will be selected.

5. Use a Center Osnap to pick the circle labeled A at the lower right of the pick area, as shown in Figure 4.33; move to the upper-left corner as shown; use a Center Osnap to pick the circle labeled B; and press ↵ twice.

    Note that only triangles visible onscreen will be removed. In other words, if you zoom or pan such that some of the triangles you selected are not visible in your drawing area, they will not be removed from the surface definition.

6. Press ↵ to finish the selection set.

FIGURE 4.33
Using a Fence selection

7. Repeat this process, removing triangles until your site resembles the image on the right in Figure 4.34.

FIGURE 4.34
Surface before removal of extraneous triangles (left) and after (right)

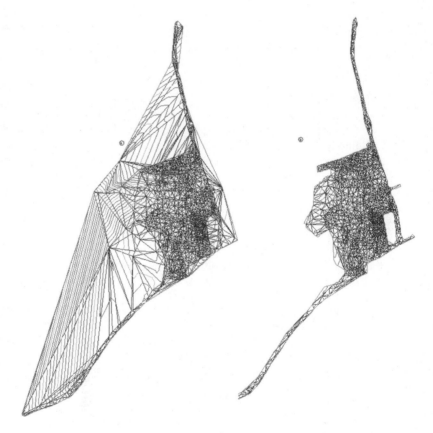

8. Zoom to the portion of your site with the red circle labeled C and you'll notice a collection of contours that seems out of place.

9. Change Surface Style to Contours And Points.

10. Select the surface. From the Tin Surface: EG contextual tab, click Edit Surface ➢ Delete Point.

11. Zoom in on the area very close.

    You will find a series of + point markers in the area with close contours. These are blown shots, and the contours are simply obeying the point elevation. In this case, there are three bad shots at 0.

12. Select each of the three markers and delete them; then notice the immediate change in the contouring.

13. Press ↵ when you are done selecting points to complete the command.

When this exercise is complete, you may save the drawing and keep it open to continue on to the next exercise.

Surface Smoothing

If you examine a surface when a style showing both contours and triangles is active, you could follow a contour and trace its path. What you would see is that contours are always a straight line across a triangle face. Civil 3D generates contours triangle by triangle.

If your goal is to create more flowing, natural looking contours, the best thing to do would be to create a surface style for which contour smoothing is turned on in the Contours tab.

If your goal is to change the surface and change the physical characteristics of the surface, the Smooth Surface tool is the way to go. Surface smoothing will interpolate additional points on the surface, forming more triangles and giving the contours a less angular appearance. Surface smoothing is often a first step before running a watershed analysis on the site because smoothing will blunt sharp ridges and create a more undulating form.

Use surface smoothing mindfully, however. Surface smoothing will smoosh breakline locations, making a retaining wall look more like a mudslide. Additionally, since this tool adds points, you are making the surface bigger from a data perspective. Similar to when applying boundaries, surface smoothing should be at/near the bottom of the list in the Definition tab of a surface. If additional changes are made to a surface, the order should be updated.

In this exercise, you'll use the NNI smoothing algorithm to reduce surface anomalies and create a more visually pleasing contour set:

1. Continue working in `0410_SurfaceEdits.dwg` file (or the `0410_SurfaceEdits_METRIC.dwg` file). It is not necessary to have completed the previous exercise to continue.

2. Select the surface.

3. From the Tin Surface: EG contextual tab, click Edit Surface ➢ Smooth Surface.

4. Expand the Smoothing Methods branch, and verify that Natural Neighbor Interpolation is the Select Method value.

5. Expand the Point Interpolation/Extrapolation branch, and click in the Select Output Region value field.

6. Click the ellipsis button.

7. At the `Select region or [rEctangle pOlygon Surface]:` prompt, pick the rectangle located in the center of the surface for smoothing and then press ↵ to return to the Smooth Surface dialog.

8. Enter **5** for the Grid X-Spacing and Grid Y-Spacing values (metric users should enter **2** for both values), and then press ↵.

   Note that Civil 3D will tell you how many points you are adding to the surface immediately below this input area by the value given in the Number Of Output Points field, as shown in Figure 4.35. It's grayed out, but it does change on the basis of your input values.

FIGURE 4.35
Smooth Surface dialog

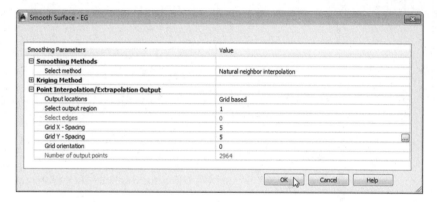

9. Click OK and the surface will be smoothed, similar to what is shown on the right in Figure 4.36.

FIGURE 4.36
Surface before NNI smoothing (left) and after (right)

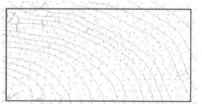 

A completed version of this drawing is available from the book's web page with the filename `0410_SurfaceEdits_FINISHED.dwg` or `0410_SurfaceEdits_METRIC_FINISHED.dwg`. Note all of the points with a circle cross symbol (verify that the surface style is set to Contours And Points to observe these symbols). These points are all new, created by the NNI surface-smoothing operation. The derived points are part of your surface, and the contours reflect the updated surface information.

When this exercise is complete, you may close the drawing.

168 | **CHAPTER 4** SURFACES

Surface Simplifying

Because of the increasing use in land development projects of GIS and other data-heavy inputs, it's critical that Civil 3D users know how to simplify the surfaces produced from these sources. In this exercise, you'll simplify the surface created from a drawing earlier in this chapter:

1. Open the 0411_SurfaceSimplify.dwg file (or the 0411_SurfaceSimplify_METRIC.dwg file). For reference, the surface statistics of the EG-GIS surface are shown in Figure 4.37.

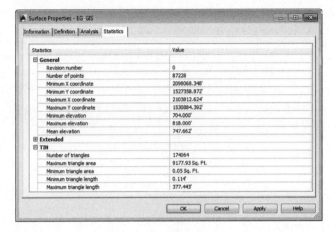

FIGURE 4.37
EG-GIS surface statistics before simplification

2. Select the surface by clicking anywhere on it in the graphic.
3. From the Tin Surface: EG-GIS contextual tab, click Edit Surface ➢ Simplify Surface.
4. Select the Point Removal radio button, as shown in Figure 4.38, and click Next to move to the Region Options page.

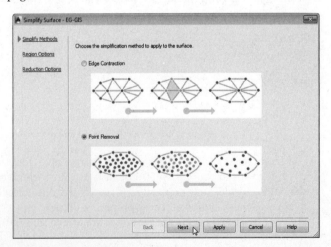

FIGURE 4.38
The Simplify Surface – Simplify Methods page

5. Accept the Region Options defaults as shown in Figure 4.39 and click Next to move to the Reduction Options page.

FIGURE 4.39
The Simplify Surface – Region Options page

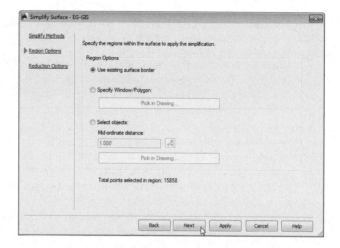

6. Set Percentage Of Points To Remove to 20 percent and then deselect the Maximum Change In Elevation option.

   This value is the maximum change allowed between the surface elevation at any point before or after the simplify process has run.

7. Click Apply.

   The program will process this calculation and display a Total Points Removed number, as shown in Figure 4.40. You can adjust the slider or place a checkbox in the Maximum Change In Elevation option to experiment with different values. Note that every time you click Apply or Finish, the number of points in the simplified surface decreases again by that percentage.

FIGURE 4.40
The Simplify Surface – Reduction Options page

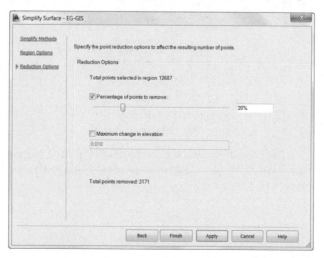

8. Click Finish to close the wizard and commit the Simplify edit.

A Word about the Simplify Build Operation

Notice that the Simplify build operation is listed twice in the preview area of Prospector under Edits. This is because the command was run once when you clicked Apply and a second time when you clicked Finish. If you want to run the Simplify Surface command only once, you should click Apply and then click Cancel or just click Finished without clicking Apply first.

When this exercise is complete, you may close the drawing. A finished copy of this drawing is available from the book's web page with the filename 0411_SurfaceSimplify_FINISHED.dwg or 0411_SurfaceSimplify_METRIC_FINISHED.dwg.

A quick visit to the Surface Properties Statistics tab shows that the number of points has been reduced, as shown in Figure 4.41. On something like an aerial topography or DEM, reducing the point count probably will not reduce the usability of the surface, but this simple point reduction will decrease the file size. Remember, you can always delete the edit or deselect the operation on the Definition tab of the Surface Properties dialog to "un-simplify" the surface.

FIGURE 4.41
EG-GIS surface statistics after simplification

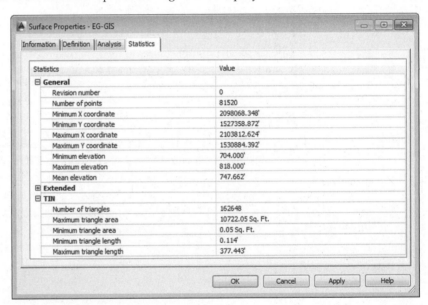

The creation of a surface is merely the starting point. Once you have a TIN to work with, you have a number of ways to view the data using analysis tools and varying styles.

Surface Analysis

Certification Objective

Once a surface is created, you can display information in a number of ways. The most common so far has been contours and triangles, but those are the basics. By using various styles, you can show a large amount of data with one single surface. Surface styles are discussed further in Chapter 19, "Object Styles." While some of the styles are used for generating plans (such as contours), others lend themselves to analyzing the surface during creation.

For a surface object in plan view, Points, Triangles, Border, Major Contour, Minor Contour, User Contours, and Gridded are standard components and are controlled like any other object component. Directions, Elevations, Slopes, Slope Arrows, and Watershed components are unique in that their display is controlled elsewhere in the surface style. Note that the Layer, Color, and Linetype fields are grayed out for these components. Each of these components has its own special coloring schemes, which we'll look at later in the section "Comparing Surfaces." In the following sections, you will explore the elevation and slope analysis styles.

Elevation Banding

Displaying surface information as bands of color is one of the most common display methods for engineers looking to make a high-impact view of the site. Elevations are a critical part of the site design process, and understanding how a site varies in terms of elevation is an important part of making the best design. Elevation analysis typically falls into two categories: showing bands of information on the basis of pure distribution of linear scales or displaying a lesser number of bands to show some critical information about the site. In this first exercise, you'll use a standard style to illustrate elevation distribution along with a prebuilt color scheme that works well for presentations:

1. Open the `0412_SurfaceAnalysis.dwg` file (or the `0412_SurfaceAnalysis_METRIC.dwg` file).

2. Select the surface on your screen to activate the contextual tab.

3. From the TIN Surface contextual tab ➢ Modify panel, choose the Surface Properties option.

4. On the Information tab, change the Surface Style field to Elevation Banding (3D).

5. Switch to the Analysis tab for the Elevations analysis type.

6. Set Create Ranges By to Number Of Ranges and set that value to **3**. Then click the Run Analysis arrow in the middle of the dialog to populate the Range Details area.

7. Click OK to close the Surface Properties dialog.

   Notice that only the border is currently showing because the surface is being shown in plan view.

[-][Top][2D Wireframe]

You may have noticed three pieces of text in the upper-left corner of your modelspace. If you click any of these viewport controls, you will find that they are drop-down lists that you can use to change what you are looking at.

The first set of bracketed text will either be [-] (denoting that one viewport is being displayed) or [+] (if two viewports are being displayed).

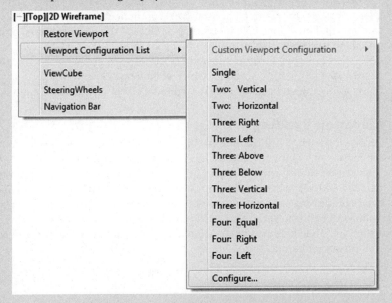

The second set of bracketed text is the view control and will list the current view within the brackets.

The third set of bracketed text is the Visual Style Control and will list the current visual style within the brackets.

All of these commands can also be changed from the View tab. If Level Of Detail is currently active, an additional small icon will appear below this line of text.

These three pieces of text provide valuable information for easy reference while you work.

8. From the View Control, select SW Isometric.
9. Zoom in if necessary to get a better view of the surface.
10. From the Visual Style Control, select the Conceptual option to see a semi-rendered view that should look something like Figure 4.42.

FIGURE 4.42
Conceptual view of the site with the Elevation Banding style

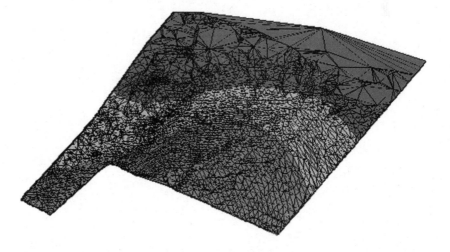

Save and keep the drawing open for the next portion of the exercise.

AutoCAD Visual Styles

When you see triangles on a surface in the object viewer, they are visible because of the view style and can be modified via the Visual Styles Manager. Turning the Edge mode off will leave you with a nicely gradated view of your site. You can edit the visual style by clicking the Visual Style Control at the upper left of your drawing screen. You can also access the Visual Style Manager on the View tab ➢ Visual Styles panel by clicking the small arrow in the lower-right corner of the panel. The Visual Styles Manager is shown here.

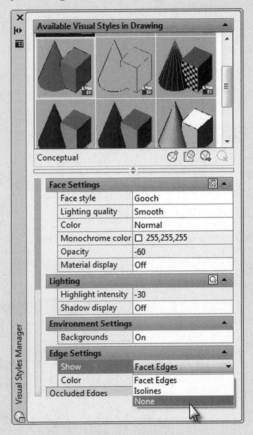

The Realistic visual style will allow you to see the render material applied to your surface—with some conditions. You must have triangles turned on in the Surface Style dialog, and you must have a render material specified in the Surface Properties dialog.

You'll use a 2D elevation to clearly illustrate portions of the site that cannot be developed. Next, you'll manually tweak the colors and elevation ranges on the basis of design constraints from outside the program.

11. Click the View Control and select Top.
12. Click the Visual Style Control and select 2D Wireframe. Zoom extents if necessary.

    This site has a limitation placed in that no development can go below the elevation of 790′ (or 240 m). Your analysis will show you the areas that are below 790′ (or 240 m), a buffer zone to 791′ (or 241 m), and then everything above that.

13. Select the surface by clicking on the border. Right-click and choose the Surface Properties option.

    The Surface Properties dialog appears.

14. On the Information tab, change the Surface Style field to Elevation Banding (2D).
15. On the Analysis tab, change Maximum Elevation for ID 1 and Minimum Elevation for ID 2 both to **790** (or **240** for metric users).
16. Change Maximum Elevation for ID 2 and Minimum Elevation for ID 3 both to **791** (or **241** for metric users).
17. Modify your color scheme to match Figure 4.43 by double-clicking on each color to open the Select Color dialog, selecting the appropriate color, and clicking OK to close the dialog. (The colors are red, yellow, and green from top to bottom, respectively.)
18. Click OK to accept the settings in the Surface Properties dialog.
19. Save the drawing for use in the next exercise.

Understanding surfaces from a vertical direction is helpful, but many times the slopes are just as important. In the next section, you'll take a look at using the slope analysis tools in Civil 3D.

FIGURE 4.43
The Surface Properties dialog after manual editing

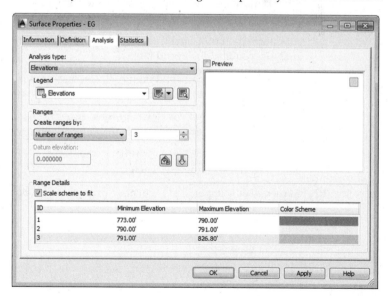

Slopes and Slope Arrows

Beyond the bands of color that show elevation differences in your models, you also have tools that display slope information about your surfaces. This analysis can be useful in checking for drainage concerns, meeting accessibility requirements, or adhering to zoning constraints. Slope is typically shown as areas of color similar to the elevation banding or as colored arrows that indicate the downhill direction and slope. In this exercise, you'll look at a proposed site grading surface and run the two slope analysis tools:

1. Continue working in the `0412_SurfaceAnalysis.dwg` file (or the `0412_SurfaceAnalysis_METRIC.dwg` file). It is not necessary to have completed the previous exercise before continuing.

2. Select the surface on your screen to activate the contextual tab.

3. From the TIN Surface contextual tab ➢ Modify panel, choose Surface Properties.

4. On the Information tab, change the Surface Style field to Slope Banding (2D).

5. Switch to the Analysis tab of the Surface Properties dialog.

6. Choose Slopes from the Analysis Type drop-down list.

7. Set the Number field in the Ranges area to **3** and click the Run Analysis arrow in the middle of the dialog to populate the Range Details area.

   The Range Details area will populate. You could change the minimum and maximum values as you did in the previous exercise, but this time you'll keep the defaults.

8. Click OK to close the dialog.

   Save the drawing and keep it open for the next portion of the exercise.

   The colors are nice to look at, but they don't mean much, and slopes don't have any inherent information that can be portrayed by color association. To make more sense of this analysis, you'll add a legend table.

9. Select the surface again, and on the TIN Surface contextual tab ➢ Labels & Tables panel, choose Add Legend.

10. At the `Enter table type [Directions Elevations Slopes slopeArrows Contours Usercontours Watersheds]:` prompt, enter **S** ↵ to select Slopes.

11. At the `Behavior [Dynamic Static]: <Dynamic>` prompt, press ↵ again to accept the default value of a Dynamic legend.

12. At the `Select upper left corner:` prompt, pick a point on screen to draw the legend, which is shown in Figure 4.44.

FIGURE 4.44
The slopes legend table

Number	Minimum Slope	Maximum Slope	Area	Color
1	0.00%	7.08%	202209.40	
2	7.08%	10.69%	182627.83	
3	10.69%	39000.33%	105351.52	

Save the drawing and keep it open for the next portion of the exercise.

By including a legend, you can make sense of the information presented in this view. Because you know what the slopes are, you can also see which way they go.

13. Select the surface, and on the TIN Surface contextual tab ➢ Modify panel, choose Surface Properties.

    The Surface Properties dialog appears.

14. On the Information tab, change the Surface Style field to Slope Arrows.

15. Switch to the Analysis tab of the Surface Properties dialog.

16. Choose Slope Arrows from the Analysis Type drop-down list.

17. Verify that the Number field in the Ranges area is set to **3** and click the Run Analysis arrow in the middle of the dialog to populate the Range Details area.

18. Click OK to close the dialog.

The benefit of arrows is in looking for "birdbath" areas that will collect water. These arrows can also verify that inlets are in the right location, as shown in Figure 4.45. Look for arrows pointing to the proposed drainage locations and you'll have a simple design-verification tool.

FIGURE 4.45
Slope arrows pointing to a proposed inlet location

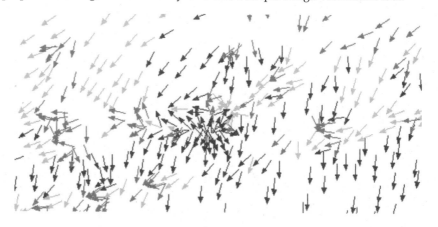

When this exercise is complete, you may close the drawing. A finished copy of this drawing is available from the book's web page with the filename `0412_SurfaceAnalysis_FINISHED.dwg` or `0412_Surface Analysis_METRIC_FINISHED.dwg`.

With these simple analysis tools, you can show a client the areas of their site that meet their constraints. Visually compelling and simple to produce, this is the kind of information that a 3D model makes available. Beyond the basic information that can be represented in a single surface, Civil 3D also contains a number of tools for comparing surfaces. You'll compare this existing ground surface to a proposed grading plan in the next section.

Visibility Checker

The Zone Of Visual Influence tool allows you to explore what-if scenarios. In this example, a 40′ (12 m) tower has been proposed for the site. A concerned neighbor wants to make sure that it won't obstruct their scenic view. You want to check it from the proposed surface:

1. Open `0413_VisibilityCheck.dwg` or `0413_VisibilityCheck_METRIC.dwg`.

2. In Prospector, expand the Surfaces branch, right-click on the FG surface, and choose the Select option.

3. On the TIN Surface contextual tab ➢ Analyze panel, choose Visibility Check ➢ Zone Of Visual Influence.

4. At the `Specify location of object:` prompt, use the Intersection Osnap to select the center of the proposed tower located on the southern portion of the site (denoted by a large white square with an X inside).

5. At the `Specify height of object:` prompt, enter **40** ↵ to set the tower height to 40′ (metric users, enter **12** ↵).

6. At the `Specify the radius of vision extent:` prompt, pan to and select the endpoint at the upper-right corner of the cyan-colored house located at the northeastern corner of the site.

   Save the drawing and keep it open for the next portion of the exercise.

   The drawing now has bands of color:

   ◆ Green near the tower location indicates that the object is completely visible.

   ◆ Yellow indicates that the object is partially visible.

   ◆ Red indicates that the object is not visible.

   So in our example, the homeowner on the upper right will be happy to know that the proposed 40′ tower will not appear in their view.

   In the next portion of this exercise, you will use the Point To Point tool.

7. Using the same drawing, zoom to the intersection of Syrah Way and Cabernet Court, where you will see a car driving on the right-hand side of the road.

   We want to check the sight distance.

8. If it is not already selected, select the FG surface.

9. On the TIN Surface contextual tab ➤ Analyze panel, choose Visibility Check ➤ Point To Point.

10. At the `Specify height of eye:` prompt, enter **3.5** ↵ (metric users enter **1** ↵).

    This sets the height of the eye of a driver sitting in a typical car.

11. At the `Specify location of eye:` prompt, click where the driver would normally be seated in the vehicle.

12. At the `Specify height of target:` prompt, enter **6** ↵ (metric users enter **1.8** ↵).

    A rubber-banding sightline ray appears.

13. Click along the path where oncoming cars would be seen.

    A sightline arrow is drawn on the screen:

    - If the arrow is green, it means that the view is unobstructed and the command line will tell you the distance from the eye.

    - If any portion of the arrow is red, it indicates that the view is obstructed and the command line will tell you the distance at which the obstruction occurs.

When this exercise is complete, you may close the drawing. A finished copy of this drawing is available from the book's web page with the filename `0413_SurfaceVisibility_FINISHED.dwg` or `0413_SurfaceVisibility_METRIC_FINISHED.dwg`.

Unfortunately, these visual tools are not dynamic; if you change the surface, you will need to rerun the visual tools.

Comparing Surfaces

Civil 3D contains a number of surface analysis tools designed to calculate earthwork quantities, and you'll look at them in the following sections. First, a simple comparison provides feedback about the volumetric difference, and then a more detailed approach enables you to perform an analysis on this difference.

For years, civil engineers have performed earthwork using a section methodology. Sections were taken at some interval, and a plot was made of both the original surface and the proposed surface. Comparing adjacent sections and multiplying by the distance between them yields an end-area method of volumes that is generally considered acceptable. The main problem with this methodology is that it ignores the surfaces in the areas between sections. These areas could include areas of major change, introducing some level of error. In spite of this limitation, this method worked well with hand calculations, trading some accuracy for ease and speed.

With the advent of full-surface modeling, more precise methods became available. By analyzing both the existing and proposed surfaces, a volume calculation can be performed. Keep in mind that the quality of the result is highly dependent on the quality of the input surfaces. At every TIN vertex in both surfaces, a distance is measured vertically to the other surface. These delta amounts can then be used to create a third dynamic surface called a volume surface, which represents the difference between the two original surfaces.

TIN Volume Surface

Using the volume utility for initial design checking is helpful, but quite often contractors and other outside users want to see more information about the grading and earthwork for their own uses. This requirement typically falls into two categories: a cut-fill analysis showing colors or contours or a grid of cut-fill tick marks.

Color cut-fill maps are helpful when reviewing your site for the locations of movement. Some sites have areas of better material or can have areas where the cost of cut is prohibitive (such as rock). In this exercise, you'll use two of the surface analysis methods to look at the areas for cut-fill on your site:

1. Open the 0414_VolumeSurface.dwg or the 0414_VolumeSurface_METRIC.dwg file.

2. From the Analyze tab ➢ Volumes And Materials panel, choose the Volumes Dashboard tool to display the Volumes Dashboard in Panorama.

3. Click the Create New Volume Surface button to display the Create Surface dialog.

   Notice that Type is already set to TIN Volume Surface. You also have the option to select Grid Volume Surface.

4. Change the name to **VOL EG-FG** and set the style to Elevation Banding (2D).

5. Click the <Base Surface> field next to Base Surface to display the ellipsis button; once it's visible, click the ellipsis to select the EG surface and then click OK.

6. Click the <Comparison Surface> next to Comparison Surface to display the ellipsis button; once it's visible, click the ellipsis to select the FG surface and click OK.

The Create Surface dialog should now look similar to Figure 4.46.

FIGURE 4.46
Creating a volume surface

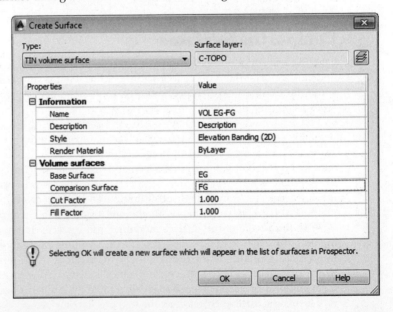

7. Click OK to accept the settings in the Create Surface dialog.

   Civil 3D will calculate the volume (Figure 4.47).

FIGURE 4.47
Composite volume calculated

Note that you can scroll right and left in Panorama to display additional information, including the ability to apply a cut or fill factor by typing directly into the cells for these values.

VOL EG-FG   This new volume surface appears in Prospector's Surfaces collection, but notice that the icon is slightly different, showing two surfaces stacked on each other. The color mapping currently shown is just a default set, though, and does not indicate much.

8. Leave Panorama open, and in Prospector, expand the Surfaces ➢ FG ➢ Definition branches.

9. Right-click Edits and select the Raise/Lower Surface option.

10. Enter **-0.25** at the command line to drop the site 3" (metric users enter **-0.075**).

    Notice that a yellow exclamation point status flag has appeared next to the volume surface in Panorama as well as in Prospector. Panorama no longer lists the volumes and instead states "Out of date."

11. Right-click the VOL EG-FG surface in Panorama or Prospector and select the Rebuild option.

12. In Prospector under the Definition branch of the FG surface, click Edits, right-click the Raise/Lower edit in the preview area, and select Delete.

13. A dialog will appear warning you that the selected definition item will be permanently removed from the surface. Click OK.

14. Return to Panorama and rebuild the volume surface again to return to the original volume calculation.

15. Close Panorama when complete.

16. In Prospector, right-click VOL EG-FG in the Surfaces branch and select the Surface Properties option.

    The Surface Properties dialog appears.

17. In the Surface Properties dialog, switch to the Statistics tab and expand the Volume branch.

    The value shown for Net Volume (Unadjusted) is the same as shown in Panorama in the first part of this exercise.

18. In the Surface Properties dialog, switch to the Analysis tab for the Elevations analysis type.
19. Verify that Create Ranges By is set to Number Of Ranges and that the value is set to **3**; then click the Run Analysis arrow in the middle of the dialog to populate the Range Of Details area.
20. Change Maximum Elevation for ID 1 and Minimum Elevation for ID 2 to **-0.5** (or **-0.15** for metric users).
21. Change Maximum Elevation for ID 2 and Minimum Elevation for ID 3 to **0.5** (or **0.15** for metric users).
22. Modify your color scheme.

    The recommended colors are red, yellow, and green, where red indicates the worst-case cut, green represents the worst-case fill, and yellow represents a balance.

    Figure 4.48 shows the completed elevation analysis settings.

FIGURE 4.48
Elevation analysis settings for earthworks

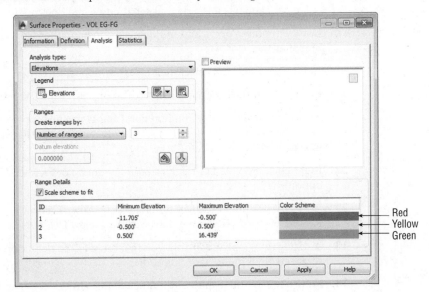

23. Click OK to close the Surface Properties dialog.

    Save the drawing and keep it open for the next portion of the exercise.

    The volume surface now indicates areas of cut, areas of fill, and areas that are nearly balanced, similar to Figure 4.49. If you leave a small range near the balance line, you can more clearly see the areas that are being left nearly undisturbed.

FIGURE 4.49
Completed elevation analysis

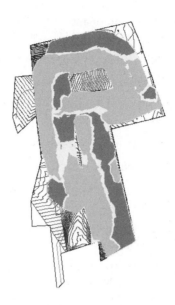

To show where large amounts of cut or fill could incur additional cost (such as compaction or excavation protection), you would simply modify the analysis range as required.

The Elevation Banding surface is great for onscreen analysis, but the color fills are too dense for most plotting purposes and can slow down the performance of the drawing. In the next steps, you use the Contour Analysis tool to prepare cut-fill contours in these same colors.

24. In Prospector, right-click VOL EG-FG in the Surfaces branch and select the Surface Properties option to display the Surface Properties dialog again.

25. On the Analysis tab, choose Contours from the Analysis Type drop-down list.

26. Verify that the Number field in the Ranges area is set to **3** and click the Run Analysis arrow in the middle of the dialog to populate the Range Details area.

27. Change the ranges to match those you entered in the previous portion of the exercise (as shown in Figure 4.48).

28. In the Major Contour column, click the small button to the far right to display the AutoCAD Select Color dialog and set a color for each ID.

Typical contour colors are as follows:

♦ Shades of red for cut

♦ A yellow for the balance line

♦ Shades of green for fill

184 | CHAPTER 4 SURFACES

29. Switch to the Information tab on the Surface Properties dialog, and change the Surface Style to Contours 1′ and 5′ (Design) or Contours 0.2 m and 1.0 m (Design).

30. Click the down arrow next to the Style field and select the Copy Current Selection option.

    The Surface Style Editor appears.

31. On the Information tab, change the Name field to **Contours 1′ and 5′ (Earthwork)** or **Contours 0.2 m and 1.0 m (Earthwork)**.

32. On the Contours tab, expand the Contour Ranges branch.

33. Change the value of the Use Color Scheme property to True.

    It's okay to ignore the range values here because you already set them in the surface properties.

34. Click OK to close the Surface Style Editor and click OK again to close the Surface Properties dialog.

When this exercise is complete, you may close the drawing. A finished copy of this drawing is available from the book's web page with the filename `0414_VolumeSurface_FINISHED.dwg` or `0414_VolumeSurface_METRIC_FINISHED.dwg`.

The volume surface can now be analyzed on a lot-by-lot basis or labeled using the surface-labeling functions to show the depths of cut and fill, which you'll look at in the next sections.

Labeling the Surface

Once the three-dimensional surface model has been created, it is time to communicate the model's information in various formats. This includes labeling contours, creating legends for the analysis you've created, adding spot labels, or labeling the slope. These exercises work through these main labeling requirements and building styles for each.

Contour Labeling

The most common requirement is to place labels on surface-generated contours. In Land Desktop, this was one of the last steps because a change to a surface required erasing and replacing all the labels. Once labels have been placed, their styles can be modified.

Contour labels in Civil 3D are created by special lines that understand their relationship with the surface. Everywhere one of these lines crosses a contour line, a label is placed. This label's appearance is based on the style applied and can be a major, minor, or user-defined contour label. Each label can have styles selected independently, so using some AutoCAD selection techniques can be crucial to maintaining uniformity across a surface. In this exercise, you'll add labels to your surface and explore the interaction of contour label lines and the labels themselves:

1. Open the `0415_SurfaceLabeling.dwg` or `0415_SurfaceLabeling_METRIC.dwg` file.

2. Select the EG surface in the drawing to display the TIN Surface contextual tab.

3. From the TIN Surface contextual tab ➢ Labels & Tables panel, choose Add Labels ➢ Contour – Single.

4. Pick any spot on a major contour to add a label.

5. Press ↵ when complete and press Esc to deselect the surface.

6. From the Annotate tab ➢ Labels & Tables panel, choose Add Labels ➢ Surface ➢ Contour – Multiple.

7. Pick a point to the west of the site and then a second point to the south of the site, crossing a number of contours in the process.

8. Press ↵ to end the picking.

9. From the Annotate tab ➢ Labels & Tables panel, click the Add Labels button instead of the drop-down to display the Add Labels dialog.

10. Set Feature to Surface and Label Type to Contour – Multiple At Interval.

11. Click the Add button.

12. Pick a point to the west of the site and then a second point to the north of the site.

13. Enter **200** (for metric **60**) at the command line for an interval value.

    Save the drawing and keep it open for the next portion of the exercise.

    You've now labeled your site in three ways to get contour labels in a number of different locations. You will need additional labels in the northeast and southwest to complete the labeling because you did not cross these contour objects with your contour label line. You can add more labels by clicking Add, but you can also use the labels created already to fill in these missing areas. By modifying the contour line labels, you can manipulate the label locations and add new labels. Next, you'll fill in the labeling to the northeast.

14. Zoom to the northeast portion of the site, and notice that some of the contours are labeled only along the boundary or not at all, as shown in Figure 4.50.

FIGURE 4.50
Contour labels applied

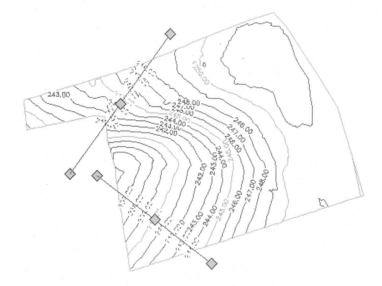

15. Zoom in to any contour label placed using the Contour – Multiple At Interval button, and pick the text.

    Three grips will appear. The original contour label lines are quite apparent, but in reality, every label has a hidden label line beneath it.

16. Grab the northernmost grip and drag across an adjacent contour northeast of the original label, as shown in Figure 4.51.

FIGURE 4.51
Grip-editing a contour label line

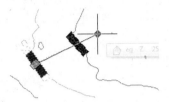

    New labels will appear everywhere your dragged line now crosses a contour.

17. Drop the grip somewhere to create labels as desired. Be sure to press Esc a few times when you are done labeling the contours to end the command.

    Save the drawing and keep it open for the next portion of the exercise.

    By using the created label lines instead of adding new ones, you'll find it easier to manage the layout of your labels.

Additional Surface Label Types

With Civil 3D's surface modeling, spot labels are dynamic and react to changes in the underlying surface. By using surface labels instead of points or text callouts, you can generate a grading plan early on in the design process and begin the process of creating sheets. In the following sections, you'll label surface slopes in a couple of ways, create a single spot label for critical information, and conclude by creating a grid of labels similar to the labels in many estimation software packages.

Labeling Slopes

Beyond the specific grade at any single point, most grading plans use slope labels to indicate some level of trend across a site or drainage area. Civil 3D can generate the following two slope labels:

One-Point Slope Labels One-point slope labels indicate the slope of an underlying surface triangle. These work well when the surface has large triangles, typically in pad or mass grading areas.

Two-Point Slope Labels Two-point slope labels indicate the slope trend on the basis of two points selected and their locations on the surface. A two-point slope label works by dividing the surface elevation distance between the points by the planar distance between the pick points. This works well in existing ground surface models to indicate a general slope direction but can be deceiving in that it does not consider the terrain between the points.

In this next exercise, you'll apply both types of slope labels:

1. Continue working in `0415_SurfaceLabeling.dwg` or `0415_Surface Labeling_METRIC .dwg`. It is not necessary to have completed the previous exercise before proceeding.

2. Select the surface to display the TIN Surface contextual tab.

3. From the TIN Surface contextual tab ➢ Labels & Tables panel, choose Add Labels ➢ Slope. (Alternately, you can go to the Annotate tab ➢ Labels & Tables panel and choose Add Labels ➢ Surface ➢ Slope.)

4. At the `Create Slope Labels or [One-point Two-point]: <One-point>` prompt, press ↵ to select the default one-point label style.

5. Zoom in on the circle drawn on the western portion of the site and use a Center Osnap to place a label at its center, similar to that shown in Figure 4.52.

FIGURE 4.52
A one-point slope label

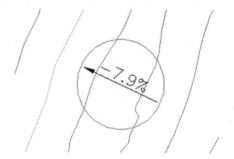

6. Press Esc or ↵ to exit the command.

7. From the TIN Surface contextual tab ➢ Labels & Tables panel, choose Add Labels ➢ Slope.

8. At the `Create Slope Labels or [One-point Two-point]: <One-point>` prompt, enter **T** ↵ to switch to a two-point label style.

9. Pan to the central portion of the site, and use an Endpoint Osnap to pick the left end of the line.

10. Use an Endpoint Osnap to select the right end of the line to complete the label.

11. Press Esc or ↵ to exit the command and view the label, shown in Figure 4.53.

FIGURE 4.53
A two-point slope label

This second label indicates the average slope of the property. By using a two-point label, you get a better understanding of the trend as opposed to a specific point.

When this exercise is complete, you may close the drawing. A finished copy of this drawing is available from the book's web page with the filename `0415_SurfaceLabeling_FINISHED.dwg` or `0415_SurfaceLabeling_METRIC_FINISHED.dwg`.

SURFACE GRID LABELS

Multiple labels placed in a grid pattern on the plan can be a huge help for grading contractors and estimators. In this exercise, you'll use the volume surface you generated earlier in this chapter to create a set of surface labels that reflect this requirement:

1. Open the `0414_SurfaceVolumeSurface_FINISHED.dwg` file (or the `0414_SurfaceVolumeSurface_METRIC_FINISHED.dwg` file).

2. From the Annotate tab ➢ Labels & Tables panel, choose Add Labels ➢ Surface ➢ Spot Elevations On Grid.

3. At the `Select a surface <or press enter key to select from list>:` prompt, press ↵ to display the Select Surface dialog.

4. Select VOL EG-FG and click OK.

5. At the `Specify a grid basepoint:` prompt, pick a point southwest of the surface to set a base point for the grid.

6. At the `Grid rotation:` prompt, enter 0 ↵.

7. At the `Grid X spacing:` prompt, enter 25 ↵ (7 ↵ for metric users).

8. At the `Grid Y spacing:` prompt, enter 25 ↵ (7 ↵ for metric users).

9. At the `Specify the upper right location for the grid:` prompt, pick a point northeast of the surface to set the area for the labels.

10. Verify that the preview window encompasses the VOL EG-FG surface and press ↵ at the command line to continue.

Wait a few moments as Civil 3D generates all the labels just specified. Your drawing should look similar to Figure 4.54.

When this exercise is complete, you may close the drawing. A finished copy of this drawing is available from the book's web page with the filename `0416_SurfaceGridLabels_FINISHED.dwg` or `0416_SurfaceGridLabels_METRIC_FINISHED.dwg`.

FIGURE 4.54
Volume surface with grid labels

Point Cloud Surfaces

A point cloud is a huge bunch of 3D points, usually collected by laser scanner or *Light Detection and Ranging (LiDAR)*. In a geographic information system (GIS), point clouds are often used as a source for a *Digital Elevation Model (DEM)*. The technology has gotten less expensive and more accurate over the last few years, allowing LiDAR to quickly take over from traditional methods of collecting photogrammetry data.

Point clouds in many formats can be imported to Civil 3D. The most common format is the Log ASCII Standard (LAS) file. This binary format is a public format and at minimum contains x, y, and z data. LAS format can also include color, classification, and other information. For more information on the LAS standard, visit www.asprs.org.

Civil 3D can import a point cloud and use it in several ways. For instance, a laser scan of a bridge can be imported and placed for reference when you're designing a road through an existing abutment. In the example that follows, you will convert LiDAR data into a Civil 3D surface. It is important to note that point clouds often contain millions of points and require a beefy computer (and a little patience on your part) to process.

Importing a Point Cloud

A typical point cloud contains millions of points. These large files are kept external to Civil 3D in a point cloud database. After the LAS has been imported, the data is passed to three files: PRMD, IATI, and ISD. The ISD file contains the points themselves and is the only file needed by CAD if the point cloud would need to be re-created or used in base AutoCAD. By default these files get created in the same directory as the DWG but the location can be changed when importing the information.

If the point cloud you are working with contains coordinate system information (as all the examples in this book do), the software will automatically convert the point cloud to the units and coordinate system of the drawing. For the exercises in this chapter, it does not matter whether you choose the Metric or the Imperial template.

Civil 3D may take a long time to process these files, and you must ensure that you have sufficient disk space to store them, as you will find in the following exercise. It is also a good idea to close the software and then reopen it to help clear some memory before starting memory-intensive tasks such as working with point clouds. The following exercise will show you how to import and work with a point cloud:

1. Start a new file by using the default Civil 3D template of your choice. Save the file before proceeding as **0417_PointCloud.dwg**.

2. Imperial users, set the coordinate system to UTM Zone 17, NAD83, US Foot (UTM83-17F). Metric users, set the coordinate system to UTM Zone 17, NAD83, Meters (UTM83-17).

3. In Prospector, right-click the Point Clouds branch and select the Create Point Cloud option to display the Create Point Cloud Wizard, shown in Figure 4.55.

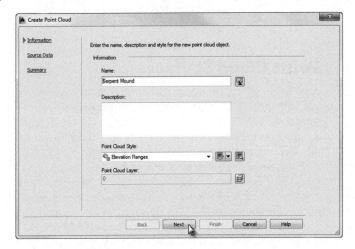

FIGURE 4.55
The Create Point Cloud – Information page

4. Set the name of the point cloud to **Serpent Mound**.

5. Set Point Cloud Style to Elevation Ranges, as shown in Figure 4.55, and click the Next button.

The Source Data page is displayed, shown in Figure 4.56.

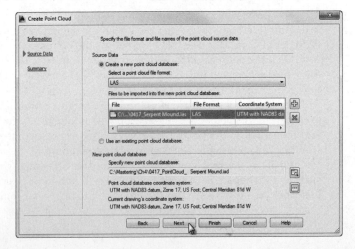

FIGURE 4.56
The Create Point Cloud – Source Data page

6. Using the white plus sign, browse to the `0417_Serpent Mound.las` file (both Imperial and metric users can use this file).

   Remember, all data and drawing files for this book can be downloaded from www.sybex.com/go/masteringcivil3d2014.

   This is a large (90 MB) file containing roughly 1.4 million points and may require a minute or two to process.

7. Click the Next button to display the Summary page, shown in Figure 4.57.

FIGURE 4.57
The Create Point Cloud – Summary page

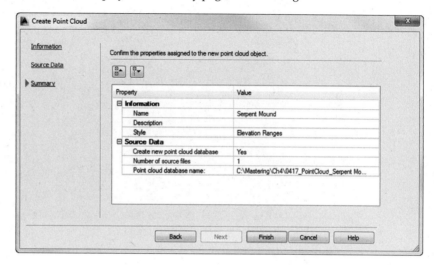

8. Accept the defaults as shown in Figure 4.57 and click Finish to process the point cloud.

9. If the New Point Cloud Database – Processing In Background dialog appears, click Close to dismiss it.

   You may notice a pop-up message in the lower-right corner of your screen indicating that the point cloud database is being created in the background. The components of the point cloud are being processed, including the graphic which will display in the drawing file.

10. Once the point cloud is created in the graphic, it will disappear and a new pop-up message will appear stating that the point cloud has been created and providing a link that says Click Here To Zoom. Click this link.

When the exercise is complete, a portion of a bounding box outlining a portion of the point cloud is displayed in the center of the screen. Save the drawing but leave it open to complete the next exercise.

Autodesk® ReCap™

Included with your installation of Civil 3D 2014 is the new Autodesk product called ReCap. ReCap allows you to look at laser-scanned data in a zippy environment outside of Civil 3D.

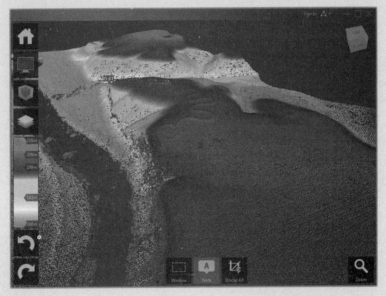

Using ReCap, you can consolidate, measure, crop, and visualize your scanned data before bringing into Civil 3D.

Working with Point Clouds

Once the point cloud is visible in your drawing, you'll want to follow a few rules of thumb to prevent performance problems. The key-in POINTCLOUDDENSITY value controls what percentage of the full point cloud displays on the screen at once. You can also access this value using a slider bar in the Point Cloud contextual tab. However, it is easier to hit the percentage you want on the first try if you use the key-in value. The lower this value, the fewer points are visible, hence the easier it will be to navigate your drawing. The POINTCLOUDDENSITY value does not have any effect on the number of points used when generating a surface model (this is similar to the Level Of Detail value used to aid surface processing).

When you are changing view directions on a point cloud, we recommend that you use preset views and named views to flip around the object. The orbit commands should not be used because they are a surefire way to max out your computer's RAM. If you used the default template, your surface will be located on the V-SITE-SCAN layer. We suggest that you freeze the layer if you do not need to see the point cloud. Use Freeze instead of Off for layer management so the point cloud is not accounted for during pan, zoom, and regen operations (this is true for all AutoCAD objects, but it makes a huge difference when working with point clouds).

Creating a Point Cloud Surface

By specifying either an entire point cloud or a small region of a point cloud, you can create a new TIN surface in your drawing. Any changes to the point cloud object will render the surface definition out-of-date. In the following exercise, a new TIN surface is created from the point cloud previously imported:

1. Continue using the `0417_PointCloud.dwg` file that you created in the previous exercise.

2. Select the bounding box representing the point cloud to display the Point Cloud: Serpent Mound contextual tab, shown in Figure 4.58.

FIGURE 4.58
The Point Cloud contextual tab

3. From the Point Cloud contextual tab ➤ Point Cloud Tools panel, choose Add Points To Surface to display the Add Points To Surface Wizard, shown in Figure 4.59.

FIGURE 4.59
The Add Points To Surface – Surface Options page

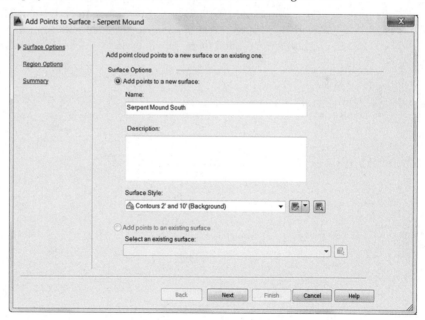

4. Name the surface **Serpent Mound South**. Leave the style set to the default.

5. Click Next and the Region Options page is displayed (Figure 4.60).

FIGURE 4.60
The Add Points To Surface – Region Options page

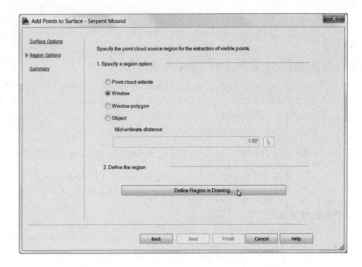

6. Choose the Window radio button, and click Define Region In Drawing.

7. Define the region by creating a window around the southern half of the point cloud.

8. Click Next to see the Summary page, shown in Figure 4.61, and click the Finish button. (Note that your number of points to be added will differ depending on the size of the window you created in the previous step.)

FIGURE 4.61
The Add Points To Surface – Summary page

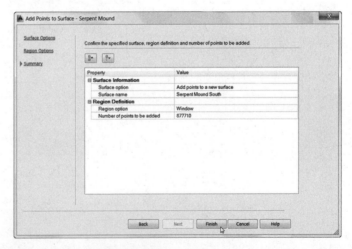

When this exercise is complete, you may close the drawing. Due to the large file size, a finished state of this drawing is not available for download.

Surfaces created from point clouds behave in the same way as any other Civil 3D surface. You will see the point cloud source information in the Definition tab of surface properties. By default, the surface's existence depends on the point cloud. Because point clouds are so large, you may wish to delete the point cloud from the drawing after the Civil 3D surface is created.

If you choose to do this, be sure to set the Copy Deleted Dependent Objects option to Yes in the surface Definition tab. That way, once the point cloud has been removed, the surface will retain the point information.

The Bottom Line

Create an existing ground surface using points.   The most common way to create a surface model is by adding point data to the definition of a surface.

Master It   Open the `MasterIt0401.dwg` or the `MasterIt0401_METRIC.dwg` file. Create a new surface called Existing. Add the point group Topo to its definition. Leave the default styles.

Modify and update a TIN surface.   TIN surface creation is mathematically precise, but sometimes the assumptions behind the equations leave something to be desired. By using the editing tools built into Civil 3D, you can create a more realistic surface model.

Master It   Continue working in the file from the previous exercise or open the `MasterIt0402.dwg` or the `MasterIt0402_METRIC.dwg` file. Use the irregular-shaped polyline and apply it to the surface as an outer boundary of the surface. Make the boundary a destructive breakline.

Prepare a slope analysis.   Surface analysis tools allow users to view more than contours and triangles in Civil 3D. Engineers working with non-technical team members can create strong meaningful analysis displays to convey important site information using the built-in analysis methods in Civil 3D.

Master It   Open the `MasterIt0403.dwg` or the `MasterIt0403_METRIC.dwg` file. Create a slope banding analysis showing slopes under and over 10 percent and insert a dynamic slope legend to help clarify the result of the analysis.

Label surface contours and spot elevations.   Showing a stack of contours is useless without context. Using the automated labeling tools in Civil 3D, you can create dynamic labels that update and reflect changes to your surface as your design evolves.

Master It   Open the `MasterIt0404.dwg` or the `MasterIt0404_METRIC.dwg` file. Label the major contours on the surface at 2′ and 10′ (Background) or 1 m and 5 m (Background).

Import a point cloud into a drawing and create a surface model.   As laser scan data collection becomes more common and replaces other large-scale data-collection methods, the ability to use point clouds in Civil 3D is critical. Intensity helps post processing software determine the ground cover type. While Civil 3D can't do post processing, you can see the intensity as part of the point cloud style.

Master It   Import an LAS format point cloud file (`MasterIt0405_Denver.las`) into the Civil 3D template (with a coordinate system) of your choice. As you create the point cloud file, set the style to Elevation Ranges. Use a portion of the file to create a Civil 3D surface model. No coordinate system needs to be set for this example.

Chapter 5

Parcels

Land development projects often involve the subdivision of large pieces of land into smaller lots. Even if your projects don't directly involve subdivisions, you're often required to show the legal boundaries of your site and the adjoining sites.

AutoCAD® Civil 3D® parcels give you a dynamic way to create, edit, manage, and annotate these legal land divisions. If you edit a parcel segment to make a lot larger, all of the affected labels will update—including areas, bearings, distances, curve information, and table information.

In this chapter, you will learn to:

- Create a boundary parcel from objects
- Create a right-of-way parcel using the right-of-way tool
- Create subdivision lots automatically by layout
- Add multiple-parcel segment labels

Introduction to Sites

In Civil 3D, a *site* is a collection of parcels, alignments, grading objects, and feature lines that share a common topology. In other words, Civil 3D objects that are in the same site are related to, as well as interact with, one another. The objects that react to one another are called *site geometry* objects.

The following objects must be placed in a site:

- Feature lines
- Grading groups
- Parcels

Feature lines and grading groups are discussed in depth in Chapter 14, "Grading." Alignments, discussed in Chapter 6, can be placed in a site but it is not a requirement.

Think Outside of the Lot

You will want to separate by site any objects you don't want interfering with one another. For example, you may have a set of parcels that represent impervious areas for drainage calculations. In the same location, you may have parcels representing property boundaries. By keeping these items on separate sites, you will be able to keep area information separate.

Dynamic area labels are useful for delineating and analyzing soil boundaries; paving, open space, and wetlands areas; and any other region enclosed with a boundary.

Like all Civil 3D objects, parcels utilize styles. With parcel styles, you can apply different layers, colors, hatch patterns, and other graphical properties to differentiate between parcel types.

It's important to understand how site geometry objects react to one another. Figure 5.1 shows a typical parcel that might represent a property boundary.

FIGURE 5.1
A typical property boundary

When an alignment is drawn and placed in the same site as the property boundary, the parcel splits into two parcels, as shown in Figure 5.2.

FIGURE 5.2
An alignment that crosses a parcel divides the parcel in two if the alignment and parcel exist in the same site.

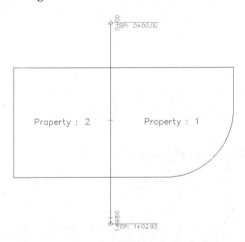

You must plan ahead to create meaningful sites based on interactions between the desired objects. For example, if you want a road centerline, a road right-of-way (ROW) parcel, and the lots in a subdivision to react to one another, they need to be in the same site (see Figure 5.3).

FIGURE 5.3
Alignments, ROW parcels, open space parcels, and subdivision lots react to one another when drawn on the same site.

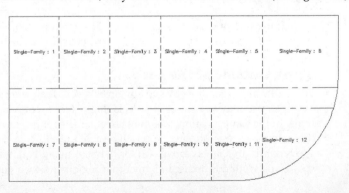

The alignment (or road centerline), ROW parcel, and lots all relate to one another. A change in the centerline of the road should prompt a change in the ROW parcel and the subdivision lots.

If you'd like to avoid the interaction between site geometry objects, place them in different sites. Figure 5.4 shows an alignment that has been placed in a different site from the boundary parcel. Notice that the alignment doesn't split the boundary parcel.

FIGURE 5.4
An alignment that crosses a parcel won't interact with the parcel if they exist in different sites.

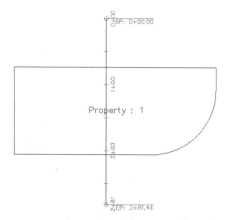

It's important that only objects that are intended to react to each other be placed in the same site. For example, in Figure 5.5 you can see parcels representing both subdivision lots and soil boundaries. Because it wouldn't be meaningful for a soil boundary parcel segment to interrupt the area or react to a subdivision lot parcel, the subdivision lot parcels have been placed in a Subdivision Lots site, and the soil boundaries have been placed in a Soil Boundaries site.

FIGURE 5.5
Parcels can be used for subdivision lots and soil boundaries as long as they're kept in separate sites.

If you didn't realize the importance of site topology, you might create both your subdivision lot parcels and your soil boundary parcels in the same site and find that your drawing looks similar to Figure 5.6. This figure shows the soil boundary segments dividing and interacting with subdivision lot parcel segments, which doesn't make any sense.

FIGURE 5.6
Subdivision lots and soil boundaries react inappropriately when placed in the same site.

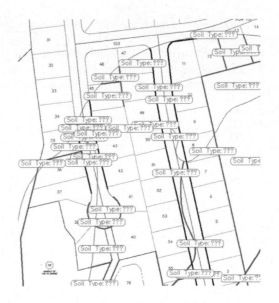

Another way to avoid site geometry problems is to do site-specific tasks in different drawings and use a combination of external references and data references to share information. For example, you could place soil boundaries in one drawing file and subdivision lots in another drawing file and then use external references to show both drawings together.

You should consider keeping your legal site plan in its own drawing. Because of the interactive and dynamic nature of Civil 3D parcels, it might be easy to accidentally modify a parcel segment when you meant to modify a manhole and unintentionally edit a portion of your plat.

You'll see additional examples and drawing divisions later in this chapter, as well as in Chapter 16, "Advanced Workflows."

If you decide to have sites in the same drawing, here are some sites you may want to create. These suggestions are meant to be used as a starting point. Use them to help find a combination of sites that works for your projects:

Roads and Lots   This site could contain road centerlines, ROWs, platted subdivision lots, open space, adjoining parcels, utility lots, and other aspects of the final legal site plan.

Grading   Feature lines and grading objects are considered part of site geometry. If you're using these tools, you must make at least one site for them. You may even find it useful to have several grading sites.

Easements   If you'd like to use parcels to manage, analyze, and annotate your easements, you may consider creating a separate site for easements.

Impervious Areas   If you'd like to use parcels to manage, analyze, and annotate paved areas, a separate site will be useful. Drainage areas can be tracked using a separate object called a catchment, so there is no need to create a site for drainage areas. You can read more about catchments in the bonus chapter, "Storm and Sewer Analysis," found on this book's web page www.sybex.com/go/masteringcivil3d2014.

Soils Many projects require knowledge of the different soil types present on the site. You will want soils in a separate site so that their boundaries don't interact with other objects in the drawing.

As you learn new ways to take advantage of alignments, parcels, and grading objects, you may find additional sites that you'd like to create at the beginning of a new project.

> **WHAT ABOUT THE "SITELESS" ALIGNMENT?**
>
> As mentioned earlier in this chapter, you have a choice whether or not to place your alignments in a site. There are many situations where having the alignment independent from other objects is desirable, and therefore the <none> site can be used.
>
> However, you can still create alignments in traditional sites, if you desire, and they will react to other site geometry objects. For example, if you wish to use the Create Right Of Way tool, the alignment you are working with must be on the same site as the main parcel.
>
> You'll likely find that best practices for most alignments are to place them in the <none> site. For example, if road centerlines, road transition alignments, swale centerlines, and pipe network alignments are placed in the <none> site, you'll save yourself quite a bit of site geometry management.
>
> If you decide you'd like to move the alignment to a site or to <none>, you can do so at any time. Click the alignment you wish to reassign. On the Alignment contextual tab ➢ Modify panel flyout, click Move To Site. See Chapter 6 for more information about alignments.
>
>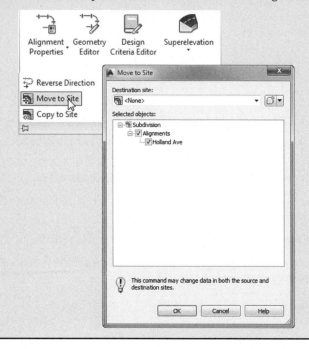

Creating a New Site

You can create a new site in Prospector. You'll find the process easier if you brainstorm potentially needed sites at the beginning of your project and create those sites right away—or, better yet, save them as part of your standard Civil 3D template. You can always add or delete sites later in the project.

You can access the Sites collection in Prospector, along with the other Civil 3D objects in your drawing.

The following exercise will lead you through creating a new site that you can use for creating subdivision lots:

1. Open the CreateSite.dwg (CreateSite_METRIC.dwg) file, which you can download from this book's web page at www.sybex.com/go/masteringcivil3d2014.

    Note that the drawing contains alignments and a boundary parcel, as shown in Figure 5.7.

FIGURE 5.7
The Create Site drawing contains alignments and a boundary parcel.

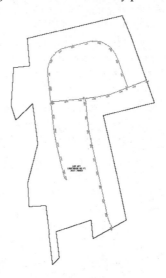

2. On the Prospector tab of Toolspace, go to Sites.

3. Right-click the Sites collection, and select New to open the Site Properties dialog.

4. On the Information tab of the Site Properties dialog, enter **Subdivision Lots** for the name of your site.

5. Confirm that the settings on the 3D Geometry tab match what is shown on Figure 5.8.

    As you create parcels, Civil 3D will automatically number them for you. The values in the Numbering tab are the starting point.

FIGURE 5.8
Confirm the settings on the 3D Geometry tab.

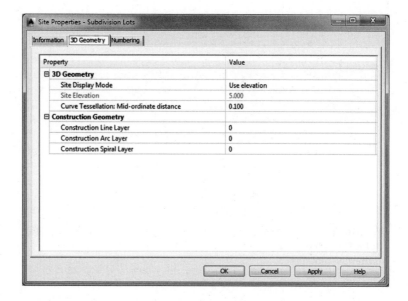

6. Confirm that both values on the Numbering tab are set to **1**. Click OK.

7. Locate the Sites collection on the Prospector tab of Toolspace, and note that your Subdivision Lots site appears on the list, as shown in Figure 5.9. You can repeat the process for all the sites you anticipate needing over the course of the project.

FIGURE 5.9
Your new site is listed in Prospector.

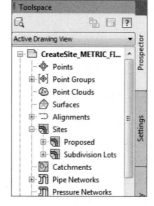

8. Save and close the drawing. If you would like to see what the drawing should look like at this point, you can open CreateSite_FINISHED.dwg (CreateSite_METRIC_FINISHED.dwg), available from the book's website.

Creating a Boundary Parcel

The Create Parcel From Objects tool allows you to create parcels by choosing AutoCAD entities in your drawing or in an XRef'd drawing. In a typical workflow, it's common to encounter a boundary created by AutoCAD entities, such as polylines, lines, and arcs.

When you're using AutoCAD geometry to create parcels, it's important that the geometry be created carefully and meet certain requirements. The AutoCAD geometry must be lines, arcs, polylines, 3D polylines, or polygons. It can't include blocks, ellipses, circles, or other entities. Civil 3D may allow you to pick objects with an elevation other than zero, but you'll find you get better results if you flatten the objects so all objects have an elevation of zero. Sometimes the geometry appears sound when elevation is applied, but you may notice this isn't the case once the objects are flattened. Flattening all objects before creating parcels can help you prevent frustration when creating parcels.

This exercise will teach you how to create a parcel from Civil 3D objects:

1. Open the CreateBoundaryParcel.dwg (CreateBoundaryParcel_METRIC.dwg) file, which you can download from this book's web page.

   This drawing has several alignments, which were created on the Proposed site, and some AutoCAD lines representing a boundary. A parcel automatically formed on the site because the alignments form a closed area.

2. On the Home tab ➢ Create Design panel, select Parcel ➢ Create Parcel From Objects.

3. At the Select lines, arcs, or polylines to convert into parcels or [Xref]: prompt, pick the red lines that represent the site boundary, and press ↵.

   The Create Parcels – From Objects dialog appears.

4. From the drop-down menus, select Proposed, Lot (Prop), and Name Square Foot & Acres (or Name Square Meter & HA if you are working in metric units) in the Site, Parcel Style, and Area Label Style selection boxes, respectively.

   Leave everything else set to the defaults, as shown in Figure 5.10. Notice that Erase Existing Entities is selected. This means that the red lines that you selected will be removed from the drawing and replaced with parcel objects.

5. Click OK to dismiss the dialog.

6. Save and close the drawing. If you would like to see what the drawing should look like at this point, you can open CreateBoundaryParcel_FINISHED.dwg (CreateBoundaryParcel_METRIC_FINISHED.dwg), available from the book's website.

The boundary polyline forms parcel segments that react with the alignments. Area labels are placed within the newly created parcels, as shown in Figure 5.11.

FIGURE 5.10
Site and Style settings for your new boundary parcel

FIGURE 5.11
The boundary parcel segments, alignments, and area labels

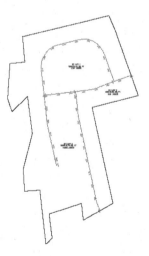

Using Parcel Creation Tools

When you don't have existing lines to work with, the best option is to draw parcel segments using the Parcel Layout tools. Figure 5.12 shows the many commands available to you.

FIGURE 5.12
Selecting parcel creation tools

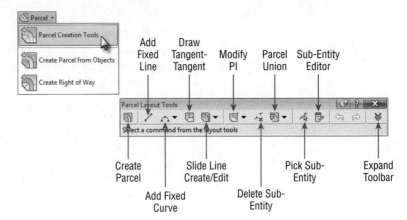

Although you may never have thought of things like wetland areas or easements as parcels in the past, you can take advantage of the parcel tools to assist in labeling, stylizing, and analyzing these features for your plans.

This exercise will teach you how to create a parcel representing wetlands using the transparent commands and Draw Tangent-Tangent With No Curves tool from the Parcel Layout Tools toolbox:

1. Open the WetlandsParcel.dwg (WetlandsParcel_METRIC.dwg) file, which you can download from this book's web page.

   Note that this drawing has several alignments and parcels and a series of points that represent a wetlands delineation.

2. Choose Parcel ➢ Parcel Creation Tools on the Create Design panel.

   The Parcel Layout Tools toolbar appears.

3. Click the Draw Tangent-Tangent With No Curves tool on the Parcel Layout Tools toolbar.

   The Create Parcels – Layout dialog appears.

4. From the drop-down menus, select Proposed, Wetland, and Name Square Foot & Acres (or Name Square Meters & HA) in the Site, Parcel Style, and Area Label Style selection boxes, respectively.

   Keep the default settings for all other options.

5. Click OK.

6. At the Specify start point: prompt, click the Point Number Transparent command (or type '**PN** ↵ at the command prompt).

7. At the Enter Point Number: prompt, enter **1-6** ↵.

   You will see a line form through the wetland boundary points in the northwest corner of the project and immediately form the parcel.

8. Press Esc once to exit the Transparent command, and press Esc a second time to complete the parcel.

9. Press Enter to start a new lot line. Then repeat steps 6 through 8 for the other wetland points near the south of the project using points **7** through **18**.

10. Press Esc to exit the command.

    Your drawing should look similar to Figure 5.13.

FIGURE 5.13
The wetlands defined on the site

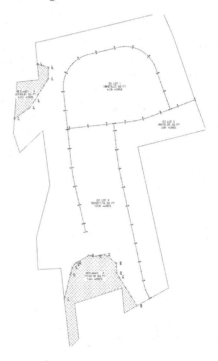

To illustrate how to change the styles that are in use on a parcel, you will next change the parcel style and the area selection label style.

11. Select the northern wetland parcel's area label. From the Parcel contextual tab ➢ Modify panel, click Parcel Properties.

    The Parcel Properties dialog appears.

12. On the Information tab, select Mitigated Wetland from the drop-down menu in the Object Style selection box, and then click Apply to observe the change.

    Remain in the Parcel Properties dialog. The parcel hatch pattern will turn blue and be applied to only the perimeter of the parcel.

13. To change the parcel area label style, switch to the Composition tab in the Parcel Properties dialog.

14. From the Area Selection Label Style pull-down, select Wetland Area Label, and click OK.

15. Select the south wetland parcel by clicking on its area label. From the Parcel contextual tab ➢ Modify panel, click Parcel Properties as you did in step 11.

16. In the Composition tab, change the Area Selection Label Style setting to Wetland Area Label and click OK.

17. Save and close the drawing. If you would like to see what the drawing should look like at this point, you can open `WetlandsParcel_FINISHED.dwg` (`WetlandsParcel _METRIC_ FINISHED.dwg`), available from the book's website.

Your parcels will look like Figure 5.14.

FIGURE 5.14
The wetlands parcels with the appropriate parcel styles and label styles applied

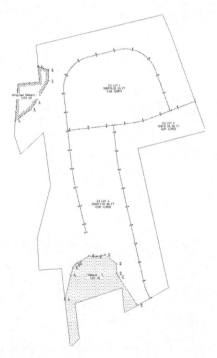

Creating a Right-of-Way Parcel

The Create ROW tool creates ROW parcels on either side of an alignment based on your specifications. The Create ROW tool can be used only when alignments are placed on the same site as the boundary parcel. The resulting ROW parcel will look similar to Figure 5.15.

FIGURE 5.15
The resulting parcels after application of the Create ROW tool

The Create ROW tool includes the following options:

- Offset distance from alignment
- Fillet or chamfer cleanup at parcel boundaries
- Alignment intersections

Figure 5.16 shows an example of chamfered cleanup at alignment intersections.

FIGURE 5.16
A ROW with chamfer cleanup at alignment intersections

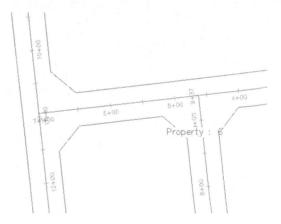

Once the ROW parcel is created, it's no different from any other parcel. It should be noted that the ROW parcel doesn't maintain a dynamic relationship with the alignment that created it. A change to the alignment will require the ROW parcel to be edited or, more likely, re-created.

This exercise will teach you how to use the Create ROW tool to automatically place a ROW parcel for each alignment on your site:

1. Open the `CreateROWParcel.dwg` (`CreateROWPARCEL_METRIC.dwg`) file, which you can download from this book's web page.

   Note that this drawing has some alignments on the same site as the boundary parcel, resulting in several smaller parcels between the alignments and boundary.

2. On the Home tab ➢ Create Design panel, choose Parcel ➢ Create Right Of Way.
3. At the Select parcels: prompt, pick EX LOT 1, EX LOT 2, and EX LOT 3 on the screen.
4. Press ↵ to stop picking parcels.

   The Create Right Of Way dialog appears.

5. Expand the Create Parcel Right Of Way branch, and enter **25′ (8 m)** as the value for Offset From Alignment.
6. Expand the Cleanup At Parcel Boundaries branch.
7. Enter **25′ (8 m)** as the value for Fillet Radius At Parcel Boundary Intersections.
8. Select Fillet from the drop-down menu in the Cleanup Method selection box.
9. Expand the Cleanup At Alignment Intersections branch.
10. Enter **35′ (10 m)** as the value for Fillet Radius At Alignment Intersections.
11. Select Fillet from the drop-down menu in the Cleanup Method selection box. Verify that the Create Right Of Way dialog matches Figure 5.17.

FIGURE 5.17
The Create Right Of Way dialog

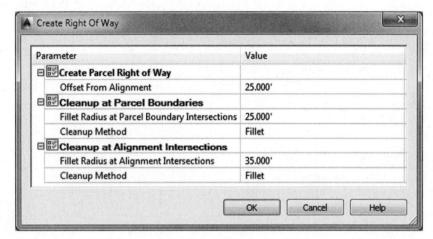

12. Click OK to dismiss the dialog and create the ROW parcels.
13. Save the drawing but keep it open for the next exercise. If you would like to see what the drawing should look like at this point, you can open `CreateROWParcel_FINISHED.dwg` (`CreateROWParcel_METRIC_FINISHED.dwg`), available from the book's website.

    Your drawing should look similar to Figure 5.18.

FIGURE 5.18
The completed ROW parcels

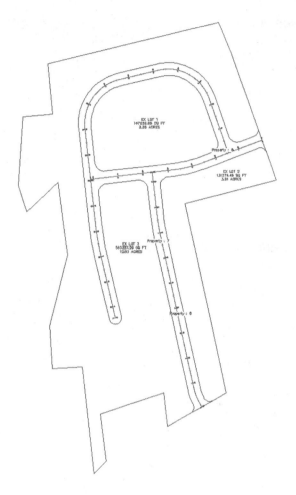

Adding a Cul-de-Sac Parcel

In this section you will add a cul-de-sac at the end of the western road. Although drawing the cul-de-sac requires only basic drafting techniques, there are numerous steps. To save time, a block will be provided so that you don't have to draw it yourself. After inserting this block, you will convert its components into a parcel. This exercise also introduces some editing tools.

You will need to have completed the previous exercise before continuing or open the "FINISHED" version of the drawing.

1. Continue working in the drawing from the previous exercise or open `CreateROWParcel_FINISHED.dwg` (`CreateROWParcel_METRIC_FINISHED.dwg`), available on the book's website.

2. Insert the `CulDeSacBlock.dwg` (`CulDeSacBlock_METRIC.dwg`) file using the settings shown in Figure 5.19.

FIGURE 5.19
Inserting the cul-de-sac block settings

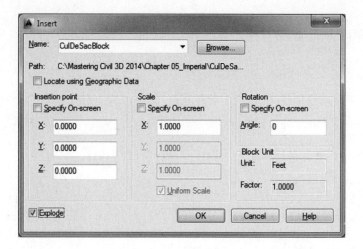

3. From the Home tab ➢ Create Design panel, expand Parcel ➢ Create Parcel From Objects.

4. At the `Select lines, arcs, or polylines to convert into parcels or [Xref]:` prompt, draw a window around all of the cul-de-sac objects, and press ↵.

5. In the Create Parcels – From Objects dialog, verify the following settings:
   - Site is set to Subdivision Lots.
   - Parcel Style is set to Property.
   - Area Label Style is set to Name Square Foot & Acres (Name Square Meter & HA).
   - Erase Existing Entities is checked.

6. Click OK.

Your drawing should look similar to Figure 5.20.

FIGURE 5.20
The cul-de-sac turned into a parcel

The cul-de-sac is now a parcel, but there are some extra lines that need to be taken care of. Let's see how to clean this up a bit.

7. Change the drawing scale to 1" = 20' (1:250). This will cause the parcel area labels to scale down, making the next few steps easier.

8. From the Home tab ➢ Create Design panel ➢ Parcel, select Parcel Creation Tools.

   The Parcel Layout Tools toolbar opens.

9. Select the Delete Sub-Entity tool.

10. At the `Select subentity to remove:` prompt, select the right of way that interferes with the cul-de-sac, as shown in Figure 5.21.

FIGURE 5.21
Delete these portions.

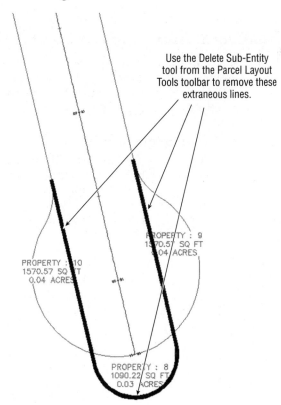

Use the Delete Sub-Entity tool from the Parcel Layout Tools toolbar to remove these extraneous lines.

11. Press Esc twice to exit the command.

12. Save and close the drawing. If you would like to see what the drawing should look like at this point, you can open CreateCulDeSac_FINISHED.dwg (CreateCulDeSac_METRIC_FINISHED.dwg), available from the book's website.

Your cul-de-sac is complete, as shown on Figure 5.22.

FIGURE 5.22
The finished cul-de-sac

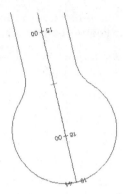

Creating Subdivision Lot Parcels Using Precise Sizing Tools

The precise sizing tools allow you to create parcels to your exact specifications. You'll find these tools most useful when you have your roadways established and understand your lot-depth requirements. These tools provide automatic, semiautomatic, and freeform ways to control frontage, parcel area, and segment direction.

Attached Parcel Segments

Before learning about the parcel sizing tools themselves, let's discus attached parcel segments. Parcel segments created with the precise sizing tools are called *attached segments*. Attached parcel segments have a start point that is attached to a frontage segment and an endpoint that is defined by the next parcel segment they encounter. Attached segments can be identified by their distinctive diamond-shaped grip at their start point and no grip at their endpoint (see Figure 5.23).

FIGURE 5.23
A series of attached parcel segments, with their endpoints at the front lot line

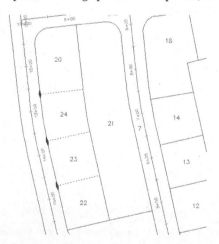

In other words, you establish their start point and their direction, but they seek another parcel segment to establish their endpoint. Figure 5.23 shows a series of attached parcel segments. You can tell the difference between their start points and endpoints because the start points have the diamond-shaped grips.

You can drag the diamond-shaped grip along the frontage to a new location and the parcel segment will maintain its angle from the frontage. If the rear lot line is moved or erased, the attached parcel segments find a new endpoint (see Figure 5.24) at the next available parcel segment.

FIGURE 5.24
The endpoints of attached parcel segments extend to the next available parcel segment if the initial parcel segment is erased.

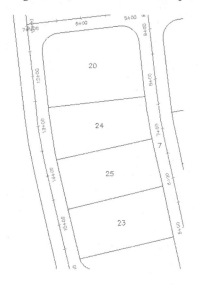

Parcel Sizing Settings

Before using the precise parcel sizing tools, you should understand that there are a number of settings that affect the behavior of these tools. These settings are found by expanding the Parcel Layout Tools toolbar as shown in Figure 5.25. Each of these settings is discussed in detail in the following sections.

FIGURE 5.25
Automated sizing options on the expanded Parcel Layout Tools toolbar

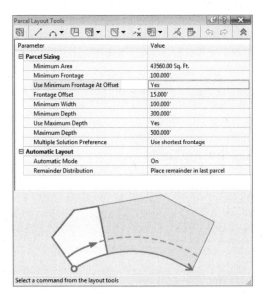

Parcel Sizing

When you create new parcels, the tools respect your default area and minimum frontage (measured from either a ROW or a building setback line). The program always uses these numbers as a minimum; it bases the actual lot size on a combination of the geometry constraints (lot depth, frontage curves, and so on) and the additional settings that follow. Keep in mind that the numbers you establish under the New Parcel Sizing option must make geometric sense. For example, if you'd like a series of 7,500-square-foot (700-square-meter) lots that have 100′ (30 m) of frontage, you must make sure that your rear parcel segment allows for at least 75′ (23 m) of depth; otherwise, you may wind up with much larger frontage values than you desire or a situation where the software can't return a meaningful result.

Automatic Layout

Automatic Layout has two parameters when the list is expanded: Automatic Mode and Remainder Distribution. The Automatic Mode parameter can have the following values:

On   Automatically follows your settings and puts in all the parcels, without prompting you to confirm each one.

Off   Allows you to confirm each parcel as it's created. In other words, this option provides you with a way to semiautomatically create parcels.

The Remainder Distribution parameter tells Civil 3D how you'd like "extra" land handled. This parameter has the following options:

Create Parcel From Remainder   Makes a last parcel with "the leftovers" once the tool has made as many parcels as it can based on the settings in this dialog. This parcel is usually smaller than the other parcels.

Place Remainder In Last Parcel   Adds the leftover area to the last parcel once the tool has made as many parcels as it can based on the settings in this dialog.

Redistribute Remainder   Takes the leftover area and distributes it across all the default-sized parcels once the tool has made as many parcels as it can based on the settings in this dialog. The resulting lots aren't always evenly sized because of differences in geometry around curves and other variables, but the leftover area is absorbed.

There aren't any rules per se in a typical subdivision workflow. Usually the goal is to create as many parcels as possible within the limits of available land. To that end, you'll use a combination of AutoCAD tools and Civil 3D tools to divide and conquer the particular tract of land with which you are working.

Parcel Sizing Tools

The precise sizing tools consist of the Slide Line, Swing Line, and Free Form Create tools (see Figure 5.26).

FIGURE 5.26
The parcel sizing tools

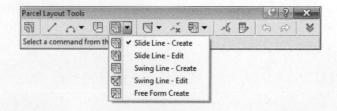

SLIDE LINE – CREATE TOOL

The Slide Line – Create tool creates an attached parcel segment based on an angle from frontage. You may find this tool most useful when your jurisdiction requires a uniform lot-line angle from the right of way.

This exercise will lead you through using the Slide Line – Create tool to create a series of subdivision lots:

1. Open the `CreateSubdivisionLots.dwg` (`CreateSubdivisionLots_METRIC.dwg`) file, which you can download from this book's web page.

   Note that this drawing has several alignments on the same site as the boundary parcel, resulting in several smaller parcels between the alignments and boundary.

2. Choose Parcel ➢ Parcel Creation Tools on the Create Design panel. The Parcel Layout Tools toolbar appears.

3. Expand the toolbar by clicking the Expand The Toolbar button.

4. In the Parcel Sizing section, change the value of the following parameters by clicking in the Value column and typing in the new values if they aren't already set. Notice how the preview window changes to accommodate your preferences:

   ♦ Minimum Area: **7500.00** sq. ft. (**700 m²**)

   ♦ Minimum Frontage: **75.000′** (**25 m**)

   ♦ Use Minimum Frontage At Offset: **Yes**

   ♦ Frontage Offset: **25.000′** (**10 m**)

   ♦ Minimum Width: **75.000′** (**25 m**)

   ♦ Minimum Depth: **50.000′** (**15 m**)

   ♦ Use Maximum Depth: **No**

   ♦ Maximum Depth: (leave default value as this will not be used)

   ♦ Multiple Solution Preference: **Use Shortest Frontage**

5. In the Automatic Layout section, change the following parameters by clicking in the Value column and selecting the appropriate option from the drop-down menu, if they aren't already set:

   ♦ Automatic Mode: On

   ♦ Remainder Distribution: Redistribute Remainder

6. Click the Slide Line – Create tool. The Create Parcels – Layout dialog appears.

7. From the drop-down menus, select Subdivision Lots, Lot (Prop), and Name Square Foot & Acres (Name Square Meter & HA) in the Site, Parcel Style, and Area Label Style selection boxes, respectively.

   Leave the rest of the options set to their defaults.

8. Click OK to dismiss the dialog.
9. At the `Select parcel to be subdivided or [Pick]:` prompt, click the EX LOT 1 area label.
10. At the `Select start point on frontage:` prompt, use your Endpoint Osnap to pick the point of curvature along the ROW parcel segment for Property 1 (see Figure 5.27).

FIGURE 5.27
Pick the point of curvature along the ROW parcel segment.

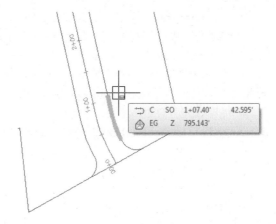

The parcel jig appears.

11. Move your mouse slowly along the ROW parcel segment, and notice that the parcel jig follows the parcel segment.
12. At the `Select end point on frontage:` prompt, use your Endpoint Osnap to pick the point of curvature along the ROW parcel segment for Property 1 (see Figure 5.28).

FIGURE 5.28
Allow the parcel-creation jig to follow the parcel segment, and then pick the point of curvature along the ROW parcel segment.

13. At the `Specify angle or [Bearing/aZimuth]:` prompt, enter 90 ↵.

Notice the preview (see Figure 5.29).

FIGURE 5.29
A preview of the results of the automatic parcel layout

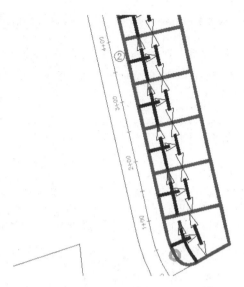

14. At the `Accept result? [Yes/No] <Yes>:` prompt, press ↵ to accept the default Yes.

15. At the `Select parcel to be subdivided or [Pick]:` prompt, press ↵.

16. Click the X to close the Parcel Layout Tools toolbar.

    Your drawing should look similar to Figure 5.30.

FIGURE 5.30
The automatically created lots

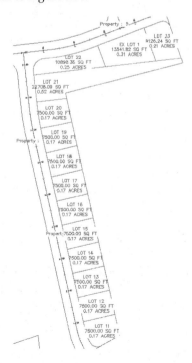

17. Save and close the drawing. If you would like to see what the drawing should look like at this point, you can open CreateSubdivisionLots_FINISHED.dwg (CreateSubdivisionLots_METRIC_FINISHED.dwg), available from the book's website.

The parcels at the north end just don't look right. We'll address that in the next section of this chapter.

> **CURVES AND THE FRONTAGE OFFSET**
>
> In most cases, the frontage along a building setback is graphically represented as a straight line behind the setback. When you specify a minimum width along a frontage offset (the building setback line) in the Parcel Layout Tools dialog, and when the lot frontage is curved, the distance you enter is measured along the curve, as shown by the dashed line here.
>
>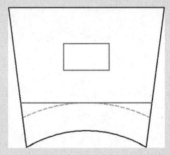
>
> In most cases, this result may be insignificant, but in a large development, the error could be the defining factor in your decision to add or subtract a parcel from the development.

SWING LINE – CREATE TOOL

The Swing Line – Create tool creates a "backward" attached parcel segment where the diamond-shaped grip appears, not at the frontage but at a different location that you specify. The tool respects your minimum frontage, and it adjusts the frontage so it is larger, if necessary, in order to respect your default area. The Swing Line – Create tool is semiautomatic because it requires your input for the swing point location.

You may find this tool most useful around a cul-de-sac or in odd-shaped corners where you must hold frontage but have a lot of flexibility in the rear of the lot.

FREE FORM CREATE TOOL

A site plan is more than just single-family lots. Areas are usually dedicated for open space, stormwater-management facilities, parks, and public utility lots. The Free Form Create tool can be useful when you're creating these types of parcels. This tool, like the precise sizing tools, creates an attached parcel segment with the special diamond-shaped grips.

NOTE The lot numbers were designed by the authors for the exercises. Your lot numbers may vary from those shown in the exercises.

In the following exercise, you'll use the Free Form Create tool to create a new parcel:

1. Open the `CreateFreeForm.dwg` (`CreateFreeForm_METRIC.dwg`) file.

   Note that this drawing contains a series of subdivision lots.

2. Pan over to LOT 20.

   You can see the lot line that was drawn automatically in the previous exercise and obviously will not work (Figure 5.31).

FIGURE 5.31
Delete the highlighted parcel line.

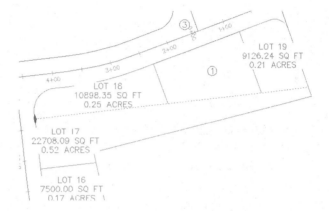

3. Use the AutoCAD Erase command to delete the parcel line highlighted in Figure 5.31.

   The parcels readjust but the resulting lot is now much larger than needed. Let's add a line using the Free Form Create tool. You will explore deleting parcel lines in more depth later in this chapter.

4. From the Home tab ➢ Create Design panel, select Parcel ➢ Parcel Creation Tools.

5. On the Parcel Layout Tools toolbar, select the Free Form Create tool. If the Parcel Layout Tools toolbar is still expanded, you can collapse it by clicking the chevron icon on the far right end of the toolbar.

   The Create Parcels – Layout dialog appears.

6. Select Subdivision Lots, Lot (Prop), and Name Square Foot & Acres (Name Square Meters & HA) from the drop-down menus in the Site, Parcel Style, and Area Label Style selection boxes, respectively.

   Keep the default values for the remaining options.

7. Click OK to dismiss the dialog.

8. Slide the Free Form Create attachment point along the frontage of the Lot: 20 area (your lot number may differ).

9. At the `Select attachment point:` prompt, pick near the point shown in Figure 5.32.

FIGURE 5.32
Use the Free Form Create tool to select an attachment point.

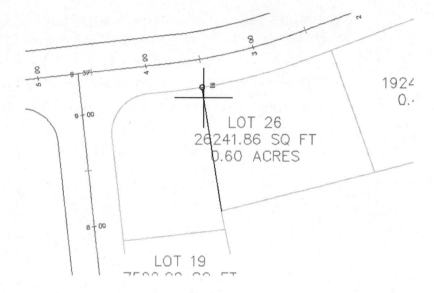

10. At the Specify lot line direction:(ENTER for perpendicular) or [Bearing/aZimuth]: prompt, press ↵ to specify a perpendicular lot line direction.

   A new parcel segment is created perpendicular to the ROW parcel segment, as shown in Figure 5.33 (your resulting lot numbers may differ). Note that a new lot parcel has formed.

   Although most parcel lines are created perpendicular to the right-of-way line, please note that you can use the Free Form Create tool to attach a new parcel segment to any parcel segment in the drawing.

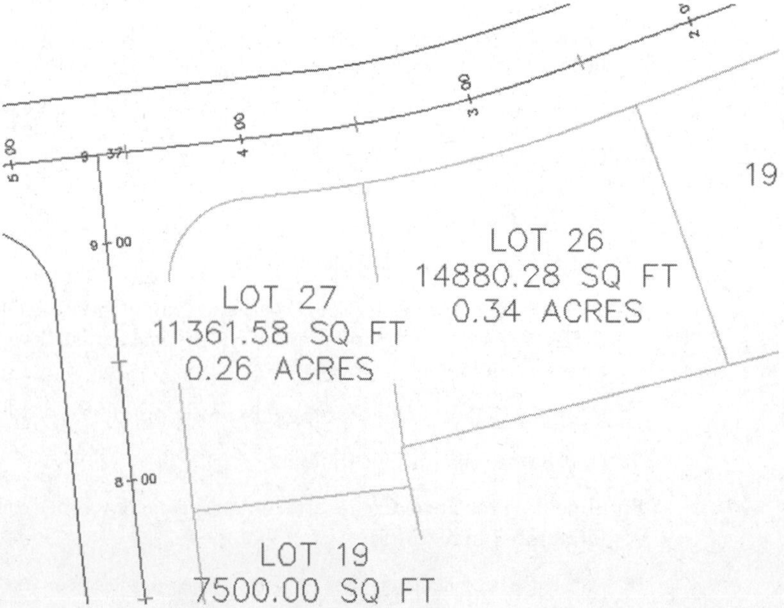

FIGURE 5.33
A new parcel created using the Free Form Create tool.

11. Press ↵ to exit the Free Form Create command.

12. Click the X or press Esc to exit the Parcel Layout Tools toolbar.

13. Pick the new parcel segment so that you see its diamond-shaped grip.

14. Grab the grip, and slide the segment along the ROW parcel segment (see Figure 5.34).

FIGURE 5.34
Sliding an attached parcel segment

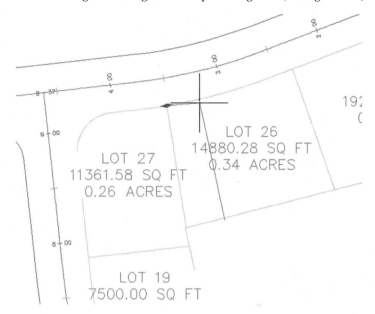

Notice that when you place the parcel segment at a new location, the segment endpoint snaps back to the rear parcel segment. This is typical behavior for an attached parcel segment.

15. Save and close the drawing. If you would like to see what the drawing should look like at this point, you can open `CreateFreeForm_FINISHED.dwg` (`CreateFreeForm_METRIC_FINISHED.dwg`), available from the book's website.

Editing Parcels by Deleting Parcel Segments

One of the most powerful aspects of Civil 3D parcels is the ability to perform many iterations of a site plan design. Typically, this design process involves creating a series of parcels and then deleting them to make room for iterations with different parameters or deleting certain segments to make room for easements, public utility lots, and more.

You can delete parcel segments using the AutoCAD Erase tool as shown in the previous exercise, or you can use the Delete Sub-Entity tool on the Parcel Layout Tools toolbar.

It's important to understand the difference between these two methods. The AutoCAD Erase tool behaves as follows:

♦ If a series of parcel segments was originally created from a polyline (or similar parcel layout tools, such as the Tangent-Tangent With No Curves tool), the AutoCAD Erase tool erases the entire series of segments (see Figure 5.35).

FIGURE 5.35
The highlighted segments will be erased after using the AutoCAD Erase tool.

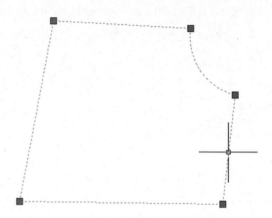

- If a series of parcel segments was originally created using lines or arcs (or similar parcel layout tools, such as the precise sizing tools), then AutoCAD Erase erases each segment individually (see Figure 5.36).

FIGURE 5.36
The highlighted segment will be erased.

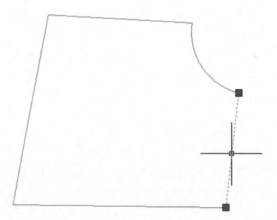

The Delete Sub-Entity tool acts more like the AutoCAD Trim tool. The Delete Sub-Entity tool erases only the parcel segments between parcel vertices. For example, if Lot 26, as shown in Figure 5.37, must be absorbed into Lot 3 to create a public utility easement, you'd want to only erase the segment at the rear of Lot 26 and not the entire segment shown previously in Figure 5.36.

As an alternate to launching the Parcel Creation tools by choosing the Home tab ➢ Create Ground Data panel ➢ Parcel ➢ Parcel Creation Tools, you can access them by selecting the area label of a parcel and then, from the Parcel contextual tab ➢ Modify panel, selecting Parcel Layout Tools. Selecting the Delete Sub-Entity tool allows you to pick only the small rear parcel segment for Lot 26. Figure 5.38 shows the result of this deletion.

FIGURE 5.37
Using the Delete Sub-Entity tool to erase the rear parcel segment for Lot 26

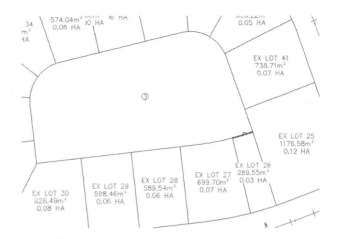

FIGURE 5.38
The rear lot line for Lot 26 was erased using the Delete Sub-Entity tool, thus creating a larger Lot 3.

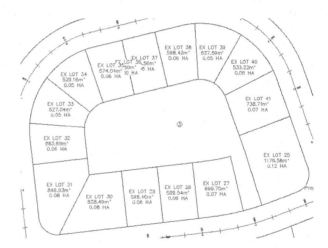

The following exercise will lead you through deleting a series of parcel segments using both the AutoCAD Erase tool and the Delete Sub-Entity tool:

1. Open the DeleteSegments.dwg (DeleteSegments_Metric.dwg) file.

   Note that this drawing contains a series of subdivision lots, along with a wetlands boundary.

   Let's say you just received word that you are allowed to build on the wetland area.

2. Use the AutoCAD Erase tool to erase the parcel segments that define the wetlands parcel.

   Note that the entire parcel disappears as soon as the first segment is removed, causing a "hole" in the parcel.

   Now the developer has decided to enlarge the lots.

3. Erase the lot lines indicated by the arrows in Figure 5.39.

FIGURE 5.39
The parcel lines to be deleted

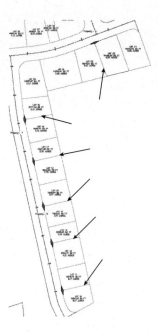

After deleting parcel segments, your lots should look similar to Figure 5.40.

FIGURE 5.40
The re-created lots

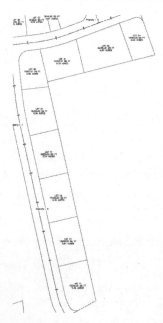

Next, you discover that Lot 26 needs to be removed and absorbed into Lot 3.

4. From the Home tab ➢ Create Ground Data panel, click Parcel ➢ Parcel Creation Tools to open the Parcel Layout Tools toolbar. On the Parcel Layout Tools toolbar, click the Delete Sub-Entity tool.
5. At the Select subentity to remove: prompt, pick the rear lot line of Lot 26.
6. Press ↵ to exit the command and then click the X to exit the Parcel Layout Tools toolbar.

The resulting parcel is shown in Figure 5.41.

FIGURE 5.41
The parcel after erasing the rear lot line

7. Save and close the drawing. If you would like to see what the drawing should look like at this point, you can open CreateFreeForm_FINISHED.dwg (CreateFreeForm_METRIC_FINISHED.dwg), available from the book's website.

Best Practices for Parcel Creation

Now that you have an understanding of how objects in a site interact and you've had some practice creating and editing parcels in a variety of ways, we'll take a deeper look at how parcels must be constructed to achieve topology stability, predictable labeling, and desired parcel interaction.

Forming Parcels from Segments

Earlier in this chapter, you saw that parcels are created only when parcel segments form a closed area (see Figure 5.42).

FIGURE 5.42
A parcel is created when parcel segments form a closed area.

Parcels must always close. Whether you draw AutoCAD lines and use the Create Parcel From Objects menu command or use the parcel segment creation tools, a parcel won't form until there is an enclosed polygon. Figure 5.43 shows four parcel segments that don't close; therefore, no parcel has been formed.

FIGURE 5.43
No parcel will be formed if parcel segments don't completely enclose an area.

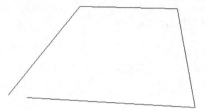

There are times in surveying and engineering when parcels of land don't necessarily close when created from legal descriptions. In this case, you must work with your surveyor to perform an adjustment or find some other solution to create a closed polygon.

You also saw that even though parcels can't be erased, if you erase the appropriate parcel segments, the area contained within a parcel is assimilated into neighboring parcels.

Parcels Reacting to Site Objects

Parcels require only one parcel segment to divide them from their neighbor (see Figure 5.44). This behavior eliminates the need for duplicate segments between parcels, and duplicate segments must be avoided.

FIGURE 5.44
Two parcels, with one parcel segment between them

As you saw in the section on site interaction, parcels understand their relationships to one another. When you create a single parcel segment between two subdivision lots, you have the ability to move one line and affect two parcels. Figure 5.45 shows the moved parcel segment from Figure 5.44 once the parcel segment between them has been shifted to the left. Note that both areas change in response.

FIGURE 5.45
Moving one parcel segment affects the area of two parcels.

A mistake that many people new to Civil 3D make is to create parcels from closed polylines, which results in a duplicate segment between parcels. Figure 5.46 shows two parcels created from two closed polylines. These two parcels may appear identical to the two seen in the previous example, because they were both created from a closed polyline rectangle; however, the segment between them is actually two segments.

FIGURE 5.46
Adjacent parcels created from closed polylines create overlapping or duplicate segments.

The duplicate segment becomes apparent when you attempt to grip-edit the parcel segments. Moving one vertex from the common lot line, as seen in Figure 5.47, reveals the second segment. Also note that a sliver parcel is formed. Duplicate site geometry objects and sliver parcels make it difficult for Civil 3D to solve the site topology and can cause unexpected parcel behavior.

FIGURE 5.47
Duplicate segments become apparent when they're grip-edited and a sliver parcel is formed.

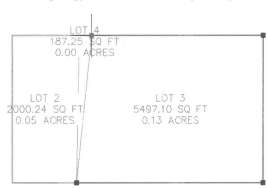

Creating a subdivision plat of parcels this way almost guarantees that your labeling won't perform properly and could lead to inaccurate data.

> **PARCELS AND LANDXML**
>
> The best method for importing lot data from other programs such as LandDesktop is LandXML. Parcels imported from LandXML will automatically clean up duplicate lines to prevent some of the pitfalls mentioned in this section.

Parcels form to fill the space contained by the original outer boundary. You should always begin a parcel-division project with an outer boundary of some sort (see Figure 5.48).

FIGURE 5.48
An outer boundary parcel

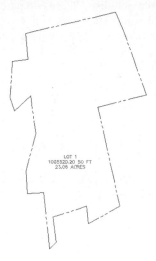

You can then add road centerline alignments to the site, which divides the outer boundary, as shown in Figure 5.49.

FIGURE 5.49
Alignments added to the same site as the boundary parcel divide the boundary parcel.

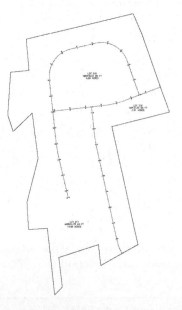

It's important to note that the boundary parcel no longer exists intact. As you subdivide this site, Parcel 1 is continually reallocated with every division. As road ROW and subdivision lots are formed from parcel segments, more parcels are created. Every bit of space that was contained in the original outer boundary is accounted for in the mesh of newly formed parcels (see Figure 5.50).

FIGURE 5.50
The total area of parcels contained within the original boundary is equal to the original boundary area.

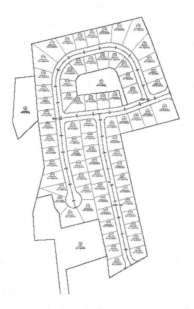

From now on, you'll consider ROW, wetlands, parkland, and open space areas as parcels, even if you didn't before. You can make custom label styles to annotate these parcels however you like, including a "no show" or none area label.

 Real World Scenario

OVERKILL IS JUST RIGHT FOR PARCELS

Frequently, parcel data comes from GIS sources or from existing plats. The quality of this data can vary, making a direct conversion to Civil 3D parcels difficult.

An excellent tool for ensuring that the conversion of these lines, arcs, and polylines goes smoothly is the Overkill command. Overkill is an AutoCAD command that will clean up drawings based on the options and tolerance that you choose.

You can find the Overkill command on the Home tab ➢ Modify panel flyout.

Once you have selected the objects you want the Overkill command to analyze, you are prompted to choose options that determine how extensive the cleanup will be. The default settings, shown here, are rather conservative.

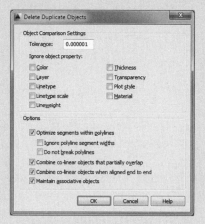

The tolerance determines how close two lines can be before they are considered to be duplicates. The higher this value, the more overlap and duplicates the command will find.

The check boxes indicate properties Overkill will disregard when performing its edits. With all of the check boxes clear, the objects are not considered duplicates unless all of the listed properties match.

Before you run the Overkill command, a selected batch of lines slated to become parcels show many extraneous vertices. The plus signs next to the grips indicate where multiple grips overlap.

Before Overkill Command

After you run the Overkill command, the selection shows fewer visible grips and no indication that there are overlapping objects.

After Overkill Command

As you can see, the Overkill command can save you valuable time cleaning up drawings.

Constructing Parcel Segments with the Appropriate Vertices

Parcel segments should have natural vertices only where necessary and split-created vertices at all other intersections. A natural vertex, or point of intersection (PI), can be identified by picking a line, polyline, or parcel segment and noting the location of the grips (see Figure 5.51).

FIGURE 5.51
Natural vertices on a parcel segment

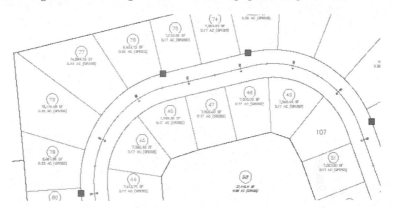

A split-created vertex occurs when two parcel segments touch or cross each other. Note that in Figure 5.52, the parcel segment doesn't show a grip even where each individual lot line touches the ROW parcel.

FIGURE 5.52
Split-created vertices on a parcel segment

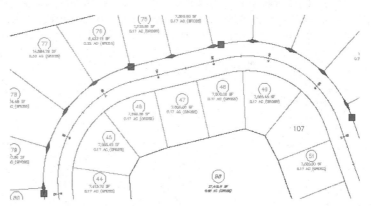

It's desirable to have as few natural vertices as possible. In the example shown in Figure 5.52, the ROW frontage line can be expressed as a single bearing and length from the end of the arc through the beginning of the next arc, as opposed to having several smaller line segments.

If the foundation geometry was drawn with a natural vertex at each lot line intersection, the resulting parcel segment won't label properly and may cause complications with editing and other functions. This subject will be discussed in more detail in the section "Labeling Spanning Segments" later in this chapter.

Parcel segments must not overhang. Spanning labels are designed to overlook the location of intersection-formed (or T-shaped) split-created vertices. However, these labels won't span a crossing-formed (X- or + [plus]-shaped) split-created vertex. Even a very small parcel segment overhang will prevent a spanning label from working and may even affect the area computation for adjacent parcels. The overhanging segment in Figure 5.53 would prevent a label from returning the full spanning length of the ROW segment it crosses.

FIGURE 5.53
Overhanging segment

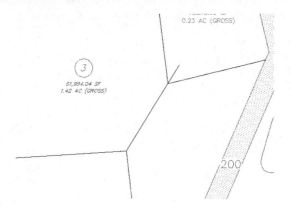

Labeling Parcel Areas

A parcel area label is placed at the parcel centroid by default, and it refers to the parcel in its entirety. It doesn't necessarily have to include the actual area of the parcel. When asked to pick a parcel, you pick the area label, however this behavior is only valid for Civil 3D commands. For example, if you pick a parcel area label and invoke the AutoCAD Erase command, the parcel will not be erased.

Area labels can be customized to suit your needs. Figure 5.54 shows a variety of customized area labels.

FIGURE 5.54
Sample area labels

Area labels often include the parcel name or number, and often the names and/or numbers need to be changed after the parcels are placed in the drawing. To accomplish this, you can select a parcel and then on the Parcel contextual tab ➢ Modify panel choose Renumber/Rename.

The following exercise will teach you how to renumber a series of parcels:

1. Open ChangeAreaLabel.dwg (ChangeAreaLabel_METRIC.dwg). Note that this drawing contains many subdivision lot parcels.

2. Near the southeast corner of the project, select Lot 25. On the Parcel contextual tab ➢ Modify panel, select Renumber/Rename. The Renumber/Rename Parcels dialog appears.

3. In the Renumber/Rename Parcels dialog, make sure Subdivision Lots is selected from the drop-down menu in the Site selection box and that the Renumber radio button is selected. Change the value of the Starting Number selection box to **1**. Click OK.

4. At the Specify start point or [Polylines/Site]: prompt, pick a point on the screen anywhere inside the Lot 25 parcel, which will become your new Lot 1 parcel at the end of the command.

5. At the End point or [Undo]: prompt, pick a point on the screen anywhere inside the Lot 35 parcel, almost as if you were drawing a line; then pick a point inside Lot 39. Press ↵ to complete choosing parcels. Press ↵ again to end the command.

Note that your parcels have been renumbered from 1 through 15.

6. Save the drawing but keep it open for the next exercise. If you would like to see what the drawing should look like at this point, you can open RenumberParcels_FINISHED.dwg (RenumberParcels_METRIC_FINISHED.dwg), available from the book's website.

Repeat the exercise with other parcels in the drawing for additional practice if desired.

The next exercise will lead you through one method of changing an area label using the Edit Parcel Properties dialog:

1. Continue working in the ChangeAreaLabel.dwg (ChangeAreaLabel_METRIC.dwg) file, or you can open RenumberParcels_FINISHED.dwg (RenumberParcels_METRIC_FINISHED.dwg), available from the book's website.

2. Select Parcel 1 and then on the Parcel contextual tab ➢ Modify panel, select Multiple Parcel Properties.

3. At the Specify start point or [Polylines/All/Site]: prompt, pick a point on the screen anywhere inside Parcel 1.

4. At the End point or [Undo]: prompt, pick a point on the screen anywhere inside Parcel 11, then Parcel 15, using the same technique that you used in steps 4 and 5 of the previous exercise.

5. Press ↵ to complete parcel selection, and press ↵ again to open the Edit Parcel Properties dialog.

6. In the Area Selection Label Styles portion of the Edit Parcel Properties dialog, use the drop-down menu to select the Parcel Number area label style, as shown in Figure 5.55.

7. Click the Apply To All Parcels button to the right of the Parcel Number listing.

FIGURE 5.55
The Edit Parcel Properties dialog

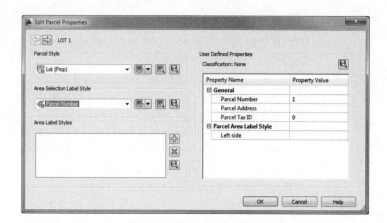

8. Click Yes in the dialog displaying the question "Apply the area selection label style to the 15 selected parcels?"

9. Click OK to exit the Edit Parcel Properties dialog.

   The 15 parcels now have parcel area labels that call out numbers only. Note that you could also use this interface to add a second area label to certain parcels by clicking the plus sign in the Area Label Styles section of the Edit Parcel Properties dialog.

10. Save the drawing but keep it open for the next exercise. If you would like to see what the drawing should look like at this point, you can open ChangeMultipleLabels_FINISHED.dwg (ChangeMultipleLabels_METRIC_FINISHED.dwg), available from the book's website.

This section's final exercise will show you how to use Prospector to change a group of parcel area labels at the same time:

1. Continue working in the previous file or open the ChangeMultipleLabels_FINISHED.dwg (ChangeMultipleLabels_METRIC_FINISHED.dwg) file, available on the book's website.

2. In Prospector, expand Sites ➤ Subdivision Lots and select the Parcels collection.

3. Hold down the Ctrl key, and select all of the lots whose names begin with LOT.

4. Release the Ctrl key, and your parcels should remain selected.

5. Slide over to the Area Label Style column, right-click the column header, and select Edit (see Figure 5.56).

FIGURE 5.56
Right-click the Area Label Style column header and select Edit.

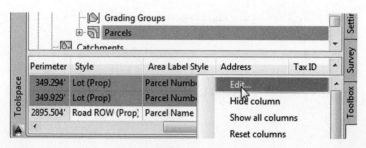

6. In the Select Label Style dialog, select Parcel Number from the drop-down menu in the Label Style selection box.

7. Click OK to dismiss the dialog.

   The drawing will process for a moment.

8. Once the processing is finished, minimize Prospector and inspect your parcels.

   All the single-family parcels should now have the Parcel Number area label style.

9. Save and close the drawing. If you would like to see what the drawing should look like at this point, you can open `ChangeAreaLabel_FINISHED.dwg` (`ChangeAreaLabel_METRIC_FINISHED.dwg`), available from the book's website.

 Real World Scenario

What If the Area Label Needs to Be Split onto Two Layers?

You may have a few different types of plans that show parcels. Because it would be awkward to have to change the parcel area label style before you plot each sheet, it would be best to find a way to make a second label on a second layer so that you can freeze the area component in sheets or viewports when it isn't needed. Here's an example where the square footage has been placed on a different layer so it can be frozen in certain viewports.

You can accomplish this by creating a second parcel area label that calls out the area only:

1. In any drawing containing parcels, change to the Annotate tab.
2. From the Labels & Tables panel, select Add Labels ➢ Parcel ➢ Add Parcel Labels.

3. Select Area from the drop-down menu in the Label Type selection box, and then select an area style label that will be the second area label.

4. Click Add, and then pick your parcel onscreen.

You'll find a second parcel area label to be a little more automatic when you place it (it already knows what parcel to reference).

You can also use the Edit Parcel Properties dialog, shown in Figure 5.55 in earlier in the section "Labeling Parcel Areas" to add a second label.

Labeling Parcel Segments

Although parcels are used for much more than just subdivision lots, most parcels you create will probably be used for concept plans, record plats, and other legal subdivision plans. These plans, such as the one shown in Figure 5.57, almost always require segment labels for bearing, distance, direction, crow's-feet, and more.

FIGURE 5.57
A fully labeled site plan

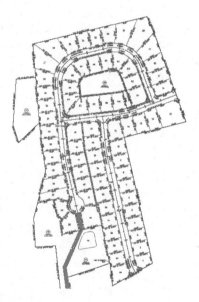

Labeling Multiple-Parcel Segments

The following exercise will teach you how to add labels to multiple-parcel segments:

1. Open the SegmentLabels.dwg (SegmentLabels_METRIC.dwg) file, which you can download from this book's web page.

   Note that this drawing contains many subdivision lot parcels.

2. On the Annotate tab ➢ Labels & Tables panel, click Add Labels.
3. From the drop-down menus in the Add Labels dialog, select Parcel, Multiple Segment, Bearing Over Distance, and Delta Over Length And Radius in the Feature, Label Type, Line Label Style, and Curve Label Style selection boxes, respectively, as shown in Figure 5.58.

FIGURE 5.58
The Add Labels dialog

4. Click Add.
5. At the `Select parcel to be labeled by clicking on area label:` prompt, pick the area label for Parcel 1.
6. At the `Label direction [CLockwise/COunterclockwise]<CLockwise>:` prompt, press ↵ to accept the default and again to exit the command.

   Each parcel segment for Parcel 1 should now be labeled.
7. Continue picking Parcels 2 through 15 in the same manner.

   Note that segments are never given a duplicate label, even along shared lot lines.
8. If time permits, label all of the parcels in the drawing.
9. Press ↵ to exit the command. Close the Add Labels dialog.
10. Save the drawing but keep it open for the next exercise. If you would like to see what the drawing should look like at this point, you can open `CreateSegmentLabels_FINISHED .dwg` (`CreateSegmentLabels_METRIC_FINISHED.dwg`), available from the book's website.

The following exercise will show you how to edit and delete parcel segment labels:

1. Continue working in the `SegmentLabels.dwg` (`SegmentLabels_METRIC.dwg`) file, or you can open `CreateSegmentLabels_FINISHED.dwg` (`CreateSegmentLabels_METRIC_FINISHED.dwg`), available from the book's website.

2. Zoom in on the label along the frontage of Parcel 8.
3. Select the label.

   You'll know your label has been picked when you see a diamond-shaped grip at the label midpoint (see Figure 5.59).

FIGURE 5.59
A diamond-shaped grip appears when the label has been picked.

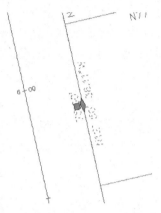

4. On the Labels – Parcel Segment Label contextual tab ➤ Modify panel, click Flip Label, shown in Figure 5.60.

FIGURE 5.60
The Parcel Segment Label contextual tab

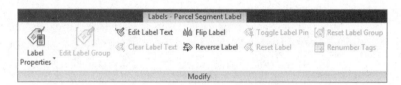

   The label flips so that the bearing component is outside the parcel and the distance component is inside.

5. Select the label again. On the Labels – Parcel Segment Label contextual tab ➤ Modify panel, click Reverse Label. Press Esc to clear your selection.

   The label reverses so that the bearing now reads SE instead of NW.

6. Repeat steps 3 through 5 for several other segment labels, and note their reactions.
7. Select one of the labels along the back lot lines of parcels 1–15. Once the label is picked, execute the AutoCAD Erase tool or press the Delete key.

   Note that the label disappears.

8. Erase the remaining back lot line labels for parcels 1–15 in preparation for the next exercise.
9. Save and close the drawing. If you would like to see what the drawing should look like at this point, you can open `EditSegmentLabels_FINISHED.dwg` (`EditSegmentLabels_METRIC_FINISHED.dwg`), available from the book's website.

Labeling Spanning Segments

Spanning labels are used where you need a label that spans the overall length of an outside segment, such as the example in Figure 5.61.

FIGURE 5.61
A spanning label

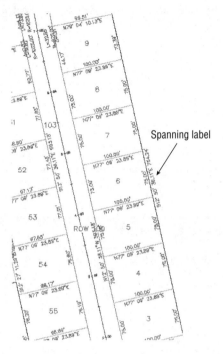

Spanning labels require that you use the appropriate vertices, as discussed in detail in the earlier section "Constructing Parcel Segments with the Appropriate Vertices." Spanning labels have the following requirements:

- Spanning labels can only span across split-created vertices. Natural vertices will interrupt a spanning length.

- Spanning label styles must be composed to span the outside segment.

- Spanning label styles must be composed to attach the desired spanning components (such as length and direction arrow) on the outside segment (as shown previously in Figure 5.61), with perhaps a small offset.

Once you've confirmed that your geometry is sound and your label is properly composed, you're set to span. The following exercise will teach you how to add spanning labels to single-parcel segments:

1. Open `SpanLabels.dwg` (`SpanLabels_METRIC.dwg`), available from the book's website.

2. Zoom in on the outer parcel segment that runs from Parcel 1 through Parcel 10.

3. Change to the Annotate tab and select Add Labels ➢ Parcel ➢ Add Parcel Labels from the Labels & Tables panel.

4. From the drop-down menus in the Add Labels dialog, select Single Segment, (Span) Bearing And Distance With Crows Feet, and Delta Over Length And Radius in the Label Type, Line Label Style, and Curve Label Style selection boxes, respectively.

5. Click Add.

6. At the Select point on entity: prompt, pick somewhere near the middle of the outer parcel segment that runs from Parcel 1 through Parcel 10.

    A label that spans the full length between natural vertices appears (see Figure 5.61).

7. As time permits, create other span labels along the other back lot lines in the project.

8. Save and close the drawing. If you would like to see what the drawing should look like at this point, you can open SpanLabels_FINISHED.dwg (SpanLabels_METRIC_FINISHED.dwg), available from the book's website.

> **FLIP IT, REVERSE IT**
>
> If your spanning label doesn't seem to work on your first try and you've followed all the spanning label guidelines, try flipping your label to the other side of the parcel segment, reversing the label, or using a combination of both flipping and reversing.

Adding Curve Tags to Prepare for Table Creation

To keep plans tidy, it is common to show labels that reference a table for curves and lines. Civil 3D parcels provide tools for creating dynamic line and curve tables. You can keep lines and curves together or create separate tables for each.

Parcel segments must be labeled before they can be used to create a table. They can be labeled with any type of label, but you'll likely find it to be best practice to create a tag-only style for segments that will be placed in a table.

The following exercise will show you how to replace curve labels with tag-only labels:

1. Open CurveTags.dwg (CurveTags_METRIC.dwg), available from the book's website.

    Note that the labels along tight curves, such as the cul-de-sac, would be better represented as curve tags.

2. Change to the Annotate tab and select Add Labels ➢ Parcel ➢ Add Parcel Labels from the Labels & Tables panel.

3. From the drop-down menus in the Add Labels dialog, select Replace Multiple Segment, Bearing Over Distance, and Curve Tag Only in the Label Type, Line Label Style, and Spanning Curve Label Style selection boxes, respectively.

4. Click Add. At the `Select parcel to be labeled by clicking on area label:` prompt, pick the area label for Parcel 1.

   Note that the line labels for Parcel 1 are reset and the curve labels convert to tags.

5. Repeat step 4 for Parcels 9 through 13, and 15.

6. Press ↵ to exit the command.

7. Save the drawing but keep it open for the next exercise. If you would like to see what the drawing should look like at this point, you can open `CurveTags_FINISHED.dwg` (`CurveTags_METRIC_FINISHED.dwg`), available from the book's website.

Now that each curve label has been replaced with a tag, it's desirable to have the tag numbers be sequential. The following exercise will show you how to renumber tags:

1. Continue working in the previous drawing or you can open `CurveTags_FINISHED.dwg` (`CurveTags_METRIC_FINISHED.dwg`), available from the book's website.

2. Zoom into the curve on the northeast corner of Parcel 15 (see Figure 5.62).

FIGURE 5.62
Curve tags on Parcel 15

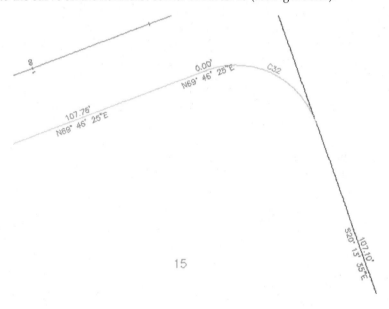

Your curve may have a different number than the one in the figure.

3. Click one of the parcel area labels. Then from the Parcel contextual tab ➢ Labels & Tables panel, select Renumber Tags.

4. At the `Select label to renumber tag or [Settings]:` prompt, type **S**, and then press ↵.

   The Table Tag Numbering dialog appears (see Figure 5.63).

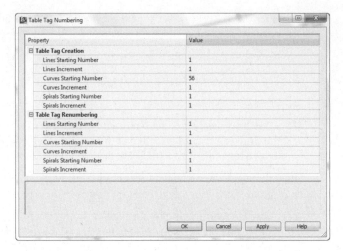

FIGURE 5.63
The Table Tag Numbering dialog

5. Within the Table Tag Creation branch, change the value in the Curves Starting Number selection box to **1**, and click OK.

6. At the `Select label to renumber tag or [Settings]:` prompt, click the curve tag label at the northeast corner of Parcel 15.

   The command line may say `Current tag number is being used, press return to skip to next available or [Create duplicate]`, in which case you should type **C** ↵ to create a duplicate.

7. Continue clicking curve tag labels working west and then south. As you do so you'll eventually resolve all of the duplicates and end up with a sequential set of curve tag numbers increasing as you move west and then south along the lot frontages. When you're finished, press ↵ to exit the command.

8. Save and close the drawing. If you would like to see what the drawing should look like at this point, you can open `Renumber_FINISHED.dwg` (`Renumber_METRIC_FINISHED.dwg`), available from the book's website.

Creating a Table for Parcel Segments

The following exercise demonstrates how to create a table from curve tags:

1. Open `SegmentTable.dwg` (`SegmentTable_METRIC.dwg`), available from the book's website.

   You should have several curves labeled with the Curve Tag Only label.

2. Select a parcel area label. From the Parcel contextual tab ➤ Labels & Tables panel, select Add Tables ➤ Add Curve.

3. In the Table Creation dialog, select Length Radius & Delta from the drop-down menu in the Table Style selection box.

4. In the Select By Label Or Style area of the dialog, click the Apply check box for the Parcel Curve: Curve Tag Only entry under Label Style Name.

5. Now that the selection rule is active for Parcel Curve: Curve Tag Only, change it to Add Existing And New.

    The Add Existing And New option will ensure that the table updates as more labels fitting this criteria are added to the drawing. Keep the default values for the remaining options. The dialog should look like Figure 5.64.

FIGURE 5.64
The Table Creation dialog

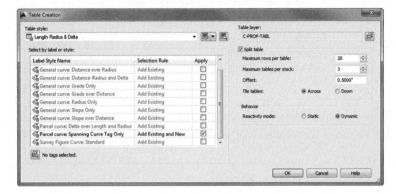

6. Click OK.
7. At the Select upper left corner: prompt, pick a location in your drawing for the table.

    A curve table appears, shown in Figure 5.65.

FIGURE 5.65
A curve table

Curve #	Length	Radius	Delta	Chord Direction	Chord Length
C1	54.98	35.00	90.00	S65° 13' 34"E	49.50
C2	35.72	525.00	3.90	N71° 43' 21"E	35.71
C3	74.38	525.00	8.12	N77° 43' 48"E	74.31
C4	17.25	525.00	1.88	N82° 43' 47"E	17.25
C5	39.27	25.00	90.00	N38° 40' 15"E	35.36
C6	78.30	975.00	4.60	N8° 37' 48"W	78.28
C7	32.55	975.00	1.91	N11° 53' 13"W	32.55
C8	32.07	150.00	12.25	N18° 58' 08"W	32.01
C9	57.21	35.00	93.65	N71° 55' 15"W	51.05

8. Save and close the drawing. If you would like to see what the drawing should look like at this point, you can open SegmentTable_FINISHED.dwg (SegmentTable_METRIC_FINISHED.dwg), available from the book's website.

The Bottom Line

Create a boundary parcel from objects.   The first step to any parceling project is to create an outer boundary for the site.

   Master It   Open the MasteringParcels.dwg (MasteringParcels_METRIC.dwg) file, which you can download from www.sybex.com/go/masteringcivil3d2014. Convert the line segments in the drawing to a parcel.

Create a right-of-way parcel using the right-of-way tool.   For many projects, the ROW parcel serves as frontage for subdivision parcels. For straightforward sites, the automatic Create ROW tool provides a quick way to create this parcel. A cul-de-sac serves as a terminal point for a cluster of parcels.

   Master It   Continue working in the Mastering Parcels.dwg (MasteringParcels_METRIC.dwg) file or you can open MasteringParcels1.dwg (MasteringParcels1_METRIC.dwg), available from the book's website. Create a ROW parcel that is offset by 25' (10 m) on either side of the road centerline with 25' (10 m) fillets at the parcel boundary and alignment ends. Then add the circles representing the cul-de-sac as a parcel. Edit the cul-de-sac area to remove unwanted parcel lines.

Create subdivision lots automatically by layout.   The biggest challenge when creating a subdivision plan is optimizing the number of lots. The precise sizing parcel tools provide a means to automate this process.

   Master It   Continue working in the previous drawing or open MasteringParcels2.dwg (MasteringParcels2_METRIC.dwg), available from the book's website. Create a series of lots with a minimum of 8,000 sq. ft. (700 m^2) and 75' (20 m) frontage. Set the Use Minimum Offset option to No. Leave all other options at their defaults.

Add multiple-parcel segment labels.   Every subdivision plat must be appropriately labeled. You can quickly label parcels with their bearings, distances, direction, and more using the segment labeling tools.

   Master It   Continue working in the previous drawing, or you can open MasteringParcels3.dwg (MasteringParcels3_METRIC.dwg), available from the book's website. Place Bearing Over Distance labels on every parcel line segment and Delta Over Length And Radius labels on every parcel curve segment using the Multiple Segment Labeling tool.

Chapter 6

Alignments

Some roads were laid out hundreds of years ago by herds of cows; others were microdesigned by planners with careful consideration for each bend and twist. Either way, when working on a linear project, whether it be the renovation of an old New England cow path or a network of new subdivision roads, horizontal layout information will need to be conveyed to the contractor. This horizontal layout is called the *alignment* and drives much of the design. This chapter shows you how alignments can be created, how they interact with the rest of the design, how to edit and analyze them, and finally, how they work with the overall project.

In this chapter, you will learn to:

- Create an alignment from an object
- Create a reverse curve that never loses tangency
- Replace a component of an alignment with another component type
- Create alignment tables

Alignment Concepts

Before you can efficiently work with alignments, you must understand two major concepts: the interaction of alignments and sites and the idea of geometry that is fixed, floating, or free.

Alignments and Sites

Prior to the AutoCAD® Civil 3D® 2008 release, alignments were always a part of a site and interacted with the topology contained in that site. This interaction led to the pickle analogy: alignments are like pickles in a Mason jar. You don't put pickles and peppers in the same jar unless you want hot pickles, and you don't put lots and alignments in the same site unless you want subdivided lots.

Civil 3D now has two ways of handling alignments in terms of sites: they can be contained in a site as before, or they can be independent of a site.

Both the alignments contained in a site and those independent of a site can be used to cut profiles or control corridors, but only the alignments contained in a site will react with and create parcels as members of a site topology.

Unless you have good reason for them to interact (as in the case of an intersection), it makes sense to create alignments outside of any site object. They can be moved later if necessary. For the purpose of the exercises in this chapter, you won't place any alignments in a site. For more information about sites, check out the section "Introduction to Sites" in Chapter 5, "Parcels."

Alignment Entities

Civil 3D recognizes five types of alignments: centerline alignments, offset alignments, curb return alignments, rail alignments, and miscellaneous alignments. Each alignment type can consist of three types of entities, or segments: lines, arcs, and spirals. These segments control the horizontal alignment of your design. Their relationship to one another is described by the following terminology:

Fixed Segments   Fixed segments are fixed in space (see Figure 6.1). They're defined by connecting points in the coordinate plane and are independent of the segments that occur either before or after them in the alignment. Fixed segments may be created as tangent to other components, but their independence from those objects lets you move them out of tangency during editing operations.

FIGURE 6.1
Alignment fixed segments

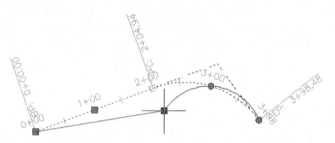

Floating Segments   Floating segments float in space but are attached to a point in the drawing and to some segment to which they maintain tangency (see Figure 6.2). Floating segments work well in situations where you have a critical point but the other points of the horizontal alignment are flexible.

FIGURE 6.2
Alignment floating segments

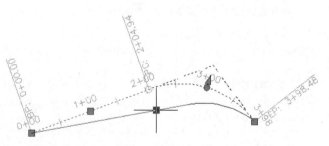

Free Segments   Free segments are dependent upon the entities that come before and after them in the alignment structure (see Figure 6.3). Unlike fixed or floating segments, a free segment must have segments that come before and after it. Free segments maintain tangency to the segments that come before and after them and move as required to make that happen. Although some geometry constraints can be put in place, these constraints can be edited and are user dependent.

FIGURE 6.3
Alignment free segments

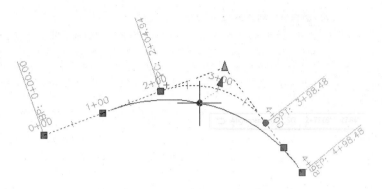

During the exercises in this chapter, you'll use a mix of these entity types to understand them better.

Creating an Alignment

Alignments can be created from AutoCAD objects (lines, arcs, or polylines) or by layout. This section looks at both ways to create an alignment and discusses the advantages and disadvantages of each. The exercise will use the street layout shown in Figure 6.4 as well as the different methods to achieve your designs.

FIGURE 6.4
Proposed street layout

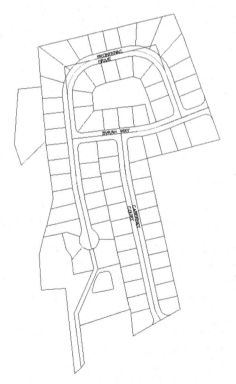

Creating from a Line, Arc, or Polyline

Most designers have used either polylines or lines and arcs to generate the horizontal control of their projects. It's common for surveyors to generate polylines to describe the center of a right-of-way or for an environmental engineer to draw a polyline to show where a new channel should be constructed. These team members may or may not have Civil 3D software, so they use their familiar friends—the line, arc, and polyline—to describe their design intent.

Although these objects are good at showing where something should go, they don't have much data behind them. To make full use of these objects, you should convert them to Civil 3D alignments that can then be shared and used for myriad purposes. Once an alignment has been created from a polyline, offsets can be created to represent rights-of-way, building lines, and so on. In this exercise, you'll convert a polyline to an alignment and create offsets:

1. Open the `AlignmentsFromPolylines.dwg` file (or for metric users, the `AlignmentsFromPolylines_METRIC.dwg` file).

   You can download this file from the book's web page at www.sybex.com/go/masteringcivil3d2014. You will see the red polylines representing the centerlines of roads, rights-of-way, and parcels.

2. From the Home tab ➢ Create Design panel, choose Alignment ➢ Create Alignment From Objects.

3. When prompted to select the first line, pick the two lines and arc labeled Syrah Way, shown previously in Figure 6.4, and press ↵.

4. Verify that the direction is from left to right and press ↵.

   The Create Alignment From Objects dialog appears.

5. Change the Name field to **Syrah Way,** and select the Centerline type.

6. Verify that Alignment Style is set to Proposed and Alignment Label Set is set to Major And Minor Only. The Create Alignment From Objects dialog should match Figure 6.5.

7. Accept the other settings, and click OK.

> **A Word about Constraints**
>
> In previous versions of the software, creating from objects held no constraints (everything was fixed). In current versions, Civil 3D tries to find tangency where a curve is between two lines and establish that tangency as a constraint. Lines will always be set to Not Constrained (Fixed), but if the curve is tangent, it will be set to Constrained On Both Sides (Free).

Keep this drawing open for the next exercise.

FIGURE 6.5
The settings used to create the Syrah Way alignment

> **THE CREATE ALIGNMENT FROM OBJECTS DIALOG**
>
> In the Create Alignment From Objects dialog (shown in Figure 6.5), there are many settings that you can adjust when creating an alignment:
>
> **Name**   This is the alignment name. No alignment name can be duplicated in a drawing.
>
> **Type**   The alignment types can be thought of as places for objects that are alike. The objects can react differently depending on which type is selected.
>
> **Centerline**   Used mainly for centers of roads, streams, or swales. Civil 3D places this type of alignment in the Alignments ➢ Centerline Alignments collection.
>
> **Offset**   Used for offset alignments. The difference between this and the centerline alignment is that you have the option in the Alignment Properties dialog to set Offset parameters, such as naming a parent alignment and offset values. Offset alignments are dynamically linked to their parent alignments and will update automatically when the parent alignment changes. Civil 3D places this type of alignment in the Alignments ➢ Offset Alignments collection.
>
> **Curb Return**   Used for curb returns, which are the radii at intersections. The difference between this and the offset alignment is that instead of offset, you have the option in the Alignment Properties dialog to set Curb Return parameters, such as setting two parent alignments and offsets. Curb return alignments are also dynamically linked to their parent alignments. Civil 3D places this type of alignment in the Alignments ➢ Curb Return Alignments collection.

Rail Used for rail design. The difference between this and the other alignments is that this alignment is set using typical rail geometry such as degree of curvature and cant. In addition, a unique tab is added to the Alignment Properties dialog called Rail Parameters, where you can specify the track width. Civil 3D places this alignment type in the Alignments ➢ Rail Alignments collection.

Miscellaneous This is a stripped-down alignment type that only contains Information, Stationing, Masking, Point Of Intersection, and Constraint Editing tabs. Civil 3D places this alignment type in the Alignments ➢ Miscellaneous Alignments collection.

Description An optional field where you can be as verbose as you want with information describing your alignment.

Starting Station This value, either positive or negative, will be the starting stationing for the alignment. This is handy if you need to start your alignment to coincide with existing stationing or if you wish to have your 0+00 stationing at an intersection of a road. If you forget to set this value here, you can change it later in the Alignment Properties dialog.

The General tab contains the following options:

Site A place to keep Civil 3D objects that you want to interact with each other. As previously mentioned, all of your alignments in these exercises will be put on the <none> site.

Alignment Style You can set different styles to visually show your alignment. For more on styles, refer to Chapter 19, "Object Styles."

Alignment Layer Overrides the layer that is specified in the drawing settings for alignments.

Alignment Label Set As with the Alignment Style, you can choose how your alignment will be labeled. For more on label styles, refer to Chapter 18, "Label Styles."

Conversion Options Depending on your selections, these will add curves or erase the original entities.

The Design Criteria tab contains these options:

Starting Design Speed Specify the design speed of the alignment for the starting station. If no additional design speeds are applied to a different section of the alignment, this speed will be used for the entire alignment. Design speed is used to calculate superelevation and to establish design criteria.

Use Criteria-Based Design When this check box is selected, the Use Design Criteria File and Use Design Check Set options can be used. These options enable the software to keep track of certain design standards that you would like your alignment to meet and notify you when those standards are not met.

Use Design Criteria File Here you can define the design criteria, such as the design manual *A Policy on Geometric Design of Highways and Streets, 2001* from the American Association of State Highway and Transportation Officials (AASHTO).

Use Design Check Set Here you can set rules or expressions for lines, curves, spirals, and tangent intersections.

Many of these settings will be discussed further throughout this chapter.

You've created your first alignment and attached stationing and geometry point labels. It is common to create offset alignments from a centerline alignment to begin to model rights-of-way. In the following exercise, you'll create offset alignments and mask them where you don't want them to be seen:

1. Using the drawing from the previous exercise, from the Home tab ➢ Create Design panel, choose Alignment ➢ Create Offset Alignment.

2. When prompted to select an alignment, pick the Syrah Way alignment to open the Create Offset Alignments dialog shown in Figure 6.6.

FIGURE 6.6
The Create Offset Alignments dialog

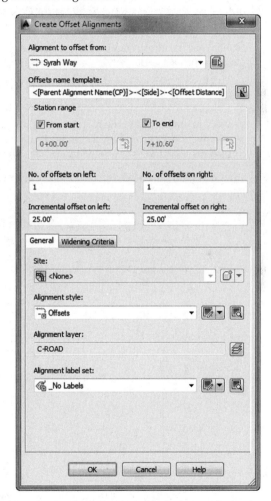

3. Change Incremental Offset On Left to **25′** (metric users, **7.5**).

4. Change Incremental Offset On Right to **25′** (metric users, **7.5**).

5. Verify that Alignment Label Set is set to _No Labels, and click OK to accept the rest of the defaults shown in Figure 6.6.

6. Select the offset alignment just created along the northerly right-of-way of Syrah Way to activate the Offset Alignment contextual tab.

7. From the Offset Alignment contextual tab ➢ Modify panel, choose Alignment Properties to open the Alignment Properties dialog.

8. Change to the Masking tab and click the Add Masking Region button.

9. Enter **0** for the first station and **50** for the second station when prompted (metric users, enter **0** and **15**), and click OK.

   Notice that the alignment is now masked at the intersection of Syrah Way and Frontenac Drive at the east end.

10. Repeat the process for the rest of the intersections, starting at the end of the arc on both right-of-way alignments on Syrah Way.

    While you entered the information using text in the previous step, notice that you could alternatively use the station picker button, which becomes visible when you click into the cell to enter a value.

    Once complete, the Masking tab of the Alignment Properties dialog should look like Figure 6.7.

FIGURE 6.7
Creating an alignment mask

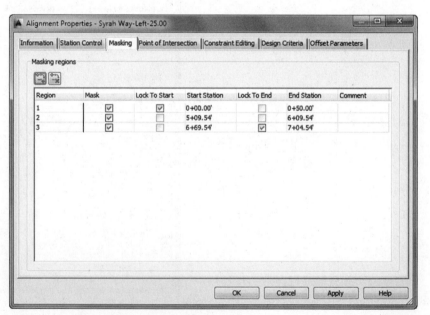

When this exercise is complete, you may close the drawing. A saved finished copy of this drawing is available from the book's web page with the filename `AlignmentsFromPolylines_FINISHED.dwg` or `AlignmentsFromPolylines_METRIC_FINISHED.dwg`. Note that when you selected the beginning and end of the offset alignment, the Lock To Start and Lock To End boxes were checked automatically.

Offset alignments are simple to create, and they are dynamically linked to a centerline alignment. To test this, grip the centerline alignment, select the endpoint grip, and stretch the alignment to the west. Notice the change, and then undo this change to return to the original state.

> **OFFSET GRIPS AND MORE**
>
> Offset alignments have two special grips: the arrow and the plus sign. The arrow is used to change the offset value, and the plus sign is used to create a transition, called a *widening*, such as a turning lane. To access the Create Widening command, from the Home tab ➢ Create Design panel, select the Alignment drop-down. Widening criteria can also be found in the Create Offset Alignments dialog and the Offset Alignment contextual tab.
>
> Even offset alignments with widening remain dynamic to their host alignment. Offset alignment objects can be found in Prospector in the Alignments collection.

Creating by Layout

Now that you've made an alignment from polylines, let's look at another creation option: Create By Layout. You'll use the same street layout (Figure 6.4) that was provided by a planner, but instead of converting from polylines, you'll trace the alignments. Although this seems like duplicate work, it will pay dividends in the relationships created between segments:

1. Open the `AlignmentsByLayout.dwg` or the `AlignmentsByLayout_METRIC.dwg` file.

2. From the Home tab ➢ Create Design panel, choose Alignment ➢ Alignment Creation Tools.

   The Create Alignment – Layout dialog appears.

3. Change the Name field to **Cabernet Court** if it is not already set.

4. Verify that Alignment Style is set to Proposed and Alignment Label Set is set to Major And Minor Only.

5. Click OK to accept the other settings shown in Figure 6.8.

FIGURE 6.8
Create Alignment – Layout dialog

The Alignment Layout Tools toolbar for Cabernet Court appears (Figure 6.9).

FIGURE 6.9
The Alignment Layout Tools toolbar for Cabernet Court

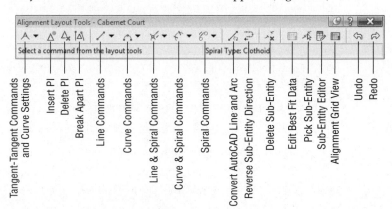

6. Click the down arrow next to the Tangent-Tangent (No Curves) tool at the far left, and select the Tangent-Tangent (With Curves) option (see Figure 6.10).

FIGURE 6.10
The Tangent-Tangent (With Curves) tool

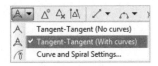

The tool places a curve automatically; you'll later adjust the curve, watching the tangents extend as needed.

7. Turn on the Endpoint Osnap and snap to the southernmost end of the Cabernet Court centerline.

8. Continue to pick the endpoints of the red lines along the Cabernet Court centerline, from south to north, to finish creating this alignment.

9. After selecting the endpoint at the intersection of Cabernet Court and Syrah Way, press ↵ to end the command.

10. Click the X button at the upper-right on the Alignment Layout Tools toolbar for Cabernet Court to close it.

   Keep this drawing open for the next portion of the exercise.

   Zoom in on the southern arc. Notice that it follows closely with the desired arc radius. Now zoom in on the northern arc, and notice that it doesn't match the arc the planner put in for you to follow. That's okay—you will fix it in a later exercise.

> **DESIGN AND THEN REFINE**
>
> It bears repeating that in dealing with Civil 3D objects, it is good practice to get something in place and *then* refine. With Land Desktop or other packages, you didn't want to define the object until it was fully designed. In Civil 3D, you design and then refine.

The alignment you just made is one of the most basic. Let's move on to some of the others and use a few of the other tools to complete your initial layout. In this exercise, you build the alignment at the north end of the site, but this time you use a floating curve to make sure the two segments you create maintain their relationship.

11. From the Home tab ➢ Create Design panel, choose Alignment ➢ Alignment Creation Tools.

    The Create Alignment – Layout dialog appears.

12. In the Create Alignment – Layout dialog, do the following:

    a. Change the Name field to **Frontenac Drive**.

    b. Set the Alignment Style field to Layout.

    c. Set the Alignment Label Set field to Major And Minor Only.

13. Click OK to display the Alignment Layout Tools toolbar for Frontenac Drive.
14. Select the Fixed Line (Two Points) tool, as shown in Figure 6.11.

FIGURE 6.11
The Fixed Line
(Two Points) tool

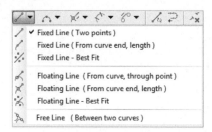

15. Using Endpoint Osnaps, select the eastern end of Frontenac Drive, and working south to north, draw the first fixed-line segment to the starting endpoint of the cyan arc.

    When you've finished, the command line will read `Specify start point:` in case you want to draw another line. You can either press ↵ to end the command or draw the next entity without ending the command.

16. Click the down arrow next to the Add Fixed Curve (Three Point) tool on the toolbar, and select More Floating Curves ➤ Floating Curve (From Entity End, Through Point), as shown in Figure 6.12.

FIGURE 6.12
Selecting the
Floating Curve tool

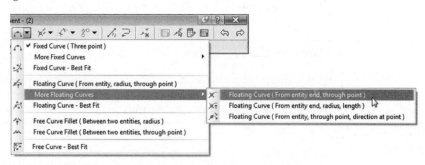

17. Select the fixed-line segment you drew in steps 14 and 15.

    Make sure you select the line segment somewhere north of the segment's midpoint in order to connect to the northern endpoint instead of the southern endpoint.

    Move your cursor around and a blue rubber band should appear, indicating that the alignment of the curve segment is being floated off the endpoint of the fixed segment.

18. Pick the endpoint of the arc of the Frontenac Drive polyline arc segment.

    Notice that you are generating the segments from low station to high station. If you perform these steps backward, you can reverse a segment using the Reverse Sub-Entity Direction button on the Alignment Layout Tools toolbar.

19. Press ↵ to end the command.
20. Close the Alignment Layout Tools toolbar.
21. Pick the Frontenac Drive alignment and then pick the grip on the upper end of the line and pull it away from its location.

    Notice that the line and the arc move in sync and tangency is maintained (see Figure 6.13).

FIGURE 6.13
Floating curves maintain their tangency.

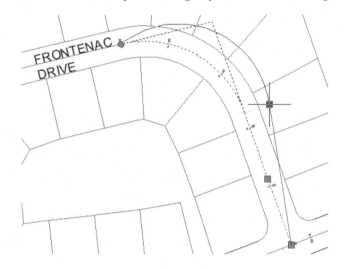

22. Press Esc to cancel the grip edit.

When this exercise is complete, you may close the drawing. A saved finished copy of this drawing is available from the book's web page with the filename `AlignmentsByLayout_FINISHED.dwg` or `AlignmentsByLayout_METRIC_FINISHED.dwg`.

ALL ABOARD! RAIL ALIGNMENTS

Rail alignments were introduced in Civil 3D 2013 to allow for the differences between rail design and other types of design. For example, railways are characterized by long sweeping curves and the curve geometry is expressed differently on railway construction plans. Rail design also has its own type of superelevation called cant, which is handled differently than road superelevation. These differences are accounted for when you use rail alignments in Civil 3D. Let's look at a quick, simple example:

1. Start a new blank drawing from the `_AutoCAD Civil 3D (Imperial) NCS` template that ships with Civil 3D. Metric users should use the `_AutoCAD Civil 3D (Metric) NCS` template.

2. From the Home tab ➤ Create Design panel, choose Alignment ➤ Alignment Creation Tools.

    The Create Alignment – Layout dialog appears.

3. Change the name to **Mastering Railway** and set Type to Rail.
4. On the General tab, make these changes:
   - Set Alignment Style to Proposed.
   - Set Alignment Label Set to All Labels.
5. Switch to the Design Criteria tab.
6. Change the starting design speed to **30** mi/hr (or **50** km/hr for metric users).
7. Select the Use Criteria-Based Design check box.

   If the design criteria file doesn't default to the one that references railway design standards, you may need to click the ellipsis next to the design criteria file to browse to `_Autodesk Civil 3D Imperial Rail Cant Design Standards.xml` or `_Autodesk Civil 3D Metric Rail Cant Design Standards.xml`.

8. Verify that Use Design Criteria File is also checked.
9. Click OK.

   The Alignment Layout Tools toolbar appears.

10. Select the Fixed Line (Two Points) tool.
11. Click anywhere on the screen to define the start point, and define the second point by entering **@330,0** on the command line (or **@100,0** for metric users).
12. Press ↵ to end the command.
13. Select Floating Curve (From Entity End, Radius, Length).
14. At the `Select entity to attach to:` prompt, select the line segment that you just drew.
15. At the `Specify curve direction [Clockwise counterclockwise] <Clockwise>:` prompt, press ↵ to accept the default, Clockwise.
16. At the `Specify radius or [Degree of curvature]:` prompt, enter **D** ↵ to switch to entering the radius by degree of curvature.
17. At the `Specify degree of curvature or [Radius]:` prompt, enter **3** ↵.
18. At the `Specify curve length or [deltaAngle Tanlen Chordlen midOrd External]:` prompt, enter **C** ↵ to define by chord length.
19. At the `Specify chord length or [curveLen deltaAngle Tanlen midOrd External]:` prompt, enter **100** ↵.
20. Press ↵ to end the command.

    Note that this will result in a different curve for imperial and metric users.

    When this exercise is complete, you may close the drawing. A saved finished copy of this drawing is available from the book's web page with the filename `Rail_FINISHED.dwg` or `Rail_METRIC_FINISHED.dwg`.

Best Fit Alignments

Often designers have to re-create an alignment for an existing road that does not have true horizontal geometry. Civil 3D has multiple tools to re-create the alignment using a best fit algorithm. You can either re-create a full best fit alignment or use best fit lines or best fit curves in an alignment. We will look at both methodologies in this section.

The Create Best Fit Alignment command can use AutoCAD blocks, AutoCAD entities, AutoCAD points, Civil 3D COGO points, or Civil 3D feature lines. You can also simply click on the screen. The Line and Curve drop-down menus on the Alignment Layout toolbar include options for Floating and Fixed Lines By Best Fit as well as Best Fit curves in all three flavors: Fixed, Float, and Free. It is similar to what we covered in Chapter 1, "The Basics," in the section "Best Fit Entities." Let's see how it works with alignments:

1. Open the AlignmentsBestFit.dwg or AlignmentsBestFit_METRIC.dwg file, which you can download from this book's web page.

2. From the Home tab ➢ Create Design panel, choose Alignment ➢ Alignment Creation Tools.

   The Create Alignment – Layout dialog appears.

3. Enter **Best Fit Lines** in the Name field.

4. Leave the rest at their defaults and click OK.

   The Alignment Layout Tools toolbar opens.

5. Click the down arrow next to the Fixed Line (Two Points) tool on the toolbar, and select the Fixed Line – Best Fit option.

   The Tangent By Best Fit dialog opens (Figure 6.14).

FIGURE 6.14
The Tangent By Best Fit dialog

Here, you can choose various methods to create a best fit line alignment:

- From COGO Points
- From Entities
- From AutoCAD Points
- By Clicking On The Screen

6. Pick the By Clicking On The Screen radio button and click OK.

7. With the running Endpoint Osnap, click on all the endpoints of the polyline in the lower left of your drawing.

   As you progress, you see a red dashed line being formed. In your selections, this line looks at all the endpoints selected in order to create the best fit line alignment (Figure 6.15).

FIGURE 6.15
The best fit line being formed

8. After selecting the last endpoint, press ↵ to open the Regression Data tab of the Panorama window.

9. On the Regression Data tab of the Panorama window, you can choose to exclude endpoints or force them to be pass-through endpoints by checking the appropriate boxes (Figure 6.16).

FIGURE 6.16
Regression Data tab on Panorama

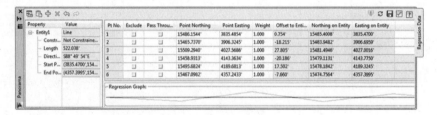

As you check and uncheck boxes in the Regression Data chart, notice the changes that occur on your best fit line alignment.

10. Click the green check mark on the upper-right side of Panorama to accept and dismiss the Panorama window.

    The best fit line alignment is complete. You can also create a similar command for a best fit curve sub-entity that can be either fixed or floating. The procedure is similar to that of the best fit line and can be performed on the polyline that vaguely resembles a curve in the exercise drawing.

    Keep this drawing open for the next portion of the exercise.

CREATING AN ALIGNMENT 263

One of the benefits to doing individual best fit segments (lines or curves) is that it is easy to exclude points. However, sometimes you will want a single "rough and dirty" alignment full of curves and lines without having to define them individually. Next you will create a full best fit alignment.

11. Zoom and pan to the area with the 13 points.

12. From the Home tab ➢ Create Design panel, choose Alignment ➢ Create Best Fit Alignment. The Create Best Fit Alignment dialog appears.

13. Set the input type to AutoCAD Points and click the selection button next to Path 1 Points.

14. In modelspace, use a crossing window to select the 13 points and press ↵.

15. Deselect the Create Spirals check box.

16. Change the name to **Best Fit Alignment**.

17. Verify that the alignment style is set to Layout and the Alignment Label Set is set to _No Labels, as shown in Figure 6.17, and click OK to accept all the other defaults.

FIGURE 6.17
The Create Best Fit Alignment dialog

The Best Fit Report dialog opens, as shown in Figure 6.18.

FIGURE 6.18
The Best Fit Report dialog

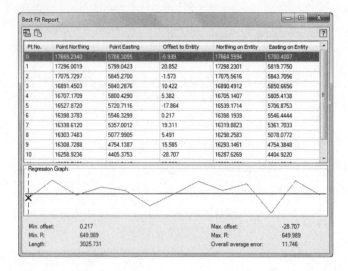

Review the results in the Best Fit Report. Notice that if you select a row, it will highlight the location on the regression graph.

18. Click the close (X) button to dismiss the dialog.

Unlike the Regression Data tab in the Panorama window, the Best Fit Report is purely informational and does not allow you to select to exclude or pass through a specific point. While the Create Best Fit Alignment procedure is fast, it may not be precise. Therefore, this command may be useful for providing a draft alignment, and then you can manually create your final alignment using results similar to those shown in the Best Fit Report that meet your specific design criteria.

The completed best fit alignment is shown in Figure 6.19. Although a best fit alignment may not give you exactly what you are looking for (especially if you like nice, whole-number radii), it generates a good starting point that you can then edit to fit your design needs.

FIGURE 6.19
The best fit alignment

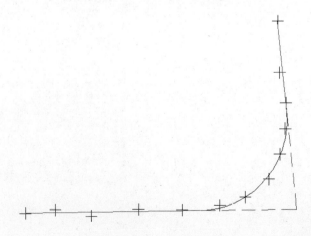

When this exercise is complete, you may close the drawing. A saved finished copy of this drawing is available from the book's web page with the filename AlignmentsBestFit_FINISHED.dwg or AlignmentsBestFit_METRIC_FINISHED.dwg.

Reverse Curve Creation

Next, let's look at a more complicated alignment construction—building a reverse curve connecting two curves:

1. Open the AlignmentReverse.dwg or AlignmentReverse_METRIC.dwg file.

2. For this exercise, confirm that your Endpoint Osnap is activated and the other Osnaps can be deactivated.

3. From the Home tab ➢ Create Design panel, choose Alignment ➢ Alignment Creation Tools.

   The Create Alignment – Layout dialog appears.

4. In the Create Alignment – Layout dialog, do the following:

   a. Change the Name field to **Reverse**.

   b. Set the Alignment Style field to Layout.

   c. Set the Alignment Label Set field to Major Minor And Geometry Points.

5. Click OK to display the Alignment Layout Tools toolbar for the reverse alignment.

6. Start by drawing a fixed line from the north end of the western portion to its endpoint using the same Fixed Line (Two Points) tool as used in an earlier exercise.

7. Use the Floating Curve (From Entity, Radius, Through Point) tool to connect a curve from the southern end of this segment.

8. When prompted for radius, enter 500 ↵ (for metric, 150 ↵).

9. At the Is curve solution angle [Greaterthan180 Lessthan180] <Lessthan180>: prompt, press ↵ to accept the default.

10. At the Specify end point: prompt, click on the other end of the arc.

11. The command repeats; at the Select entity to attach to: prompt, select the curve that you just created.

12. At the Specify radius or [Degree of curvature]: prompt, enter 400 ↵ (for metric, 120 ↵).

13. At the Is curve solution angle [Greaterthan180 Lessthan180] <Lessthan180>: prompt, press ↵ to accept the default.

14. The program detects that you are attaching a curve to a curve; at the Is curve compound or reverse to curve before? [Compound Reverse] <Compound>: prompt, enter **R** ↵ to specify that it is a reverse curve.

15. At the Specify end point: prompt, click the endpoint of the southern arc.

16. Press ↵ to end the command.
17. Use the Floating Line (From Curve, Through Point) tool to connect a line from the second curve and to finish the alignment by selecting the two-point line (Figure 6.20).

FIGURE 6.20
Segment layout for the reverse curve alignment

The alignment now contains a perfect reverse curve. Move any of the pieces and you'll see the other segments react to maintain the relationships shown in Figure 6.21. The flexibility of the Civil 3D tools allows you to explore an alternative solution (the reverse curve) as opposed to the basic solution. Flexibility is one of the strengths of Civil 3D.

FIGURE 6.21
Curve relationships during a grip edit

You've completed your initial reverse curve layout. Unfortunately, the reverse curve may not be acceptable to the designer, but you'll look at those changes later in the section "Component-Level Editing."

When this exercise is complete, you may close the drawing. A saved finished copy of this drawing is available from the book's web page with the filename AlignmentReverse_FINISHED.dwg or AlignmentReverse_METRIC_FINISHED.dwg.

Creating with Design Constraints and Check Sets

Civil 3D allows you to use design constraints and design check sets during the process of creating alignments and design profiles. Typically, these constraints check for things like curve radius, length of tangents, and so on. Design constraints use information from AASHTO or other design manuals to set curve requirements. Check sets allow users to create their own criteria to match local requirements, such as subdivision or county road design. First, you'll make one quick set of design checks:

1. Open the CreatingChecks.dwg or CreatingChecks_METRIC.dwg file.
2. On the Settings tab in Toolspace, expand the Alignment ➢ Design Checks branch.
3. Right-click the Line folder, and select New to display the New Design Check dialog.
4. Change the name to **Subdivision Tangent**.

5. Click the Insert Property drop-down menu, and select Length.
6. Click the greater-than/equals symbol (>=) button, and then enter **100** (for metric, **30**) in the Expression field as shown in Figure 6.22.

   When complete, your dialog should look like Figure 6.22. Click OK to accept the settings in the dialog.

FIGURE 6.22
The completed Subdivision Tangent design check

7. Right-click the Curve folder, and select New to display the New Design Check dialog.
8. Change the name to **Subdivision Radius**.
9. Click the Insert Property drop-down menu, and select Radius.

10. Click the greater-than/equals symbol (>=) button, and then enter **200** (for metric, **60**) in the Expression field, as shown in Figure 6.23.

FIGURE 6.23
The completed Subdivision Radius design check

11. Click OK to accept the settings in the dialog.
12. Right-click the Design Check Sets folder, and select New to display the Alignment Design Check Set dialog.
13. On the Information tab, change the name to **Mastering Subdivision**, and then switch to the Design Checks tab.
14. Choose Line from the Type drop-down list and select the Subdivision Tangent line check that you just created.
15. Click the Add button to add the Subdivision Tangent check to the set.
16. Choose Curve from the Type drop-down list and select the Subdivision Radius curve check that you just created.
17. Click the Add button again to complete the set, shown in Figure 6.24.

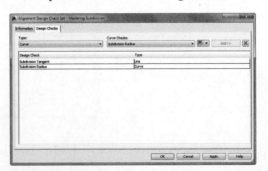

FIGURE 6.24
The completed Mastering Subdivision design check set

18. Click OK to accept the settings in the dialog.

You may keep this drawing open to continue to the next exercise or use the saved finished copy of this drawing available from the book's web page (CreatingChecks_FINISHED.dwg or CreatingChecks_METRIC_FINISHED.dwg).

Once you've created a number of design checks and design check sets, you can apply them as needed during the design and layout stage of your projects.

> **DESIGN CHECKS VS. DESIGN CRITERIA**
>
> What's the difference between design checks and design criteria? A design check uses basic properties such as radius, length, grade, and so on to check a particular portion of an alignment or profile. These constraints are generally dictated by a governing agency based on the type of road involved. Design criteria use speed and related values from design manuals such as AASHTO to establish these geometry constraints. Think of design criteria as a suite of check sets with different sets for each city, type of street, design speed, and so on.

In the next exercise, you'll see the results of your Mastering Subdivision check set in action:

1. If not still open from the previous exercise, open the CreatingChecks_FINISHED.dwg or CreatingChecks_METRIC_FINISHED.dwg file.

2. Change to the Prospector tab in Toolspace, and expand the Alignments ➢ Centerline Alignments branch.

3. Right-click on Frontenac Drive and select Properties.

4. Change to the Design Criteria tab, and set the start station design speed to 30 mi/h (or 50 km/h), as shown in Figure 6.25.

FIGURE 6.25
Setting up design checks from Alignment Properties

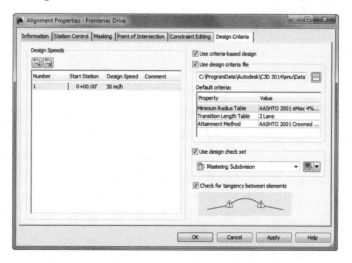

5. Verify that all of the check boxes are selected and that Use Design Check Set is set to Mastering Subdivision as shown in Figure 6.25.

   Note that the Use Criteria-Based Design check box must be selected to activate the other two.

6. Click OK to accept the settings in the dialog.

   You could have alternatively set the design criteria when you were originally creating the alignment.

   The curve radius on the northernmost part of Frontenac Drive is less than the Subdivision Radius design check, as illustrated by the design check failure indicators.

7. In Prospector, right-click on the alignment named Frontenac Drive and click Select to activate the Alignment contextual tab.

8. From the Alignment contextual tab ➢ Modify panel, choose Geometry Editor to display the Alignment Layout Tools toolbar.

9. Select the Tangent-Tangent With Curves option in the drop-down list on the left side of the Alignment Layout Tools toolbar.

10. Starting at the end of the northeastern Frontenac Drive curve, connect the endpoints of the tangent lines to finish creating the Frontenac Drive alignment.

    Notice that the two curves generated do not match the ones that the planner had originally laid out; you will fix them in the next exercise.

    If you hover over the exclamation-point symbol, as shown in Figure 6.26, it will indicate which design criteria and design checks have been violated.

FIGURE 6.26
Completed alignment layout with design criteria and design checks failure indicator

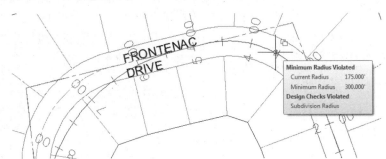

11. Close the Alignment Layout Tools toolbar.

When this exercise is complete, you may close the drawing. A saved finished copy of this drawing is available from the book's web page with the filename `AlignmentsChecked_FINISHED.dwg` or `AlignmentsChecked_METRIC_FINISHED.dwg`.

Now that you know how to create an alignment that doesn't pass the design checks, let's look at different ways of modifying alignment geometry. As you correct and fix alignments that violate the assigned design checks, the warning symbols indicating those violations will disappear.

Editing Alignment Geometry

The general power of Civil 3D lies in its flexibility. The documentation process is tied directly to the objects involved, so making edits to those objects doesn't create hours of work in updating the documentation. With alignments, there are three major ways to edit the object's horizontal geometry without modifying the underlying construction:

Graphical Grip Editing Select the object, and use the various grips to move critical points. This method works well for realignment, but precise editing for things like a radius or direction can be difficult without the ability to enter values.

Tabular Design Use Panorama to view all the alignment segments and their properties; type in values to make changes. This approach works well for modifying lengths or radius values, but setting a tangent perpendicular to a line in the drawing or placing a control point in a specific location is better done graphically.

Component-Level Editing Use the Alignment Layout Parameters dialog to view the properties of an individual piece of the alignment. This method makes it easy to modify one piece of an alignment that is complicated and that consists of numerous segments, whereas picking the correct field in a Panorama view can be difficult.

In addition to these methods, you can use the Alignment Layout Tools toolbar to make edits that involve removing components or adding to the underlying component count. The following exercises look at the three simple edits and then explain how to add and remove components of an alignment without redefining it.

Grip Editing

You already used graphical editing techniques when you created alignments from polylines, but those techniques can also be used with considerably more precision than shown previously. The alignment object has a number of grips that reveal important information about the elements' creation (see Figure 6.27).

FIGURE 6.27
Alignment grips

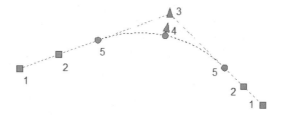

You can use the grips in Figure 6.27 to do the following actions:

Grip 1 The square pass-through grip at the beginning of the alignment that indicates a segment point that can be moved at will to change the length and angle of the line. This grip doesn't attach to any other components.

Grip 2 The square pass-through grip in the middle of the tangents that allows the element to be translated while maintaining the length and angle. Other components attempt to hold their respective relationships, but moving the grip to a location that would break the alignment isn't allowed.

Grip 3 The triangular grip at the intersection of tangents that indicates a Point of Intersection (PI) relationship. The curve shown is a function of these two tangents and is free to move on the basis of incoming and outgoing tangents while still holding a radius.

Grip 4 The triangular and circular radius grips near the middle of the curve that allow the user to modify the radius directly. The tangents must be maintained, so any selection that would break the alignment geometry isn't allowed.

Grip 5 The circular pass-through grip on the end of the curve that allows the radius of the curve to be indirectly changed by changing the point of the Point of Curvature (PC) of the alignment. You make this change by changing the curve length, which in effect changes the radius.

In the following exercise, you'll use grip edits to make one of your alignments match the planner's intent more closely:

1. Open the `AlignmentsChecked_FINISHED.dwg` or `AlignmentsChecked_METRIC_FINISHED.dwg` file.

2. In Prospector, expand the Alignments ➢ Centerline Alignments branch.

3. Right-click on Cabernet Court and select Zoom To.

4. Zoom in to the northern curve of the Cabernet Court alignment.

   This curve was inserted in a previous exercise using the default settings and doesn't match the guiding polyline well.

5. Select the Cabernet Court alignment to activate the grips.

6. Select the circular grip shown in Figure 6.28, and use an Endpoint Osnap to place it on the original cyan arc. Doing so changes the radius without changing the PI.

FIGURE 6.28
Grip-editing the Cabernet Court curve

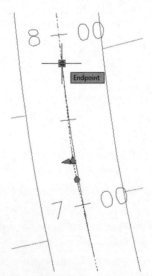

Keep this drawing open for the next exercise.

EDITING ALIGNMENT GEOMETRY

Your alignment of Cabernet Court now follows the planned layout. With no knowledge of the curve properties or other driving information, you've quickly reproduced the design's intent.

Tabular Design

When you're designing on the basis of governing requirements, one of the most important elements is meeting curve radius requirements. It's easy to work along an alignment in a tabular view, verifying that the design meets the criteria. In this exercise, you'll verify that your curves are suitable for the design:

1. Using the drawing from the previous exercise, zoom to the Frontenac Drive alignment, and select it to activate the Alignment contextual tab.

2. From the Alignment contextual tab ➢ Modify panel, choose Geometry Editor.

   The Alignment Layout Tools toolbar opens.

3. Select the Alignment Grid View tool.

   Panorama appears as shown in Figure 6.29, with all the elements of the alignment listed along the left. You can use the scroll bar along the bottom to review the properties of the alignment if necessary. Note that the columns can be resized as well as toggled off by right-clicking the column headers. The segment selected in the alignment grid view is also highlighted in the model, which can also make identifying the segment easier. Note that the design check failure indicators that are shown on the plan also appear in the table.

FIGURE 6.29
Alignment Entities vista for Frontenac Drive

Creating and Saving Custom Panorama Views

If you right-click a column heading in Panorama and select the Customize Columns option at the bottom of the menu, you're presented with a Customize Columns dialog. This dialog allows you to set up any number of column views, such as Road Design or Stakeout, to show a set of different columns. These views can be saved, allowing you to switch between views easily.

Notice that you can click many of the fields to edit them.

4. Change the radius value of segment No. 6 to **1000** (or **300** for metric users).

5. Click the close X button (notice that in the Panorama window, there is no green check mark to accept the changes) to dismiss Panorama, and then close the toolbar.

   Keep this drawing open for the next exercise.

Panorama allows for quick and easy review of designs and for precise data entry, if required. Grip editing is commonly used to place the line and curve of an alignment in an approximate working location, but then you use the tabular view in Panorama to make the values more reasonable—for example, to change a radius of 292.56 to 300.00.

Component-Level Editing

Once an alignment gets more complicated, the tabular view in Panorama can be hard to navigate, and deciphering which element is which can be difficult. In this case, reviewing individual elements by picking them onscreen can be easier:

1. Using the drawing from the previous exercise, zoom to the Frontenac Drive alignment, and select it to activate the Alignment tab if not still active from the previous exercise.

2. If the Alignment Layout Tools toolbar for Frontenac Drive is not still open from the previous exercise, from the Alignment contextual tab ➢ Modify panel, choose Geometry Editor.

3. Select the Sub-Entity Editor tool to open the Alignment Layout Parameters dialog for Frontenac Drive.

4. Select the Pick Sub-Entity tool on the Alignment Layout Tools toolbar.

5. Pick the first line of Frontenac Drive on the east side of the site to display its properties in the Alignment Layout Parameters dialog (see Figure 6.30).

FIGURE 6.30
The Alignment Layout Parameters dialogs for the first line (on the left) and the first curve (on the right) on Frontenac Drive

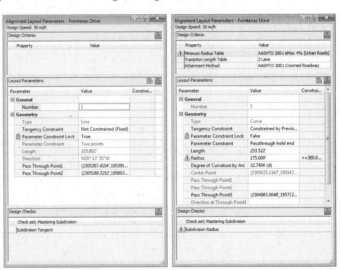

6. Zoom in, and pick the first curve.

   Notice that the Tangency Constraint value now reports Constrained By Previous (Floating).

7. Change Parameter Constraint Lock to False.

   Notice that this will enable the ability to set many of the values that previously were grayed out and could not be set.

8. Change the value in the Radius field to **300** (for metric, **100**), and watch the screen update.

   This value is too far from the original design intent to be a valid alternative and the tangency with the next line was not maintained. We will reinvestigate the tangency constraints in the next exercise.

9. Change the value in the Radius field back to **175** (for metric, **55**), and again watch the update.

   This value is closer to the design and is acceptable despite being less than the minimum radius for the 30 mi/hr design speed.

10. Change the radius of the second curve to match the first curve by using the Pick Sub-Entity tool and changing the value in the Alignment Layout Parameters dialog.

11. Close the Alignment Layout Parameters dialog and the Alignment Layout Tools toolbar.

You may keep this drawing open to continue on to the next exercise or use the saved finished copy of this drawing available from the book's web page (EditingAlignments_FINISHED.dwg or EditingAlignments_METRIC_FINISHED.dwg).

By using the Alignment Layout Parameters dialog, you can concisely review all the individual parameters of a component. In each of the editing methods discussed so far, you've modified the elements that were already in place. We will look at how to change the makeup of the alignment itself, not just the values driving it. But first, let's look at some of the constraints and understand how they work.

Understanding Alignment Constraints

In the previous exercise you were exposed to constraints. The various constraints will help keep geometry together to maintain tangency or to maintain the radius.

When an alignment is created from objects, the lines are not constrained (Fixed). This is the same if you select the Fixed Lines (Two Points) tool from the Alignment Layout Tools toolbar. The curves are all constrained on both sides (Free). When you grip-edit a line or arc, the lines maintain tangency to the arc, but the arc loses its original radius, as shown in the figures at the beginning of this chapter. You may have noticed in Panorama the Tangency Constraint field that we looked at briefly in the previous exercise. You can click on any segment and change constraints (Figure 6.31). You can also change the constraints in the Sub-Entity Editor.

FIGURE 6.31
The tangency constraints in Panorama

No.	Type	Tangency Constraint
1	Line	Not Constrained (Fixed)
2	Curve	Constrained by Previous (Floating)
3	Line	Not Constrained (Fixed)
4	Curve	Constrained on Both Sides (Free)
5	Line	Not Constrained (Fixed)
6	Curve	Constrained on Both Sides (Free)
7	Line	Not Constrained (Fixed)

In this exercise you will experiment with constraints and their effect on the behavior of an alignment.

1. If your file is not still open from the previous exercise, open the EditingAlignments_FINISHED.dwg or EditingAlignments_METRIC_FINISHED.dwg file.
2. Select the Frontenac Drive alignment to activate the Alignment contextual tab.
3. From the Alignment contextual tab ➢ Modify panel, choose Geometry Editor.
4. In the Alignment Layout Tools toolbar, select the Alignment Grid View.
5. Click in the Tangency Constraint field for the first curve and change it to Not Constrained (Fixed).

   Notice that when you change the first curve from Constrained By Previous (Floating) to Not Constrained (Fixed), the first line changes from Not Constrained (Fixed) to Constrained By Next (Floating).

6. Grip-edit the curve and notice how it does not maintain tangency or radius with the second line but does maintain tangency with the first line because of the first line's tangency constraint (Figure 6.32). Also notice that the angle of the first line (set to Floating) does change.

FIGURE 6.32
Gripping on an alignment with the first line set to Constrained By Next (Floating) and the first curve set to Not Constrained (Fixed)

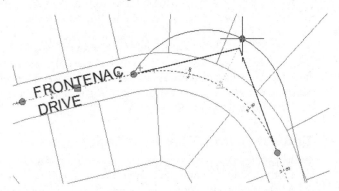

You may need to move Panorama out of the way to do this, but don't close it yet.

7. Press Esc on your keyboard to cancel the grip edit. If you inadvertently moved the curve, you can use the Undo command.
8. In Panorama, click in the Tangency Constraint field for the first curve and change it back to Constrained By Previous (Floating). The first line will change back to Not Constrained (Fixed).
9. Grip-edit the curve again and notice that the curve still maintains its tangency with the previous line but does not for the following line, just as it did in the previous steps (Figure 6.33). But unlike in the previous steps, the first line (set to Fixed) does not change its angle.

FIGURE 6.33
Gripping on an alignment with the first line set to Not Constrained (Fixed) and the first curve using Constrained By Previous (Floating)

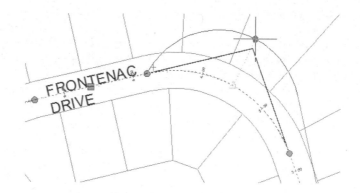

10. Press Esc on your keyboard to cancel the grip edit. If you inadvertently moved the curve, you can use the Undo command.

11. Click in the Tangency Constraint field for the first curve and change it to Constrained By Next (Floating).

    The lines before it and after it will be set to Not Constrained (Fixed).

12. Grip-edit the curve and notice that the curve now maintains its tangency with the following line but not the previous line (Figure 6.34).

FIGURE 6.34
Gripping on an alignment with the first curve set to Constrained By Next (Floating) and the following line set to Not Constrained (Fixed)

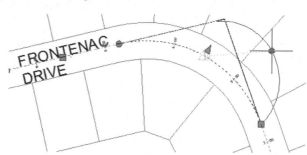

13. Press Esc on your keyboard to cancel the grip edit action. If you inadvertently moved the curve, you can use the Undo command.

14. Click in the Tangency Constraint field for the first curve and change it to Constrained On Both Sides (Free).

15. Grip-edit the curve. See that the curve now maintains its tangency with both the previous and the following lines (Figure 6.35).

FIGURE 6.35
Gripping on an alignment with the first curve set to Constrained On Both Sides (Free) with both adjoining lines set to Not Constrained (Fixed)

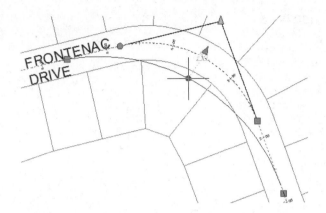

16. Press Esc on your keyboard to cancel the grip edit. If you inadvertently moved the curve, you can use the Undo command.

17. Close the Alignment Layout Tools toolbar.

When you click on an alignment that contains curves and select Alignment Properties, some additional options become available that were not available when the alignment was originally created. In the Point Of Intersection tab, you can select whether you want to visually show points of intersection by a change in alignment direction or by individual curves and curve group. You can also choose to not display any implied points of intersection (Figure 6.36).

FIGURE 6.36
The Point Of Intersection tab

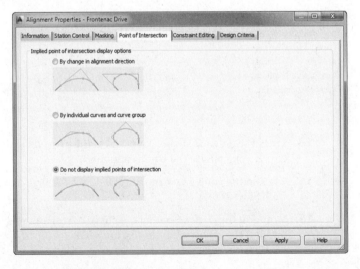

In the Constraint Editing tab, you can select if you wish to always perform any implied tangency constraint swapping and whether to lock all parameter constraints (Figure 6.37). The Always Perform Implied Tangency Constraint Swapping check box is what caused the swapping seen in this exercise.

FIGURE 6.37
The Constraint Editing tab

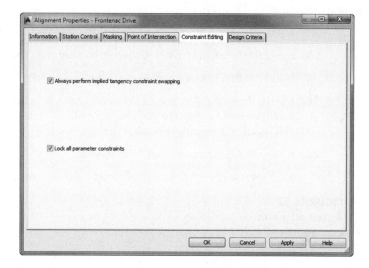

When this exercise is complete, you may close the drawing. A saved finished copy of this drawing is available from the book's web page with the filename `AlignmentConstraints_FINISHED.dwg` or `AlignmentConstraints_METRIC_FINISHED.dwg`.

Changing Alignment Components

One of the most common changes is adding a curve where there was none before or changing the makeup of the curves and tangents already in place in an alignment. Other design changes can include swapping out curves for tangents or adding a second curve to smooth a transition area.

It turns out that your perfect reverse curve isn't allowed by the current ordinances for subdivision design! In this example, you'll go back to the design the planner gave you and place a minimum length tangent between the curves:

1. Open the `AlignmentReverse_FINISHED.dwg` or `AlignmentReverse_METRIC_FINISHED.dwg` file from a previous exercise in this chapter.

2. Select the alignment named Reverse to activate the Alignment contextual tab.

3. From the Alignment contextual tab ➤ Modify panel, choose Geometry Editor.

   The Alignment Layout Tools toolbar appears.

4. Select the Delete Sub-Entity tool.

5. When prompted to select a sub-entity, pick the northern curve to remove it and press ↵ to end the command.

   Note that the last curve and tangent is still part of the alignment—it just isn't connected.

6. Select the Floating Line (From Curve End, Length) option.

7. When prompted to select an entity for the start point, click the northern end of the southern curve, and then specify a length of **100** ↵ (for metric, **30** ↵).

If you inadvertently select the polyline instead of the alignment, you may want to turn on selection cycling by entering **Selectioncycling** on the command line and entering **2** ↵.

8. Select the Free Curve Fillet (Between Two Entities, Through Point) option.

9. Select the first line segment to the north and then the line just drawn.

10. For the through point, use an Endpoint Osnap to select the free end of the tangent line just drawn and press ↵ to end the command. Notice how the stationing is now shown the full length of the completed alignment.

11. Close the Alignment Layout Tools toolbar.

Your reverse curve with tangent section is complete, as shown in Figure 6.38.

FIGURE 6.38
Reverse curve with tangent segment

When this exercise is complete, you may close the drawing. A saved finished copy of this drawing is available from the book's web page with the filename `AlignmentReverseEdit_FINISHED.dwg` or `AlignmentReverseEdit_METRIC_FINISHED.dwg`.

So far in this chapter, you've created and modified the horizontal alignments, adjusted them onscreen to look like what your planner delivered, and tweaked the design using a number of different methods. Now let's look beyond the lines and arcs and get into the design properties of the alignment.

Alignments As Objects

Beyond the simple nature of lines and arcs, alignments represent other things such as highways, streams, sidewalks, or even flight patterns. All these items have properties that help define them, and many of these properties can also be part of your alignments. In addition to obvious properties like names and descriptions, you can include functionality such as superelevation, station equations, reference points, and station control. The following sections will look at other properties that can be associated with an alignment and how to edit them.

Alignment Properties

While the properties of the alignment were originally assigned during creation, later in design there are often changes that are required. In this exercise, you'll learn an easy way to change the object style and how to add a description.

Most of an alignment's basic properties can be modified in Prospector. In this exercise, you'll change the name in a couple of ways:

1. Open the `AlignmentConstraints_FINISHED.dwg` or `AlignmentConstraints_METRIC_FINISHED.dwg` file, and make sure Prospector is open.

2. In Prospector, expand the Alignments ➢ Centerline Alignments branch.

   Notice that Cabernet Court, Frontenac Drive, and Syrah Way are listed as members.

3. Click the Centerline Alignments branch, and the individual alignments appear in the Toolspace preview area (see Figure 6.39).

FIGURE 6.39
The Alignments collection listed in the preview area of Prospector

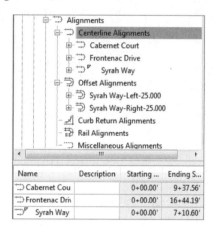

4. Down in the Prospector preview area, click in the Name field for Cabernet Court, and pause briefly before clicking again.

   The text highlights for editing.

5. Click in the Description field, and enter a description (such as **North South**). Press ↵.

6. Click in the Style field (you may have to scroll left/right to find the Style field), and the Select Label Style dialog appears.

7. Select Basic from the drop-down menu, and click OK.

   The screen updates. Keep this drawing open for the next portion of the exercise.

   That's one method to change alignment properties. The next is to use the Properties palette:

8. Open the Properties palette by using the Ctrl+1 keyboard shortcut or some other method.

9. Select the Syrah Way alignment in the drawing.

   The Properties palette looks like Figure 6.40.

FIGURE 6.40
Syrah Way in the Properties palette

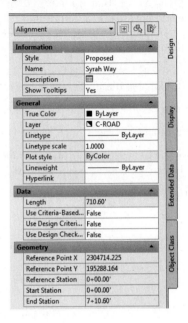

10. Click in the Description field and the Description dialog will open. Enter a description (such as **Central East West**), and click OK.

11. Press Esc on your keyboard to deselect all objects, and close the Properties palette if you'd like.

    Keep this drawing open for the next portion of the exercise.

    The final method involves getting into the Alignment Properties dialog, your access point to information beyond the basics:

12. In Prospector, right-click Frontenac Drive and select Properties. This is the alignment at the north end of the site.

    The Alignment Properties dialog for Frontenac Drive opens.

13. Change to the Information tab if it isn't selected.

14. Enter a description in the Description field (such as **Loop Cul-de-sac**).

15. Set Object Style to Existing.

16. Click the Apply button.

    Notice that the display style in the drawing updates immediately.

17. Click OK to accept the settings in the dialog.

Keep this drawing open for the next portion of the exercise.

Now that you've updated your alignments, let's make them all the same style for ease of viewing. The best way to do this is in the Prospector preview area.

18. In Prospector, expand the Alignments ➢ Centerline Alignments branch, and select the Centerline Alignments text.

19. Select one of the alignments in the preview area.

20. Press Ctrl+A to select them all, or pick the top and then Shift+click the bottom item.

    The idea is to pick *all* of the alignments.

21. Right-click the Style column header and select Edit (see Figure 6.41).

FIGURE 6.41
Editing alignment styles en masse via Prospector

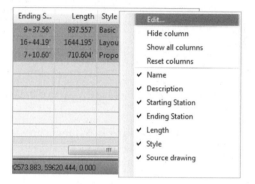

22. Select Layout from the drop-down list in the Select Label Style dialog that appears, and click OK.

23. Press Esc to deselect all of the alignments.

    Notice that all alignments pick up this style. Although the dialog is named Edit Label Style, you are actually assigning styles to the objects.

You may keep this drawing open to continue on to the next exercise or use the saved finished copy of this drawing available from the book's web page (`AlignmentProperties_FINISHED.dwg` or `AlignmentProperties_METRIC_FINISHED.dwg`).

The alignments now look the same, and they all have a name and description. Let's look beyond these basics at the other properties you can modify and update.

> ### Don't Forget This Technique
>
> This technique of selecting all, right-clicking the column heading in the preview area, and editing the value works on every object that displays in the Toolspace preview: parcels, pipes, corridors, assemblies, and so on. It can be painfully tedious to change a large number of objects from one style to another using any other method. If you left-click the column header instead of right-click, the alignments will sort alphabetically based on the values in that column.

The Right Station

At the end of the process, every alignment has stationing applied to help locate design information. This stationing often starts at zero, but it can also tie to an existing object and may start at some arbitrary value. Stationing can also be fixed in both directions, requiring station equations that help translate between two disparate points that are the basis for the stationing in the drawing.

One common problem is an alignment that was drawn in the wrong direction. Thankfully, Civil 3D has a quick edit command to fix that:

1. Open the AlignmentStations.dwg file or the AlignmentStations_METRIC.dwg file, and make sure Prospector is open.

2. Pick the Syrah Way alignment to activate the Alignment contextual tab.

3. From the Alignment contextual tab ➢ Modify drop-down panel, choose Reverse Direction.

    A warning dialog appears, reminding you of the consequences of such a change.

4. Click OK to dismiss the warning dialog.

    The stationing reverses, with 0+00 (or 0+000 for metric users) now at the east end of the street.

    Notice that the masking is no longer applied to the offset alignments. This is one of the reasons that warning message was given. You can reapply the masking if you would like by going to the Station Control tab of the Alignment Properties dialog.

5. Press Esc to deselect the alignment.

    Keep this drawing open for the next portion of the exercise.

    This technique allows you to reverse an alignment almost instantly. The warning that appears is critical, though! When an alignment is reversed, the information that was derived from its original direction may not translate correctly, if at all. One prime example of this is design profiles: they don't reverse themselves when the alignment is reversed, and this can lead to serious design issues if you aren't paying attention.

    Beyond reversing, it's common for alignments to not start with zero. For example, the Cabernet Court alignment may be a continuation of an existing street, and it makes sense to make the starting station for this alignment the end station from the existing street. In this next portion of the exercise, you'll set the beginning station.

6. Select the Cabernet Court alignment to activate the Alignment contextual tab.

7. From the Alignment contextual tab ➢ Modify panel, choose Alignment Properties.

8. Switch to the Station Control tab of the Alignment Properties dialog.

    This tab controls the base stationing and lets you create station equations.

9. In the Reference Point section of the dialog, enter **315.62** (for metric, **96.20**) in the Station field (see Figure 6.42).

FIGURE 6.42
Setting a new starting station on the Cabernet Court alignment

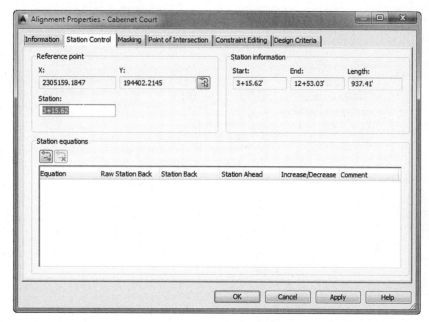

10. Click OK to dismiss the warning message that appears, and click OK to dismiss the dialog.

11. Press Esc to deselect the alignment.

    The Station Information section in the top right will update. These options can't be edited but provide a convenient way for you to review the alignment's length and station values.

    Keep this drawing open for the next portion of the exercise.

    In addition to changing the value for the start of the alignment, you could use the Pick Reference Point button, as shown in Figure 6.42, to select another point as the stationing reference point.

    Station equations can occur multiple times along an alignment. They typically come into play when plans must match existing conditions or when the stationing has to match other plans, but the lengths in the new alignment would make that impossible without some translation. In this last portion of the exercise, you'll add a station equation on Frontenac Drive at the intersection with Syrah Way.

12. Select the Frontenac Drive alignment to activate the Alignment contextual tab.

13. From the Alignment contextual tab ➢ Modify panel, choose Alignment Properties.

14. On the Station Control tab of the Alignment Properties dialog, click the Add Station Equations button.

15. Use an Endpoint Osnap to select the intersection at the west end of Syrah Way.

16. Change the Station Ahead value to **5000** (for metric, **1500**).

17. Click the Apply button, and notice the change in the Station Information section as shown in Figure 6.43.

FIGURE 6.43
Frontenac Drive station equation in place

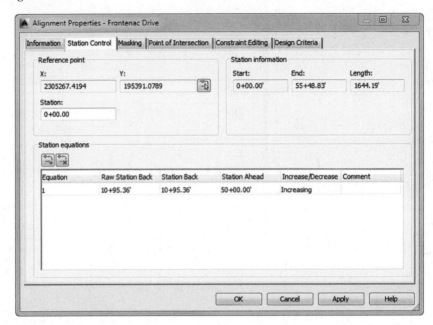

18. Click OK to accept the settings in the dialog, and review the stationing that has been applied to the alignment.

You may keep this drawing open to continue on to the next exercise or use the saved finished copy of this drawing available from the book's web page (`AlignmentStations_FINISHED.dwg` or `AlignmentStations_METRIC_FINISHED.dwg`).

Stationing constantly changes as alignments are modified during the initial stages of a development or as late design changes are pushed back into the plans. With the flexibility shown here, you can reduce the time you spend dealing with minor changes that seem to ripple across an entire plan set.

Assigning Design Speeds

One driving part of transportation design is the design speed. Civil 3D considers the design speed a property of the alignment, and it can be used in labels or calculations as needed. In this exercise, you'll add a series of design speeds to Frontenac Drive. Later in the chapter, you'll label these sections of the road using a label set:

1. If it's not still open from the previous exercise, open the `AlignmentStations_FINISHED.dwg` file or the `AlignmentStations_METRIC_FINISHED.dwg` file.

2. Bring up the Alignment Properties dialog for Frontenac Drive to the north using any of the methods discussed previously.

3. Switch to the Design Criteria tab.

4. Verify that the Design Speed field for Number 1 is 30 mi/hr (or 50 km/hr for metric users).

   This speed is typical for a subdivision street.

5. Click the Add Design Speed button.

6. Select the new row, then click in the Start Station field for Number 2.

   A small Pick On Screen button appears to the right of the Start Station value, as shown in Figure 6.44.

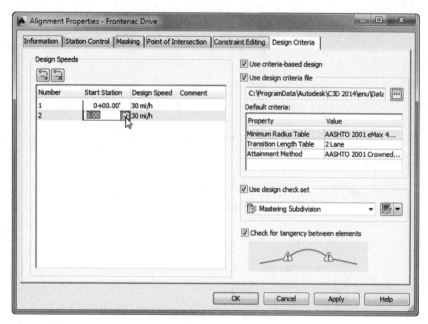

FIGURE 6.44
Setting the design speed for a Start Station field

7. Click the Pick On Screen button, and then use an Osnap to pick the PC on the east portion of the site, near station 2+25.88 (or station 0+067.38 for metric users).

8. Enter a value of 20 (for metric, 40) in the Design Speed field for the line you just added.

9. Click the Add Design Speed button again.

   Notice that the additional row is added to the top of the list.

10. Select the new row. Then click in the Start Station field for the new row, click the Pick On Screen button again to add one more design speed portion, and enter 900 ↵ (for metric, 275 ↵).

288 | **CHAPTER 6** ALIGNMENTS

11. Enter a value of **30** (for metric, **50**) for this design speed.

    When complete, the tab should look like Figure 6.45.

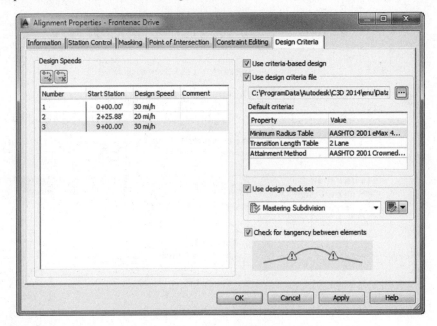

FIGURE 6.45
The design speeds assigned to Frontenac Drive

12. Click OK to accept the settings in the Alignment Properties dialog.

When this exercise is complete, you may close the drawing. A saved finished copy of this drawing is available from the book's web page with the filename `DesignSpeed_FINISHED.dwg` or `DesignSpeed_METRIC_FINISHED.dwg`.

In a subdivision, these values can be inserted for labeling purposes. In a highway design, they can be used to drive the superelevation calculations that are critical to a working design. Chapter 11, "Superelevations," looks at this subject.

Labeling Alignments

Labeling in Civil 3D is one of the program's strengths, but it's also an easy place to get lost. There are myriad options for every type of labeling situation under the sun, and keeping them straight can be difficult.

The Power of Label Sets

When you think about it, any number of items can be labeled on an alignment. These include major and minor stations, geometry points, design speeds, and profile information. Each of these objects can have its own style. Keeping track of all these individual labeling styles and options would be burdensome and uniformity would be difficult, so Civil 3D features the concept of *label sets*.

A label set lets you build up the labeling options for an alignment, picking styles for the labels of interest, or even multiple labels on a point of interest, and then save them as a set. These sets are available during the creation and labeling process, making the application of individual labels less tedious. Out of the box, a number of sets are available, primarily designed for combinations of major and minor station styles along with geometry information.

You'll learn how to create individual label styles in Chapter 18. We will use the out-of-the-box styles and label sets over the next couple of exercises.

Major Station

Major station labels typically include a tick mark and a station callout.

Geometry Points

Geometry points reflect the PC, PT, and other points along the alignment that define the geometric properties.

In this exercise, you'll apply a label set to all of your alignments and then see how an individual label can be changed from the set:

1. Open the `AlignmentLabels.dwg` file or the `AlignmentLabels_METRIC.dwg` file.

2. Select the Cabernet Court alignment on the screen.

3. Right-click, and select Edit Alignment Labels to display the Alignment Labels dialog shown in Figure 6.46.

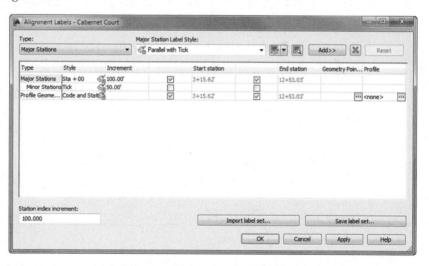

FIGURE 6.46
The Alignment Labels dialog for Cabernet Court

4. Click the Import Label Set button near the bottom of this dialog to display the Select Label Set dialog.

5. In the Select Label Set drop-down list, select All Labels and click OK.

   The Alignment Labels dialog now populates with the additional labels imported from the label set, as shown in Figure 6.47.

FIGURE 6.47
The All Labels label set

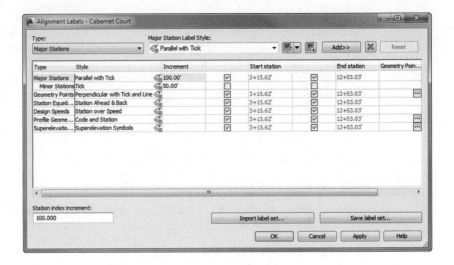

6. Click OK to accept the settings in the Alignment Labels dialog.
7. Repeat this process to import the All Labels set for the rest of the centerline alignments.
8. When you've finished, zoom in on any of the major station labels (for example, on Frontenac Drive, the 1+00 label, or the 0+020 label for metric users).
9. Select the label and notice how all of the label set group labels are selected.
10. Press Esc to deselect.
11. Hold down the Ctrl key, and select the label.

    Notice that a single label is selected, not the label set group, and the Labels contextual tab is activated.

> **CTRL+CLICK? WHAT'S THAT ABOUT?**
>
> Prior to AutoCAD Civil 3D 2008, clicking an individual label picked the label and the alignment. Because labels are part of a label set object now, Ctrl+click is the *only* way to access the Flip Label and Reverse Label functions!

12. With the individual major station selected, right-click and select Label Properties.

    The Properties palette for the label appears.

13. Use the Major Station Label Style drop-down list to change from <default> to Perpendicular With Line.

14. Change the Flipped Label value to True to move the label from the left side of the alignment to the right side.

    This command can be helpful if your plans become busy and you need to move a label to the other side for better visibility.

15. Close the Properties palette and press Esc to deselect the label item.

    Keep this drawing open for the next exercise.

By using alignment label sets, you'll find it easy to standardize the appearance of labeling and stationing across alignments. Building label sets can take some time, but it's one of the easiest, most effective ways to enforce standards. Building label sets will be discussed further in Chapter 18.

Station Offset Labeling

Beyond labeling an alignment's basic stationing and geometry points, you may want to label points of interest in reference to the alignment. Station offset labeling is designed to do just that. In addition to labeling the alignment's properties, you can include references to other object types in your station offset labels. The objects available for referencing are as follows:

- Alignments
- COGO points
- Parcels
- Profiles
- Surfaces

In this last portion of this exercise, you'll use an alignment reference to create a label suitable for labeling the station information for the center point of the cul-de-sac at the end of Frontenac Drive:

1. Using the drawing from the previous exercise, from the Annotate tab ➢ Labels & Tables panel ➢ Add Labels drop-down, choose Alignment ➢ Add Alignment Labels. The Add Labels dialog appears with Feature set to Alignment.

2. In the Label Type drop-down list, select Station Offset.

    The Station Offset label type creates a label that floats with the alignment. If the alignment is edited graphically, the label will move with it and the station and offset information that it displays will remain constant. The counterpart label type is Station Offset – Fixed Point. This type of label will hold its position if the alignment is modified and the station and offset information will change.

3. In the Station Offset Label Style drop-down, select Station And Offset.

4. Leave the Marker Style field at the default setting, but remember that you could use any of these styles to mark the selected point.

5. Click the Add button.

6. At the `Select Alignment:` prompt, select the Frontenac Drive alignment.
7. Use the Center Osnap to snap to the center of the right-of-way arc at the cul-de-sac of Frontenac Drive.
8. At the `Specify station offset:` prompt, enter **0** ↵ for the offset amount, and press ↵ to end the command.
9. Close the Add Labels dialog.
10. Select the label and use the square grip to drag the label to a convenient location (Figure 6.48). Label grips will be discussed further in Chapter 18.

FIGURE 6.48
The alignment station offset label in use

You may keep this drawing open to continue on to the next exercise or use the saved finished copy of this drawing available from the book's web page (`AlignmentLabels_FINISHED.dwg` or `AlignmentLabels_METRIC_FINISHED.dwg`).

Using station offset labels and their reference object ability, you can label most site plans quickly with information that dynamically updates. Because of the flexibility of labels in terms of style, you can create "design labels" that are used to aid in modeling yet never plot and aren't seen in the final deliverables. More advanced alignment labels are discussed in Chapter 18.

Alignment Tables

There isn't always room to label alignment objects directly on top of them. Sometimes doing so doesn't make sense, or a reviewing agency wants to see a table showing the radius of every curve in the design. Documentation requirements are endlessly amazing in their disparity and seeming randomness. Beyond labels that can be applied directly to alignment objects, you can also create tables to meet your requirements.

You can create four types of tables:

- Line
- Curve
- Spiral
- Segment

Each of these is self-explanatory except perhaps the segment table. That table generates a mix of all the lines, curves, and spirals that make up an alignment, essentially re-creating the alignment in a tabular format. In this section, you'll generate a new line table and segment table.

All the tables work in a similar fashion. To add a table:

1. From the Annotate tab ➢ Labels & Tables panel, choose the drop-down from the Add Tables button.

2. From the Add Tables drop-down, choose the object type (such as Alignment).

3. Pick a table type (such as Add Segment) that is relevant to your work.

   The Alignment Table Creation dialog appears (see Figure 6.49).

FIGURE 6.49
The Alignment Table Creation dialog for the Add Segment table type

4. Select a table style from the drop-down list or create a new one.

The Select By Label Or Style area determines how the table is populated. All the label style names for the selected type of component are presented, with a check box to the right of each one. Applying one of these styles enables the Selection Rule setting, which has the following two options:

Add Existing   Any label using this style that currently exists in the drawing is converted to a tag format, substituting a key number such as L1 or C27, and added to the table. Any labels using this style created in the future will *not* be added to the table.

Add Existing And New   Any label using this style that currently exists in the drawing is converted to a tag format and added to the table. In addition, any labels using this style created in the future will be added to the table.

To the right of the Select section is the Split Table section, which determines how the table is stacked up in modelspace once it's populated. You can modify these values after a table is generated, so it's often easier to leave them alone during the creation process.

Finally, the Behavior section provides two options for Reactivity Mode: Static and Dynamic. This section determines how the table reacts to changes in the labeled objects. In some cases in surveying, this disconnect is used as a safeguard to the platted data, but in general, the point of a 3D model is to have live labels that dynamically react to changes in the object. Be sure to cancel out of the Alignment Table Creation dialog before proceeding to the next exercise.

Before you draw any tables, you need to apply labels so the tables will have data to populate. In this exercise, you'll place some labels on your alignments, and then you'll move on to drawing tables:

1. If it's not still open from the previous exercise, open the AlignmentLabels_FINISHED.dwg or the AlignmentLabels_METRIC_FINISHED.dwg file.

2. From the Annotate tab ➢ Labels & Tables panel, choose Add Labels to open the Add Labels dialog.

3. In the Feature field, select Alignment.

4. In the Label Type field, select Multiple Segment from the drop-down list.

   With these options, you'll click each alignment one time, and every subcomponent will be labeled with the style selected here.

5. Verify that the Line Label Style field is set to Line Label Style: Bearing Over Distance (not General Line Label Style: Bearing Over Distance).

   You won't be left with these labels—you just want them for selecting elements later.

6. Verify that the Curve Label Style field is set to Delta Over Length & Radius.

   Since you have no spirals, no Spiral label style needs to be specified.

7. Click the Add button, and select each of the three alignments; then press ↵ to end the command.

8. Click the Close button to dismiss the Add Labels dialog.

Keep this drawing open for the next exercise.

Now that you've got labels to play with, let's build some tables.

Creating a Line Table

Most line tables are simple: a line tag, a bearing, and a distance. You'll also see how Civil 3D can translate units without having to change anything at the drawing level:

1. Using the drawing from the previous exercise, from the Annotate tab ➢ Labels & Tables panel, click the Add Tables drop-down arrow and choose Alignment ➢ Add Line to open the Table Creation dialog.

2. Verify that a check mark appears in the Apply column for the Alignment Line: Bearing Over Distance label, as shown in Figure 6.50.

FIGURE 6.50
Creating an alignment line table

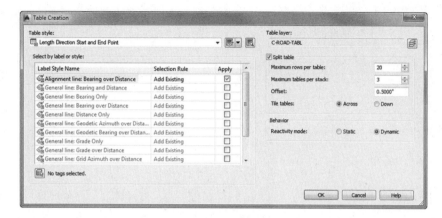

3. Click the Pick On-Screen button and make a crossing window across the entire project.
4. Press ↵ to complete the selections set.
5. When presented with the Create Table – Convert Child Styles warning dialog, select Convert All Selected Label Styles To Tag Mode.

   Notice that there are now 15 tags selected.
6. Click OK to accept the settings in the Table Creation dialog.
7. At the `Select upper left corner:` prompt, select an insertion point onscreen and the table will generate.

Keep this drawing open for the next exercise.

Pan back to your drawing, and you'll notice that the line labels have turned into tags on the line segments. After you've made one table, the rest are similar. Be patient as you create tables—a lot of values must be tweaked to make them look just right. By drawing one onscreen and then editing the style, you can quickly achieve the results you're after.

An Alignment Segment Table

An individual segment table allows a reviewer to see all the components of an alignment. In this exercise, you'll draw the segment table for Frontenac Drive:

1. Using the drawing from the previous exercise, from the Annotate tab ➢ Labels & Tables panel, choose the Add Tables drop-down arrow, and choose Alignment ➢ Add Segment to open the Alignment Table Creation dialog.
2. In the Select Alignment field, choose the Frontenac Drive alignment from the drop-down list (Figure 6.51), and click OK.

FIGURE 6.51
Creating an alignment segment table

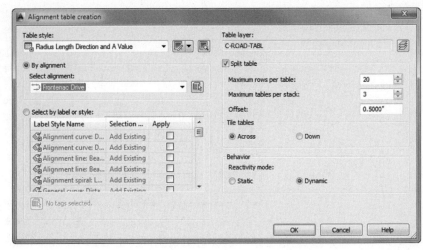

3. At the Select upper left corner: prompt, select an insertion point on the screen and the table will generate.

When this exercise is complete, you may close the drawing. A saved finished copy of this drawing is available from the book's web page with the filename AlignmentTables_FINISHED.dwg or AlignmentsTables_METRIC_FINISHED.dwg.

The Bottom Line

Create an alignment from an object.   Creating alignments based on polylines is a traditional method of building engineering models. With built-in tools for conversion, correction, and alignment reversal, it's easy to use the linework prepared by others to start your design model. These alignments lack the intelligence of crafted alignments, however, and you should use them sparingly.

Master It   Open the MasteringAlignments-Objects.dwg or MasteringAlignments-Objects_METRIC.dwg file, and create alignments from the linework found there with the All Labels label set.

Create a reverse curve that never loses tangency.   Using the alignment layout tools, you can build intelligence into the objects you design. One of the most common errors introduced to engineering designs is curves and lines that aren't tangent, requiring expensive revisions and resubmittals. The free, floating, and fixed components can make smart alignments in a large number of combinations available to solve almost any design problem.

Master It   Open the MasteringAlignments-Reverse.dwg or the MasteringAlignments-Reverse_METRIC.dwg file, and create an alignment using the linework on the right as a guide. Create a reverse curve with both radii equal to 200 (or 60 for metric users) and with a pass-through point at the intersection of the two arcs.

Replace a component of an alignment with another component type. One of the goals in using a dynamic modeling solution is to find better solutions, not just the first solution. In the layout of alignments, this can mean changing components out along the design path or changing the way they're defined. The ability of Civil 3D to modify alignments' geometric construction without destroying the object or forcing a new definition lets you experiment without destroying the data already based on an alignment.

> **Master It** Convert the reverse curve indicated in the `MasteringAlignments-Rcurve.dwg` or the `MasteringAlignments-Rcurve_METRIC.dwg` file to a floating arc that is constrained by the next segment. Then change the radius of the curves to 150 (or 45 for metric users).

Create alignment tables. Sometimes there is just too much information that is displayed on a drawing, and to make it clearer, tables are used to show bearings and distances for lines, curves, and segments. With their dynamic nature, these tables are kept up to date with any changes.

> **Master It** Open the `MasteringAlignments-Table.dwg` or `MasteringAlignments-Table_METRIC.dwg` file, and generate a line table, a curve table, and a segment table. Use whichever style you want to accomplish this.

Chapter 7

Profiles and Profile Views

Profile information is the backbone of vertical design. The AutoCAD® Civil 3D® software takes advantage of sampled data, design data, and external input files to create profiles for a number of uses. Profiles will be an integral part of corridors, as we'll discuss in Chapter 9, "Basic Corridors." In this chapter we'll look at using profile creation tools, editing profiles, and generating and editing profile views, and you'll learn ways to get your labels just so.

In this chapter, you will learn to:

- Sample a surface profile with offset samples
- Lay out a design profile on the basis of a table of data
- Add and modify individual entities in a design profile
- Apply a standard band set

The Elevation Element

The whole point of a three-dimensional model is to include the elevation element that's been missing for years on two-dimensional plans. But to get there, designers and engineers still depend on a flat 2D representation of the vertical dimension as shown in a profile view (see Figure 7.1).

FIGURE 7.1
A typical profile view of the surface elevation along an alignment

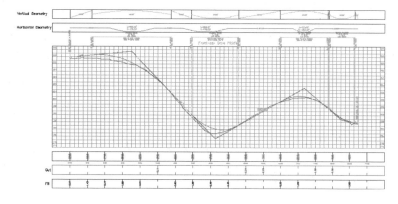

A profile is nothing more than a series of data pairs in a station, elevation format. There are basic curve and tangent components, but these are purely the mathematical basis for the paired data sets. In AutoCAD Civil 3D, you can generate profile information in one of the following four ways:

Sampling from a Surface Sampling from a surface involves taking vertical information from a surface object every time the sampled alignment crosses a TIN line of the surface. This is perfect for generating a profile for the existing ground.

Using a Layout to Create a Profile Using a layout to create a profile allows you to input design information, setting critical station and elevation points, calculating curves to connect linear segments, and typically working within design requirements laid out by a reviewing agency.

Creating a Profile from a File Creating from a file lets you reference a specially formatted text file to pull in the station and elevation pairs. Doing so can be helpful in dealing with other analysis packages or spreadsheet tabular data.

Creating a Best Fit Profile Similar to the ability to generate a best fit alignment that we discussed in Chapter 6, "Alignments," you can also create a best fit profile. You may find yourself using this method when you are trying to generate defined geometry for an existing road.

The following sections look at all four methods of creating profiles:

Surface Sampling

Working with surface information is the most elemental method of creating a profile. This information can represent any of the surfaces already in your drawing, such as an existing surface or any number of other surface-derived data sets. Surfaces can also be sampled at offsets, as you'll see in the next series of exercises. Follow these steps:

1. Open the `ProfileSampling.dwg` file (or the `ProfileSampling_METRIC.dwg` file for metric users) shown in Figure 7.2. Remember, all data files can be downloaded from this book's web page at www.sybex.com/go/masteringcivil3d2014.

FIGURE 7.2
The drawing you'll use for this exercise

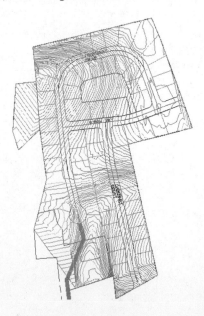

2. From the Home tab ➢ Create Design panel, choose Profile ➢ Create Surface Profile to display the Create Profile From Surface dialog (Figure 7.3).

FIGURE 7.3
The Create Profile From Surface dialog

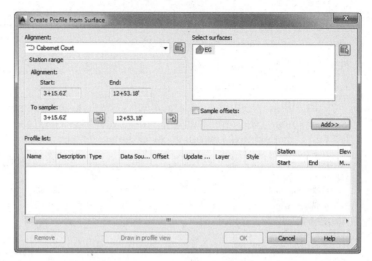

A Quick Tour around the Create Profile From Surface Dialog

This dialog has a number of important features, so take a moment to see how it breaks down:

- The upper-left quadrant is dedicated to information about the alignment. You can select the alignment from a drop-down list, or you can use the pick button to select it in the drawing. The Station Range area shows the starting and ending stations of the alignment and sets the To Sample range automatically to run from the beginning to the end of the alignment. You can control it manually by entering the station ranges in the To Sample text boxes or graphically in the drawing by using the pick buttons provided.

- The upper-right quadrant controls the selection of the surface that will be sampled and allows you to add sample offsets. You can select a surface from the list, or you can click the Pick On Screen button. Beneath the Select Surfaces box is a Sample Offsets check box. The offsets aren't applied in the left and right direction uniformly. You must enter a negative value to sample to the left of the alignment or a positive value to sample to the right. You can add multiple offsets in a comma-delimited list here, such as, for example, -50, -25, -10, -5, 0, 5, 10, 25, 50. In all cases, whether or not you are sampling offsets, the profile isn't generated until you click the Add button.

- In the bottom half, the Profile List box displays all profiles associated with the alignment currently selected in the Alignment drop-down menu. This area is generally static (it won't change), but you can modify the Update Mode, Layer, and Style columns by clicking the appropriate cells in this table. You can stretch and rearrange the columns to customize the view. The columns can only be modified in this location as they are created. Upon returning to the Create Profile From Surface dialog, the profiles previously created will have static values and can be changed in the Profile Properties dialog instead.

3. Select Syrah Way from the Alignment drop-down list if it isn't already selected.
4. In the Select Surfaces box, select EG.
5. Click Add to add the centerline profile to the Profile List section.
6. Check the box next to Sample Offsets.
7. Enter **-25, 25** (or **-7.5, 7.5** for metric users) to sample at the left and right right-of-way lines and click Add again.
8. In the profile list, select the cell in the Style column that corresponds to the negative (left offset) value (see Figure 7.4) to activate the Pick Profile Style dialog. If you need to widen the columns, you can do so by double-clicking the line between the column headings.

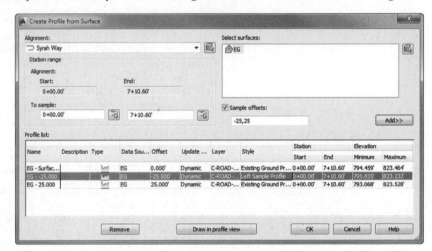

FIGURE 7.4
The Create Profile From Surface dialog with styles assigned on the basis of the Offset value

9. Select the Left Sample Profile option from the drop-down list, and click OK to dismiss the Pick Profile Style dialog.

    The style changes from Existing Ground Profile to Left Sample Profile in the table.

10. Select the cell in the Style column that corresponds to the positive (right offset) value to activate the Pick Profile Style dialog.
11. Select the Right Sample Profile option from the drop-down list, and click OK to dismiss the Pick Profile Style dialog.
12. Click Draw In Profile View to dismiss this dialog and open the Create Profile View Wizard, as shown in Figure 7.5.

FIGURE 7.5
The Create Profile View – General wizard page

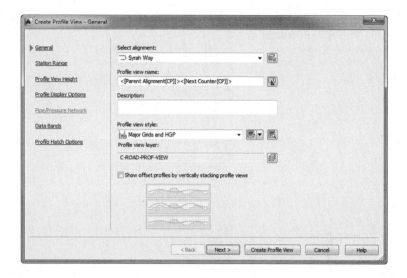

13. Verify that the Select Alignment drop-down list shows Syrah Way and that Profile View is selected in the Profile View Style drop-down list. Click Next.

14. On the Create Profile View – Station Range wizard page, verify that the Automatic option has been selected (Figure 7.6). Click Next.

FIGURE 7.6
The Create Profile View – Station Range wizard page

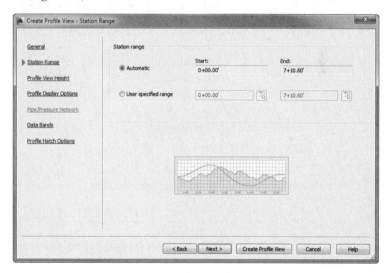

15. On the Create Profile View – Profile View Height wizard page, verify that the Automatic option has been selected (Figure 7.7). Click Next.

FIGURE 7.7
The Create Profile View – Profile View Height wizard page

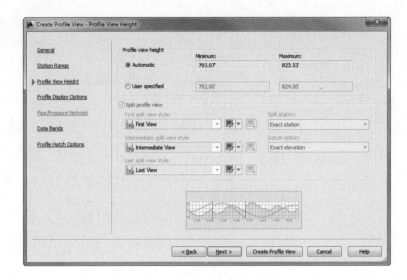

We will examine split profile views in a later exercise.

16. On the Create Profile View – Profile Display Options wizard page, look at the settings but do not make any changes (Figure 7.8). Click Next.

FIGURE 7.8
The Create Profile View – Profile Display Options wizard page

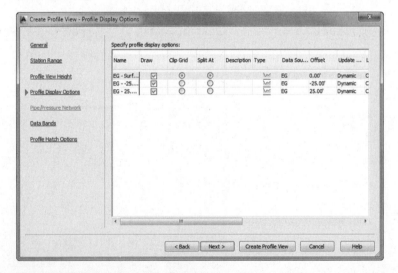

17. On the Create Profile View – Data Bands wizard page, verify that the band set is set to EG-FG Elevations And Stations (Figure 7.9). Notice in the Set Band Properties area that the Profile1 and Profile2 columns are both set to EG - Surface (1). We will look at data bands in greater detail a bit later.

FIGURE 7.9
The Create Profile View – Data Bands wizard page

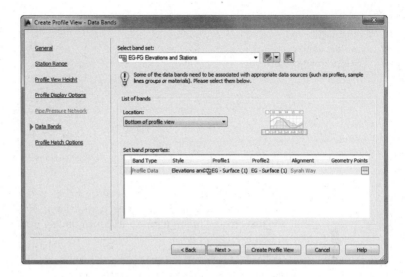

Notice that you could continue to click Next to step through the remainder of the wizard; however, you have no need to adjust further options at this time.

18. Click the Create Profile View button to dismiss the wizard.
19. Pick a point on the screen somewhere to the right of the site to draw the profile view, as shown in Figure 7.10.

FIGURE 7.10
The complete profile view for Syrah Way

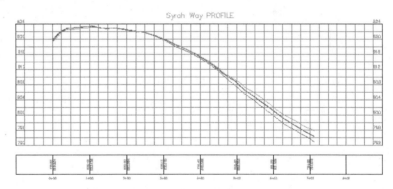

If the Events tab in Panorama appears, telling you that you've sampled data or if an error in the sampling needs to be fixed, then click the green check mark or the X to dismiss Panorama.

Keep the drawing open for the next portion of the exercise.

Profiles are dependent on the alignment they're derived from, so they're stored as profile branches under their parent alignment on the Prospector tab, as shown in Figure 7.11.

FIGURE 7.11
Alignment profiles on the Prospector tab

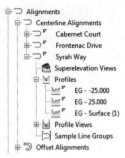

By maintaining the profiles under the alignments, you make it simpler to review what has been sampled and modified for each alignment. Note that the profiles from surface that you just created are dynamic and continuously update, as you'll see in the next portion of this exercise.

20. From the View tab ➢ Model Viewports panel, choose the drop-down list on the Viewport Configuration button and select Two: Horizontal.
21. Click in the top viewport to activate it.
22. On the Prospector tab, expand the Alignments ➢ Centerline Alignments branch.
23. Right-click Syrah Way, and select the Zoom To option.
24. Click in the bottom viewport to activate it.
25. Expand the Alignments ➢ Centerline Alignments ➢ Syrah Way ➢ Profile Views branch.
26. Right-click the profile view named Syrah Way1, and select Zoom To.

    Your screen should now look similar to Figure 7.12.

FIGURE 7.12
Splitting the screen for plan and profile editing

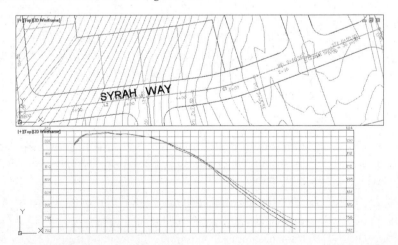

27. Click in the top viewport.

28. Zoom out so you can see more of the plan view.

29. Pick the alignment to activate the grips, and stretch the western end grip to lengthen and/or move the alignment, as shown in Figure 7.13.

FIGURE 7.13
Grip-editing the alignment

30. Click to complete the edit.

    The profile view automatically adjusts to reflect the change in the starting point of the alignment. Note that the offset profiles move dynamically as well.

31. Press Ctrl+Z enough times to undo the movement of your alignment and return it to its original location.

32. Select the top viewport and then switch back to a single viewport by once again clicking the viewport controls in the upper-left corner of one of the modelspace viewports and selecting Viewport Configuration List ➢ Single.

    Keep the drawing open for the next brief exercise.

By maintaining the relationships between the alignment, the surface, the sampled information, and the offsets, the software creates a much more dynamic feedback system for designers. This system can be useful when you're analyzing a situation with a number of possible solutions, where the surface information will be a deciding factor in the final location of the alignment. Once you've selected a location, you can use this profile view to create a vertical design, as you'll see in the next section.

> **LEFT TO RIGHT AND RIGHT TO LEFT?**
>
> You may have noticed that the alignment for Syrah Way is drawn right to left but the profile shows it left to right. It is often desirable to have the plan and profile go in the same direction.
>
> One option is to rotate the plan view 180 degrees. If you have your labeling all set to be plan-readable, it will follow along nicely.
>
> Another option is to generate a profile view object style that is set to read right to left instead of the default left to right.

In the following short exercise you will generate a profile view that displays right to left:

1. Using the file from the previous exercise, select the profile view (grid), and from the Profile View contextual tab ➤ Modify View panel, choose Profile View Properties.

2. On the Information tab, click the drop-down edit button to the right of the current object style (Profile View) and select Copy Current Selection.

3. On the Information tab, change the Name to **Profile View: Right to Left**.

   You may revise the description if desired.

4. In the Profile View Style dialog, on the Graph tab, change the profile view direction to Right To Left.

5. Click OK to dismiss the Profile View Style dialog.

6. Click OK to dismiss the Profile View Properties dialog.

You may need to move the profile view since the insert point of the profile view will now be in the lower-right corner instead of the lower-left corner as it was previously.

When this exercise is complete, you may close the drawing. A finished copy of this drawing is available from the book's website with the filename `ProfileSampling_FINISHED.dwg` or `ProfileSampling_METRIC_FINISHED.dwg`.

Changing a profile view style is straightforward, but because of the large number of settings in play with a profile view style, the changes can be dramatic. A profile view style includes information such as labeling on the axis, vertical scale factors, grid clipping, and component coloring.

Using various styles lets you make changes to the view to meet requirements without changing any of the design information associated with the profile. To learn more about editing and creating profile styles, refer to Chapter 19, "Object Styles."

Layout Profiles

Certification Objective

Working with sampled surface information is dynamic, and the improvement over previous generations of Autodesk civil design software is profound. Moving into the design stage, you'll see how these improvements continue as you look at the nature of creating design profiles. By working with layout profiles as a collection of components that understand their relationships with each other as opposed to independent finite elements, you will realize the power of the AutoCAD Civil 3D software as a design tool in addition to being a drafting tool.

You can create layout profiles in two basic ways:

PVI-Based Layout PVI-based layouts are the most common, using tangents between points of vertical intersection (PVIs) and then applying curve parameters to connect them. PVI-based editing allows editing in a more conventional tabular format.

Entity-Based Layout Entity-based layouts operate like horizontal alignments in the use of free, floating, and fixed entities. The PVI points are derived from pass-through points and other parameters that are used to create the entities. Entity-based editing allows for the selection of individual entities and editing in an individual component dialog.

You'll work with both methods in the next series of exercises to illustrate a variety of creation and editing techniques. First, you'll focus on the initial layout, and then you'll edit the various layouts.

Layout by PVI

PVI layout is the most common methodology in transportation design. Using long tangents that connect PVIs by derived parabolic curves is a method most engineers are familiar with, and it's the method you'll use in the first exercise:

1. Open the `LayoutByPVI.dwg` file or the `LayoutByPVI_METRIC.dwg` file.
2. From the Home tab ➢ Create Design panel, choose Profile ➢ Profile Creation Tools.
3. At the `Select profile view to create profile:` prompt, pick the Syrah Way profile view by clicking one of the grid lines to display the Create Profile – Draw New dialog.
4. Set Name to **Syrah Way FG**.
5. On the General tab, set Profile Style to Design Profile and Profile Label Set to Complete Label Set, as shown in Figure 7.14.

FIGURE 7.14
The General tab of the Create Profile – Draw New dialog

6. Switch to the Design Criteria tab to examine the options provided.

    Criteria-based design for profiles is similar to what you learned in Chapter 6 for alignments in that the software compares the design speed to a selected design table (typically AASHTO 2004 in North America) and sets minimum values for curve K values. This can be helpful when you're laying out long highway design projects, but most site and subdivision designers have other criteria to design against. We won't be using design criteria in this exercise, so you can leave everything unchecked.

7. Click OK to display the Profile Layout Tools toolbar shown in Figure 7.15.

FIGURE 7.15
Profile Layout Tools toolbar

Notice that the toolbar is *modeless*, meaning it stays open even if you do other AutoCAD operations such as Pan or Zoom.

8. On the Profile Layout Tools toolbar, click the drop-down arrow next to the Draw Tangents button on the far left.

9. Select the Curve Settings option.

    The Vertical Curve Settings dialog opens (Figure 7.16).

FIGURE 7.16
The Vertical Curve Settings dialog

The Select Curve Type drop-down menu should be set to Parabolic, and the Length values in both the Crest Curves and Sag Curves areas should be **150.000′** (or **45.000** m for metric users), as shown in Figure 7.16. Selecting a Circular or Asymmetric curve type activates the other options in this dialog.

> **TO K OR NOT TO K**
>
> You don't have to choose. Realizing that users need to be able to design using both, the software lets you modify your design based on what's important. You can enter a K value to see the required length and then enter a length with a nice round value that satisfies the K. The choice is up to you.

10. Click OK to dismiss the Vertical Curve Settings dialog.

11. On the Profile Layout Tools toolbar, click the drop-down arrow next to the Draw Tangents button on the far left again. This time, select the Draw Tangents With Curves option.

12. Use a Center Osnap to pick the center of the circle at the far right in the profile view.

    Remember, the profile is reversed so that station 0+00 is lined up with the plan. Therefore, station 0+00 is on the right of the profile. You need to draw your profile from low station to high station, which in this case is right to left.

13. Continue working your way across the profile view, picking the center of each circle right to left with a Center Osnap.

14. Right-click or press ↵ after you select the center of the last circle.

15. The profile labels will default to a location; however, you can click any of the profile labels and use the grips to move them to a more legible location.

    Your drawing should look similar to Figure 7.17.

FIGURE 7.17
A completed layout profile with labels

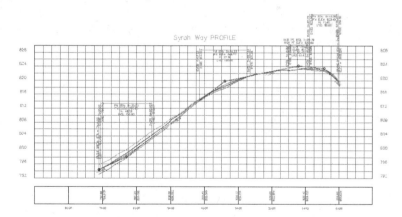

16. Close the Profile Layout Tools toolbar.

When this exercise is complete, you may close the drawing. A finished copy of this drawing is available from the book's web page with the filename LayoutByPVI_FINISHED.dwg or LayoutByPVI_METRIC_FINISHED.dwg.

The layout profile is labeled with the complete label set you selected in the Create Profile dialog. As you'd expect, this labeling and the layout profile are dynamic. If you select the profile and then zoom in on this profile line, not the labels or the profile view, you'll see something like Figure 7.18.

FIGURE 7.18
The types of grips on a layout profile

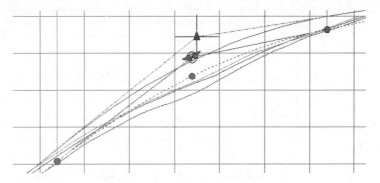

The PVI-based layout profiles include the following unique grips:

Vertical Triangular Grip   The vertical triangular grip at the PVI point is the PVI grip. Moving this alters the inbound and outbound tangents, but the curve remains in place with the same design parameters of length and type.

Angled Triangular Grips   The angled triangular grips on either side of the PVI are sliding PVI grips. Selecting and moving either moves the PVI, but movement occurs only along the tangent of the selected grip. The curve length isn't affected by moving these grips, but the PVI station and elevation will be, as well as the grade of the other tangent.

Circular Pass-Through Grips   The circular pass-through grips near the PVI and at each end of the curve are curve grips. Moving any of these grips makes the curve longer or shorter without adjusting the inbound or outbound tangents or the PVI point.

Although this simple pick-and-go methodology works for preliminary layout, it lacks a certain amount of control typically required for final design. For that, you'll use another method of creating PVIs:

1. Open the LayoutByPVITransparent.dwg or LayoutByPVITransparent_METRIC.dwg file.

2. Verify that the Transparent Commands toolbar (Figure 7.19) is displayed somewhere on your screen. If it is not shown, from the View tab ➢ User Interface panel, choose Toolbars ➢ CIVIL ➢ Transparent Commands.

FIGURE 7.19
The Transparent Commands toolbar

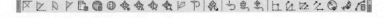

3. From the Home tab ➢ Create Design panel, choose Profile ➢ Profile Creation Tools.

4. Pick the Cabernet Court profile view (located at the lower right, inside a rectangle) by clicking on one of the grid lines to display the Create Profile – Draw New dialog.

5. Set the name to **Cabernet Court FG**.

THE ELEVATION ELEMENT | 313

6. On the General tab, set Profile Style to Design Profile and Profile Label Set to Complete Label Set. Click OK to display the Profile Layout Tools toolbar.

7. On the Profile Layout Tools toolbar, click the drop-down arrow next to the Draw Tangents button on the far left, and select the Draw Tangents With Curves option, as in the previous exercise.

8. Use a Center Osnap to snap to the center of the circle near the left edge of the profile view.

   Unlike in the previous exercise, this profile view is set up left to right.

9. On the Transparent Commands toolbar, select the Profile Station Elevation transparent command.

   For those who prefer using the command line, the key-in command for this transparent command is **'PSE**.

10. When prompted to select a profile view, click a grid line on the Cabernet Court profile view to select it.

    If you move your cursor within the profile grid area, a vertical red line appears. Notice that the tooltip currently shows the station value of the cursor.

11. When prompted for a station, enter **365** ↵ (or **112.8** ↵ for metric users) at the command line.

    If you move your cursor within the profile grid area, a horizontal line appears (see Figure 7.20), but it can only move vertically along the station just specified.

FIGURE 7.20
Using the Profile Station Elevation transparent command

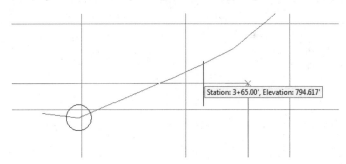

12. When prompted to specify an elevation, enter **795** ↵ (or **242.5** ↵ for metric users) at the command line to set the elevation for the second PVI.

13. Press Esc only once.

    The Profile Station Elevation transparent command is no longer active, but the Draw Tangents With Curves button that you previously selected on the Profile Layout Tools toolbar continues to be active.

14. When prompted to specify a point, select the Profile Grade Station transparent command on the Transparent Commands toolbar.

    For those who prefer using the command line, the key-in command for this transparent command is **'PGS**.

Notice that you did not need to select a profile view this time; that's because you are still in the same command (Draw Tangents With Curves in this case). The transparent command will default to the same profile view that was previously selected.

15. When prompted to specify the grade, enter **10** ↵ at the command line for the profile grade.

16. When prompted for the station, enter **430** ↵ (or **131** ↵ for metric users) at the command line.

17. Press Esc only once to deactivate the Profile Grade Station transparent command.

18. When prompted to specify the grade, select the Profile Grade Length transparent command on the Transparent Commands toolbar.

    For those who prefer using the command line, the key-in command for this transparent command is **'PGL**.

19. When prompted to specify the grade, Enter **-1.5** ↵ at the command line for the profile grade.

20. Enter **120** ↵ (or **36.5** ↵ for metric users) for the profile grade length.

21. Press Esc only once to deactivate the Profile Grade Length transparent command.

22. Continue defining the profile using the Profile Grade Station transparent command and the following input:

    a. **-3.5** ↵, **850** ↵ (**-3.5** ↵, **259** ↵ for metric users)

    b. **10** ↵, **1035** ↵ (**10** ↵, **315.5** ↵ for metric users)

    c. **3.5** ↵, **1165** ↵ (**3.5** ↵, **355** ↵ for metric users)

23. Press Esc only once to deactivate the Profile Grade Station transparent command and to continue using the Draw Tangent With Curves command.

24. Use a Center Osnap to select the center of the circle along the far-right side of the profile view.

25. Press ↵ to complete the profile.

    Your profile should look like Figure 7.21.

26. Close the Profile Layout Tools toolbar.

FIGURE 7.21
A layout profile created using the Transparent Commands toolbar

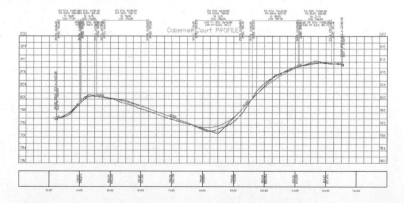

When this exercise is complete, you may close the drawing. A finished copy of this drawing is available from the book's web page with the filename `LayoutByPVITransparent_FINISHED.dwg` or `LayoutByPVITransparent_METRIC_FINISHED.dwg`.

Using PVIs to define tangents and fitting curves between them is the most common approach to create a layout profile, but you'll look at an entity-based design in the next section.

Layout by Entity

In this exercise you will lay out a design profile using the concepts of fixed, floating, and free entities in much the same way that you used them for laying out alignments in Chapter 6:

1. Open the `LayoutByEntity.dwg` file or the `LayoutByEntity_METRIC.dwg` file.
2. From the Home tab ➢ Create Design panel, choose Profile ➢ Profile Creation Tools.
3. Pick the Frontenac Drive profile view (located at the lower right, inside a rectangle) by clicking one of the grid lines to display the Create Profile – Draw New dialog.
4. For Name enter **Frontenac Drive FG**.
5. On the General tab, set Profile Style to Design Profile and Profile Label Set to Complete Label Set; then click OK to display the Profile Layout Tools toolbar.

> **Oops, You Closed the Profile Layout Tools Toolbar!**
>
> If at any point you inadvertently close the Profile Layout Tools toolbar, have no fear. You can reopen it by selecting the profile that you were editing and selecting Geometry Editor from the Profile contextual tab. If you created the profile but there aren't any entities to select, you can select it in Prospector by expanding the Alignments ➢ Centerline Alignments ➢ *Alignment Name* ➢ Profiles branch. Right-click the profile and choose Delete to start again from scratch. Unfortunately there isn't a Select option as you have with alignments, so deleting the profile is the only option.

6. On the Profile Layout Tools toolbar, click the drop-down arrow next to the Tangent Creation button, and select the Fixed Tangent (Two Points) option.
7. Using a Center Osnap, pick the circle at the left edge of the profile view labeled A.

    A rubber-band line appears.

8. Using a Center Osnap, pick the circle labeled B.

    A tangent is drawn between these two circles.

9. Using a Center Osnap, pick the circle labeled B again as the start point and the circle labeled C as the endpoint.

A tangent is drawn between these two circles. Notice that the tangent does not automatically continue from the previous two-point fixed tangent; therefore, you have to select the B circle again.

10. Using a Center Osnap, pick the circle labeled D as the start point and the circle labeled E as the endpoint.

   A tangent is drawn between these two circles.

11. On the Profile Layout Tools toolbar, click the drop-down arrow next to the Vertical Curve Creation button and select the More Free Vertical Curves ➤ Free Vertical Parabola (PVI Based) option.

   Notice that the image shown to the left of the drop-down arrow for the Tangent Creation button and Vertical Curve Creation button will match the last type of entity you selected from the drop-down menu.

12. At the `Pick point near PVI or curve to add curve:` prompt, pick the circle labeled B as the PVI.

13. At the `Specify curve length or [Passthrough K]:` prompt, enter **450** ↵ (or **137** ↵ for metric users) as the curve length.

14. Press ↵ to end the command.

   Your drawing should now look similar to Figure 7.22. Notice that although you have added the tangent between D and E, it is not yet labeled since it is not connected with the main portion of the profile created up until this point.

FIGURE 7.22
Some tangent and vertical curve entities placed on Frontenac Drive

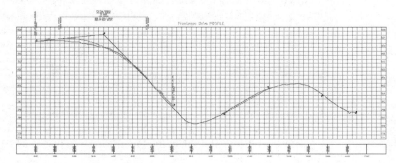

15. On the Profile Layout Tools toolbar, click the drop-down arrow next to the Vertical Curve Creation button and select the Free Vertical Curve (Parameter) option.

16. When prompted to select the first entity, click the tangent between B and C. Then click the tangent between D and E as the next entity.

   Remember to pick the tangent line and not an end circle.

17. At the `Specify curve length or [Radius K]:` prompt, enter **300** ↵ (or **90** ↵ for metric users) as the curve length and press ↵ again to end the command.

Notice that with this command the tangents do not have to meet at a PVI, unlike the previous Free Vertical Curve (PVI Based) curve.

18. On the Profile Layout Tools toolbar, click the drop-down arrow next to the Vertical Curve Creation button and select the More Fixed Vertical Curves ➢ Fixed Vertical Curve (Entity End, Through Point) option.

19. At the `Select entity to attach to:` prompt, select the tangent between D and E to attach the fixed vertical curve.

Remember to pick the tangent line and not the end circle. Also, you have to select the tangent between the midpoint and the endpoint of the tangent at the circle labeled E. Selecting too close to the endpoint at the circle labeled D will give a result of `End of selected entity already has an attachment`. A rubber-band curve appears. If you move the cursor on the wrong side of the tangent endpoint, it will become a large red circle with an X across it indicating that you cannot select that point.

20. At the `Specify end point:` prompt, using a Center Osnap, pick the circle labeled F and press ↵ to end the command.

21. On the Profile Layout Tools toolbar, click the drop-down arrow next to the Tangent Creation button and select the Float Tangent (Through Point) option.

22. At the `Select entity to attach to:` prompt, select the curve between E and F to attach the floating tangent.

A rubber-band line appears.

23. At the `Select through point:` prompt, using a Center Osnap, select the circle labeled G.

24. Press ↵ or right-click to end the Fixed Tangent (Through Point) command; then close the Profile Layout Tools toolbar.

Your drawing should look like Figure 7.23.

FIGURE 7.23
Completed profile built using entities

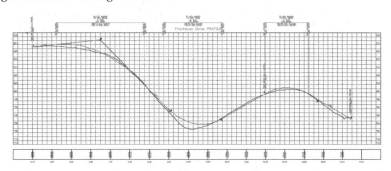

When this exercise is complete, you may close the drawing. A finished copy of this drawing is available from the book's web page with the filename `LayoutByEntity_FINISHED.dwg` or `LayoutByEntity_METRIC_FINISHED.dwg`.

With the entity-creation method, grip editing works in a similar way to other layout methods based on the fixed, floating, and free constraints.

Profile Layout Tools

Certification Objective

Although we have touched on many of the available tools in the Profile Layout Tools toolbar, shown in Figure 7.24, there are still many that we have not.

FIGURE 7.24
Profile Layout Tools toolbar

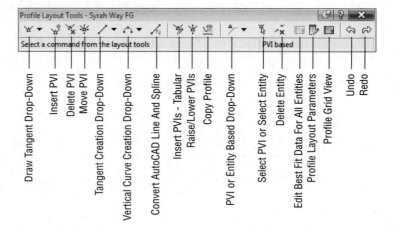

Draw Tangent Drop-Down The Draw Tangent drop-down button contains three options: Draw Tangents, Draw Tangents With Curves, and Curve Settings. Draw Tangents lays out a profile point to point with no curves. Draw Tangent With Curves lays out a profile from point to point with the curve type and length determined from the Curve Settings options. Curve Settings sets the type of curve (parabolic, circular, asymmetric) along with specific geometric properties that define each type of curve.

Insert PVI The Insert PVI button adds a new PVI at the specified location, consequently breaking an existing tangent and generating two new tangents connected to the new PVI.

Delete PVI The Delete PVI button removes an existing PVI at the specified location, consequently taking two tangents and replacing them with a single tangent.

Move PVI The Move PVI button allows you to select an existing PVI and relocate it to a specified location while keeping the two existing tangents. You can get the same result by grip-editing the PVI with the vertical triangular grip, as described earlier. In fact, one might argue that the grip-editing approach is better because it shows you a preview of your edit as you make it. The Move PVI command does not.

Tangent Creation Drop-Down The Tangent Creation Drop-down button contains six tools:

- Fixed Tangent (Two Points)
- Fixed Tangent – Best Fit
- Float Tangent (Through Point)
- Float Tangent – Best Fit
- Free Tangent
- Solve Tangent Intersection

The fixed, float, and free tools are consistent with those discussed in Chapter 6 when we were generating alignments, and therefore many of these options should be self-explanatory. The Solve Tangent Intersection option extends two tangents that do not currently connect to form a PVI.

Vertical Curve Creation Drop-Down The Vertical Curve Creation Drop-down button contains 14 options:

- Fixed Vertical Curve (Three Points)
- Fixed Vertical Curve (Two Points, Parameter)
- Fixed Vertical Curve (Entity End, Through Point)
- Fixed Vertical Curve (Two Points, Grade At Start Point)
- Fixed Vertical Curve (Two Points, Grade At End Point)
- Fixed Vertical Curve – Best Fit
- Floating Vertical Curve (Parameter, Through Point)
- Floating Vertical Curve (Through Point, Grade)
- Floating Vertical Curve – Best Fit
- Free Vertical Curve (Parameter)
- Free Vertical Parabola (PVI Based)
- Free Asymmetrical Parabola (PVI Based)
- Free Circular Curve (PVI Based)
- Free Vertical Curve – Best Fit

Once again the fixed, floating, and free terminology should be familiar from Chapter 6. By trying these various options, you will become comfortable with their capabilities and you will find the ones that best fit your design needs.

Convert AutoCAD Line And Spline The Convert AutoCAD Line And Spline button takes a singular line/spline and converts it into a profile object, either a tangent or a three-point vertical curve, as applicable.

Insert PVIs – Tabular The Insert PVIs – Tabular button allows you to enter PVI station and elevation information in a table-like dialog, which is helpful if you want to create multiple PVIs at once using station and elevation information. This table allows you to insert PVIs and curves anywhere geometrically possible in the profile. You are not required to enter the PVIs in any specific order when using this method of entry.

Raise/Lower PVIs The Raise/Lower PVIs button allows you to raise or lower the entire profile or a subset of PVIs within a specified station range. This button will be discussed in a later exercise in the section "Other Profile Edits."

Copy Profile The Copy Profile button allows you to copy either the entire profile or a portion of the profile within a specified station range. This button will be discussed later in the section "Other Profile Edits."

PVI or Entity Based Drop-Down The PVI or Entity Based drop-down button allows you to choose to select and display profile layout parameters based on either PVI or entity.

Select PVI or Select Entity The Select PVI or Select Entity button opens the Profile Layout Parameters dialog for the selected PVI or Entity.

Delete Entity The Delete Entity button removes a selected curve or tangent.

Edit Best Fit Data For All Entities The Edit Best Fit Data For All Entities button turns on the display of a table of the regression data for a profile that was created by best fit. A discussion of best fit profiles is provided in the next section.

Profile Layout Parameters The Profile Layout Parameters button opens the Profile Layout Parameters dialog, which shows numeric data for editing the selected entity or PVI.

Profile Grid View The Profile Grid View button opens the Profile Entities tab in Panorama, showing information about all the entities and PVIs in the profile. This is where you have access to make edits on all the entities of the profile.

Undo/Redo The Undo button reverses that last command and the Redo button reverses the last undo operation. This includes commands and operations that are not part of creating or editing a profile.

The Best Fit Profile

You've surveyed along a centerline, and you need to closely approximate the tangents and vertical curves as they were originally designed and constructed.

The Create Best Fit Profile option is found in the Home tab ➢ Create Design panel, on the Profile drop-down. Once you select a profile view, the Create Best Fit Profile dialog appears (Figure 7.25).

FIGURE 7.25
Create Best Fit Profile dialog

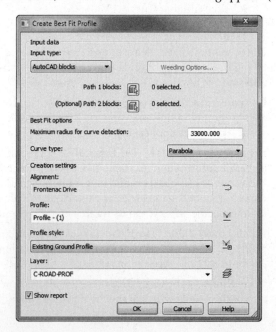

Similar to the best fit alignment we discussed in Chapter 6, a best fit profile can be based on an Input Type setting of AutoCAD Blocks, AutoCAD 3D Polylines, AutoCAD Points, COGO Points, Surface Profile, or Feature Lines. The most common option is Surface Profile.

The command attempts to run a complex algorithm to determine the best fit profile, including both tangents and vertical curves. However, the only best fit option available for determining vertical curves is the maximum curve radius. The maximum curve radius does not apply in a design that uses parabolic curves, the most common curves found in roadway design. Once the analysis is run, a Best Fit Report is provided; however, unlike the Best Fit command for lines and curves as seen in Chapter 1, "The Basics," this command has no options for selecting or deselecting points.

Creating a Profile from a File

Working with profile information in the AutoCAD Civil 3D environment is nice, but it isn't the only place where you can create or manipulate this sort of information. Many programs and analysis packages generate profile information. One common case is the plotting of a hydraulic grade line against a stormwater network profile of the pipes. When information comes from outside the program, it is often output in a variety of formats. If you convert this data to a text file in the format required by AutoCAD Civil 3D, the profile information can be imported directly.

There is a specific format that is required for creating a profile from a text file. Each line is a PVI definition (station and elevation) listed in ascending order. The station should not include the plus character (use 100, not 1+00 or 0+100). Curve information is an optional third bit of data on any line except for the first and last lines in the file. The vertical curve that is created will be a parabolic curve, which is the most popular type of vertical curve. Note that each line is space delimited. Here's one example of a profile text file:

```
0 550.76
127.5 552.24
200.8 554 100
256.8 557.78 50
310.75 561
```

In this example, the third and fourth lines include the curve length as the optional third piece of information. The only inconvenience of using this input method is that the information in Civil 3D doesn't directly reference the text file. Once the profile data is imported, no dynamic relationship exists with the text file, but other methods can be used to edit the profile once imported.

In this exercise, you'll import a small text file to see how the function works:

1. Open the `ProfilefromFile.dwg` file (or the `ProfilefromFile_METRIC.dwg` file).

2. From the Home tab ➢ Create Design panel, choose Profile ➢ Create Profile From File.

    The Import Profile From File – Select File dialog appears.

3. Browse to and select the `ProfileFromFile.txt` file (or the `ProfileFromFile_METRIC.txt` file for metric users), and click Open to display the Create Profile – Draw New dialog.

4. For Alignment, choose Frontenac Drive, and set Name to **Frontenac Drive FG**.

5. On the General tab, set Profile Style to Design Profile and Profile Label Set to Complete Label Set; then click OK.

Your drawing should look like Figure 7.26. The Frontenac profile view is located at the lower right, inside a rectangle.

FIGURE 7.26
Completed profile created from a file

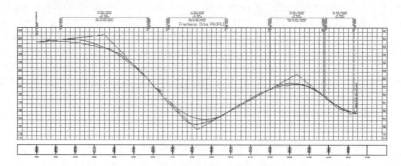

When this exercise is complete, you may close the drawing. A finished copy of this drawing is available from the book's web page with the filename `ProfileFromFile_FINISHED.dwg` or `ProfileFromFile_METRIC_FINISHED.dwg`.

Now that you've tried the three main ways of creating profiles, you'll edit a profile.

Editing Profiles

The methods just reviewed let you quickly create profiles. You saw how sampled profiles reflect changes in the surface along the parent alignment and how to lay out a design profile using a few different techniques. You also imported a text file with profile information. In all these cases, you just left the profile as originally designed with no analysis or editing.

In the following sections, you will begin to look at some of the profile editing methods available. The most basic is a more precise grip-editing methodology, which you'll learn about first. Then you'll see how to modify the PVI-based layout profile, how to change out the components that make up a layout profile, and how to use some other miscellaneous editing functions.

Grip-Editing Profiles

Once a profile layout is in place, sometimes a simple grip edit will suffice. But for precision editing, you can use the grips along with transparent commands or dynamic input, as in this short exercise:

1. Open the `GripEditingProfiles.dwg` file (or the `GripEditingProfiles_METRIC.dwg` file).

2. Zoom to the Cabernet Court profile view (located inside a rectangle) and pick the Cabernet Court FG profile (the blue line) to activate its grips.

3. Locate the PVI around Sta. 8+50 (or 0+260 for metric users) and pick the vertical triangular grip on the vertical sag curve to begin a grip stretch of the PVI, as shown in Figure 7.27.

FIGURE 7.27
Grip-editing a PVI

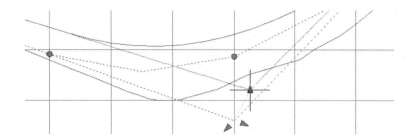

The command line states Specify stretch point or [Base point Copy Undo eXit]:.

4. On the Transparent Commands toolbar, select the Profile Station Elevation command. That's **'PSE** for the command-line users.

5. At the Select a profile view: prompt, pick a grid line on the Cabernet Court profile view to select.

6. At the Specify station: prompt, enter **835** ↵ (or **255** ↵ for metric users).

7. At the Specify elevation: prompt, enter **788** ↵ (or **240.5** ↵ for metric users).

8. Click the vertical triangular grip for the PVI near station 10+35 (or 0+315 for metric users).

9. If dynamic input is not turned on already, click the Dynamic Input icon at the bottom of your screen.

   You should see two editable tooltips on your screen, one for station and one for elevation. You may need to zoom out to see them.

10. Press the Tab key to change from editing the station tooltip to editing the elevation tooltip and enter **810** ↵ (or **247** ↵ for metric users).

When this exercise is complete, you may close the drawing. A finished copy of this drawing is available from the book's web page with the filename GripEditingProfiles_FINISHED.dwg or GripEditingProfiles_METRIC_FINISHED.dwg.

The grips can go from quick-and-dirty editing tools to precise editing tools when you use them in conjunction with the transparent commands or dynamic input. They lack the ability to precisely control a curve length, though, so you'll look at editing a curve next.

Editing Profiles using Profile Layout Parameters

Beyond the simple grip edits, but before changing out the components of a typical profile, you can modify the values that generate an individual component. In this exercise, you'll use the Profile Layout Parameters dialog to modify the curve properties on your design profile:

1. Open the ParameterEditingProfiles.dwg file (or the ParameterEditingProfiles_METRIC.dwg file).

2. Zoom to the Cabernet Court profile view (located inside a rectangle) and pick the Cabernet Court FG profile (the blue line) to activate the Profile contextual tab.

3. From the Profile contextual tab ➢ Modify Profile panel, choose Geometry Editor.

4. On the Profile Layout Tools toolbar, click the Profile Layout Parameters button to open the Profile Layout Parameters dialog and place the dialog somewhere on your screen so that you can still see the profile view.

5. On the Profile Layout Tools toolbar, click the Select PVI button.

   If the Select Entity button is showing on the toolbar instead, from the PVI or Entity Based drop-down button select the PVI Based option.

6. Zoom in to click near the PVI at station 11+65 (or 0+355 for metric users) to populate the Profile Layout Parameters dialog (Figure 7.28).

FIGURE 7.28
The Profile Layout Parameters dialog

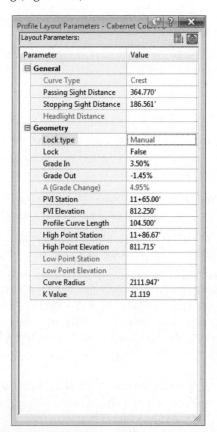

Values that can be edited are in black; the rest, shown grayed out, are mathematically derived and can be of some design value but can't be directly modified. The two buttons at the top of the dialog adjust how much information is displayed. The one on the left is the Show More/Show Less button and the one on the right is the Collapse All Categories/Expand All Categories button.

7. In the Profile Layout Parameters dialog, change the K value to **19** ↵ (or **7** ↵ for metric users). Notice that the curve changes but the label does not update. This is because you are still in the command. Once you end the command, all appropriate labels will update.

8. Click near the PVI at station 10+35 (or 0+315.5 for metric users) to repopulate the Profile Layout Parameters dialog with a different curve.

9. In the Profile Layout Parameters dialog, change Profile Curve Length to **130** ↵ (or **42** ↵ for metric users) and press ↵ or right-click to end the command and apply the changes to the profile. Since you've ended the command, the labels will now update.

10. Close the Profile Layout Parameters dialog by clicking the X in the upper-right corner.

Keep the file open for the next exercise.

Editing Profiles using Profile Grid View

In this exercise, you'll use the Profile Grid View command to view and modify the profile within Panorama:

1. Using the file from the previous exercise, on the Profile Layout Tools toolbar, click the Profile Grid View tool to activate the Profile Entities tab in Panorama.

   If you have closed the Profile Layout Tools toolbar, click the Cabernet Court FG profile, then from the Profile: Cabernet Court contextual tab ➤ Modify Profile panel, click Geometry Editor.

   Panorama allows you to view all the profile components at once, in a compact form.

2. Scroll right in Panorama until you see the Profile Curve Length column.

   You can show and hide columns by right-clicking one of the column headings. You can also resize the columns by dragging on the breaks between the columns or by double-clicking the break between two columns to auto-size to the column contents.

3. Double-click the Profile Curve Length value for Entity No. 5 (see Figure 7.29) and change the value from 150′ to **250′** (or from 45 to **78** for metric users).

FIGURE 7.29
Direct editing of the curve length in Panorama

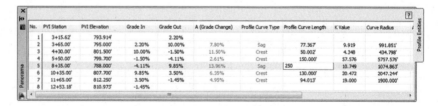

4. Close Panorama and the Profile Layout Tools toolbar, and zoom out to review your edits.

   Your complete profile should now look like Figure 7.30.

FIGURE 7.30
The completed editing of the curve length in the layout profile

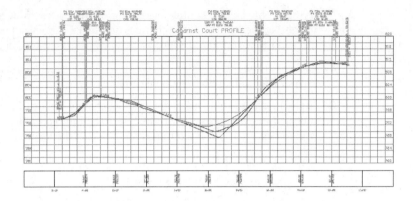

When this exercise is complete, you may close the drawing. A finished copy of this drawing is available from the book's web page with the filename `ParameterEditingProfiles_FINISHED.dwg` or `ParameterEditingProfiles_METRIC_FINISHED.dwg`.

You can use these tools to modify the PVI points or tangent parameters, but they won't let you add or remove an entire component. You'll do that in the next section.

> **PVIs in Lockdown**
>
> You may have noticed while editing the profile in Panorama that the last column is named Lock. You can also lock a PVI at a specific station and elevation in the Profile Layout Parameters dialog. PVIs that are locked cannot be moved with edits to adjacent entities. However, it's important to note that a PVI can be unlocked by simply clicking the lock icon in the profile view.

Component-Level Editing

In addition to editing basic parameters and locations, sometimes you have to add or remove entire components. In this exercise, you'll delete a PVI, remove a curve from an area in order to revise a nearby PVI location, and insert a new curve into the layout profile:

1. Open the `ComponentEditingProfiles.dwg` file (or the `ComponentEditingProfiles_METRIC.dwg` file).

2. Zoom to the Cabernet Court profile view (located inside a rectangle) and pick the Cabernet Court FG profile (the blue line) to activate the Profile contextual tab.

3. From the Profile contextual tab ➢ Modify Profile panel, choose the Geometry Editor to display the Profile Layout Tools toolbar.

4. On the Profile Layout Tools toolbar, click the Delete PVI button.

5. Pick a point near the 4+30 station (or the 0+131 station for metric users) and right-click or press ↵ to end the command.

The profile is adjusted accordingly. The vertical curves that were in place are also modified to accommodate this new geometry.

6. On the Profile Layout Tools toolbar, click the Delete Entity button.

7. Zoom in, and pick the curve entity near station 3+65 (or 0+112.8 for metric users) to delete it and right-click or press ↵ to end the command.

   Notice that the incoming and outgoing tangents remain.

8. Using one of the methods discussed previously, change the PVI currently at station 5+50 (or 0+167.5 for metric users) to PVI Station **4+70** and PVI Elevation **802.30** (or PVI Station **0+143** and PVI Elevation **244.650** for metric users).

9. On the Profile Layout Tools toolbar, click the drop-down arrow next to the Vertical Curve Creation button.

10. Select the More Free Vertical Curves ➢ Free Vertical Parabola (PVI Based) option.

11. When prompted to pick a point, click near the PVI where you removed the vertical curve in step 8 to re-add a curve.

12. At the Specify curve length or [Passthrough K]: prompt, enter **50** ↵ (or **25** ↵ for metric users) at the command line to set the curve length.

13. Right-click or press ↵ to end the command and update the profile display.

14. Close the Profile Layout Tools toolbar.

    Your drawing should look similar to Figure 7.31.

FIGURE 7.31
The completed editing of the curve using component-level editing

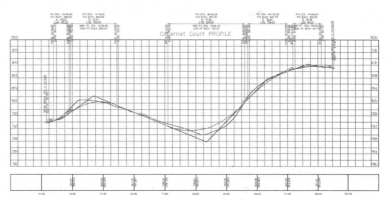

When this exercise is complete, you may close the drawing. A finished copy of this drawing is available from the book's web page with the filename ComponentEditingProfiles_FINISHED.dwg or ComponentEditingProfiles_METRIC_FINISHED.dwg.

Editing profiles using any of these methods gives you precise control over the creation and layout of your vertical design. In addition to these tools, some of the tools on the Profile Layout Tools toolbar are worth investigating and somewhat defy these categories. You'll look at them next.

Other Profile Edits

Some handy tools exist on the Profile Layout Tools toolbar for performing specific actions. These tools aren't normally used during the preliminary design stage, but they come into play as you're working to create a final design for grading or corridor design. They include raising or lowering a whole layout in one shot, as well as copying profiles. Try this exercise:

1. Open the OtherProfileEdits.dwg file (or the OtherProfileEdits_METRIC.dwg file).

2. Zoom to the Cabernet Court profile view (located inside a rectangle) and pick the Cabernet Court FG profile (the blue line) to activate the Profile contextual tab.

3. From the Profile contextual tab ➢ Modify Profile panel, choose the Geometry Editor to display the Profile Layout Tools toolbar.

4. On the Profile Layout Tools toolbar, click the Copy Profile button to display the Copy Profile Data dialog, shown in Figure 7.32.

FIGURE 7.32
The Copy Profile Data dialog

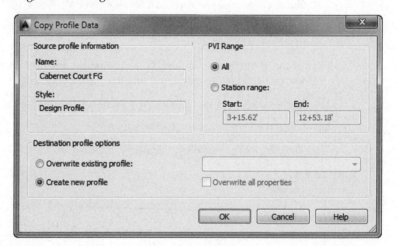

5. Click OK to create a new layout profile directly on top of Cabernet Court FG.

6. In Prospector, expand the Alignments ➢ Centerline Alignments ➢ Cabernet Court ➢ Profiles branch to see that a profile named Cabernet Court FG [Copy] has been added.

7. Press Esc to clear the selection of the original profile and then click the profile again to select it. This time the Cabernet Court FG [Copy] profile is selected because it is on top. The Profile Layout Tools toolbar now references the new profile.

8. On the Profile Layout Tools toolbar, click the Raise/Lower PVIs button to display the Raise/Lower PVI Elevation dialog, shown in Figure 7.33.

9. Set Elevation Change to **-1** (or **-0.3** for metric users).

FIGURE 7.33
The Raise/Lower PVI Elevation dialog

10. Click the Station Range radio button, set the Start value to **3+16** (or **0+097** for metric users), and leave the default End value to modify all the PVIs after the starting PVI.

11. Click OK to dismiss the Raise/Lower PVI Elevation dialog.

12. Close the Profile Layout Tools toolbar.

When this exercise is complete, you may close the drawing. A finished copy of this drawing is available from the book's web page with the filename OtherProfileEdits_FINISHED.dwg or OtherProfileEdits_METRIC_FINISHED.dwg.

Generating a copy is useful if you want to remember a conceptual profile layout but would like to experiment with a different layout. The copies do not stay dynamically related to one another.

Using the layout and editing tools discussed in these sections, you should be able to create profiles for many different types of designs.

Up to now, you have learned how to use some of the available tools for modifying profiles, but you might be wondering about intersecting roads and how their profiles will interact with one another. Have no fear; we are going to discuss those later in Chapter 10, "Advanced Corridors, Intersections, and Roundabouts," when we discuss corridor intersections.

Profile Views

Certification Objective

Working with vertical data is an integral part of building the model. Once profile information has been created in any number of ways, displaying it to make sense is another task. It can't be stated enough that profiles and profile views are not the same thing. The profile view displays the profile data. A single profile can be shown in an infinite number of views, with different grids, exaggeration factors, labels, or linetypes. In the following sections, you'll look at the various methods available for creating profile views.

Creating Profile Views during Sampling

The easiest way to create a profile view is to draw it as an extended part of the surface sampling procedure as shown in the first exercise in this chapter. By combining the profile sampling step with the creation of the profile view, you have avoided one more trip to the menus. This is the most common method of creating a profile view, but we'll look at a manual creation in the next section.

Creating Profile Views Manually

Once an alignment has profile information associated with it, any number of profile views might be needed to display the proper information in the right format. To create a 2nd, 3rd, or even 10th profile view once the sampling is done, you must use a manual creation method. In this exercise, you'll create a profile view manually for an alignment that already has a surface-sampled profile associated with it:

1. Open the `ProfileViews.dwg` file (or the `ProfileViews_METRIC.dwg` file).

2. From the Home tab ➢ Profile & Section Views panel, choose Profile View ➢ Create Profile View to display the Create Profile View Wizard.

   This is the same wizard that was discussed in the surface sampling example.

3. In the Select Alignment text box, select Cabernet Court from the drop-down list.

4. Set the Profile View name as **Cabernet Court Full Grid**.

5. In the Profile View Style drop-down list, select the Full Grid style.

6. Click the Create Profile View button and pick a point onscreen to draw the profile view, as shown in Figure 7.34.

FIGURE 7.34
The completed profile view of Cabernet Court using the Full Grid profile view style

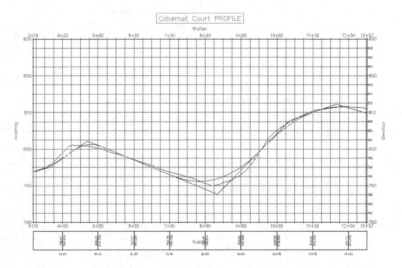

When this exercise is complete, you may close the drawing. A finished copy of this drawing is available from the book's web page with the filename `ProfileViews_FINISHED.dwg` or `ProfileViews_METRIC_FINISHED.dwg`.

Using this creation method, you've made a short, simple profile view, but in the next exercise we will look at a longer alignment as well as some more of the options available in the Create Profile View Wizard.

> **SO YOU WANT TO DELETE A PROFILE VIEW?**
>
> Getting rid of a profile view is easy, but be careful not to inadvertently delete your profiles at the same time. If you use a crossing window when selecting the objects to delete, then you will also be deleting your profiles that hold the data shown in your profile view.
>
> The easiest way to get rid of a profile view is to click one of the grid lines to activate the profile view object and press the Delete key on your keyboard.
>
> Alternatively, you can delete the profile view by expanding the branches in Prospector until you see the profile view you want to get rid of and then right-clicking it and selecting the Delete option.

Splitting Views

Dividing up the data shown in a profile view can be time consuming. The Profile View Wizard is used for simple profile view creation, but the wizard can also be used to create manually limited profile views, staggered (or stepped) profile views, multiple profile views with gaps between the views, and stacked profiles (aka three-line profiles). You'll now look at these variations on profile view creation.

CREATING MANUALLY LIMITED PROFILE VIEWS

Continuous profile views like you made in the exercises prior to this point work well for design purposes, but they are often unusable for plotting or documentation purposes. In this exercise, you'll use the Profile View Wizard to create a manually limited profile view. This variation will allow you to control how long and how high each profile view will be, thereby making the views easier to plot:

1. Open the `ProfileViewsSplit.dwg` file (or the `ProfileViewsSplit_METRIC.dwg` file).

2. From the Home tab ➤ Profile & Section Views panel, choose Profile View ➤ Create Profile View to display the Create Profile View Wizard.

3. Verify that the Select Alignment drop-down list shows Frontenac Drive, Profile Name is set to **Frontenac Drive Limited Full Grid**, and Full Grid is selected in the Profile View Style drop-down list; then click Next.

4. On the Station Range wizard page, select the User Specified Range radio button.

5. Enter **0** for the start station and **800** for the end station (or **0** and **245** for metric users), as shown in Figure 7.35. It isn't necessary to include the + when entering station data.

FIGURE 7.35
The start and end stations for the user-specified profile view

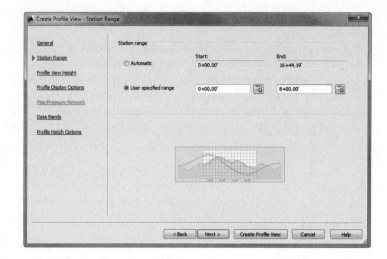

Notice that the preview picture now shows a clipped portion of the total profile.

6. Click Next.

7. On the Profile View Height wizard page, select the User Specified radio button.

8. Set the minimum height to **780** and the maximum height to **820** (or **238** and **250** for metric users). It isn't necessary to include the foot mark (') or m for meters when entering elevations.

9. Click the Create Profile View button and pick a point onscreen to draw the profile view.

Your screen should look similar to Figure 7.36. When this exercise is complete, you may save the drawing and keep it open to continue on to the next exercise. If you would like to view the result of this exercise, it is included in the finished drawing (along with the finished portions of the next two exercises) available from the book's web page (ProfileViewsSplit_FINISHED.dwg or ProfileViewsSplit_METRIC_FINISHED.dwg).

FIGURE 7.36
Applying user-specified station and height values to a profile view

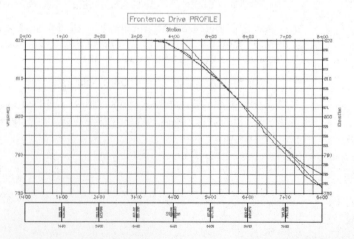

Creating Staggered Profile Views

When large variations occur in profile height, the profile view must often be split just to keep from wasting much of the page with empty grid lines. In this exercise, you use the Profile View Wizard to create a staggered, or stepped, view:

1. You may continue using the file from the previous exercise or start this exercise with the ProfileViewsSplit.dwg file (or the ProfileViewsSplit_METRIC.dwg file).

2. From the Home tab ➢ Profile & Section Views panel, choose Profile View ➢ Create Profile View to display the Create Profile View Wizard.

3. Verify that the Select Alignment drop-down list shows Syrah Way, Profile Name is set to **Syrah Way Staggered Full Grid**, and Full Grid is selected in the Profile View Style drop-down list; then click Next.

4. Verify that Station Range is set to Automatic to allow the view to show the full length, and click Next.

5. In the Profile View Height field, select the User Specified option and set the values to **790.00′** and **805.00′** (or **240** and **246** for metric users), as shown in Figure 7.37. These heights are only important in that they set the height of the profile view (15′ or 6 m in this case).

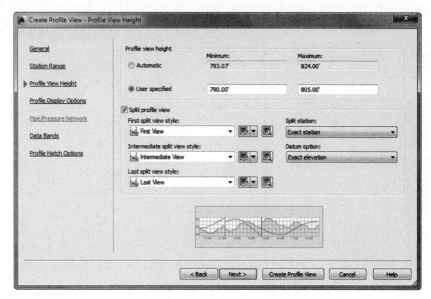

FIGURE 7.37
Split Profile View settings

6. Check the Split Profile View option and set the view styles to First View, Intermediate View, and Last View, as shown in Figure 7.37.

7. Click the Create Profile View button and pick a point onscreen to draw the staggered display, as shown in Figure 7.38.

FIGURE 7.38
A staggered (stepped) split profile view created via the wizard

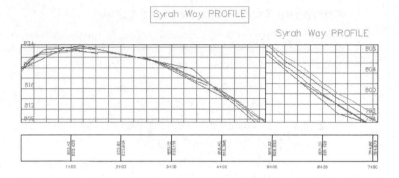

The profile view is split into views according to the settings that were selected in the Create Profile View Wizard in step 6. The first portion of the profile view shows the profile from 0 to the station where the elevation change of the profile exceeds the limit for height. If more splits were required, the command would have created them accordingly. Each of these portions is part of the same profile view and can be adjusted via the Profile View Properties dialog. In fact, the locations where the profile view splits can be controlled manually on the Elevations tab of the Profile View Properties dialog box. Here you can add and remove splits, control the station at which a given split occurs, and control the datum elevation at each split station (see Figure 7.39).

FIGURE 7.39
The Elevations tab of the Profile View Properties dialog showing manual control of a split profile view

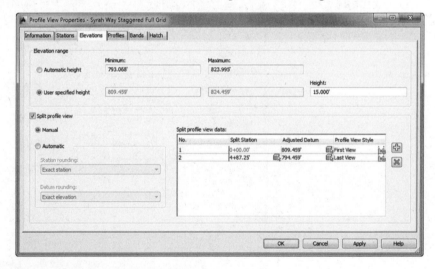

When this exercise is complete, you may save the drawing and keep it open to continue on to the next exercise. If you would like to view the result of this exercise, it is included in the finished drawing (along with the finished portions of the previous exercise and next exercise), available from the book's web page (ProfileViewsSplit_FINISHED.dwg or ProfileViewsSplit_METRIC_FINISHED.dwg).

Creating Gapped Profile Views

Profile views must often be limited in length and height to fit a given sheet size. Gapped views are a way to show the entire length and height of the profile, by breaking the profile into different sections with "gaps," or spaces, between each view.

When you are using the Plan Production tools (covered in Chapter 15, "Plan Production"), the gapped profile views are automatically created.

In this exercise, you will use a variation of the Create Profile View Wizard called the Create Multiple Profile Views Wizard to create gapped views automatically:

1. You may continue using the file from the previous exercise or start this exercise with the `ProfileViewsSplit.dwg` file (or the `ProfileViewsSplit_METRIC.dwg` file).

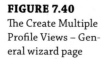

2. From the Home tab ➢ Profile & Section Views panel, choose Profile View ➢ Create Multiple Profile Views to display the Create Multiple Profile Views Wizard.

3. Verify that the Select Alignment drop-down list shows Frontenac Drive, Profile Name is set to **Frontenac Drive Gapped Full Grid**, and Full Grid is selected in the Profile View Style drop-down list, as shown in Figure 7.40; then click Next.

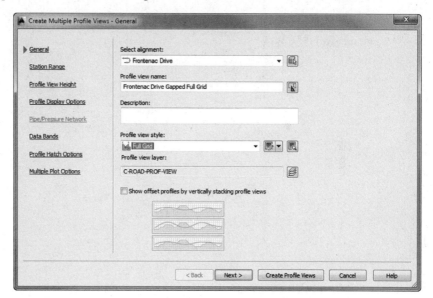

FIGURE 7.40
The Create Multiple Profile Views – General wizard page

4. On the Station Range wizard page, verify that the Automatic option is selected.

5. Set the length of each view to **650** (or **200** for metric users), and click Next.

6. On the Profile View Height wizard page, verify that the Automatic option is selected.

    Note that you could use the Split Profile View options from the previous exercise here as well if you use the User Specified profile view height.

7. Click Next.

8. On the Profile Display Options wizard page, scroll across until you get to the Labels column and verify that Style is set to _No Labels on both profiles; then click Next.

9. On the Data Bands wizard page, verify that the band set is EG-FG Elevations And Stations, and click the Multiple Plot Options link in the left sidebar of the wizard to jump ahead to that wizard page. We will look at the data bands in further depth a little later in this chapter.

   The Multiple Plot Options wizard page shown in Figure 7.41 is unique to the Create Multiple Profile Views Wizard. This wizard page controls whether the gapped profile views will be arranged in a column, a row, or a grid. The Frontenac Drive alignment is fairly short, so the gapped views will be aligned in a row. However, it could be prudent with longer alignments to stack the profile views in a column or a compact grid, thereby saving screen space.

FIGURE 7.41
The Create Multiple Profile Views – Multiple Plot Options wizard page

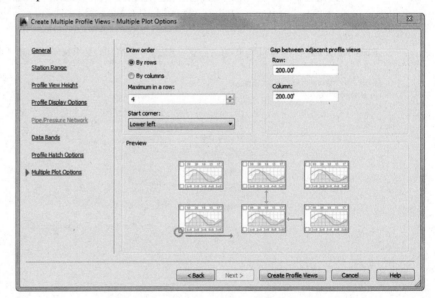

10. Click the Create Profile Views button and pick a point onscreen to create a view similar to Figure 7.42.

FIGURE 7.42
The staggered and gapped profile views of the Frontenac alignment

The gapped profile views are the three profile views on the bottom of the screen, and just like the staggered profile view, they show the entire alignment from start to finish. Unlike the staggered view, however, the gapped view is separated by a "gap" creating three individual profile views. In addition, the gapped profile views are independent of each other so they can be modified to have their own styles, properties, and labeling. This is also the primary way to create divided profile views for sheet production.

When this exercise is complete, you may close the drawing. A finished copy of this drawing showing the results of the previous three exercises is available from the book's web page with the filename ProfileViewsSplit_FINISHED.dwg or ProfileViewsSplit_METRIC_FINISHED.dwg.

Creating Stacked Profile Views

In some parts of the United States, a three-line profile view is a common requirement. In this situation, the centerline is displayed in a central profile view, with left and right offsets shown in profile views above and below the centerline profile view. These are then typically used to show top-of-curb design profiles in addition to the centerline design. In this exercise, you look at how the Create Profile View Wizard makes generating these stacked views a simple process:

1. Open the StackedProfiles.dwg file (or the StackedProfiles_METRIC.dwg file). This drawing has sampled profiles for the Syrah Way alignment at center as well as left and right offsets.

2. From the Home tab ➢ Profile & Section Views panel, choose Profile View ➢ Create Profile View to display the Create Profile View Wizard.

3. Verify that the Select Alignment drop-down list shows Syrah Way, Profile Name is set to **Syrah Way Stacked Full Grid**, and Full Grid is selected in the Profile View Style drop-down list.

4. Check the Show Offset Profiles By Vertically Stacking Profile Views option on the General wizard page.

    Notice that when you check this box, an additional link named Stacked Profile is added to the left sidebar of the wizard.

5. Click Next.

6. On the Create Profile View – Station Range wizard page, verify that Station Range is set to Automatic, and click Next.

7. On the Create Profile View – Profile View Height wizard page, verify that Profile View Height is set to Automatic, and click Next.

8. On the Create Profile View – Stacked Profile wizard page, set the gap between views to **15** (or **4** for metric users).

9. Set the view styles to Top Stacked View, Middle Stacked View, and Bottom Stacked View, as shown in Figure 7.43. Click Next.

FIGURE 7.43
The Create Profile View – Stacked Profile wizard page

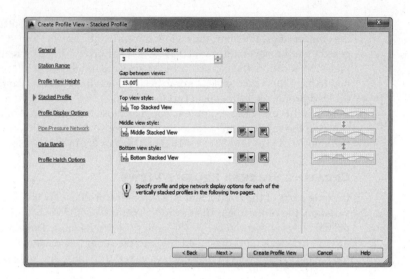

10. On the Create Profile View – Profile Display Options wizard page, select Top View in the Select Stacked View To Specify Options For list box.

11. Toggle the Draw option for the left offset profile (EG - Syrah Way -25.000 or EG - Syrah Way -7.500), as shown in Figure 7.44. If you need to widen the columns, you can do so by double-clicking the line between the column headings.

FIGURE 7.44
Setting the stacked view options for each view

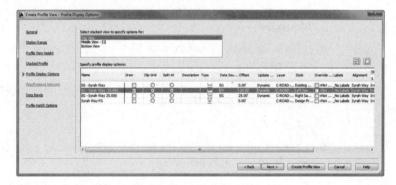

Remember that the negative offset denotes a left profile whereas a positive offset denotes a right profile.

12. Select Middle View - [1] in the Select Stacked View To Specify Options For list box.

13. Toggle the Draw option for the sampled centerline profile (EG - Syrah Way) as well as the layout centerline profile (Syrah Way FG).

14. Select Bottom View in the Select Stacked View To Specify Options For list box.

15. Toggle the Draw option for the right offset profile (EG - Syrah Way 25.000 or EG - Syrah Way 7.500), and click Next.

16. On the Data Bands wizard page, verify that the band set is set to EG-FG Elevations And Stations.

17. Click the Create Profile View button and pick a point on the screen to draw the stacked profiles, as shown in Figure 7.45.

FIGURE 7.45
Completed stacked profiles

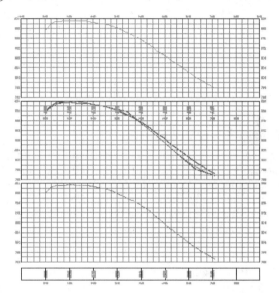

When this exercise is complete, you may close the drawing. A finished copy of this drawing is available from the book's web page with the filename StackedProfiles_FINISHED.dwg or StackedProfiles_METRIC_FINISHED.dwg.

Like the gapped profile views that you generated in a previous exercise, the profile views are independent of one another so they can be modified to have their own styles, properties, and labeling associated with them. The stacking here simply automates a process that many users previously found tedious. At this point you do not have finished grade information at the offsets, but you can add it to these views later by editing the Profile View Properties for those profile views.

When you create a profile, that profile will appear in any profile views that reference the same alignment. In the Profile View Properties dialog, you can always turn the Draw option off for any profile that should not appear in a given profile view.

> **STYLES: WHERE TO LOOK**
>
> The exercises in this chapter have many different styles created to show variety. You'll learn more about styles in Chapter 18, "Label Styles," and Chapter 19, "Object Styles." It's okay to take a peek ahead once in a while.

Editing Profile Views

The profile view is one of the most sophisticated and flexible objects in the AutoCAD Civil 3D package. After a profile view is created, many modifications can be made to it without modifying the style or assigning a different style. In this series of exercises, you'll look at a number of changes that can be applied to any profile view in a given drawing.

> **SUPERELEVATION VIEWS**
>
> Although not the focus of this chapter, superelevation views behave much like profile views. Once design speeds have been assigned to an alignment and superelevation has been calculated, you'll find the Create Superelevation View command on the Alignment contextual tab ➢ Modify panel. You can access their properties via a contextual tab after selecting a view.

Profile View Properties

Picking a profile view and then selecting Profile View Properties from the Profile View contextual tab ➢ Modify View panel yields the dialog shown in Figure 7.46. The properties of a profile include the style applied, station and elevation limits, the number of profiles displayed, the bands associated with the profile view, and any hatching that has been included. If a pipe network is displayed, a tab labeled Pipe Networks will appear. These tabs should look very similar to the links in the sidebar of the Create Profile View Wizard.

FIGURE 7.46
Typical Profile View Properties dialog

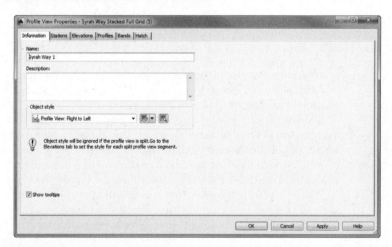

ADJUSTING THE PROFILE VIEW STATION LIMITS

There are often times when a profile view needs to be manually adjusted. For example, the most common change is to limit the length of the profile view that is being shown so it fits on a specific size of paper or viewport. You can make some of these changes during the initial creation of a profile view (as shown in a previous exercise), but you can also make changes after the profile view has been created.

One way to do this is to use the Profile View Properties dialog to make changes to the profile view. The profile view is an AutoCAD Civil 3D object, so it has properties and styles that can be adjusted in this dialog to make the profile view look like you need it to.

1. Open the ProfileViewProperties.dwg file (or the ProfileViewsProperties_METRIC.dwg file).

2. Zoom to the Cabernet Court Full Grid profile view (located inside a rectangle labeled "Cabernet Court Adjust Stations").

3. Pick a grid line, and from the Profile View contextual tab ➢ Modify View panel, choose Profile View Properties to display the Profile View Properties dialog.

4. On the Stations tab, click the User Specified Range radio button, and set the value of the end station to **700** (or **200** for metric users), as shown in Figure 7.47. Note that you do not need to type the + symbol.

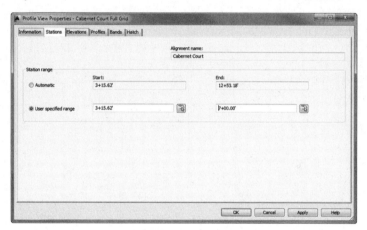

FIGURE 7.47
Adjusting the end station values for Cabernet Court

5. Click OK to dismiss the dialog.

   The profile view will now reflect the updated end station value.

   One of the nice things about Civil 3D is that copies of a profile view retain the properties of that view, making a gapped view easy to create manually if it was not created with the wizard.

6. Press F8 on your keyboard to enable Ortho mode.

7. Enter **Copy** ↵ on the command line. Pick the Cabernet Court profile view you just modified. Make sure you are selecting the grid representing the profile view and not the linework that represents the profile. Press Enter to complete the selection.

8. Pick a base point and move the crosshairs to the right.

9. When the crosshairs reach a point where the two profile views do not overlap, pick that as your second point, and press ↵ to end the Copy command.

10. Pick a grid line on the copy just created, and from the Profile View contextual tab ➢ Modify View panel, choose Profile View Properties to display the Profile View Properties dialog.

11. On the Stations tab, set the Start field to **700** (or **200** for metric users) and the End field to **1253.18** (or **381.94** for metric users).

12. Click OK to dismiss the dialog.

    The total length of the alignment will now be displayed on the two profile views, with a gap between the two views at station 7+00 (or 0+200 for metric users).

    You may want to move the copied profile view since it held the station location, thus shifting it to the right.

    Once this exercise is complete, your drawing will look like Figure 7.48.

FIGURE 7.48
A manually created gap between profile views

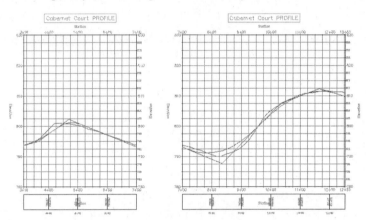

When this exercise is complete, you may save the drawing and keep it open to continue on to the next exercise. If you would like to view the result of this exercise, it is included in the finished drawing (along with the finished portions of the next two exercises) available from the book's web page (`ProfileViewProperties_FINISHED.dwg` or `ProfileViewProperties_METRIC_FINISHED.dwg`).

In addition to creating gapped profile views by changing the profile view properties, you could show phase limits by applying a different style to the profile in the second view.

ADJUSTING THE PROFILE VIEW ELEVATIONS

Another common issue is the need to control the height of the profile view. Civil 3D automatically sets the datum and the top elevation of profile views on the basis of the data to be displayed. In most cases this is adequate, but in others, this simply creates a view too large for the space allocated on the sheet or does not provide the adequate room for layout PVIs to be placed.

1. You may continue using the file from the previous exercise or start this exercise with the `ProfileViewProperties.dwg` file (or the `ProfileViewProperties_METRIC.dwg` file).

2. Zoom to the Syrah Way Full Grid profile view (located inside a rectangle labeled "Syrah Way: Adjust Elevation Step 2").

3. Pick a grid line, and from the Profile View contextual tab ➢ Modify View panel, choose Profile View Properties to display the Profile View Properties dialog.

4. On the Elevations tab, in the Elevation Range section, check the User Specified Height radio button and set the maximum height to **830** (or **255** for metric users), as shown in Figure 7.49.

FIGURE 7.49
Modifying the height of the profile view

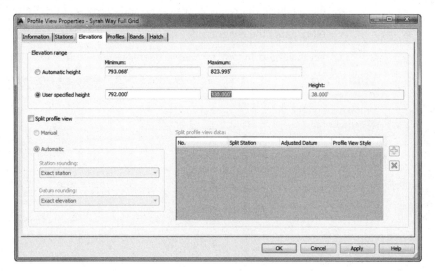

5. Click OK to dismiss the dialog.

The profile view of Syrah Way should reflect the updated elevations, as shown in Figure 7.50.

FIGURE 7.50
The updated profile view with the heights manually adjusted

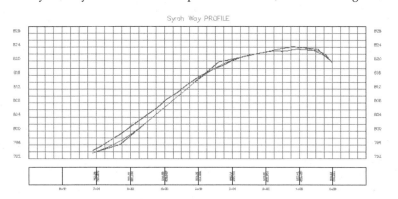

The Elevations tab can also be used to split the profile view and create the staggered view that you previously created with the wizard.

6. Pick the Syrah Way Full Grid (1) profile view (located inside a rectangle labeled "Syrah Way: Adjust Elevations Step 6").

7. Select a grid line, then click Profile View Properties on the Modify View panel of the Profile View contextual tab.
8. On the Profile View Properties dialog, switch to the Elevations tab.
9. In the Elevations Range area, click the User Specified Height radio button.
10. Check the Split Profile View option, and verify that the Automatic radio button is selected.

    Notice that the Height field is now active.
11. Set Height to **16** (or **7** for metric users), as shown in Figure 7.51.

FIGURE 7.51
Defining a split profile view on the Elevations tab

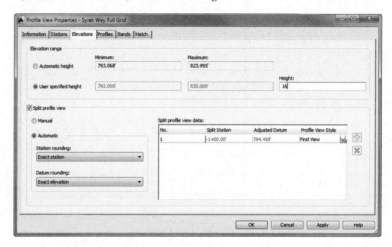

12. Click OK to exit the dialog.
13. Enter **REGEN** on the command line.

    The profile view should look similar to Figure 7.52.

FIGURE 7.52
A split profile view for the Syrah Way alignment

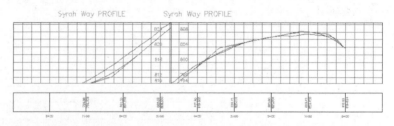

When this exercise is complete, you may save the drawing and keep it open to continue on to the next exercise. If you would like to view the result of this exercise, it is included in the finished drawing (along with the finished portions of the previous exercise and next exercise) available from the book's web page (`ProfileViewProperties_FINISHED.dwg` or `ProfileViewProperties_METRIC_FINISHED.dwg`).

Automatically creating split views is a good starting point, but you'll often have to tweak them as you've done here. The selection of the proper profile view styles is an important part of the Split Profile View process. We'll look at object styles in Chapter 19.

USING PROFILE DISPLAY OPTIONS

AutoCAD Civil 3D allows the creation of literally hundreds of profiles for any given alignment, which makes it easy to evaluate multiple design solutions, but it can also mean that profile views get very crowded. In this exercise, you'll look at some profile display options that allow the toggling of various profiles within a profile view:

1. You may continue using the file from the previous exercise or start this exercise with the ProfileViewProperties.dwg file (or the ProfileViewProperties_METRIC.dwg file).

2. Pick the left Cabernet Court Full Grid profile view from the adjusting stations exercise (located inside a rectangle labeled "Cabernet Court Adjust Stations") to activate the Profile View contextual tab.

3. From the Profile View contextual tab ➤ Modify View panel, choose Profile View Properties to display the Profile View Properties dialog.

4. Switch to the Profiles tab.

5. Uncheck the Draw option in the EG - Cabernet Court row and click OK.

Your profile view should look similar to Figure 7.53.

FIGURE 7.53
The Cabernet Court profile view with the Draw option toggled off for the EG profile

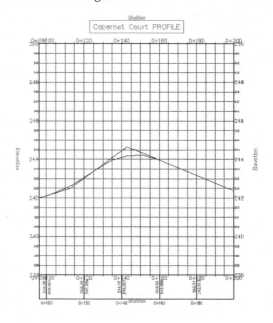

Toggling off the Draw option for the EG surface has created a profile view in which a profile of the existing ground surface will not be drawn.

The sampled profile from the EG surface still exists under the Cabernet Court alignment; it simply isn't shown in the current profile view.

When this exercise is complete, you may close the drawing. A finished copy of this drawing showing the results of the previous three exercises is available from the book's web page with the filename ProfileViewProperties_FINISHED.dwg or ProfileViewProperties_METRIC_FINISHED.dwg.

Now that you've modified profile views using a number of methods, let's look at another option that is available on the Profile View Properties dialog: Profile View Bands.

Adding Profile View Bands

Data bands are horizontal elements that display additional graphical and numerical information about the profile or alignment that is referenced in a profile view. Bands can be applied to both the top and bottom of a profile view, and there are six different band types:

Profile Data Bands   Display information about the selected profile. This information can include simple elements such as elevation or more complicated information such as the cut-fill between two profiles at a given station.

Vertical Geometry Bands   Create a schematic view of the elements making up a profile. Typically used in reference to a design profile, vertical data bands make it easy for a designer to see the locations of tangents and vertical curves along an alignment as well as information about the tangents and vertical curves.

Horizontal Geometry Bands   Create a schematic view of the horizontal alignment elements, giving the designer or reviewer information about line, curve, and spiral segments and their relative location to the profile data being displayed.

Superelevation Bands   Display the various options for superelevation values at the critical points along the alignment.

Sectional Data Bands   Can display information about the sample line locations and the distance between them as well as other section-related information.

Pipe Data Bands   Can display specific information such as part, offset, elevation, or direction about each pipe or structure being shown in the profile view.

In this exercise, you'll add bands to give feedback on the EG and layout profiles as well as horizontal and vertical geometry:

1. Open the ProfileViewBands.dwg file (or the ProfileViewBands_METRIC.dwg file).

2. Zoom to the Frontenac Drive profile view (the one enclosed by a rectangle).

3. Pick a grid line, and from the Profile View contextual tab ➢ Modify View panel, choose Profile View Properties to display the Profile View Properties dialog.

4. On the Bands tab (Figure 7.54), in the List Of Bands area, verify that the Location drop-down is set to Bottom Of Profile View and notice that an Elevations And Stations band has already been set during the creation of this profile view.

FIGURE 7.54
The Bands tab of the Profile View Properties dialog

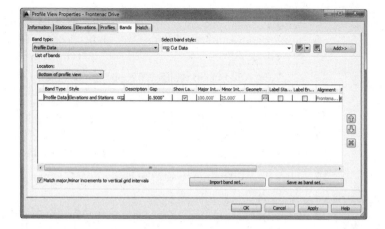

Selecting the type of band from the Band Type drop-down menu changes the Select Band Style drop-down menu so that it includes only styles that are available for the band type you select. Next to the Select Band Style drop-down menu are the usual Style Edit/Copy button and a preview button. Once you've selected a style from the Select Band Style drop-down, clicking the Add button places it on the profile. The Location drop-down list in the List Of Bands section of this dialog allows you to switch between the bands shown at the bottom of the profile view or the top of the profile view; you'll look at that in a moment.

5. Change the Band Type drop-down to the Profile Data option and choose the Cut Data option from the Select Band Style drop-down.

6. Click the Add button. The Geometry Points To Label In Band dialog will open (Figure 7.55).

FIGURE 7.55
The Geometry Points To Label In Band dialog showing the Alignment Points tab (left) and the Profile Points tab (right)

 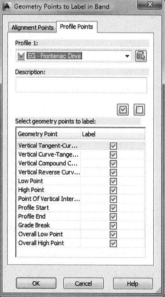

348 | **CHAPTER 7** PROFILES AND PROFILE VIEWS

   7. Click OK to accept the defaults in the Geometry Points To Label In Band dialog.

   8. Leave Band Type set to Profile Data, and choose the Fill Data option from the Select Band Style drop-down.

   9. Click the Add button which will open the Geometry Points To Label In Band dialog once again. Click OK to accept the defaults and dismiss that dialog.

   10. Change the Location drop-down to Top Of Profile View.

   11. Change the Band Type drop-down to the Horizontal Geometry option and choose the Geometry option from the Select Band Style drop-down.

   12. Click the Add button to add the Horizontal Geometry band to the table in the List Of Bands area.

   13. Change the Band Type drop-down to the Vertical Geometry option.

   Do not change the Select Band Style field from its current selection (Geometry).

   14. Click the Add button to also add the Vertical Geometry band to the table in the List Of Bands area.

   15. Click OK to exit the dialog.

   Your profile view should look like Figure 7.56.

FIGURE 7.56
Applying bands to a profile view

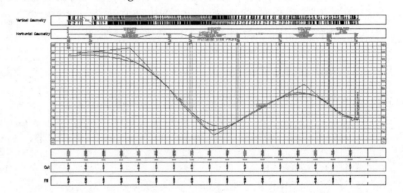

There are obviously problems with the bands. The Vertical Geometry band is a mess and is located above the title of the profile view, whereas the Horizontal Geometry band overwrites the title. In addition, the elevation information has only the existing ground profile being referenced. Next, you'll fix those issues:

   16. Pick a grid line on Frontenac Drive profile view.

   17. From the Profile View contextual tab ➢ Modify View panel, choose Profile View Properties to display the Profile View Properties dialog.

   18. On the Bands tab, verify that the Location drop-down in the List Of Bands area is set to Bottom Of Profile View.

19. Verify that the "Match major/minor increments to vertical grid intervals" option at the bottom of the page is selected.

    Checking this option ensures that the major/minor intervals of the profile data band match the major/minor profile view style's major/minor grid spacing.

    Three Profile Data bands are listed in the table in the List Of Bands area (Elevations And Stations, Cut Data, and Fill Data). If you need to widen the columns, you can do so by double-clicking the line between the column headings.

20. Scroll right and notice the two columns labeled Profile1 and Profile2.

21. For all three rows change the value of Profile2 to Frontenac Drive FG, as shown in Figure 7.57.

FIGURE 7.57
Setting the profile view bands to reference the Frontenac Drive FG profile

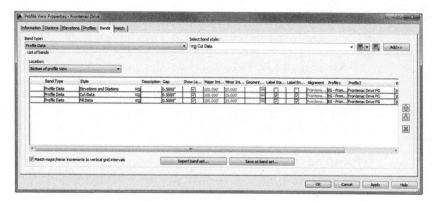

22. Change the Location drop-down to Top Of Profile View.

    The Horizontal Geometry and the Vertical Geometry bands are now listed in the table.

23. Scroll to the right again, and set the value of Profile1 in the Vertical Geometry band to Frontenac Drive FG.

    Notice that some of the Profile1 and Profile2 boxes are not available for editing, such as those in the horizontal geometry band in the Top Of Profile View; this is because profile information isn't needed for this band.

24. Scroll back to the left and set the Gap value for the Horizontal Geometry band to **1.5"** (or **35 mm** for metric users).

    This value controls the distance from one band to the next or to the edge of the profile view itself and will move the bands above the profile view title.

25. Click OK to dismiss the dialog.

    Your profile view should now look like Figure 7.58.

FIGURE 7.58
Completed profile view with the Bands set appropriately

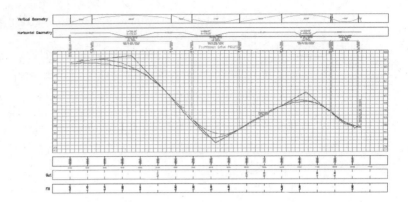

When this exercise is complete, you may close the drawing. A finished copy of this drawing is available from the book's web page with the filename `ProfileViewBands_FINISHED.dwg` or `ProfileViewBands_METRIC_FINISHED.dwg`.

Bands use the Profile1 and Profile2 designation as part of their style construction. By changing the profile referenced as Profile1 or Profile2, you change the values that are calculated and displayed (e.g., existing versus proposed elevations). These bands are just more items that are driven by object styles, which you will learn more about in Chapter 19.

In the next section we'll look at using band sets to make working with bands more efficient.

Band Sets

Band sets are simply collections of bands, much like the profile label sets or alignment label sets. In this exercise, you'll save a band set and then apply it to a second profile view:

1. Open the `ProfileViewBandSets.dwg` file (or the `ProfileViewBandSets_METRIC.dwg` file).

2. Pick a grid line in the Frontenac Drive profile view (at the lower right located in a rectangle), and from the Profile View contextual tab ➢ Modify View panel, choose Profile View Properties to display the Profile View Properties dialog.

3. On the Bands tab, click the Save As Band Set button to display the Band Set – New Profile View Band Set dialog.

4. On the Information tab, in the Name field, enter **Cut Fill Elev Station and Horiz Vert Geometry**, as shown in Figure 7.59.

5. Click OK to dismiss the Band Set – New Profile View Band Set dialog.

6. Click OK to dismiss the Profile View Properties dialog.

7. Press Esc to clear your selection, then pick a grid line in the Syrah Way profile view (at the upper right located in a rectangle), and from the Profile View contextual tab ➢ Modify View panel, choose Profile View Properties to display the Profile View Properties dialog.

8. On the Bands tab, click the Import Band Set button, and the Band Set dialog opens.

FIGURE 7.59
The Information tab for the Band Set – New Profile View Band Set dialog

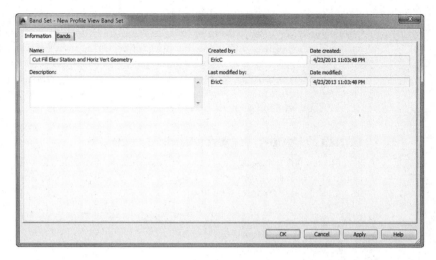

9. Select the Cut Fill Elev Station And Horiz Vert Geometry option from the drop-down list and click OK.
10. Select Top Of Profile View from the Location drop-down list.
11. Scroll over on the Vertical Geometry row and set Profile1 to Syrah Way FG.
12. Select Bottom Of Profile View from the Location drop-down list.
13. Scroll over and change Profile2 to Syrah Way FG for all three rows.
14. Click OK to exit the Profile View Properties dialog.

Your Syrah Way profile view (Figure 7.60) now looks like the Frontenac Drive profile view.

FIGURE 7.60
Completed profile view after importing the band set and matching properties

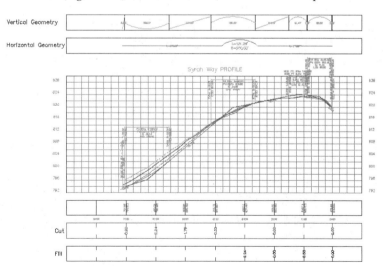

Band sets allow you to create uniform labeling and callout information across a variety of profile views. By using a band set, you can apply a collection of settings and styles that you've assigned to a single profile view to a number of profile views. The simplicity of enforcing standard profile view labels and styles makes using profiles and profile views simpler than ever.

When this exercise is complete, you may close the drawing. A finished copy of this drawing is available from the book's web page with the filename `ProfileViewBandSets_FINISHED.dwg` or `ProfileViewBandSets_METRIC_FINISHED.dwg`.

Understanding Profile View Hatch

Sometimes it is necessary to hatch cut/fill areas in a profile view. The settings on the Hatch tab of the Profile View Properties dialog are used to specify upper and lower cut/fill boundary limits for associated profiles (see Figure 7.61).

FIGURE 7.61
Shape style selection on the Hatch tab of the Profile View Properties dialog

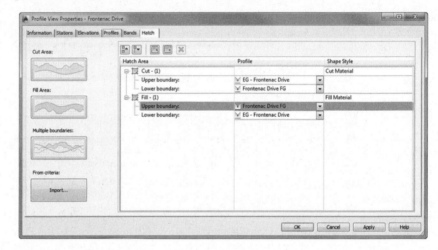

Shape styles from the General Multipurpose Styles collection found on the Settings tab of Toolspace can also be selected here. These settings include the following:

Cut Area   Click this button to add hatching to a profile view in areas of cut (the layout profile is at a lower elevation than the sampled surface profile).

Fill Area   Click this button to add hatching to a profile view in areas of fill (the layout profile is at a higher elevation than the sampled surface profile).

Multiple Boundaries   Click this button to add hatching to a profile view in areas of a cut/fill where the area must be averaged between two existing profiles (for example, finished ground at the centerline vs. the left and right top of a curb).

From Criteria   Click this button to add hatch in areas where quantity takeoff criteria are used to define a hatch region.

Figure 7.62 shows a cut and fill hatched profile based on the criteria shown previously in Figure 7.61.

FIGURE 7.62
A portion of the Frontenac Drive profile shown with cut and fill shading

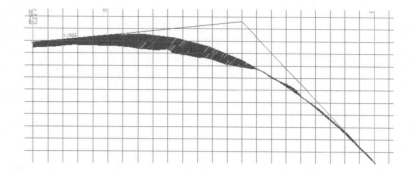

Mastering Profiles and Profile Views

One of the most difficult concepts to master in AutoCAD Civil 3D is the notion of which settings control which display property. Although the following two rules may sound overly simplistic, they are important to remember and will help you work through your understanding of the software:

- Every object has a label and an object style.
- Every label has a label style.

Furthermore, if you can remember that there is a distinct difference between a profile object and the profile view object you place it in, you'll be well on your way to mastering profiles and profile views. When in doubt, select an object and observe the options that become available on the contextual ribbon tab. Label styles and object styles will be discussed further in Chapters 18 and 19, respectively.

Profile View Labeling Styles

Now that the profile view is created, the profile view grid spacing is set, and the titles all look good, it's time to add some specific callouts and detail information. Civil 3D uses profile view labels and bands for annotating. The specific label styles will be discussed further in Chapter 18, but for now we will just discuss how to apply the labels.

View Annotation

Profile view annotations label individual points in a profile view, but they are not tied to a specific profile object. Profile view labels can be station elevation labels, depth labels, or projection labels. Station elevation labels can be used to label a single point or the depth between two points in a profile while recognizing the vertical exaggeration of the profile view and applying the scaling factor to label the correct depth. In this exercise, you'll use both the station elevation label and the depth label:

1. Open the ProfileViewLabels.dwg file (or the ProfileViewLabels_METRIC.dwg file).
2. Zoom to the Cabernet Court profile view located in the rectangle.

3. From the Annotate tab ➤ Labels & Tables panel, choose Add Labels (not the drop-down list) to open the Add Labels dialog.

4. In the Feature drop-down, select Profile View.

5. In the Label Type drop-down, verify that Station Elevation is selected.

6. In the Station Elevation Label Style drop-down, verify that Station And Elevation is selected.

7. Verify that the marker style is set to Basic Circle With Cross.

8. Click the Add button.

9. When prompted to select a profile view, click a grid line in the Cabernet Court profile view and a vertical red line will appear.

10. Zoom in around station 7+85 (or 0+235 for metric users) so that you can see the point where the EG and layout profiles cross.

11. Turn off your running Osnaps and pick this profile crossover point visually; then pick the same point to set the elevation and press ↵.

    Your label should look like Figure 7.63 although the actual values will vary slightly.

FIGURE 7.63
An elevation label for a profile station

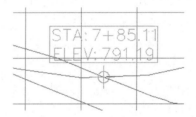

> **SNAPPING TO PROFILES AND PROFILE VIEWS**
>
> For a number of releases now, users have been asking for the ability to simply snap to the intersection of two profiles. We mention this because you'll try to snap and wonder if you've lost your mind. You haven't—it just doesn't work. If you are after a solution (it isn't elegant), you can draw lines on top of the profiles.

12. In the Add Labels dialog, change both Label Type and Depth Label Style to the Depth option.

13. Click the Add button.

14. Pick the Cabernet Court profile view by clicking one of the grid lines.

15. Pick a point along the layout profile; then pick a point along the EG profile and press ↵.

    The depth between the two profiles will be measured, as shown in Figure 7.64. Your value may vary slightly from what is shown.

FIGURE 7.64
A depth label applied to the Cabernet Court profile view

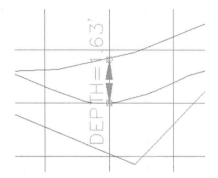

16. Close the Add Labels dialog.

When this exercise is complete, you may close the drawing. A finished copy of this drawing is available from the book's web page with the filename `ProfileViewLabels_FINISHED.dwg` or `ProfileViewLabels_METRIC_FINISHED.dwg`.

Depth labels can be handy in earthworks situations where cut and fill become critical, and individual spot labels are important to understanding points of interest, but most design documentation is accomplished with labels placed along the profile view axes in the form of data bands and band sets discussed earlier in this chapter.

Profile Labels

It's important to remember that the profile and the profile view aren't the same thing. The labels discussed in the following sections are those that relate directly to the profile but are visible for a specific profile view. This usually means station-based labels, individual tangent and curve labels, or grade breaks.

Applying Labels

As with alignments, you can apply labels as a group of objects separate from the profile. In this portion of the exercise, you'll learn how to add labels along a profile object:

1. Open the `ApplyingProfileLabels.dwg` file (or the `ApplyingProfileLabels_METRIC.dwg` file).

2. Zoom to the Cabernet Court profile view in the rectangle and pick the Cabernet Court FG profile (the blue line) to activate the Profile contextual tab.

3. From the Profile contextual tab ➤ Labels panel, choose Edit Profile Labels to display the Profile Labels dialog (see Figure 7.65).

FIGURE 7.65
An empty Profile Labels dialog

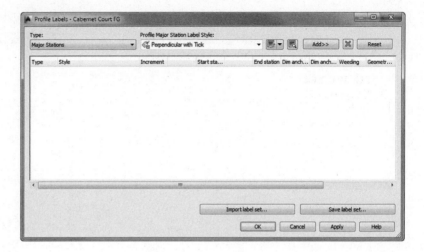

Selecting the type of label from the Type drop-down menu changes the style drop-down menu to include styles that are available for that label type. Next to the Style drop-down menu are the usual Style Edit/Copy button and preview button. Once you've selected a style from the Style drop-down menu, clicking the Add button places it on the profile. The middle portion of this dialog displays information about the labels that are being applied to the profile selected; you'll look at that in a moment.

4. Choose the Major Stations option from the Type drop-down menu.

   The name of the second drop-down menu changes to Profile Major Station Label Style to reflect this option.

5. Verify that Perpendicular With Tick is selected in this menu.

6. Click the Add button to apply this label to the profile.

7. Choose Horizontal Geometry Points from the Type drop-down menu.

   The name of the style drop-down menu changes to Profile Horizontal Geometry Point.

8. Select the Horizontal Geometry Station option, and click the Add button again to display the Geometry Points dialog.

   This dialog lets you apply different label styles to different geometry points if necessary.

9. Deselect the Alignment Beginning and Alignment End rows, as shown in Figure 7.66, and click OK to dismiss the dialog.

10. On the Profile Labels dialog, click the Apply button.

11. Drag the dialog out of the way to view the changes to the profile, as shown in Figure 7.67.

FIGURE 7.66
The Geometry Points dialog appears when you apply labels to horizontal geometry points.

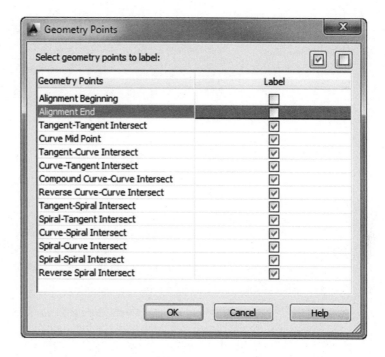

FIGURE 7.67
Labels applied to major stations and alignment geometry points

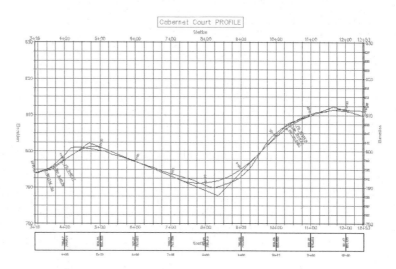

12. In the middle of the Profile Labels dialog, change the Increment value in the Major Stations row to **50** (or **10** for metric users), as shown in Figure 7.68.

FIGURE 7.68
Modifying the major station labeling increment

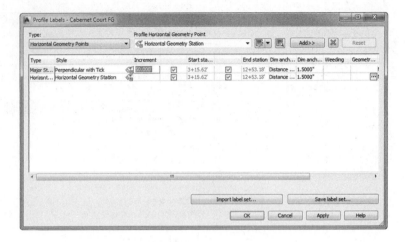

This modifies the labeling increment only, not the grid or other values.

13. Click OK to dismiss the Profile Labels dialog.

When this exercise is complete, you may close the drawing. A finished copy of this drawing is available from the book's web page with the filename ApplyingProfileLabels_FINISHED.dwg or ApplyingProfileLabels_METRIC_FINISHED.dwg.

Using Profile Label Sets

Applying labels to both crest and sag curves and to tangents, grade breaks, and geometry with the label style selection and various options can be monotonous. Thankfully, Civil 3D gives you the ability to use label sets, as in alignments, to make the process quick and easy. Just as you saved and imported band sets in a previous example, you can save and import label sets as well.

In all of the previous examples, you used either the complete label set or the _No Labels label set, which you specified on the Profile Display Options wizard page.

> **SOMETIMES YOU DON'T WANT TO SET EVERYTHING**
>
> Resist the urge to modify the beginning or ending station values in a label set. If you save a specific value, that value will be applied when the label set is imported. For example, if you set a station label to end at 15+00 because the alignment is 15+15 long, that label will always stop at 15+00, even if the target profile is 5,000' long!

Label sets are the best way to apply profile labeling uniformly. When you're working with a well-developed set of styles and label sets, it's quick and easy to go from sketched profile layout to plan-ready output. We will discuss profile label sets in further detail and go over an example in Chapter 18.

Profile Utilities

One common requirement is to compare profile data for objects that are aligned similarly but not parallel. Another is the ability to project objects from a plan view into a profile view. The abilities to superimpose profiles and project objects are both discussed in this section.

Superimposing Profiles

In a profile view, a profile is sometimes superimposed to show one profile adjacent to another (e.g., a ditch adjacent to a road centerline). In this brief exercise, you'll superimpose one of your street designs onto the other to see how they compare over a certain portion of their length:

1. Open the SuperimposeProfiles.dwg file (or the SuperimposeProfiles_METRIC.dwg file).
2. From the Home tab ➢ Create Design panel, choose Profile ➢ Create Superimposed Profile.
3. At the Select source profile: prompt, zoom to the Frontenac Drive profile view at the bottom right and pick the Frontenac Drive FG profile (the blue line).
4. At the Select destination profile view: prompt, click one of the grid lines in the Cabernet Court profile view at the middle right to display the Superimpose Profile Options dialog shown in Figure 7.69.

   On the Limits tab you can control the start and end station of the superimposed profile, in the event that you do not want the entire profile to be superimposed. On the Accuracy tab, you can control the handling of curves through the mid-ordinate settings. Superimposed curves aren't curves at all; they are lots of small segments that approximate a curve. The smaller the mid-ordinate distance values, the smaller and more numerous the segments will be.

FIGURE 7.69
The Superimpose Profile Options dialog

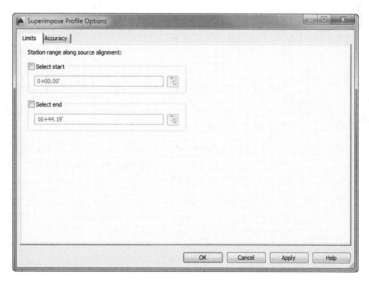

5. Click OK to dismiss the dialog, accepting the default settings.

6. Zoom in on the Cabernet Court profile view to see the superimposed data, as shown in Figure 7.70.

FIGURE 7.70
The Frontenac Drive layout profile superimposed on the Cabernet Court profile view

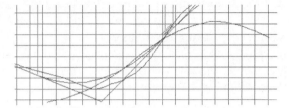

Note that the vertical curve in the Frontenac Drive layout profile has been approximated on the Cabernet Court profile view, using a series of PVIs. Superimposing works by projecting a line from the target alignment (Cabernet Court) to a perpendicular intersection with the other source alignment (Frontenac Drive).

The target alignment is queried for an elevation at the intersecting station and a PVI is added to the superimposed profile. Note that this superimposed profile is still dynamic! A change in the Frontenac Drive layout profile will be reflected on the Cabernet Court profile view.

When this exercise is complete, you may close the drawing. A finished copy of this drawing is available from the book's web page with the filename SuperimposeProfiles_FINISHED.dwg or SuperimposeProfiles_METRIC_FINISHED.dwg.

Projecting Objects in Profile View

Some AutoCAD and some AutoCAD Civil 3D objects can be projected from a plan view into a profile view. The list of available AutoCAD objects includes points, blocks, 3D solids, and 3D polylines. The list of available AutoCAD Civil 3D objects includes COGO points, feature lines, and survey figures. These objects can be projected to the object's elevation, a manually selected elevation, a surface, or a profile. In the following exercise, you'll project a 3D object into a profile view:

1. Open the ObjectProjection.dwg file (or the ObjectProjection_METRIC.dwg file).

2. From the Home tab ➢ Profile & Section Views panel, choose Profile View ➢ Project Objects To Profile View.

3. At the Select objects to add to profile view: prompt, select the Fire Hydrant object located in the center of the circle near the eastern intersection of Syrah Way and Frontenac Drive and press ↵.

4. At the Select a profile view: prompt, select the Syrah Way profile view (located at the upper right, inside a rectangle) by clicking one of the grid lines to display the Project Objects To Profile View dialog.

5. Verify that Style is set to Fire Hydrant, Elevation Options is set to Surface ➢ EG, and Label Style is set to Projection Dimension Below, as shown in Figure 7.71.

6. Click OK to dismiss the dialog, and review your results as shown in Figure 7.72.

FIGURE 7.71
A completed Project Objects To Profile View dialog

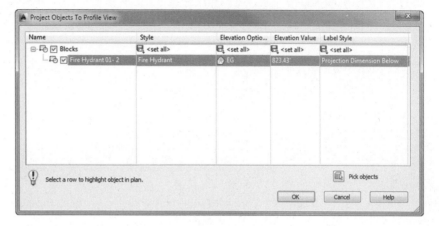

FIGURE 7.72
The COGO point object projected into a profile view

We actually wanted the fire hydrant to show on the proposed surface. No problem. Follow the next steps.

7. Click the fire hydrant in profile view to activate the Projected Object contextual tab.

8. From the Projected Object contextual tab ➢ Modify Projected Object panel, choose Projection Properties to display the Profile View Properties dialog.

9. In the Profile View Properties dialog, click the <set all> option in the Elevation Options column and select Syrah Way FG, as shown in Figure 7.73. Click OK.

FIGURE 7.73
Selecting the Syrah Way FG elevation

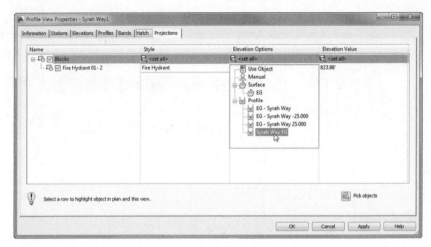

The projected fire hydrant is now adjusted to match the elevation of the Syrah Way FG profile, as shown in Figure 7.74.

FIGURE 7.74
The COGO point object projected onto the Syrah Way FG profile

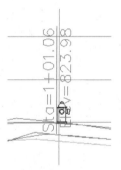

When this exercise is complete, you may close the drawing. A finished copy of this drawing is available from the book's web page with the filename `ObjectProjection_FINISHED.dwg` or `ObjectProjection_METRIC_FINISHED.dwg`.

Once an object has been projected into a profile view, the Profile View Properties dialog will display a new Projections tab. Projected objects will remain dynamically linked with respect to their plan placement. If you move them manually after placing them dynamically, a warning will appear to confirm that you want to break the dynamic setting. Because profile views and section views are similar in nature, objects can be projected into section views in the same fashion. However, projecting a feature line onto a profile view will give you a different result than projecting it onto a section view. On a profile view, it looks more like a superimposed profile. On a section view, it's more like a pipe crossing since it appears only where it intersects the section line.

Creating a Quick Profile

There are going to be times when all you want is to quickly look at a profile and not keep it for later use. When this is the case, instead of creating an alignment, a profile, and a profile view, you can create a quick profile. A quick profile is a temporary object that will not be saved with the drawing. You can create a quick profile for 2D or 3D lines or polylines, lot lines, feature lines, survey figures, and even a series of points.

From the Home tab ➢ Create Design panel, choose Profile ➢ Quick Profile. The command line will state Select object or [by Points]:. Once you select your object (or points), the Create Quick Profiles dialog is displayed (Figure 7.75).

FIGURE 7.75
The Create Quick Profiles dialog

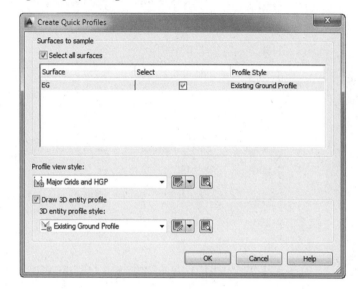

You can select which surface you want to sample as well as what profile view style and 3D entity profile style to use.

The Bottom Line

Sample a surface profile with offset samples. Using surface data to create dynamic sampled profiles is an important advantage of working with a three-dimensional model. Quick viewing of various surface centerlines and grip-editing alignments makes for an effective preliminary planning tool. Combined with offset data to meet review agency requirements, profiles are robust design tools in Civil 3D.

Master It Open the MasteringProfiles.dwg file (or MasteringProfiles_METRIC.dwg file) and sample the ground surface along Alignment A, along with offset values at 15′ left and 15′ right (or 4.5 m left and 4.5 m right) of the alignment. Generate a profile view showing this information using the Major Grids profile view style with no data band sets.

Lay out a design profile on the basis of a table of data. Many programs and designers work by creating pairs of station and elevation data. The tools built into Civil 3D let you input this data precisely and quickly.

Master It Continue in the MasteringProfiles.dwg file (or the MasteringProfiles_METRIC.dwg file) or open MasteringProfiles_SolutionA.dwg (MasteringProfiles_SolutionA_METRIC.dwg). Create a layout profile on Alignment A using the Layout profile style and a complete label set with the following information for Imperial users:

Station	PVI Elevation	Curve Length
0+00	822.00	
1+80	825.60	300'
6+50	800.80	

Or the following information for metric users:

Station	PVI Elevation	Curve Length
0+000	250.400	
0+062	251.640	100 m
0+250	244.840	

Add and modify individual entities in a design profile. The ability to delete, modify, and edit the individual components of a design profile while maintaining the relationships is an important concept in the 3D modeling world. Tweaking the design allows you to pursue a better solution, not just a working solution.

Master It Continue in the MasteringProfiles.dwg file (or the MasteringProfiles_METRIC.dwg file) or you can open MasteringProfiles_SolutionB.dwg (MasteringProfiles_SolutionB_METRIC.dwg), available from the book's website. For the layout profile created in the previous exercise, modify the curve so that it is 200' (or 60 m for metric users). Then insert a PVI at Station 4+90, Elevation 794.60 (or at Station 0+150, Elevation 242.840 for metric users) and add a 300' (or 96 m for metric users) parabolic vertical curve at the newly created PVI.

Apply a standard band set. Standardization of appearance is one of the major benefits of using styles in labeling. By applying band sets, you can quickly create plot-ready profile views that have the required information for review.

Master It Continue in the drawing you have open from the previous exercise or open MasteringProfiles_SolutionC.dwg (MasteringProfiles_SolutionC_METRIC.dwg). Apply the Cut And Fill band set to the layout profile created in the previous exercise with the appropriate profiles referenced in each of the bands.

Chapter 8

Assemblies and Subassemblies

Roads, ditches, trenches, and berms usually follow a predictable pattern known as a *typical section*. Assemblies are how you tell the AutoCAD® Civil 3D® software what these typical sections look like. Assemblies are made up of smaller components called *subassemblies*. For example, a typical road section assembly contains subassemblies such as lanes, sidewalks, and curbs.

In this chapter, the focus will be on understanding where these assemblies come from and how to build and manage them.

In this chapter, you will learn to:

- Create a typical road assembly with lanes, curbs, gutters, and sidewalks
- Edit an assembly
- Add daylighting to a typical road assembly

Subassemblies

A *subassembly* is a building block of a typical section, known as an *assembly*. Examples of subassemblies include lanes, curbs, sidewalks, channels, trenches, daylighting, and any other component required to complete a typical corridor section.

An extensive selection of subassemblies has been created for use in Civil 3D. More than a hundred subassemblies are available in the tool palettes, and each subassembly has a list of adjustable *parameters*. There are also about a dozen generic links you can use to further refine your most complex assembly needs. From ponds and berms to swales and roads, the design possibilities are almost infinite.

To expand the possibilities even more, you can use the Subassembly Composer to create custom subassemblies from scratch. Subassembly Composer is a separate program whose sole purpose is to build custom subassemblies. You can learn more about Subassembly Composer in the bonus chapter named "Custom Subassemblies" that is available on the book's website, www.sybex.com/go/masteringcivil3d2014.

The Tool Palettes

You will add subassemblies to a design by clicking on them from the subassembly tool palette, as you'll see later in this chapter. By default, Civil 3D has several tool palettes created for corridor modeling.

You can access these tool palettes from the Home tab by clicking the Tool Palettes button on the Palettes panel or by pressing Ctrl+3.

When Civil 3D is installed, you have an initial set of the most commonly used assemblies and subassemblies ready to go. The Tool Palettes window consists of multiple customizable tabs that categorize the assemblies and subassemblies so that they are easy to find.

The top, default tab in the Tool Palettes window is the Assemblies tab. On this tab you will find a selection of predefined, completed assemblies (Figure 8.1). These are a great starting point for beginners who are looking for examples of how subassemblies are put together into an assembly. There are examples of simple roadway sections as well as more advanced items, such as intersections and roundabouts. To use one, click the desired assembly tool on the Tool Palettes window, and then click within the drawing area and press ↵ to end the command.

FIGURE 8.1
Tool Palettes predefined assemblies

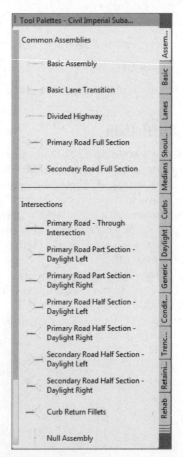

> **GETTING TO THE TOOL PALETTES**
>
> The exercises in this chapter depend heavily on the use of the Tool Palettes window of AutoCAD®. To avoid redundancy, we will assume that you have the Tool Palettes window open for each exercise and not include instructions to open it. In case you need a reminder, the easiest way to open the Tool Palettes window is either from the Home tab ➢ Palettes panel or by pressing Ctrl+3 on your keyboard.

The Corridor Modeling Catalogs

If the default set of subassemblies in the Tool Palettes window is not adequate for your design situation, check the Corridor Modeling Catalogs for one that will work.

The Corridor Modeling Catalogs are installed by default on your local hard drive. You access it on the Home tab by expanding the Palettes panel and clicking the Content Browser button to open a content browser interface.

Choose either the metric or imperial corridor catalogs to explore the entire collection of subassemblies available in each category (see Figure 8.2).

FIGURE 8.2
The front page of the Corridor Modeling Catalog

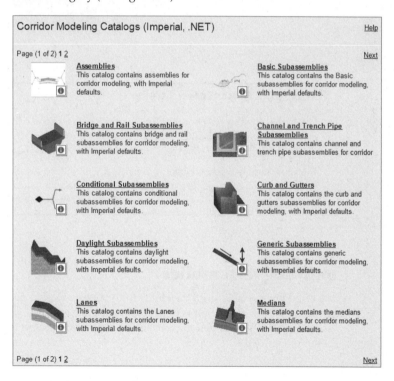

Adding Subassemblies to a Tool Palette

If you'd like to add additional subassemblies to your Tool Palettes window, you can use the i-drop to grab subassemblies from the catalog and drop them onto the Tool Palettes window. To use the i-drop:

1. Click the small blue *i* next to any subassembly, and continue to hold down your left mouse button until you're over the desired tool palette.

2. Release the button, and your subassembly should appear on the tool palette (see Figure 8.3).

FIGURE 8.3
Using the i-drop to add the RailSingle subassembly to a tool palette

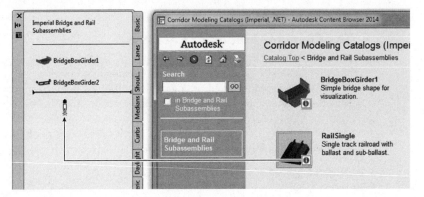

> ### Accessing Subassembly Help
>
> Later, this chapter will point out other shortcuts to access the extensive subassembly documentation. You can get quick access to information by right-clicking any subassembly entry on the Tool Palettes window or the Corridor Modeling Catalog page and selecting the Help option.
>
> The Subassembly Reference page in the help file provides a detailed breakdown of each subassembly, examples for its use, its parameters, a coding diagram, and more. While you're searching the catalog for the right subassemblies to use, you'll find the Subassembly Reference page quite useful.

Building Assemblies

You build an assembly from the Home tab ➢ Create Design panel by choosing Assembly ➢ Create Assembly. The result is the main assembly baseline marker. This is the point on the assembly that gets attached to your design alignment and profile. A typical assembly baseline is shown in Figure 8.4.

FIGURE 8.4
Creating an assembly (left); an assembly baseline marker (right)

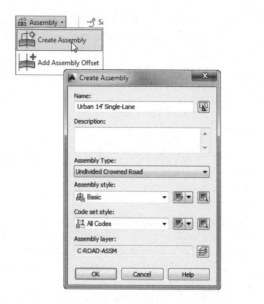

When an assembly is created, you have the option of telling Civil 3D what type of assembly this will be:

- Undivided Crowned Road
- Undivided Planar Road
- Divided Crowned Road
- Divided Planar Road
- Railway
- Other

These categories will help the software determine the axis of rotation options in superelevation, if needed.

Once an assembly is created and assigned a type, you start piecing it together using various subassemblies to meet your design intent. In the next section we will look at how you can create the most common assembly type, an undivided crowned road.

Creating a Typical Road Assembly

The process for building an assembly requires the use of the Tool Palettes window (accessible using Ctrl+3) and the AutoCAD Properties palette (accessible using Ctrl+1), both of which can be docked. You'll quickly learn how to best orient these palettes with your limited screen real estate. If you run dual monitors, you may find it useful to place both of these palettes on your second monitor.

The exercise that follows builds a typical assembly, as shown in Figure 8.5, using LaneSuperelevationAOR, UrbanCurbGutterGeneral, UrbanSidewalk, and DaylightMaxOffset subassemblies.

FIGURE 8.5
A typical road assembly

Let's have a more detailed look at each component you'll use in the following exercise. A quick peek into the subassembly help file will give you a breakdown of attachment options; input parameters; target parameters; output parameters; behavior; layout mode operation; and the point, link, and shape codes.

The LaneSuperelevationAOR Subassembly The LaneSuperelevationAOR subassembly is an all-purpose subassembly for lanes. It can superelevate for an inside or outside lane if needed and allows for up to four layers of materials. The input parameters available are Side, Width, Default Slope, Pave1 Depth, Pave2 Depth, Base Depth, Subbase Depth, Use Superelevation, Slope Direction, Potential Pivot, Inside Point Code, and Outside Point Code. The default width of 12′ (3.6 m) can be adjusted in the parameters or can be used with an offset alignment to control its width. Figure 8.6 shows the image provided in the subassembly help file for this subassembly.

FIGURE 8.6
The LaneSuper-elevationAOR subassembly help diagram

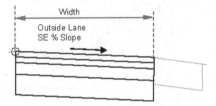

The UrbanCurbGutterGeneral Subassembly The UrbanCurbGutterGeneral subassembly is another standard component that creates an attached curb and gutter. Looking into the subassembly help file, you'll see a diagram of UrbanCurbGutterGeneral with input parameters for Side, Insertion Point, Gutter Slope Method, Gutter Slope, Gutter Slope Direction, Subbase Depth, Subbase Extension, Subbase Slope Method, Subbase %Slope, and the subassembly's seven dimensions. You can adjust these parameters to match many standard curb-and-gutter configurations. Figure 8.7 shows the image provided in the subassembly help file for this subassembly.

FIGURE 8.7
The UrbanCurb-GutterGeneral subassembly help diagram

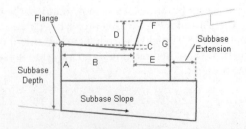

The UrbanSidewalk Subassembly   The UrbanSidewalk subassembly creates a sidewalk and terrace buffer strips. The help file lists the following six input parameters for the UrbanSidewalk subassembly: Side, Inside Boulevard Width, Sidewalk Width, Outside Boulevard Width, %Slope, and Depth. These input parameters let you adjust the sidewalk width, material depth, and buffer widths to match your design specification. Figure 8.8 shows the image provided in the subassembly help file for this subassembly.

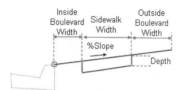

FIGURE 8.8
The UrbanSidewalk subassembly help diagram

The UrbanSidewalk subassembly can return quantities of concrete (or other sidewalk construction material) but not gravel bedding or other advanced material layers. If you needed to track these materials, you could use generic subassemblies (covered further on in this chapter) or build a custom subassembly using Subassembly Composer (covered in the bonus chapter, "Custom Subassemblies").

The DaylightMaxOffset Subassembly   The DaylightMaxOffset subassembly is a nice "starter" for creating simple, single-slope daylight instructions for your corridor. In Civil 3D, an offset dimension is measured from the baseline, and a width is measured from the attachment point. Therefore, the maximum offset in our example is measured from the centerline of the road, which is the baseline. The slope will attempt a default of 4:1, but it will adjust if it needs to in order to keep inside your specified maximum offset (such as a right-of-way line). Options are also available for rounding. Figure 8.9 shows the image provided in the subassembly help file for this subassembly.

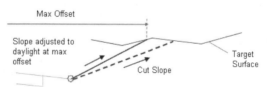

FIGURE 8.9
The Daylight-MaxOffset subassembly help diagram for the cut scenario

---

> **What's With the Funny Names?**
>
> You'll notice that all subassemblies have names with no spaces. This is because of the underlying .NET coding that makes up a subassembly. When you place one of these in your project, it will retain the name from the tool palette.
>
> Prior to Civil 3D 2013, each subassembly was required to have a unique name; therefore, it was suffixed with a number. This is no longer the case. Nonetheless, later in this chapter you'll see how to rename them to something more user friendly if you so desire.

In the following exercise, you'll build a typical road assembly using the subassemblies discussed previously. Follow these steps:

1. Start a new blank drawing from the _AutoCAD Civil 3D (Imperial) NCS template that ships with Civil 3D. Metric users can use the _AutoCAD Civil 3D (Metric) NCS template.

2. Confirm that your Tool Palettes window is showing the subassembly set appropriate for your drawing units (or you may end up with monster 12 meter lanes!).

   If you need to change your active Tool Palettes window from metric to Imperial or vice versa, right-click the Tool Palettes control bar located at the top of the window, as shown in Figure 8.10.

FIGURE 8.10
Right-click the Tool Palettes control bar to change assembly sets if needed.

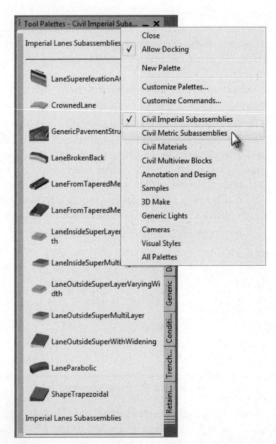

3. Verify that your drawing scale is set to 1≤=10′ (1:50 for metric users).

4. From the Home tab ➢ Create Design panel, choose Assembly ➢ Create Assembly.

   The Create Assembly dialog opens.

5. In the Create Assembly dialog:

   a. Enter **Urban 14′ Single-Lane** (or **Urban 4.5m Single-Lane**) in the Name text box.

   b. Set Assembly Type to Undivided Crowned Road.

   c. Confirm that Assembly Style is set to Basic and Code Set Style is set to All Codes, and click OK.

6. Pick a location in your drawing to place your red assembly baseline marker.

7. Locate and select the Lanes tab on the Tool Palettes window. Position the palette on your screen so that you can clearly see the assembly baseline.

8. Click the LaneSuperelevationAOR button on the Tool Palettes window.

   The AutoCAD Properties palette appears (if it is not already open).

9. Locate the Advanced Parameters section on the Design tab of the AutoCAD Properties palette (Figure 8.11).

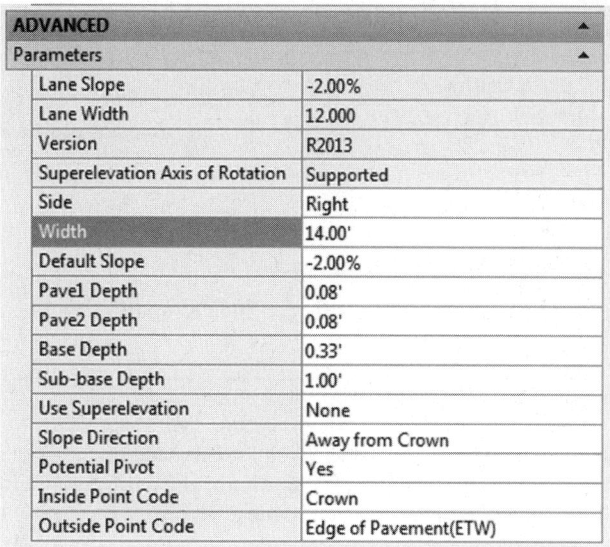

FIGURE 8.11
Advanced Parameters on the Properties palette

ADVANCED	
Parameters	
Lane Slope	-2.00%
Lane Width	12.000
Version	R2013
Superelevation Axis of Rotation	Supported
Side	Right
Width	14.00′
Default Slope	-2.00%
Pave1 Depth	0.08′
Pave2 Depth	0.08′
Base Depth	0.33′
Sub-base Depth	1.00′
Use Superelevation	None
Slope Direction	Away from Crown
Potential Pivot	Yes
Inside Point Code	Crown
Outside Point Code	Edge of Pavement(ETW)

This section lists the LaneSuperelevationAOR parameters.

10. Change the Width parameter to **14′ (4.5 m)**.

    Your Properties palette should resemble Figure 8.11.

11. At the `Select marker point within assembly or [Insert Replace Detached]:` prompt, select anywhere on the red assembly baseline marker to place the first lane.

    Note that it is placed on the right side as specified in the Advanced Parameters section.

12. Before ending the command, click the red assembly baseline marker again to place the left lane.

    Civil 3D has the intelligence to automatically detect the side and place a subassembly on the appropriate side regardless of what is specified in the Advanced Parameters.

13. Press ↵ to end the command.

14. Switch to the Curbs tab in the Tool Palettes window.

15. Click the UrbanCurbGutterGeneral button on the Tool Palettes window.

16. You will accept the parameter defaults, so no changes are needed. Remember that the Side parameter will automatically be detected so there is no need to change it.

17. At the Select marker point within assembly or [Insert Replace Detached]: prompt, select the circular marker located at the top right of the right LaneSuperelevationAOR subassembly.

    This marker represents the top-right edge of pavement (see Figure 8.12).

FIGURE 8.12
The UrbanCurbGutterGeneral subassembly placed on the Lane Superelevation AOR subassembly

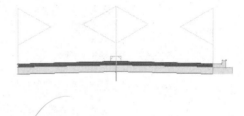

18. Press ↵ to end the command.

    You will add the left curb later. Keep this drawing open for the next portion of the exercise.

IF YOU GOOF...

Often, the first instinct when a subassembly is misplaced is to Undo or erase the wayward piece. However, if you have spent a lot of time diligently tweaking parameters, there is a way to fix things without redoing the subassembly.

MOVING A SUBASSEMBLY

Select the errant subassembly component and use the Move option from the Subassembly contextual tab ➢ Modify Subassembly panel. Use this instead of the base AutoCAD Move tool to get the best results. Using regular AutoCAD Move may cause unexpected results in the corridor.

INSERTING A SUBASSEMBLY

Sometimes you forget to place a subassembly component, or your design changes and you want to include a subassembly that wasn't there before. Prior to Civil 3D 2013, you had to delete the subassembly components from the outside in until you got to where you wanted to insert your missing subassembly and then had to re-create the deleted subassembly pieces.

In this version of Civil 3D, you may have noticed that every time you place an assembly the command line states Select marker point within assembly or [Insert Replace Detached]:. If you enter I, the command line will state Select the subassembly to insert after or [Before]:. The ability to insert a subassembly will come in useful when the planner decides to add a sidewalk at the shoulder of your road.

REPLACING A SUBASSEMBLY

Similar to the insert operation, you can also replace a subassembly component with another component. Again, you will find this very helpful when the planner decides to make changes to your design.

DELETING A SUBASSEMBLY

To delete a subassembly component, you can simply select the subassembly component and press the Delete key. The assembly will connect the subassemblies on either side at the connection points previously used with the deleted component.

CHANGING SUBASSEMBLY PARAMETERS

If you placed everything correctly but forgot to change a parameter or two, there's an easy fix for that, too. Cancel out of any active subassembly placement and select the subassembly you wish to change. Most subassembly parameters can be changed from the AutoCAD Properties palette. For more heavy-duty modifications (such as specifying the side), you will want to get into the Subassembly Properties discussed later in this chapter.

19. In the Curbs tab, click the UrbanSidewalk button on the Tool Palettes window.

20. In the Advanced section of the Design tab on the AutoCAD Properties palette, change the following parameters, leaving all other parameters at their default values:

    Sidewalk Width: **5' (1.5 m)**

    Inside Boulevard Width: **2' (0.7 m)**

    Outside Boulevard Width: **2' (0.7 m)**

    It may be hard to ignore the Side parameter, but setting it is not required with Civil 3D side autodetection.

21. At the Select marker point within assembly or [Insert Replace Detached]: prompt, select the circular marker on the UrbanCurbGutterGeneral subassembly that represents the top rear of the curb to attach the UrbanSidewalk subassembly (see Figure 8.13).

FIGURE 8.13
The BasicSidewalk subassembly placed on the Urban-CurbGutterGeneral subassembly

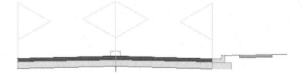

376 CHAPTER 8 ASSEMBLIES AND SUBASSEMBLIES

22. Switch to the Daylight tab on the subassemblies Tool Palettes window, and select the DaylightMaxOffset subassembly.

23. In the Advanced Parameters area of the AutoCAD Properties Palette, change Max Offset From Baseline to **50′ (17 m)**, leaving all other parameters at their default values.

24. At the Select marker point within assembly or [Insert Replace Detached]: prompt, select the circular marker on the outermost point of the sidewalk subassembly. Press Enter or Esc to complete the command.

Your drawing should now resemble Figure 8.14.

FIGURE 8.14
The complete right side of the assembly with DaylightMaxOffset

To complete the left side, you will use the Mirror Subassemblies command.

25. Select the curb, sidewalk, and daylight subassemblies on the right side of the baseline.

The Subassembly contextual tab will show a variety of tools, including Mirror (Figure 8.15).

FIGURE 8.15
The Subassembly contextual tab with subassembly modification tools

ASSEMBLY LABELS

You may notice the 4.00:1 label shown in Figure 8.14, which may or may not show up in your drawing as you work through this exercise. These labels are governed by the code set style, which in this exercise is set to All Codes. If you would like to add this label or other labels, follow these simple steps:

1. Switch to the Settings tab in Toolspace.
2. Expand the General ➢ Multipurpose Styles ➢ Code Set Styles branch.
3. Right-click All Codes and select Edit to display the Code Set Style – All Codes dialog.
4. On the Codes tab, expand the Link branch.
5. In the Label Style column of the Daylight link row, click the Style button to display the Pick Style dialog.

You may need to widen the column headings in order to view the full names.

6. Use the drop-down list to select Steep Grades.
   7. Click OK to dismiss the Pick Style dialog.
   8. Click OK to dismiss the Code Set Style – All Codes dialog.

   Now all Daylight links in any assembly that uses the All Codes code-set style will display a grade label. You may find it helpful to provide other labels on your subassemblies to be able to easily differentiate visually between the assemblies that are similar. For example, now you can tell the difference between the assembly that uses the 4:1 daylight and the one that uses the 5:1 daylight.

26. From the Subassembly contextual tab ➢ Modify Subassembly panel, choose Mirror, and then click the circular point marker located at the top left of the left LaneSuperelevationAOR subassembly.
27. Your assembly should now resemble Figure 8.5 from earlier in the chapter.

   You have now completed a typical road assembly.

You may keep this drawing open to continue on to the next exercise or use the saved copy of this drawing available from the book's web page (`TypicalRoadAssembly_FINISHED.dwg` or `TypicalRoadAssembly_METRIC_FINISHED.dwg`).

Subassembly Components

A subassembly is made up of three basic parts: *links*, *marker points*, and *shapes*, as shown in Figure 8.16. Each piece plays a role in your design and is used for different purposes at each stage of the design process.

FIGURE 8.16
Schematic showing parts of a subassembly

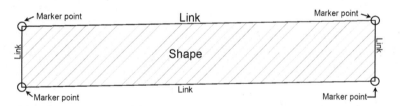

Links

Links are the linear components to your assembly. A link usually represents the top or bottom of a material but can also be used as a spacer between subassemblies.

Links can have codes assigned to them that Civil 3D uses to build the design. Think of these codes as nicknames. In the example assembly you created in the previous exercise, each of the subassembly components contained numerous coded links. As shown in Figure 8.17, on the sidewalk the topmost link has the codes Top and Sidewalk and on the lane subassembly the topmost codes are Top and Pave.

FIGURE 8.17
Link codes on the UrbanSidewalk subassembly (top) and link codes on the LaneSuperelevationAOR subassembly (bottom)

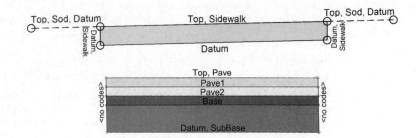

Coded links will be your primary source of data when creating proposed surfaces from your corridors.

Marker Points

Marker points are located at the endpoints of every link and usually are represented by the circles you see on the subassemblies, as shown in Figure 8.18. As you experienced in the previous exercise, the markers are used in assembly creation to "click" subassemblies together and will also "hook" to attach to alignments and/or profiles, known as *targets*.

FIGURE 8.18
Point codes on the UrbanSidewalk subassembly (top) and point codes on the LaneSuperelevationAOR subassembly (bottom)

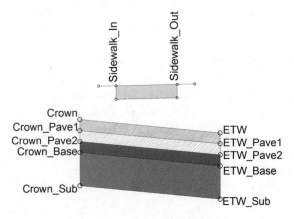

Coded markers are the starting point for *feature lines* generated by the corridor, which are used for a variety of purposes that we will discuss in the upcoming chapters.

Shapes

Shapes are the areas inside a closed formation of links. For example, Figure 8.19 shows different subassemblies with shape codes labeled. Shapes are used in end-area material quantity calculations. At the time an assembly is created, you do not need to consider what material these shapes represent. After your corridor is complete, you will specify what materials the codes represent upon computing materials.

BUILDING ASSEMBLIES | 379

FIGURE 8.19
Shape codes on the UrbanSidewalk subassembly (top) and shape codes on the LaneSuperelevationAOR subassembly (bottom)

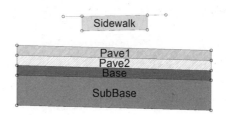

Jumping into Help

Each subassembly is capable of accomplishing different tasks in your design. There is no way to tell just by looking at the icon all the acrobatics that an assembly can do. For a detailed rundown of each parameter, and what can be done with a subassembly, you will need to pop into the help files.

Subassembly Help is extremely—well, helpful! There are many doors into the help files, including from the Corridor Modeling Catalog as you saw earlier. Another way to access the help files is to right-click any subassembly in the tool palette and select Help, as shown in Figure 8.20.

FIGURE 8.20
Getting to the subassembly help file for UrbanCurbGutterGeneral

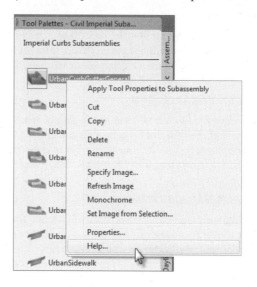

ATTACHMENT AND INPUT PARAMETERS

When you access Subassembly Help using one of the methods described in the sidebar "Accessing Subassembly Help," it will take you to the help file specific to the subassembly you are working with. At the top, you will see a diagram showing the location of the numeric parameters that can be edited in the Properties palette, as shown in Figure 8.21.

FIGURE 8.21
The top portion of Subassembly Help shown with subassembly parameters

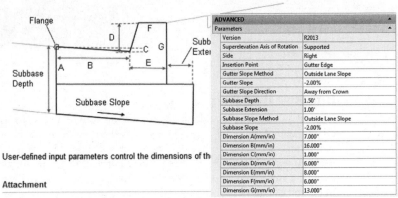

For most subassemblies, the default attachment point will be the topmost-inside marker point. The help file will tell you if this differs for the subassembly you are looking at. Scroll further down to see detailed explanations of each input parameter.

Target Parameters

The Target Parameters section is a listing of what attachments can be set for a subassembly. There are three types of targets: a target surface, a target elevation, and a target offset. The help file will also tell you whether the target is optional or required. We will look at target parameters and setting targets in Chapter 9, "Basic Corridors."

Output Parameters

Output parameters are values calculated when the corridor is built, such as the cross-slope of a lane. In several subassemblies, there is an advanced option called Parameter Reference that can use an output parameter from a previous subassembly in the assembly instead of using the value entered in the subassembly properties. We discuss this concept further in the bonus chapter, "Custom Subassemblies," available on this book's web page.

Reading a Coding Diagram

The coding diagram gives you a list of all the codes used on the subassembly you are working with. Every coded point, link, and shape is listed here. Not all subassembly components have explicit names, such as L9 shown in Figure 8.22. If the point, shape, or link is not included in the table, it is considered uncoded.

FIGURE 8.22
Coding diagram and name table for UrbanCurbGutter-General

Point / Link	Code	Description
P1	Flange	Flange point of the gutter
P2	Flowline_Gutter	Gutter flowline point
P3	TopCurb	Top-of-curb
P4	BackCurb	Back-of-curb
L1 – L3	Top, Curb	Finish grade on the curb and gutter
L7	Subbase Datum	
S1	Curb	Curb-and-gutter concrete area
S2	Subbase	

Coding Diagram

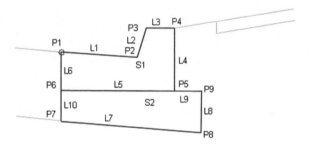

Commonly Used Subassemblies

Once you gain some skills in building assemblies, you can explore the subassembly tool palettes to find subassemblies that have more advanced parameters so that you can get more out of your corridor model. For example, if you must produce detailed schedules of road materials such as asphalt, coarse gravel, fine gravel, subgrade material, and so on, the catalog includes lane subassemblies that allow you to specify those thicknesses for automatic volume reports.

The following sections include some examples of different components you can use in a typical road assembly. Many more alternatives are available within the subassembly tool palettes provided in the software. The help file provides a complete breakdown of each subassembly provided; you'll find this useful as you search for your perfect subassembly.

Each of these subassemblies can be added to an assembly using the same process specified in the first exercise in this chapter. Choose your alternative subassembly instead of the basic parts specified in the exercise, and adjust the parameters accordingly.

Common Lane Subassemblies

The LaneSuperelevationAOR subassembly is suitable for many roads, including undivided roads as shown in the previous example, and divided roads as shown in Figure 8.23. However, you may need different road lane configurations for your locality or design situation:

LaneParabolic The LaneParabolic subassembly (Figure 8.24) is used for road sections that require a parabolic lane in contrast to the linear grade of LaneSuperelevationAOR. The LaneParabolic subassembly also adds options for four material depths. This is useful in jurisdictions that require two lifts of asphalt, base material, and sub-base material; taking advantage of these additional parameters gives you an opportunity to build corridor models that can return more detailed quantity takeoffs and volume calculations.

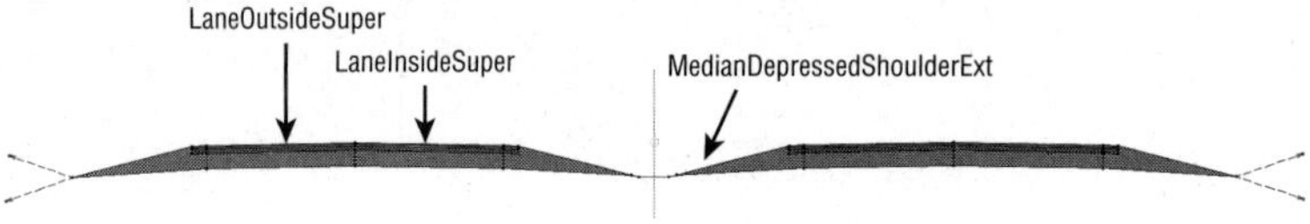

FIGURE 8.23
Use of LaneSuperelevationAOR in a divided highway

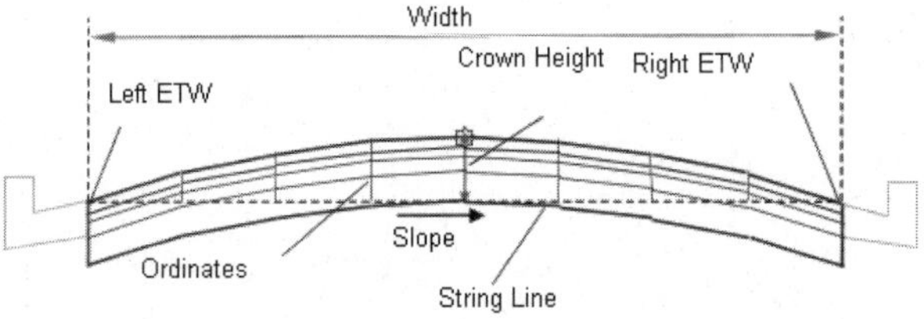

FIGURE 8.24
The LaneParabolic subassembly help diagram

Note that the LaneParabolic subassembly doesn't have a Side parameter. The parabolic nature of the component results in a single attachment point that would typically be the assembly centerline marker.

LaneBrokenBack For designs that call for two lanes, and those lanes must each have a unique slope, the LaneBrokenBack subassembly (Figure 8.25) can be used. This subassembly provides parameters to change the road-crown location and specify the width and slope for each lane. Like LaneParabolic, the LaneBrokenBack subassembly provides parameters for additional material thicknesses.

FIGURE 8.25
The LaneBrokenBack subassembly and parameters

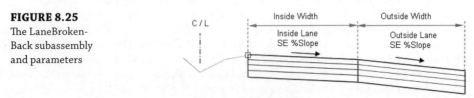

The LaneBrokenBack subassembly, like LaneSuperelevationAOR, allows for the use of target alignments and profiles to guide the subassembly horizontally and/or vertically for both of the lanes.

Common Shoulder and Curb Subassemblies

There are many types of curbs, and the UrbanCurbGutterGeneral subassembly can't model them all. Sometimes you may need a mountable curb, or perhaps you need a shoulder instead. In those cases, the Curbs tool palette provides many alternatives:

UrbanCurbGutterValley (1, 2, or 3) The UrbanCurbGutterValley subassemblies are great if you need mountable curbs. UrbanCurbGutterValley 1, 2, and 3, shown in Figure 8.26, Figure 8.27, and Figure 8.28, respectively, vary slightly in how they handle the sub-base slope. UrbanCurbGutterValley 1 also differs because it comes to a point instead of offering a width at the top of curb.

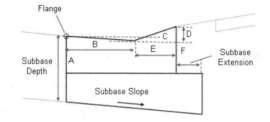

FIGURE 8.26
The UrbanCurbGutterValley1 subassembly help diagram

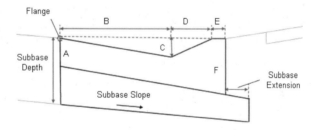

FIGURE 8.27
The UrbanCurbGutterValley2 subassembly help diagram

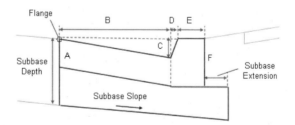

FIGURE 8.28
The UrbanCurbGutterValley3 subassembly help diagram

BasicShoulder BasicShoulder (see Figure 8.29) is another simple yet effective subassembly for use with road sections that require a shoulder. The predefined shape for this subassembly is Pave1, which is good if you are planning to treat this as a paved shoulder and include it with other pavement material in a quantity calculation.

FIGURE 8.29
The BasicShoulder subassembly help diagram

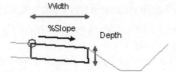

ShoulderExtendSubbase and ShoulderExtendAll Shoulders that can work with your lanes in a superelevation situation, as these two do, are extremely helpful. These two subassemblies, shown in Figure 8.30, will "play nice" with your breakover-removal settings, as you will see in Chapter 11, "Superelevation."

FIGURE 8.30
ShoulderExtendSubbase subassembly help diagram (top) and ShoulderExtendAll subassembly help diagram (bottom)

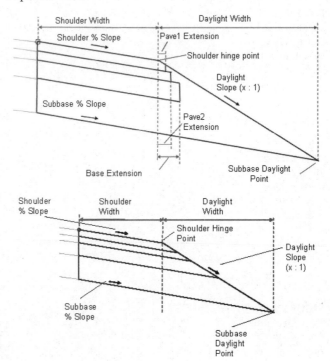

Editing an Assembly

As you saw earlier in this chapter, the AutoCAD Properties palette is an option for changing subassembly parameters for one or more subassemblies of the same type. However, there are a handful of settings that can only be controlled in the Civil 3D Subassembly Properties dialog. For example, the side (left or right) is a parameter that must be changed in the Subassembly Properties dialog or on the Construction tab of the Assembly Properties dialog.

Editing a Single Subassembly's Parameters

Once your assembly is created, you can edit individual subassembly components as follows:

1. Pick the subassembly component you'd like to edit.

   This will bring up the Subassembly contextual tab.

Subassembly Properties

2. From the Subassembly contextual tab ➢ Modify Subassembly panel, choose the Subassembly Properties option.

   The Subassembly Properties dialog appears.

3. Switch to the Parameters tab, shown in Figure 8.31, to access the same parameters you saw in the AutoCAD Properties palette when you first placed the subassembly.

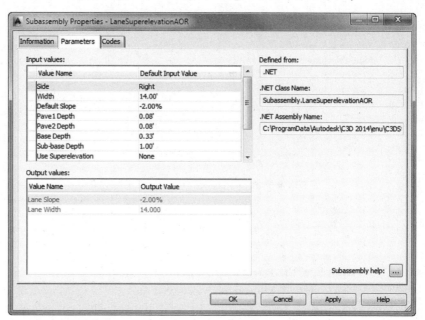

FIGURE 8.31
Subassembly Properties – Parameters tab

4. Click the Subassembly Help ellipsis button at the bottom right of the dialog if you want to access the help page that gives detailed information about the use of this particular subassembly.

   Do not confuse the Subassembly Help ellipsis button with the plain Help button, which will just give you help on the Subassembly Properties dialog.

5. Close the help file when you've finished viewing it.

6. On the Parameters tab of the Subassembly Properties dialog, click inside any field in the Default Input Value column to make changes.

Editing the Entire Assembly

Assembly Properties

Sometimes it's more efficient to edit all the subassemblies in an assembly at once. To do so, pick the assembly baseline marker or any subassembly that is connected to the assembly you'd like to edit. This time, select the Assembly Properties option from the Modify Assembly panel of either the Subassembly or Assembly contextual tab to display the Assembly Properties dialog, shown in Figure 8.32.

FIGURE 8.32
Assembly Properties – Information tab

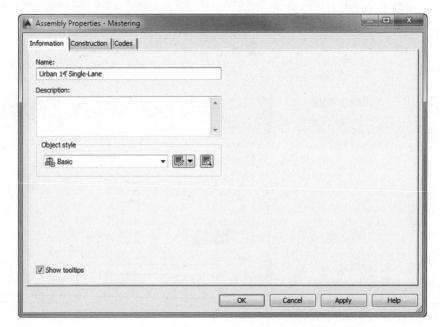

Renaming the Assembly

The Information tab on the Assembly Properties dialog shown in Figure 8.32 gives you an opportunity to rename your assembly and provide an optional description. It is good practice to be consistent and detailed in your assembly names (for example, **Divided 4-Lane 12′ w Paved Shoulder**). With informative assembly names, you will eliminate much of the guesswork when it comes to building corridors later on.

Changing Parameters

The Construction tab in the Assembly Properties dialog, as shown in Figure 8.33, houses each subassembly and its parameters. At the top of the dialog you can change the Assembly Type setting using the drop-down list. In addition, you can change the parameters for individual subassemblies by selecting the subassembly in the Item pane on the left side of the Construction tab and changing the desired parameter in the Input Values pane on the right side.

FIGURE 8.33
Assembly
Properties –
Construction tab

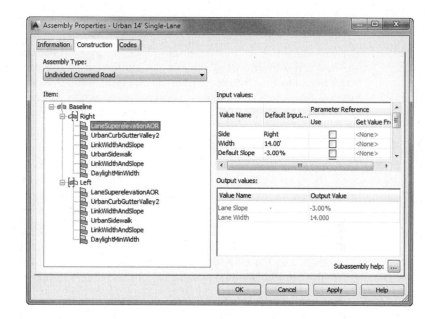

Renaming Groups and Subassemblies

Note that the left side of the Construction tab displays a list of groups. Under each group is a list of the subassemblies in use in your assembly. A new group is formed every time a subassembly is connected directly to the assembly baseline marker.

With side autodetection, you will notice that the groups have already been named Right and Left with the appropriate symbol next to the group name, as shown in Figure 8.33. The subassemblies in each group appear in the same order in which they were originally placed, usually from the inside out. The first subassembly under the Right group is LaneSuperelevationAOR. If you dig into its parameters on the right side of the dialog, you'll learn that this lane is attached to the right side of the assembly marker, the UrbanCurbGutterGeneral is attached to right side of the LaneSuperelevationAOR, and the UrbanSidewalk is attached to the right side of the UrbanCurbGutterGeneral. In this example, the next group, Left, is identical but attached to the left side of the assembly baseline marker.

> **RENAMING TO INCLUDE LEFT OR RIGHT?**
>
> The automatic naming conventions are somewhat simple but usually provide enough information. In previous editions of Civil 3D, many users would change the subassembly names to reference what side they were on; this approach was convenient so users did not have to dig into the subassembly parameters to determine which side of the assembly a certain group was on when it came time to attach targets to a corridor. However, since AutoCAD Civil 3D 2013, both the subassembly and the assembly group are listed in the Target Mapping dialog, as you will see in Chapter 9, "Basic Corridors," so there is no longer any need to add the left or right information to the subassembly name.

If you want, you can rename any of the groups or subassemblies on the Construction tab of the Assembly Properties dialog by right-clicking on the group or subassembly you wish to rename and choosing Rename. From this same right-click menu you can also delete the group or subassembly.

There is no official best practice for renaming your groups and subassemblies, but you may find it useful to keep the designation of what type of subassembly it is or other distinguishing features. For example, if a lane is to be designated as a transition lane or a generic link used as a ditch foreslope, it would be useful to name them descriptively.

Creating Assemblies for Non-road Uses

There are many uses for assemblies and their resulting corridor models aside from road sections. The Corridor Modeling Catalog also includes components for retaining walls, rail sections, bridges, channels, pipe trenches, and much more. In Chapter 9, you'll use a channel assembly and a pipe trench assembly to build corridor models. Let's investigate how those assemblies are put together by building a channel assembly for a stream section:

1. Start a new blank drawing from the `_AutoCAD Civil 3D (Imperial) NCS` template that ships with Civil 3D. Metric users can use the `_AutoCAD Civil 3D (Metric) NCS` template. You can also continue working in your drawing from the first exercise in this chapter.

2. Confirm that your Tool Palettes window is showing the subassembly set (Imperial or metric) appropriate for your drawing units.

3. From the Home tab ➢ Create Design panel, choose Assembly ➢ Create Assembly.

   The Create Assembly dialog opens.

4. In the Create Assembly dialog:

   a. Enter **Channel** in the Name text box.

   b. Set Assembly Type to Other.

   c. Confirm that Assembly Style is set to Basic and that Code Set Style is set to All Codes, and click OK.

5. Pick a location in your drawing to place your red assembly baseline marker.

6. Locate the Trench Pipes tab on the Tool Palettes window.

7. Click the Channel button on the Tool Palettes window.

   The AutoCAD Properties palette appears.

8. Locate the Advanced Parameters section of the Design tab on the AutoCAD Properties palette.

   You'll place the channel with its default parameters and make adjustments through the Assembly Properties dialog, so don't change anything for now. Note that there is no Side parameter. This subassembly will be centered on the assembly baseline marker.

9. At the `Select marker point within assembly or [Insert Replace Detached]:` prompt, select the red assembly baseline marker, and a channel is placed on the assembly (see Figure 8.34).

FIGURE 8.34
The Channel subassembly with default parameters

10. Press Esc to leave the assembly insertion command and dismiss the Properties palette.
11. Click the assembly baseline marker, and then on the Assemblies contextual tab ➢ Modify Assembly palette, click Assembly Properties.

    The Assembly Properties dialog appears.
12. Switch to the Construction tab.

    Notice that while the typical road assembly in the previous exercise generated a Left group and a Right group, the Channel subassembly generated a Centered group.
13. Select the Channel entry on the left side of the dialog (under the Centered group).
14. Click the Subassembly Help ellipsis button located at the bottom right on the dialog's Construction tab.

    The Subassembly Reference page of the AutoCAD Civil 3D 2014 help file appears.
15. Familiarize yourself with the diagram, shown in Figure 8.35, and the input parameters for the Channel subassembly.

FIGURE 8.35
The Channel subassembly help diagram

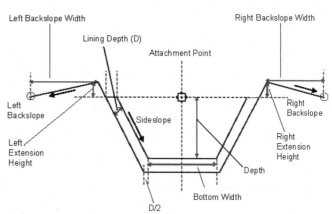

Especially note the Attachment Point, Bottom Width, Depth, and Sideslope parameters. The attachment point indicates where your baseline alignment and profile will be applied.

16. Minimize or close the help file.

    To match the engineer's specified design, you need a stream section 6′ (2 m) deep with a bottom that's 3′ (1 m) wide, 1:1 sideslopes, and no backslopes.

17. Change the following parameters in the Assembly Properties dialog, leaving all other parameters at their default values:

Depth: **6′ (2 m)**

Bottom Width: **3′ (1 m)**

Sideslope: **1** (This value will automatically change to be displayed as 1.00:1.)

Left and Right Backslope Width: **0′ (0 m)**

> **ZERO SUBASSEMBLY VALUES**
>
> There are some subassemblies that do not like zero values, so you may be taught to enter 0.001 or some other value that is so small that it is as if you enter 0. However, there are also some subassemblies that like zero values. If you look in the help file for the Channel subassembly, you will notice that in the Behavior section it explains that if a zero value is specified for left or right extensions and backslope widths, those links are omitted or are not drawn. So in this case a zero is what you want, but be sure to check the help files before using a zero in your subassemblies!

18. Click OK, and confirm that your completed assembly looks like Figure 8.36.

FIGURE 8.36
The channel assembly with customized parameters

You may keep this drawing open to continue on to the next exercise or use the finished copy of this drawing available from the book's web page (ChannelAssembly_FINISHED.dwg or ChannelAssembly_METRIC_FINISHED.dwg).

 Real World Scenario

A PIPE TRENCH ASSEMBLY

Projects that include piping, such as sanitary sewers, storm drainage, gas pipelines, or similar structures, almost always include trenching. The trench must be carefully prepared to ensure the safety of the workers placing the pipe as well as to provide structural stability for the pipe in the form of bedding and compacted fill.

The corridor is an ideal tool for modeling pipe trenching. With the appropriate assembly combined with a pipe-run alignment and profile, you cannot only design a pipe trench but also use cross-section tools to generate section views, materials tables, and quantity takeoffs. The resulting corridor model can also be used to create a surface for additional analysis.

The following exercise will lead you through building a pipe trench corridor based on an alignment and profile that follow a pipe run and a typical trench assembly:

1. Start a new blank drawing from the _AutoCAD Civil 3D (Imperial) NCS template that ships with Civil 3D. Metric users can use the _AutoCAD Civil 3D (Metric) NCS template. You can also continue working in your drawing from the previous exercise.
2. Confirm that your Tool Palettes window is showing the subassembly set (Imperial or metric) appropriate for your drawing units.
3. From the Home tab ➢ Create Design panel, choose Assembly ➢ Create Assembly.
4. In the Create Assembly dialog:
    a. Enter **Pipe Trench** in the Name text box to change the assembly's name.
    b. Set Assembly Type to Other.
    c. Confirm that Assembly Style is set to Basic and Code Set Style is set to All Codes, and click OK.
5. Pick a location in your drawing for the assembly; somewhere in the center of your screen where you have room to work is fine.
6. Locate the Trench Pipes tab on the Tool Palettes window.
7. Click the TrenchPipe1 button on the Tool Palettes window.

   The AutoCAD Properties palette appears.
8. Locate the Advanced section of the Design tab on the AutoCAD Properties palette.

   This section lists the TrenchPipe1 parameters. You'll place TrenchPipe1 with its default parameters and make adjustments through the Assembly Properties dialog, so don't change anything for now. Note that similar to the Channel subassembly, there is no Side parameter. This subassembly will be placed centered on the assembly baseline marker.
9. At the Select marker point within assembly or [Insert Replace Detached]: prompt, select the assembly baseline marker. A TrenchPipe1 subassembly is placed on the assembly as shown here.

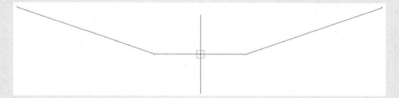

10. Press Esc to leave the assembly creation command and dismiss the AutoCAD Properties palette.
11. Select the assembly baseline marker to activate the Assembly contextual tab.
12. From the Assembly contextual tab ➢ Modify Assembly panel, choose Assembly Properties.

    The Assembly Properties dialog appears.
13. On the Construction tab, select the TrenchPipe1 assembly entry on the left side of the dialog.

14. Click the Subassembly Help ellipsis button located at the bottom right.

    The Subassembly Reference page of the AutoCAD Civil 3D 2014 help file appears.

15. Familiarize yourself with the diagram, shown in the following image, and with the input parameters for the TrenchPipe1 subassembly.

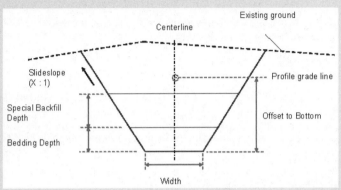

In this case, the profile grade line will attach to a profile drawn to represent the pipe invert. Because the trench will be excavated deeper than the pipe invert to accommodate gravel bedding, you'll want to provide information for the bedding depth parameter. Also note under the Target Parameters section in Help that this subassembly requires a surface target to determine where the sideslopes terminate.

16. Minimize or close the help file.

    To match the engineer's specified design, the pipe trench should be 3′ (1 m) deep and 4′ (1.3 m) wide with 2:1 sideslopes and 1′ (0.3 m) of gravel bedding.

17. In the Assembly Properties dialog – Construction tab, change the following parameters, leaving all other parameters at their default values:

    Width: **4′ (1.3 m)**
    Sideslope: **2** (This value will automatically change to be displayed as 2:1.)
    Bedding Depth: **1′ (0.3 m)**
    Offset To Bottom: **3′ (1 m)**

18. Click OK.

19. Confirm that your completed assembly looks like the image shown here.

> You may keep this drawing open to continue on to the next exercise or use the finished copy of this drawing available from the book's web page (PipeTrenchAssembly_FINISHED.dwg or PipeTrenchAssembly_METRIC_FINISHED.dwg).
>
> This assembly will be used to build a pipe trench corridor in Chapter 9.

Specialized Subassemblies

Despite the more than 100 subassemblies available in the Corridor Modeling Catalog, sometimes you may not find the perfect component. Perhaps none of the channel assemblies exactly meet your design specifications and you'd like to make a more customized assembly, or neither of the sidewalk subassemblies allows for the proper boulevard slopes. Maybe you'd like to try to do some preliminary lot grading using your corridor, or mark a certain point on your subassembly so that you can extract important features easily.

You can handle most of these situations by using subassemblies from the Generic Subassembly Catalog (see Figure 8.37). These simple yet flexible components can be used to build almost anything, although they lack the coded intelligence of some of the more intricate subassemblies (such as knowing if they're paved or grass and understanding things like subbase depth and so on).

FIGURE 8.37
The Generic Subassembly tool palette

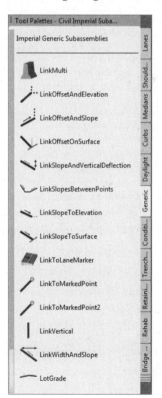

Using Generic Links

Let's look at two examples where you might take advantage of generic links.

The first example involves the typical road section you built in the first exercise in this chapter. You saw that UrbanSidewalk doesn't allow for differing cross-slopes for the inside boulevard (terrace), sidewalk, and outside boulevard (buffer strip). If you need a terrace that's 3' (1 m) wide with a 3 percent slope, and then a 5' (1.5 m) sidewalk with a 2 percent slope, followed by another buffer strip that is 6' (2 m) wide with a slope of 5 percent, you can use generic links to assist in the construction of the proper assembly.

In this exercise we will be creating a new assembly based on the typical road assembly made in the first exercise. Therefore, any of the previously saved files (which you can download from this book's web page) can be used, if you do not have one open from a previous exercise.

1. Continue from the previous exercise or open `ChannelAssembly_FINISHED.dwg` or `ChannelAssembly_METRIC_FINISHED.dwg`, available on the book's website.
2. In Prospector, locate and expand the Assemblies group.
3. Right-click Urban 14' Single-Lane or Urban 4.5 m Single-Lane and select Zoom To.
4. Select the lane and curb subassemblies as well as the assembly baseline marker; then on the Home tab ➢ Clipboard panel, click Copy Clip.
5. On the Home tab ➢ Clipboard panel, click Paste. Pick a location directly under the Urban Single-Lane assembly to paste the copied assembly.

   While you could place the subassembly anywhere, you will find that as you gather more and more assemblies in a drawing, having them organized in a logical manner with similar assemblies in a common area makes them easier to manage.

> ### Labeling Assemblies
>
> When you start getting multiple similar assemblies in your drawing, you may find it helpful to add Mtext next to the assembly with the assembly's name so that you know which assembly is which. By using a field within Mtext, these labels will remain dynamic to their associated object (i.e., if you change the name of the assembly, the Mtext will change as well). You can do this using the following simple steps:
>
> 1. Enter **MTEXT** on the command line.
> 2. Specify the location of your Mtext box.
> 3. From the Text Editor contextual tab ➢ Insert panel, click Field to display the Field dialog.
> 4. Verify that Field Category is set to Objects and Field Names is set to Object.
> 5. Click the Select Object button next to Object Type.
> 6. At the `Select object:` prompt, select the assembly baseline marker.
> 7. Set Property to Name and click OK to dismiss the Field dialog.
> 8. On the Text Editor tab ➢ Close panel, click Close Text Editor.
>
> You now have a dynamic field that will maintain the name of the associated assembly; however, you may need to run a REGEN in order for the field to update. It's also helpful to know that if you copy an assembly and its label at the same time, the new label will reference the new assembly.

6. Select the assembly baseline marker, then on the Assembly contextual tab ➢ Modify Assembly Panel, click Assembly Properties.

   The Assembly Properties dialog appears.

7. On the Information tab, change Name to **Urban 14' Single-Lane with Terraced Sidewalk** (or **Urban 4.5 m Single-Lane with Terraced Sidewalk**) and click OK.

   If you added the Mtext label containing a field, you may want to run a REGEN to update the label.

8. Locate the Generic tab on the Tool Palettes window.

9. Click the LinkWidthAndSlope subassembly (you may need to scroll down to find it), and the AutoCAD Properties palette appears.

10. Scroll down to the Advanced Parameters section of the Properties palette and change the parameters as follows to create the first buffer strip, leaving all other parameters at their default values:

    Width: **3' (1 m)**

    Slope: **3%**

11. At the Select marker point within assembly or [Insert Replace Detached]: prompt, select the circular marker on the right UrbanCurbGutterGeneral subassembly as well as the circular marker on the left UrbanCurbGutterGeneral subassembly, both of which represent the top back of the curb.

12. Switch to the Curbs tab of the Tool Palettes window, and click the UrbanSidewalk button.

13. In the Advanced Parameters area of the Properties palette, change the parameters as follows to create the sidewalk, leaving all other parameters at their default values:

    Inside Boulevard Width: **0' (0 m)**

    Sidewalk Width: **5' (1.5 m)**

    Outside Boulevard Width: **0' (0 m)**

    Slope: **2%**

14. At the Select marker point within assembly or [Insert Replace Detached]: prompt, select the upper-right circular marker on the right LinkWidthAndSlope subassembly and the upper-left circular marker on the left LinkWidthandSlope subassembly.

15. Switch to the Generic tab of the Tool Palettes window, and click the LinkWidthAndSlope button.

    The AutoCAD Properties palette appears.

16. In the Advanced Parameters area, change the parameters as follows to create the second buffer strip, leaving all other parameters at their default values:

    Width: **6' (2 m)**

    Slope: **5%**

17. At the Select marker point within assembly or [Insert Replace Detached]: prompt, select the upper-right circular marker on the right UrbanSidewalk subassembly, as well as the upper-left circular marker on the left UrbanSidewalk subassembly, both of which represent the outside edge of the sidewalk. When complete, press Esc to end the command.

18. Select the right daylight subassembly from the Urban Single-Lane assembly. On the Subassembly contextual tab ➢ Modify Subassembly panel, click Copy.

19. Select the outermost marker on the right side of the new assembly that you are working on. Do the same for the left daylight.

   The completed assembly should look like Figure 8.38 (shown with the typical road assembly from the first exercise for comparison).

FIGURE 8.38
The completed Urban Single-Lane assembly from the first exercise (top) and the Urban Single-Lane with Terraced Sidewalks assembly (bottom)

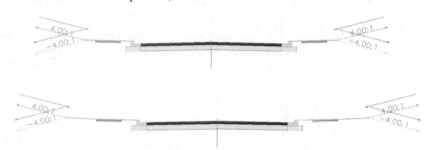

You may keep this drawing open to continue on to the next exercise, or use the finished copy of this drawing available from the book's web page (GenericLinksAssembly_FINISHED.dwg or GenericLinksAssembly_METRIC_FINISHED.dwg).

You've now created a custom sidewalk terrace for a typical road.

Daylighting with Generic Links

The next example involves the channel section you built earlier in this chapter. This exercise will lead you through using the LinkSlopetoSurface generic subassembly, which will provide a surface target to the channel assembly that will seek the target assembly at a 25 percent slope. For more information about surface targets, see Chapter 9.

In this exercise, you will be creating another new assembly based on the channel assembly made in the second exercise.

1. Open ChannelAssembly_FINISHED.dwg or ChannelAssembly_METRIC_FINISHED.dwg, available on the book's website.

2. In Prospector, locate and expand the Assemblies group.

3. Right-click Channel and select Zoom To.

4. Locate the Generic tab on the Tool Palettes window.

5. Click the LinkSlopetoSurface button.

6. In the Advanced Parameters area of the Properties palette, change the Slope parameter to 25%, leaving all other parameters at their default values.

7. At the Select marker point within assembly or [Insert Replace Detached]: prompt, select the circular marker at the upper right on the channel subassembly as well as the circular marker on the upper left on the channel subassembly.

A surface target link appears. Press Esc to end the command.

The completed assembly should look like Figure 8.39.

FIGURE 8.39
The completed channel assembly

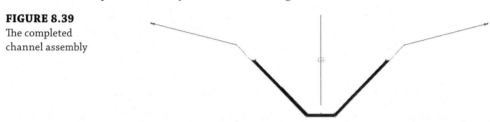

You may keep this drawing open to continue on to the next exercise or use the finished copy of this drawing available from the book's web page (ChannelLinkDaylight_FINISHED.dwg or ChannelLinkDaylight_METRIC_FINISHED.dwg).

Adding a surface link to the channel assembly provides a surface target for the assembly. Now that you've added the LinkSlopetoSurface, you will be able to specify your existing ground as the surface target for a corridor and the subassembly will grade between the top of the bank and the surface for you. You can achieve additional flexibility for connecting to existing ground with the more complex daylight subassemblies, as discussed in the next section.

Working with Daylight Subassemblies

In previous examples, we worked with a generic daylight subassembly, but now let's take a closer look at what they can do for you.

A daylight subassembly tells Civil 3D how to extend a link to a target surface. The instructions might specify that a ditch or berm be inserted before looking for existing ground. Others provide a straight shot but with contingencies for certain design conditions. Figure 8.40 shows the many options you have for adding a daylight subassembly to an assembly.

FIGURE 8.40
Daylight subassemblies in the Tool Palettes window

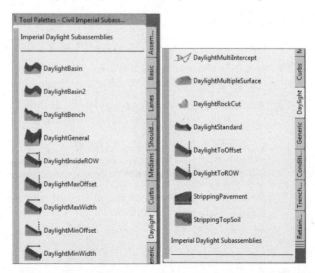

In the following exercise, you'll use the DaylightInsideROW subassembly. This subassembly contains parameters for specifying the maximum distance from the centerline or offset alignments. If the 4:1 slope hits the surface inside the right-of-way (ROW), no adjustment is made to the slope. If 4:1 causes the daylight to hit outside of the ROW, the slope adjusts to stay inside the specified location.

In this exercise you will be creating a new assembly based on the typical road assembly you made in the first exercise.

1. Open `TypicalRoadAssembly_FINISHED.dwg` or `TypicalRoadAssembly_METRIC_FINISHED.dwg`, available from the book's website.

2. In Prospector, locate and expand the Assemblies group.

3. Right-click Urban 14′ Single-Lane or Urban 4.5 m Single-Lane, and select Zoom To.

4. Select the lane, curb, and sidewalk subassemblies as well as the assembly baseline marker. On the Home tab ➢ Clipboard panel, click Copy Clip.

5. Pick a location directly under the Urban Single-Lane assemblies to paste the copied assembly.

6. Click the assembly baseline marker, and then on the Assemblies contextual tab ➢ Modify Assembly palette, click Assembly Properties.

   The Assembly Properties dialog appears.

7. On the Information tab, change the name to **Urban 14′ Single-Lane Daylight ROW** (or **Urban 4.5 m Single-Lane Daylight ROW**), and click OK.

8. Locate the Daylight tab on the Tool Palettes window.

9. Right-click the DaylightInsideROW button on the Tool Palettes panel and select Help.

   The Subassembly Reference page opens in a new window.

10. Familiarize yourself with the options for the DaylightInsideROW subassembly, especially noting the optional parameters for a lined material, a mandatory daylight surface target, and an optional ROW offset target that can be used to override the ROW offset specified in the parameters.

11. Close the Help dialog. Click the DaylightInsideROW button on the Tool Palettes window.

12. In the Advanced Parameters area of the Properties palette, change the parameter ROW Offset From Baseline to **33′** (**10 m**), leaving all other parameters at their default values.

13. At the `Select marker point within assembly or [Insert Replace Detached]:` prompt, select the circular marker on the farthest-right link.

14. Click the DaylightInsideROW button on the Tool Palettes window again, but this time in the Advanced Parameters area of Properties, change the parameter ROW Offset From Baseline to **-33′** (**-10 m**) before placing the left side.

    Notice that there is no Left or Right parameter. The negative value in the ROW Offset From Baseline parameter is what tells Civil 3D the daylight is to the left.

15. You can now dismiss the Properties palette.

The completed assembly should look like Figure 8.41.

FIGURE 8.41
An assembly with the Daylight-InsideROW subassembly attached to each side

You may keep this drawing open to continue on to the next exercise or use the finished copy of this drawing available from the book's web page (`DaylightROWAssembly_FINISHED.dwg` or `DaylightROWAssembly_METRIC_FINISHED.dwg`).

> ### When to Ignore Daylight Input Parameters
>
> The first time you attempt to use many daylight subassemblies, you may become overwhelmed by the sheer number of parameters.
>
> The good news is that many of these parameters are unnecessary for most uses. For example, many daylight subassemblies, such as DaylightGeneral (shown here), include multiple cut-and-fill widths for complicated cases where the design may call for test scenarios. If your design doesn't require this level of detail, leave those parameters set to zero.
>
> | ADVANCED | | | |
> |---|---|---|---|
> | Parameters | | | |
> | Version | R2013 | Fill 1 Slope | Horizontal |
> | Side | Left | Fill 2 Width | 0.00' |
> | Daylight Link | Include Daylight link | Fill 2 Slope | Horizontal |
> | Cut Test Point Link | 3 | Fill 3 Width | 0.00' |
> | Cut 1 Width | 0.00' | Fill 3 Slope | Horizontal |
> | Cut 1 Slope | Horizontal | Flat Fill Slope | 6.00:1 |
> | Cut 2 Width | 0.00' | Flat Fill Max Height | 5.00' |
> | Cut 2 Slope | Horizontal | Medium Fill Slope | 4.00:1 |
> | Cut 3 Width | 0.00' | Medium Fill Max Height | 10.00' |
> | Cut 3 Slope | Horizontal | Steep Fill Slope | 2.00:1 |
> | Cut 4 Width | 0.00' | Guardrail Width | 2.00' |
> | Cut 4 Slope | Horizontal | Guardrail Slope | -2.00% |
> | Cut 5 Width | 0.00' | Include Guardrail | Omit Guardrail |
> | Cut 5 Slope | Horizontal | Width to Post | 1.000 |
> | Cut 6 Width | 0.00' | Rounding Option | None |
> | Cut 6 Slope | Horizontal | Rounding By | Length |
> | Cut 7 Width | 0.00' | Rounding Parameter | 1.50' |
> | Cut 7 Slope | Horizontal | Rounding Tessellation | 6 |
> | Cut 8 Width | 0.00' | Place Lined Material | None |
> | Cut 8 Slope | Horizontal | Slope Limit 1 | 1.00:1 |
> | Flat Cut Slope | 6.00:1 | Material 1 Thickness | 1.00' |
> | Flat Cut Max Height | 5.00' | Material 1 Name | Rip Rap |
> | Medium Cut Slope | 4.00:1 | Slope Limit 2 | 2.00:1 |
> | Medium Cut Max Height | 10.00' | Material 2 Thickness | 0.50' |
> | Steep Cut Slope | 2.00:1 | Material 2 Name | Rip Rap |
> | Fill 1 Width | 0.00' | Slope Limit 3 | 4.00:1 |
> | | | Material 3 Thickness | 0.33' |
> | | | Material 3 Name | Seeded Grass |
>
> Some daylight subassemblies include guardrail options. If your situation doesn't require a guardrail, leave the default parameter set to Omit Guardrail and ignore it from then on. Another common, confusing parameter is Place Lined Material, which can be used for riprap or erosion-control matting. If your design doesn't require this much detail, ensure that this parameter is set to None, and ignore the thickness, name, and slope parameters that follow.
>
> If you're ever in doubt about which parameters can be omitted, investigate the help file for that subassembly.

Alternative Daylight Subassemblies

Over a dozen daylight subassemblies are available, varying from a simple cut-fill parameter to a more complicated benching or basin design. Your engineering requirements may dictate something more challenging than the exercise in the previous section. Here are some alternative daylight subassemblies and the situations where you might use them. For more information on any of these subassemblies and the many other daylighting choices, see the AutoCAD Civil 3D 2014 Subassembly Reference page in the help file.

DaylightToROW and DaylightInsideROW The DaylightToROW subassembly differs slightly from the DaylightInsideROW, as shown in Figure 8.42. DaylightToROW constantly adjusts the slope to stay a certain distance away from your ROW, as specified by the Offset Adjustment input parameter. For example, you can have a ROW alignment specified but use this subassembly to tell Civil 3D to always stay 3′ inside the ROW line. The DaylightInsideROW uses the typical slope but adjusts up to a maximum slope in order to stay inside of the ROW. In both subassemblies, you must specify an offset value or an offset target to use as the ROW.

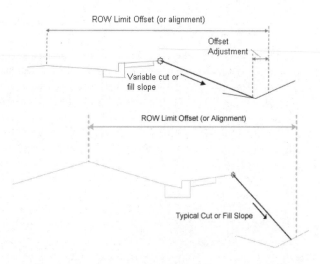

FIGURE 8.42
DaylightToROW subassembly help diagram (top) and DaylightInsideROW subassembly help diagram (bottom)

BasicSideSlopeCutDitch In addition to including cut-and-fill parameters, the BasicSideSlopeCutDitch subassembly (see Figure 8.43) creates a ditch in a cut condition. This is most useful for road designs that require a roadside ditch through cut sections but omit it when passing through areas of fill. If your corridor model is revised in a way that changes the location of cut-and-fill boundaries, the ditch will automatically adjust. Note that this subassembly is located on the Basic tab whereas the other subassemblies in this section are located on the Daylight tab.

When you insert this subassembly you will notice that it does not look anything like the help diagram and instead will display the "LayoutMode" text on the design assembly, as shown in Figure 8.44. This will not display on the completed corridor. There are several subassemblies where this will occur—which is another good reason to always check the help file for an accurate representation of what the final product will look like.

FIGURE 8.43
The BasicSideSlopeCutDitch subassembly help diagram

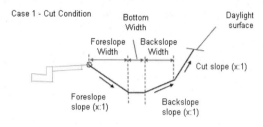
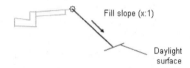

FIGURE 8.44
The BasicSideSlopeCutDitch in layout mode

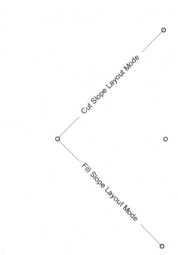

DaylightBasin Many engineers must design berms to contain roadside swales when the road design is in the fill condition. The process for determining where these berms are required is often tedious. The DaylightBasin subassembly (see Figure 8.45) provides a tool for automatically creating these "false berms." The subassembly contains parameters for the specification of a basin (which can be easily adapted to most roadside ditch cross sections as well) and parameters for containment berms that appear only when the subassembly runs into areas of roadside cut.

FIGURE 8.45
The DaylightBasin subassembly help diagram

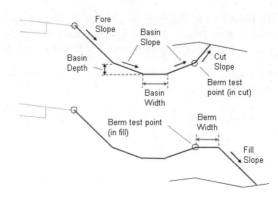

Advanced Assemblies

As you get to know Civil 3D better, you will want it to do more for you. With the tools you are given and your own creativity and problem-solving skills, you can use Civil 3D to create some complex designs. Offset assemblies and marked point assemblies are powerful tools you have at your fingertips.

Offset Assemblies

Offset assemblies are an advanced option when you want to model a coordinating component of the design whose cross section is related to the main assembly. An example of where an offset assembly would be helpful is a main road adjacent to a meandering bike path. The bike path generally follows the main road, but its alignment is not always parallel and the profile might be altogether different. Figure 8.46 shows what the assembly for a bike path to the left of a road would look like.

FIGURE 8.46
An example of an assembly with an offset to the left representing a bike path

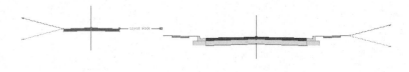

To use an offset assembly, from the Home tab ➢ Create Design panel, choose Assembly ➢ Add Assembly Offset. You will be prompted to select the main assembly and place the offset in the graphic. The location of the offset assembly in relation to the main assembly will have no effect on the final design.

Once the offset assembly is placed, the construction of the offset assembly is identical to any other assembly. We will use an example of an assembly with an offset in Chapter 10, "Advanced Corridors, Intersections, and Roundabouts."

Marked Points and Friends

The marked point assembly is a small but powerful subassembly found in the Generic palette. It consists of a single marker, and you can place it on an assembly to flag a location. You can use the marked point by itself to generate a feature line where no coded marker currently exists, say in the midpoint of a lane link. Where marked points really shine are when used with one of the subassemblies designed to look for a marked point.

When using a marked point, name it right away, and make note of that name for using it with its "friends" (Figure 8.47).

FIGURE 8.47
Name the marked point in the Advanced Parameters area of the Properties palette

Linking to a Marked Point

In the example shown in Figure 8.48, a LinkToMarkedPoint2 subassembly is placed on the right side of the bike path pavement. The LinkToMarkedPoint2 subassembly has been created to look for the marked point on the left side of the sidewalk buffer.

FIGURE 8.48
Add the name of the marked point before you place it on the assembly

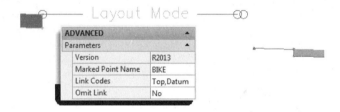

Before placing a marked point subassembly, change the name of the marked point in the Advanced Parameters section. Before you create a subassembly that references a marked point, make certain that you have properly defined the marked point that it references. Be sure to reference this marked point by name with the subsequent subassemblies that reference it before you place it on your assembly.

At this stage, the geometry for a subassembly using a marked point is not known. The final geometry will be determined when the corridor is built. All subassemblies that use the marked point will appear with the "Layout Mode" placeholder. The following subassemblies are designed to look for a marked point:

- Channel
- ChannelParabolicBottom
- LinkToMarkedPoint
- LinkToMarkedPoint2
- LinkSlopesBetweenPoints
- MedianDepressed
- MedianRaisedConstantSlope
- MedianRaisedWithCrown

- OverlayBrokenBackBetweenEdges
- OverlayBrokenBackOverGutters
- OverlayParabolic
- UrbanReplaceCurbGutter (1 and 2)
- UrbanReplaceSidewalk

> **MAKING SURE YOUR MARKED POINT PROCESSES**
>
> Always place the marked point before the links that use it to avoid having to reorder subassemblies in the Construction tab of Assembly Properties. If the marked point is listed below the subassembly that needs it in the Construction tab, Civil 3D will not process it.
>
>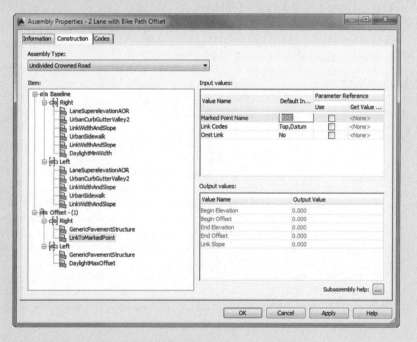
>
> To reorder subassemblies in this dialog, right-click the subassembly and select Move Up or Move Down as needed.

Organizing Your Assemblies

The more geometry changes that occur throughout your corridor, the more assemblies you will have. Civil 3D offers several tools to keep your assemblies organized and available for future use.

Storing a Customized Subassembly on a Tool Palette

Customizing subassemblies and creating assemblies are both simple tasks. However, you'll save time in future projects if you store these items for later use.

A typical jurisdiction usually has a finite number of allowable lane widths, curb types, and other components. It would be extremely beneficial to have the right subassemblies with the parameters already available on your Tool Palettes window.

The following exercise will lead you through storing a customized subassembly on a tool palette.

In this exercise you will be storing some of the subassemblies you made in earlier exercises; therefore, any of the previously saved files (which you can download from this book's web page) can be used, if you do not have one open from a previous exercise.

You can only add a tool from a saved drawing, so make sure you save the drawing you are working in before following these steps:

1. Be sure your Tool Palettes window is displayed.
2. Right-click the Tool Palettes control bar located at the top of the window, and select New Palette to create a new tool palette.
3. Enter **My Road Parts** in the Name text box.
4. Select the sidewalk subassembly from the Urban 14' Single-Lane assembly.

    You'll know it's selected when you can see it highlighted and the grip appears.

5. Click on the dashed portion of the subassembly (i.e., a subassembly marker, *not* the grip point) and drag the assembly into the Tool Palettes window.

    It may take you several tries to get the click-and-drag timing correct, but it will work. You'll know it is working when the cursor appears with a plus sign in the tool palette.

    When you release the mouse button, an entry appears on your tool palette with the name of the subassembly component that you are adding as well as a graphic of the subassembly.

6. Select and right-click this entry, and select the Properties option.

    The Tool Properties dialog appears (see Figure 8.49).

7. If desired, change the image, description, and other parameters in the Tool Properties dialog, and click OK.

8. Try this process for several lanes and curbs in the drawing.

    The resulting tool palette looks similar to Figure 8.50.

Note the tool palette entries for each subassembly point to the location of the Subassembly .NET directory and not to this drawing. If you share this tool palette, make sure the subassembly directory is either identical or accessible to the person with whom you're sharing.

Storing a Completed Assembly on a Tool Palette

In addition to storing individual subassemblies on a tool palette, it's often useful to store entire completed assemblies. Many jurisdictions have several standard road cross sections; once each standard assembly has been built, you can save time on future projects by pulling in a prebuilt assembly.

FIGURE 8.49
The Tool Properties dialog

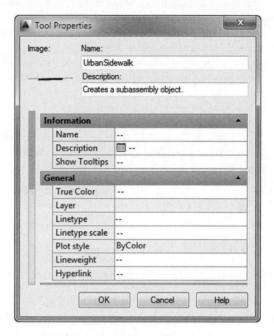

FIGURE 8.50
A tool palette with three customized subassemblies

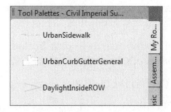

The process for storing an assembly on a tool palette is nearly identical to the process of storing a subassembly. Simply select the assembly baseline, hover your cursor over the assembly baseline, left-click, and drag to a palette of your choosing.

It's usually a good idea to create a library drawing in a shared network location for common completed assemblies and to create all assemblies in that drawing before dragging them onto the tool palette. By using this approach, you'll be able to test your assemblies for validity before they are rolled into production. Alternatively, you can right-click the new palette name and choose Import Subassemblies to display the Import Subassemblies dialog. Here you can choose a source file and then specify whether you want the subassemblies from that source file to import into the palette (optional) and/or the Catalog Library/My Imported Tools.

> **ORGANIZING ASSEMBLIES WITHIN PROSPECTOR**
>
> There are multiple features that help you keep a drawing with many assemblies organized. In Prospector, you will see your listing of assemblies and an Unassigned Subassemblies entry.
>
> Unassigned subassemblies are orphaned parts that are not attached to any main assembly. They may be left over from some assembly customization or they may just be a mistake. In either case, you will want to clean them out. Right-click on the Assemblies collection and select Erase All Unreferenced Assemblies.
>
>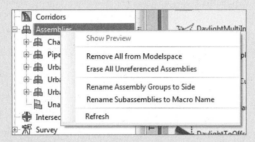
>
> In this same right-click menu, you can also choose to remove the display of the assemblies from modelspace. This hides the display of the assembly but retains its definition in the drawing.
>
> You can still use a hidden assembly in a corridor. If you need it visible again for editing purposes, right-click on the assembly and select Insert To Modelspace.
>
> The niftiest part of this method of organizing assemblies is that they can now be part of your Civil 3D template without being visible.

The Bottom Line

Create a typical road assembly with lanes, curbs, gutters, and sidewalks. Most corridors are built to model roads. The most common assembly used in these road corridors is some variation of a typical road section consisting of lanes, curb, gutter, and sidewalk.

> **Master It** Create a new drawing from the DWT of your choice. Build a symmetric assembly using LaneSuperelevationAOR, UrbanCurbGutterValley2, and LinkWidthAndSlope for terrace and buffer strips adjacent to the UrbanSidewalk. Use widths and slopes of your choosing.

Edit an assembly. Once an assembly has been created, it can be easily edited to reflect a design change. Often, at the beginning of a project you won't know the final lane width. You can build your assembly and corridor model with one lane width and then change the width and rebuild the model immediately.

> **Master It** Working in the drawing from the preceding exercise, edit the width of each LaneSuperelevationAOR to 14' (4.3 m), and change the cross slope of each LaneSuperelevationAOR to -3.00%.

Add daylighting to a typical road assembly. Often, the most difficult part of a designer's job is figuring out how to grade the area between the last engineered structure point in the cross section (such as the back of a sidewalk) and existing ground. An extensive catalog of daylighting subassemblies can assist you with this task.

Master It Working in the drawing from the preceding exercise, add the DaylightMinWidth subassembly to both sides of your typical road assembly. Establish a minimum width between the outermost subassembly and the daylight offset of 10′ (3 m).

Chapter 9

Basic Corridors

The corridor object is a three-dimensional model that combines the horizontal geometry of an alignment, the vertical geometry of a profile, and the cross-sectional geometry of an assembly.

Corridors range from extremely simple roads to complicated highways and interchanges, but they aren't limited to just road travel ways. Corridors can be used to model many linear designs. This chapter focuses on building several simple corridors that can be used to model and design roads, channels, and trenches.

In this chapter, you will learn to:

- Build a single baseline corridor from an alignment, profile, and assembly
- Use targets to add lane widening
- Create a corridor surface
- Add an automatic boundary to a corridor surface

Understanding Corridors

In its simplest form, a corridor combines an alignment, a profile, and an assembly (see Figure 9.1).

FIGURE 9.1
A corridor shown in 3D view

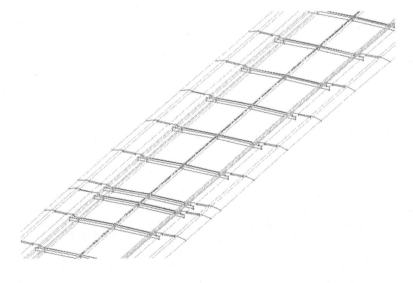

You can also build corridors with complex combinations of alignments, profiles, and assemblies to make complicated intersections, interchanges, or branching streams (see Figure 9.2).

FIGURE 9.2
An intersection modeled with a corridor

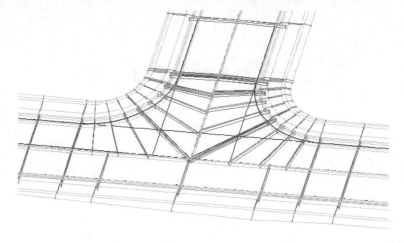

The horizontal properties of the alignment, the vertical properties of the profile, and the cross-sectional properties of the assembly are merged together to form a dynamic model that can be used to build surfaces, sample cross sections, generate quantities, and much more.

Most commonly, corridors are thought of as being used to model roads, but they can also be adapted to model berms, streams (see Figure 9.3), trails, and even parking lots.

FIGURE 9.3
A stream modeled with a corridor

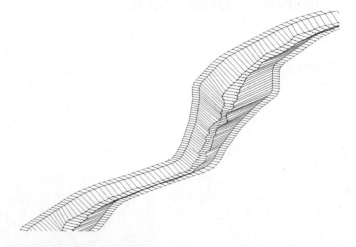

Recognizing Corridor Components

Certification Objective

First, let's look at some important corridor components you will want to become familiar with before proceeding. *Baseline, region, assembly, frequency,* and *target* are all parts of a corridor that you will encounter even on your first design.

Baseline

The first component for any corridor is a *baseline*. The baseline is actually composed of two AutoCAD® Civil 3D® objects, an alignment providing the horizontal layout and a profile providing the vertical layout. The baseline generates the backbone skeleton on which the assembly can hang. Corridors can contain more than one baseline. Therefore, a network of roads can be modeled using one corridor.

In most of the examples in this chapter, the baseline will correspond to the centerline alignment with a profile representing the elevation at the crown of a proposed road. However, this is not always the case, as we will explore in more depth in Chapter 10, "Advanced Corridors, Intersections, and Roundabouts." As your designs become more detailed, you may have corridors with multiple baselines.

Regions

When the geometry along a baseline changes enough to warrant a new assembly, a new *region* is needed. Regions specify the station range where a specific assembly is applied to the design. There may be many regions along a baseline to accommodate design geometry, but the regions may not overlap. Therefore, each region has a start station and an end station, with the end station of one region often matching the start station of the next region.

Assemblies

Marker points, *links*, and *shapes* are coded into the subassemblies that the assembly comprises, as you saw in Chapter 8, "Assemblies and Subassemblies." Assemblies are the third Civil 3D object that is required to generate the corridor by providing cross-sectional information to be applied along some or all of the length of the baseline. Figure 9.4 shows a symmetric roadway assembly that consists of lanes, curbs, sidewalks, and daylight links.

FIGURE 9.4
Typical roadway assembly

Frequency

Frequency refers to how often the assembly is applied to the corridor design. You can set the frequency distance for the corridor as a whole, but in most cases, you should apply it at the region level.

The frequency value will vary depending on the situation. The default frequencies of the stock Civil 3D templates are 25′ in Imperial units and 20 m for metric units. But like many of the settings in Civil 3D, these defaults can be changed in your drawing or template settings by using the following simple steps:

1. On the Settings tab of Toolspace, expand the Corridor ➢ Commands branch.

2. Right-click the CreateCorridor command and choose Edit Command Settings to display the Edit Command Settings – CreateCorridor dialog.

3. Expand the Assembly Insertion Defaults.

4. Change Frequency Along Tangents, Frequency Along Curves, Frequency Along Spirals, Frequency Along Profile Curves, or any other of the multitude of default settings associated with creating a corridor.

In addition to the frequencies based on the alignment or profile entity, Civil 3D can place frequency lines at special stations such as horizontal geometry stations, superelevation critical stations, profile geometry stations, profile high/low stations, and offset target geometry stations. You can also manually create additional frequency stations for things like driveways or culvert crossings.

Targets

As you learned in Chapter 8, there are three types of targets that can be configured in a corridor: a target surface, a target elevation, and a target offset. Targets can be used in lieu of additional assemblies to change corridor geometric characteristics such as cross slope (elevation targets) and lane width (offset target). Surface targets can be used in daylighting or roadway rehabilitation scenarios. Some targets are optional, whereas others are required. For a detailed explanation on targeting for each stock assembly, refer to the help files.

Corridor Feature Lines

When a corridor is created, corridor feature lines are generated. These feature lines can represent back of curb, top of curb, flow lines, edge of pavements, crowns, and any other breakline that would be produced based on your proposed typical roadway section. Corridor feature lines are drawn along the corridor, connecting marker points of identical codes in between assembly frequencies as shown in Figure 9.5. This takes traditional roadway design to another level: a collection of cross sections occurring at specified frequencies has just become a dynamic three-dimensional object model.

FIGURE 9.5
The anatomy of a corridor

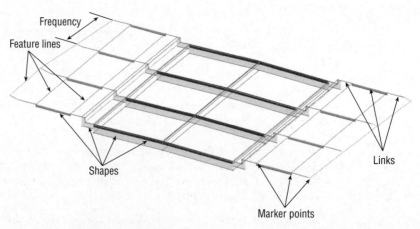

Later on in this chapter, we will take a closer look at corridor feature lines.

This exercise gives you hands-on experience in building a corridor model from an alignment, a profile, and an assembly:

1. Open the `RoadCorridor.dwg` file or `RoadCorridor_METRIC.dwg` file. You can download files from www.sybex.com/go/masteringcivil3d2014.

   Note that the drawing has an alignment for Cabernet Court, a profile view containing the existing and design profiles for Cabernet Court, an assembly, and an existing ground surface.

2. From the Home tab ➢ Create Design panel, choose Corridor to display the Create Corridor dialog.

3. In the Name text box, name the corridor **Cabernet Court Corridor**.

   Keep the default values for Corridor Style and Corridor Layer.

4. Verify that Alignment is set to Cabernet Court and Profile is set to Cabernet Court FG.

   Notice that you can use the small green selection button to select the object on the screen instead of using the drop-down list if you prefer.

5. Verify that Assembly is set to Urban 14′ Single-Lane (or Urban 4.5 m Single-Lane for metric users).

6. Verify that Target Surface is set to the EG surface.

7. Verify that the Set Baseline And Region Parameters check box is selected.

   The Create Corridor dialog should now look similar to Figure 9.6.

FIGURE 9.6
The Create Corridor dialog

8. Click OK to accept the settings and to display the Baseline And Region Parameters dialog, as shown in Figure 9.7.

FIGURE 9.7
The Baseline And Region Parameters dialog

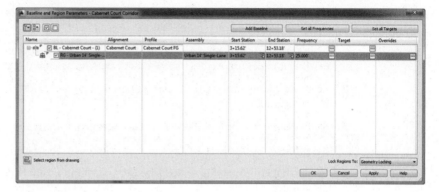

Notice that there is currently one baseline containing one region. The Start Station and End Station values of the region match that of the baseline.

> **Understanding the Locking Regions**
>
> At the bottom of the Baseline And Region Parameters dialog is a drop-down list for locking regions. In versions prior to Civil 3D 2013, all regions were locked to the station. However, now there are two options for locking regions: Geometry Locking and Station Locking.
>
> By default on new drawing files, this drop-down will be set to Geometry Locking. If you open a drawing created in a previous version, the drop-down will be set to Station Locking, but you can change it to Geometry Locking if you desire. You can update this default in your existing templates as needed by using the following steps:
>
> 1. On the Settings tab of Toolspace, expand the Corridor ➢ Commands branch.
> 2. Right-click the CreateCorridor command and choose Edit Command Settings to display the Edit Command Settings – CreateCorridor dialog.
> 3. Expand the Assembly Insertion Defaults.
> 4. Change the Lock Region To value to the desired setting.
>
> You can experiment with either option to see how it will affect your corridor as you make changes to the corridor's baseline.

Most of the other settings have already been set from the information you provided in the Create Corridor dialog, but let's take a few minutes to look at some of the settings in the Baseline And Region Parameters dialog.

9. Click the ellipsis button in the Frequency column in the first row associated with the baseline (BL) to display the Frequency To Apply Assemblies dialog.

   By clicking on the ellipsis in the first row that is associated with the baseline, you will be setting the frequency for all of the regions within that baseline. In this instance you have only one region, but setting them all at once is a good habit to get into when applicable. You could also click the Set All Frequencies button, which would set the frequency for all of the baselines and all of the baseline regions.

10. Examine the settings in the Frequency To Apply Assemblies dialog, as shown in Figure 9.8.

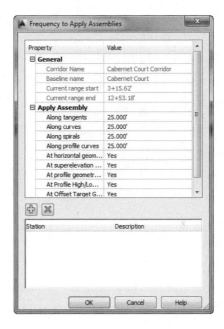

FIGURE 9.8
The Frequency To Apply Assemblies dialog

You can vary the default frequency for the portions of the region along tangents, curves, spirals, and profile curves. In addition, you can add frequency lines at various geometry points. At the bottom of the dialog you can add a user-defined station.

11. Click OK to accept the settings in the Frequency To Apply Assemblies dialog.

12. Click the ellipsis button in the Targets column in the baseline row to display the Target Mapping dialog.

   The ellipsis buttons in this column behave the same way as the ellipsis buttons in the Frequency column. Notice that there is also a Set All Targets button, which can be used to set the targets for all of the baselines and all of the baseline regions.

13. Examine the settings in the Target Mapping dialog, as shown in Figure 9.9.

FIGURE 9.9
The Target Mapping dialog

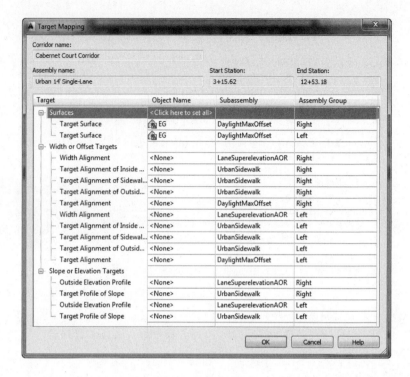

The information in this dialog will vary depending on the subassemblies assigned to the assembly used for this corridor baseline but will always be broken into three categories: Surfaces, Width Or Offset Targets, and Slope Or Elevation Targets. Notice that because you selected EG as the target surface when creating the corridor, the EG surface is already set as the target surface for the left and right subassemblies targeting a surface.

14. Click OK to accept the settings in the Target Mapping dialog.

15. Click OK to accept the settings in the Baseline And Region Parameters dialog.

   If at any point you want to return to the information shown on this dialog, you can do so on the Parameters tab of the Corridor Properties dialog, which we will look at a little later in this chapter.

16. You will see a dialog warning you that the corridor definition has been modified and giving you two options: Rebuild The Corridor or Mark The Corridor As Out-Of-Date. Select the Rebuild The Corridor option.

   If you click the check box at the bottom of this warning dialog that says, "Always perform my current choice," you will not see this warning dialog again.

> **UNDERSTANDING THE LOCKING REGIONS**
>
> If you select the check box (or any others) to hide a message and later want to receive the warning dialog again, you can reactivate it in the future by entering **Options** on the command line. On the System tab, click the Hidden Message Settings button.

You will receive several error messages in Panorama, as shown in Figure 9.10, that read, "Intersection with target could not be computed," "Intersection Point doesn't exist," or something similar. You will rectify this issue in the following steps.

FIGURE 9.10
Corridor errors in Panorama

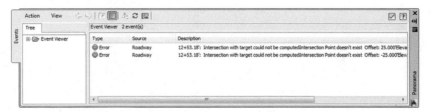

If Panorama did not automatically display, from the Home tab ➢ Palettes expanded panel, choose Event Viewer.

17. Dismiss Panorama.

Your corridor should look similar to Figure 9.11.

FIGURE 9.11
A portion of the nearly completed corridor

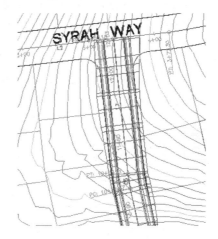

Now, let's try to figure out why those messages appeared in Panorama. In plan view, everything looks normal. However, a look at the corridor in the Object Viewer tells a different story (see Figure 9.12).

FIGURE 9.12
A "waterfall" at the end of the alignment viewed in the Object Viewer

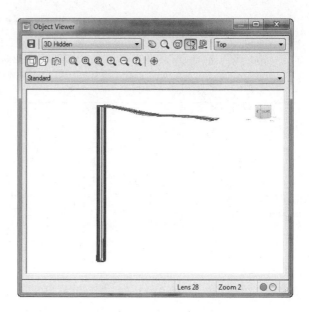

> ### Viewing the Corridor in the Object Viewer
>
> If you'd like to view the corridor in the Object Viewer, do the following:
>
> **1.** Select one of the corridor lines to activate the Corridor contextual tab.
>
> **2.** From the Corridor contextual tab ➢ General Tools panel, choose Object Viewer.
>
>
>
> **3.** Use the View Control drop-down from the top of the Object Viewer to view the corridor from various isometric views, or press and hold the left button on your mouse and move the object around to view it from various custom angles.
>
> **4.** After you have examined the corridor, click the X in the upper-right corner to dismiss the Object Viewer.

18. Select the corridor to activate the Corridor contextual tab.

19. From the Corridor contextual tab ➢ Modify Corridor panel, choose Corridor Properties to display the Corridor Properties dialog.

20. In the Corridor Properties dialog, switch to the Parameters tab.

Notice in Figure 9.13 that the end station for the region is 12+53.18 (or 0+381.94 for metric users), which is based on the full length of the alignment. However, the design profile ends at 12+50 (or 0+381 for metric users). Even though there is no elevation assigned at 12+53.18 (or 0+381.94 for metric users), the corridor assumes an elevation of zero at this station, which creates this waterfall effect. To fix this, you need to tell the corridor to end the region at the final station of the design profile. Note that even though the display value shows only two (or three) decimal places, the corridor examines up to eight decimal places of precision, which means that the value being displayed as the final station of the design profile may be a rounded-up value. If this is the case, this too will create a waterfall effect.

FIGURE 9.13
Corridor Properties dialog, Parameters tab

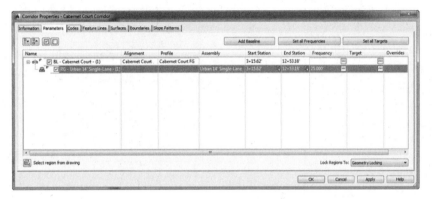

21. Click into the End Station field for the region and enter **1250** (or **381** for metric users).

Entering the value with station notation (the plus sign) is not needed.

Notice that you could alternatively click the station picker button in this cell to select the station on the plan. This may be preferable if you do not know the exact station at which your profile starts or ends or if the precision of the value may be an issue.

22. Click OK to accept the settings in the Corridor Properties dialog.

23. If you didn't click the check box at the last warning dialog, you will receive the same warning again. If so, select the Rebuild The Corridor option to allow the corridor to rebuild.

No new errors will appear in Panorama.

A quick look in the Object Viewer should reveal that the waterfall is gone. When this exercise is complete, you may close the drawing. A finished copy of this drawing is available from the book's website with the filename `RoadCorridor_FINISHED.dwg` or `RoadCorridor_METRIC_FINISHED.dwg`.

Rebuilding Your Corridor

A corridor is a *dynamic* model—which means that if you modify any of the objects used to create the corridor, the corridor must be rebuilt to reflect those changes. For example, if you make a change to the design profile, you must rebuild the corridor to bring it up to date. The same principle applies to changes to alignments, assemblies, target surfaces, and any other corridor ingredients or parameters.

You can access the Rebuild command by right-clicking the corridor name in Prospector, as shown in Figure 9.14.

FIGURE 9.14
Right-click the corridor name in the Corridor collection in Prospector to rebuild it.

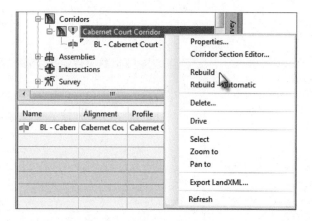

You can also rebuild the corridor or rebuild all corridors by selecting the corridor object and choosing Rebuild Corridor or Rebuild All Corridors from the Corridor contextual tab ➢ Modify Corridor panel, as shown in Figure 9.15.

FIGURE 9.15
Rebuilding the corridor from the Corridor contextual tab

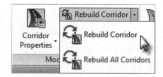

Avoiding Lengthy Rebuilds

For large, complex corridors, you may want to consider leaving the option for Rebuild Automatic unchecked. Every time a change is made that affects a region in your corridor, those regions that have been modified will go through the rebuilding process. This is an improvement from previous versions of Civil 3D, which rebuilt the full corridor even if a modification affected only one region. Nonetheless, the larger or more complex the corridor is and the more regions you are modifying, the longer the rebuild process will take.

Tweaking Corridors

Your corridor may not be perfect on your first iteration of the design. Get comfortable getting into and working with the Parameters tab of the Corridor Properties dialog. You will spend the majority of your time modifying your corridor on this tab.

Whether you are building your first corridor or your five hundredth, odds are good that you will run into one of the following common issues:

Problem   The waterfall effect: Your corridor seems to fall off a cliff, meaning the beginning or ending station of your corridor drops down to elevation zero.

Typical Cause   The range of your corridor is longer than the design profile.

Fix   This is exactly what you ran into in the first exercise. The corridor takes the initial station range from your alignment. However, most designers don't tie into existing ground at the exact alignment start and end stations, so you need to adjust the corridor stations accordingly.

If you need to check the exact start and end stations of the design profile, the best place to do so is the Profile Data tab in the Profile Properties dialog (as shown in Figure 9.16). Edit your corridor region to begin and end at the design profile station.

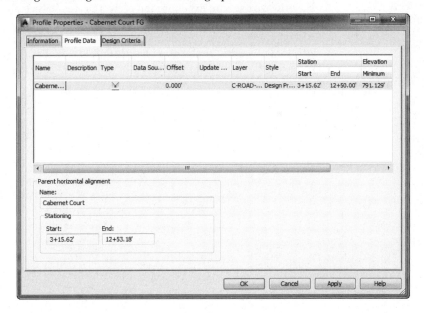

FIGURE 9.16
Check your Profile Properties dialog to verify the station range of the design profile.

Alternately, you can use the station picker button to select the first and last corridor region stations by snapping to the first and last frequency lines that correspond to your profile geometry.

Problem   Your corridor seems to take longer to build and has irregular frequency stations. Also, your daylighting may not extend out to where you expect it (see Figure 9.17).

FIGURE 9.17
An example of unexpected corridor frequency

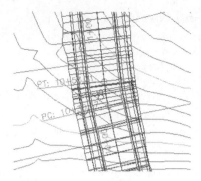

Typical Cause You accidentally chose the existing profile instead of the design profile for your baseline profile. Most corridors are set up to place a frequency line at every vertical geometry point, and a profile, such as the existing profile, has many more vertical geometry points than a layout profile (in this case, the design profile). These additional points on the existing profile are the cause of the unexpected frequency lines.

Fix Always use care to choose the correct profile. Either physically select the profile onscreen or make sure your naming conventions clearly define your finished grade as finished grade. If your corridor is already built, select the corridor, right-click, and choose Corridor Properties, as shown in Figure 9.18. On the Parameters tab of the Corridor Properties dialog, change the profile from the existing profile to the design profile.

FIGURE 9.18
The right-click menu available on the Corridor object

Adding a surface target throws another variable into the mix. Here is a list of some of the most typical problems new users face and how to solve them:

Problem Your corridor doesn't show daylighting even though you have a daylight subassembly on your assembly. You may also get an error message in Event Viewer, as shown in Figure 9.19.

FIGURE 9.19
Error messages associated with absence of daylighting

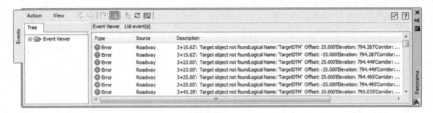

Typical Cause You forgot to set the surface target when you created your corridor.

Fix Select the corridor, right-click, and choose Corridor Properties. On the Parameters tab of the Corridor Properties dialog, click the Set All Targets button. The Target Mapping dialog opens, and its first category is Surfaces. Click the <Click Here To Set All> text in the Object Name column field to display the Pick A Surface dialog. In this dialog, you can choose a surface for the daylight subassembly to target.

Problem Your corridor seems to be missing areas of daylighting. You may also get an error message in Event Viewer, as shown in Figure 9.20.

FIGURE 9.20
Error messages associated with missing areas of daylighting

Typical Cause The Daylight link cannot find the target surface within the parameters you've set in the subassembly properties. This could also mean that the target surface doesn't fully extend the full length of your corridor or your target surface is too narrow at certain locations.

Fix Add more data to your target surface. The daylighting subassemblies in your corridor need a surface to tie into at the slope, grade, or distance specified in subassembly properties.

You can also revisit your daylight subassembly settings to give the program a narrower offset or steeper slope or grade. Alternatively, you can adjust your alignment and/or profile to require less cut or fill, which would cause daylighting to occur over a shorter distance.

If this is not possible, omit daylighting through those specific stations, and once your corridor is built, do hand-grading using feature lines or grading objects. You can also investigate other subassemblies such as LinkOffsetAndElevation that will meet your design intention without requiring a surface target.

Working with Corridor Feature Lines

Corridor feature lines are first drawn connecting marker points of the same code in between assembly frequencies. For example, a feature line will work its way down the corridor and connect all the Daylight points occurring from assembly to assembly. If there are Daylight points on the entire length of your corridor, then the feature line is drawn for the entire length of your corridor. If a region is defined along that same corridor without Daylight points, then the daylight feature line will end at that region.

The Feature Lines tab of the Corridor Properties dialog has a drop-down menu called Branching (see Figure 9.21) with two options: Inward and Outward.

FIGURE 9.21
The Feature Lines tab of the Corridor Properties dialog

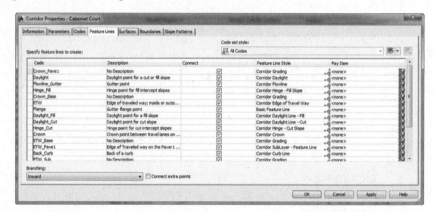

In the case of a two-lane road transitioning to a four-lane road, you would have two lane edges transitioning to four lane edges in the corridor. When branching inward, the feature lines representing those two lane edges will be drawn to the innermost lane edges of the four-lane road on their respective sides. When branching outward, the feature lines representing the two lane edges will be drawn to the outermost lane edges on their respective sides. This setting is particularly important for surface building purposes. Also on this tab is a check box labeled Connect Extra Points. If checked, this will cause the feature lines representing those two lane edges to branch out and connect to both the inner and outer lane edges on their respective sides (see Figure 9.22).

FIGURE 9.22
Feature line branching and connectivity options

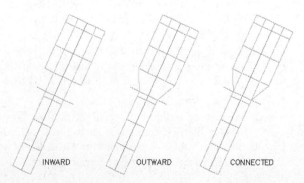

As mentioned earlier, a feature line will connect the same point codes in between assembly frequencies by default. However, the Feature Lines tab of the Corridor Properties dialog allows you to eliminate certain feature lines on the basis of the point code. For example, if for some reason you did not want your Daylight points connected with a feature line, you could toggle that feature line off using the check box in the Connect column.

Corridor feature lines represent the linear features of your corridor like crown, pavement edge, flow line, back and top of curb, and sidewalk edges. Feature lines can also be generated for the edges of all pavement layers, base layers, and sub-base layers. You will be using feature lines in Chapter 14, "Grading," for grading purposes. The difference between the two types of feature line is that corridor feature lines live in the Corridor object.

There are several types of objects you can extract from the Corridor object. If you select the corridor to activate the Corridor contextual tab, you will notice that the Launch Pad panel, shown in Figure 9.23, offers tools for extracting points, alignments, profiles, polylines, and feature lines. The latter four objects are created from the corridor feature lines.

FIGURE 9.23
Launch Pad panel of the Corridor contextual tab

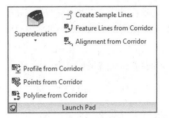

A step-by-step explanation of how to extract a corridor feature line is given in the next exercise.

In the case of extracting a feature line, you have the choice of keeping the line dynamically linked to the corridor geometry or making it an entirely separate entity. When the feature line is linked, it automatically updates when the corridor updates. It cannot be modified. It can be used for grading, but it cannot be used for targeting in that corridor. If the feature line is not linked, it will not update with the corridor. It will be a duplicate of the corridor feature line, but it can be modified. It can also be used for targeting in that corridor. Both types of feature lines can be used as breaklines in surface models.

Once an alignment, profile, or 3D polyline has been extracted, its geometry can be adjusted independently. In this case, the profile, alignment, or polyline no longer retains a link to the corridor from which it originated. Once a profile, alignment, or 3D polyline has been extracted, you can use it as a target back in the corridor that formed it. In the following exercise, a corridor feature line is extracted to produce an alignment and profile:

1. Open the CorridorFeatureLine.dwg or CorridorFeatureLine_METRIC.dwg file.

2. From the Home tab ➢ Create Design panel, choose Alignment ➢ Create Alignment From Corridor.

   Alternatively you can select the corridor by selecting the corridor feature line that you want to create an alignment from to activate the Corridor contextual tab. From the Launch Pad, choose Alignment From Corridor. *If you do this, then skip the next step.*

3. At the Select a corridor feature line: prompt, select the feature line labeled in the drawing as East ETW Feature Line (ETW stands for Edge of Traveled Way).

This opens a Select A Feature Line dialog similar to the one shown in Figure 9.24.

FIGURE 9.24
Selecting the ETW feature line to be extracted as an alignment

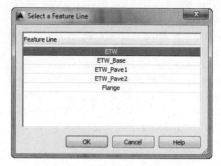

4. Select the ETW feature line from the Select A Feature Line dialog.
5. Click OK to accept the settings in the Select A Feature Line dialog and display the Create Alignment From Objects dialog.
6. In the Name text box of the Create Alignment From Objects dialog, name the alignment **Cabernet Court East ETW**.
7. Verify that Type is set to Offset.
8. On the General tab of the Create Alignment From Objects dialog, verify the following settings:
   - Alignment Style is set to Basic.
   - Alignment Label Set is set to _No Labels.
   - The Create Profile check box is selected.

   The dialog should match Figure 9.25.
9. Click OK to accept the settings in the Create Alignment From Objects dialog and display the Create Profile – Draw New dialog.
10. In the Name text box of the Create Profile – Draw New dialog, name the profile **Cabernet Court East ETW Profile**.
11. On the General tab of the Create Profile – Draw New dialog, verify that the profile style is set to Design Profile and that the profile label set is set to _No Labels.

    The dialog should match Figure 9.26.
12. Click OK to accept the settings in the dialog and then press ↵ to exit the command.

FIGURE 9.25
The completed Create Alignment From Objects dialog

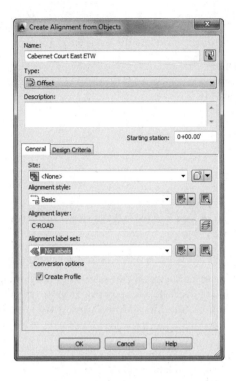

FIGURE 9.26
The Create Profile – Draw New dialog

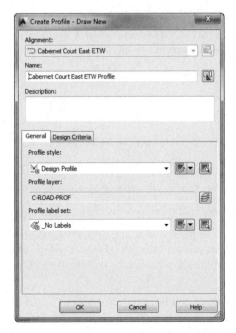

13. In Prospector, expand Alignments ➢ Offset Alignments ➢ Cabernet Court East ETW ➢ Profiles to review the completed alignment and profile (see Figure 9.27).

FIGURE 9.27
A completed alignment and profile shown on the Prospector tab of Toolspace

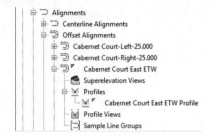

Next, you will extract a corridor feature line and keep it dynamically linked to the corridor.

14. From the Home tab ➢ Create Design panel, choose Feature Line From Corridor.

    Alternatively, you can select the corridor by selecting the corridor feature line that you want to create a feature line from to activate the Corridor contextual tab and then selecting Feature Lines From Corridor from the Launch Pad. *If you do this, then skip the next step.*

15. At the Select a corridor feature line: prompt, select the feature line labeled in the drawing as East Hinge Feature Line.

    This opens the Select A Feature Line dialog.

16. Select the Hinge feature line from the Select A Feature Line dialog.

17. Click OK to close the Select A Feature Line dialog and display the Create Feature Line From Corridor dialog.

18. Select the Name check box and name the feature line **Cabernet Court Hinge East**.

19. Verify that Style is set to Corridor Grading.

20. Verify that the Create Dynamic Link To The Corridor check box is selected.

    All other settings can be left at their defaults. The dialog should match Figure 9.28.

21. Click OK to accept the settings in the dialog and then press ↵ to exit the command.

22. Select the newly created auto corridor feature line.

    Note that it can be selected but does not have grips.

23. From the Feature Line contextual tab ➢ Modify panel, click the Edit Geometry and Edit Elevations buttons to display the hidden panels. Note which options are grayed out and which are not, as shown in Figure 9.29.

These alignments and feature lines can now be used for further design. Remember that feature lines created from the corridor but not dynamically linked to the corridor are no longer connected to the corridor and will not update as you make modifications to the corridor.

When this exercise is complete, you may close the drawing. A finished copy of this drawing is available from the book's web page with the filename CorridorFeatureLine_FINISHED.dwg or CorridorFeatureLine_METRIC_FINISHED.dwg.

FIGURE 9.28
The Create Feature Line From Corridor dialog with the option to create a dynamic link turned on

FIGURE 9.29
Auto Corridor Feature Line contextual tab

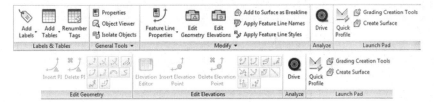

Understanding Targets

Every subassembly is programmed with parameters. Some of these parameters are fixed values like pavement thickness. A lot of subassemblies are programmed with targeting parameters. Examples of targeting parameters would be lane width and cross slope. Lane width and cross slope can be variable. Therefore, if this is the case, you could use alignment and profile targets to define the variable lane width and cross slope.

Alignments, profiles, feature lines, survey figures, and polylines can be configured as targets for subassemblies with width or cross slope. Surface targets are used with daylighting and roadway rehabilitation subassemblies. Now in Civil 3D 2014, you can select these targets through external references. Not all subassemblies contain targeting parameters. You can refer to the help file for target information regarding any subassembly you plan to use.

When you built the corridor in the first exercise, the opening dialog prompted you for a surface target. Next we will take a look at linear targets.

Using Target Alignments and Profiles

So far, all of the corridor examples you looked at have a constant cross section. In this section, we'll take a look at what happens when a portion of your corridor needs to transition to a wider section and then transition back to normal.

Many subassemblies have been programmed to allow for not only a baseline attachment point but also additional outer attachment points that can target alignments, feature lines, and profiles. Be sure to check the subassembly help file to make sure the subassembly you are using will accept targets if you need them. In Chapter 8, you learned that you can right-click any subassembly in the Tool Palettes window to enter the help file. For instance, the BasicLane subassembly will show None in the Target Parameters area of the help file, but LaneSuperElevationAOR lists Lane Width and Outside Elevation.

Think of a subassembly as a rubber band that is attached both to the baseline of the corridor (such as the road centerline) and the target alignment. As the target alignment, such as a lane widening, gets further from the baseline, the rubber band is stretched wider. As that target alignment transitions back toward the baseline, the rubber band changes to reflect a narrower cross section.

Figure 9.30 shows what happens to a cross section when various targets are set for the left edge of a traveled way for a lane subassembly. Figure 9.30a shows the assembly as it was originally placed in the drawing. The width from the original assembly is 14′ with a cross slope of 2 percent to the edge of pavement. Figure 9.30b shows how the geometry changes if an alignment target is set for the edge of pavement. Notice that because there is no profile specified to change the elevation, the 2 percent cross slope is held and the lane width is the only geometry that changes. Figure 9.30c shows that if both an alignment and profile are specified for an edge of pavement alignment, the design cross slope and width both change. Last, you see the assembly with just a profile target assigned to the edge of pavement in Figure 9.30d. In this case, the width stays at 14′ but the elevation of the edge of pavement is dictated by the profile.

FIGURE 9.30
How geometry changes with a target on the left: original assembly geometry (a), assembly with width alignment target only (b), assembly with both alignment and profile target (c), and assembly with only profile target set (d).

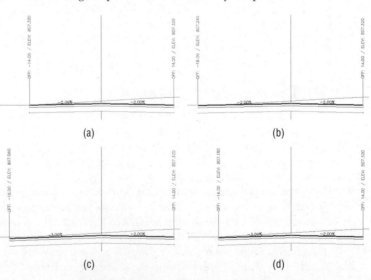

UNDERSTANDING TARGETS | **431**

The following exercise shows you how to set targets using alignments and profiles for a corridor lane widening:

1. Open the `TargetPractice.dwg` or `TargetPractice_METRIC.dwg` file. Note that this drawing has an incomplete corridor.

   You will be modifying the alignment generated from the East ETW corridor feature line for some on-street parking and using the Target Mapping dialog to add the offset alignments.

2. From the Home tab ➢ Create Design panel, choose Alignment ➢ Create Widening.

3. At the `Select an alignment:` prompt, select the Cabernet Court East ETW alignment.

4. At the `Select start station:` prompt, enter **690** ↵ (or **210** ↵ for metric users).

5. At the `Select end station:` prompt, enter **900** ↵ (or **275** ↵ for metric users).

6. At the `Enter widening offset:` prompt, enter **12** ↵ (or **3.75** ↵ for metric users).

7. At the `Specify side: [Left Right]:` prompt, enter **R** ↵.

   The Offset Alignment Parameters palette appears as shown in Figure 9.31. Though you won't be making any changes to the defaults at this time, observe that you can change the transition parameters at the entry and exit if desired.

FIGURE 9.31
Offset Alignment Parameters palette

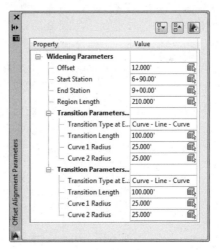

8. Dismiss the Offset Alignment Parameters palette by clicking the X in the upper corner.

   Notice that a new alignment was created named Cabernet Court East ETW-Left-0.000. This is because Add Widening is actually an offset alignment so it uses the Offset Alignment Name template. Since the widening is created on top of the original alignment, it is named Cabernet Court East ETW-Left-0.000 regardless of whether you add the widening to the left or to the right. You could rename this alignment, but for the purposes of this exercise leave it as is.

9. Select the corridor to activate the Corridor contextual tab.

10. From the Corridor contextual tab ➢ Modify Corridor panel, choose Corridor Properties.

11. On the Parameters tab, in the Target column for region RG – Urban 14′ Single-Lane – (1) for imperial units or RG – Urban 4.5 m single Lane – (1) for metric units row, click the ellipsis button to display the Target Mapping dialog. You may need to make the Name column wider to see the row name.

12. In the Width Or Offset Targets ➢ Width Alignment branch for Lane SuperelevationAOR for the Right Assembly Group, click <None> to display the Set Width Or Offset Target dialog.

13. With the Select Object Type To Target drop-down set to Alignments, use the Select From The Drawing button to select the alignment that you just created. Press Enter when finished selecting to return back to the Set Width Or Offset Target dialog, or select Cabernet Court East ETW-Left-0.000 from the list.

14. Click the Add button.

    The alignment should appear in the listing of selected entities to target at the bottom of the dialog, as shown in Figure 9.32.

FIGURE 9.32
The Set Width Or Offset Target dialog

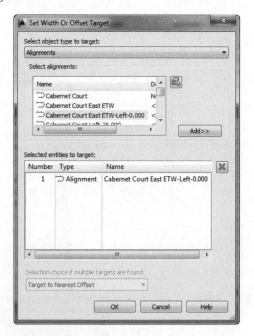

15. Click OK to accept the settings in the Set Width Or Offset Target dialog.

    The Target Mapping dialog should now look like Figure 9.33.

FIGURE 9.33
Targets set for surface and width for the right side

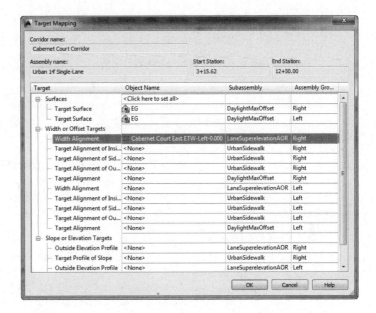

16. Click OK to accept the settings in the dialog.

17. Click OK to accept the settings in the Corridor Properties dialog and allow the corridor to rebuild.

    The corridor should now be following the ETW alignment and resemble Figure 9.34.

FIGURE 9.34
The completed exercise in plan view

You may keep this drawing open to continue on to the next exercise or use the finished copy of this drawing available from the book's web page (`TargetPractice_FINISHED.dwg` or `TargetPractice_METRIC_FINISHED.dwg`).

Corridor Properties Modifications Made Easy

It can be tedious to modify alignments, profiles, starting stations, ending stations, frequencies, and targets along a corridor. Each of these items can be modified in the item view at the bottom of Prospector as shown here.

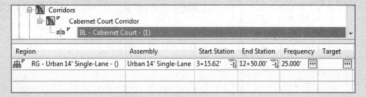

In the following quick exercise, you will use Toolspace to access and modify Frequency values for the Cabernet Court Corridor region:

1. If not still open from the previous exercise, open the TargetPractice_FINISHED.dwg or TargetPractice_FINISHED_METRIC.dwg file.
2. In Prospector, expand and highlight the Cabernet Court Corridor.
3. Click the Frequency ellipsis button for the BL-Cabernet Court -(1) region to display the Frequency To Apply Assemblies dialog.
4. Set the Frequency values for Along Curves and Along Profile Curves to **5′** (or **1.5** m for metric users).
5. Click OK.

The corridor will rebuild and the additional frequency lines should be visible in the plan as shown here.

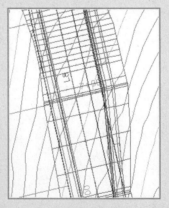

When this exercise is complete, you may close the drawing. A finished copy of this drawing is available from the book's web page with the filename CorridorFrequency_FINISHED.dwg or CorridorFrequency_METRIC_FINISHED.dwg.

Editing Sections

Once your corridor is built, chances are you will want to examine it section by section, make some adjustments, and check for problems. For a station-by-station look at a corridor, select the corridor and from the Corridor contextual tab ➢ Modify Corridor Sections panel, choose Section Editor (see Figure 9.35).

FIGURE 9.35
Some of the many tools, including the Section Editor, available on the Corridor contextual tab

Once you are in the Section Editor, you are in a purely data-driven view. That means that this is a live, editable section of the corridor and is not for plotting purposes. We will discuss plotting cross sections in Chapter 12, "Cross Sections and Mass Haul."

The Section Editor allows for multiple viewport configurations so that you can see the plan, profile, and section all at the same time, as seen in the following exercise:

1. From the Section Editor contextual tab ➢ View Tools panel, choose Viewport Configuration to display the Corridor Section Editor: Viewport Configuration dialog, as shown in Figure 9.36.

FIGURE 9.36
The Corridor Section Editor: Viewport Configuration dialog

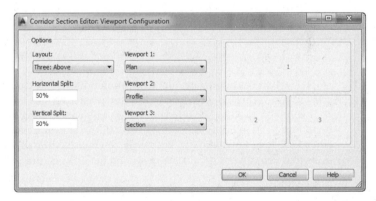

2. Set Layout to Three: Above and Viewport 1 to Plan, Viewport 2 to Profile, and Viewport 3 to Section with a 50% horizontal split and a 50% vertical split.

3. Click OK to accept the settings in the Viewport Configuration dialog. Your screen should now look similar to Figure 9.37.

FIGURE 9.37
The Corridor Section Editor with Viewport Configuration set to Three: Above

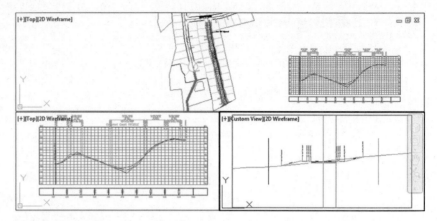

4. You may receive a warning dialog asking if you would like to turn on Viewport Configuration; if so, select Yes.

The Section Editor contextual tab offers many commands other than just Viewport Configuration, as shown in Figure 9.38.

FIGURE 9.38
The Corridor Section Editor contextual tab

The Station Selection panel on the Section Editor contextual tab allows you to move forward and backward through your corridor to see what each section looks like.

If you wish to edit a section, you may do so geometrically in the viewport showing the section in the Section Editor or through the Parameter Editor palette, but not both. To graphically edit a link, hold down the Ctrl key on the keyboard while selecting the item. This will activate grips that you can use to relocate or stretch the link (Figure 9.39).

FIGURE 9.39
A Daylight link ready for grip-editing in the Section Editor

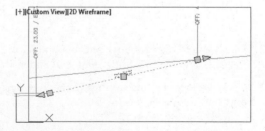

If you would rather use the Parameter Editor to make changes that are more precise to your section, do the following:

1. From the Section Editor contextual tab ➢ Corridor Edit Tools panel, choose Parameter Editor.

2. Look through the listing of subassemblies and their current parameter values.

3. When you find a value you wish to edit, click on the value field to override the design value, as shown in Figure 9.40.

FIGURE 9.40
The Corridor Parameter Editor

When you change the value, a check mark will appear in the Override column. You will find that some of the values are uneditable.

The changes you make can be applied to just the section you are viewing or to a range of stations in the region you are working in (use the Apply To A Station Range button in the Corridor Edit Tools panel to apply the changes to a region).

To exit the Section Editor, click the Close button on the Close panel.

> ### Ack! Stuck in Bizarro Coordinate System
>
> If you accidentally exit your drawing without exiting the Section Editor, you will return to a drawing in a rotated coordinate system as shown here. It may be difficult to see your design, but don't panic.
>
>
>
> At the command line, enter **Plan ↵ W ↵**. You will see your project, but there is still another step.
>
> From the View tab ➢ Coordinates panel, set the current UCS to World, as shown in the following graphic. The UCS icon will return to normal and you can continue working.
>
>
>
> If your viewport configuration was set to three views, you can return to a single view on the View tab ➢ Model Viewports panel. Click the drop-down from the Viewport Configuration button and select Single. Whichever viewport is current when you select Single will be the only one visible after you complete the command.

Creating a Corridor Surface

A corridor provides the raw components for surface creation. Just as you would use points, breaklines, and boundaries to make a surface, a corridor surface uses corridor points as point data, feature lines and links as breaklines, and various commands for using corridor geometry as an outer boundary.

The Corridor Surface

Civil 3D does not automatically build a corridor surface when you build a corridor. From examining subassemblies, assemblies, and the corridors you built in the previous exercises, you have probably noticed that there are many "layers" of points, links, and feature lines. Some represent the very top of the finished ground of your road design, some represent subsurface gravel or concrete thicknesses, and some represent subgrade, among other possibilities. You can choose to build a surface from any one of these layers or from all of them. Figure 9.41 shows an example of a TIN surface built from the links that are coded Top, which would represent final finished ground.

When you create a surface from a corridor, the surface is dependent on the Corridor object. This means that if you change something about your corridor and then rebuild the corridor, the surface will also update.

FIGURE 9.41
A surface built from Top code links

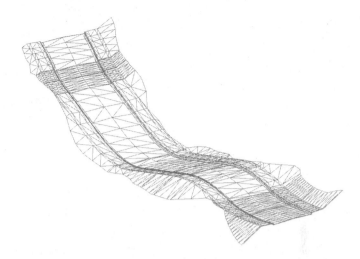

A corridor surface shows up as a surface under the Surfaces branch in Prospector with a slightly modified icon that denotes it is related to the corridor.

After you create the initial corridor surface, you can create a static export of the surface by changing to the Home tab ➢ Create Ground Data panel and choosing Surfaces ➢ Create Surface From Corridor. The Create Corridor Surfaces dialog will appear. If you have created multiple surfaces inside multiple corridors, all of the surfaces will be listed. Fill the check box next to each surface to be exported and then click OK. A detached surface will not react to corridor changes and can be used to archive a version of your surface. To remind you that this surface is not related to a corridor, the icon in the Surfaces branch in Prospector will not show the corridor icon.

Corridor Surface Creation Fundamentals

You create corridor surfaces on the Surfaces tab in the Corridor Properties palette using the following three steps (which are examined in detail later):

1. Click the Create A Corridor Surface button to add a surface item.

2. Choose the data type, either links or feature lines.

3. Choose the data to add, and click the Add Surface Item button.

You can choose to create your corridor surface on the basis of links, feature lines, or a combination of both.

Creating a Surface from Link Data

Most of the time, you will build your corridor surface from links. As discussed earlier, each link in a subassembly is coded with a name such as Top, Pave, Datum, and so on. Choosing to build a surface from Top links will create a surface that triangulates between the points at the link vertices that represent the final finished grade.

The most commonly built link-based surfaces are Top for contours and Datum for earthworks; however, you can build a surface from any link code in your corridor. Figure 9.42 shows a schematic of how links are used to form the most common surfaces—in this case a top surface, as shown in the top image of Figure 9.42, and a datum surface, as shown in the bottom image of Figure 9.42.

FIGURE 9.42
Schematic of Top links connecting to form a surface (top) and schematic of Datum links connecting to form a surface (bottom)

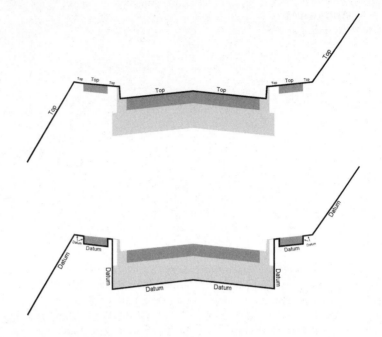

OVERHANG CORRECTION FOR CONFUSED DATUM SURFACES

A common situation with assemblies is a material that juts out past another material, such as curb sub-base. Triangulated surfaces cannot contain caves or perfectly vertical faces because both of those scenarios result in two elevation points for a given (x,y) coordinate pair, so Civil 3D needs to go around the material in a logical manner. What the software sees as logical and what you actually want from the surface may not always be the same thing, as shown here.

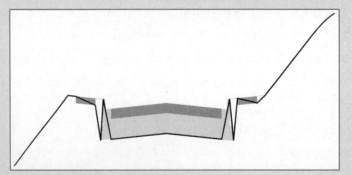

If you have a datum surface doing an unexpected zigzag, change the Overhang Correction setting to Bottom Links. There is also an option to set Overhang Correction to Top Links. This Overhang Correction setting is in the Corridor Properties dialog on the Surfaces tab. After your corridor rebuilds, the result will resemble the bottom image of Figure 9.42.

When building a surface from links, you have the option of checking a box in the Add As Breakline column. Checking this box will add the actual link lines themselves as additional breaklines to the surface. In most cases, especially in intersection design, checking this box forces better triangulation.

CREATING A SURFACE FROM FEATURE LINES

There might be cases where you would like to build a simple surface from your corridor—for example, by using just the crown and edge-of-travel way. If you build a surface from feature lines only or a combination of links and feature lines, you have more control over what Civil 3D uses as breaklines for the surface.

If you added all the topmost corridor feature lines to your surface item and built a surface, you would get a result that's very similar to the result you would get if you had added the Top link codes.

CREATING A SURFACE FROM BOTH LINK DATA AND FEATURE LINES

A link-based surface can be improved by the addition of feature lines. A link-based surface does not automatically include the corridor feature lines but instead uses the link vertex points to create triangulation. Therefore, the addition of feature lines ensures that triangulation occurs where desired along ridges and valleys. This is especially important for intersection design, curves, and other corridor surfaces where triangulation around tight corners is critical. Figure 9.43 shows the Surfaces tab of the Corridor Properties dialog where a Top link surface will be improved by the addition of Back_Curb, ETW, and Top_Curb feature lines.

FIGURE 9.43
The Surfaces tab indicates that the surface will be built from Top links as well as from several feature lines.

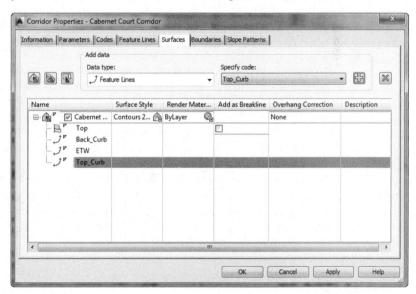

If you are having trouble with triangulation or contours not behaving as expected, experiment with adding a few feature lines to your corridor surface definition.

Creating a Corridor Surface for Each Link

To the right of the Create Corridor Surface button is the Create A Corridor Surface For Each Link button. Clicking this button populates the Surface List area with a multitude of corridor surfaces; as the button name would suggest, you will now have a corridor surface for each of the link codes.

This may not be desirable for many people because you rarely need a surface for every link, but once they are created, you can always remove the unwanted corridor surfaces using the Delete Surface Item button.

Using the Corridor Surface Name Template

The third button at the upper left of the Surfaces tab of the Corridor Properties dialog is the Surface Name Template button. When you click this button, the Name Template dialog shown in Figure 9.44 appears.

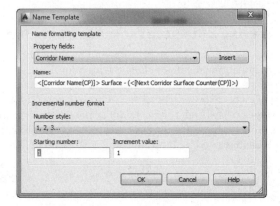

FIGURE 9.44
The Name Template dialog for corridor surfaces

Here you can set the formatting for the corridor name as well as the number style, starting number, and increment value. The Property fields are Corridor Name and Next Corridor Surface Counter. Based on the name shown in Figure 9.44, the next surface created for the corridor named Cabernet Court Corridor will be named Cabernet Court Corridor Surface – (1). The Number style can be set to 1, 2, 3 … or 01, 02, 03 … or a multitude of other styles based on the number of leading zeros.

Completing Other Surface Tasks

You can do several other tasks on the Surfaces tab. For each corridor surface, you can set a surface style, revise the default name assigned by the Name Template, and provide a description for your corridor surface. Alternatively, you can do all those things once the corridor surface appears in the Surface collection through Prospector.

Adding a Surface Boundary

Surface boundaries are critical to any surface, but especially so for corridor surfaces. Tools that automatically and interactively add surface boundaries, using the corridor intelligence, are available. Figure 9.45 shows a corridor surface before and after the addition of a boundary. Notice how the extraneous contours have been eliminated along the line of intersection between the existing ground and the proposed ground (the daylight line), thereby creating a much more accurate surface.

CREATING A CORRIDOR SURFACE | 443

FIGURE 9.45
A corridor surface before the addition of a boundary (left) and after the addition of a boundary (right)

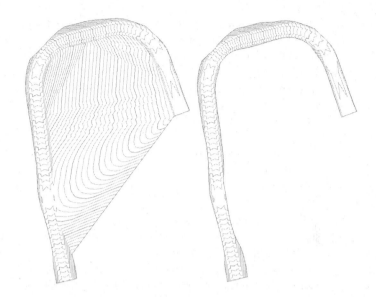

You can create corridor surface boundaries using the Boundaries tab of the Corridor Properties dialog. Each corridor surface will be listed. To add the boundary, right-click on the corridor surface and select the desired boundary type.

Boundary Types

There are several tools to assist you in corridor surface boundary creation. They can be automatic, semiautomatic, or manual in nature, depending on the complexity of the corridor.

You access these options on the Boundaries tab of the Corridor Properties dialog by right-clicking the name of your surface item, as shown in Figure 9.46.

FIGURE 9.46
Corridor surface boundary options for a corridor containing a single baseline

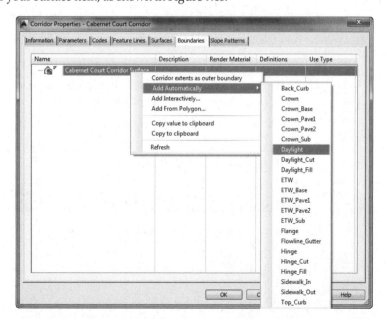

The following corridor boundary methods are listed in order of desirability. Corridor Extents As Outer Boundary is the most user-friendly, whereas Add From Polygon is fast but needs constant updating because it is not dynamically linked to the corridor.

Corridor Extents As Outer Boundary With this selection, Civil 3D will shrink-wrap the corridor, taking into account intersections and various daylight options on different alignments. Corridor Extents As Outer Boundary will probably be your most-used boundary option unless you are modeling other parts of your roadway network in separate corridors. This topic will be covered in more detail in Chapter 10.

Add Automatically The Add Automatically boundary tool allows you to pick a point code and use the associated feature lines as your corridor boundary. This tool is available only for single-baseline corridors. This tool is automatic, easy to apply, and will remain dynamically linked to the corridor.

Add Interactively The Add Interactively boundary tool allows you to work your way around a corridor and choose which corridor feature lines you would like to use as part of the boundary definition.

Choosing this option is better than using Add From Polygon if Add Automatically and Corridor Extents are not available. It takes a bit of patience to trace the corridor, but the result is a dynamically linked boundary that changes when the corridor changes. Using this method, once you select a feature line, a thick line will trace around the corridor following your mouse; to switch to a different feature line, simply click the new feature line at the transition location. When complete, you can close the boundary just as you would a polyline.

Add From Polygon The Add From Polygon tool allows you to choose a closed 2D or 3D polyline or polygon in your drawing that you would like to add as a boundary for your corridor surface. This method is quick, but unlike the other methods, the resulting boundary is not dynamic to your design.

The next exercise leads you through creating a corridor surface with an automatic boundary:

1. Open the CorridorBoundary.dwg or CorridorBoundary_METRIC.dwg file.

2. Select the corridor named Frontenac Drive Corridor to activate the Corridor contextual tab.

3. From the Corridor contextual tab ➢ Modify Corridor panel, choose Corridor Properties.

4. On the Surfaces tab of the Corridor Properties dialog, click the Create A Corridor Surface button in the upper-left corner of the dialog.

   You should now have a surface item in the bottom half of the dialog.

5. Click the surface item under the Name column and change the default name of your surface to **Frontenac – Top Surface**.

6. Verify that Links has been selected from the drop-down list in the Data Type selection box.

7. Verify that Top has been selected from the drop-down list in the Specify Code selection box.

8. Click the Add Surface Item button to add Top Links to the Surface Definition.

9. Click OK to accept the settings in this dialog. Choose Rebuild The Corridor when prompted and examine your surface.

   The road surface should look fine; however, because you have not yet added a boundary to this surface, undesirable triangulation is occurring outside your corridor area.

10. Expand the Surfaces branch in Prospector.

    Note that you now have a corridor surface listed in addition to the EG surface that was already in the drawing.

11. Select the corridor named Frontenac Drive Corridor to activate the Corridor contextual tab.

12. From the Corridor contextual tab ➢ Modify Corridor panel, choose Corridor Properties.

    If you do not see the Corridor Properties button on the Modify Corridor panel, you may have inadvertently chosen the corridor surface and may be viewing the Surface contextual ribbon.

13. On the Boundaries tab of the Corridor Properties dialog, right-click Frontenac – Top Surface in the listing.

14. Hover over the Add Automatically menu flyout, and select Daylight as the feature line that will define the outer boundary of the surface.

15. Verify that the Use Type column says Outside Boundary to ensure that the boundary definition will be used to define the desired extreme outer limits of the surface.

16. Click OK to accept the settings in the Corridor Properties dialog and choose Rebuild The Corridor when prompted.

17. Examine your surface, and note that the triangulation terminates at the Daylight point all along the corridor model.

18. (Optional) Experiment with making changes to your finished grade profile, assembly, or alignment geometry and rebuilding both your corridor and finished ground surface to see the boundary in action.

> **REBUILD: LEAVE IT ON OR OFF?**
>
> Once you rebuild your corridor, your corridor surface will need to be updated. Typically, the best practice is to leave Rebuild – Automatic off for corridors and keep it on for corridor surfaces. The corridor surface will want to rebuild only when the corridor is rebuilt. For very large corridors, this may become a bit of a memory lag, so try it both ways to see what you like best.

When this exercise is complete, you may close the drawing. A finished copy of this drawing is available from the book's web page with the filename `CorridorBoundary_FINISHED.dwg` or `CorridorBoundary_METRIC_FINISHED.dwg`.

Common Surface Creation Problems

Here are some common problems you may encounter when creating surfaces:

Problem Your corridor surface does not appear or seems to be empty.

Typical Cause You might have created the surface item but not added any data. Another cause could be the surface style is set to No Display or the surface was created on a frozen layer.

Fix Open the Corridor Properties dialog and switch to the Surfaces tab. Select a data type from the drop-down menus in the Data Type and Specify Code selection boxes, and click the Add Surface Item button. Make sure your dialog shows both a surface item and a data type, as shown in Figure 9.47.

FIGURE 9.47
A surface cannot be created without both a surface item and a data type.

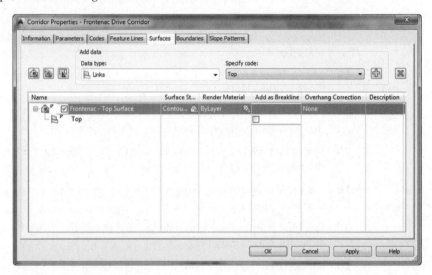

Problem Your corridor surface does not seem to respect its boundary after a change to the assembly or surface-building data type (in other words, you switched from link data to feature lines).

Typical Cause Automatic and interactive boundary definitions are dependent on the codes used in your corridor. If you remove or change the codes used in your corridor, the boundary needs to be redefined.

Fix Open the Corridor Properties dialog and switch to the Boundaries tab. Remove any boundary definitions that are no longer valid (if any) by right-clicking the boundary and selecting Remove Boundary. Once the outdated boundary has been removed, you can redefine the corridor surface boundaries using any of the applicable boundary types.

Problem Your corridor surface seems to have gaps at points of curvature (PCs) and points of tangency (PTs) near curb returns.

Typical Cause You may have encountered an error in rounding station values at these locations and as a result created gaps in your corridor. This is commonly caused by osnapping to start and end region stations using two-dimensional linework as a guide when some segments of that linework do not touch.

PERFORMING A VOLUME CALCULATION | 447

Fix Be sure your corridor region definitions produce no gaps. You might consider using the PEDIT command to join lines and curves representing corridor elements that will need to be modeled later. You might also consider setting a COGO point at these locations (PCs, PTs, and so on) and using the Node object snap instead of the Endpoint object snap to select the same location each time you are required to do so.

Performing a Volume Calculation

One of the most powerful aspects of Civil 3D is having instant feedback on your design iterations. Once you create a preliminary road corridor, you can immediately compare a corridor surface to existing ground and get a good understanding of the earthwork magnitude. When you make an adjustment to the finished grade profile and then rebuild your corridor, you can see the effect that this change has on your earthwork within minutes, if not sooner.

Even though volumes were covered in detail in Chapter 4, "Surfaces," it is worth revisiting the subject here in the context of corridors.

This exercise uses a TIN-to-TIN composite volume calculation to compare the existing ground surface and the datum corridor surface; average end area and other section-based volume calculations are covered in Chapter 12.

1. Open the `CorridorVolume.dwg` or `CorridorVolume_METRIC.dwg` file.

   Note that this drawing has a completed Frontenac Drive corridor, as well as a top corridor surface and a datum corridor surface for Frontenac Drive.

2. From the Analyze tab ➢ Volumes And Material panel, choose Volumes Dashboard to display the Volumes Dashboard tab in the Panorama.

3. Click the Create New Volume Surface button to display the Create Surface dialog.

4. Change the name to **VOL-Frontenac-EG-Datum** and set Style to Elevation Banding (2D).

5. Click the <Base Surface> field to display the ellipsis button; once it's visible, click the ellipsis button to select EG. Then click OK.

6. Click the <Comparison Surface> field to display the ellipsis button; once it's visible, click the ellipsis button to select Frontenac – Datum Surface. Then click OK.

7. Click OK to accept the settings in the Create Surface dialog.

   A Cut/Fill breakdown should appear in the Volumes Dashboard tab of Panorama, as shown in Figure 9.48.

FIGURE 9.48
Panorama showing an example of a volume surface and the cut/fill results

8. Make a note of these numbers.

9. Leave Panorama open on your screen (make it smaller, if desired), and pan over to the proposed profile for Frontenac Drive.

10. Select the Finished Ground profile and move the vertical triangular grip on the first PVI to the center of the circle shown in the profile view. Notice that the volume calculations have changed to Out Of Date because the corridor isn't set to rebuild automatically.

11. In Prospector, expand Corridors, select Frontenac Drive Corridor, right-click, and choose Rebuild. If any errors appear in Panorama, they can be ignored for this exercise.

    Notice that the corridor changes, and therefore the corridor surfaces (which are both set to Rebuild Automatic) change as well. However, the volume surface did not automatically rebuild.

12. On the Volumes Dashboard tab of Panorama, right-click the volume surface and select Rebuild.

    Notice the new values for cut and fill.

13. Close Panorama using the X in the upper corner.

When this exercise is complete, you may close the drawing. A finished copy of this drawing is available from the book's web page with the filename CorridorVolume_FINISHED.dwg or CorridorVolume_METRIC_FINISHED.dwg.

Building Non-Road Corridors

As discussed in the beginning of this chapter, corridors are not just for roads. Once you have the basics about corridors down, your ingenuity can take hold.

Corridors can be used for far more than just road designs. You will explore some more advanced corridor models in Chapter 10, but there are plenty of simple, single-baseline applications for alternative corridors such as channels, berms, retaining walls, and more. You can take advantage of several specialized subassemblies or build your own custom assembly using the Subassembly Composer. Figure 9.49 shows an example of a channel corridor.

FIGURE 9.49
A simple channel corridor viewed in 3D built from the channel subassembly and a generic link subassembly

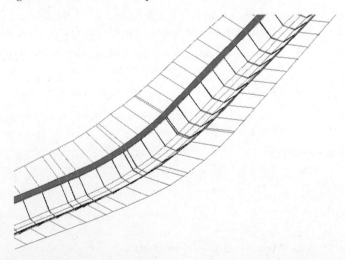

One of the subassemblies discussed in Chapter 8 is the channel subassembly. The following exercise shows you how to apply this subassembly to design a simple drainage channel:

1. Open the `CorridorChannel.dwg` or `CorridorChannel_METRIC.dwg` file.

   Note that there is an alignment that represents a drainage channel centerline, a profile that represents the drainage channel normal water line, and an assembly created using the Channel and LinkSlopetoSurface subassemblies.

2. From the Home tab ➢ Create Design panel, choose Corridor to display the Create Corridor dialog.

3. In the Name text box, name your corridor **Drainage Channel**.

   Keep the default values for Corridor Style and Corridor Layer.

4. Verify that Alignment is set to Channel CL and Profile is set to Channel NWL.

   NWL stands for *normal water level*.

5. Verify that Assembly is set to Project Channel.

6. Verify that Target Surface is set to Existing Ground.

7. Verify that the Set Baseline And Region Parameters check box is checked.

8. Click OK to accept the settings in the Create Corridor dialog and to display the Baseline And Region Parameters dialog.

9. Click the Set All Frequencies button.

10. Change the values for Along Tangents and Along Curves to **10'** (**3 m** for metric users).

11. Click OK to accept the settings in the Frequency To Apply Assemblies dialog. Click OK to accept the settings in the Baseline And Region Parameters dialog.

12. You may receive a dialog warning that the corridor definition has been modified. If you do, select the Rebuild The Corridor option.

13. Select the corridor to activate the Corridor tab.

14. From the Corridor contextual tab ➢ Modify Corridor Sections panel, choose Section Editor.

15. From the Section Editor contextual tab ➢ Station Selection panel, you can navigate through the drainage channel cross sections by clicking the forward or backward arrows.

    The cross section should look similar to Figure 9.50.

FIGURE 9.50
The completed Drainage Channel corridor viewed in the Section Editor

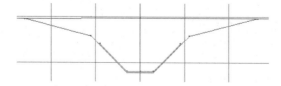

This corridor can be used to build a surface for a TIN-to-TIN volume calculation or to create sections and generate material quantities, cross-sectional views, and anything else that can be done with a more traditional road corridor.

When you are finished viewing the sections, dismiss the dialog by clicking the X on the Section Editor ➤ Close panel. When this exercise is complete, you may close the drawing. A finished copy of this drawing is available from the book's web page with the filename `CorridorChannel_FINISHED.dwg` or `CorridorChannel_METRIC_FINISHED.dwg`.

Real World Scenario

CREATING A PIPE TRENCH CORRIDOR

Another use for a corridor is a pipe trench. A pipe trench corridor is useful for determining quantities of excavated material, limits of disturbance, trench-safety specifications, and more. This graphic shows a completed pipe trench corridor.

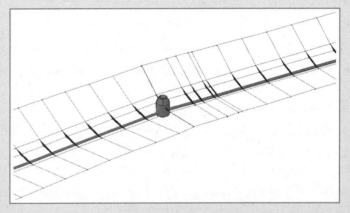

One of the subassemblies discussed in Chapter 8 is the TrenchPipe1 subassembly. The following exercise leads you through applying this subassembly to a pipe trench corridor:

1. Open the `CorridorPipeTrench.dwg` or `CorridorPipeTrench_METRIC.dwg` file.

   Note that there is a pipe network with a corresponding alignment, profile view, and pipe trench assembly. Also note that there is a profile drawn that corresponds with the inverts of the pipe network.

2. From the Home tab ➤ Create Design panel, choose Corridor.

3. In the Name text box, name the corridor **Pipe Trench**.

   Keep the default values for Corridor Style and Corridor Layer.

4. Verify that Alignment is set to Pipe Centerline and Profile is set to Bottom Of Pipe Profile.
5. Verify that Assembly is set to Pipe Trench.
6. Verify that Target Surface is set to Existing Ground.
7. Verify that the Set Baseline And Region Parameters check box is selected.
8. Click OK to accept the settings in the Create Corridor dialog and display the Baseline And Region Parameters dialog.
9. Click the Set All Frequencies button.
10. Change the values for Along Tangents and Along Curves to **10′** (**3** m for metric users).
11. Click OK to accept the settings in the Frequency To Apply Assemblies dialog and click OK again to accept the settings in the Baseline And Region Parameters dialog.
12. You may receive a dialog warning that the corridor definition has been modified. If you do, select the Rebuild The Corridor option.

    The corridor will build.
13. Select the corridor to activate the Corridor contextual tab.
14. From the Corridor contextual tab ➢ Modify Corridor Sections panel, choose Section Editor. Browse the cross sections through the trench.
15. When you are finished viewing the sections, dismiss the Section Editor by clicking the X on the Section Editor ➢ Close panel.

You may notice that at the sharp bends in the pipe alignment the corridor frequency lines cross one another. This is called a bow-tie and will be discussed further in the next chapter.

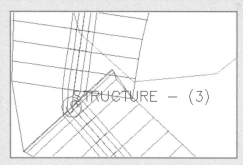

When this exercise is complete, you may close the drawing. A finished copy of this drawing is available from the book's web page with the filename `CorridorPipeTrench_FINISHED.dwg` or `CorridorPipeTrench_METRIC_FINISHED.dwg`.

The Bottom Line

Build a single baseline corridor from an alignment, profile, and assembly. Corridors are created from the combination of alignments, profiles, and assemblies. Although corridors can be used to model many things, most corridors are used for road design.

Master It Open the MasteringCorridors.dwg or MasteringCorridors_METRIC.dwg file. Build a corridor named Corridor A on the basis of the Alignment A alignment, the FG profile, and the Basic Assembly. Set all frequencies to 10′ (or 3 m for metric users).

Use targets to add lane widening. Targets are an essential design tool used to manipulate the geometry of the road.

Master It Open the MasteringCorridorTargets.dwg or MasteringCorridorTargets_METRIC.dwg file. Set Right Lane to target Alignment A-Right.

Create a corridor surface. The corridor model can be used to build a surface. This corridor surface can then be analyzed and annotated to produce finished road plans.

Master It Open the MasteringCorridorSurface.dwg or MasteringCorridorSurface_METRIC.dwg file. Create a corridor surface for the Alignment A corridor from Top links. Name the surface Corridor A-Top.

Add an automatic boundary to a corridor surface. Surfaces can be improved with the addition of a boundary. Single-baseline corridors can take advantage of automatic boundary creation.

Master It Open the MasteringCorridorBoundary.dwg or MasteringCorridorBoundary_METRIC.dwg file. Use the Automatic Boundary Creation tool to add a boundary using the Daylight code.

Chapter 10

Advanced Corridors, Intersections, and Roundabouts

This chapter focuses on taking your corridor-modeling skills to a new level by introducing more tools to your corridor-building toolbox, such as intersecting roads, cul-de-sacs, advanced techniques, and troubleshooting. You will use advanced corridor targets and work with conditional subassemblies.

This chapter assumes that you've worked through the examples in the chapters on alignments, profiles, profile views, assemblies, and basic corridors. Without a strong knowledge of the foundational skills, many of the tasks in this chapter will be difficult.

In this chapter, you will learn to:

- Create corridors with non-centerline baselines
- Add alignment and profile targets to a region for a cul-de-sac
- Create a surface from a corridor and add a boundary

Using Multiregion Baselines

In the previous chapter, you modeled corridors with one baseline and one region. A question many people ask when working with corridors is, "At what point do I need another region?" The answer is simple: If you need a different assembly, you need a different region.

In the following example, you will step through adding an additional region to an existing baseline:

1. Open the drawing 1001_MultiRegionCorr.dwg (1001_MultiRegionCorr_METRIC.dwg).

    This drawing has been split into two modelspace viewports so that you can observe the results of your efforts in 3D.

2. Select the corridor and, from the Corridor contextual tab ➢ Modify Corridor panel, select Corridor Properties.

    On the Parameters tab of the Corridor Properties dialog, notice there is a baseline containing a single region.

3. Right-click on the region and select Split Region, as shown in Figure 10.1.

FIGURE 10.1
Right-click on a region to access region modification tools.

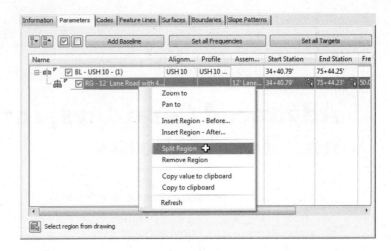

4. At the command line, enter **4000** ↵ (**1220** ↵ for metric users) to create a split at 40+00 (1+220 for metric users), and create a second split at 55+00 (1+680 for metric users) by entering **5500** ↵ (**1680** ↵).

5. Press ↵ again to return to the Corridor Parameters dialog.

   If you see a warning dialog referring to 0+00 being outside of the station limits, click OK and continue.

   You should have three regions at this step, as shown in Figure 10.2.

FIGURE 10.2
New regions with different assemblies applied

6. Change the assembly for the middle region to use the Road w Guardrail assembly, and then click OK.

7. In the same row, scroll over and click the Target ellipsis button.

8. Verify that the targets have been maintained for this region. If necessary, set the targets as shown in Figure 10.3.

9. Click OK when target mapping is complete, and click OK again to close the Corridor Properties dialog.

10. Select Rebuild The Corridor when prompted.

    You should now see additional feature lines in the plan view representing the top of the guardrail. In the right viewport, it will resemble Figure 10.4.

11. Save and close the drawing.

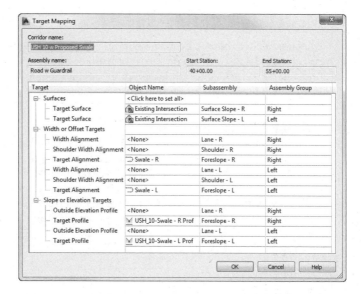

FIGURE 10.3
Target alignments and target profiles for the center region

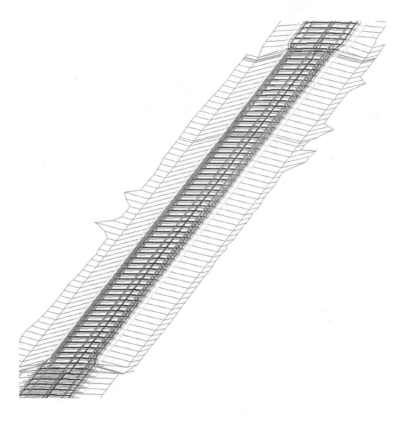

FIGURE 10.4
The completed corridor with guardrails

Modeling a Cul-de-Sac

Even if you never plan to design one in real life, understanding what is going on in a cul-de-sac corridor model will set you on the right path for building more complex models. If you truly understand the principles explained in the section that follows, then expanding your repertoire to include intersections and roundabouts will become much easier.

Using Multiple Baselines

Up to this point, every corridor we've examined has had a single baseline. In our examples, the centerline alignment has been the main driving force behind the corridor design. In this section, the training wheels are coming off! You've seen the last of single baseline corridors in this book. A cul-de-sac by itself can be modeled in two baselines, as shown in Figure 10.5. The procedures that follow will work for most cul-de-sacs, symmetrical, asymmetrical, and hammerhead styles. You will need the centerline alignment and design profile as well as an edge of pavement alignment and design profile.

FIGURE 10.5
Example cul-de-sac alignment setup

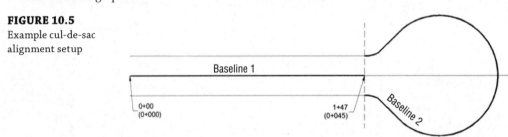

In the section leading to the cul-de-sac bulb, the corridor can be modeled using the tools you learned to use in Chapter 9, "Basic Corridors." In the example shown in Figure 10.5, the centerline of the road is the baseline, with an assembly using two lanes and with the crown as the baseline marker from a beginning station of 0+00 to the station 1+47 (for metric users 0+045). However, once the curvature of the bulb starts at 1+47 (for metric users 0+045), the centerline is no longer an acceptable baseline.

Assemblies are always applied to a baseline, and the assembly geometry will be perpendicular to the baseline. To get correct pavement grades in the bulb, you'll need to hop over to the edge of pavement alignment for a baseline. This second baseline will use a partial assembly that is built with the main assembly marker at the outside edge of pavement (Figure 10.6). The crown of the road will stretch to meet the centerline alignment and profile as its targets.

FIGURE 10.6
Assembly used for designing off the edge of pavement

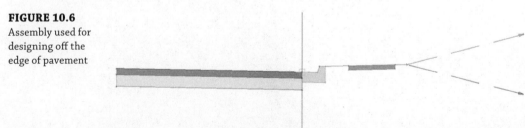

It helps to think of the assembly as radiating away from the baseline, from the assembly base outward, toward a target (Figure 10.7). Because the assemblies are applied to the baseline in a perpendicular manner, using the edge of pavement for a baseline in curved areas (such as cul-de-sac bulbs or curb returns) will result in a smooth, properly graded pavement surface.

FIGURE 10.7
It helps to think of the assemblies radiating away from the baseline toward the targets.

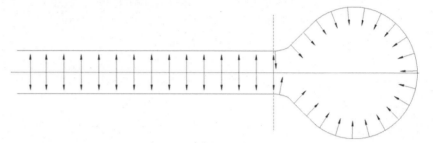

Establishing EOP Design Profiles

One of the most challenging parts of any "fancy" corridor (i.e., cul-de-sac, intersection, or roundabout) is establishing design profiles for non-centerline alignments. You must have design profiles for both the centerline and edge of pavement alignments, but it does not have to be a painful process to obtain them.

Using a simple, preliminary corridor and the profile creation tools you learned about in Chapter 7, "Profiles and Profile Views," you'll find that establishing an edge of pavement (EOP) profile can go quickly.

In the exercise that follows, you will work through the steps of creating an EOP profile:

1. Open the 1002_EOPProfile.dwg (1002_EOPProfile_METRIC.dwg) file, which you can download from this book's web page, www.sybex.com/go/masteringcivil3d2014.

   This drawing contains several alignments and an existing surface whose style is set to show the border only.

2. From the Home tab ➢ Create Design panel, click Corridor, and do the following:

   a. Name the corridor **PRELIM**.

   b. Set Alignment to Frontenac Drive.

   c. Set Profile to Frontenac Drive – FG.

   d. Set Assembly to PRELIM.

   e. Clear the check box for Set Baseline And Region Parameters.

3. No targets are needed in this preliminary corridor, so click OK to complete the corridor.

4. Select the new corridor, and from the Corridor contextual tab ➢ Modify Panel, click Corridor Properties.

5. On the Surfaces tab, click the leftmost button to start a corridor surface.

6. With Data Type set to Links and Specify Code set to Top, click the plus sign to add data to the surface.

7. On the Boundaries tab, right-click the PRELIM surface, select Corridor Extents As Outer Boundary.

8. Click OK and rebuild the corridor.

   You should now see contours in your drawing representing the 2 percent crossfall from the centerline.

9. Press Esc to clear the selection, if needed. Select the red EOP alignment for the cul-de-sac, and in the Profile contextual tab ➤ Launch Pad panel, click Surface Profile.

10. In the Create Profile From Surface dialog, highlight PRELIM Surface - (1) and click Add.

11. Click the Draw In Profile View button.

12. Leave all the defaults in the Create Profile View dialog and click Create Profile View.

13. Click in the graphic to the right of the surface to place the profile view.

    The profile you are seeing will have gaps in the middle that need grading information. However, the bulb portion of the profile only needs to exist between 16+87.00 (0+514.2 for metric users) and 19+42.89 (0+592.2 for metric users). They are the PC and PT stations for the EOP alignment in plan. In this example, you will create a design profile that starts slightly before, and ends slightly after, these critical stations.

14. Select the profile view. (Hint: Click a grid line rather than the profile itself.) From the Profile View contextual tab ➤ Modify panel, click Profile Creation Tools.

15. Click OK to accept the defaults in the Create Profile dialog.

16. In the Profile Layout Tools toolbar, select Draw Tangents, and snap to the intersection of the preliminary profile and the grid line at 16+50 (0+500 for metric users). The elevation is also determined by this location.

17. Next, snap to the endpoint where the preliminary surface trails off at station 17+02.38 (0+517.13 for metric users), elevation 792.88′ (241.737 m).

18. The next VPI will be the peak of the preliminary profile in the center of the view; station 18+14.94 (0+553.20 for metric users), elevation 790.35′ (240.898 m).

19. Snap to the endpoint where the preliminary profile picks up again; station 19+27.51 (0+589.26 for metric users), elevation 792.88′ (241.737 m).

20. Finally, snap to the grade break at station 19+78.30 (0+602.99 for metric users), elevation 794.68′ (242.219 m).

21. Press ↵ to complete the command and close the Profile Layout Tools toolbar.

    You now have a proposed profile that is acceptable to use in the cul-de-sac corridor. Check your work against `1002_EOPProfile_ FINISHED.dwg` or `1002_EOPProfile_ METRIC_FINISHED.dwg` to see how your stations and elevations compare.

Putting the Pieces Together

You have all the pieces in place to perform the first iterations of this cul-de-sac design.

The following exercise will walk you through the steps to put the cul-de-sac together. You will complete several steps and let the corridor build to observe what is happening at each stage. This exercise will also encourage you to get comfortable using the Corridor Properties dialog to make design modifications.

1. Open the `1003_Cul-de-SacDesign.dwg` (`1003_Cul-de-SacDesign_METRIC.dwg`) file, which you can download from this book's web page.

   This drawing contains the cul-de-sac centerline alignment and profile, the EOP alignment and profile, and the assemblies needed to complete the process. The PRELIM corridor layer is frozen. The view of this drawing is twisted 90° to better fit most monitors.

2. From the Home tab ➢ Create Design panel, click Corridor, and do the following:

   a. Name the corridor **Cul-de-Sac**.
   b. Set Alignment to Frontenac Drive.
   c. Set Profile to Frontenac Drive FG.
   d. Set Assembly to Urban.
   e. Set Target Surface to EG.
   f. Keep the check mark next to Set Baseline And Region Parameters.
   g. Click OK.

3. In the Baseline And Region Parameters dialog, click the Pick Station button for the start station of the first region.

4. At the `Specify station along alignment:` prompt, click (using the Endpoint Osnap) to the start of the EOP alignment on the west side, as shown in Figure 10.8 (top).

   This will result in a start station of 11+42.86 (0+348.34 for metric users) in the Create Corridor dialog.

5. At the `Specify station along alignment:` prompt, click the Pick Station icon for the end station of the first region.

6. Snap to the PT station on the east side of the cul-de-sac bulb, as shown in Figure 10.8 (bottom).

   This will result in the region end station of 15+35.41 (0+467.99 for metric users).

7. Click OK to dismiss the Baseline And Region Parameters dialog. Click Rebuild The Corridor if necessary.

   Examine the corridor you just created. It should start south of the intersection with the Syrah Way alignment and end before the curvy part of the cul-de-sac bulb.

Corridor Properties

8. Select the corridor. From the Corridor contextual tab ➢ Modify Corridor panel, click Corridor Properties.

   Next you will switch to the Parameters tab, the corridor designer's best friend. This is a screen very similar to the one you saw when you set the baseline regions and stations initially. You can change the region stations here if you missed this step earlier in the process.

FIGURE 10.8
Setting the start station (top) and end station (bottom) for the centerline region

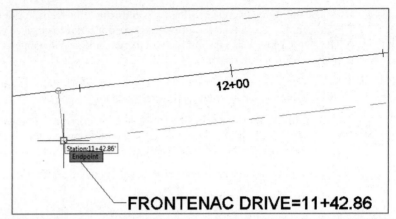

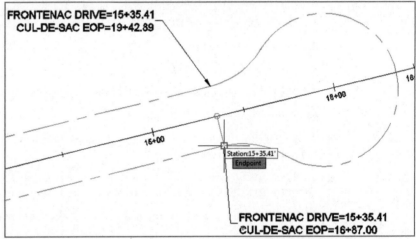

9. In the Parameters tab of the Corridor Properties dialog, click Add Baseline.

10. In the Create Corridor Baseline dialog, select Cul-de-Sac EOP as the alignment, and click OK.

11. In the Profile column of the newly added baseline, click in the field containing <Click here...>, and in the Select A Profile dialog, select the Cul-de-Sac EOP – FG profile and click OK.

    This is the profile that you learned how to develop in the previous exercise.

12. Right-click on the newly added baseline, and select Add Region, as shown in Figure 10.9.

13. In the Create Corridor Region dialog, select the Curb Right assembly, and click OK.

14. Expand the newly added Cul-de-Sac baseline and then click the Pick Station button for the Curb Right region start station.

15. Select the PC station of the corridor bulb (station 16+87.00 in the Imperial drawing and 0+514.20 in the metric drawing).

FIGURE 10.9
Right-click the baseline to add a region.

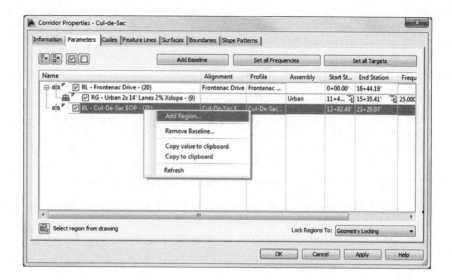

16. Click the Pick Station icon for the end station of the new region.
17. Select the PT station of the corridor bulb.

    This will result in an end station of 19+42.89 (0+592.19 for metric users).

18. Click OK and let the corridor rebuild once again.

    Have a look at the corridor in its current state (see Figure 10.10). Even if you are not exactly sure what you are looking at, you should at least see that a few things are amiss with the corridor so far.

FIGURE 10.10
The cul-de-sac corridor several steps away from completion

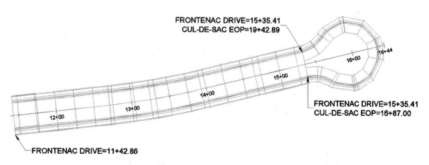

There are three things that need to be modified before the cul-de-sac is complete. The lane needs to extend to the center of the cul-de-sac bulb and the daylight surface needs to be set, both of which can be corrected in the Target Mapping area. Finally, you will want to increase the frequency around the curvy area to get a smoother, more precise design.

In the next steps, you will correct these issues and complete the cul-de-sac.

19. Select the corridor and return to Corridor Properties (last time in this exercise, we promise).

20. Scroll over and click the ellipsis in the Target column to enter the Target Mapping dialog for the region named RG – Curb Right – (36). Note that the number following the region name may vary because these are added automatically by the AutoCAD® Civil 3D® software.

    In the Target Mapping dialog, do the following:

    a. Set Target Surface to EG, and then click OK.

    b. Set Width Alignment for Lane – L to Frontenac Drive. Click Add and then click OK.

    c. Set Outside Elevation Profile to Frontenac Drive – FG for Lane – L. Click Add and then click OK.

    d. Click OK again to dismiss the Target Mapping dialog.

21. In the Cul-de-Sac region row, click the frequency ellipsis next to the current value of 25′ (20 m), and in the Frequency To Apply Assemblies dialog, set the frequency for both tangents and curves to 5′ (1 m). Set the At Offset Target Geometry Points to Yes.

22. Click OK to dismiss the Frequency To Apply Assemblies dialog.

23. Click OK and let the corridor rebuild one last time.

    The completed corridor will look like Figure 10.11.

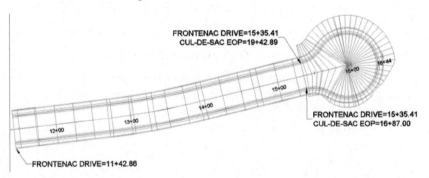

FIGURE 10.11
The completed cul-de-sac corridor. Gorgeous!

24. Once you are satisfied with your work, save and close the file.

    The files `1003_Cul-de-SacDesign_FINISHED.dwg` and `1003_Cul-de-SacDesign_METRIC_FINISHED.dwg` are available for you to check your work.

Troubleshooting Your Cul-de-Sac

People make several common mistakes when modeling their first few cul-de-sacs:

Your cul-de-sac appears with a large gap in the center. If your curb line seems to be modeling correctly but your lanes are leaving a large empty area in the middle (see Figure 10.12), chances are pretty good that you forgot to assign targets or perhaps assigned the incorrect targets.

Fix this problem by opening the Target Mapping dialog for your region and checking to make sure you assigned the road centerline alignment and FG profile for your transition lane. If you have a more advanced lane subassembly, you may have accidentally set the targets for another subassembly somewhere in your corridor instead of the lane for the cul-de-sac transition, especially if you have poor subassembly-naming conventions. Poor naming conventions become especially confusing if you use the Map All Targets button.

FIGURE 10.12
A cul-de-sac without targets

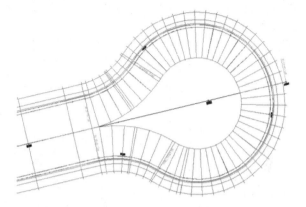

To pinpoint the location of errors, use the Select Region From Drawing button in the Parameters tab of Corridor Properties to select the region you wish to modify. Edit targets one region at a time to avoid confusion.

Your cul-de-sac appears to be backward.   Occasionally, you may find that your lanes wind up on the wrong side of the EOP alignment, as shown in Figure 10.13. The direction of your alignment will dictate whether the lane should be on the left or right side. In the example from the previous exercise, the alignment was running counterclockwise around the cul-de-sac bulb; therefore, the lane was on the left side of the assembly.

You can fix this problem by changing the assembly applied to the region to one that was created for the correct side.

Your cul-de-sac drops down to 0.   A common problem when you first begin modeling cul-de-sacs, intersections, and other corridor components is that one end of your baseline drops down to 0. You probably won't notice the problem in plan view, but once you build your surface (see Figure 10.14, left) or rotate your corridor in 3D (see Figure 10.14, right), you'll see it. This problem will always occur if your region station range extends beyond the proposed profile.

To fix this problem, make sure your region station range corresponds with the design profile length. You may need to extend the design profile in some cases, but usually restricting the station range of the region will do the trick.

Your cul-de-sac seems flat.   When you're first learning the concept of targets, it's easy to mix up baseline alignments and target alignments. In the beginning, you may accidentally choose your EOP alignment as a target instead of the road centerline. If this happens, your cul-de-sac will look similar to Figure 10.15.

You can fix this problem by opening the Target Mapping dialog for this region and making sure the target alignment is set to the road centerline and the target profile is set to the road centerline FG profile.

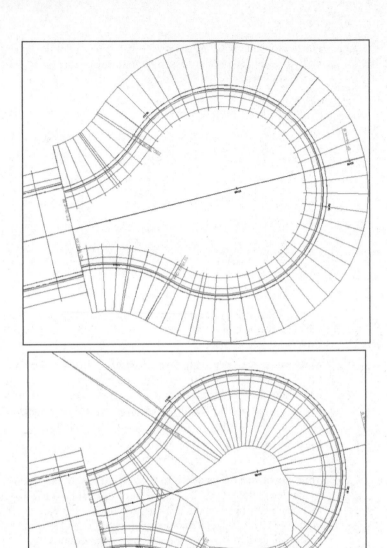

FIGURE 10.13
A cul-de-sac with the lanes modeled on the wrong side without targets (top) and with targets (bottom)

FIGURE 10.14
Contours indicating that the corridor surface drops down to 0 (top), and a corridor viewed in 3D showing a drop down to 0 (bottom)

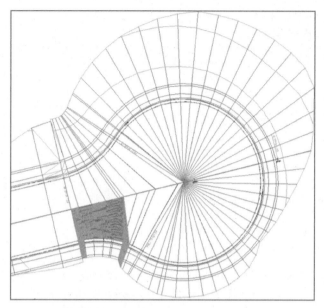

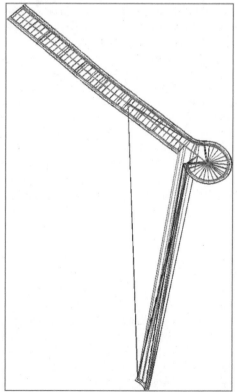

FIGURE 10.15
A flat cul-de-sac with the wrong lane target set

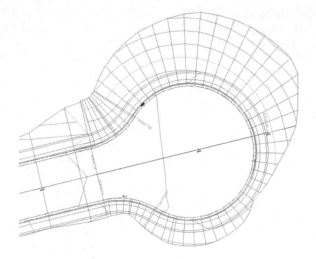

Moving Up to Intersections

Ask yourself the following question: "Did I understand why we did what we did to build the cul-de-sac in the previous section?" If the answer is "Not really," you may want to review a few topics before proceeding. If the answer is "Yeah, mostly," then you are ready to move to the next level of corridor complexity: intersections.

The steps that follow apply to all intersections, regardless of whether it is a T-shaped intersection, a four-way intersection, perfectly perpendicular, or skewed at an angle.

Certification Objective

Corridor modeling is an iterative process. The more advanced your model, the more iterations it may take to get to the correct design. You will often not know the final design parameters until you see how the model relates to existing conditions or ties into other pieces of the design. Get comfortable jumping in and out of corridor properties and identifying regions within your corridor.

Plan what alignments, profiles, and assemblies you'll need to create the right combination of baselines, regions, and targets to model an intersection that will interact the way you want. It helps to create a simple sketch, as shown in Figure 10.16.

FIGURE 10.16
Plan your intersection model in sketch form.

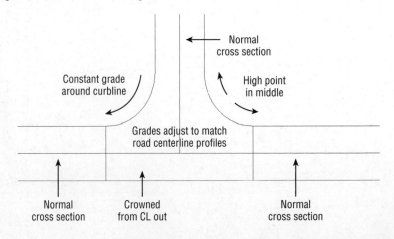

Figure 10.17 shows a sketch of required baselines. As you saw in the previous example, *baselines* are the horizontal and vertical foundations of a corridor. Each baseline consists of an alignment and its corresponding finished ground (FG) profile. You may never have thought of edge of pavement (EOP) in terms of profiles, but after you build a few intersections, thinking that way will become second nature. The Intersection tool on the Create Design panel of the Home tab will create EOP baselines as curb return alignments for you, but it will rely on your input for curb return radii.

FIGURE 10.17
Required baselines for modeling a typical intersection

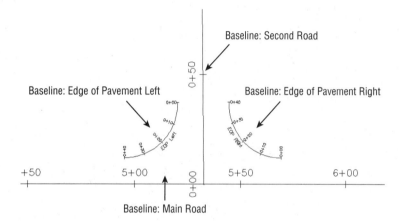

Figure 10.18 breaks each baseline into regions where a different assembly or different target will be applied. Once the intersection has been created, target mapping as well as other particulars can be modified as needed.

FIGURE 10.18
Required regions for modeling an intersection created by the Intersection tool

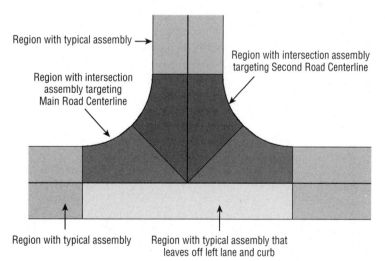

Using the Intersection Wizard

Certification
Objective

All the work of setting baselines, creating regions, setting targets, and applying the correct frequencies can be done manually for an intersection. However, Civil 3D contains an automated Intersection tool that can handle many types of intersections.

On the basis of the schematic you drew of your intersection, your main road will need several assemblies to reflect the different road cross sections. Figure 10.19 shows the full range of potential assemblies you may need in an intersection and the design situations in which they may arise.

FIGURE 10.19
Various assembly schematics and applications

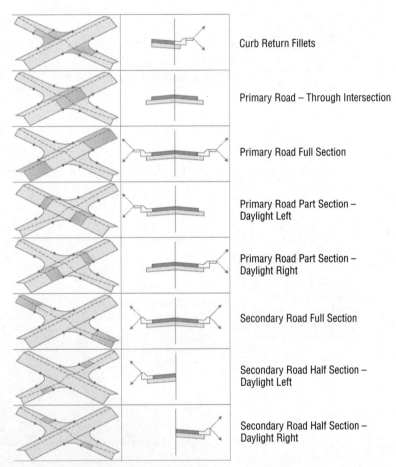

This exercise will take you through building a typical peer-road intersection using the Create Intersection Wizard:

1. Open the 1004_Intersection.dwg (1004_Intersection_METRIC.dwg) file, which you can download from this book's web page.

    The drawing contains two centerline alignments (USH 10 and Mill Creek Drive) that are part of the same corridor.

2. From the Home tab ➢ Create Design panel, choose the Intersections ➢ Create Intersection tool.

3. At the `Select intersection point:` prompt, choose the intersection of the two existing alignments. (Hint: Use the Intersection Osnap.)

4. At the `Select main road alignment <or press enter key to select from list>:` prompt, click the USH 10 alignment that runs vertically in the project.

   The Create Intersection – General dialog will appear (Figure 10.20).

Assembly Sets

When you are ready to create an intersection, you do not need to have all the special assemblies created ahead of time. On the Corridor Regions page of the Create Intersection Wizard, you will see a list of the assemblies Civil 3D plans to use. If the assemblies are not already part of the drawing, they will get pulled in automatically when you click Create Intersection.

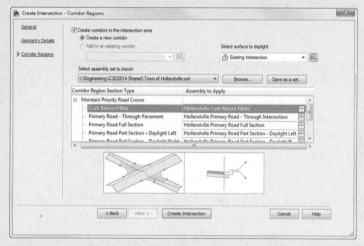

The default intersection assemblies are general and may not work for your design situation. You will want to create and save an assembly set of your own.

1. In a file that contains all of your desired assemblies, work through the Create Intersection Wizard to get to the Corridor Regions page.

2. Click the ellipsis to select the appropriate assembly to use for each corridor region section type.

3. Once the listing is complete, click the Save As A Set button.

Civil 3D creates an XML file that stores the listing of the assemblies. It also creates a copy of each assembly as a separate DWG file. Save the set in a network shared location so your office colleagues can use the set as well.

The next time an intersection is created, you can use the assembly set by clicking Browse and selecting the XML file. Civil 3D will pull in your assemblies, saving lots of time!

FIGURE 10.20
The General page of the Create Intersection Wizard

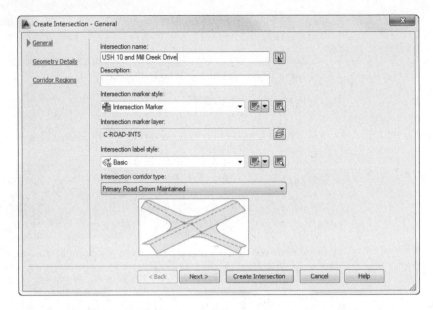

5. Name the intersection **USH 10 and Mill Creek Drive**. Set the intersection corridor type to Primary Road Crown Maintained, as shown in Figure 10.20, and click Next.

6. In the Geometry Details page (Figure 10.21), verify that USH 10 is the primary road by looking at the Priority listing. If USH 10 is not at the top of the list, use the arrow buttons on the right side to reorder the roads.

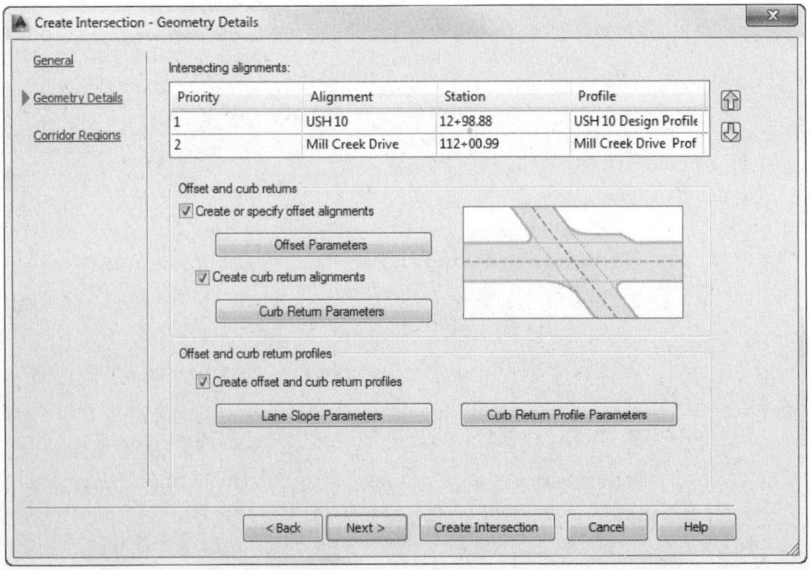

FIGURE 10.21
The Geometry Details page of the Create Intersection Wizard

7. Click the Offset Parameters button, and do the following in the Intersection Offset Parameters dialog:

   a. Set the offset values for both the left and right sides to **24′ (8 m)** for USH 10.

   b. Set the left and right offset values for Mill Creek Drive to **18′ (5 m)**.

   c. Select the check box Create New Offsets From Start To End Of Centerlines.

      At this step, the screen should resemble Figure 10.22.

FIGURE 10.22
Intersection Offset Parameters dialog

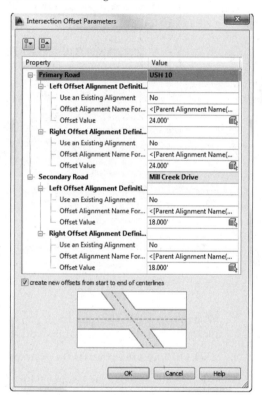

   d. Click OK to close the Intersection Offset Parameters dialog.

8. On the Geometry Details page, click the Curb Return Parameters button to enter the Intersection Curb Return Parameters dialog.

   a. For all four quadrants of the intersection, place a check mark next to Widen Turn Lane For Incoming Road and Widen Turn Lane For Outgoing Road.

   b. Click the Next button at the top of the dialog to move from quadrant to quadrant.

      You will see a schematic name of the quadrant listed at the top of the dialog that corresponds to the glyph. As shown in Figure 10.23, the temporary glyph will help you determine which quadrant you are currently modifying.

FIGURE 10.23
Adding lane widening to the SW – Quadrant of the intersection

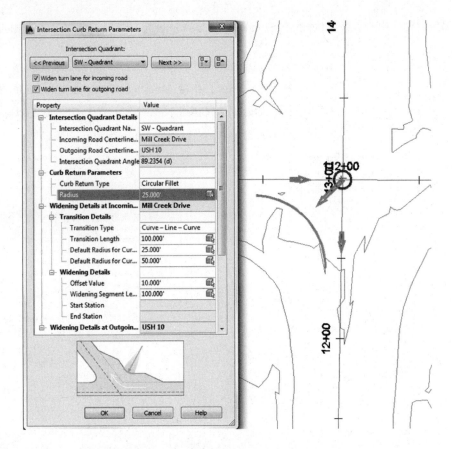

When you reach the last quadrant, you will see that the Next button is grayed out. This means that you have successfully worked through all four curb returns.

   c. Click OK to return to the wizard's Geometry Details page.

Some locales require that lane slopes flatten out to a 1% cross-slope in an intersection. If this is the case for you, you can change the lane slope parameters in the Intersection Lane Slope Parameters dialog (Figure 10.24). In this exercise you will leave this as -2%.

Civil 3D is performing the task of generating the curb return profile. The profile will be at least as long as the rounded curb plus the turn lanes that are added in this exercise. If you wish to have Civil 3D generate even more than the length needed, you can specify that in the Intersection Curb Return Profile Parameters dialog (Figure 10.25).

You will be keeping all default settings in both the Lane Slope Parameters area (Figure 10.24) and the Curb Return Parameters area (Figure 10.25).

9. Click Next to continue to the Corridor Regions page of the Create Intersection Wizard (Figure 10.26).

FIGURE 10.24
Lane slope parameters control the cross-slope in the intersection.

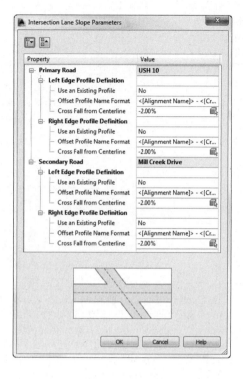

FIGURE 10.25
Intersection Curb Return Profile Parameters options extend the Civil 3D–generated profile beyond the curb returns by the value specified.

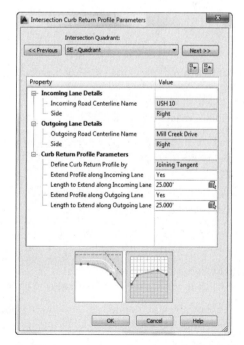

FIGURE 10.26
The Corridor Regions page of the Create Intersection Wizard drives the assemblies used in the intersection.

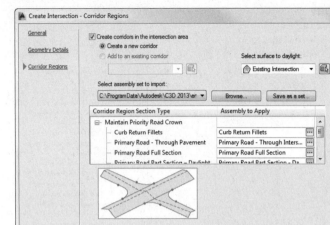

The Corridor Regions page is where you control which assemblies are used for the different design locations around the intersection. Clicking each entry in the Corridor Region Section Type list will give you a clear picture of which assemblies you should use and where (as shown at the bottom of Figure 10.26). If your assemblies have the same names as the default assemblies, as is the case in this example, they will be pulled from the current drawing. Alternately, you can click the ellipsis to select any assembly from the drawing.

If you don't have all the necessary assemblies at this point, you can still create your intersection. Civil 3D will pull in the default set of assemblies. You can always modify these assemblies after they are brought in.

10. Click through the Corridor Region Section Type list to view the schematic preview for each.

   Do not make any changes to the assembly listing.

11. Click Create Intersection.

   After a few moments of processing, you will see a corridor appear at the intersection of the roads. Use the REGEN command if you do not see the frequency lines. Your corridor should now resemble Figure 10.27.

12. Select the corridor, and from the Corridor contextual tab ➢ Modify Corridor panel, select Corridor Properties.

13. Switch to the Parameters tab of the Corridor Properties dialog and highlight one of the regions in the listing, as shown in Figure 10.28.

   Notice how the region highlighted in the Parameters tab is outlined in the graphic. This will help you determine which region to edit, even in the largest of corridors.

   To move from region to region without searching through the somewhat daunting list, use the Select Region From Drawing button.

   In the next steps you will leverage the work Civil 3D has done for you and extend the intersection in all four directions.

FIGURE 10.27
The nearly completed intersection

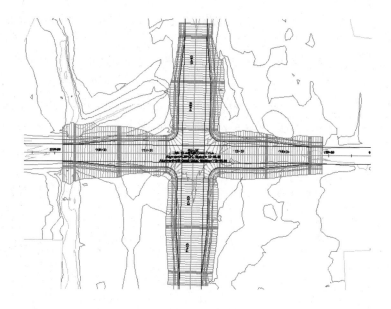

FIGURE 10.28
The best way to modify regions is to use the Corridor Properties. The selected region will highlight graphically.

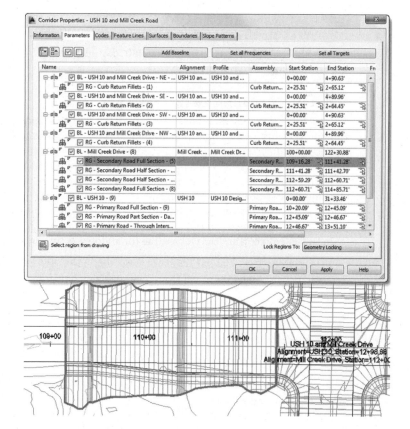

14. Change the region start station on the west side of Mill Creek Road to 101+00 (3+200 for metric users).
15. Change the east side end station to 121+00 (3+700 for metric users).
16. Change the start station of the south end of USH 10 to 10+00 (0+300 for metric users).
17. Change the north side end station to 16+50 (0+500 for metric users).
18. Click OK and rebuild the corridor.

    When the corridor is complete, it will look like Figure 10.29.

FIGURE 10.29
The completed intersection

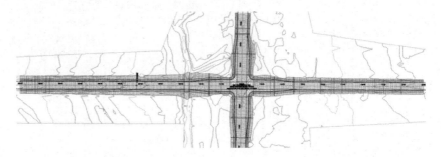

19. Save and close the drawing when you have completed the exercise.

    The files `1004_Intersection_FINISHED.dwg` and `1004_Intersection_METRIC_FINISHED.dwg` are available for your review.

Manually Modeling an Intersection

In some cases, you may find it necessary to model an intersection manually. Five-way intersections and intersections containing superelevated curves can't be created with the automated tools alone. You need to understand what the intersection tool is doing behind the scenes before you can be a true corridor guru. The next example will take you through an overview of the manual steps.

At the point where the example begins, a few pieces are already in place: the centerline alignments and profiles, EOP alignments and profiles, and the full road assemblies. The corridor for the intersection is started for you, but it contains only one baseline. For simplicity's sake, daylight subassemblies have been omitted from this example.

In this example, you'll add a baseline for an intersecting road to your corridor:

1. Open the `1005_ManualIntersection.dwg` (`1005_ManualIntersection_METRIC.dwg`) file, which you can download from this book's web page.
2. Select the corridor, and from the Corridor contextual tab ➢ Modify Corridor panel, select Corridor Properties.
3. Switch to the Parameters tab and click Add Baseline.

    The Create Corridor Baseline dialog opens.

4. In the Create Corridor Baseline dialog, pick the Syrah Way alignment, and then click OK to dismiss the dialog.

5. In the row containing the newly added baseline, click in the Profile column on the Parameters tab of the Corridor Properties dialog.

   The Select A Profile dialog opens.

6. Pick the Syrah Way FG profile, and then click OK to dismiss the dialog.

   You now have a new baseline, but the region still needs to be added. Note that your baseline and region numbers may vary; these are added automatically and increase every time you make a new baseline or region. For this reason, these numbers have been omitted from the exercise steps.

7. In the Corridor Properties dialog, right-click BL – Syrah Way and select Add Region.

8. In the Create Corridor Region dialog, select Full Road Section, and then click OK to dismiss the dialog.

9. Expand BL – Syrah Way by clicking the small + sign, and see the new region you just created.

10. Click OK to dismiss the Corridor Properties dialog and select the option to rebuild the corridor.

    Once the corridor finishes rebuilding, it will look like Figure 10.30.

FIGURE 10.30
The Syrah Way baseline added

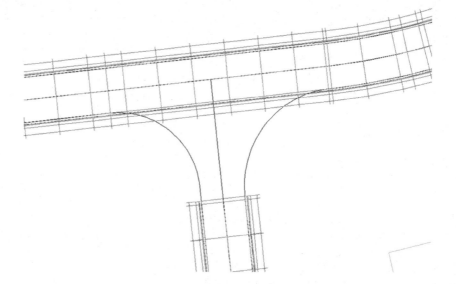

Notice that the region for the second baseline extends all the way through the intersection. You must now split the Syrah region to accommodate the curb returns.

11. Open the Corridor Properties dialog and switch to the Parameters tab. Select the region RG – Full Road Section under the BL – Syrah Way baseline (number may vary), and then right-click and select Split Region.

12. Enter **377.3 (115 m)** and press ↵ to create the first region.

13. Enter 537.21 (163.74 m) and press ↵ to form the second split.
14. Press ↵ again to return to the Corridor Parameters dialog.
15. Switch the middle assembly to Daylight Right and then click OK.

    Your corridor parameters will now match Figure 10.31.

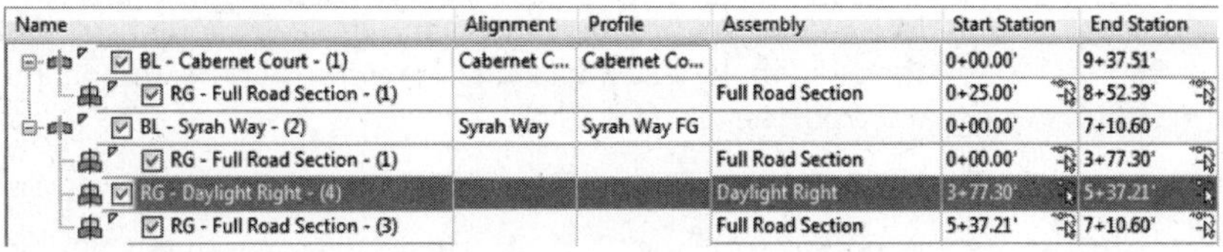

FIGURE 10.31
Split the Syrah Way baseline into three regions to accommodate curb returns.

Name	Alignment	Profile	Assembly	Start Station	End Station
BL - Cabernet Court - (1)	Cabernet C...	Cabernet Co...		0+00.00'	9+37.51'
RG - Full Road Section - (1)			Full Road Section	0+25.00'	8+52.39'
BL - Syrah Way - (2)	Syrah Way	Syrah Way FG		0+00.00'	7+10.60'
RG - Full Road Section - (1)			Full Road Section	0+00.00'	3+77.30'
RG - Daylight Right - (4)			Daylight Right	3+77.30'	5+37.21'
RG - Full Road Section - (3)			Full Road Section	5+37.21'	7+10.60'

16. Click OK and, as always, choose to rebuild the corridor.

    You will see that the way has been cleared for the curb assemblies to be applied, as shown in Figure 10.32.

FIGURE 10.32
The intersection corridor…so far

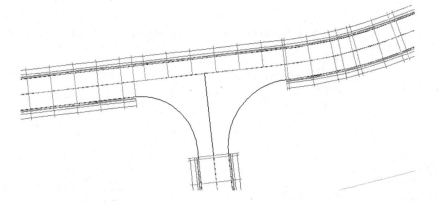

17. Save the drawing for use in the next exercise.

Creating an Assembly for the Intersection

You built several assemblies in Chapter 8, "Assemblies and Subassemblies," but most of them were based on the paradigm of using the assembly marker along a centerline. This next exercise leads you through building an assembly that attaches at the EOP.

This exercise uses the LaneSuperelevationAOR subassembly (see Figure 10.33), but you're by no means limited to this subassembly in practice. You can use any lane subassembly that allows for a width and elevation target such as BasicLaneTransition, GenericPavementStructure, or LaneOutsideSuperWithWidening.

To create an assembly suitable for use on the filleted alignments of the intersection, follow these steps:

1. Continue working in your drawing from the previous exercise (1005_ManualIntersection .dwg or 1005_ManualIntersection_METRIC.dwg). Be sure the previous exercise is complete before proceeding.

FIGURE 10.33
An assembly for an intersection curb return

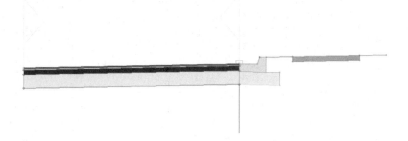

2. From Prospector, expand the Assemblies group and locate Curb Return Fillets.
3. Right-click on this assembly and select Zoom To.

   The assembly base has been placed in the drawing for you, but it does not contain any subassemblies.

4. Open your subassemblies tool palette if it is not already open (hint: use Ctrl+3), and switch to the Lanes tab.
5. Click the LaneSuperelevationAOR subassembly, and on the Properties palette, change the side to Left.
6. Click on the main baseline assembly to add the LaneSuperelevationAOR subassembly to the left side of the assembly.
7. Switch to the Curbs tab of the subassemblies tool palette, and select UrbanCurbGutterGeneral.
8. In the Properties palette, change the side to Right and leave all other values at their defaults.
9. Click anywhere on the baseline to add the curb and gutter to the right side of the assembly.
10. Also on the Curbs tab of the tool palette, click the UrbanSidewalk assembly.
11. On the Properties palette, set the boulevard widths to **2′ (0.5 m)** for both inside and outside, and click to place the subassembly at the top back of the curb.

    An alternate method (which would replace steps 7–10) would be to use the Copy To Assembly option after selecting the curb and gutter and sidewalk from the Daylight Right Assembly.

    Your assembly should now look like Figure 10.33.

> **Take the Pebble from My Hand, Grasshopper**
>
> If you are a little worried about the downward slope of the lane in the subassembly you just created, you needn't be. Remember that the slope and length of the lane will be controlled by the alignment and profile target in the corridor. The geometry that you see in the assembly is purely preliminary. If this concept makes sense to you, you are ready for bigger and better corridors!
>
> And of course, if you'd rather see that slope go +2% in the subassembly, you can change it in the good old AutoCAD properties, but keep in mind it won't make one bit of difference to the corridor.

Adding Baselines, Regions, and Targets for the Intersections

The Intersection assembly attaches to alignments created along the EOP. Because a baseline requires both horizontal and vertical information, you also need to make sure that every EOP alignment has a corresponding finished ground (FG) profile.

It may seem awkward at first to create alignment and profiles for things like the EOP, but after some practice you'll start to see things differently. If you've been designing intersections using 3D polylines or feature lines, think of the profile as a vertical representation of a feature line and the profile grid view as the feature-line elevation editor. If you've designed intersections by setting points, think of alignment PIs and profile PVIs as points. If you need a low point midway through the EOP, as indicated in the sketch at the beginning of this section, you'll add a PVI with the appropriate elevation to the EOP FG profile.

To avoid getting confused by multiple targets, baselines, and regions, here are a few tips:

Stay Organized   Onscreen, it is helpful to group similar objects. For example, many people find it helpful to group assemblies in the same area of the drawing. As you see in many of the examples in this book, the names of the assemblies can be added with a base AutoCAD Mtext label.

Group profile views in a way that helps you identify with which alignment they are associated. For example, place your predominant alignment in the drawing with cross-street alignments placed below it in station order.

Naming Conventions   If you let Civil 3D defaults have their way, you'll end up with objects that have generic names whose only distinguishing feature is the number indicating how many tries you've done before. Be sure to give your alignments, profiles, and assemblies names that can be easily identified. Rename the subassemblies to have more user-friendly names, as you learned in Chapter 8, "Assemblies and Subassemblies."

Working at the Region Level   There are several buttons you can click to get into the Target Mapping dialog. You can set targets for the entire corridor, at the baseline level, or at the region level. The simplest place to go is the Targets button for the region you are working with. Use Set All Targets for setting the target surface, but you'll get lost in a long list of subassemblies if you use it for much else.

You'll finalize the corridor intersection in the next exercise. Along the way, you'll examine the corridor in various stages of completion so that you'll understand what each step accomplishes. In practice, you'll likely continue working until you build the entire model. Make sure you've completed the last two exercises, and then follow these steps:

1. Continue working in your drawing from the previous exercise (1005_ManualIntersection.dwg or 1005_ManualIntersection_METRIC.dwg).

2. (Optional) Zoom to the area where the profile views are located. Notice that existing and proposed ground profiles are created for both EOP alignments. The step of developing the design profiles has been completed for you.

3. Select the corridor, and then open the Corridor Properties dialog and switch to the Parameters tab.

4. Click Add Baseline.

    The Create Corridor Baseline dialog opens.

Using Multiple Targets

Civil 3D allows you to select more than one item in Width Or Offset Target and Slope Or Elevation Target areas. You can use any number of targets. You can even mix and match the types of targets you use, such as alignments with polylines, survey figures, and feature lines.

In the example that follows, you need two alignment targets and two profile targets in each curb region. To help Civil 3D figure out what you want, you need to verify that it will use the target that it finds *first* as it radiates out from the baseline. In most cases, you can click the Target To Nearest Offset option found in the target selection dialogs.

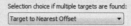

In the following image, the arrows represent the assemblies seeking out a target. From station 1+73.40 (0+052.85 for metric users) to 2+19.56 (0+066.92 for metric users), the Syrah Way alignment is closer to the west curb baseline; therefore, Syrah Way's geometry is used as the target. From station 2+19.56 (0+066.92 for metric users) to 2+76.07 (0+084.15 for metric users), Cabernet Court's alignment is closer and is used as the target. When multiple targets are used, Civil 3D automatically adds a frequency line at the intersection of the two alignments.

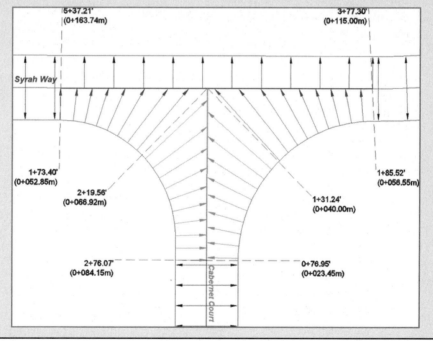

5. In the Create Corridor Baseline dialog, select the Cab-Syrah EOP W alignment, and click OK.

6. Click in the Profile field for the new baseline on the Parameters tab of the Corridor Properties dialog where it says <Click here...>.

7. In the Select A Profile dialog, select Cab-Syrah EOP W – FG, and click OK.
8. Right-click baseline BL – Cab-Syrah EOP W and select Add Region. (Your numbering may differ from that shown in Figure 10.34.)
9. In the Create Corridor Region dialog, select Curb Return Fillets, and then click OK to dismiss the dialog.
10. Click the small plus sign to expand the baseline, and see the new region you just created.

    The Parameters tab of the Corridor Properties dialog will look like Figure 10.34.

FIGURE 10.34
The Parameters tab with a new baseline and region

Name	Alignment	Profile	Assembly	Start Station	End Station
BL - Cabernet Court - (1)	Cabernet C...	Cabernet Co...		0+00.00'	9+37.51'
RG - Full Road Section - (1)			Full Road Section	0+25.00'	8+52.39'
BL - Syrah Way - (2)	Syrah Way	Syrah Way FG		0+00.00'	7+10.60'
RG - Full Road Section - (1)			Full Road Section	0+00.00'	3+77.30'
RG - Daylight Right - (4)			Daylight Right	3+77.30'	5+37.21'
RG - Full Road Section - (3)			Full Road Section	5+37.21'	7+10.60'
BL - Cab-Syrah EOP W - (3)	Cab-Syrah ...	Cab-Syrah E...		0+00.00'	3+50.00'
RG - Curb Return Fillets - (5)			Curb Return Fillets	1+73.40'	2+76.07'

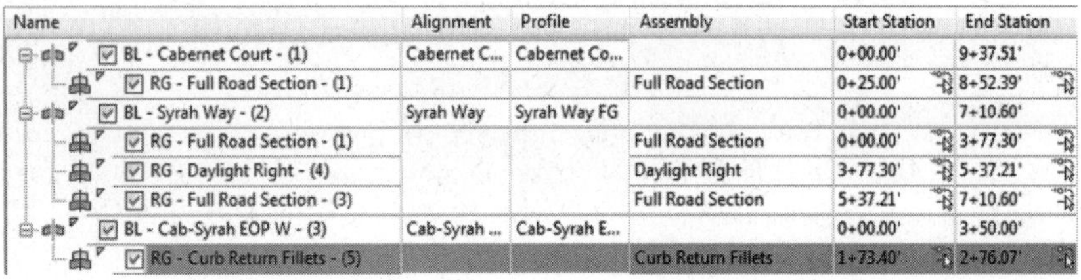

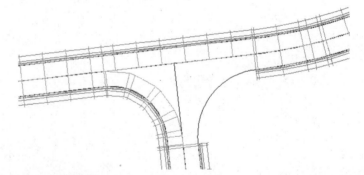

11. Change the region start station to 1+73.40 (0+052.85 for metric users) and the region end station to 2+76.07 (0+084.15 for metric users).

    These stations represent the PC and PT stations of the curb alignment. You could also have selected these start and end stations in the drawing.

12. Click OK to dismiss the Corridor Properties dialog and rebuild the corridor.

    Your corridor should now look similar to Figure 10.35.

FIGURE 10.35
Your corridor after applying the Curb Return Fillets assembly

What is missing from the corridor? You need to set targets, and add more frequency lines along the curves. Of course, you will repeat the process on the east side of the road as well.

13. Select the corridor, and from the Corridor contextual tab ➢ Modify Corridor panel, click Corridor Properties.
14. On the Parameters tab, set the frequency by clicking the ellipsis button in the Frequency column of the RG – Curb Return Fillets region row. Set both tangents and curves to **5′** (**1** m). Set the At Offset Target Geometry Points to Yes and click OK.

15. Click the ellipsis for targets of this region, and do the following:

    a. Click the Object Name field for the Width Alignment target for LaneSuperelevationAOR. In the Set Width Or Offset Target dialog, set Select Object Type To Target to Alignments.

    b. Highlight both Syrah Way and Cabernet Court centerline alignments. (Hint: Press the Ctrl key as you click to highlight both alignments in the listing.)

    c. In the Set Width Or Offset Target dialog, click Add. Both alignments will appear in the listing, as shown on the left in Figure 10.36.

FIGURE 10.36
In the Set Width Or Offset Target dialog, pick both Syrah Way and Cabernet Court centerlines (left). In the Set Slope Or Elevation Target dialog, pick an alignment first, and then add the FG profile (right).

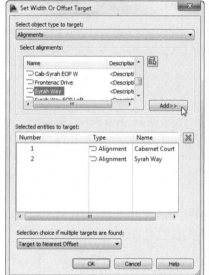 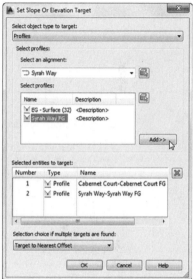

    d. Be sure Target To Nearest Offset is set at the bottom, and click OK when complete.

16. In the Target Mapping dialog, click the Outside Elevation Profile target for LaneSuperelevationAOR, and do the following:

    a. Change Select Object Type To Target to Profiles, and then from the Select An Alignment drop-down, select Syrah Way.

    b. Highlight the Syrah Way FG profile and click Add.

    c. Change the selected alignment to Cabernet Court.

    d. Highlight the Cabernet Court FG profile and again click Add. Both profiles will appear in the listing, as shown in Figure 10.36 on the right.

    e. Keep the default option for Target To Nearest Offset and click OK.

17. Click OK to dismiss the Target Mapping dialog, and click OK in the Corridor Properties dialog. Click Rebuild The Corridor if prompted to do so.

    Your intersection should now look like Figure 10.37.

FIGURE 10.37
Three baselines completed—one to go

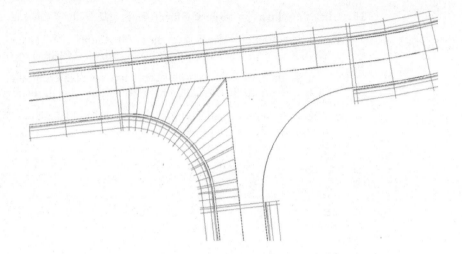

18. Repeat steps 4 through 17 for the Cab-Syrah EOP E.

    Here are some hints to get you going: The region should go from station 0+76.95 (0+023.45 for metric users) to 1+85.51 (0+056.54 for metric users). The frequency should be 5′ (1 m) for both tangents and curves. The target mapping will be identical to the region you created in steps 15 and 16.

19. Let the corridor rebuild one last time and admire your work.

    The completed intersection should look like Figure 10.38.

FIGURE 10.38
The completed intersection

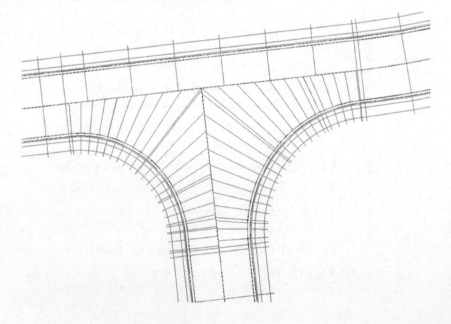

Troubleshooting Your Intersection

The best way to learn how to build advanced corridor components is to go ahead and build them, make mistakes, and try again. This section provides some guidelines on how to "read" your intersection to identify what steps you may have missed.

Your lanes appear to be backward. Occasionally, you may find that your lanes wind up on the wrong side of the EOP alignment, as in Figure 10.39. The most common cause is that your assembly is backwards from what is needed based on your alignment direction.

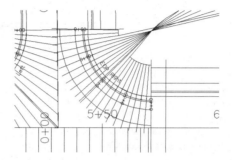

FIGURE 10.39
An intersection with the lanes modeled on the wrong side

Fix this problem by editing your subassembly to swap the lane to the other side of the assembly. If the assembly is used in another region that is correct, just make a new assembly that is the mirror image of the other assembly and apply the new one to the alignment.

Since so many design elements rely on the alignment as their base, it is better to add a new assembly rather than reversing the direction of the alignment.

Your intersection drops down to zero. A common problem when modeling corridors is the cliff effect, where a portion of your corridor drops down to zero. You probably won't notice in plan view, but if you rotate your corridor in 3D using the Object Viewer (see Figure 10.40), you'll see the problem. The most common cause for this phenomenon is incorrect region stationing.

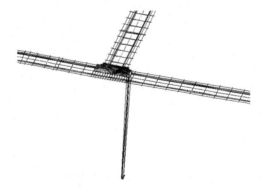

FIGURE 10.40
A corridor viewed in 3D, showing a drop down to zero

Fix this problem by making sure your baseline profile exists where you need it, and make note of the station range. Set your station range in the corridor to be within the correct range.

Your lanes extend too far in some directions. There are several variations on this problem, but they all appear similar to Figure 10.41. All or some of your lanes extend too far down a target alignment, or they may cross one another, and so on.

FIGURE 10.41
The intersection lanes extend too far down the main road alignment.

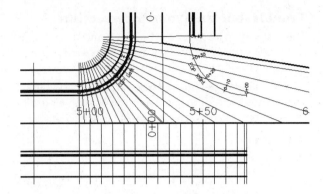

This occurs when a target alignment and profile have been omitted for one or more regions. In the case of Figure 10.41, the EOP Left baseline region was only set to one alignment. In an intersection, you need two targets in a corner region to model the road correctly. You would see a similar issue appear if the Selection Choice If Multiple Targets Are Found option is set to Target Farthest From Offset.

Your lanes don't extend far enough. If your intersection or portions of your intersection look like Figure 10.42, you neglected to set the correct target alignment and profile.

FIGURE 10.42
Intersection lanes don't extend out far enough.

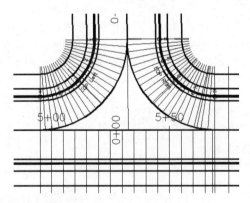

You can fix this problem by opening the Target Mapping dialog for the appropriate regions and double-checking that you assigned targets to the right subassembly. It's also common to accidentally set the target for the wrong subassembly if you use Map All Targets or if you have poor naming conventions for your subassemblies.

Checking and Fine-Tuning the Corridor Model

Recall that the EOP profiles are developed by creating a preliminary corridor surface model. To create the EOP elevations, you filled in the gaps by creating the EOP design profile where the preliminary surface stops and then picks back up again on the adjacent road (or in the case of the cul-de-sac, the opposite side of the street).

You will want to check the elevations of the corridor model against the profiles you created. The preliminary surface model was a best guess for elevations; now you need to reconcile your corridor geometry with those best-guess profiles. Figure 10.43 shows a common elevation problem that arises in the first iteration of intersection design.

FIGURE 10.43
A potential elevation problem in the intersection

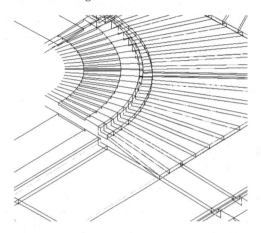

The first step in correcting this is creating a corridor surface model out of Top links. The surface will show you exactly how Civil 3D is interpreting your design, elevation-wise.

Before beginning this exercise, review the basic corridor surface creation in Chapter 9. When you think you are ready, follow these steps:

1. Open the 1006_ManualFineTune.dwg (1006_ManualFineTune_METRIC.dwg) file, which you can download from this book's web page.

2. Select the corridor in plan view, and from the Corridor contextual tab ➢ Modify Corridor panel, select Corridor Properties.

3. On the Surfaces tab, click the Create A Corridor Surface button.

   A Corridor Surface entry appears.

4. Rename the surface **Corridor Top** by clicking on the surface name.

5. Ensure that Links is selected as the data type and that Top appears under Specify Code, and then click the white plus button.

   An entry for Top appears under the Corridor Top entry.

6. Switch to the Boundaries tab.

7. Right-click Corridor Top and select Corridor Extents As Outer Boundary.

8. Click OK, and select Rebuild The Corridor.

9. Save the drawing for use in the next exercise.

   You should now see a surface model in addition to the corridor.

The next exercise will lead you through placing some design labels to assist in perfecting your model and then show you how to easily edit your EOP FG profiles to match the design intent on the basis of the draft corridor surface built in the previous section.

Don't forget a few basics that will help you navigate this exercise more easily. If you wish to see the drawing without the corridor, then freeze the corridor layer, which is C-ROAD-CORR. Don't forget to thaw this layer when you need it later! You can change the view of your surface around by changing the surface style. Also, remember that display order may be an issue. You will have corridor feature lines overlapping alignments, so it may be best to use the Send To Back option on the corridor.

1. Continue working in 1006_ManualFineTune.dwg (1006_ManualFineTune_METRIC.dwg) from the previous exercise.

2. For the next few steps, either freeze the corridor layer or send your corridor display order to the back.

3. From the Annotate tab ➢ Labels & Tables panel, click the Add Labels button.

   The Add Labels dialog will appear.

4. Set Feature to Alignment and Label Type to Station Offset.

5. Set Station Offset Label Style to Intersection Centerline Label, as shown in Figure 10.44.

FIGURE 10.44
Add alignment labels to identify potential problem areas.

6. Set Marker Style to Basic X, and click Add.

   This label references two alignments and two profiles.

7. The command line will read `Select Alignment:`. Click Syrah Way.

8. The command line will read `Specify station along alignment:`. Use object snaps to specify the intersection of Syrah Way and Cabernet Court.

9. At the `Specify station offset:` prompt, type **0** (zero) and press ↵.

10. At the `Select profile for label style component Profile 1:` prompt, right-click to bring up a list of profiles and choose Syrah Way FG.

11. Click OK to dismiss the dialog.

12. At the Select alignment for label style component Alignment 2: prompt, pick the Cabernet Court alignment.
13. At the Select profile for label style component Profile 2: prompt, right-click to open a list of profiles, and choose Cabernet Court FG.
14. Click OK to dismiss the profile listing, and press Esc to exit the labeling command.
15. Pick the label, and use the square-shaped grip to drag the label somewhere out of the way.

    Keep the Add Labels dialog open for the next part of the exercise.

    This label shows you that the crown elevations of the Cabernet Court FG and Syrah Way FG are both equal to 811.204' (247.255 m), which is great! If they differed at all, you'd have some adjustments to make on the FG profiles where the roads meet.

    The next part of the exercise guides you through the process of adding a label to help determine what elevations should be assigned to the start and end stations of the EOP Right and EOP Left alignments.

16. In the Add Labels dialog do the following:
    a. Change the feature to Surface.
    b. Set Label Type to Spot Elevation.
    c. Verify that Spot Elevation Label Style is set to Compare Elevations and click Add.

    This label compares proposed surface elevation and the corresponding profile elevation.

17. At the Select a Surface <or press enter to select from list>: prompt, pick any contour from the corridor surface.
18. At the Select a Point: prompt, use your Endpoint Osnap to pick the PC station of the right curb alignment.
19. At the Select surface for label style component Surface2: prompt, press ↵.
20. Highlight Prelim, and click OK.
21. At the Select alignment for label style component Alignment: prompt, press ↵.
22. Highlight Cab-Syrah EOP E and click OK. Remain in the command.

    The question marks that initially came in with the label should now show the elevation difference between proposed and Prelim elevation at the station of interest.

23. Place another label at the PT station.

    Because you have already specified the surface and profile, the label will pop right in.

24. Press Esc and click Add in the Add Labels dialog.

    You need to stop and start the command to switch the alignment used in the label.

25. Using the same techniques you used on the west side of the road, add labels at the PC and PT stations of Cab-Syrah EOP W.

26. Select the labels. Click and drag the square grip and add leaders to make them easier to read.

    Your labels should look like the ones in Figure 10.45.

FIGURE 10.45
The corridor with all labels placed

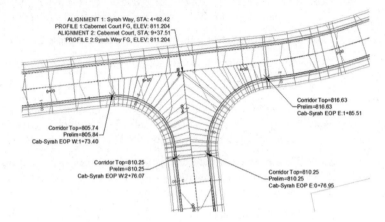

27. Save the drawing for use in the next exercise.

    As you can observe from the labels, the profiles and the surface model are perfect except for station 1+85.51 (0+056.54 for metric users) along Cab-Syrah EOP E. In fact, if the elevation difference doesn't bother you, you can skip the next steps. If you want your design to be as perfect as it can be, carry on:

    1. Continue working in 1006_ManualFineTune.dwg (1006_ManualFineTune_METRIC.dwg) from the previous exercise.

    2. Split your model space into two views for easier working: Go to the View tab ➢ Model Viewports panel and select Viewport Configuration ➢ Two: Vertical.

        Pan in the drawing to show the profile view for CAB-Syrah EOP E. Your screen should look similar to Figure 10.46.

FIGURE 10.46
Use a split screen to see the plan and profile simultaneously.

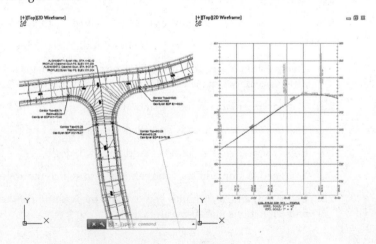

MOVING UP TO INTERSECTIONS | **491**

3. Pick the Cab-Syrah EOP E – FG profile (the blue line in the profile view), and from the Profile contextual tab ➢ Modify Profile panel, click Geometry Editor.

4. On the Profile Layout Tools toolbar, click Insert PVIs-Tabular.

5. In the Insert PVIs dialog, set the vertical curve type to None.

6. Type **185.51** (**56.54** m) for the station. (Station notation is not needed.) Type **816.63** (**248.907** m) for the elevation. Click OK.

7. Thaw the layer C-ROAD-CORR if you froze it in the previous exercise. Select the corridor and choose Rebuild Corridor from the contextual tab.

   Your first curb return label will now report a perfect match between the Corridor Surface elevation and the FG Profile elevation (Figure 10.47).

FIGURE 10.47
Add PVI stations and elevations to match the desired FG elevations.

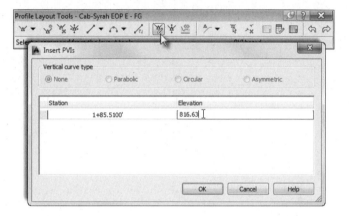

8. Pick the Corridor Top surface, and use the Object Viewer to study the TIN in the intersection area.

Study the contours in the intersection area. You may wish to add feature lines as breaklines to the surface model, as you learned in Chapter 9, "Basic Corridors."

9. Save and close the drawing.

Completed versions of the drawings (`1006_ManualFineTune_FINISHED.dwg` and `1006_ManualFineTune_METRIC_FINISHED.dwg`) are available for your reference.

🌐 Real World Scenario

WHEN AUTOMATIC SURFACE BOUNDARIES ARE NOT AVAILABLE

You will come to a point in the corridor modeling where you need to add a corridor surface boundary, but the best options (Add Automatically ➢ Daylight or Corridor Extents As Outer Boundary) are not available. There will also be situations where Corridor Extents As Outer Boundary does not go where you want it to go. In those situations, you will need to use the Add Interactively tool.

Add Interactively allows you to direct Civil 3D to use the correct feature line as the boundary. You will trace your desired outer bounds with a temporary graphic called a *jig*.

1. Open the `1007_InteractiveBoundary.dwg` (`1007_InteractiveBoundary_METRIC.dwg`) file, which you can download from the book's web page.

   You will see a surface in this file that needs some serious reining in.

2. Select the corridor. From the Corridor contextual tab ➢ Modify Corridor panel, click Corridor Properties. Select the Boundaries tab.

3. Right-click on the Corridor Top surface and select Corridor Extents As Outer Boundary. Click OK and rebuild the corridor.

4. Pan around the drawing.

   The contours are tidier, but the surface inside the loop formed by Frontenac Drive has not been cleaned up. In the next steps of the exercise, you will clean up the surface using the Add Interactively tool.

5. Set your AutoCAD object snaps to use only Endpoint and Nearest.

   These will be the most helpful Osnaps to steer Civil 3D in the right direction.

6. Select the corridor. From the Corridor contextual tab ➢ Modify Corridor panel, click Corridor Properties.

7. In the Corridor Properties dialog ➢ Boundaries tab, right-click Concord Commons Corridor Top and select Add Interactively.

8. Use the Endpoint object snap to click station 0+00 (0+000 for metric users) on the inside of the loop and click the Sidewalk_Out feature line. Trace it with your cursor.

9. When the jig is going to the correct location, click to commit the boundary.

   The following image shows what the jig will look like.

As you move around, you will see that the jig tends to move unexpectedly at region boundaries. Steer the jig where you want it to go by clicking and tracing. If you made a mistake, type **U** for undo at the command line.

If you click a location where there is more than one feature line, or you are zoomed out to where your pickbox encompasses two lines, you will see a dialog like this one asking you to pick the line you are after.

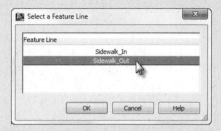

10. After you have finally come back around to where you started the boundary, type **C** (for close) at the command line, and press ↵.

11. When you complete the inner loop and return to the Corridor Properties dialog, set the Corridor Boundary(2) boundary type to Hide Boundary, as shown here. Click OK and rebuild the corridor.

Yes, this process can be tedious. However, your work will pay off when changes are made to your design. These boundaries are dynamic to the design, unlike a polyline boundary.

Using an Assembly Offset

In Chapter 9, you completed a road-widening example with a simple lane transition. Earlier in this chapter, you worked with intersections and cul-de-sacs. These are just a few of the techniques for adjusting your corridor to accommodate a widening, narrowing, interchange, or similar circumstances. There is no single method for building a corridor model; every method discussed so far can be combined in a variety of ways to build a model that reflects your design intent.

Another tool in your corridor-building arsenal is the assembly offset. In Chapter 8, "Assemblies and Subassemblies," you had your first glimpse of an offset assembly, but in the example that follows you will have a chance to use one for a bike path design.

Notice in Figure 10.48 how the frequency lines in the corridor are running perpendicular to the main alignment. The bike path is an alignment that is not a constant offset through the length of the corridor. In this scenario, the cross section of the bike path itself is skewed. This could prove problematic when computing end area volumes for the bike path pavement. This is the result of using an assembly where all of the design is based on one main baseline assembly.

FIGURE 10.48
A bike path modeled with a traditional assembly

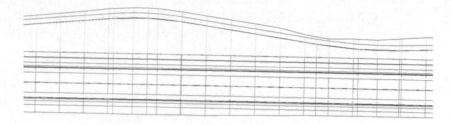

There are several advantages to using an offset assembly instead of creating an additional, separate assembly. The offset assembly requires its own alignment and profile for design. In the corridor that results, a secondary set of frequency lines is generated perpendicular to the offset alignment, as shown in Figure 10.49. Additionally, you can use a marked point assembly to model the ditch between the bike path and the main road.

FIGURE 10.49
Modeling a bike path with an assembly offset

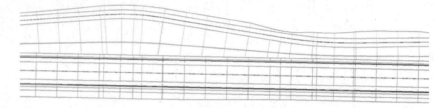

There are many uses of offset assemblies besides bike paths. Typical examples of when you'll use an assembly offset include transitioning ditches, divided highways, and interchanges. The assembly in Figure 10.50, for example, includes two assembly offsets.

FIGURE 10.50
An assembly with two offsets representing roadside swale centerlines

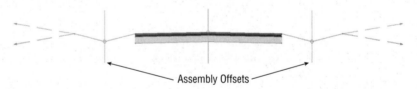

When you use an assembly with an offset in your corridor, you must assign an alignment and profile to it. The only restriction to the offset assembly is that it can't use the same alignment or profile as the main part of the assembly. In the case where you want your offsets to follow the same elevation, you will need to use the Superimposed Profile tool to effectively make a copy of the desired profile.

In this exercise, you will model a bike path with an assembly offset:

1. Open the 1008_BikePath.dwg (1008_BikePath_METRIC.dwg) file, which you can download from this book's web page.

2. Zoom to the area of the drawing where the assemblies are located.

   You'll see an incomplete assembly called Road With GR And Bikepath.

3. Click the main assembly marker, and then from the Assembly contextual tab ➤ Modify Assembly panel, click Add Offset.

4. At the `Specify offset location:` prompt, click to the left of the Road With GR And Bikepath assembly, leaving enough room for the bike lane and ditch.

   Your result should look like Figure 10.51. The warning symbols indicate that you can no longer use the assembly for roads where superelevation occurs at points other than the centerline. Superelevation at the crown will still work, which is the situation used here.

FIGURE 10.51
Road With GR And Bikepath assembly, so far

5. Use Ctrl+3 to open the subassembly tool palette.
6. Switch to the Basic tab, and select the BasicLane subassembly.
7. In the advanced parameters in the Properties palette, set Side to Right and Width to **5′ (1 m)**.
8. Click to place the subassembly on the offset assembly, and click the offset assembly again to form the left side.

   Your assembly should now look like Figure 10.52.

FIGURE 10.52
Road With GR And Bikepath assembly with the BasicLane subassembly as a bike path

Next, you will use a MarkPoint assembly to set the stage for building a ditch between the bike path and the main road.

9. Switch to the Generic tab in the subassembly tool palette, and click the MarkPoint subassembly.
10. Change Point Name to **BIKE** (use all capital letters).
11. Click on the outermost point of the left shoulder subassembly.

   Your marker will look like Figure 10.53. Note that the point code defaults to MarkedPoint in the subassembly parameters but can be renamed as needed.

FIGURE 10.53
A close-up of the MarkPoint subassembly

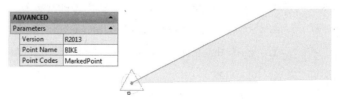

12. From the Generic tab of the subassembly tool palette, click the LinkSlopesBetweenPoints subassembly, and do the following:

    a. Set Marked Point Name to **BIKE** (again, use all capital letters).

    b. Set Ditch Width to **0.5' (0.15 m)**.

    c. Click on the right side of the bike path.

    The offset assembly will now look like Figure 10.54.

FIGURE 10.54
The LinkSlopes BetweenPoints subassembly in layout mode

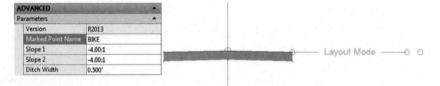

13. Add a LinkSlopeToSurface generic link subassembly, and do the following:

    a. Set Side to Left and Slope to 25%.

    b. Place the subassembly on the left side of the bike path. Press Esc to complete adding subassemblies and save the drawing.

    The completed assembly will look like Figure 10.55.

FIGURE 10.55
The completed assembly with offset

Next, you will create a corridor using this new assembly. You need to have completed the previous exercise before proceeding:

1. Continue working in your drawing from the previous exercise.

2. From the Home tab ➢ Create Design panel, click Corridor and do the following:

    a. Name the corridor **Bike Path**.

    b. Set Alignment to USH 10.

    c. Set Profile to USH 10 Roadway CL Prof.

    d. Set Assembly to Road W GR And Bikepath.

    e. Set Target Surface to Existing Intersection.

    f. Verify that there is a check next to Set Baseline And Region Parameters.

    g. Click OK.

In the Baseline And Region Parameters dialog, notice that the Offset – (1) is not associated with an alignment (your numbers may vary).

3. Click the alignment field for Offset – (1), select Bike Path from the drop-down list, and click OK.

4. Click the profile field for Offset – (1) and do the following:

   a. From the Select An Alignment drop-down list, select Bike Path.

   b. From the Select A Profile drop-down list, select Bike Path FG, and click OK.

   The Bike Path alignment is slightly shorter than the main USH 10 alignment, which would cause the "waterfall" effect explained in Chapter 9.

5. To prevent corridor errors, do the following:

   a. Set the start station for the Offset – (1) region to **0+25.00** (**0+010.0** for metric users).

   b. Set the end station to **40+00** (**1+220** for metric users).

   c. Click OK and rebuild the corridor.

   Your completed corridor will resemble the example shown earlier in Figure 10.49.

6. Select the corridor by clicking on one of the frequency lines anywhere near the middle of the alignment, and click Section Editor.

   You may want to change your annotation scale to 1" = 1' (1: 1 for metric users) to get an unobstructed view of your masterpiece.

7. Explore the finished design by clicking through the Corridor Editor.

   At each station, the offset assembly ties back to the main assembly because of the use of the LinkSlopeBetweenPoints subassembly. Your design in the Section Editor should resemble Figure 10.56.

FIGURE 10.56
Inside the Corridor Section Editor

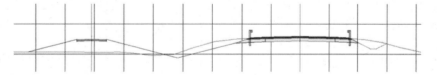

8. On the Section Editor contextual tab ➤ Close panel, click Close.

   Completed versions of these drawings (1008_BikePath_FINISHED.dwg and 1008_BikePath_METRIC_FINISHED.dwg) are located with the rest of the dataset for your review.

9. Save and close the drawing.

The Trouble with Bowties

In your adventures with corridors, chances are pretty good that you'll create an overlapping link or two. These overlapping links are known not so affectionately as *bowties*. Here's an example.

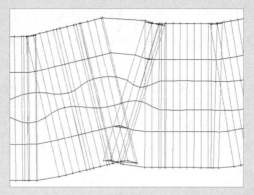

Bowties are problematic for several reasons. In essence, the corridor model has created two or more points at the same x and y locations with a different z, making it difficult to build surfaces, extract feature lines, create a boundary, and apply codeset styles that render or hatch.

When your corridor surface is created, the TIN has to make some assumptions about crossing breaklines that can lead to strange triangulation and incoherent contours, such as in the following image:

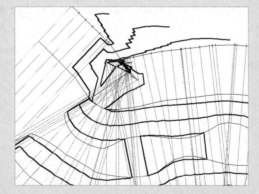

When you create a corridor that produces bowties, the corridor won't behave as expected. Using the Corridor contextual tab ➤ Launch Pad panel to extract polylines or feature lines from overlapping corridor areas yields an entity that is difficult to use for additional grading or manipulation because of extraneous, overlapping, and invalid vertices. If the corridor contains many overlaps, you may have trouble even executing the extraction tools. The same concept applies to extracted alignments, profiles, and COGO points.

If you try to add an automatic or interactive boundary to your corridor surface, either you'll get an error or the boundary jig will stop following the feature line altogether, making it impossible to create an interactive boundary.

To prevent these problems, the best plan is to try to avoid link overlap. Be sure your baseline, offset, and target alignments don't have redundant or PI locations that are spaced excessively close.

If you initially build a corridor with simple transitions that produce a lot of overlap, try using an assembly offset and an alignment besides your centerline as a baseline. Another technique is to split your assembly into several smaller assemblies and to use your target assemblies as baselines, similar to using an assembly offset. This method was used to improve the river corridor shown in the previous images. The following image shows the two assemblies that were created to attach at the top of bank alignments instead of the river centerline.

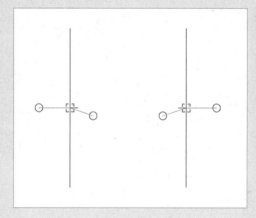

The resulting corridor is shown here.

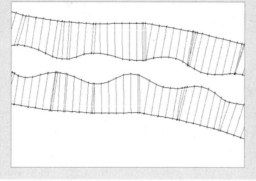

The TIN connected the points across the flat bottom and modeled the corridor perfectly, as you can see in the following image.

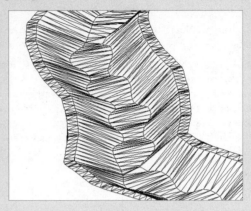

Another method for eliminating bowties is to notice the area where they seem to occur and then adjust the regions. If your daylight links are overlapping, perhaps you can create an assembly that doesn't include daylighting and create a region to apply that new assembly.

If overlap can't be avoided in your corridor, don't panic. If your overlaps are minimal, you should still be able to extract a polyline or feature line—just be sure to weed out vertices and clean up the extracted entity before using it for projection grading. You can create a boundary for your corridor surface by drawing a regular polyline around your corridor and adding it as a boundary to the corridor surface under the Surfaces branch in Prospector. The surface-editing tools, such as Swap Edge, Delete Line, and Delete Point, can also prove useful for the final cleanup and contour improvement of your final corridor surface.

As you gain more experience building corridors, you'll be able to prevent or fix most overlap situations, and you'll also gain an understanding of when they aren't having a detrimental effect on the quality of your corridor model and resulting surface.

Understanding Corridor Utilities

You'll now take advantage of some of the corridor utilities found in the Launch Pad panel (Figure 10.57) by selecting the corridor.

FIGURE 10.57
A bounty of corridor utilities on the Launch Pad panel

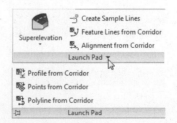

The utilities on this panel are as follows:

Superelevation   This button will jump you to the alignment superelevation parameters. See Chapter 11, "Superelevation," for a detailed look at how Civil 3D creates banked curves for your design speed.

Create Sample Lines   Corridors and sample lines are both linked to alignments. Civil 3D gives you a shortcut to the sample line creation tool. Chapter 12, "Cross Sections and Mass Haul," explores the creation and uses of sample lines.

Feature Lines From Corridor   This utility extracts a grading feature line from a corridor feature line. This grading feature line can remain dynamic to the corridor, or it can be a static extraction. Typically, this extracted feature line will be used as a foundation for some feature-line grading or projection grading. If you choose to extract a dynamic feature line, it can't be used as a corridor target due to possible circular references.

> **DYNAMIC FEATURE LINES CANNOT BE USED AS TARGETS**
>
> It's important to note that dynamic feature lines extracted using the Feature Lines From Corridor tool can't be used as targets.

Alignment From Corridor   This utility creates an alignment that follows the horizontal path of a corridor feature line. By default, the alignment is categorized as an offset alignment, but it is not tied to the baseline. You can use this alignment to create target alignments, profile views, special labeling, or anything else for which a traditional alignment could be used. Extracted alignments are not dynamic to the corridor. If you place a check box next to the Create Profile option, the resulting profile will be related to the new alignment rather than the baseline.

Profile From Corridor   This utility creates a profile that follows the vertical path of a corridor feature line. This profile appears in Prospector under the baseline alignment and is drawn on any profile view that is associated with that baseline alignment. This profile is typically used to extract edge of pavement (EOP) or swale profiles for a finished profile view sheet or as a target profile for additional corridor design. Extracted profiles are not dynamic to the corridor.

Points From Corridor   This utility creates Civil 3D points that are based on corridor point codes. You select which point codes to use as well as a range of corridor stations. A Civil 3D point is placed at every point-code location in that range. These points are a static extraction and don't update if the corridor is edited. For example, if you extract COGO points from your corridor and then revise your baseline profile and rebuild your corridor, your COGO points won't update to match the new corridor elevations.

Polyline From Corridor   This utility extracts a 3D polyline from a corridor feature line. The extracted 3D polyline isn't dynamic to the corridor. You can use this polyline as is or flatten it to create road linework.

Using Corridor Utilities in Practice

There are many uses for the utilities outlined in this section. Once you get the hang of using some of the corridor utilities, you should find that they are straightforward. In the exercise that follows, you will dabble in the corridor utilities.

1. Open the file 1009_CorrUtils.dwg (1009_CorrUtils_METRIC.dwg).

2. Select the corridor by clicking one of the frequency lines (these are the magenta lines that are perpendicular to the alignment).

3. From the Corridor contextual tab ➢ Launch Pad panel, click Feature Lines From Corridor.

4. At the Select a Corridor Feature Line: prompt, click the south daylight line (this will be the line that is not a constant offset from the centerline alignment).

   The Select A Feature Line dialog will appear if you click the daylight line in a cut or fill region. This is because Civil 3D makes two distinct feature lines in these areas. Recall from Chapter 9 that feature lines are formed as a result of marker points with the same name in the assembly connecting together at frequency stations. In other words, why do we have two feature lines here? Because the daylight subassembly creates two marker points at the catch point.

   You want the Daylight feature line because it is continuous through the length of the corridor. Daylight_Cut only appears in the cut areas (red) and Daylight_Fill only appears in the fill areas (green). Where the corridor transitions from cut to fill (or fill to cut), you will see a yellow line. Only the Daylight feature line will appear in the transition regions.

5. If needed, highlight Daylight and click OK.

6. In the Create Feature Line From Corridor dialog (Figure 10.58), clear the check box labeled Create Dynamic Link To The Corridor.

FIGURE 10.58
Creating a feature line from a corridor without the dynamic link to corridor

7. Leave all other options at their defaults and click OK.

8. You should still be in the Create Feature Line From Corridor command and the command line should return to the `Select a Corridor Feature Line:` prompt. If you accidentally exited the Feature Lines From Corridor command, start it again from the Launch Pad panel. Click the daylight line on the north side of the road.

9. Repeat steps 5–7 to create a second feature line.

10. Press Esc to end the command.

11. Select the corridor again if it is not already selected.

12. From the Corridor contextual tab ➢ Launch Pad panel flyout, click Points From Corridor.

13. In the Create COGO Points dialog, do the following:

    a. Toggle on the option For Entire Corridor Range.

    b. Name the new point group **Corridor Stakeout**.

    c. In the Select column, clear all the check boxes except Daylight and ETW, as shown in Figure 10.59. (Hint: Press the Shift key as you click to unselect multiple items at once.)

    d. Click OK.

FIGURE 10.59
Creating points for stakeout along corridor feature lines

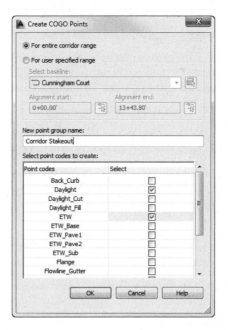

14. Save and close the drawing.

Completed versions of this exercise are available with the rest of the dataset.

Using a Feature Line as a Width and Elevation Target

You've gained some hands-on experience using alignments and profiles as targets in an intersection and in a cul-de-sac design. Civil 3D adds options for corridor targets beyond alignments and profiles. You can use grading feature lines, survey figures, or polylines to drive horizontal and/or vertical aspects of your corridor model.

Imagine using an existing polyline that represents a curb for your lane-widening projects without duplicating it as an alignment, or grabbing a survey figure to assist with modeling an existing road for a rehabilitation project. Better yet, what if the object you are targeting is visible to the corridor drawing only as an XRef? The next exercise will lead you through an example where a lot-grading feature line is integrated with a corridor model through an external reference:

1. Open the 1010_FeatureLineTarget.dwg (1010_FeatureLineTarget_METRIC.dwg) file, which you can download from this book's web page. Note that the file 1010_XREF.dwg (or 1010_XREF_METRIC.dwg) must be extracted to the same folder as the main file in order to see it for use in this exercise.

   This drawing includes a completed assembly and a partially completed corridor. Your task will be to use the yellow feature lines that run through the project as targets in the corridor. These lines are in an external reference file.

2. Zoom to the corridor and select it. From the Corridor contextual tab ➢ Modify Corridor panel, choose Corridor Properties.

3. Switch to the Parameters tab in the Corridor Properties dialog.

4. Click the ellipsis button in the Targets field in the RG – Subdivision region. Note that the number following the region will vary. The higher the number, the more previous attempts at building the corridor the authors have made before handing the drawing over to you.

   The Target Mapping dialog appears.

5. In the Target Mapping dialog, locate the Width Or Offset Target in the Object Name column. Click the <None> field next to Target Alignment for the Slope-Left subassembly.

   The Set Width Or Offset Target dialog appears.

6. Choose Feature Lines, Survey Figures And Polylines from the Select Object Type To Target drop-down list (see Figure 10.60).

7. In the Set Width Or Offset Target dialog, click the Select From Drawing button. At the Select feature lines, survey figures or polylines to target: prompt, select the yellow feature line to the north of the alignment and then press ↵.

   The Set Width Or Offset Target dialog reappears, with an entry in the Selected Entries To Target area.

8. Click OK to return to the Target Mapping dialog.

   If you stopped at this point, the horizontal location of the feature line would guide the Slope-Left subassembly, and the vertical information would be driven by the slope set in the subassembly properties. Although this has its applications, most of the time you'll want the feature line elevations to direct the vertical information. The next few steps will teach you how to dynamically apply the vertical information from the feature line to the corridor model.

FIGURE 10.60
The Select Object Type To Target drop-down list at the top of the Set Width Or Offset Target dialog

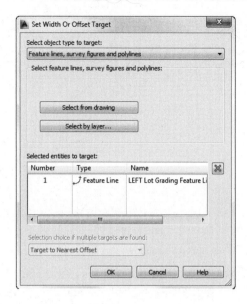

9. In the Object Name column in the Slope Or Elevation Targets section, click the <None> field next to Target Profile for the Slope-Left subassembly.

10. Make sure Feature Lines, Survey Figures And Polylines is selected in the Select Object Type To Target drop-down.

11. Click the Select From Drawing button. At the Select feature lines, survey figures or 3D polylines to target: prompt, select the yellow feature line to the north again and then press ↵.

The Set Slope Or Elevation Target dialog reappears, with an entry in the Selected Entries To Target area.

12. Click OK to return to the Target Mapping dialog.

13. Click OK to return to the Corridor Properties dialog.

14. Repeat steps 5–11 for Slope-Right on the right side of the corridor using the feature line to the south of the alignment.

15. Click OK to exit the Corridor Properties dialog and choose Rebuild Corridor.

The corridor will rebuild to reflect the new target information and should look similar to Figure 10.61.

Once you've linked the corridor to these feature lines, any edits to the feature lines will be incorporated into the corridor model. You can establish this feature line at the beginning of the project and then make horizontal edits and elevation changes to perfect your design. The next few steps will lead you through making some changes to this feature line and then rebuilding the corridor to see the adjustments.

FIGURE 10.61
The corridor now uses the grading feature lines as width and elevation targets.

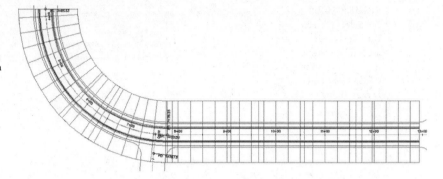

16. Select the text that resides in the XRef. From the External Reference context tab ➢ Edit panel, click Open Reference.

    This opens the external reference so you can modify the feature lines.

17. Select one of the feature lines so that you can see its grips. Experiment with the feature lines by moving several grips.

18. When you have edited several areas in the external reference file (1010_XREF.dwg or 1010_XREF_METRIC.dwg), save and close the drawing.

    After the external reference closes, you should be back in the corridor drawing. A message will appear in the lower-right corner of your screen indicating External Reference File Has Changed.

19. In the bubble message, click the link to reload 1010_XREF.dwg (for metric users, 1010_XREF_METRIC.dwg).

20. Select the corridor. From the Corridor contextual tab ➢ Modify Corridor panel, click Rebuild Corridor.

    The corridor will rebuild to reflect the changes to the target feature lines.

See 1010_FeatureLineTarget_FINISHED.dwg (1010_FeatureLineTarget_METRIC_FINISHED.dwg) to view a completed version of this exercise.

Edits to targets—whether they're feature lines, alignments, profiles, or other Civil 3D objects—drive changes to the corridor model, which in turn drives changes to any corridor surfaces, sections, section views, associated labels, and other objects that are dependent on the corridor model.

 Real World Scenario

CONDITIONAL LOVE

Driveways are a common source of grief for designers. They pop up along corridors at irregular intervals, making them difficult to model with a traditional assembly. Luckily, we have the powerful *conditional subassemblies* available to us. Conditional subassemblies allow designers to test scenarios and have the assembly react to different conditions. In programmer-speak, the conditional subassemblies allow "if-then" statements within an assembly.

Using a Feature Line as a Width and Elevation Target

There are two conditional subassemblies available to Civil 3D users: ConditionalCutOrFill and ConditionalHorizontalTarget.

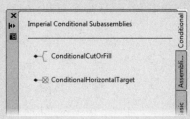

The ConditionalCutOrFill subassembly will check for a range of depth of cut or fill and apply the desired subassembly accordingly. For example, you may have a shoulder subassembly with daylight for cut scenarios, but after a cut of more than 10′, you want a retaining wall subassembly to be placed instead.

The ConditionalHorizontalTarget subassembly checks for the presence of a specific offset object or objects. For example, you will use the existence of a driveway to trigger the assembly to apply geometry in the following example:

1. Open the drawing 1011_Conditional.dwg (or 1011_Conditional_METRIC.dwg).

   There are five total assemblies in this drawing. The one that is modified in this exercise will be the DRIVEWAY CONDITIONAL. The other four assemblies in the drawing are there to illustrate the assemblies that would be needed to accomplish the corridor modeling task without the use of conditional subassemblies. In other words, one assembly has the power of four!

2. In the right viewport, zoom in to the assembly labeled DRIVEWAY CONDITIONAL.

3. Open the Subassembly tool palette if it is not already open. Switch to the Conditional tab.

4. Click the ConditionalHorizontalTarget subassembly. In the Properties palette, set the side to Right. Change the type to **Target Found**. Leave all other parameters as default.

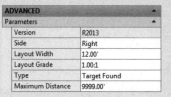

5. Click on the marker point at the top-right edge of the lane to place the subassembly in the drawing.

6. Again, click the ConditionalHorizontalTarget subassembly. In the Properties palette, set the side to Right. Change Layout Grade to **-1.00:1** and change the type to **Target Not Found**.

The Layout Width and Layout Grade values do not have any effect on the resulting corridor. Changing Layout Grade only helps to visually separate the different conditions so they do not overlap graphically.

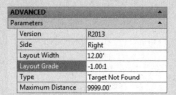

7. Click on the same marker point at the top-right edge of the lane to place the subassembly in the drawing.
8. Press Esc to clear any previous selections. Select one of the new subassemblies. From the Subassembly contextual tab, click Assembly Properties.
9. In the Assembly Properties dialog, switch to the Construction tab.
10. Right-Click the first ConditionalHorizontalTarget in the listing and select Rename. Set the name to **DRIVEWAY**.
11. Right-click the second ConditionalHorizontalTarget in the listing and select Rename. Change the name to **NO DRIVEWAY**.
12. Click OK to close the Assembly Properties dialog. Press Esc to clear the selection.
13. Select the shoulder and the daylight subassemblies from the right side of the Driveway Right assembly.
14. From the Subassemblies contextual tab ➢ Modify Subassembly panel, click Copy.
15. At the Select marker point within assembly for the copied subassemblies: prompt, click the marker point at the end of the DRIVEWAY subassembly. This will be the geometry that is created when the driveway target is found.
16. Press Esc to clear the selection.
17. Select the shoulder and the daylight subassemblies from the right side of the Driveway Left assembly.
18. From the Subassemblies contextual tab ➢ Modify Subassembly panel, click Copy.
19. At the Select marker point within assembly for the copied subassemblies: prompt, click the marker point at the end of the NO DRIVEWAY subassembly, completing the right side of the assembly.
20. Select all the subassemblies that you have added in this exercise.
21. From the Subassemblies contextual tab ➢ Modify Subassembly panel, click Mirror.

22. Click the top-left edge of pavement marker point.

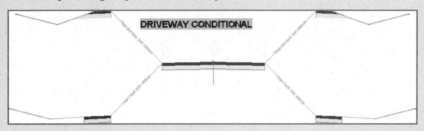

23. In Toolspace ➢ Prospector, expand Corridors. Right-click USH 10 – w Driveways and select Properties.

24. In the Corridor Properties dialog, click the ellipsis in the Target column of the DRIVEWAY CONDITIONAL region row.

25. In the Corridor Properties dialog, switch to the Parameters tab. Click the <Click Here To Set All> option in the Object Name column in the Surfaces row. Highlight the surface Existing Intersection and click OK.

26. Click the <None> field next to Target Offset for the DRIVEWAY subassembly (in the assembly group Right).

27. Change Select Object Type To **Target to Feature Lines, Survey Figures And Polylines**, and then click Select By Layer.

> **28.** Place a check mark next to the C-TOPO-FEAT layer and click OK. Click OK again to close the Set Width Or Offset Target dialog.
>
> **29.** Repeat steps 24–26 for NO DRIVEWAY on the right and again for both DRIVEWAY and NO DRIVEWAY on the left.
>
> **30.** Click OK to close the Target Mapping dialog and then click OK (click Rebuild The Corridor if prompted to do so) to close the Corridor Properties dialog.

Tackling Roundabouts: The Mount Everest of Corridors

If you really understand what went on earlier in this chapter, you are almost ready for roundabout design. You may want to wait to tackle your first roundabout until after reading Chapter 14, "Grading."

The same concepts apply to a roundabout as for a standard road junction, but you will have several more regions, baselines, and corresponding profiles.

The following sections will help you prepare files for roundabout design. We will not take you through every detail of corridor creation, but once you master the topics of intersection design, a roundabout is an extension of the same concepts.

A roundabout is best done in several corridors:

Preliminary Corridor for Circulatory Road This is a corridor used to determine the elevations of the approach road profiles. The circular portion of the roadway will set the elevations for all of the alignments leading into it. Using similar techniques to those used earlier in the chapter, you create a corridor surface from this corridor and use it as a tie-in for approach road profiles.

Main Corridor with Approaches and Circulatory Road This corridor is the main part of your design. You will spend lots of time in the corridor properties tweaking stations, adding baselines and regions, and targeting the appropriate locations.

Curb Island Corridors (Optional) There are many different philosophies about the best way to show curb islands. If showing them in a section is not important, you can omit this altogether. Some people prefer to use grading feature lines (as discussed in Chapter 14). In the following sections, you will go "all out" and use the dynamic capabilities of the corridor model to make curb islands.

Drainage First

Based on your existing ground surface, determine the general direction that you want water to flow away from the center of the roundabout. Use grading tools and a feature line to create the general drainage direction of the roundabout.

Chapter 14 will go into much more depth on creating grading. You will certainly want to have an understanding of grading basics before you tackle a roundabout.

Create a feature line that represents the highest elevations. In the example shown in Figure 10.62, the feature line slopes downward and acts as a ridge to separate water flow. The grading tools are then used to create grading objects and a corresponding surface model called Roundabout Grading.

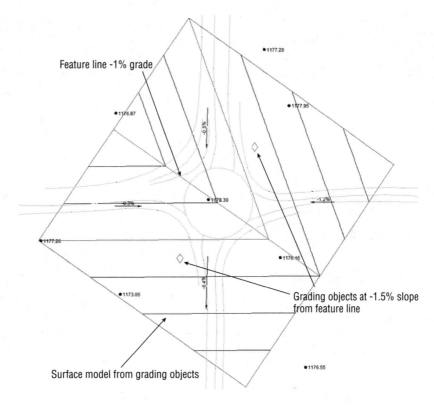

FIGURE 10.62
Feature lines and grading create a preliminary surface to ensure proper drainage through the roundabout.

The Roundabout Grading surface will be the basis for our profile elevations through the rest of the design process.

Roundabout Alignments

Roundabouts need alignments to guide the design for the same reasons that an intersection needs them. Alignments will be baselines and targets for the approaches and rotary. Create alignments manually with the tools you learned to use earlier in this chapter, or start with the handy roundabout layout tool.

The roundabout layout tool creates horizontal data based on the location of the center of the roundabout and the approach alignments.

In the exercise that follows, you will create the Civil 3D alignments needed to create a roundabout:

1. Open the 1012_RoundaboutLayout.dwg (1012_RoundaboutLayout_METRIC.dwg) file, which you can download from this book's web page.

2. From the Home tab ➢ Create Design panel, select Intersections ➢ Create Roundabout, as shown in Figure 10.63.

3. At the Specify roundabout center point: prompt, use your Intersection Osnap to select the point where the alignments intersect.

   The alignments leading into the roundabout are often referred to as *approaches*.

FIGURE 10.63
Access the Create Roundabout tool from the Home tab.

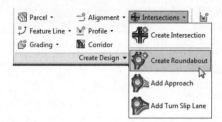

4. At the Select approach road: prompt, select all four alignments leading into the roundabout, and press ↵ when you are done.

    You now see the Create Roundabout – Circulatory Road page of the wizard, shown in Figure 10.64.

FIGURE 10.64
The first roundabout layout screen for designing the main circulatory road

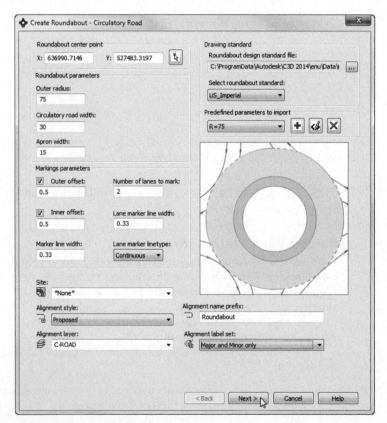

5. Verify that you are using the correct standards for your units. Click the ellipsis next to the Roundabout Design Standard File field, and verify that you are using the correct XML standards file. Imperial users should be using Autodesk Civil 3D Imperial

Roundabouts Presets.xml. Metric users should be using `Autodesk Civil 3D Metric Roundabouts Presets.xml`.

   a. If Roundabout Design Standard File is set correctly, click Cancel and continue to step 6.

   b. If it is not set correctly, browse to the `C:\ProgramData\Autodesk\C3D 2014\enu\Data\Corridor Design Standards\` folder. Browse to either the Imperial or Metric units folder and select the correct roundabout presets XML file. Click Open to return to the Create Roundabout dialog.

6. From the Predefined Parameters To Import drop-down, choose **R = 75 (Rg = 25m)**.

   This will be the radius from the center of the roundabout to the outermost circular edge of pavement.

7. Set Alignment Layer to C-ROAD and Alignment Label Set to Major And Minor Only. Click Next.

   Now, you'll design the approach road exit and entry geometry. The options in the Create Roundabout – Approach Roads page of the wizard (see Figure 10.65) can be set independently for each approach, or you can click Apply To All, which will set the geometry for all four approaches.

FIGURE 10.65
Approach road widths at entry and exit

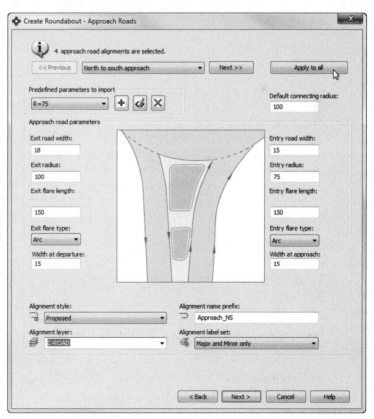

8. On the Create Roundabout – Approach Road page, do the following:
   a. Set Predefined Parameters To Import to **R = 75 (Rg = 25m)**.
   b. Change the alignment layer to C-ROAD. Hint: You can type in the layer name or pick it from the drop-down.
   c. For Alignment Label Set, choose Major And Minor Only.
   d. Leave all other settings at their defaults.
   e. Click the Apply To All button, and click Next.
9. In the Create Roundabout – Islands screen (see Figure 10.66), again set Predefined Parameters To Import to **R = 75 (Rg = 25m)**.

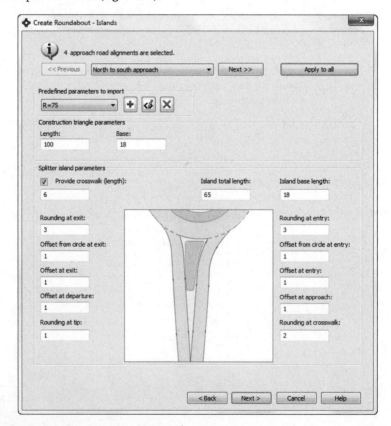

FIGURE 10.66
Roundabout Islands parameters

10. Click Apply To All, and then click Next.

    The final screen of the Create Roundabout Wizard deals with pavement markings and signage. Notice that you can specify your own blocks for the signs that will be placed in this process.

Everything created in this last step is an AutoCAD polyline or block. The polylines have a global width set to indicate pavement marking thicknesses. These thicknesses are set in the Markings And Signs page (Figure 10.67).

FIGURE 10.67
Pavement markings galore!

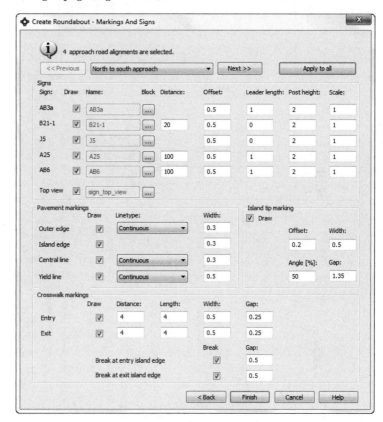

11. Leave all defaults in the Create Roundabout – Markings And Signs page (Figure 10.67), and click Finish.

    Your roundabout should resemble Figure 10.68. Since standards vary by region, the metric drawing will have slightly different default pavement markings.

    Finally, you will add a turn lane in the NW quadrant of the roundabout. When you're creating slip turn lanes, remember that the turn radius must be large enough to fillet the exit and entry roads without overlapping the other alignments.

    When selecting the approach entry and exit alignments, you need to click the shorter approach alignments created by Civil 3D rather than the original approach road. For this reason, the exercise has you select inside the islands, just to be sure.

12. From the Home tab ➢ Create Design panel, select Intersections ➢ Add Turn Slip Lane.

13. When prompted to select the entry approach, select the north approach alignment inside the curb island.

FIGURE 10.68
Completed roundabout alignment layout

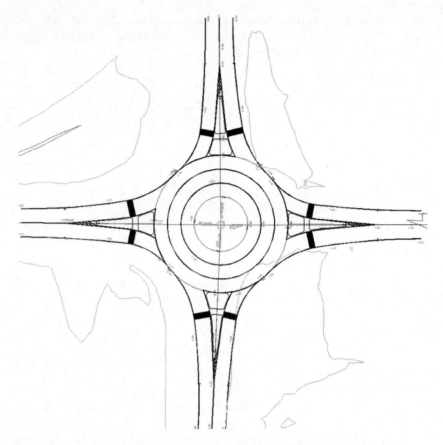

14. When prompted for the exit approach, select the west approach alignment inside the curb island, as shown in Figure 10.69.

15. In the Draw Slip Lane dialog shown in Figure 10.69, set the lane width to **14′ (4 m)** and the radius to **150′ (45 m)**.

16. For the alignment layer, choose C-ROAD and for Alignment Label Set, choose Major And Minor Only. Click OK.

    Your roundabout will now look like Figure 10.71.

17. Save and close the drawing.

The completed files, 1012_RoundaboutLayout_FINISHED.dwg and 1012_RoundaboutLayout_METRIC_FINISHED.dwg, are available for your review if desired.

You now have all the alignments you need to start your roundabout design. At this point, you can add geometry to the alignments and modify what Civil 3D has created for you.

The horizontal layout is complete, but the roundabout design is far from done. No vertical data has been created; that is up to you.

FIGURE 10.69
Entry and exit approach alignments for the slip lane

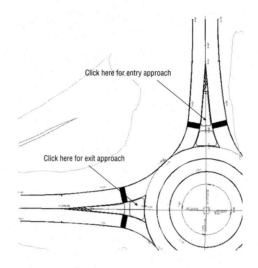

FIGURE 10.70
Adding a slip lane

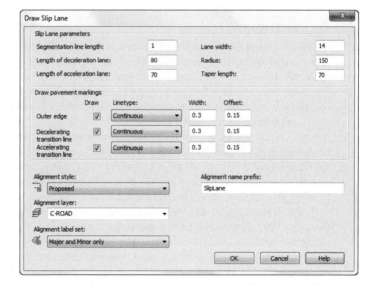

FIGURE 10.71
The completed slip lane alignments

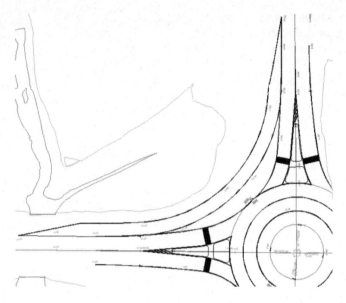

This is as far as we will take you in this book in regard to building a roundabout step by tedious step. Rest assured, however, that if you have truly mastered corridors, the technique for completing a roundabout is similar to that used for any intersection.

The remainder of this chapter gives you an overview of how to accomplish the rest on your own.

Center Design

All profiles need to meet at the elevations inside the traveled way in the circular pavement area of the roundabout. Therefore, the main circle design comes first.

To see an example of a completed roundabout corridor, open the drawing 1013_RoundaboutExample_FINISHED.dwg or 1013_RoundaboutExample_METRIC_FINISHED.dwg. Use these examples as a guide to "reverse engineer" your own roundabouts when you've mastered other forms of intersections.

You can use any of the circular alignments created by Civil 3D as the basis for this step, as long as your assembly works with the design. Remember to make note of which direction the alignment goes, to ensure that the assembly you create is not backwards.

Extract profiles for the main circle design from the Existing Intersection and Roundabout Grading surfaces, as shown in Figure 10.72.

The assembly you create for this preliminary design will also be used in the main design. Decide which alignment will be used as the circular design basis, and create an assembly based on your alignment location and desired geometry, as shown in Figure 10.73.

This assembly can be used in several steps of the process. First it is used in a preliminary corridor called RAB Center. It can also be recycled to be the centerpiece of your main corridor. In the example, the main circle alignment created by Civil 3D, Roundabout_OUTER_EDGE, was used as the baseline for this initial corridor. There are no targets or frequencies set in this corridor. Like the cul-de-sac example earlier in this chapter, this is a preliminary corridor, used to ensure that profiles from the approach roads tie in at the correct elevations.

In the example drawings, a Top link surface was created from the RAB Center corridor. At this point, the roundabout will resemble Figure 10.74.

FIGURE 10.72
Extract surface profiles around the main circular alignment.

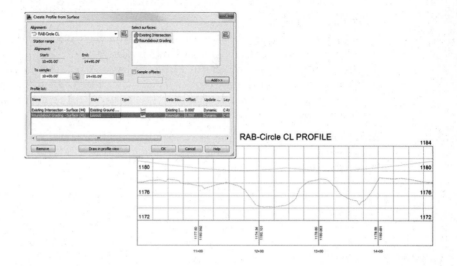

FIGURE 10.73
Center assembly for roundabout

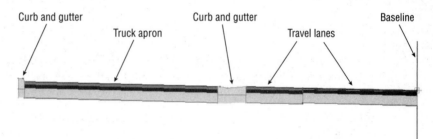

FIGURE 10.74
Preliminary center corridor and surface

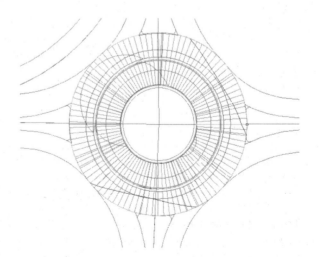

Profiles for All

You have all the preliminary surfaces in place, and you have all the alignments you need, so it is time to extract profiles from your various surfaces.

In the `1013_RoundaboutExample_FINISHED.dwg` and `1013_RoundaboutExample_METRIC_FINISHED.dwg` files, the surface profiles for all the approach alignments were created by sampling the existing ground, drainage surface, and preliminary RAB Center corridor surface. These profiles will look something like Figure 10.75.

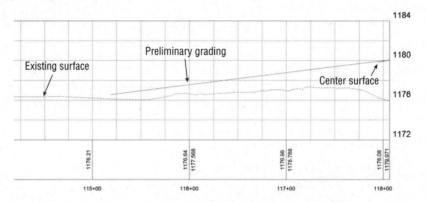

FIGURE 10.75
Surface profiles needed for design

When you create your design, you will see all your design considerations in the profile views. No matter how you decide to tie into existing ground and slope upward toward the surface, your design must tie into the preliminary center surface, as shown in Figure 10.76.

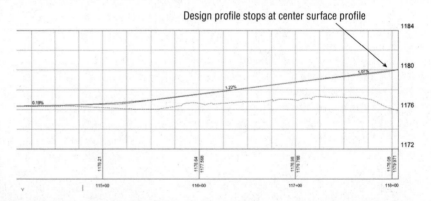

FIGURE 10.76
Design profiles must tie into the center.

Use techniques you learned earlier in this chapter to assist you. Labels are an especially valuable tool for roundabouts. Keep your profile views organized because you will have at least three for each approach. If you have a slip turn lane, you will have a profile for that as well.

Tie It All Together

Stretch your legs and go for more coffee. It is time to put this thing together into a completed corridor. When you model the corridor initially, ignore curb islands—you will add them as individual corridors in a later step.

A simple roundabout can be completed using as few as three assemblies. In our example, however, the slip turn lane necessitates a total of four assemblies. In addition to the RAB Center assembly you saw in Figure 10.73, you will need three more assemblies, as shown in Figure 10.77.

FIGURE 10.77
Assemblies needed for the main roundabout corridor

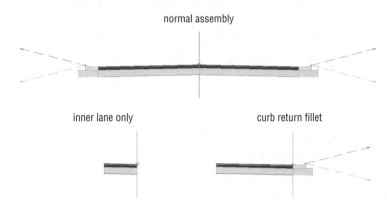

These assemblies will be tied to the EOP alignments and profiles as baselines, similar to a traditional intersection. Each quadrant of the roundabout will target at least two alignments and profiles, as shown in Figure 10.78.

FIGURE 10.78
Roundabout corridor regions and targets

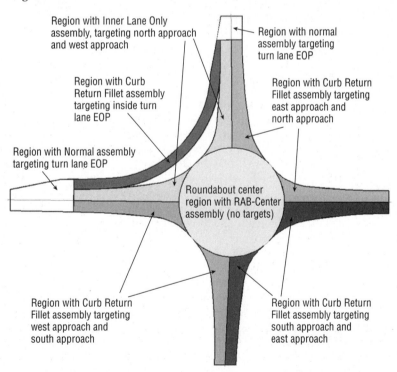

Keep in mind the direction of your alignments. If you build the corridor in stages, check the corridor periodically to make sure it is building correctly.

Create a corridor surface from your completed roundabout lanes. You will likely have to use the Add Interactively tool to add the boundary correctly.

Finishing Touches

The median islands are the last parts to go on the corridor. You can create them using simple grading objects, but since this is a book about mastering skills, and this is a chapter about corridors, you should examine the dynamic way.

Create a simple assembly containing the curb and gutter for the curb islands. This will be the assembly that you use with the median corridors (Figure 10.79).

FIGURE 10.79
Simple assembly containing just the curb and gutter on the median islands

Each median island will need its own alignment. Take note of the direction of the alignments to make sure they are compatible with the curb island assembly. If necessary, change the direction of the alignments using the Reverse Direction tool in the Modify panel of the contextual Ribbon tab. Figure 10.80 shows the bypass island and north island with directions.

The good news is that the elevation data for the medians is already complete. Your main corridor's surface will act as the profile for each individual median. This also means that once the curb return corridors are created, they will be dynamic to the main corridor. After these little corridors are created and surface model information has been obtained, you can set them to Rebuild – Automatic and forget all about them.

Extract a profile for all the curb return alignments from the Top link surface model from the main roundabout corridor, as shown in Figure 10.81. You do not need to see this profile in a view, so you can click OK to extract.

When the design roundabout corridors are complete and surfaces are made, the next step is to merge the surfaces. Create a final surface model and paste the main roundabout design in first. After the main corridor is pasted in, paste the smaller median corridor surfaces as shown in Figure 10.82. The center median is already taken care of by the first baseline.

Your next step is to use the corridor fine-tuning techniques you learned earlier in this chapter to ensure that your grades are correct and the design is correct. To see an example of a completed corridor using these steps, take a look at 1013_RoundaboutExample_FINISHED.dwg (1013_RoundaboutExample_FINISHED_METRIC.dwg), which you can download from this book's web page.

TACKLING ROUNDABOUTS: THE MOUNT EVEREST OF CORRIDORS | 523

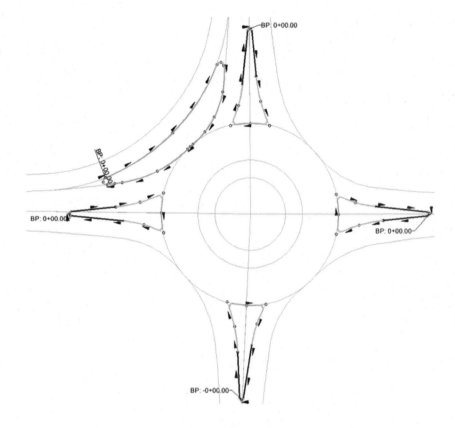

FIGURE 10.80
Several median island alignments with direction shown

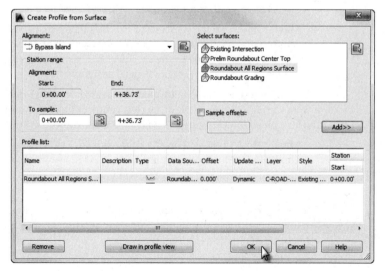

FIGURE 10.81
Extract the profile for medians from your main roundabout surface.

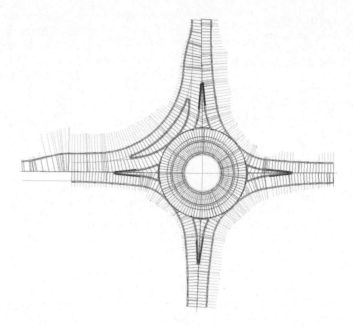

FIGURE 10.82
The completed roundabout

The Bottom Line

Create corridors with non-centerline baselines. Although for simple corridors you may think of a baseline as a road centerline, other elements of a road design can be used as a baseline. In the case of a cul-de-sac, the EOP, the top of curb, or any other appropriate feature can be converted to an alignment and profile and used as a baseline.

>**Master It** Open the MasterIt1001.dwg (MasterIt1001_METRIC.dwg) file, which you can download from www.sybex.com/go/masteringcivil3d2014. Add the cul-de-sac alignment and profile to the corridor as a baseline. Create a region under this baseline that applies the Intersection Typical assembly.

Add alignment and profile targets to a region for a cul-de-sac. Adding a baseline isn't always enough. Some corridor models require the use of targets. In the case of a cul-de-sac, the lane elevations are often driven by the cul-de-sac centerline alignment and profile.

>**Master It** Continue working in the MasterIt1001.dwg (MasterIt1001_METRIC.dwg) file. You need to have completed the previous exercise before continuing. Add the Second Road alignment and Second Road FG profile as targets to the cul-de-sac region. Adjust Assembly Application Frequency to **5′ (1 m)** for tangents and curves.

Create a surface from a corridor and add a boundary. Every good surface needs a boundary to prevent bad triangulation. Bad triangulation creates inaccurate contours and can throw off volume calculations later in the process. Civil 3D provides several tools for creating corridor surface boundaries, including an Interactive Boundary tool.

>**Master It** You need to have completed the previous exercise before continuing. Keep working in the MasterIt1001.dwg (MasterIt1001_METRIC.dwg) file. Create an interactive corridor surface boundary for the entire corridor model.

Chapter 11

Superelevation

Superelevation and cant are methods for changing the cross section of a design to keep cars and trains safely and comfortably on their paths when driving along a curve or series of curves. Superelevation tools also provide a convenient way to force the slope of a road for drainage purposes, without the need for additional assemblies.

Once you have a good grasp of alignments, assemblies, and corridors, you can add a level of sophistication to your design with the superelevation and cant tools within the AutoCAD® Civil 3D® software. Whether you are trying to match in to an existing road's superelevation or creating new data, you will find the tools you need.

In this chapter, you will learn to:

- Add superelevation to an alignment
- Create a superelevation assembly
- Create a rail corridor with cant
- Create a superelevation view

Preparing for Superelevation

Several pieces need to be in place before superelevation can be applied to the design. You will need a *design criteria file* appropriate for your locale, design speeds applied to an alignment, and an assembly that consists of subassemblies that recognize superelevation.

There are a lot of abbreviations and terminology thrown around when it comes to superelevation, so let's take a look at that first.

When an assembly is applied without superelevation, the geometry comes directly from the original design, as shown in Figure 11.1.

FIGURE 11.1
An example assembly without superelevation

As the assembly begins its entrance into a curve with superelevation applied to it, it will first start to lose its normal crown. Figure 11.2 shows the same assembly at the *End Normal Crown (ENC)* station.

FIGURE 11.2
At the End Normal Crown (ENC) station, the default lane slope starts to change.

When the assembly has one side flattened out, as shown in Figure 11.3, it is called *Level Crown (LC)*.

FIGURE 11.3
Level Crown entering a left-hand turn

When the assembly straightens out into a plane that matches the inside lane, it is called *Reverse Crown (RC)*, as shown in Figure 11.4.

FIGURE 11.4
Reverse Crown (RC)

If accommodations have been made for shoulder slope *rollover* and *breakover removal*, the shoulder will shift as well. The left image of Figure 11.5 shows where the superelevated lane becomes steep enough to cause an outside rollover problem with the outside shoulder. The shoulder begins to adjust to increase the safety of the road. The right image of Figure 11.5 shows the change in the lower shoulder.

FIGURE 11.5
(Left) Begin Shoulder Rollover (BSR) and (right) Low Shoulder Match (LSM)

Finally, the lane gets to its maximum slope at the *Begin Full Super (BFS)* station. As you can see in Figure 11.6, all the geometry adjustment has taken place.

FIGURE 11.6
Begin Full Super (BFS)

On the way out of the curve, you will see the geometry transitioning back to its original design. It will pass through *End Full Super (EFS)*, *Low Shoulder Match (LSM)*, *End Shoulder Rollover (ESM)*, *Reverse Crown (RC)*, *Level Crown (LC)*, *Begin Normal Crown (BNC)*, and finally back to *Begin Normal Shoulder (BNS)*.

Now that you are familiar with the terminology and abbreviations Civil 3D uses, let's get started on some design.

Design Criteria Files

Having the correct design criteria file in place is the first step to applying superelevation to your corridor. These XML-based files contain instructions to the software on when to flag your design for geometry problems both horizontally and vertically. Design criteria files are the brains behind how your road behaves when superelevation is applied to the design.

Several design criteria files are supplied with Civil 3D upon installation. The out-of-the-box standards include AASHTO 2001 and AASHTO 2004 for both metric and Imperial units. Several of the country kits include design criteria files for your locality if you are outside of the United States. If country or state kits do not exist for your situation, you can create your own, user-defined files.

To create your own design criteria:

1. Select any alignment.

2. From the Alignment context tab ➢ Modify panel, click the Design Criteria Editor.

   It is easiest to modify an existing table in your desired units rather than starting from an empty file.

3. Be sure to click the Save As icon before making any changes.

> ### User-Defined Criteria Files
>
> If you create design criteria files for your region or design scenario, you will need to be able to share the file with anyone who will be working with your design.
>
> Inside your organization, the best way to handle the design criteria file is to move it to a shared network location. If the design criteria file is stored in a shared location, be sure to have your IT personnel set the file to read-only to prevent inexperienced users from modifying it.
>
> The default location for road design standards is
>
>   C:\ProgramData\Autodesk\C3D 2014\enu\Data\Corridor Design Standards\
>
> The default location for rail design standards is
>
>   C:\ProgramData\Autodesk\C3D 2014\enu\Data\Railway Design Standards\
>
> Once you re-path to the design criteria file for an alignment, the location will be saved with the alignment.
>
> If you are collaborating with someone without direct access to your design criteria files, you will need to send that person the XML file with your DWG. The XML file will not automatically go along for the ride if you use the eTransmit command.

Inside the Design Criteria Editor (Figure 11.7), you will see three headings: Units, Alignments, and Profiles. The Units page tells Civil 3D what type of values it will be using in the file. The Alignments page is used for checking design, creating superelevation, and widening outside curves. The Profiles page provides tabular data for minimum K-value used to check vertical design.

FIGURE 11.7
Inside the Design Criteria Editor

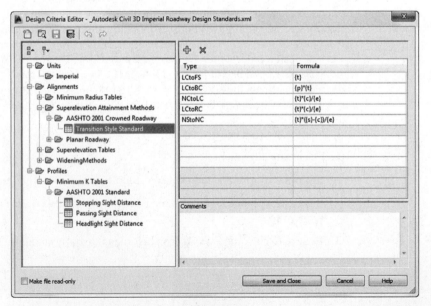

Civil 3D will graphically flag alignments when the design speed specified in the alignment properties has a radius less than the value specified in the Minimum Radius Table. The Minimum Radius Tables from AASHTO use superelevation rates in the table names, but this does not lock you into that rate for applying superelevation to the corridor. In other words, just because you use a more conservative value in your radius check, that doesn't mean you can't superelevate at a steeper rate. The tables are independent of one another.

Also in the Alignments branch you will find the superelevation attainment equations. These equations determine the distance between superelevation critical stations. Familiarize yourself with the terminology and locations represented by these stations, as shown in Figure 11.8.

FIGURE 11.8
Superelevation critical stations and regions calculated by Civil 3D

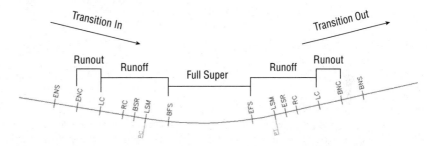

In the following exercise, you will modify an example design criteria file and save it:

1. Open the `1101_Criteria.dwg` (`1101_Criteria_METRIC.dwg`) file, which you can download from this book's web page at www.sybex.com/go/masteringcivil3d2014.

2. Select the USH 10 alignment that already exists in the drawing.

Design Criteria Editor

3. On the Alignment contextual tab ➢ Modify panel, select Design Criteria Editor.

4. Click the Open button at the top of the dialog, and browse to the `1101_CriteriaExample.xml` (`1101_CriteriaExample_METRIC.xml`) file. Click Open.

5. Expand the Alignments branch, and expand the Superelevation Tables branch.

   There is only one superelevation table in this example for 4% maximum slope.

6. Right-click Superelevation Tables, and select New Superelevation Table.

7. Double-click the new table and rename it **Example 6% Super**.

8. Right-click Example 6% Super, and select New SuperelevationTypeByTable.

9. Expand the new branch, right-click SuperelevationTypeByTable, and select New Superelevation Design Speed, as shown in Figure 11.9.

FIGURE 11.9
Adding a design speed to the design criteria file

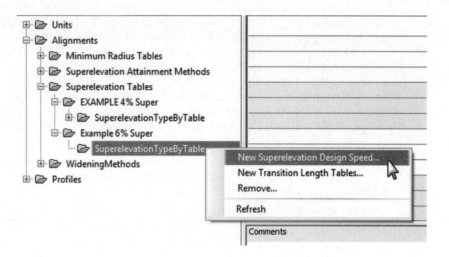

When you expand the SuperelevationTypeByTable branch, you will see that Civil 3D has placed a new design speed with a default of 10 mi/hr in the listing.

10. Double-click Design Speed 10 and change the design speed to **30** mi/hr (**50** km/hr).

    Notice that you only need the numeric value; Civil 3D fills in the rest of the table name for you.

11. Highlight your example design speed.

    The right side of the dialog will have an empty table containing columns labeled Radius and Superelevation Rate.

12. Select the first row of the table. Click the first field in the Radius column to start entering data. Add several radius and superelevation values, as follows:

Curve radius (feet)	Curve radius (meters)	Superelevation %
300	90	6
1000	300	4.5
1500	450	3.2
2000	600	2.6
4000	1250	NC

When you have completed your data entry from the table, the US units Design Criteria Editor will resemble Figure 11.10.

FIGURE 11.10
Adding example data using the Design Criteria Editor

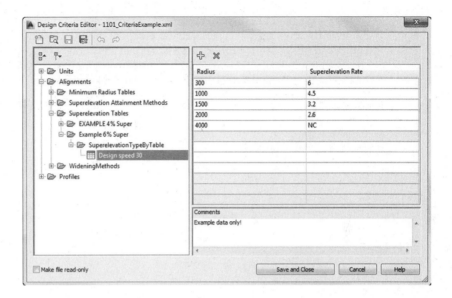

13. Add a note in the Comment field that reads **Example Data Only!**
14. Click the Save And Close button at the bottom of the editor.
15. Click Save Changes and exit when prompted.
16. Save and close the current drawing.

Ready Your Alignment

Superelevation stations are connected to alignment curves (unless you create a user-defined curve in a tangent section). The design speed from the alignment properties is needed at each curve to specify which superelevation rate tables to use from the design criteria. The design speed has an effect on the distance between superelevation critical stations and the cross-slope used when the road is at full-super.

It is a good idea to get your alignment geometry and design speed locations finalized before attaching superelevation. If a change is made to your alignment, the superelevation stationing will be marked as out of date.

Super Assemblies

As a general rule, if the lane subassembly has the word *super* somewhere in its name, it will respond to superelevation. If you want to verify that the lane you are choosing will behave the way you want it to in a superelevation situation, you can right-click it from the tool palette and access the subassembly help.

As long as you stay away from the Basic tab, all of the shoulder and curb subassemblies have parameters you can set to dictate how the assembly is to behave when an adjacent lane superelevates.

Most subassemblies that are capable of superelevating are intended for use where the pivot point for the cross section is at the center crown of the road. When the pivot point is at the center of the road, the baseline profile dictates the final elevation of the crown of the road. Figure 11.11 shows an example of a two-lane highway (top image) and a four-lane divided highway (bottom image) that are designed to be used with the superelevation tools.

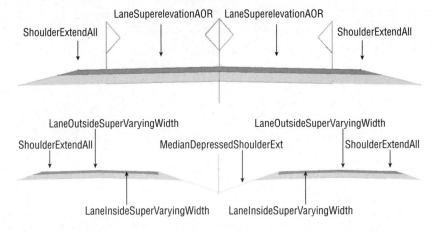

FIGURE 11.11
Two-lane road ready for super; four-lane road used in super

Axis of Rotation (AOR)

The following is a list of the limitations and other factors to be aware of if you decide to take these new assemblies for a spin:

- Don't use an offset assembly with an axis of rotation superelevation assembly. The result in the corridor may not come out as expected.

- When working with curbs, medians, and shoulders, keep an eye out for the parameter Superelevation Axis Of Rotation. If this parameter reads Unsupported, it means that the subassembly will not adjust for superelevation other than at the center. None of the curb and gutter subassemblies will adjust for breakover or rollover, but most of the shoulder assemblies do. This is a "hard-coded" parameter that cannot be changed by the end user.

Axis of Rotation Support

The axis of rotation (AOR) subassembly can be used when the centerline of the road is not the pivot point for superelevation. The "flag" symbols (as shown in Figure 11.12) indicate potential pivot points on the assembly.

FIGURE 11.12
AOR subassemblies used on an undivided, crowned roadway

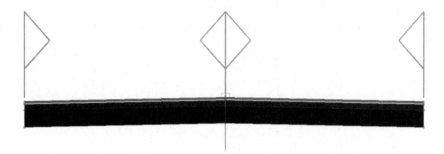

The flag symbols on LaneSuperelevationAOR indicate where the lane can be pinned down and used as a pivot point. When the axis of rotation is not the centerline of the road, the lane geometry is used to determine the change in elevation that will occur as a result.

When building assemblies with LaneSuperelevationAOR you may see warnings appear, as shown in Figure 11.13.

FIGURE 11.13
A warning symbol on an assembly using LaneSuperelevationAOR

Here are some of the warnings you may encounter:

Center Pivots Not Applied When Only One Group   This usually occurs when you start your assembly with a median at the center. The construction of the assembly then does not have a distinct left and right group. Check the Construction tab of the assembly Properties dialog to verify. The fix is to build your assembly with left and right sides and add the median last.

Unsupported Subassemblies   This warning will appear when you're attempting to use an assembly that has the Superelevation Axis Of Rotation parameter absent from its parameters.

Unsupported in Assemblies Containing Offsets   Using an offset assembly will interfere with the software's ability to calculate the correct slope and curve widening on a superelevated road. Therefore, offset assemblies are not recommended for use with the AOR subassembly. You can still use offsets in traditional center-pivot-based superelevation.

No Center Pivots Found   Make sure your assembly properties list the assembly as the correct type. For example, if you accidentally set Assembly Type to Divided Crowned Road when it is actually an undivided crowned road, you will receive this message. Check the Construction tab of the assembly Properties dialog to verify.

You can still add assemblies with warnings to a corridor; however, the superelevation may not behave as expected.

Real World Scenario

THE TIPPING POINT

In some design situations that use superelevation, the crown of the road is not the ideal pivot point. In the following example, you will create a four-lane divided highway that pivots inside the curve rather than at the crowns during superelevation.

1. Open the file 1102_AORAssembly.dwg (1102_AORAssembly_METRIC.dwg).

   This drawing contains an assembly that was started for you. Note that a generic link has been placed on each side of the main assembly marker as spacers for the subassemblies that you will be adding in the next steps.

2. Open your subassembly tool palette (Ctrl+3 will work) and select the Lanes tab.
3. Select the LaneSuperelevationAOR subassembly.
4. In the Parameters section of the AutoCAD Properties palette, change Side to Right and set Use Superelevation to Right Lane Outside (as shown in the following screen shot).

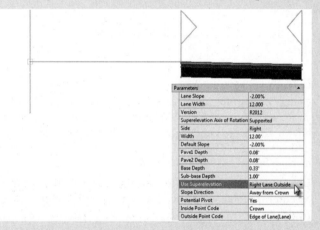

5. Click the circular marker point on the right to place the subassembly, using the generic link marker as its "hanger." Press Esc to complete the placement of the generic link.
6. Select the newly placed subassembly.
7. From the Subassembly: LaneSuperelevationAOR contextual tab ➢ Modify Subassembly, click Mirror.
8. Click the right circular marker point (the same one you clicked in step 5) to attach the mirrored subassembly to the left side of the subassembly you placed in step 5.

   A dialog will appear asking you to confirm that you want to place the mirrored subassembly on the same side. Click OK.

9. Select the newly placed subassembly and view its AutoCAD properties.

10. Change the Use Superelevation parameter to Right Lane Inside, as shown here.

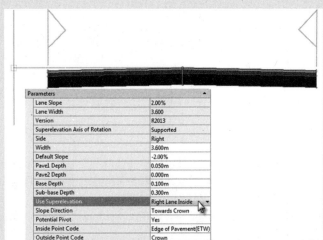

11. Select both of the new LaneSuperelevationAOR assemblies and the generic link. From the Subassemblies contextual tab ➢ Modify Subassembly panel, click Mirror.

12. Click the baseline assembly marker to mirror all three subassemblies to the left side of the assemblies.

    Be sure to set the subassembly parameters to the correct superelevation lane and side property. Note that the inside lane must have Use Superelevation set to Left Lane Inside and the outside lane must be set to Left Lane Outside.

13. On the tool palette, switch to the Generic tab. Select the MarkPoint subassembly, change Point Name to **MEDIAN**, and place it by clicking the red circle on the inside right edge of the pavement (as shown below).

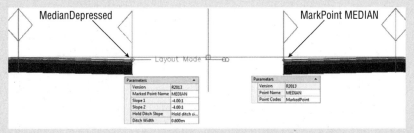

14. On the Medians tab, select Median Depressed, and change Marked Point Name to **MEDIAN**.

    Note that the MarkedPoint name is case sensitive and must match the point name exactly or the median will not form correctly.

15. Leave all other parameters at their defaults and place the median on the inside left of the subassembly as shown in the image above. Press Esc when complete.

16. As the final step in building this assembly, select both of the original generic link spacers and set the Omit Link property to Yes.

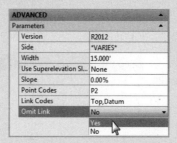

If time permits, add the ShoulderExtendAll subassembly to the outermost edges using the default settings. Press Esc when complete.

17. Select the Route 66 alignment that runs through the plan view of the project (Hint: Zoom out to see it).

18. On the Alignment contextual tab ➢ Modify panel, click Superelevation ➢ Calculate/Edit Superelevation, and click the Calculate Superelevation Now option when notified that no data exists.

19. On the Calculate Superelevation – Roadway Type page, set Roadway Type to Divided Crown With Median.

20. Set the pivot method to Inside Of Curve.

21. Set the median treatment to Distorted Median, and click Next.

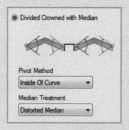

22. On the Calculate Superelevation Lanes page, set the normal lane width to **12′** (**4** m) and the normal lane slope to **-2.00%**.

23. Place a check mark next to Symmetric Roadway, and click Next.

24. On the Calculate Superelevation Shoulder Control page, leave the settings at their defaults, and click Next.

25. On the Calculate Superelevation Attainment page, set the design criteria file to `Autodesk Civil 3D Imperial (2004) Roadway Design Standards.xml` (metric users use `Autodesk Civil 3D Metric (2004) Roadway Design Standards.xml`).

26. Set the superelevation rate table to AASHTO 2004 US Customary eMax 6% (AASHTO 2004 Metric eMax 6%) using the 4 Lane Transition Length table.

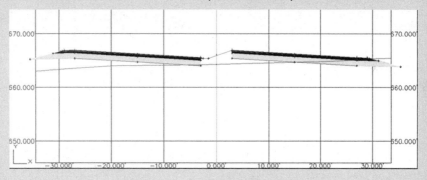

27. Place a check mark next to Automatically Resolve Overlap, and click Finish.

    The Panorama palette will pop up containing the results of your superelevation calculation. Because the Automatically Resolve Overlap option was turned on, no warnings should appear in the Superelevation Tabular Editor.

28. Click the green check mark to dismiss Panorama.

29. In the Prospector tab of Toolspace, expand the Corridors branch, select the corridor Route 66, right-click, and select Rebuild.

30. Select the corridor in plan and enter the Section Editor to examine your corridor. (Hint: Review "Editing Sections" in Chapter 9, "Basic Corridors.")

    You should observe that superelevation is occurring based on our specified pivot point inside the curves. Close the Section Editor when you can't take any more awesomeness.

Check out `1102_AORAssembly_FINISHED.dwg` and `1102_AORAssembly_METRIC_FINISHED.dwg` to see a complete version of this example.

Applying Superelevation to the Design

Civil 3D takes into account other factors such as curve station locations and assembly geometry. Superelevation information is associated with the alignment but is handled in a separate calculation area. In this section, you will put all the pieces in place that are needed for the software to dynamically apply superelevation or cant to your design.

Start with the Alignment

To begin applying superelevation to the design, select your alignment:

1. Open the 1103_Super.dwg (1103_Super_METRIC.dwg) file, which you can download from this book's web page.

2. Select the USH 10 alignment that already exists in the drawing.

3. From the Alignment contextual tab, select Superelevation ➢ Calculate/Edit Superelevation.

4. When prompted by the dialog, click Calculate Superelevation Now.

5. On the Calculate Superelevation – Roadway Type page, select the Undivided Crowned radio button.

6. From the Pivot Method drop-down, choose Center Baseline, as shown in Figure 11.14, and click Next.

FIGURE 11.14
Roadway type specification for superelevation

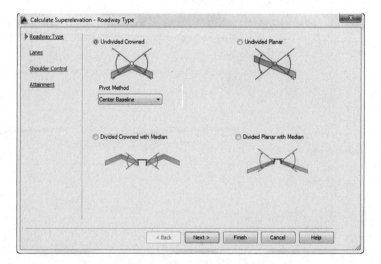

7. On the Calculate Superelevation – Lanes page of the wizard, verify that the Symmetric Roadway check box is selected.

8. Set Normal Lane Width to **12′ (4 m)**, and set Normal Lane Slope to **-2.00%**, as shown in Figure 11.15, and click Next.

9. On the Calculate Superelevation – Shoulder Control page, make sure the Calculate check box is selected on the right side of the page.

   Only the Outside Edge Shoulders options should be active. Since this is an undivided road, the options for the inside median shoulders are grayed out.

FIGURE 11.15
Lane information

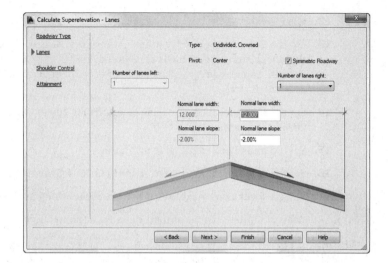

10. Set Normal Shoulder Width to **6′ (2 m)**, and set Normal Shoulder Slope to **-5.00%**.

11. For Shoulder Slope Treatment, set the following:
    - Set the Low Side option to Breakover Removal.
    - Set the High Side option to Default Slopes.
    - Place a check mark next to Maximum Shoulder Rollover and set the value to **8.00%**.

    Your Shoulder Control screen should look like Figure 11.16.

FIGURE 11.16
Shoulder Control and Breakover Removal parameters

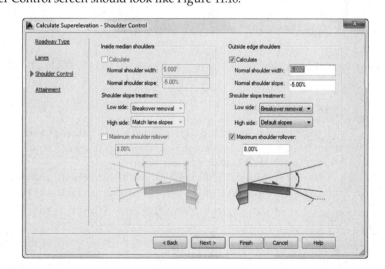

12. Click Next.

13. On the Calculate Superelevation – Attainment screen, click the ellipsis button to verify that the design criteria file is set to Autodesk Civil 3D Imperial (2004) Roadway Design Standards.xml (Autodesk Civil 3D Metric (2004) Roadway Design Standards.xml).

14. Set the superelevation rate table to **AASHTO 2004 US Customary eMax 4% (AASHTO 2004 Metric eMax 4%)**.

15. Set the transition length table to **2 Lane**.

16. Set the Attainment method to **AASHTO 2004 Crowned Roadway**.

17. Place a check mark next to Apply Curve Smoothing and set Curve Length to **50′ (20 m)**.

    Leave all other settings at their defaults, as shown in Figure 11.17.

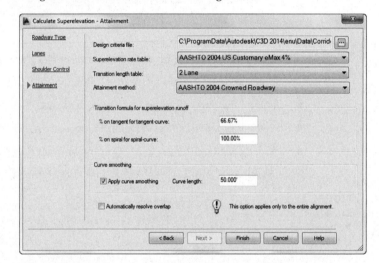

FIGURE 11.17
Finalizing the superelevation on the Attainment screen

18. Click Finish.

You should now see the superelevation table appear inside Panorama with the data resulting from the wizard. Examine your alignment; you should now have labels showing the superelevation critical stations created by the wizard.

As you click in the table, you will see helpful glyphs showing you which superelevation station and corresponding curve you are editing, as shown in Figure 11.18.

FIGURE 11.18
Superelevation table with glyphs in the graphic

Superelevation Curve	Start Station	End Station	Length	Overl...	Left Outsi...	Curve Smooth
Level Crown	13+44.10'				-5.00%	50.000
Runout	13+44.10'	13+85.71'	41.613'			
Level Crown	13+44.10'				-5.00%	50.000
Begin Normal Crown	13+85.71'				-5.00%	50.000
Begin Normal Shoulder	14+48.13'				-5.00%	50.000
Curve.2						
Transition In Region	18+60.80'	20+25.85'	165.045'			
End Normal Shoulder	18+60.80'				-5.00%	50.000
Runout	19+22.81'	19+64.15'	41.340'			
End Normal Crown	19+22.81'				-5.00%	50.000
Level Crown	19+64.15'				-5.00%	50.000
Runoff	19+64.15'	20+25.85'	61.695'			
Level Crown	19+64.15'				-5.00%	50.000
Begin Curve	20+05.28'					
Reverse Crown	20+05.49'				-5.00%	50.000
Begin Full Super	20+25.85'				-5.00%	50.000
Transition Out Region	21+99.81'	23+64.85'	165.045'			
Runoff	21+99.81'	22+61.50'	61.695'			
End Full Super	21+99.81'				-5.00%	50.000
Reverse Crown	22+20.16'				-5.00%	50.000
End Curve	22+20.37'					
Level Crown	22+61.50'				-5.00%	50.000
Runout	22+61.50'	23+02.84'	41.340'			
Level Crown	22+61.50'				-5.00%	50.000
Begin Normal Crown	23+02.84'				-5.00%	50.000
Begin Normal Shoulder	23+64.85'				-5.00%	50.000

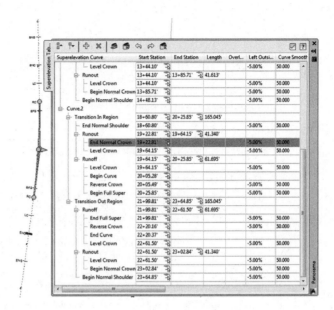

Transition Station Overlap

It is not uncommon to have overlap warnings in your superelevation table. You should resolve the transition station overlap before you continue your design.

Overlap occurs when there is not enough room between curves to fully transition out of one curve and back into the next. Transition station overlap will always occur when a reverse curve or compound curve exists in your alignment. As you can see in Figure 11.19, Curve 1 does not complete its transition out until station 16+82.56, but according to the attainment calculations, Curve 2 will begin affecting the shoulder starting at station 15+35.24.

FIGURE 11.19
Superelevation table showing overlap between two curves

Transition Out Region	15+26.92'	16+82.56'	155.643'			
Runoff	15+26.92'	15+78.53'	51.612'			
End Full Super	15+26.92'				-5.00%	50.000
Reverse Crown	15+36.92'			⚠	-5.00%	50.000
End Curve	15+44.12'			⚠		
Level Crown	15+78.53'			⚠	-5.00%	50.000
Runout	15+78.53'	16+20.14'	41.613'			
Level Crown	15+78.53'			⚠	-5.00%	50.000
Begin Normal Crown	16+20.14'			⚠	-5.00%	50.000
Begin Normal Shoulder	16+82.56'			⚠	-5.00%	50.000
Curve.2						
Transition In Region	15+35.24'	17+00.29'	165.045'			
End Normal Shoulder	15+35.24'			⚠	-5.00%	50.000
Runout	15+97.25'	16+38.59'	41.340'			
End Normal Crown	15+97.25'			⚠	-5.00%	50.000
Level Crown	16+38.59'			⚠	-5.00%	50.000
Runoff	16+38.59'	17+00.29'	61.695'			
Level Crown	16+38.59'			⚠	-5.00%	50.000
Begin Curve	16+79.73'			⚠		
Reverse Crown	16+79.93'			⚠	-5.00%	50.000
Begin Full Super	17+00.29'				-5.00%	50.000

You have several options for fixing superelevation overlap:

- You can choose to have Civil 3D rectify the overlap for you.
- You can manually modify the stations in the table.
- You can change the stationing for superelevation by modifying the superelevation view, which we will discuss later in this chapter.

IMPORT OR EXPORT SUPERELEVATION DATA TO CSV

You may spend hours getting your superelevation stations to work out correctly. However, it just takes one wrong button click to blow away all your time-consuming edits to the tabular input. To save you from yourself (or the click-happy intern), back up your superelevation tables by exporting them to a file. You'll find the Export Superelevation Data button at the top of the tabular input.

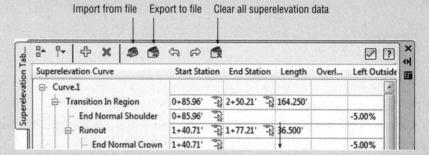

The nice thing about a CSV file is that it can easily be edited in an external spreadsheet program, such as Microsoft Excel. If you choose to modify your superelevation stations outside of Civil 3D, there are a few things to keep in mind:

- You must keep the same number of columns and column names.
- It is OK to add or delete rows.
- Blank shoulder or lane slope fields will be treated as 0.00%. However, if all of your lane and shoulder slope fields are blank, the row will be omitted from the import.
- It is OK to rename the superelevation critical stations.
- If your CSV file has more curves in it than the alignment (i.e. superelevation regions), you must create a user-defined curve before importing the data. More information on user-defined curves can be found later in the section, "Applying Superelevation to the Design."
- Importing data from a CSV file will override any data for existing stations in the file.

When you need to reimport the superelevation data, it is a best practice to clear any data before clicking Import From File.

To have Civil 3D clear the overlap for you, click the warning symbol that appears in the Superelevation Tabular Editor. Civil 3D resolves overlap by omitting noncritical stations and/or by compressing the transition length between certain stations. In the case of a reverse curve, Civil 3D will pivot the road from full-super to full-super, without transitioning back to normal crown. Be sure to verify that the software has made the update that meets the requirements of your locale.

USER-DEFINED CURVES: FORCING SUPERELEVATION

By default, superelevation stations are calculated per curve. In most situations, the curves on the alignment will correspond to the needed superelevation regions. However, there are several situations where you want to force superelevation to occur on a tangent. In those situations, you can create a user-defined curve.

A user-defined curve starts in the same place it starts with traditional superelevation, by selecting the alignment along which you want to create superelevation. From the Alignment context tab ➢ Modify panel, click Superelevation ➢ Calculate/Edit Superelevation.

If your alignment already has curves with superelevation data attached to them, Civil 3D will open the Superelevation Curve Manager palette right away. If no superelevation data exists for the alignment, you will be prompted to either calculate superelevation now or open the superelevation curve manager. If your alignment has no curves, click the Open The Superelevation Curve Manager option.

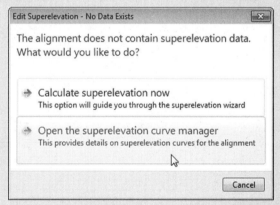

Once you are in the Superelevation Curve Manager, you will have access to any curves—if they exist. If you are working with an alignment that contains no curves, this palette will be blank.

To create a user-defined curve, click the Create User-defined Curve button. You will be prompted to select an entity to associate to the user-defined curve.

The entity (or entities) you select must be a tangent segment that does not already have a user-defined curve associated with it. The segment geometry will be used as the Begin Curve and End Curve stations. If you pick more than one entity, they must be adjacent to each other.

After you pick your elements, you will have the ability to add "fake" curve information. If you plan to use Civil 3D to calculate superelevation, you must have a design speed assigned to the alignment. You can specify a radius and curve direction in the Superelevation Curve Manager palette (as shown below) so that the Superelevation Wizard knows which way you intend to tip the road.

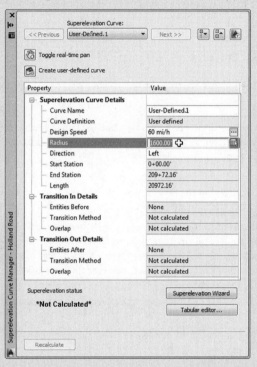

After a user-defined curve has been assigned a design speed and radius, the Superelevation Wizard can be used to calculate superelevation similar to a traditional, "real" curve.

If you are planning to import all your stations from a CSV file or manually create stations and cross-slopes in the tabular editor, then you may omit design speed and radius.

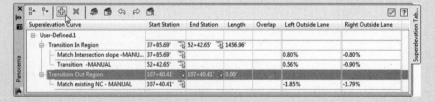

As shown in the image above, you can manually add superelevation stations by clicking the plus sign in the Superelevation Tabular Editor. Initially, the lane and shoulder slopes will be blank in a manually added station. Use the Superelevation Tabular Editor to rename the special regions and set the slopes as needed.

Oh Yes, You Cant

In the civil engineering industry, the terms *superelevation* and *cant* are often used interchangeably. Inside Civil 3D, the terms have distinct meanings (Figure 11.20).

FIGURE 11.20
Example cant curve data table

Cant Curve	Start Station	End Station	Length	Applied Cant	Equ
Curve.1					
Transition In Region	4+75.44'	8+42.44'	367.000'		
End Level Rail	4+75.44'			0.000"	0.000
Begin Curve	6+58.94'				2.047
Begin Full Cant	8+42.44'			0.400"	2.047
Transition Out Region	13+86.97'	15+22.24'	135.264'		
End Full Cant	13+86.97'			0.400"	2.047
Begin Level Rail	15+22.24'			0.000"	2.047
Curve.2					
Transition In Region	15+22.24'	17+91.31'	269.074'		
End Level Rail	15+22.24'			0.000"	0.000

Superelevation tools in Civil 3D are used for roads where the cross-slope changes within a curve are expressed by a percentage. Unlike superelevation, cant is expressed as a difference in height between the outer and inner rails.

Workin' on the Railroad

Cant tools were introduced in Civil 3D 2013 and are specifically designed for rail. In order to work with cant, the following must be in place:

Alignment Type Set to Rail Normally, you would set the alignment type when you first define the alignment. If you forget to set this initially, you can change it at any time in the alignment properties on the Information tab.

Alignment Design Speed Like motorways and superelevation, rail requires a design speed in order to apply cant. Cant design standards are provided with the software and can be edited in the same manner as other design criteria files.

Cant Calculation Like superelevation, cant is attached to an alignment and its curves. As seen in Figure 11.21, the icon's location in the Alignment contextual tab should be reminiscent of the superelevation button.

FIGURE 11.21
Accessing the Cant Calculation tools from the Rail Alignment contextual tab

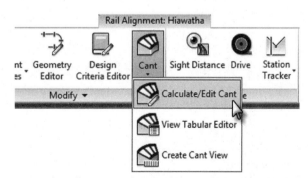

Rail Assembly When working with an assembly for rail, the type must be set to Railway. You can set this on assembly creation or in the assembly properties after the fact. There is one subassembly that is purpose-built for cant, the Rail subassembly, shown in Figure 11.22. If you are working with a drawing created in a release previous to 2013, make sure you replace the old assemblies with this new rail subassembly to ensure cant takes place. You can verify the version by checking Advanced Parameters ➢ Version in the AutoCAD Properties palette.

FIGURE 11.22
The new Rail subassembly

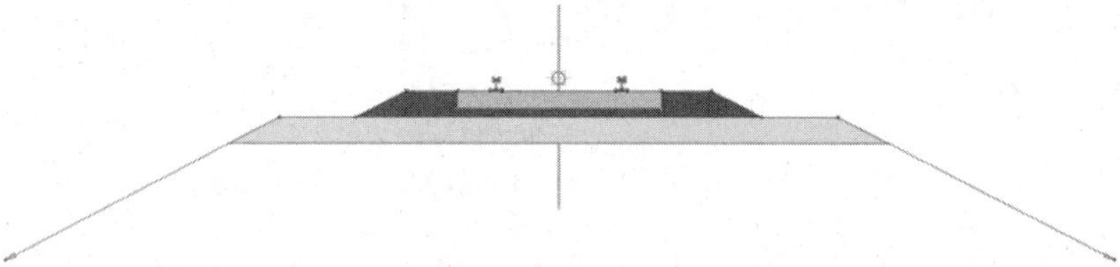

Creating a Rail Assembly

You will find the Rail assembly in the Bridge And Rail tab of the tool palettes. The following exercise will walk you through creating a typical rail bed design:

1. Open the drawing 1104_Rail.dwg (1104_Rail_METRIC.dwg), which you can download from this book's website.

   This drawing contains an alignment and design profile.

2. On the Home tab ➢ Create Design panel ➢ Assembly, click Create Assembly.

3. Name the assembly **Rail w Service Road**.

4. Set Assembly Type to Railway, and click OK.

5. Click to place the assembly anywhere in the graphic.

6. Open the subassembly tool palettes if they are not already open. In the subassembly tool palettes, locate the Bridge And Rail palette, and click Rail Single. In the Properties palette, make the following changes:

   a. Set the subballast width to **20′ (6 m)**.

   b. Set the subballast side slope to **0.001:1**.

   Leave all other parameters at their defaults.

7. Click the assembly to place the RailSingle subassembly.

8. Switch to the Basic tab of the subassembly tool palettes, and click GenericPavementStructure.

   This will be our service road, constructed out of the same material as the subballast.

9. Enter the following data into the Properties palette, remembering that the codes are case sensitive:

   a. Set Side to **Left**.

   b. Set Width to **8′ (2.5 m)**.

c. Set Depth to **1′ (0.3 m)**.
   d. Set DeflectOuterVerticalFace to **Yes**.
   e. Set Outer Edge Slope to **2.00:1**.
   f. Set TopLink Codes to **Top**, and set BottomLink Codes to **Datum**.
   g. Set Shape Codes to **Subballast**.
10. When your parameters resemble Figure 11.23, click to place the GenericPavementStructure to the left of the subballast rail subassembly.

FIGURE 11.23
Placing an assembly to represent additional subballast for a service road next to the rail bed

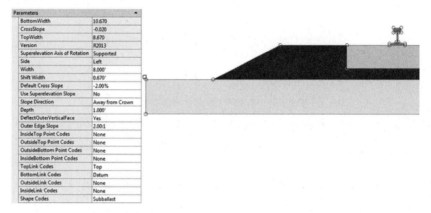

11. Remain in the GenericPavementStructure tool, but change the width to **0.1′ (0.03 m)**.
12. Click to place the structure on the right side of the rail subassembly.
13. Switch to the Generic palette, and click LinkSlopeToSurface.
14. Set Slope to **-50** and click the "tip" of the generic pavement structure on both sides of the assembly.

   Your completed assembly will look like Figure 11.24.

FIGURE 11.24
Your completed rail assembly

15. Save and close the drawing.

   Compare your work with 1104_Rail_FINISHED.dwg or 1104_Rail_METRIC_FINISHED.dwg to see how you fared.

Applying Cant to the Alignment

Like superelevation in a roadway alignment, *cant* is related to a rail alignment. The following exercise will walk you through applying cant to the alignment. You should experience a distinct feeling of *déjà-vu* if you completed earlier exercises involving applying superelevation to the alignment.

1. Open the drawing 1105_RailAlignment.dwg (1105_RailAlignment_METRIC.dwg), which you can download from this book's website.

   This drawing contains a corridor with rail; your task is to update the alignment to apply cant. Observe the section views in the drawing (shown in the lower modelspace viewport in the file). Each is created at a location on the curve that should have cant applied. However, these views show flat rails.

2. Select the alignment from the upper viewport. From the Rail Alignment contextual tab ➢ Modify panel, click Cant ➢ Calculate/Edit Cant.

3. Click Calculate Cant Now.

4. Keep the pivot method as Low Side Rail, and click Next.

5. Place a check mark next to Automatically Resolve Overlap.

6. Set Applied Cant Table to Freight Train Applied Cant Table (metric users, use Mixed Passenger And Freight Cant).

7. Leave all other settings at their defaults, and click Finish.

8. In Prospector, rebuild the Hiawatha corridor.

9. Save and close the drawing.

   You should observe in the lower viewport that your section views show the cant applied to the design. Check your design against 1105_RailAlignment.dwg or 1105_RailAlignment_METRIC.dwg if desired.

Superelevation and Cant Views

Superelevation and cant views are a graphic representation of the roadway or rail superelevation. Grip edits to the graphical view will also edit the superelevation stations. The view itself is not intended for plotting. The superelevation view plots station against lane slope to form a graph of the left and right edges of the pavement.

In the following exercise, you will create a superelevation view:

1. Open the drawing 1106_SuperView.dwg (1106_SuperView_METRIC.dwg), which you can download from this book's website.

   The alignment in this file already has superelevation calculated for it. Your task is to create the superelevation view.

2. Select the Route 66 alignment by clicking it in the graphic.

3. From the Alignment context tab of the ribbon ➢ Modify panel, click Superelevation ➢ Create Superelevation View as shown in Figure 11.25.

FIGURE 11.25
Creating a super-elevation view

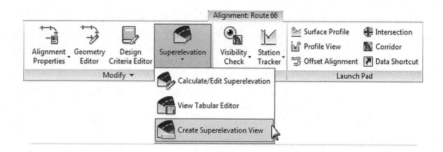

4. In the Create Superelevation View dialog, set the colors for the lane and shoulder slopes as shown in Figure 11.26 by double-clicking the ByBlock field and picking the colors indicated.

FIGURE 11.26
Set colors for different assembly components to easily differentiate them in the super-elevation view

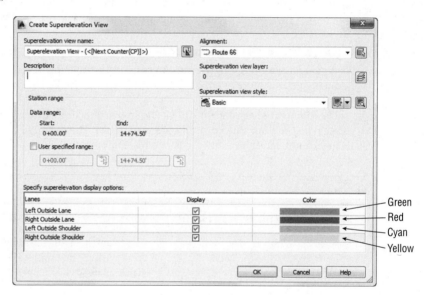

5. Click OK and then click in the graphic to place the superelevation view. Your result should look like Figure 11.27.

FIGURE 11.27
The superelevation view for the Route 66 alignment

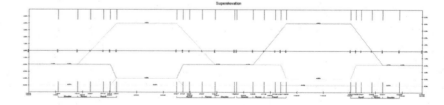

6. Save the file for use in the next exercise.

At first glance, the superelevation view may seem difficult to read, but with a little explanation it can shed a lot of light on what is going on with your lane and shoulder slopes. The superelevation graphic plots the station value against the percent cross-slope of each edge of pavement and edge of shoulder. The upper line shows the behavior of the right edge of the pavement, and the lower line shows the left edge of the pavement.

Where no superelevation is applied, the graph data for the lanes remains at -2% while the shoulders are shown at their default cross-slope of -5%. As the assembly twists into position during superelevation, the distances between the lines become greater as the right edge slopes up to a maximum superelevation of 4%.

There are a few more observations you can make about your superelevation view. No overlap exists between the two curves' superelevation data. You can tell this by observing the center portion of the graph; the superelevation lines go back to the default cross-slope. The very astute observer can ascertain by looking at this graph that a maximum breakover slope of 8% was used on the shoulder. How can one tell? By seeing that when the lanes are in max super (i.e., the lane is at +4% slope), the shoulder slope jumps up to -4%.

Using a Superelevation View to Edit Data

Next, you will use the superelevation view to edit superelevation data. Editing superelevation data by the superelevation view is an alternative to editing the data in tabular form as you learned about earlier.

The diamond-shaped grips can be slid in one axis to modify stationing (the horizontally oriented grips) or slope (the vertically oriented grips). The rectangular grip can be moved to reduce the maximum lane slope when it is in a full-super state.

TABLE 11.1: Superelevation view grips

SUPERELEVATION VIEW SYMBOL	MEANING
(vertical diamond)	Grip (blue) is at a superelevation critical station and a grade break occurs at that station. Vertically oriented grips can be moved up or down to change the slopes associated with them.
(horizontal diamond)	Grip is at a superelevation station. Horizontally oriented grips change the value of superelevation stations.
(rectangle)	Grip will appear at locations of constant slope. These can be moved up or down to change the superelevation cross-slope.
(gray diamond)	Grip (gray) is at a superelevation critical station, but no grade breaks occur at the location. Vertically oriented grips can be moved up or down to change the slopes associated with them.
(diamond with +)	The plus sign next to any grip indicates that more than one item is the same slope at that station.

In the following exercise, you will use the superelevation view to remove the normal crown area in the middle of the alignment. In other words, you will force the curves to transition directly from one to the other. You will also adjust the rate of maximum superelevation.

1. Continue working in the drawing 1106_SuperView.dwg (1106_SuperView_METRIC.dwg). You need to have completed the previous exercise before continuing.

2. Isolate the right outside lane line by holding down Ctrl as you click the red line. (Hint: If you are not sure which line represents the various superelevation slopes, pause your cursor over the line to get a tooltip showing the line's information.)

3. Pause your cursor (without clicking) over the blue grip at station 5+73.47 (0+175.06 in the metric drawing).

4. Select Remove Grade Break from the grip menu as shown in Figure 11.28.

FIGURE 11.28
The grip menu at the superelevation critical station

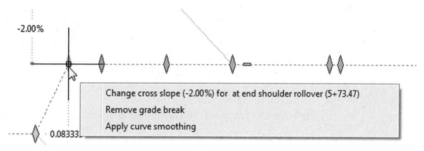

5. Repeat steps 3 and 4 to remove the grade break at 8+35.56 (0+258.53 in the metric drawing).

6. Press Esc to clear your selection.

7. Hold down Ctrl and select the green line representing the left outside lane.

8. Using the same technique as before, remove the grade break at station 6+93.47 (0+207.56 in the metric drawing) and 9+39.31 (0+286.03 in the metric drawing). Press Esc to clear the selection when complete.

9. Turn on dynamic input on your status bar (F12 on your keyboard) if it is not already on.

10. Click your superelevation view to select it. All of the grips should be available.

11. Click the flat blue grip in the first left outside shoulder region, just after station 1+09.41 (0+039.60 in the metric drawing). The grip will be associated with the cyan line.

12. In the dynamic input box, key-in -4.5 as shown in Figure 11.29, and then press Enter.

FIGURE 11.29
Changing the slope using grips and dynamic input

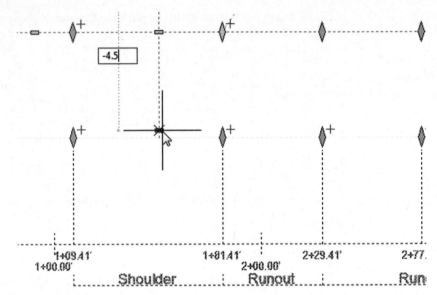

The shoulder slope has been changed slightly to match existing conditions.

13. Save the file.

    Compare your work to `1106_SuperView_FINISHED.dwg` or `1106_SuperView_METRIC_FINISHED.dwg` if desired.

The Bottom Line

Add superelevation to an alignment. Civil 3D has convenient and flexible tools that will apply safe, correct superelevation to an alignment curve.

Master It Open the `MasterIt1101.dwg` (`MasterIt1101_METRIC.dwg`) file, which you can download from www.sybex.com/go/masteringcivil3d2014. Verify that the design speed of the road is 20 miles per hour (35 km per hour) and apply superelevation to the entire length of the alignment. Use AASHTO 2004 design criteria with an eMax of 6% 2-Lane. Use the option to automatically resolve overlap. For the remainder of the options, leave default settings unless otherwise directed.

Create a superelevation assembly. For superelevation to happen, you need to have an assembly that is capable of superelevation.

Master It Continue working in `MasterIt1101.dwg` (`MasterIt1101_METRIC.dwg`). Create an assembly similar to the one in the top image shown earlier in the chapter in Figure 11.11. Set each lane to be **14′** (**4.5 m**) wide and each shoulder to be **6′** (**2 m**) wide. Leave all other options at their defaults. If time permits, build a corridor based on the alignment and assembly.

Create a rail corridor with cant. Cant tools allow users to create corridors that meet design criteria specific to rail needs.

> **Master It** In the drawing `MasterIt1102.dwg` (`MasterIt1102_METRIC.dwg`), create a Railway assembly with the RailSingle subassembly using the default parameters for width and depth. Add a LinkSlopetoSurface generic link with 50% slope to each side. Add cant to the alignment in the drawing using the default settings for attainment. Create a corridor from these pieces.

Create a superelevation view. Superelevation views are a great place to get a handle on what is going on in your roadway design. You can visually check the geometry as well as make changes to the design.

> **Master It** Open the drawing `MasterIt1103.dwg` (`MasterIt1103_METRIC.dwg`). Create a superelevation view for the alignment. Show only the left and right outside lanes as blue and red, respectively.

Chapter 12

Cross Sections and Mass Haul

Cross sections are used in the AutoCAD® Civil 3D® program to allow the user to have a graphic confirmation of design intent as well as to calculate the quantities of materials used in a design. All that is needed for section creation is an alignment and a surface. Other objects, such as pipes, structures, and corridor components, can be sampled in a sample line group, which is used to create the graphical section displayed in a section view. These section views and sections remain dynamic throughout the design process, reflecting any changes made to the sampled information. The result is a plot-worthy set of section views and accurate end area volume information.

In this chapter, you will learn to:

- Create sample lines
- Create section views
- Define and compute materials
- Generate volume reports

Section Workflow

When the time comes that you wish to see how the information along your alignment will appear plotted, you can create *sample lines*. If your goal is to show your completed design, at this point you should have a completed corridor, corridor top surface, and corridor datum surface.

Sample Lines vs. Frequency Lines

Sample lines are created at any stations where you wish to create a *section view*. Sample lines are also used to compute end area volumes.

A common point of confusion with new users is the difference between *frequency lines* and *sample lines*. Table 12.1 explains the differences.

TABLE 12.1: Sample lines vs. frequency lines

SAMPLE LINES & SECTION VIEWS	FREQUENCY LINES & SECTION EDITOR
Sample lines can be created without a corridor present (e.g., when you wish to see existing surface sections along the alignment).	Frequency lines are always part of a corridor.
Sample lines occur at any station where a section view or end area is needed.	Frequency lines occur anywhere the design needs to be calculated or modified (e.g., at certain station intervals, a driveway).
Sample lines are used for end area volume computation.	Frequency lines are used to apply assembly calculations to the design (e.g., locating slope-intercept).
Sample lines can be skewed at an angle other than 90° from the baseline.	Frequency lines are always perpendicular to the baseline.
Sample line swath width is usually uniform and dependent on user plotting needs.	Frequency lines' length from baseline depends on assembly and will vary from station to station.
Section views are read-only reflections of the design.	Design can be modified in the Section Editor at each frequency line.
Section views are readily adapted to plotting.	Plotting should never occur from the Section Editor.

When you create your sample line group, you will have the option to sample any surface in your drawing, including corridor surfaces, the corridor assembly itself, and pipes. The sections are then sampled along the alignment with the left and right *swath widths* specified and at the intervals specified. After you create sample lines, you can create section views or define materials for end area volume calculations.

Creating Sample Lines

A *sample line* is a powerful tool needed for both section view creation and end area material computations. Sample lines are created in batches and stored in *sample line groups*. A sample line group is always associated with an alignment and can be found under the associated alignment in Prospector, as shown in Figure 12.1.

FIGURE 12.1
A view of Prospector; sample lines are stored in sample line groups and are dependent on an alignment.

You can also see in Figure 12.1 that quite a few additional items are dependent on sample lines, such as *sections, section views, mass haul lines,* and *mass haul views.* If you click a sample line, you will see it has three types of grips, as shown in Figure 12.2.

FIGURE 12.2
The three grips on a typical sample line

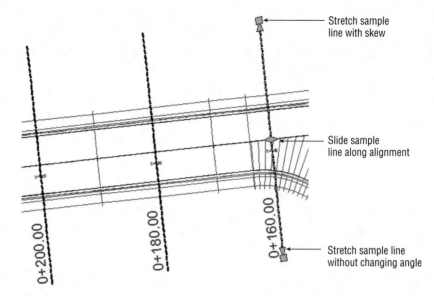

The grips are intended for changes to a sample line that cannot be accomplished through the sample line group properties. The diamond-shaped grip at the alignment location will allow you to slide the sample line to a different station. The triangular grips at the ends of a sample line allow you to extend the sample line swath width while maintaining its angle from the alignment. The square grips on the sample line will allow you to change the length and angle of the sample line. If you have used the square grip to skew a sample line but wish to return it to its original perpendicular state, right-click the grip and then click Make Orthogonal from the menu as shown in Figure 12.3.

FIGURE 12.3
Removing the skew with the Make Orthogonal command

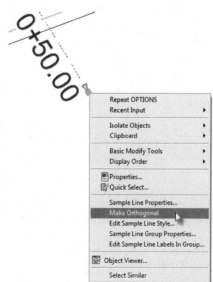

To create a sample line group, change to the Home tab ➢ Profile & Section Views panel ➢ Sample Lines. After you select the appropriate alignment, the Create Sample Line Group dialog, shown in Figure 12.4, will appear. You should name the sample line group and verify the sample line and label style.

FIGURE 12.4
The Create Sample Line Group dialog

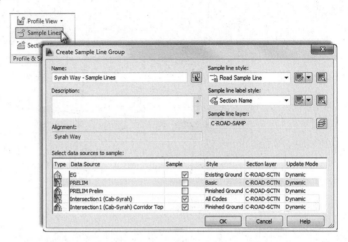

Every source object that is available will be displayed at the bottom of this box. If you wish to omit specific data from the section view, you can clear the check box. Set the applicable style for each item by clicking in the column to the right of the object. The section layer should be preset as specified in your template Object Layers settings.

Once you've selected the sample data and clicked OK, the Sample Line Tools toolbar will appear (Figure 12.5).

FIGURE 12.5
From the Sample Line Tools toolbar, choose By Range Of Stations.

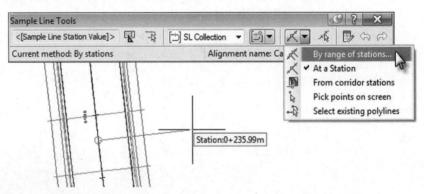

The By Range Of Stations option is used most often. You can use At A Station to create one sample line at a specific station. From Corridor Stations will insert a sample line at the same locations as corridor frequency stations. Pick Points On Screen allows you to pick any two (or more) points to define a sample line. This option can be useful in special situations, such as sampling a pipe on a skew. The last option, Select Existing Polylines, lets you define sample lines from existing polylines.

To define sample lines, you need to specify a few settings. Figure 12.6 shows these settings in the Create Sample Lines – By Station Range dialog.

FIGURE 12.6
Create Sample Lines – By Station Range dialog

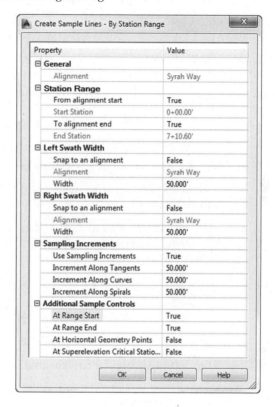

Station Range   Station Range controls where on your alignment sample lines are created. By default, the dialog picks up the start and end station of the alignment. Change the station range by changing From Alignment Start or To Alignment End to False and setting the desired stations.

Swath Width   Swath Width is the offset from the alignment along which you create sample lines. When using sample lines for end area volume calculations, be sure that the swath width is large enough to encompass the design but not so large that they overlap in curve areas. You can change the Snap To An Alignment option to True if you wish to force your sample lines to stop at an alignment that is not a constant offset from the centerline.

Sampling Increments   Sampling Increments allows you to choose the interval at which sample lines are created. Notice that you can control the sample line interval separately along tangents, curves, and spirals.

Additional Sample Controls   Additional Sample Controls adds a sample line at the beginning, end, and other special stations, such as horizontal geometry (PC, PT, and so on) and superelevation critical stations.

In the following exercise, you'll create sample lines for Cabernet Court:

Certification Objective

1. Open the `1201_SampleLines.dwg` (`1201_SampleLines_METRIC.dwg`) file, which you can download from this book's web page at www.sybex.com/go/masteringcivil3d2014.

2. From the Home tab ➢ Profile & Section Views panel, click Sample Lines.

3. At the `Select an alignment, <or press enter key to select from list>:` prompt, press ↵ to display the Select Alignment dialog.

4. Select the Syrah Way alignment and click OK.

   The Create Sample Line Group dialog opens.

5. In the Create Sample Line Group dialog:

   a. Name the group **Syrah Way - Sample Lines**.

   b. Clear the check boxes to the right of the PRELIM corridor and its surface named PRELIM Prelim.

   c. Set the surface EG style to Existing Ground by clicking in the Style column to the right. (Hint: The first click will activate the cell, and the second click will give you the list of styles.)

   d. Set the style for the surface Intersection1 (Cab-Syrah) Corridor Top to Finished Ground.

      The column is too narrow to view the full names of the items, so to see the name of the item, pause your cursor over the name or expand the column.

   e. Set the corridor Intersection1 (Cab-Syrah) code set style to All Codes.

6. Leave the default settings for Sample Line Style and the Sample Line Label Style.

7. When your dialog looks similar to Figure 12.4 (your sampled items may be listed in a different order), click OK.

8. On the Sample Line Tools toolbar, click the Sample Line Creation Methods drop-down arrow and then select By Range Of Stations.

9. In the Create Sample Lines – By Station Range dialog, leave the swath widths and sampling increments at their defaults (this will be 50′ for Imperial units and 20 m for metric units).

10. Change both the At Range Start and At Range End options to True, as shown earlier in Figure 12.6.

11. Click OK, and press ↵ to end the command.

12. If you receive a Panorama view telling you that your corridor is out of date and may require rebuilding, dismiss it by clicking the green check box.

    You should now have dashed lines at even station intervals; these are your new sample lines.

13. Save the drawing for use in the next exercise.

Editing the Swath Width of a Sample Line Group

There may come a time when you need to show information outside the limits of your section views or not show as much information. To edit the width of a section view, you will have to change the swath width of a sample line group. These sample lines can be edited manually on an individual basis, or you can edit the entire group at once.

> ### SAMPLE LINE AND SECTION VIEW WORDS OF WISDOM
>
> It is extremely rare to encounter situations where you need to touch each sample line or section view individually. From the contextual tab, select a sample line or section view to access many tools that will save time (and your sanity).
>
>
>
> ### YOU WANT TO CHANGE ALL THE SAMPLE LINE LENGTHS
>
> To change many swath widths at once:
>
> 1. Open the Sample Line Group Properties ➢ Sample Lines tab, as shown here.
>
>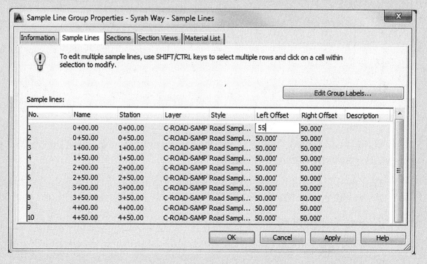
>
> 2. Select the first station you wish to change and hold down Shift as you click the last row.
> 3. Click on the offset distance to change the swath widths as desired.
>
>   You will need to change both the left and the right because they are independent from each other.

YOU WANT TO ADD OR REMOVE SECTION DATA IN MANY VIEWS

A common situation with sections is that surface or corridor data you've created after you created sample lines is not appearing in your cross sections. This is easily rectified:

1. Click Sample More Sources. (Hint: You can get to this directly from the contextual tab or from the Sample Line Group Properties ➢ Sections tab.)

    Data that exists but that is not recognized by the sample lines will appear to the left, as shown here.

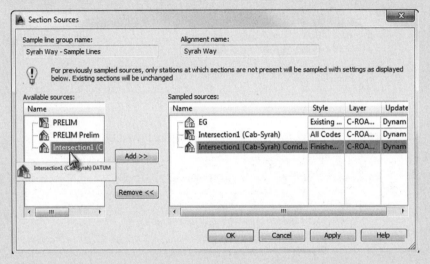

2. Highlight the data you wish to add, and click Add.

In this exercise, you'll edit the widths of an entire sample line group. You must complete the previous exercise before proceeding.

1. Continue working on the `1201_SampleLines.dwg` (`1201_SampleLines_METRIC.dwg`) file. You need to have completed the previous exercise to continue.
2. Select a sample line.
3. From the Sample Line contextual tab ➢ Modify panel, click Group Properties.

4. Switch to the Sample Lines tab.
5. Click to select the first sample line in the listing at station 0+00.00 (0+000.00).
    a. Scroll down to the bottom of the list.
    b. While holding the Shift key, select the last station in the listing. This will highlight all rows.
    c. Click your cursor in the Left Offset column and change it to **100′** (**30** m).
    d. Press ↵ to complete the edit.

The left offsets will change in the listing. If you have a long list of sample lines in a project, it may take a moment to update.

   e. With all of the rows still highlighted, make the same change to the Right Offset column.

6. Click OK.

   After a moment, the sample lines will resize to reflect your change.

7. Save the drawing.

Check your completed drawings against `1201_SampleLines.dwg` or `1201_SampleLines_METRIC.dwg` if desired.

Creating Section Views

Once the sample line group is created, it is time to create views. You can create a single view or many views arranged together (Figure 12.7).

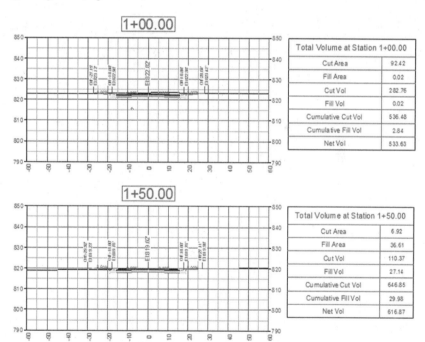

FIGURE 12.7
Section views arranged to plot by page

A section view is a reflection of the design and can be used for plotting purposes. No edits to the corridor, surface, or other design elements can be made from a section view. The view contains horizontal and vertical grids, tick marks for axis annotation, the axis annotation itself, and a title. Views can also be configured to show horizontal geometry, such as the centerline of the section, edges of pavement, and right of way. Tables displaying end areas or volumes can also be shown with the sections.

564 | **CHAPTER 12** CROSS SECTIONS AND MASS HAUL

Creating a Single-Section View

Certification Objective

There are occasions when all section views are not needed. In these situations, a single-section view can be created. In this exercise, you'll create a single-section view of station 0 + 00.00 (0+000.00) from sample lines:

1. Open the `1202_SectionViews.dwg` (`1202_SectionViews_METRIC.dwg`) file, which you can download from this book's web page.

   You will want to switch to this drawing because there are a few steps completed for you that you will learn about later in the chapter.

2. Select the sample line at 0+50.00 (0+020.00).

3. From the Sample Line contextual tab ➢ Launch Pad panel, click Create Section View ➢ Create Section View.

4. Verify that your alignment and sample line group name are correct on the General page of the wizard, shown in Figure 12.8.

FIGURE 12.8
The General page of the Create Section View Wizard

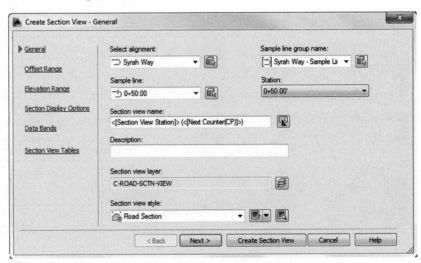

You can navigate from one page to another by either clicking Next at the bottom of the screen or clicking the links on the left side of the screen.

5. Click Next to view the Offset Range page.

   The top of Figure 12.9 shows the Offset Range page, which should match your sample line swath width. In this case, you will leave this set to Automatic. No action is needed on the Offset Range page.

6. Click Next to view the Elevation Range page.

   The bottom of Figure 12.9 shows the Elevation Range page. The values shown here are taken from your design max and min elevations. In this case, and in most cases, you will leave this set to Automatic. No action is needed on the Elevation Range page.

FIGURE 12.9
The Offset Range page (top) and Elevation Range page (bottom) of the Create Section View Wizard

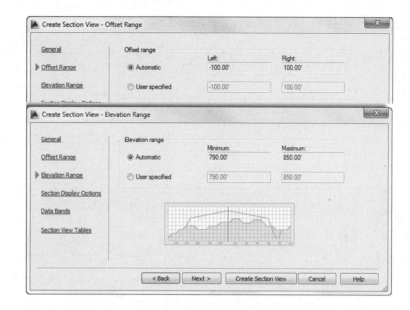

7. Click Next to view the Section Display Options page. Change all three surface label styles to _No Labels by clicking in the Change Labels column for each surface item.

   The fourth page contains the section display options, as shown in Figure 12.10. This page reflects the styles and data you selected when creating your sample lines. If you forgot to set a style or wish to omit additional data, you can change your options here.

FIGURE 12.10
The Section Display Options page of the Create Section View Wizard

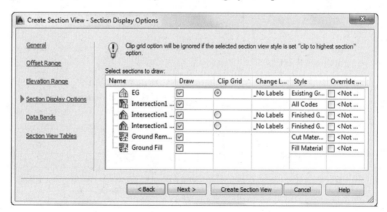

8. Click Next to view the Data Bands page.

   The fifth page, shown in Figure 12.11, lets you specify the data band options. Here, you can select band sets to add to the section view, pick the location of the band, and choose the surfaces to be referenced in the bands. The data bands used in this example show only offset distance (rather than offset and elevation), and therefore no action is needed on the Data Bands page.

FIGURE 12.11
The Data Bands page of the Create Section View Wizard

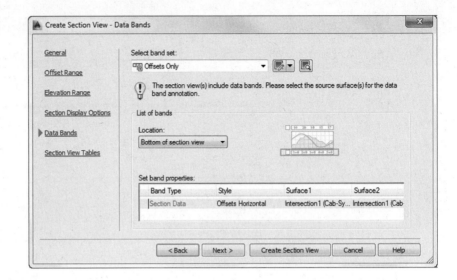

9. Click Next to view the Section View Tables page.

   The sixth and last page, shown in Figure 12.12, is where you set up the section view tables. Note that this screen will be available only if you have already computed materials for the sample line group. On this page, you can select the type of table and the table style and select the position of the table relative to the section view. The graphic on the lower-right side of the window will help to illustrate the table placement and changes as you update these settings.

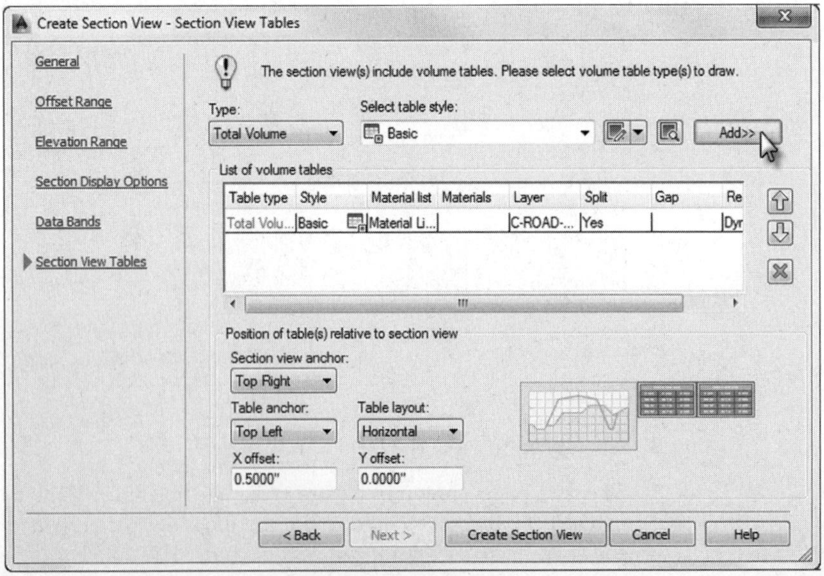

FIGURE 12.12
The Section View Tables page of the Create Section View Wizard

10. With the table type set to Total Volume and Select Table Style set to Basic, click Add.

    As shown in Figure 12.12, you will have a row of data indicating that a material table will come in to the right of your view.

11. Click Create Section View.

12. Pick any point in the drawing area to place your section view.

13. Examine your section view.

    The display should resemble Figure 12.13.

FIGURE 12.13
The finished section view

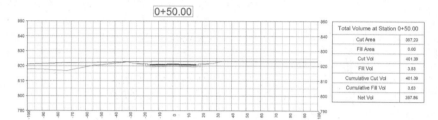

14. Save the drawing for use in the next exercise.

Creating Multiple Section Views

Section views belong in packs. In the exercise that follows, you will create section views intended to plot together on a sheet:

1. Continue working in 1202_SectionViews.dwg (1202_SectionViews_METRIC.dwg). You do not need to have completed the previous exercise to continue.

2. Select any sample line.

3. From the Sample Line contextual tab ➤ Launch Pad panel, click Create Section View ➤ Create Multiple Section Views.

    Alternately, you can access this tool from the Home tab ➤ Profile & Section Views panel ➤ Section Views ➤ Create Multiple Views.

4. On the General page, set the Section View Style option to Road Section – No Grids, and click Next.

5. On the Section Placement page, click the ellipsis next to the path for Template For Cross Section Sheet.

6. From the default cross section sheet template, select ARCH D Section 20 Scale (ISO A1 Section 1 to 500), as shown in Figure 12.14, and click OK.

FIGURE 12.14
When you're creating multiple views, the scale and spacing of the final product depend on layouts from sheet templates.

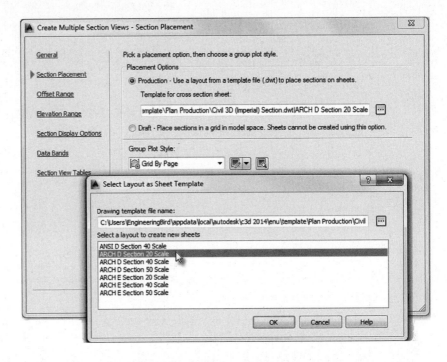

The Civil 3D default cross section template should be `Civil 3D (Imperial) Section.dwt` or `Civil 3D (Metric) Section.dwt`. Each scale listed in the dialog box (Figure 12.14) relates to a layout tab in the section template file.

7. Set Group Plot Style to Grid By Page, and click Next.

   Figure 12.14 is a page you did not see when placing a single view. Setting the Production radio button allows you to use the Create Section Sheets tool from the Output tab. The Draft option forces Civil 3D to behave like version 2010 and prior. Do not use the Draft option if you intend to run the Create Section Sheets command with the section views.

   Group Plot Style controls how the views are arranged on a page. In this example, the grid will come from the group plot style rather than the section view style.

   You can skip the rest of the wizard because you will keep the default input for the remainder of the settings.

8. Click Create Section Views.

9. Click anywhere off to the right of the graphic to place the views.

   You should see views arranged on the screen (Figure 12.15). Metric users will see one page of section views, and Imperial unit users will see three pages of section views. Notice that the scale chosen for the Select Layout As Sheet dialog box matches the annotative scale of the drawing, creating a neat, coherent set of section views.

FIGURE 12.15
One of the pages of cross-section views

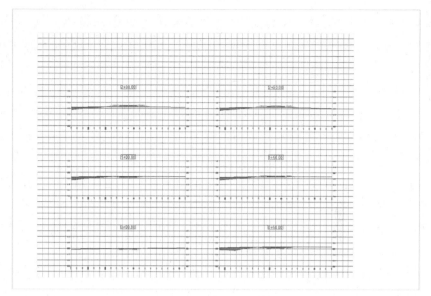

10. Save the drawing for use in the next exercise.

> **SECTION VIEW OBJECT PROJECTION**
>
> Civil 3D has the ability to project AutoCAD® points, blocks, 3D solids, 3D polylines, Civil 3D COGO points, feature lines, and survey figures into section and profile views. Each of the objects listed can be projected to a section view and labeled appropriately.
>
> In the following exercise, you will project feature lines, several COGO points, and an AutoCAD cylinder solid into section view:
>
> 1. Open the drawing 1203_SectionProjection.dwg (1203_SectionProjection_METRIC.dwg).
>
> 2. On the Home tab ➢ Profile & Section Views panel, click Project Objects To Multiple Section Views.
>
>
>
> 3. When prompted at the command line Select a sample line or a section view:, select any sample line in the drawing.
>
> 4. In the Project Objects To Multiple Section Views dialog, switch the projection rule to the By Distance option.

5. Change Distance Before and Distance After to **10′** (**3.5** m).

   The distance before and distance after tells Civil 3D how far away from the sample line to look for objects for projection.

   This tool will pick up all of the object types listed in the Name column that it encounters. If you do not wish a certain type of object to be projected into section view, clear the check box. For Cogo Points, Feature Lines, and Survey Figures, the default behavior for the tool is to use the object style section settings. You can override this by clicking in the style field and choosing a projection style. For all other objects listed in this dialog box, a project style must be specified. The elevation for the projected object can be taken from the object or a surface or set to Manual, in which case you will be able to edit the Elevation Value field.

   Many other Civil 3D objects, such as pipes, structures, and surfaces, do not appear in this listing because they can be picked up by sample lines for display in section views.

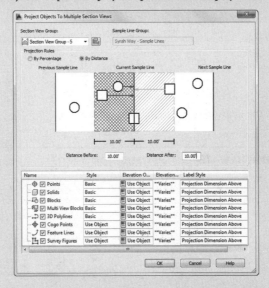

6. Leave the default settings for all other options and click OK.
7. Examine your section views. You should now see additional information in your section views representing objects that would have not been visible in section view without this command.
8. Save and close the drawing.

Section Views and Annotation Scale

In the last section, you created section views using a cross-section sheet template. You chose a scale for the section sheet layout at the same time that you picked a page size. Coincidentally, the scale you used in the exercise was already set as the annotation scale. Since both scales agreed with each other, everything came out nicely.

Depending on what portion of the design you are working on, the scale of the sections and the scale you are comfortable working with may not agree. Additionally, you may just forget to set the scale ahead of placing section views.

Ideally, your section views and their sheets should be in a drawing separate from the corridor and the rest of the design. In Chapter 16, "Advanced Workflows," you will learn how to do this using data shortcuts.

Delete Section Views the Pain-Free Way

The best way to remove an unwanted batch of section views from your graphic is to delete them through Prospector. In the following image you can see that the intrepid user tried (and apparently failed) seven times before keeping the view group.

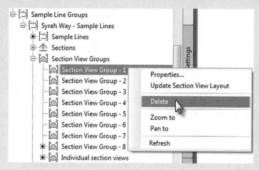

If you delete views graphically, you'll create more work for yourself. First, the section view group stays behind in Prospector, as you see in the following image. As indicated by the dot in front of the group in the first image, only group 8 and the individual section views group contain information.

The second problem with graphically deleting your section views is that Civil 3D interprets this maneuver as removing data from the sample lines. The next time you create sections, not all of the data you expect will be there. In this case, you would need to use the Sample More Sources tool to bring back the missing data.

The best way to delete section view groups is by right-clicking the group in Prospector and selecting Delete, as shown in the first image in this sidebar.

It is helpful to learn how to work with annotation scale and section views. In the following exercise, you will go through a brief lesson in reorganizing sections. You need to have completed the previous exercise before continuing.

1. Continue working in 1202_SectionViews.dwg (1202_SectionViews_METRIC.dwg).
2. Select the Annotation scale in the lower-right portion of the screen.
3. Change the scale to **1"=10′ (1:250 mm)**.

   The views probably look pretty funky—and not in a good way. The page has gotten smaller, but the views have not updated.

4. Select any section view by clicking on the station value.
5. From the Section View contextual tab ➢ Modify View panel, click Update Group Layout.
6. Save and close the drawing.

Update Group Layout

   The views should now be reorganized to a better-looking state.

The results of the exercises can be found in 1202_SectionViews.dwg and 1202_SectionViews_METRIC.dwg. Feel free to examine the section views created in these drawings to compare your work.

Real World Scenario

Showing ROW Lines in Cross Section Views

The trick you are about to learn will come in handy for more than just showing right-of-way locations in cross section. Any item that is alignment-based can work the same way.

1. Open the drawing 1204_SectionROW.dwg (1204_SectionROW_METRIC.dwg).
2. Select any sample line, and then from the Sample Line contextual tab ➢ Modify panel, click Group Properties.
3. Switch to the Section Views tab.
4. Locate the Profile Grade column for the section view group. You may need to expand the columns to read what they are.
5. Select the Section View Group row and then click the ellipsis, as shown here.

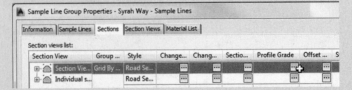

The offset alignments represent the ROW location. An existing ground profile has been created for you as well. You will use this information to place the ROW symbol in the cross-section graphic.

6. Set Syrah Way Right 33.000 (Syrah Way Right 11.000) as the active alignment in the listing, and click Add.
7. Repeat step 6 for Syrah Way Left 33.000 (Syrah Way Left 11.000).
8. Set the Marker Style option for both left and right alignments to ROW Marker.

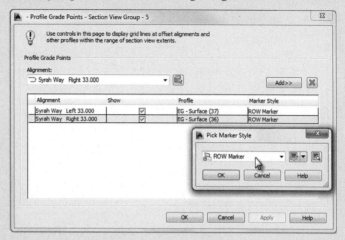

9. Click OK several times until you are out of all dialog boxes.

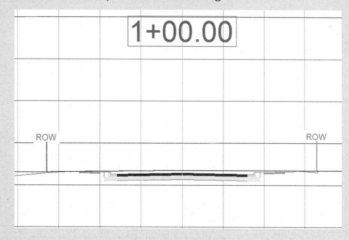

You should now have a ROW symbol whose insertion offset was determined by the alignment and whose elevation was from the surface profile.

It's a Material World

Once alignments are sampled, volumes can be calculated from the sampled surface or from the corridor section shape. These volumes are calculated in a materials list and can be displayed as a label on each section view or in an overall volume table, as shown in Figure 12.16.

FIGURE 12.16
A total volume table inserted into the drawing

Total Volume Table
SYRAH WAY

Station	Fill Area	Cut Area	Fill Volume	Cut Volume	Cumulative Fill Vol	Cumulative Cut Vol
0+00.00	3.81	66.27	0.00	0.00	0.00	0.00
0+50.00	0.00	162.07	2.82	253.71	2.82	253.71
1+00.00	0.02	92.42	0.02	282.76	2.84	536.48
1+50.00	36.61	6.92	27.14	110.37	29.98	646.85
2+00.00	126.37	0.00	120.73	7.68	150.70	654.53
2+50.00	124.23	0.00	185.63	0.00	336.33	654.53
3+00.00	73.59	0.00	146.04	0.00	482.38	654.53
3+50.00	26.71	30.59	73.65	34.14	556.03	688.67
4+00.00	4.17	74.57	22.61	117.32	578.64	805.98
5+00.00	17.59	88.75	32.23	362.94	610.87	1168.93
5+50.00	23.72	44.92	30.60	148.52	641.47	1317.45
6+00.00	30.73	37.47	40.33	91.55	681.80	1409.00
6+50.00	30.41	41.93	45.29	88.22	727.09	1497.22
7+00.00	21.98	67.39	38.81	121.47	765.89	1618.69
7+10.60	19.38	77.52	6.50	34.15	772.39	1652.84

The volumes can also be displayed in an XML report, as shown in Figure 12.17.

FIGURE 12.17
A total volume XML report shown in Microsoft Internet Explorer

Volume Report

Alignment: Syrah Way
Sample Line Group: Syrah Way - Sample Lines
Start Sta: 0+00.000
End Sta: 7+10.604

Station	Cut Area (Sq.ft.)	Cut Volume (Cu.yd.)	Reusable Volume (Cu.yd.)	Fill Area (Sq.ft.)	Fill Volume (Cu.yd.)	Cum. Cut Vol. (Cu.yd.)	Cum. Reusable Vol. (Cu.yd.)	Cum. Fill Vol. (Cu.yd.)	Cum. Net Vol. (Cu.yd.)
0+00.000	66.27	0.00	0.00	3.81	0.00	0.00	0.00	0.00	0.00
0+50.000	162.07	253.71	253.71	0.00	2.82	253.71	253.71	2.82	250.89
1+00.000	92.42	282.76	282.76	0.02	0.02	536.48	536.48	2.84	533.63
1+50.000	6.92	110.37	110.37	36.61	27.14	646.85	646.85	29.98	616.87
2+00.000	0.00	7.68	7.68	126.37	120.73	654.53	654.53	150.70	503.83
2+50.000	0.00	0.00	0.00	124.23	185.63	654.53	654.53	336.33	318.20
3+00.000	0.00	0.00	0.00	73.59	146.04	654.53	654.53	482.38	172.15
3+50.000	30.59	34.14	34.14	26.71	73.65	688.67	688.67	556.03	132.64
4+00.000	74.57	117.32	117.32	4.17	22.61	805.98	805.98	578.64	227.35
5+00.000	88.75	362.94	362.94	17.59	32.23	1168.93	1168.93	610.87	558.06
5+50.000	44.92	148.52	148.52	23.72	30.60	1317.45	1317.45	641.47	675.98
6+00.000	37.47	91.55	91.55	30.73	40.33	1409.00	1409.00	681.80	727.20
6+50.000	41.93	88.22	88.22	30.41	45.29	1497.22	1497.22	727.09	770.13
7+00.000	67.39	121.47	121.47	21.98	38.81	1618.69	1618.69	765.89	852.80
7+10.604	77.52	34.15	34.15	19.38	6.50	1652.84	1652.84	772.39	880.45

Once a materials list is created, it can be edited to include more materials or to make modifications to the existing materials. For example, soil expansion (fluff or swell) and shrinkage factors can be entered to make the volumes more accurately match the true field conditions. This can make cost estimates more accurate, which can result in fewer surprises during the construction phase of any given project.

Creating a Materials List

Certification Objective

Materials can be created from surfaces or from corridor shapes. Surfaces are great for earthwork because you can add cut or fill factors to the materials, whereas corridor shapes are great for determining quantities of asphalt or concrete. In this exercise, you practice calculating earthwork quantities for the Syrah Way corridor:

1. Open the 1205_Materials.dwg (1205_Materials_METRIC.dwg) file, which you can download from this book's web page.

2. From the Analyze tab ➢ Volumes And Materials panel, click Compute Materials.

   The Select A Sample Line Group dialog appears.

3. In the Select Alignment field, verify that the alignment is set to Syrah Way, and then click OK.

   The Compute Materials dialog appears.

4. In the Quantity Takeoff Criteria drop-down box, select Earthworks from the drop-down menu.

5. Click the Object Name cell for the Existing Ground surface, and select EG from the drop-down menu.

6. Click the Object Name cell for the Datum surface, and select Intersection1 (Cab-Syrah) DATUM (this is the name of the corridor followed by the name of the surface) from the drop-down menu.

7. Verify that your settings match those shown in Figure 12.18, and then click OK.

FIGURE 12.18
The settings for the Compute Materials dialog

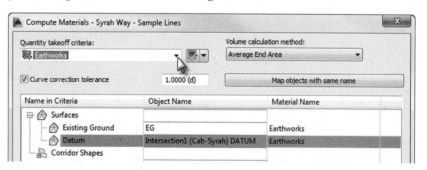

Graphically, nothing will happen. However, in the background Civil 3D has computed material data.

8. Save the drawing for use in the next exercise.

Creating a Volume Table in the Drawing

In the preceding exercise, materials were created that represent the total dirt to be moved or used in the sample line group. In the next exercise, you insert a table into the drawing so you can inspect the volumes. You need to have successfully completed the previous exercise before proceeding.

1. Continue working in 1205_Materials.dwg (1205_Materials_METRIC.dwg). You need to have successfully completed the previous exercise before continuing.

2. From the Analyze tab ➤ Volumes And Materials panel, click Total Volume Table.

   The Create Total Volume Table dialog appears.

3. Verify that your settings match those shown in Figure 12.19.

FIGURE 12.19
The Create Total Volume Table dialog settings

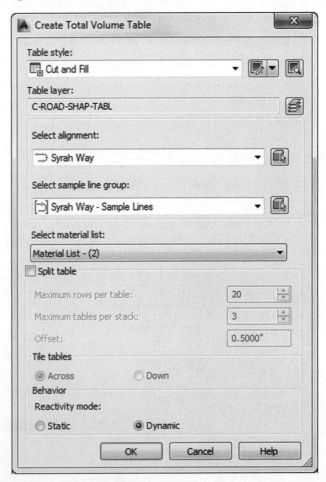

4. Verify that Reactivity Mode at the bottom of the dialog is set to Dynamic.

5. Clear the check box next to split table and click OK.

6. Pick a point in the drawing to place the volume table.

    The table indicates a cumulative fill volume of 772.39 cubic yards (572.16 cubic meters) and a cumulative cut volume of 1652.84 cubic yards (1081.87 cubic meters).

7. Save the drawing for use in the next exercise.

Adding Soil Factors to a Materials List

Certification Objective

Civil 3D allows for more accurate earthwork computations by providing entry of cut factors, fill factors, and refill factors.

Cut Factor   Sometimes known as expansion factor, or "fluff" factor, the *cut factor* is always expressed as a number greater than or equal to 1.0. Cut volume is multiplied by this value to determine the volume of soil after excavation has taken place. For example, soil that expands 5% after excavation would use a cut factor of 1.05.

Fill Factor   Commonly known as compaction factor, the *fill factor* is used to determine the volume of material after it has been mechanically compacted. Counter to most people's intuition, this value is also expressed as a value greater than or equal to 1.0. A material that compacts to 90% of its original volume would use a fill factor of 1.1. The fill volume is divided by the fill factor to determine the final amount of fill needed on a site.

Refill Factor   The *refill factor* dictates what amount of the cut material can be reused as fill. This value is expressed as a percentage of the cut volume (with cut factor applied). This factor is only used when the material quantity type is set to Cut and Refill.

In the following exercise, the materials need to be modified to bring them closer in line with true field numbers. For this exercise, the shrinkage factor will be assumed to be 80%, which is entered into Civil 3D as 1.2. The expansion on cut will be 115%, or as entered into Civil 3D, 1.15. In addition to these numbers (which Civil 3D represents as cut factor for swell and fill factor for shrinkage), for this exercise, assume a Refill Factor value of 1.00.

1. Continue working in 1205_Materials.dwg (1205_Materials_METRIC.dwg). Be sure you have completed the previous exercises that use this file before continuing.

2. From the Analyze tab ➢ Volumes And Materials panel, click Compute Materials.

    The Select A Sample Line Group dialog appears.

3. Select the Syrah Way alignment and the Syrah Way - Sample Lines group, and click OK.

    The Edit Material List dialog appears.

4. In the row labeled Earthworks, enter a cut factor of **1.15** and a fill factor of **1.2**.

5. Verify that all other settings are the same as in Figure 12.20, and click OK.

FIGURE 12.20
The Edit Material List dialog

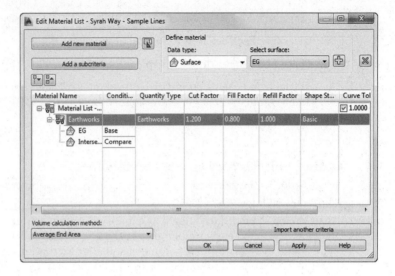

6. Examine the total volume table again.

   Notice that the new Cumulative Fill Volume value is 772.39 cubic yards (572.16 cubic meters) and the new Cumulative Cut Volume value is 1652.84 cubic yards (1081.87 cubic meters).

7. Save the drawing for use in the next exercise.

Generating a Volume Report

Civil 3D provides you with a way to create a report that is suitable for printing or for transferring to a word processing or spreadsheet program. In this exercise, you'll create a volume report for the Intersection1 (Cab-Syrah) corridor. You need to have successfully completed the previous exercises before proceeding.

1. Continue working in 1205_Materials.dwg (1205_Materials_METRIC.dwg).

2. From the Analyze tab ➢ Volumes And Materials panel, click Report.

   The Report Quantities dialog appears.

3. Verify that Material List–(2) is selected in the dialog, and click OK.

4. You may get a warning message that says, "Scripts are usually safe. Do you want to allow scripts to run?" Click Yes.

   Civil 3D temporarily takes over your web browser to display the volume report. This is the same information that you placed in your drawing, but since it's in this form, you can more readily copy and paste it into Excel, Word, or another program.

5. Note the cut-and-fill volumes and compare them to your volume table in the drawing.

6. Close the report when you are done viewing it.

7. Save and close the drawing.

Compare your work with 1205_Materials_FINISHED.dwg (1205_Materials_METRIC_FINISHED.dwg) if desired.

Section View Final Touches

Before you move away from section views, a few last touches are needed in your sections. First you will add last-minute data to the sections. You will also add labels to the sections.

Sample More Sources

It is a common occurrence that data is created from the design after sample lines have been generated. For example, you may need to add surface data or pipe network data to existing section views. To accomplish this, you need to add that data to the sample line group using the Sample More Sources command.

In this exercise, you'll add a pipe network to a sample line group and inspect the existing section views to ensure that the pipe network was added correctly:

1. Open the 1206_FinalTouches.dwg (1206_FinalTouches_METRIC.dwg) file, which you can download from this book's web page.

2. Select one of the section views by clicking the station label.

3. From the Section View contextual tab ➢ Modify Section panel, click Sample More Sources.

4. In the Section Sources dialog, click Sanitary Network on the left side of the dialog to highlight it, as shown in Figure 12.21, and then click Add.

FIGURE 12.21
Adding sewer data to the cross-section view via Section Sources

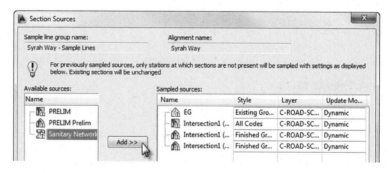

The Sanitary network will now appear on the right side of the dialog.

5. Click OK to dismiss the Section Sources dialog.

Take a look at the sections. You should now see pipes and structures in the various section views.

6. Save the drawing for use in the next exercise.

Cross-Section Labels

The best way to label cross sections that contain corridor data is to use the code set style. Using the code set style, you can control which parts of the corridor are labeled. You can learn more about creating code set styles in Chapter 18, "Label Styles." In the meantime, you will look at how to change the active code set style on a cross-section view group.

In the following exercise, you will use the view group properties to add labels. It is *not* necessary to have completed the previous exercise before continuing.

1. Continue working in the 1206_FinalTouches.dwg (1206_FinalTouches_METRIC.dwg) file.
2. Select one of the section views by clicking its station label.
3. From the Section View contextual tab ➤ Modify View panel, click View Group Properties.
4. On the Sections tab, locate the Style column for the corridor and click the field that currently reads All Codes.

   This is the code set style that is current in the views.

5. Click the field again to switch the style to All Codes w Labels, as shown in Figure 12.22, and click OK.

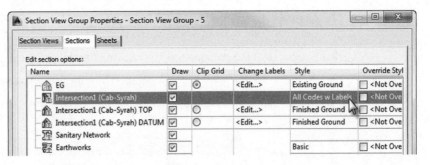

FIGURE 12.22
Changing the active code set style for all views in the group

You should now have lovely little labels on all of the views, as shown in Figure 12.23.

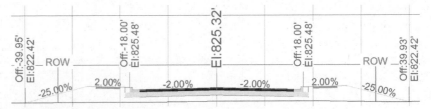

FIGURE 12.23
Slope, elevation, and offset labels from the code set style

6. Save and close the drawing.

Check your work against the file 1206_FinalTouches.dwg or 1206_FinalTouches_METRIC.dwg if time permits and you have the inclination.

Mass Haul

Mass haul diagrams help designers and contractors gauge how far and how much soil needs to be moved around a site. Figure 12.24 shows the mass haul diagram for Syrah Way. The free haul area is material the contractor has agreed to move at no extra charge. The fact that the mass haul line is always above 0 indicates that the project is in a net cut situation through the length of the Syrah Way alignment.

FIGURE 12.24
Syrah Way Mass Haul diagram (Note that the legend was added for illustration purposes only.)

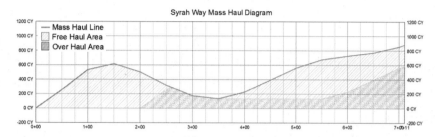

Taking a Closer Look at the Mass Haul Diagram

In an ideal design situation, there is no leftover cut material and no extra material needs to be brought in. This would mean that net volume = 0. When the line appears above the zero volume point, it is showing net cut values.

As the mass haul diagram continues, it shows the cumulative effect of net cut and fill for the alignment. When the net cut and net fill converge at the zero volume, the earthwork along the alignment is balanced. When the line appears below the zero volume point, it is showing net fill values, as you can see in Figure 12.25.

FIGURE 12.25
The volume, net cut, and net fill on an idealized mass haul diagram shown with profile

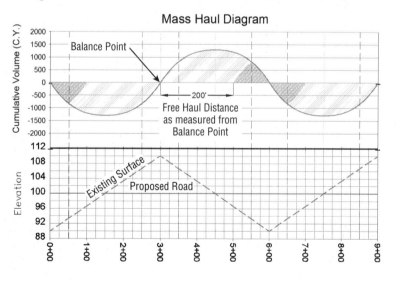

Here is some of the terminology you will encounter:

Balanced The state where the cumulative cut and fill volumes are equal.

Origin Point The beginning of the mass haul diagram, typically at station 0+00, but it can vary depending on your stationing.

Borrow A negative value, typically at the end of the mass haul diagram, that indicates fill material that will need to be brought into the site.

Waste A positive value, typically at the end of the mass haul diagram, that indicates cut material that will need to be hauled out of the site.

Free Haul Earthwork that a contractor has contractually agreed to move. This typically specifies a contracted distance.

Over Haul Earthwork that the contractor has not contractually agreed to move. This excess can be used for borrow pits or waste piles.

Create a Mass Haul Diagram

Now, let's put it all together and build a mass haul diagram in Civil 3D for Syrah Way and you'll see how easy it is:

1. Open the `1207_MassHaul.dwg` (`1207_MassHaul_METRIC.dwg`) file.

   Remember, you can download all the data files from this book's web page.

2. From the Analyze tab ➢ Volumes And Materials panel, select Mass Haul.

   The Create Mass Haul Diagram wizard opens.

3. On the General page:

   a. Verify that you are creating a mass haul diagram for the Syrah Way alignment.

   b. Verify that you are using the Syrah Way – Sample Lines group.

   c. Leave all other options at their defaults, as shown in Figure 12.26, and click Next.

FIGURE 12.26
The General options of the Create Mass Haul Diagram Wizard

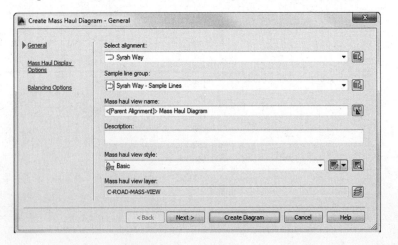

On the Mass Haul Display Options page, no changes are needed.

4. Click Next.
5. In the Balancing Options page, place a check box next to Free Haul Distance.
6. Set the distance to **200'** (**60 m**), as shown in Figure 12.27, and click Create Diagram.

FIGURE 12.27
The Balancing Options page of the Create Mass Haul Diagram dialog

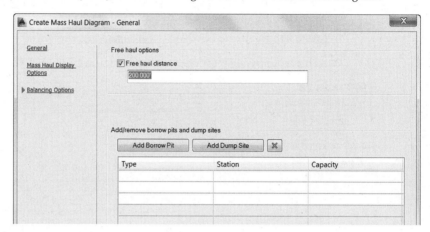

7. Find a clear spot on your drawing to place the mass haul diagram.

   The diagram should look similar to Figure 12.24 (shown previously). Note that the legend was added for explanation purposes and is not generated by Civil 3D. We created this using AutoCAD text and hatching tools.

8. Save the drawing and keep it open for the next exercise.

Editing a Mass Haul Diagram

When you create a mass haul diagram, you can easily modify parameters and get instant feedback on it. Follow these steps to see how (you must have completed the previous exercise before continuing):

1. Continue working in the 1207_MassHaul.dwg (1207_MassHaul_METRIC.dwg) file.
2. Select anywhere on the mass haul object or grid.

   If you look at your mass haul diagram, you can see that nearly all the earthwork for Syrah Way involves net cut, which means hauling away dirt.

   The Balancing options give you a chance to change or add waste and borrow pits. You will now make a few changes and observe what happens to the mass haul diagram.

Balancing Options

3. From the Mass Haul View (or Mass Haul Line) contextual tab ➢ Modify panel, click Balancing Options.

   The Mass Haul Line Properties dialog opens with the settings you used when initially generating the mass haul diagram.

4. Free Haul Distance is presently set for 200′ (60 m). Change it to **500′ (150 m)**.

5. Drag the dialog away from the screen to see the changes and then click Apply.

You can further tweak this amount to cut down on the net cut values by adding a dump site.

6. Click the Add Dump Site button.

  a. Set the station to **1+31.00 (0+040.00)**.

  b. Set the capacity to **850** cubic yards (**650** cubic meters).

7. When your dialog resembles the top of Figure 12.28, click OK.

FIGURE 12.28
The Mass Haul Line Properties dialog: adding a dump site (top). The mass haul diagram adjusted for the dump site (bottom).

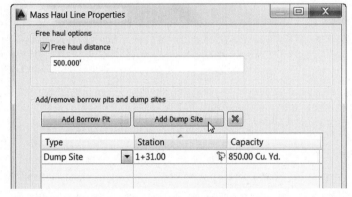

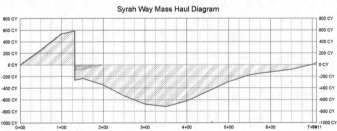

The mass haul diagram will immediately update to reflect the changes, as shown at the bottom of Figure 12.28.

8. Save and close the drawing.

If you'd like to see how you did, open the file 1207_MassHaul_FINISHED.dwg or 1207_MassHaul_METRIC_FINISHED.dwg.

For this stretch of the project, the mass haul diagram ends near zero. The changes you made also decrease the amount of overhaul in the project, potentially decreasing the earthwork cost.

The Bottom Line

Create sample lines. Before any section views can be displayed, sections must be created from sample lines.

>   **Master It** Open `MasterIt1201.dwg` (`MasterIt1201_METRIC.dwg`) and create sample lines along the USH 10 alignment every 50′ (20 m). Sample all data, and set the left and right swath widths to **50′ (20** m).

Create section views. Just as profiles can be shown only in profile views, sections require section views to be displayed. Section views can be plotted individually or all at once. You can break them up into groups for plotting into sheets.

>   **Master It** In the previous exercise, you created sample lines. In that same drawing, create section views for all the sample lines. For US units, use a cross section scale of 1″=20′ on an Arch D size layout sheet. For metric units, use a cross section scale of 1:500 on an ISO A0 size sheet. For all other options, use the default settings and styles. You need to have completed the previous exercise before continuing.

Define and compute materials. Materials are required to be defined before any quantities can be displayed. You learned that materials can be defined from surfaces or from corridor shapes. Corridors must exist for shape selection, and surfaces must already be created for comparison in materials lists.

>   **Master It** Using `MasterIt1201.dwg` (`Master It1201_METRIC.dwg`), create a materials list that compares Existing Intersection with HWY 10 DATUM Surface. Use the Earthworks Quantity takeoff criteria. You need to have completed the previous exercise before continuing.

Generate volume reports. Volume reports give you numbers that can be used for cost estimating on any given project. Typically, construction companies calculate their own quantities, but developers often want to know approximate volumes for budgeting purposes.

>   **Master It** Continue using `MasterIt1201.dwg` (`MasterIt1201_METRIC.dwg`). Be sure you have completed all the previous Master It exercises before continuing. Use the materials list created earlier to generate a volume report. Create a web browser-based report and a total volume table that can be displayed on the drawing.

Chapter 13

Pipe Networks

In this chapter, you'll look at two different types of pipe systems in AutoCAD® Civil 3D® . *Pipe networks* refer to the gravity-based object type in Civil 3D that works best for sewer systems. *Pressure networks* are a separate object type that works best for systems such as water and gas.

First, you will take an in-depth look at pipe networks. In the latter part of this chapter, you'll learn about pressure networks.

In this chapter, you will learn to:

- ♦ Create a pipe network by layout
- ♦ Create an alignment from network parts and draw parts in profile view
- ♦ Label a pipe network in plan and profile
- ♦ Create a dynamic pipe table

Pipe Network Setup

Before you can draw pipes in your project, some setup is needed. The setup discussed in this first section should be done in your Civil 3D template so it can be applied to multiple projects. In this section, you'll examine what is needed for an example sanitary sewer network.

Pipe networks contain the following object types:

Pipes   *Pipes* are components of a pipe network that primarily represent underground pipes or culverts. The standard catalog has pipe shapes that are circular, elliptical, egg-shaped, and rectangular and are made of materials that include PVC, RCP, DI, and HDPE. You can use Part Builder (discussed later in this chapter) to create your own shapes and materials if the default shapes and dimensions can't be adapted for your design.

Structures   *Structures* are the components of a pipe network that represent manholes, catch basins, inlets, joints, and any other type of junction between two pipes. The standard catalog includes inlets, outlets, junction structures with frames (such as manholes with lids or catch basins with grates), and junction structures without frames (such as simple cylinders and rectangles).

Null Structures   *Null structures* are needed when two pipes are joined without a structure; they act as a placeholder for a pipe endpoint. They have special properties, such as allowing pipe cleanup at pipe intersections. Most of the time, you'll create a style for them that doesn't plot or is invisible for plotting purposes. (See Chapter 19, "Object Styles," for more information on creating structure styles.)

Parts List—Sewer Systems

A *parts list* contains the pieces needed to complete a pipe design. Both pipe networks and pressure networks use parts lists to help you organize design elements and determine how they will appear in a drawing. For example, you'll want to have different parts available when working with sanitary sewers than you'll want when working with storm sewers.

Pipe Network parts lists (for gravity systems) contain pipes, structures, pipe rules, structure rules, styles, and the ability to associate a quantity takeoff pay item number with each item. You'll learn more about Pressure Network parts lists later in this chapter.

Examples of Pipe Network parts lists include:

- Storm sewer
  - Catch basin/inlet structures
  - Manhole structures
  - Concrete pipe
  - HDPE pipe
- Sanitary sewer (gravity) as shown in Figure 13.1:
  - Manhole structures
  - HDPE pipe
  - Ductile iron pipe

FIGURE 13.1
A Sanitary Sewer parts list

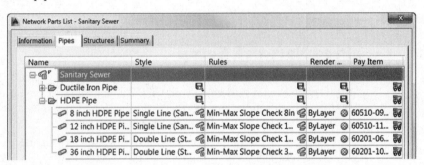

In the upcoming section, you will explore planning and creating a parts list. You'll start by examining your local requirements, and then you will compare those needs with what is available in the software.

Planning a Typical Pipe Network

Let's look at a typical sanitary sewer design. You'll start by going through the sewer specifications for the jurisdiction in which you're working. The following is an example of a completed checklist for the example's jurisdiction, Sample County:

Sanitary Sewers in Sample County

- Recommended Structures: Standard concentric manhole, small-diameter cleanout.
- Structure Behavior: All structures have 1.5′ (0.46 m) sump, rims, a 0.10′ (0.03 m) invert drop across all structures. All structures designed at finished road grade.
- Structure Symbology: Manholes are shown to scale in plan view as a circle with an S inside. Cleanouts are shown to scale as a hatched circle. (See Figure 13.2.)

FIGURE 13.2
Sanitary sewer manhole in plan view (left) and a cleanout in plan view (right)

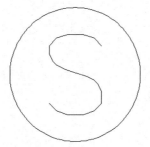 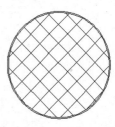

Manholes are shown in profile view with a coned top and rectangular bottom. Cleanouts are shown as a rectangle (see Figure 13.3).

FIGURE 13.3
Profile view of a sanitary sewer manhole (left) and a cleanout (right)

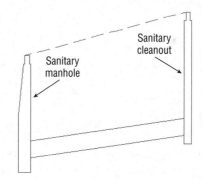

- Recommended Pipes: 8″ (200 mm), 10″ (250 mm), and 12″ (300 mm) PVC pipe, per manufacturer specifications.
- Pipe Behavior: Pipes must have cover of 4′ (1.22 m) to the top of the pipe; the maximum slope for all pipes is 10 percent, although minimum slopes may be adjusted to optimize velocity as follows:

Sewer size	Minimum slope
8″ (200 mm)	0.40%
10″ (250 mm)	0.28%
12″ (300 mm)	0.22%

- Pipe Symbology: In plan view, pipes are shown with a CENTER2 linetype line that has a thickness corresponding to the inner diameter of the pipe. In profile view, pipes show both inner and outer walls, with a hatch between the walls to highlight the wall thickness (see Figure 13.4).

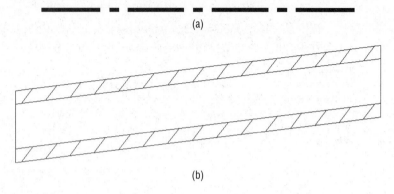

FIGURE 13.4
Sanitary pipe in plan view (a) and in profile view (b)

Now that you know your requirements for Sample County, the next step is to begin entering this information into your Civil 3D template file.

Part Rules

Rules define the constraints on things like minimum slope, sump depths, and pipe-invert drops across structure. Rules are assigned to parts in the parts list. Depending on the type of network and the complexity of your design, you may have many rules. Civil 3D allows you to establish structure and pipe rules that will assist in respecting these constraints during initial layout and edits.

Rules don't restrict you from drawing the location and lengths you want. As you draw your pipe network in plan view, Civil 3D tries to adhere to as many rules as it can. If the constraints defined in the rules conflict with each other (for example, maximum slope may be violated to maintain minimum depth), rules are violated even in the layout stage.

Furthermore, rules will never change your design without your direct guidance. For example, if the surface tied to a structure changes, the rim elevation *will* change, but the pipe invert elevations *will not* change.

To see where the design needs to be altered, you will need to view the violations in the Status column of Panorama. Panorama is the only place where rule violations are flagged; there is no graphic representation of rule violations in plan view.

Structures and pipes have separate rule sets. When creating rules, don't be thrown off by the fact that the category always reads Storm Sewer, as you'll see in Figure 13.6. You can use these rules regardless of the type of parts list you are creating. You can then add these rule sets to specific parts in your parts list, which you'll build later in this chapter.

STRUCTURE RULES

Structure rule sets are located on the Settings tab of Toolspace, under the Structure branch as shown in Figure 13.5.

FIGURE 13.5
On the Settings tab of Toolspace, right-click a structure to edit it.

Click the Add Rule button on the Rules tab in the Structure Rule Set dialog. The Add Rule dialog will appear, which will allow you to access all the various structure rules (see Figure 13.6). Although it looks like you can, you won't be able to change the values until you finalize adding the rule.

FIGURE 13.6
In the Add Rule dialog, the category always shows Storm Sewer, but you can use rules for whatever type you want.

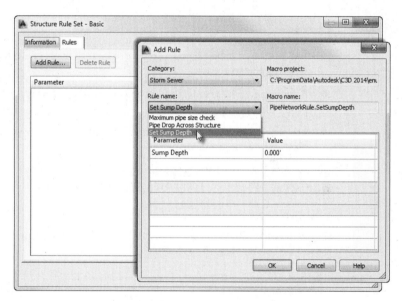

You will have a chance to work with this firsthand in the upcoming exercise.

Maximum Pipe Size Check

The Maximum Pipe Size Check rule (see Figure 13.7) examines all pipes connected to a structure and flags a violation in Prospector if any pipe is larger than your rule. This is a violation-only rule—it won't change your pipe size automatically.

FIGURE 13.7
The Maximum Pipe Size Check rule option

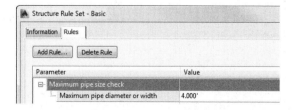

Pipe Drop Across Structure

The Pipe Drop Across Structure rule (see Figure 13.8) tells any connected pipes how their inverts (or alternatively, their crowns or centerlines) must relate to one another.

FIGURE 13.8
The Pipe Drop Across Structure rule options

When a new pipe is connected to a structure that has the Pipe Drop Across Structure rule applied, the following checks take place:

- A pipe drawn to be exiting a structure has an invert equal to or lower than the lowest pipe entering the structure.

- A pipe drawn to be entering a structure has an invert equal to or higher than the highest pipe exiting the structure.

- Minimum specified drop distance is calculated by measuring between the lowest entering pipe and the highest exiting pipe.

In the hypothetical sanitary sewer example, you're required to maintain a 0.10' (3 cm) invert drop across all structures. You'll use this rule in your structure rule set in the next exercise.

Set Sump Depth

Sump depth is additional structure depth below the lowest pipe invert. The Set Sump Depth rule (Figure 13.9) establishes sump depth for structures.

FIGURE 13.9
The Set Sump Depth rule option

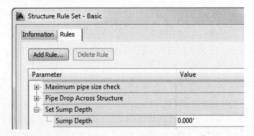

It's important to add a sump depth rule to all of your structure rule sets. If no sump rule is used, Civil 3D will assume a 2' sump for English units and 2 meters in metric units! If you forget to set this before placing structures, you will need to edit the individual structure properties or make the rule and retroactively apply it to the network.

In the hypothetical sanitary sewer example, all the structures have a 1.5' (0.5 m) sump depth. You'll use this rule in your structure rule set in the next exercise.

Pipe Rules

Pipe rule sets are located on the Settings tab of Toolspace, under the Pipe branch. For a detailed breakdown of pipe rules and how they're applied, including images and illustrations, please see the *Civil 3D User's Guide* (located online at http://usa.autodesk.com/adsk/servlet/index?siteID=123112&id=21329326).

After you right-click on a pipe rule set and click Edit (as shown in Figure 13.10), you can access all the pipe rules by clicking the Add Rule button on the Rules tab of the Pipe Rule Set dialog.

FIGURE 13.10
Accessing the pipe rules from Settings

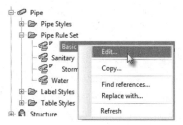

Cover and Slope Rule

The Cover And Slope rule (Figure 13.11) allows you to specify your desired slope range and cover range. As you place your pipe network, Civil 3D tries to use the minimum and maximum depth and minimum and maximum slope to set the initial pipe depth and slope.

FIGURE 13.11
The Cover And Slope rule options

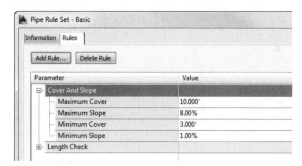

Depending on your site conditions, applying this rule to every pipe may not be feasible. In this situation, the rule becomes what is referred to as *violation only*. You will still be able to place pipes, and the rule will cause a violation message to show in the Pipe Network panel of Panorama.

If part of your design changes and you'd like Civil 3D to make another attempt to enforce the Cover And Slope rule, you can use the Apply Rules feature, which is discussed later in this chapter.

Cover Only Rule

The Cover Only rule (Figure 13.12) is designed for use with pipe systems where slope can vary or isn't a critical factor. Like Cover And Slope, this rule is used on first placement. Manual edits can cause rules to be violated. The rule will show as a violation in the Pipe Network panel, but no changes to your design take place until you use the Apply Rules command.

FIGURE 13.12
The Cover Only rule options

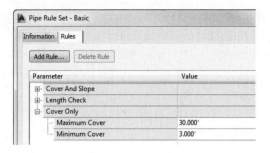

Length Check

Length Check is a violation-only rule; it won't change your pipe length size automatically. The Length Check options (see Figure 13.13) allow you to specify a minimum and maximum pipe length.

FIGURE 13.13
The Length Check rule options

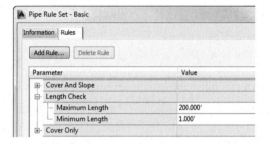

Pipe To Pipe Match Rule

The Pipe To Pipe Match rule (Figure 13.14) is also designed for use where there are no true structures (only null structures), including situations where pipe is placed to break into an existing pipe. This rule determines how pipe inverts are assigned when two pipes come together, similar to the Pipe Drop Across Structure rule.

FIGURE 13.14
The Pipe To Pipe Match rule options

Set Pipe End Location Rule

Without the Set Pipe End Location rule (Figure 13.15), Civil 3D assumes you are measuring pipes from center of structure to center of structure. With the rule in place, you have the capability to determine where the pipe end is located on the structure. The options are Structure Center (the default without the rule), Structure Inner Wall, or Structure Outer Wall.

FIGURE 13.15
The Set Pipe End Location rule options

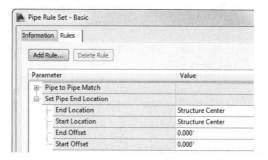

End Offset and Start Offset are used if you would like to have your pipes protrude past the inner or outer wall on the respective ends. The offset is ignored if the location for the end is set to the center of the structure. The value must be positive and will be ignored if the additional length causes the pipe end to be located past the center of the structure.

Graphically, you probably won't be able to tell that this setting is active until you add labels. The Set Pipe End Location will make a difference in the length and slope labels on pipes.

Creating Structure and Pipe Rule Sets

In this exercise, you'll create one structure rule set and three pipe rule sets for a hypothetical sanitary sewer project:

1. Open the 1301_RulesTemplate.dwg (1301_RulesTemplate_METRIC.dwg) drawing, which you can download from this book's web page at www.sybex.com/go/masteringcivil3d2014.

2. On the Settings tab of Toolspace, expand the Structure branch, right-click Structure Rule Set, and click New.

3. Switch to the Information tab and enter **Sanitary Structure Rules** in the Name text box.

4. Switch to the Rules tab, and click the Add Rule button.

5. In the Add Rule dialog, choose Pipe Drop Across Structure in the Rule Name drop-down. Click OK.

6. Expand the new rule and confirm that the parameters in the Structure Rule Set dialog are the following:

Drop Reference Location	Invert
Drop Value	0.1' (0.03 m)
Maximum Drop Value	3' (1 m)

These parameters establish a rule that will match your hypothetical municipality's standard for the drop across sanitary sewer structures.

7. Click the Add Rule button again.

8. In the Add Rule dialog, choose Set Sump Depth in the Rule Name drop-down. Click OK.

9. Expand the new rule. Change the Sump Depth parameter to **1.5′ (0.5 m)** in the Structure Rule Set dialog to meet the hypothetical municipality's standard for sump in sanitary sewer structures, and click OK.

10. On the Settings tab of Toolspace, expand the Pipe branch, right-click the pipe rule set, and choose New.

11. Switch to the Information tab and enter **8 Inch Sanitary Pipe Rule** (for metric, **200 mm Sanitary Pipe Rule**) for the name.

12. Switch to the Rules tab. Click Add Rule.

13. In the Add Rule dialog, choose Cover And Slope in the Rule Name drop-down. Click OK.

14. Expand the Cover And Slope rule and then modify the parameters to match the constraints established by your hypothetical municipality for 8″ (200 mm) pipe, as follows:

Maximum Cover	10′ (3m)
Maximum Slope	10%
Minimum Cover	4′ (1.5m)
Minimum Slope	0.40%

15. Click OK.

16. In the Settings tab of Toolspace, expand Pipe Rule Set and select the rule set you just created. Right-click, and choose Copy.

17. Switch to the Information tab and enter **10 Inch Sanitary Pipe Rule** (for metric, **250 mm Sanitary Pipe Rule**) in the Name text box.

18. Switch to the Rules tab, expand the Cover And Slope rule, and then modify the parameters to match the constraints established by your hypothetical municipality for a 10″ (250 mm) pipe, as follows:

Maximum Cover	10′ (3 m)
Maximum Slope	10%
Minimum Cover	4′ (1.5 m)
Minimum Slope	0.28%

19. Click OK when you have finished modifying the rule set. Repeat the process to create a rule set for the 12″ (300 mm) pipe using the following parameters:

Maximum Cover	10′ (3 m)
Maximum Slope	10%
Minimum Cover	4′ (1.5 m)
Minimum Slope	0.22%

20. You should now have one structure rule set and three pipe rule sets.

21. Save your drawing.

For your reference, completed versions of the drawing (1301_RulesTemplate.dwg_FINISHED and 1301_RulesTemplate_METRIC_FINISHED.dwg) are available with the rest of this book's download.

Putting Your Parts List Together

Everything you've done in this chapter up to this point is leading up to the creation of the parts list. Your parts lists should reside in your Civil 3D template file. Having this information in your template will prevent you from needing to re-create the parts list for every project. You will have multiple parts lists for each type of system you are creating and possibly for each jurisdiction you work in.

1. Open the 1302_PartsListTemplate.dwg (1302_PartsListTemplate_METRIC.dwg) drawing.

2. From the Settings tab, expand the Pipe Network branch, and expand Parts Lists.

   In the drawing, there are currently two parts lists: Standard and Storm Sewer.

3. Right-click Parts Lists and select Create Parts List.

4. In the Network Parts List dialog, switch to the Information tab and name the parts list **Example County San**.

5. Switch to the Pipes tab.

6. Right-click New Parts List and select Add Part Family, as shown in Figure 13.16.

FIGURE 13.16
Add a part family for your new parts list.

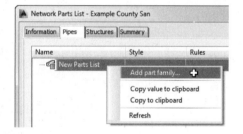

The Part Catalog dialog will appear.

7. As shown in Figure 13.17, place a check mark next to PVC Pipe (PVC Pipe SI in the metric catalog) and click OK.

   At this step, the Parts List name should appear at the top of the Pipe list.

8. Expand the Pipe List name to see the new PVC Pipe family you just added.

9. Right-click PVC Pipe and select Add Part Size, as shown in Figure 13.18.

10. In the Part Size Creator dialog, click the drop-down in the Inner Pipe Diameter field, as highlighted in Figure 13.19.

11. Select 8 Inch (metric: 200 mm).

FIGURE 13.17
Choosing a part family to add

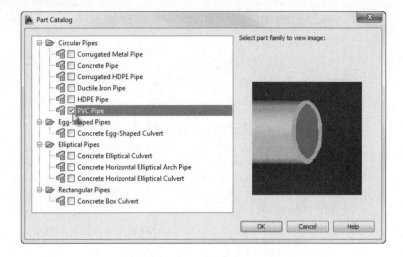

FIGURE 13.18
Add a new part size to the PVC Pipe part family.

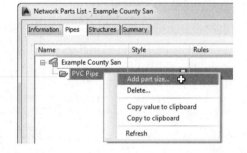

FIGURE 13.19
Set the diameter and material for all the needed pipes.

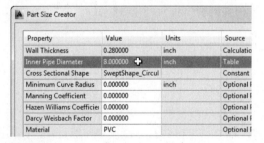

12. In the Material field, set the material to PVC and click OK.

   For Material, you can use the drop-down to select from several preexisting material types, or you can type in your own material name. The material name is used frequently in labels.

13. Expand the new PVC category to examine the result. Repeat steps 9 through 12 for 10" (250 mm) and 12" (300 mm) pipes.

    In this example, the pipes will share the same style. Using the disk icon in the PVC Pipe part family row of the table, you will apply your style choice to the entire PVC family.

    For the columns Render Material and Pay Item, leave the defaults. Render materials and pay items do not affect the design portion of the pipe network. Render materials are used if you want to give a realistic material to the object for visualization purposes. You will take an in-depth look at assigning pay items to a parts list in Chapter 17, "Quantity Takeoff."

14. In the PVC Pipe part family row, click the disk icon in the Style column.
15. Set Pipe Style to Single Line (Sanitary) and click OK.
16. For each pipe size, click the Pipe Rule Set icon in the Rules column, choose the respective rule in the Pipe Rule Set dialog, and then click OK.

    At the end of the process, your Pipes tab will look like Figure 13.20.

FIGURE 13.20
The completed Pipes tab

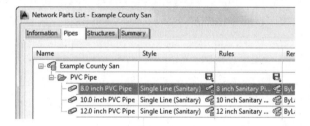

17. Switch to the Structures tab and expand New Parts List.

    Notice there is already a null structure in the listing.

18. Expand the Null Structure part family and click the Structure Style icon for the null structure.
19. Change the Null Structure style to Null and click OK.
20. Right-click on the main heading New Parts List and select Add Part Family.
21. In the Part Catalog dialog, locate the Junction Structures With Frames grouping and place a check mark next to Concentric Cylindrical Structure and Cylindrical Structure Slab Top Circular Frame, as shown in Figure 13.21. Click OK.

    Note that in the metric drawings, the structure descriptions end with SI.

22. Right-click Concentric Cylindrical Structure and select Add Part Size.
23. In the Part Size Creator, select an Inner Structure diameter of 48" (1,200 mm).
24. Leave all other size options at their defaults, and click OK.
25. Repeat steps 23 through 24 to add the 60" (1,500 mm) structure.
26. Use the disk icon to set the style for both concentric cylindrical structures to Sanitary Sewer Manhole.

FIGURE 13.21
Adding structure part families

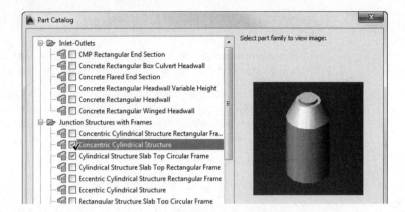

27. Right-click Cylindrical Structure Slab Top Circular Frame, and select Add Part Size.
28. Set Inner Structure Diameter to 15″ (450 mm), and click OK.
29. Using the same process you used in step 26, set Structure Style to Cleanout.
30. Use a similar procedure to set the rules for the three new structures to Sanitary Structure Rules.

    If you expand all the structure part families, your Network parts list will look like Figure 13.22.

FIGURE 13.22
The completed Structures tab in your new parts list

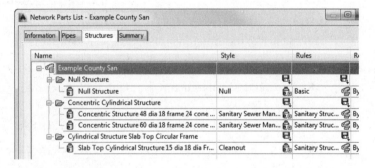

31. Click OK to close the Network parts list.
32. Save the drawing.

Creating a Sanitary Sewer Network

Earlier, you prepared a parts list for a typical sanitary sewer network. This chapter will lead you through several methods for using that parts list to design, edit, and annotate a pipe network.

A pipe network, such as the one in Figure 13.23, can have many branches. In most cases, the pipes and structures in your network will be connected to each other; however, they don't necessarily have to be physically touching to be included in the same pipe network.

FIGURE 13.23
A typical Civil 3D pipe network

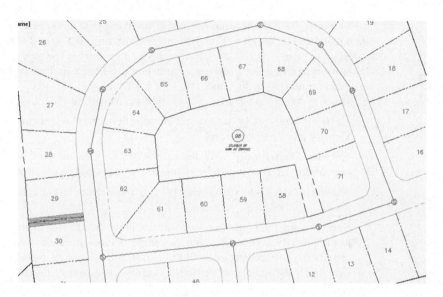

There are several ways to create pipe networks. You can do so using the Civil 3D pipe layout tools. You can also create pipe networks from certain AutoCAD and Civil 3D objects, such as lines, polylines, alignments, and feature lines.

Establishing Pipe Network Parameters

This section will give you an overview of establishing pipe network parameters. Use this section as a reference for the exercises in this chapter. When you're ready to create a pipe network, select the Home tab ➢ Create Design panel and choose Pipe Network ➢ Pipe Network Creation Tools. The Create Pipe Network dialog appears (see Figure 13.24), and you can establish your settings.

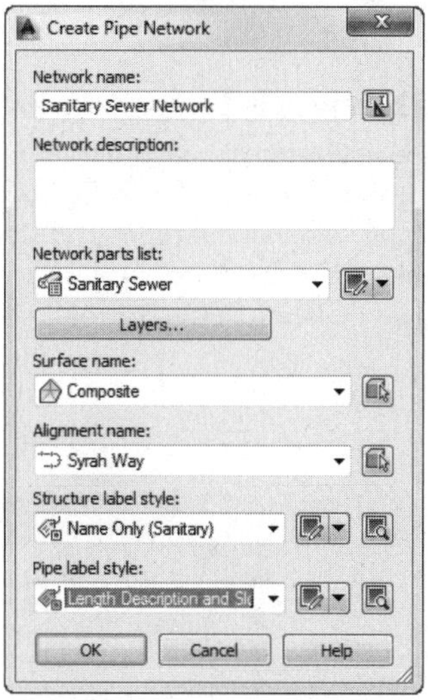

FIGURE 13.24
The Create Pipe Network dialog

Before you can create a pipe network, you must give your network a name, but more important, you need to assign a parts list for your network. As you saw earlier, the parts list provides a toolkit of pipes, structures, rules, and styles to automate the pipe network design process. It's also important to select a reference surface in this interface. This surface will be used for rim elevations and rule application.

When creating a pipe network, you'll be prompted for the following options:

Network Name Choose a name for your network that is meaningful and will help you identify it in Prospector and other locations.

Network Description The description of your pipe network is optional. You might make a note of the date, the type of network, and any special characteristics.

Network Parts List Choose the parts list that contains the parts, rules, and styles you want to use for this design.

Surface Name Choose the surface that will provide a basis for applying cover rules as well as provide an insertion elevation for your structures (in other words, rim elevations). You can change this surface later or for individual structures. For proposed pipe networks, this surface is usually a finished ground surface or a composite surface of the existing and proposed surfaces.

Alignment Name Choose an alignment that will provide station and offset information for your structures in Prospector as well as any labels that call for alignment stations and/or offset information. Because most pipe networks have several branches, it may not be meaningful for every structure in your network to reference the same alignment. Therefore, you may find it better to leave your Alignment option set to None in this dialog and set it for individual structures later using the layout tools or Structure list in Prospector.

Structure Label Style and Pipe Label Style As you create the network, you'll have the option to add labels as you go. If you choose to use a label style that displays text (i.e., not the <None> option), these labels will apply only to the plan view. Section and profile pipe network labels are added in a separate area. You will learn about adding labels later in this chapter. For more information about creating label styles, see Chapter 18, "Label Styles."

Using the Network Layout Creation Tools

Creating a pipe network with layout tools is much like creating other Civil 3D objects. After naming and establishing the parameters for your pipe network, you'll be presented with a special toolbar that you can use to lay out pipes and structures in plan, which will also drive a vertical design.

Certification Objective

After establishing your pipe network parameters in the Create Pipe Network dialog (shown earlier in Figure 13.24), click OK; the Network Layout Tools toolbar will appear (see Figure 13.25).

FIGURE 13.25
The Network Layout Tools toolbar

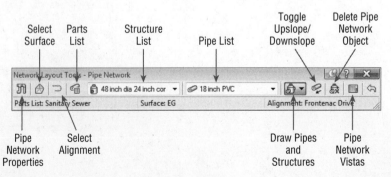

Clicking the Pipe Network Properties tool displays the Pipe Network Properties dialog, which contains the settings for the entire network. If you mistyped any of the parameters in the original Create Pipe Network dialog, you can change them here. In addition, you can set the default label styles for the pipes and structures in this pipe network.

The Pipe Network Properties dialog contains the following tabs:

Information On this tab, you can rename your network, provide a description, and choose whether you'd like to see network-specific tooltips.

Layout Settings Here you can change the default label styles, Network parts list, reference surface and alignment, master object layers for plan pipes and structures, as well as name templates for your pipes and structures (see Figure 13.26).

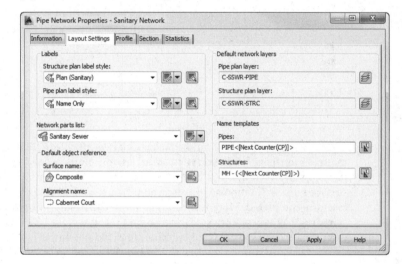

FIGURE 13.26
The Layout Settings tab of the Pipe Network Properties dialog

Profile On this tab, you can change the default profile label styles and master object layers for profile pipes and structures (see Figure 13.27).

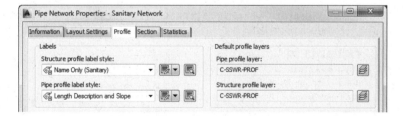

FIGURE 13.27
The Profile tab of the Pipe Network Properties dialog

Section Here you can change the master object layers for network parts in a section (see Figure 13.28).

Statistics This tab gives you a snapshot of your pipe network information, such as minimum and maximum elevation information, pipe and structure quantities, and references in use such as alignments and surfaces (see Figure 13.29).

FIGURE 13.28
The Section tab of the Pipe Network Properties dialog

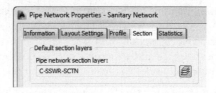

FIGURE 13.29
The Statistics tab of the Pipe Network Properties dialog

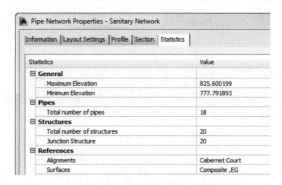

The Select Surface tool on the Network Layout Tools toolbar allows you to switch between reference surfaces while you're placing network parts if needed. However, you should create a merged surface containing your existing surface plus your final design. In these examples, the surface named "Composite" was created by using the Paste command after the proposed surface was completed. More information about the Paste command can be found in Chapter 14, "Grading." Whenever possible, you should use a surface that incorporates the entire site (e.g., existing and proposed pasted together) to avoid needing to switch between surfaces.

The Select Alignment tool on the Network Layout Tools toolbar lets you switch between reference alignments while you're placing network parts, similar to the Select Surface tool.

The Parts List tool allows you to switch parts lists for the pipe network.

The Structure drop-down (the image on the left in Figure 13.30) lets you choose which structure you'd like to place next, and the Pipes drop-down (the image on the right in Figure 13.30) allows you to choose which pipe you'd like to place next. Your choices come from the active Network parts list.

FIGURE 13.30
The Structure drop-down (left), and the Pipes drop-down (right)

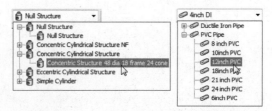

The options for the Draw Pipes and Structures category let you choose what type of parts you'd like to lay out next. You can choose Pipes And Structures, Pipes Only, or Structures Only.

Real World Scenario

PRE-COOKED PIPE NETWORKS IN YOUR TEMPLATE

The following tip is a revelation from a client—as many of the coolest tricks are. Even book authors have more to discover every day!

If you know that you will most likely have pipe networks in your project, you can create a few preset pipe networks in your template.

These "dummy" networks should be empty, but you can set the layout settings ahead of time. This will save you the steps of setting the correct parts catalog, object layers, object styles, and label styles.

PLACING PARTS IN A NETWORK

You place parts much as you do other Civil 3D objects or AutoCAD objects such as polylines. You can use your mouse, transparent commands, dynamic input, object snaps, and other drawing methods when laying out your pipe network.

If you choose Pipes And Structures, a structure is placed wherever you click, and the structures are joined by pipes. If you choose Pipes Only, you can connect previously placed structures. If you have Pipes Only selected and there is no structure where you click, a null structure is placed to connect your pipes.

 Use the Structures Only option when you want to add a structure along an existing pipe run. Watch for the "boxing glove" glyph to appear, indicating that the pipe network recognizes the connection. Clicking to connect to the pipe breaks the pipe in two pieces and places a structure (or null structure) at the break point.

 While you're actively placing pipes and structures, you may want to connect to a previously placed part. For example, there may be a service or branch that connects into a structure along the main trunk. Begin placing the new branch. When you're ready to tie into a structure, you'll get a circular connection marker as your cursor comes within connecting distance of that structure. If you click to place your pipe when this marker is visible, a structure-to-pipe connection will form.

 As you create pipes, the default behavior is to draw them upstream to downstream. The Toggle Upslope/Downslope tool changes the flow direction of your pipes as they're placed. In Figure 13.31, structure 9 was placed before structure 10.

 Click Delete Pipe Network Object to delete pipes or structures of your choice. AutoCAD Erase can also delete network objects, but be careful that you don't accidentally remove more objects than you intend.

FIGURE 13.31
Using the Downslope toggle (a) and the Upslope toggle (b) to create a pipe network leg

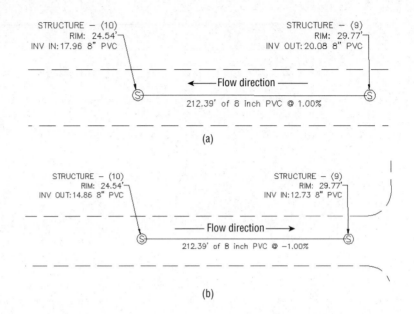

Clicking Pipe Network Vistas brings up Panorama (see Figure 13.32), where you can make tabular edits to your pipe network while the Network Layout Tools toolbar is active.

FIGURE 13.32
Pipe Network Vistas via Panorama

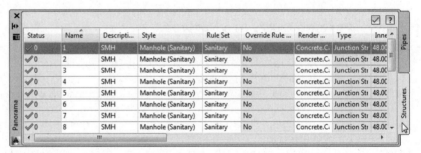

The Pipe Network Vistas interface is similar to what you encounter in the Pipe Networks branch of Prospector. The advantage of using Pipe Network Vistas is that you can make tabular edits without leaving the Network Layout Tools toolbar. You can edit pipe properties, such as Invert and Slope, on the Pipes tab, and you can edit structure properties, such as Rim and Sump, on the Structure tab.

Creating a Sanitary Sewer Network

This exercise will apply the concepts taught in this section and give you hands-on experience using the Network Layout Tools toolbar:

1. Open 1303_Pipes.dwg (1303_Pipes_METRIC.dwg), which you can download from this book's web page.

2. Expand the Surfaces branch in Prospector.

This drawing has several surfaces that have a _No Display style applied to simplify the drawing. The surface you will be working from is a composite of the existing conditions, corridor surfaces, and grading surfaces.

OPTIMIZING THE COVER BY STARTING UPHILL

If you're using the Cover And Slope rule for your pipe network, you'll achieve better cover optimization if you begin your design at an upstream location and work your way down to the connection point.

The Cover And Slope rule prefers to hold minimum slope over optimal cover. In practice, this means that as long as minimum cover is satisfied, the pipe will remain at minimum slope. If you start your design from the upstream location, the pipe is forced to use a higher slope to achieve minimum cover. The following graphic shows a pipe run that was created starting from the upstream location (right to left):

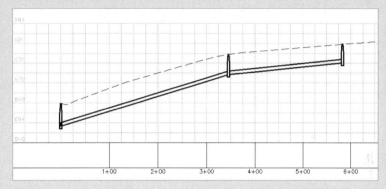

When you start from the downhill side of your project, the Minimum Slope rule is applied as long as minimum cover is achieved. The following graphic shows a pipe run that was created starting from the downstream location (left to right):

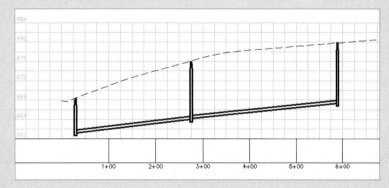

Notice how the slope remains constant even as the pipe cover increases. Maximum cover is a violation-only rule, which means it never forces a pipe to increase slope to remain within tolerance; it provides only a warning that maximum cover has been violated.

3. Expand the Alignments and Centerline Alignments branches, and notice that there are several road alignments (no action is required).

4. On the Home tab ➢ Create Design panel ➢ Pipe Network drop-down, select Pipe Network Creation Tools.

5. In the Create Pipe Network dialog (shown previously in Figure 13.24), give your network the following information:

    ♦ Network Name: **Sanitary Sewer Network**

    ♦ Network Parts List: **Sanitary Sewer**

    ♦ Surface Name: **Composite**

    ♦ Alignment Name: **Syrah Way**

    ♦ Structure Label Style: **Data with Connected Pipes (Sanitary)**

    ♦ Pipe Label Style: **Length Description and Slope**

6. Click OK.

    The Network Layout Tools toolbar will appear.

7. From the structure list, expand the Concentric Cylindrical Structure branch, choose Concentric Structure 48 Dia 18 Frame 24 Cone (Metric: Concentric Structure 1,200 Dia 450 Frame 600 Cone).

8. From the pipe list, expand the PVC pipe branch and choose 8 Inch PVC (200 mm PVC).

9. Click the Draw Pipes And Structures tool. Working right to left, click the X labeled 1 in the drawing to place the first structure. Click the X labeled 2 to place the second structure.

10. Without exiting the command, go back to the Network Layout Tools toolbar and change the Pipe drop-down from 8 Inch PVC to 10 Inch PVC (200 mm PVC to 250 mm PVC) and then place structures at the Xs labeled 3, 4, and 5.

    The labels show that the diameter of the pipe between these structures is 10" (250 mm).

11. Press ↵ to exit the command.

    Next, you'll add a branch of the network from Frontenac Drive. You may want to use the label grip to drag the label off to the side. This will form the leaders as shown in Figure 13.33, making the next step easier.

12. Go back to the Network Layout Tools toolbar, and select 8 Inch PVC (200 mm PVC) from the Pipe list.

13. Click the Draw Pipes And Structures tool button again.

14. Working north to south, click to place the next structure at the X labeled 6.

15. Tie into the Syrah Way branch by moving your cursor near structure 2. You will be ready to click when you see the "sunshine" glyph, indicating that the pipe will tie into the structure.

You'll know you're about to connect when you see the connection marker appear next to your previously inserted structure.

FIGURE 13.33
The connection marker appears when your cursor is near the existing structure.

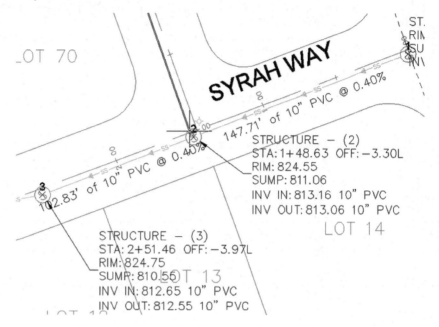

16. Press ↵ to exit the command.

    Observe your pipe network, including the labeling that automatically appeared as you drew the network.

17. Expand the Pipe Networks branch in Prospector in Toolspace, and locate your Sanitary Sewer network.

18. Click the Pipes branch.

    The list of pipes appears in the preview pane.

19. Click the Structures branch; the list of structures appears in the preview pane.

20. Save the drawing.

    The 1303_Pipes_FINISHED.dwg and 1303_Pipes_METRIC_FINISHED.dwg files are available for reference.

Creating a Storm Drainage Pipe Network from a Feature Line

Certification Objective

If you already have an object in your drawing that represents a pipe network (such as a polyline, an alignment, or a feature line), you can take advantage of the Create Pipe Network From Object command in the Pipe Network drop-down.

This option can be used for applications such as converting surveyed pipe runs into pipe networks and bringing forward legacy drawings that used AutoCAD linework to represent pipes. The Create Pipe Network From Object option creates a pipe for every linear segment of your object and places a structure at every vertex of your object. Each object you convert will initially go to a separate network. However, once you have several networks you can easily merge them if needed.

This exercise will give you hands-on experience building a pipe network from a feature line with elevations and merging it with an existing network:

1. Open the 1304_PipesFromObject.dwg (1304_PipesFromObject_METRIC.dwg) file.

   This drawing contains the feature line you will use to generate a storm network.

2. Expand the Surfaces branch in Prospector.

   This drawing has several surfaces that have a _No Display style applied to simplify the drawing (no action is required).

3. Expand the Alignments and Centerline Alignments branches, and notice that there are several road alignments.

   In the drawing, a cyan feature line runs through Syrah Way (the horizontal road) and then goes onto Frontenac Drive. This feature line represents utility information for a storm-drainage line. The elevations of this feature line correspond with invert elevations that you'll apply to your pipe network.

4. Choose Create Pipe Network From Object from the Pipe Network drop-down.

5. At the Select Object or [Xref]: prompt, select the cyan feature line near the eastern side of the line.

   You'll see a preview (see Figure 13.34) of the pipe-flow direction that is based on where you selected the line. In this case, the east end of the feature line is considered the upstream end.

FIGURE 13.34
The flow-direction preview

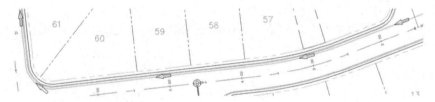

6. At the Flow Direction [OK Reverse] <Ok>: prompt, press ↵ to choose OK.

   The Create Pipe Network From Object dialog appears as shown in Figure 13.35.

7. In the dialog, give your pipe network the following information:

   ♦ Network Name: **Storm Network**

   ♦ Network Parts List: **Storm Sewer**

   ♦ Pipe To Create: **12 Inch Concrete Pipe (300 mm Concrete Pipe)**

   ♦ Structure To Create: From the Rectangular Junction Structure NF collection **2 × 2 (750 × 750 mm Rectangular Structure)**

- Surface Name: **Composite**
- Alignment Name: **<none>**
- Erase Existing Entity check box: Selected
- Use Vertex Elevations check box: Selected
- Set Vertex Elevation Reference to **Invert**

If you select Use Vertex Elevations, the pipe rules for your chosen parts list will be ignored. If it becomes necessary later, you can use the Apply Rules tool as discussed in the next section of this chapter.

FIGURE 13.35
Converting a feature line to a pipe network

8. Click OK. A pipe network will be created.

   Next, you will merge this network with another network that exists in the drawing.

9. Select a pipe or structure from the newly created network. From the Pipe Networks contextual tab ➢ Modify panel, select Merge Networks.

10. In the Select Pipe Network to be merged into another network dialog, highlight Storm Network and click OK.

11. In the Select Destination Pipe Network dialog, highlight Existing Storm Network and click OK.
12. Save the drawing.

This exercise combined all your object-created networks into a more manageable single network. Completed versions of these drawings (1304_PipesFromObject.dwg and 1304_PipesFromObject_METRIC.dwg) are available for your reference.

Editing a Pipe Network

You can edit pipe networks in several ways:

- Using drawing layout edits such as grip, move, and rotate
- Grip-editing the pipe size
- Using vertical movement edits using grips in profile (see the "Vertical Movement Edits Using Grips in Profile" section later in this chapter)
- Using tabular edits in the Pipe Networks branch in Prospector, or from the Pipe Network Vista (via Panorama) from the Pipe toolbar
- Right-clicking a network part to access tools such as Swap Part or Pipe/Structure Properties
- Returning to the Network Layout Tools toolbar by right-clicking the object and choosing Edit Network
- Selecting the pipe network in Toolspace ➢ Prospector, right-clicking, and choosing Edit Network
- Selecting a network part to access the Pipe Networks contextual tab

You will have the chance to explore most of these methods in the following sections.

Changing Flow Direction

By default, Civil 3D pipes are drawn upstream to downstream. It is easy to forget the Toggle Upslope/Downslope tool located on the Network Layout Tools. If you forget to use it, don't fret; you can change the flow direction after the pipes are placed.

To change flow direction, select any part from a pipe network to open the Pipe Networks contextual tab. Select Change Flow Direction from the drop-down portion of the Modify panel. Change Flow Direction allows you to reverse the pipe's understanding of which direction it flows, which comes into play when you're using the Apply Rules command and when you're annotating flow direction with a pipe label–slope arrow.

Changing the flow direction of a pipe doesn't make any changes to the pipe's invert. By default, a pipe's flow direction depends on how the pipe was drawn and how the Toggle Upslope/Downslope tool was set when the pipe was drawn:

- If the toggle was set to Downslope, the pipe flow direction is set to Start To End, which means the first endpoint you placed is considered the start of flow and the second endpoint is established as the end of flow.

♦ If the toggle was set to Upslope when the pipe was drawn, the pipe flow direction is set to End To Start, which means the first endpoint placed is considered the end for flow purposes and the second endpoint the start.

After pipes are drawn, you can set four additional Flow Methods: Start To End, End To Start, Bi-Directional, and By Slope. These settings can be found in Pipe Properties or by going to Prospector ➢ Pipe Networks ➢ *Network Name* ➢ Pipe Branch:

Start To End   A pipe label–flow arrow shows the pipe direction from the first pipe endpoint drawn to the second endpoint drawn, regardless of invert or slope.

End To Start   A pipe label–flow arrow shows the pipe direction from the second pipe endpoint drawn to the first pipe endpoint drawn, regardless of invert or slope.

Bi-Directional   Typically, this is a pipe with zero slope that is used to connect two bodies that can drain into each other, such as two stormwater basins, septic tanks, or overflow vessels. The direction arrow is irrelevant in this case.

By Slope   A pipe label–flow arrow shows the pipe direction as a function of pipe slope. For example, if End A has a higher invert than End B, the pipe flows from A to B. If B is edited to have a higher invert than A, the flow direction flips from B to A.

Editing Your Network in Plan View

When selected, a structure has two types of grips, shown in Figure 13.36. The first is a square grip located at the structure insertion point. You can use this grip to grab the structure and stretch/move it to a new location using the insertion point as a base point. Stretching a structure results in the movement of the structure as well as any connected pipes. You can also scroll through Stretch, Move, and Rotate by using your spacebar once you've grabbed the structure by this grip.

FIGURE 13.36
Two types of structure grips

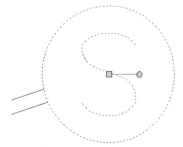

The second structure grip is a rotational grip that you can use to spin the structure about its insertion point. This is useful for aligning eccentric structures, such as rectangular junction structures.

In plan view, many common AutoCAD Modify commands work with structures. You can execute the following commands normally: Move, Copy, Rotate, Align, and Mirror. (Scale doesn't have an effect on structures.) You can use the AutoCAD Erase command to erase network parts. Note that erasing a network part in plan completely removes that part from the network. Once erased, the part disappears from plan, profile view, Prospector, and so on.

When selected, a pipe end has two types of grips (see Figure 13.37). The first is a square endpoint-location grip. Using this grip, you can change the location of the pipe end without constraint. You can move it in any direction; make it longer or shorter; and take advantage of Stretch, Move, Rotate, and Scale by using your spacebar.

FIGURE 13.37
Two types of pipe-end grips

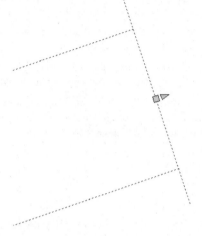

The second grip is a pipe-length grip. This grip lets you extend a pipe along its current bearing and slope.

A pipe midpoint also has two types of grips (see Figure 13.38). The first is a square location grip that lets you move the pipe using its midpoint as a base point.

FIGURE 13.38
Two types of pipe midpoint grips

The second grip is a triangular-shaped pipe-diameter grip. Stretching this grip gives you a tooltip showing allowable diameters for that pipe, which are based on your parts list. Use this grip to make quick visual changes to the pipe diameter.

> **A Word about Pay Items**
>
> Be careful about using the grip to change the pipe diameter. Note that if a pay item is associated with a pipe in your parts list, changing the pipe diameter graphically does not change the associated pay item. For more information on associating pipe parts to pay items, see Chapter 17.

Using the Pipe Network Vista Effectively

To access the full tabular version of your pipe network, the best place to go is to the Pipe Networks contextual tab ➢ Modify panel ➢ Edit Pipe Network button.

Clicking Edit Pipe Network will display the Network Layout toolbar, from which you can click the Pipe Network Vistas (aka Panorama) tool. This will display the grid view shown in Figure 13.39.

FIGURE 13.39
Selecting multiple rows and right-clicking the column name allows you to edit multiple items at once.

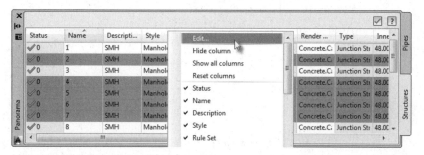

Note that pipes and structures are listed on separate tabs. Many items can be edited from this view. Subtle shading of the columns indicates which rows can be modified. If a column has a light gray background, this means the value is calculated or for information only.

The Status column indicates any rule violations that exist in the network. When a rule violation exists, a red symbol will appear with the number of violations. When you pause your cursor over one of these symbols, a tooltip will inform you which rule has been violated.

You can adjust many things in this interface, but you'll find it cumbersome for some tasks. The interface is best used for the following:

Batch Changes to Styles, Render Materials, Reference Surfaces, Reference Alignments, Rule Sets, and so on   Use your Shift or Ctrl key to select the desired rows, and then right-click the column header of the property you'd like to change. Choose Edit, and then select the new value from the drop-down. If you find yourself doing this on every project for most network parts, confirm that the correct values are set in your parts list and in the Pipe Network Properties dialog.

Batch Changes to Pipe Description   Use your Shift key to select the desired rows, and then right-click the Description column header. Choose Edit, and then enter your new description. If you find yourself doing this on every project for most network parts, check your parts list. If a certain part will always have the same description, you can add it to your parts list and prevent the extra step of changing it here.

Changing Pipe or Structure Names   You can change the name of a network part by typing in the Name field. If you find yourself doing this on every project for every part, check that you're taking advantage of the Name templates in your Pipe Network Properties dialog (which can be further enforced in your Pipe Network command settings).

This interface can be useful for changing pipe inverts, crowns, and centerline information. Keep in mind that Civil 3D will allow you to make changes that violate your pipe rules. Edits to a single pipe or structure are dynamic to parts that are directly connected, but Civil 3D will not reapply the rules for you.

Pipe Network Contextual Tab Edits

You can perform many edits at the individual part level by selecting it. The Pipe Network contextual tab will appear for the object you have selected.

If you realize you placed the wrong part at a certain location—for example, if you placed a catch basin where you need a drainage manhole—use the Swap Part option on the Pipe Network contextual tab (see Figure 13.40). You'll be given a list of all the relevant parts from the parts list associated with the Pipe Network.

FIGURE 13.40
Selecting a network part brings up the Pipe Network contextual tab with many options, including Swap Part.

The same properties listed in Prospector (or Pipe Network Vista) can be accessed on an individual part level by selecting a pipe or structure and choosing Pipe Properties or Structure Properties from the contextual tab. A dialog like the Structure Properties dialog in Figure 13.41 will open, with several tabs that you can use to edit that particular part.

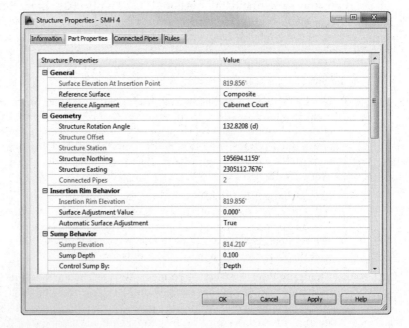

FIGURE 13.41
The Part Properties tab in the Structure Properties dialog gives you the opportunity to perform many edits and adjustments.

The Labels & Tables Panel

Many objects have a Labels & Tables panel in their contextual tabs. This panel (Figure 13.42) is where you do the most annotation labeling for pipes.

FIGURE 13.42
The Labels & Tables panel

Add Labels   For the best control over what styles are used in your label, use the Add Pipe Network Labels option. From there, you can add labels for the selected pipe network using the sub-options such as Entire Network Plan, Entire Network Profile, Entire Network Section, Single Part Plan, Single Part Profile, Single Part Section, Spanning Pipes Plan, and Spanning Pipes Profile.

Add Tables   You can add a structure or pipe table.

Reset Labels   This Reset Labels is different from the Reset Label option you see when clicking directly on a label. In this version of the command, pipe networks that are in the drawing via data reference will be labeled using styles from the source drawing. For more information on source drawings and data references, see Chapter 16, "Advanced Workflows."

The General Tools Panel

The General Tools panel (Figure 13.43) is common to all Civil 3D objects contextual tabs.

FIGURE 13.43
The General Tools panel in the Pipe Network contextual tab

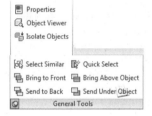

Properties   Toggles the Properties palette on and off.

Object Viewer   Allows you to view the selected object or objects in 3D via a separate viewer.

Isolate Objects   Selected objects will be the only objects visible on the screen. This is useful if you are working in a tight area and do not want to see extraneous objects, and it is much quicker than clicking to turn off or freeze layers.

These tools are available via the General Tools panel drop-down:

Select Similar   This is the best tool ever invented! Select an example object or objects you would like to select, and then click Select Similar. All objects of the same type and layer will be selected. This works great on base AutoCAD objects, not just Civil 3D objects.

Quick Select or QSelect   Opens the Quick Select dialog, which allows custom filtering.

Draw Order Icons   Allow you to move objects either to the front or back of other objects, or above or behind a specific object.

The Modify Panel

The Modify panel (shown previously in Figure 13.40) is where you edit an existing pipe network. The Modify panel contains the following tools:

Network Properties   Opens the Pipe Network Properties dialog.

Pipe Properties   Opens the Pipe Properties dialog. If you are seeing the Modify panel because you selected a structure, you will be prompted to select a pipe.

Edit Pipe Style   This tool resides in the Pipe Properties drop-down and opens the Pipe Style dialog. If you are seeing the Modify panel because you have selected a structure, you will be prompted to select a pipe. For more on pipe styles, see Chapter 19.

Structure Properties   Opens the Structure Properties dialog. If you are seeing the Modify panel because you have selected a pipe, you will be prompted to select a structure.

Edit Structure Style   This tool resides in the Structure Properties drop-down and opens the Structure Style dialog. For more on structure styles, see Chapter 19, "Object Styles."

Edit Pipe Network   Opens the Network Layout Tools toolbar.

Connect and Disconnect Part   Allow you to connect pipes to structures that may have been disconnected and to disconnect pipes from structures.

Swap Part   Allows you to replace a structure or pipe type with another one from a Swap Part Size dialog.

Split and Merge Network   Allow you to take an existing network and split it into two networks, or take an existing pipe network and merge it into another pipe network.

The Modify panel has a flyout that contains the following tools:

Rename Parts   Opens the Rename Pipe Network Parts dialog. Here, you can rename pipes and structures, modify pipe numbering, and decide how you want to handle conflicting names or numbers.

Apply Rules   Apply Rules is an important yet easy-to-miss tool that will help you recalculate pipe slopes if changes need to be made. The easiest way to use this tool is to key in **APPLYRULES** with no objects selected. The reason for this is that it will give you a chance to pick an upstream part and a downstream part and automatically select items between. If you have items selected at the time you click Apply Rules, the rules will be rechecked only for the selected items, often leading to unexpected results.

Change Flow Direction   Changes the path of the selected objects. It is very important to have this option set correctly if you are going to use any of the analysis programs.

The Network Tools Panel

The Parts List drop-down on the Network Tools panel (Figure 13.44) contains the following tools:

Create Parts List   Allows you to create a new parts list via the Network Parts List – New Parts List dialog.

Create Full Parts List   Takes all the parts available in the parts catalog and creates a list called Full Catalog.

FIGURE 13.44
The Network Tools panel shown with the Parts List menu expanded

Edit Parts List   Opens the Parts List dialog, where you can select the network with which you want to work. A common mistake people make when getting to the parts lists this way is clicking OK. To edit, copy, or make a new list, use the drop-down to the right of the Parts List name.

Set Network Catalog   Opens the Pipe Network Catalog Settings dialog, where you can indicate whether you want to use Imperial or metric parts, and also sets the location of the `Catalog` folder.

Part Builder   The last item in the Parts List drop-down is the Part Builder tool. This will open the Getting Started screen of the Part Builder tool. For more information about Part Builder, see the section "Part Builder" later in this chapter.

You'll also find Draw Parts In Profile on the Network Tools panel. This tool adds the selected pipe network objects into an existing profile view.

The Analyze Panel

The Analyze panel (Figure 13.45) contains the following tools for performing various checks on a pipe network:

FIGURE 13.45
The Analyze panel

Interference Check Properties   Allows you to create an interference check between parts, whether or not they are on the same network. The following tools are available via the submenu:

Create Interference Check   When you select the Interference Check Properties tool, you will be prompted to select Interference and then the Interference Check Properties dialog will open:

- The Information tab displays basic information such as the name of the interference check set, style, render material, and layer.

- The Criteria tab contains specific editable items, such as whether to use 3D proximity check, distance, and scale factors.

- The Statistics tab is a combined listing of the other tabs but also allows you to see the networks used for the interference check.

Interference Properties   The Interference Properties tool will give you information about a single interference location in your drawing. This differs from the Interference Check Properties in that you can't change the criteria from here. You see only the following information about a specific instance of interference:

- Information tab (where you can name the individual interference objects, and change the style and render material)
- Statistics tab (which contains information about the networks causing the selected interference and its *xyz* location)

Edit Interference Style   This tool will prompt you to select an interference object (by default a brown, circular block). After selecting the interference object, this opens the Interference Style dialog, where you can change the visual parameters for the interference object. For more on styles, see Chapter 19.

Storm Sewers   These commands interact with the Hydraflow Storm Sewers Extension. Hydraflow is a separate program that is included with Civil 3D. An in-depth discussion of Hydraflow is beyond the scope of this book; see the Hydraflow Storm Sewers Extension help files for more information.

Edit In Storm And Sanitary Analysis   Opens the Storm and Sanitary Analysis (SSA) program. For more on SSA, see this book's website for a bonus chapter, "Storm and Sewer Analysis."

The Launch Pad Panel

The Launch Pad panel (Figure 13.46) contains the following tools:

FIGURE 13.46
The Launch Pad panel

Alignment From Network   Allows you to create an alignment from pipe network parts.

Storm Sewers   Launches the Hydraflow Storm Sewers program. Hydraflow Storm Sewers is a separate program that is included with Civil 3D. An in-depth discussion of Hydraflow Storm Sewers is beyond the scope of this book; see the Hydraflow Storm Sewers Extension help files for more information.

Hydrographs   Launches the Hydraflow Hydrographs program. Hydraflow Hydrographs is a separate program that is included with Civil 3D. An in-depth discussion of Hydraflow Hydrographs is beyond the scope of this book; see the Hydraflow Hydrographs Extension help files for more information.

Express   Opens the Hydraflow Express program. Hydraflow Express is a separate program that is included with Civil 3D. An in-depth discussion of Hydraflow Express is beyond the scope of this book; see the Hydraflow Express help files for more information.

Editing with the Network Layout Tools Toolbar

You can also edit your pipe network by retrieving the Network Layout Tools toolbar. This is accomplished by selecting a pipe network object and choosing Edit Network from the contextual tab. You can also go to the Modify tab and click Pipe Network on the Design panel.

Once the toolbar is up, you can continue working exactly the way you did when you originally laid out your pipe network.

This exercise will give you hands-on experience in making a variety of edits to a sanitary and storm-drainage pipe network:

1. Open the 1305_PipeEditing.dwg (1305_PipeEditing_Metric.dwg) file.

    This drawing includes a sanitary sewer network and a storm drainage network, as well as some surfaces and alignments. For metric users, the structure family names end with SI.

2. Select the structure STM STR 3 in the drawing. It is labeled with the structure name showing.

3. From the Pipe Network contextual tab ➢ Modify panel, click Swap Part.

4. Select the 2 × 4 (1,000 × 750 mm Rectangular Junction Structure) structure from the Rectangular Junction Structure NF, and click OK.

5. Select the newly replaced catch basin so that you see the two structure grips.

6. Use the rotational grip and your nearest Osnap to align the catch basin to the centerline of the pipe, as shown in Figure 13.47.

FIGURE 13.47
Rotate the catch basin into place along the curb.

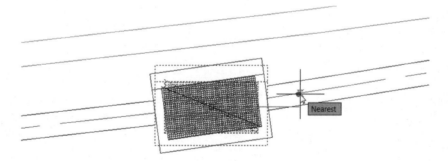

7. Press Esc to clear the selection.

8. In the Prospector tab of Toolspace ➢ Pipe Networks ➢ Networks ➢ Sanitary Network, click the Structures branch.

    At the bottom of Toolspace, you should see a list of structures present in the network.

9. Scroll through the Structures list and locate SAN STR 7. (You will probably need to expand the width of the Name column to view the structure names.)

10. Right-click SAN STR 7 and click Pan To.

11. Again, right-click SAN STR 7 in the listing. This time, click Delete.

    This is one way to remove a pipe network object. You can also delete items graphically by using the AutoCAD Erase command.

    Next, you will use the AutoCAD dynamic input tool to force pipe lengths to the desired value.

12. At the command line, type **DYNMODE** and press ↵.

13. Type **3** and press ↵.

    Your DYNMODE setting may already be set to a value of 3, meaning dimensional input will appear. This step ensures that dynamic input will allow you to enter a total length for the pipe.

14. If dynamic input is not already on, press F12 to enable it. Select the sanitary sewer pipe that is labeled in the north of the drawing.

15. Click the triangular Endpoint grip on the right end of the pipe.

    Notice the tooltips at your cursor with information about the length of the pipe.

16. Press the Tab key to switch the input to the overall length of the pipe. Enter **200'** (**60 m**), as shown in Figure 13.48.

FIGURE 13.48
Lengthen the pipe to 200' (60.96 m).

Note that this is the 3D Center To Center Pipe length. Next, you will add a structure to the end of the pipe you just modified.

17. Select any pipe in the sanitary network. In the Pipe Networks contextual tab ➢ Modify panel, click Edit Pipe Network.

18. From the Structures list, set the structure to SMH from the Concentric Cylindrical Structures NF family.

19. From the Draw Pipes and Structures drop-down, set the option to Structures Only. Place the structure in the drawing at the end of the pipe you lengthened previously. Make sure you can see the connection glyph before you click to add the structure.

20. Press Esc to clear the selection.

21. Select STM STR 3 in the drawing (this is the structure you modified at the beginning of this exercise), and from the Pipe Network contextual tab ➢ Modify panel, click Structure Properties.

22. Switch to the Part Properties tab, scroll down to the Sump Depth field, and change the value to **0'** (0 m).
23. Click OK to exit the Structure Properties dialog. Press Esc to clear the selection.
24. Select any pipe or structure from the Sanitary Network.
25. From the Pipe Network contextual tab ➢ Modify panel, click Edit Pipe Network.
26. From the Network Layout Tools, click Pipe Network Vistas.

    This will bring up the Panorama interface, which will allow you to rename the structures.
27. Make sure that the active Panorama tab is set to Structures.
28. In the Name column heading, widen the column so you can see the full name of the structures in the list. (You can widen a column in all spreadsheet-like views in Civil 3D, much as you would in Microsoft Excel.)
29. Double-click the cell in the Name column for SAN STR 1.
30. Clear the text that is there and type **MH 1**.
31. Repeat steps 29 and 30 for SAN STR 2 through SAN STR 4, renaming each but keeping the number.
32. Save the drawing.

Completed versions of these drawings are available for your review at `1305_PipeEditing.dwg` and `1305_PipeEditing_FINISHED.dwg` (`1305_PipeEditing_Metric_FINISHED.dwg`).

Real World Scenario

I Thought I Took Care of That Sump?

In steps 13 through 15 of the exercise, you were instructed to make the Sump Depth 0. So why isn't the bottom of the structure even with the pipe in the Profile Graphic? Structures are created using the outermost edges, and in the case of structures, this includes a 6" (150 mm) sump.

After you learn about Part Builder at the end of the chapter, come back to this section and add a zero value to the Floor Thickness list, as follows:

1. From the Pipe Networks contextual tab ➢ Network Tools panel ➢ Parts List drop-down, click Part Builder.
2. Change the active part catalog to Structure.
3. Expand the Junction Structures with Frames family.
4. Select the Concentric Cylinder part and click Modify Part Sizes.
5. In the Content Builder area on the left side of the screen, right-click Size Parameters and select Edit Values.
6. In the Edit Part Sizes dialog, click your cursor in the FTh field. This will enable the Edit button.

> 7. Click the Edit button. In the Edit Values dialog, click Add.
> 8. Type **0** for the new value and click OK.
> 9. Click OK again to close the Edit Part Sizes dialog.
> 10. Move your cursor to the top of the Content Builder palette until you see the X appear. Click the X to close Part Builder.
> 11. When you are prompted to Save Changes to Concentric Cylindrical Structure, click Yes.
> 12. Back in Toolspace, switch to the Settings tab and go to Pipe Network ➢ Parts Lists ➢ Storm Sewer.
> 13. Right-click Storm Sewer and click Edit.
> 14. On the Structures tab of the Network Parts List dialog, expand the Concentric Cylindrical Structure family.
> 15. Right-click on Concentric Structure 48 Dia 24 Frame 24 Cone, and choose Edit.
> 16. Notice the row called Floor Thickness. Verify that the Floor Thickness is set to 0 and click OK. Click OK to close the Network Parts List dialog.
>
> Note that this thickness will be applied only to parts going forward. You are stuck with the floor thickness at the time you placed the structure unless you use the Swap Part tool to force the parts to "refresh" from the parts list.

Creating an Alignment from Network Parts

On some occasions, certain legs of a pipe network require their own stationing. Perhaps most of your pipes are shown on a road profile, but the legs that run offsite or through open space require their own profiles. Whatever the reason, it's often necessary to create an alignment from network parts. To do so, follow these steps:

1. Open the 1306_Alignment.dwg (1306_Alignment_METRIC.dwg) file.
2. Select the CB1 structure, which will be the first structure and will be station 0+00 (0+000) on the alignment.

Alignment from Network

3. On the Pipe Networks contextual tab ➢ Launch Pad panel, select Alignment From Network.

   The command line will read Select next Network Part or [Undo].

4. Select the STM STR 7 structure, which will be the last structure on the alignment.
5. Press ↵, and a Create Alignment – From Pipe Network dialog will appear that is almost identical to the one you see when you create an alignment from the Alignments menu.

   ♦ Name your alignment **Storm CL**.

   ♦ Notice the Create Profile And Profile View check box on the last line of the dialog. Leave the box selected and click OK.

   The Create Profile From Surface dialog will appear (see Figure 13.49).

FIGURE 13.49
The Create Profile From Surface dialog

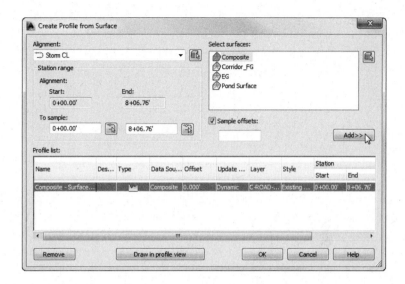

6. Click to highlight the Composite surface, and click Add.
7. Click Draw In Profile View.

   The Create Profile View Wizard will appear (see Figure 13.50).

FIGURE 13.50
The Create Profile View Wizard

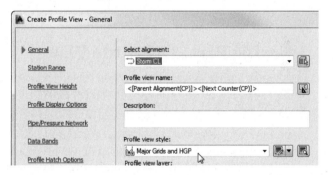

8. Click Pipe/Pressure Network Display to jump to that page.

   You should see a list of pipes and structures in your drawing.
9. Verify that Yes is selected for each pipe and structure in the Storm Network only.
10. Click Create Profile View, and place the profile view to the right of the site plan.

Seven structures and six pipes will be drawn in a profile view, which is based on the newly created alignment (see Figure 13.51).

FIGURE 13.51
The completed profile view

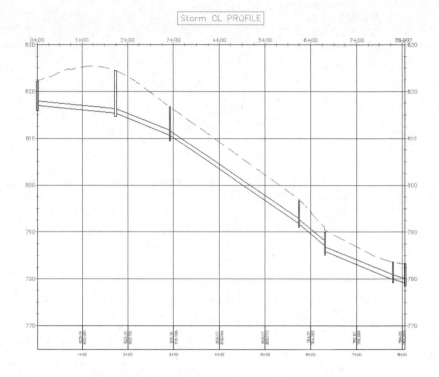

Drawing Parts in Profile View

To add pipe network parts to an existing profile view, select a network part, and choose Draw Parts In Profile from the Network Tools panel on the Pipe Networks contextual tab. When you're using this command, it's important to note that only selected parts are drawn in your chosen profile view.

Profiles and profile views are always cut with respect to an alignment. Therefore, pipes are shown in the profile view on the basis of how they appear along that alignment, or how they cross that alignment. Unless your alignment *exactly follows the centerline of your network parts,* your pipes will likely show some drafting distortion.

Let's look at Figure 13.52 as an example. This particular jurisdiction requires that all utilities be profiled along the road centerline. There's a road centerline, a storm network that jogs across the road to connect with another catch basin.

At least two potentially confusing elements show up in your profile view. First, the distance between structures (2D Length – Center To Center) isn't the same between the plan and the profile (see Figure 13.53) because the storm pipe doesn't run parallel to the alignment. Because the labeling reflects the network model, all labeling is true to the 2D Length – Center To Center or any other length you specify in your label style.

The second potential issue is that the invert of your crossing storm pipe is shown at the point where the storm pipe *crosses the alignment,* and not at the point where it crosses the sanitary pipe (see Figure 13.54).

DRAWING PARTS IN PROFILE VIEW | 627

FIGURE 13.52
These pipe lengths will be distorted in profile view.

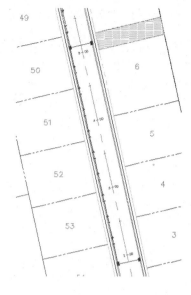

FIGURE 13.53
Pipe labels in plan view (top) and profile view (bottom)

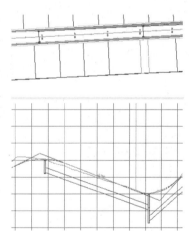

FIGURE 13.54
The invert of a crossing pipe is drawn at the location where it crosses the alignment.

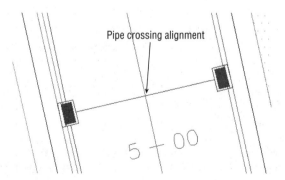

Vertical Movement Edits Using Grips in Profile

Although you can't make changes to certain part properties (such as pipe length) in profile view, pipes and structures both have special grips for changing their vertical properties in profile view.

When selected, a structure has two grips in profile view (see Figure 13.55). The first is a triangular-shaped grip representing a rim insertion point. This grip can be dragged up or down, and it affects the model structure-insertion point.

FIGURE 13.55
A structure has two grips in profile view.

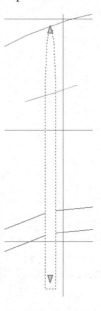

Moving this grip can affect your structure insertion point in two ways, depending on how your structure properties were established:

- If your structure has Automatic Surface Adjustment set to True, grip-editing this Rim Insertion Point grip changes the surface adjustment value. If your reference surface changes, your rim will change along with it, plus or minus the surface adjustment value.

- If your structure has the Automatic Surface Adjustment set to False, grip-editing this Rim grip modifies the insertion point of the rim. No matter what happens to your reference surface, the rim will stay locked in place.

Typically, you'll use the Rim Insertion Point grip only in cases where you don't have a surface for your rims to target, or if you know there is a desired surface adjustment value. It's tempting to make a quick change instead of making the improvements to your surface that are fundamentally necessary to get the desired rim elevation. One quick change often grows in scope. Making the necessary design changes to your target surface will keep your model dynamic and, in the long run, will make editing your rim elevations easier.

The second grip is a triangular grip located at the sump depth. This grip doesn't represent structure invert. In Civil 3D, only pipes truly have invert elevation. The structure uses the connected pipe information to determine how deep it should be. When the sump has been set at a depth of 0, the sump elevation equals the invert of the deepest connected pipe.

This grip can be dragged up or down. It affects the modeled sump depth in one of two ways, depending on how your structure properties are established:

Control Sump By Depth   If your structure is set to control sump by depth, editing with the Sump grip changes the sump depth.

The depth is measured from the structure insertion point. For example, if the original sump depth was 0, grip-editing the sump 0.5′ (15.24 cm) lower would be the equivalent of creating a new sump rule for a 0.5′ (15.24 cm) depth and applying the rule to this structure. This sump will react to hold the established depth if your reference surface changes, your connected pipe inverts change, or something else is modified that would affect the invert of the lowest connected pipe. This triangular grip is most useful in cases where most of your pipe network will follow the sump rule applied in your parts lists, but selected structures need special treatment.

Control Sump By Elevation   If your structure is set to control sump by elevation, adjusting the Sump grip changes that elevation.

When sump is controlled by elevation, sump is treated as an absolute value that will hold regardless of the structure insertion point. For example, if you grip-edit your structure so its depth is 8.219′ (2.51 m), the structure will remain at that depth regardless of what happens to the inverts of your connected pipes. The Control Sump By Elevation parameter is best used for existing structures that have surveyed information of absolute sump elevations that won't change with the addition of new connected pipes.

When selected, a pipe end has three grips in profile view (see Figure 13.56). You can grip-edit the invert, crown, and centerline elevations at the structure connection using these grips, resulting in the pipe slope changing to accommodate the new endpoint elevation.

FIGURE 13.56
Three grips for a pipe end in profile view

When selected, a pipe in profile view has one grip at its midpoint (see Figure 13.57). You can use this grip to move the pipe vertically while holding the slope of the pipe constant.

FIGURE 13.57
Use the midpoint grip to move a pipe vertically.

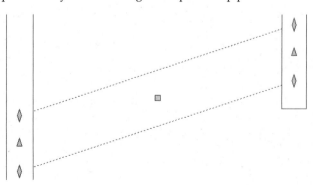

You can access pipe or structure properties by choosing a part, right-clicking, and choosing Pipe or Structure Properties.

Removing a Part from Profile View

If you have a part in profile view that you'd like to remove from the view without deleting it from the pipe network entirely, you have a few options.

AutoCAD Erase can remove a part from profile view; however, that part is then removed from every profile view in which it appears. If you have only one profile view, or if you're trying to delete the pipe from every profile view, this is a good method to use.

Be careful when using the Del key or the Erase command on objects in plan view. Keep in mind that deleting any object from plan view removes the object outright, which includes pipe network parts. Use your Esc key liberally before selecting items to remove; this will help you avoid unintended deletion of items. Of course, if you accidentally blow away something, Undo will bring it back.

A better way to remove parts from a particular profile view is through the Profile View Properties. You can access these properties by selecting the profile view, right-clicking, and choosing Profile View Properties.

The Pipe Networks tab of the Profile View Properties dialog (see Figure 13.58) provides a list of all pipes and structures that are shown in that profile view. You can deselect the check boxes next to parts you'd like to omit from this view.

FIGURE 13.58
Deselect parts to omit them from a view.

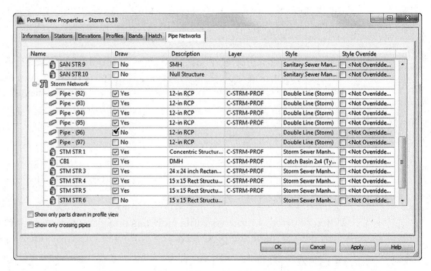

At the bottom of the Profile View Properties dialog is a check box for Show Only Parts Drawn In Profile View. This check box is off by default so that you can see every possible pipe and structure. When it is checked on, the Pipe Network tab will hide any pipes that are not visible in the profile view you are examining.

New to Civil 3D 2014 is a check box for Show Only Crossing Pipes. When this option is on, the listing will narrow down to only pipes that cross the alignment related to the profile view. This is a great help when you need to override the style of crossing pipes to show them as ellipses rather than linear pipes. Style Overrides are discussed in more detail in the section of this chapter called "Adding Pipe Network Labels."

Showing Pipes That Cross the Profile View

If you have pipes that cross the alignment related to your profile view, you can show them with a crossing style. A pipe must cross the parent alignment to be shown as a crossing in the profile view. The location of a crossing pipe is always shown *at the elevation where it crosses the alignment* (see Figure 13.59).

FIGURE 13.59
A pipe crossing a profile

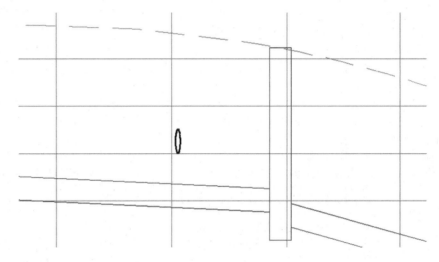

When pipes enter directly into profiled structures, they can be shown as ellipses through the Display tab of the Structure Style dialog (see Figure 13.60). See Chapter 19 for more information about creating structure styles.

FIGURE 13.60
Pipes that cross directly into a structure can be shown as part of the structure style.

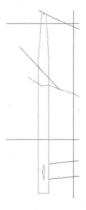

The first step to display a pipe crossing in profile is to add the pipe that crosses your alignment to your profile view by either selecting the pipe, right-clicking, and selecting Draw Parts In Profile from the Network Tools panel and selecting the profile view, or by checking the appropriate boxes on the Pipe Network tab of the Profile View Properties dialog. When the pipe is added, it's distorted when it's projected onto your profile view—in other words, it's shown as if you wanted to see the entire length of pipe in profile (see Figure 13.61).

FIGURE 13.61
The pipe crossing is distorted.

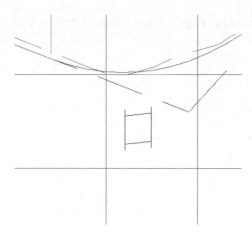

The next step is to override the pipe style *in this profile view only*. Changing the pipe style through pipe properties won't give you the desired result, because that will affect the visibility of every single instance of the pipe. You must override the style on the Pipe Networks tab of the Profile View Properties dialog (see Figure 13.62).

FIGURE 13.62
Use Style Override to display correctly a crossing in the Profile View Properties dialog.

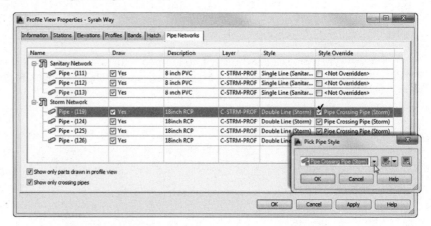

Locate the pipe you just added to your profile view and scroll to the last column on the right (Style Override). Select the Style Override check box and choose your pipe crossing style. Click OK. Your pipe should appear as an ellipsis.

If your pipe appears as an ellipsis but suddenly seems to have disappeared in the plan and other profiles, chances are good that you didn't use the Style Override but accidentally changed the pipe style. Go back to the Profile View Properties dialog and make the necessary adjustments; your pipes will appear as you expect.

Adding Pipe Network Labels

Once you've designed your network, it's important to annotate the design. This section focuses on pipe network–specific label components in plan and profile views (see Figure 13.63).

FIGURE 13.63
Typical pipe network labels in plan view (top) and in profile view (bottom)

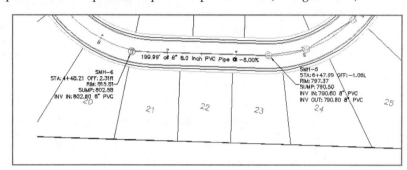

Like all Civil 3D objects, the Pipe and Structure label styles can be found in the Pipe and Structure branches of the Settings tab in Toolspace and are covered in Chapter 18.

Creating a Labeled Pipe Network Profile with Crossings

This exercise will apply several of the concepts in this chapter to give you hands-on experience producing a pipe network profile that includes pipes that cross the alignment. You will have another chance to practice creating an alignment from a pipe network and associating the new alignment to the parts. You will also practice overriding pipe display in the Profile View Properties. There are storm pipes that cross the centerline nearly perpendicular to the alignment. You will show these as ellipses in the profile view.

1. Open the 1307_PipeLabels.dwg (1307_PipeLabels_METRIC.dwg) file. (It's important to start with this drawing rather than use the drawing from an earlier exercise.) Note that surface numbering may differ between drawings so these numbers have been omitted from the exercise.

2. Locate sanitary structure 16 (near station 5+15 or 0+160 of Cabernet Court), select it to open the Pipe Networks contextual tab, and then select Alignment From Network on the Launch Pad panel.

3. When prompted to Select Next Network Part, select structure 18 and press ↵ to create an alignment from the sanitary sewer network.

4. Choose the following options in the Create Alignment From Network Parts dialog:

    - Site: <none>
    - Name: **SMH16 to SMH18 Alignment**
    - Alignment Type: Miscellaneous
    - Alignment Style: _None
    - Alignment Label Set: _No Labels
    - Create Profile And Profile View check box: Selected

5. Click OK.

6. In the Create Profile From Surface dialog, highlight the EG and Corridor FG surfaces, and click Add.

7. Change the Style for Corridor FG – Surface to Design Profile.

8. Click Draw In Profile View to open the Create Profile View dialog.

9. Click the Next button in the Create Profile View Wizard until you reach the Pipe/Pressure Network Display page.

    You should see a list of pipes and structures in your drawing. Make sure Yes is selected for each pipe and structure under Sanitary Sewer Network.

10. Click Create Profile View. At the Select profile view origin: prompt, click to place the profile view to the right of the site plan.

11. Press the Esc key to complete the command. Select either a pipe or a structure in the profile view to open the Pipe Network contextual tab. On the Labels & Tables panel, select Add Labels ➢ Entire Network Profile.

    The alignment information is missing from your structure labels because the alignment was created after the pipe network.

12. Select one of the pipes in the drawing. It does not matter if you select it from plan or profile view.

13. From the Pipe Network contextual tab ➢ Modify panel, click Edit Pipe Network.
14. From the Network Layout Tools, click Pipe Network Vistas.
15. Verify that you are on the Structures tab of Panorama.
16. Select structures 16 through 18 in Panorama using Shift+click. You may need to left-click on the Name column header to sort by name.
17. Scroll to the right until you see the Reference Alignment column.
18. Right-click the column header and select Edit. Choose SMH16 To SMH18 Alignment from the Select Alignment dialog and click OK.
19. Close Panorama by clicking the green check mark. You may need to enter **REGENALL** at the command line to see your updated structure labels.
20. Pan your drawing until you see the Storm layout located on Syrah Way in the plan view.

    The sanitary pipe network has already been laid out on the Syrah Way profile view. You want to show the storm where it crosses the sanitary so you can adjust the pipe elevations if necessary.

21. Select the four pipes that cross perpendicular to the Syrah Way alignment.
22. In the Pipe Networks contextual tab ➢ Network Tools panel, select Draw Parts In Profile.
23. Select the Syrah Way profile view.
24. Press Esc to ensure no extra objects are selected, and then select the profile view and pick Profile View Properties from the Profile View contextual tab.
25. On the Pipe Networks tab of Profile View Properties, place a check mark next to Show Only Parts Drawn In Profile View.
26. For all four Storm Network pipes shown, override the pipe style in this profile view by placing a check mark next to <Not Overridden> in the Style Override column. You will be prompted to pick the pipe style as shown previously in Figure 13.62. Click OK when you have selected the style.
27. When overrides have been set for all four pipes, click OK.

    Your crossing pipes will resemble Figure 13.64.

28. Adjust the elevations of the crossing pipes as needed. Save and close the drawing.

Completed versions of the drawings are available for your enlightenment (1307_PipeLabels.dwg and 1307_PipeLabels_METRIC.dwg).

Pipe and Structure Labels

In earlier exercises, you had a sneak peek at adding labels for structures. Civil 3D makes no distinction between a plan label and a profile label. The same label style can be used in both places. In this chapter, you will use label styles that are already part of the drawing.

To create your own pipe labels from scratch, see Chapter 18.

FIGURE 13.64
The completed exercise with crossing pipes

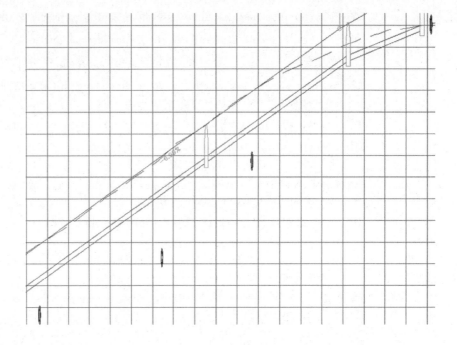

Spanning Pipe Labels

In addition to single-part labels, pipes shown in either plan or profile view can be labeled using the Spanning Pipes option. This feature allows you to choose more than one pipe; the length that is reported in the label is the cumulative length of all pipes you choose.

Unlike Parcel spanning labels, no special label-style setting is required to use this tool. The Spanning Pipes option is on the Annotate tab. Select Add Labels ➢ Pipe Network ➢ Spanning Pipes Profile or Spanning Pipes Plan to access the command.

Creating an Interference Check

When designing, you must make sure pipes and structures are appropriately separated. You can perform some visual checks by rotating your model in 3D and plotting pipes in profile and section views (see Figure 13.65). Civil 3D also provides a tool called Interference Check that makes a 3D sweep of your pipe networks and lets you know if anything is too close for comfort.

The following exercise will lead you through creating a pipe network and using Interference Check to scan your design for potential pipe network conflicts:

1. Open the 1308_Interference.dwg (1308_Interference_METRIC.dwg) file. The drawing includes a sanitary sewer pipe network and a storm drainage network.

2. Select a part from either network, and choose Interference Check from the Analyze panel, as shown in Figure 13.66.

FIGURE 13.65
Two pipe networks may interfere vertically where crossings occur (left). Viewing your pipes in profile view can also help identify conflicts (right).

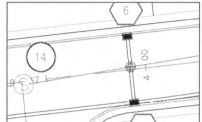
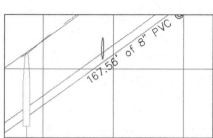

FIGURE 13.66
Creating an interference check

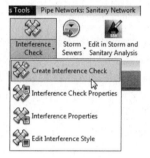

You'll see the `Select a part from the same network or different network:` prompt.

3. Select a part from the network that hasn't already been chosen.

The Create Interference Check dialog will appear (Figure 13.67).

FIGURE 13.67
The Create Interference Check dialog

4. Name the Interference Check **Collision**, and confirm that Sanitary Network and Storm Network appear in the Network 1 and Network 2 boxes.

5. Click the 3D Proximity Check Criteria button, and the Criteria dialog will appear (see Figure 13.68).

FIGURE 13.68
Criteria for the 3D proximity check

6. You're interested in finding all network parts that are within a certain tolerance of one another, so enter **1.5′ (0.5 m)** in the Use Distance box.

   This setting creates a buffer to help find parts in all directions that might interfere. If you forget to check Apply 3D Proximity Check, only direct, physical collisions will be listed as collisions.

7. Click OK to exit the Criteria dialog, and click OK to run the Interference Check.

   You should see a dialog that alerts you to three interferences.

8. Click OK to dismiss this dialog.

9. On the Prospector tab of Toolspace ➤ Pipe Networks, expand the Interference Checks branch. Right-click Exercise and select Zoom To.

   A small marker will appear at each location where interference occurs, as shown in Figure 13.69.

FIGURE 13.69
The interference marker in plan view

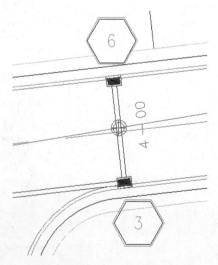

10. Select any one of the interference markers, the pipes that intersect, and the nearby inlets.
11. From the Multiple contextual tab ➢ General Tools panel, click Object Viewer.
12. In the upper-left corner of Object Viewer, click the Visual Style drop-down. Change the Visual Style to Conceptual.

    The interference marker will appear in 3D, as shown in Figure 13.70.

FIGURE 13.70
The interference marker in 3D

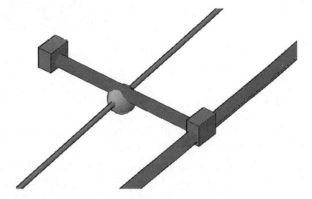

13. Use the ViewCube in Object Viewer to navigate in 3D.
14. Once you have examined the 3D objects, close Object Viewer. Save the drawing.

    Note that each instance of interference is listed in the preview pane for further study.

The `1308_Interference_FINISHED.dwg` and `1308_Interference_METRIC_FINISHED.dwg` files are available for your perusal.

Editing your pipe network will flag the interference check as "out of date." You can rerun Interference Check by right-clicking Interference Check in Prospector. You can also access the Interference Check Properties to edit your criteria in this right-click menu.

Creating Pipe Tables

Just as with parcels and labels, the process of labeling pipes can turn into a mess when all the labels are set on the plan (see Figure 13.71). In this section, you will explore the process of creating tables for pipes and structures.

Exploring the Table Creation Dialog

Because the Structure Table Creation dialog and Pipe Table Creation dialog are similar (see Figure 13.72), we will cover both of them in this section:

- The Table Style option allows you to select a table look or style. You can select the available styles from the arrow to the right of the Table Style name. You can also create new, copy, edit, or pick a table style from an existing table in the drawing. For more on table styles, see Chapter 18.

- The Table Layer option will be the layer where the table is placed.

FIGURE 13.71
Crowded pipes
and structure labels
on a plan

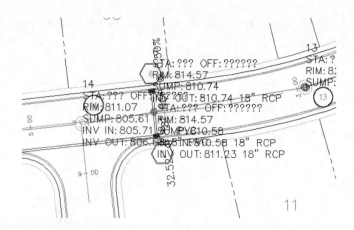

- With the By Network radio button selected, you can select the network to create a table from the drop-down, or use the Pick icon to select the network from the drawing.

- With the Multiple Selection radio button selected and by using the Pick icon, you can select structures or pipes (depending on which table type is selected) from the drawing. You can pick pipes or structures regardless of the network.

- The Split Table check box will allow you to split the table if it becomes too large. You can specify the maximum number of rows per table and the maximum number of tables per stack. Additionally, you can set the offset distance between the stacked tables.

- You can choose whether you want the split tables tiled across or down.

- The last option to choose defines the behavior you want for the table: Static or Dynamic. A static table will not update if any changes are made to the pipe network, such as swapping a part. Dynamic will update the table to those changes.

In the following exercise, you will create a pipe network table for the sanitary sewer structures:

1. Open the 1309_PipeTable.dwg (1309_PipeTable_METRIC.dwg) file.

2. Click on the MH1. This is a structure from the sanitary sewer network.

3. From the Pipe Networks contextual tab ➤ Labels & Tables panel, select Add Tables ➤ Add Structure.

   The Structure Table Creation dialog opens.

4. Verify that the By Network option is set to Sanitary Network. Click OK to accept the default settings (shown in Figure 13.72).

5. Place the table to the right of your plan.

   The table should look similar to Figure 13.73.

FIGURE 13.72
The Structure Table Creation dialog box

FIGURE 13.73
The finished structure table

STRUCTURE TABLE			
STRUCTURE NAME:	DETAILS:	PIPES IN:	PIPES OUT
MH1	48" RIM = 823.35 SUMP = 814.8 INV IN = 814.92 INV OUT = 814.92	Pipe – (98), 8" PVC INV IN =814.92	Pipe – (99), 8" PVC INV OUT =814.92
MH2	48" RIM = 824.50 SUMP = 813.6 INV IN = 813.68 INV OUT = 820.81	Pipe – (107), 4" INV IN =813.68	Pipe – (98), 8" PVC INV OUT =820.81
MH3	48" RIM = 811.05 SUMP = 806.1 INV IN = 806.21 INV OUT = 806.21	Pipe – (99), 8" PVC INV IN =806.21	Pipe – (100), 8" PVC INV OUT =806.21
SAN STR 4	48" RIM = 797.37 SUMP = 776.8 INV IN = 776.90 INV OUT = 776.90	Pipe – (105), 8" PVC INV IN =776.90	Pipe – (106), 8" PVC INV OUT =776.90
SAN STR 5	48" RIM = 793.30 SUMP = 788.7 INV IN = 788.83 INV OUT = 789.61	Pipe – (100), 8" PVC INV IN =788.63	Pipe – (101), 8" PVC INV OUT =789.61
SAN STR 6	48" RIM = 791.47 SUMP = 777.4 INV IN = 777.47 INV OUT = 777.47	Pipe – (104), 8" PVC INV IN =777.47	Pipe – (105), 8" PVC INV OUT =777.47
SAN STR 8	48" RIM = 785.43 SUMP = 778.2 INV IN = 778.28 INV OUT = 778.28	Pipe – (103), 8" PVC INV IN =778.28	Pipe – (104), 8" PVC INV OUT =778.28
SAN STR 9	48" RIM = 783.54 SUMP = 779.0 INV IN = 779.09 INV OUT = 779.09	Pipe – (102), 8" PVC INV IN =779.09	Pipe – (103), 8" PVC INV OUT =779.09
SAN STR 10	48" RIM = 815.82 SUMP = 775.0 INV IN = 775.06	Pipe – (106), 8" PVC INV IN =775.06	

The process of creating tables for pipes is similar to the process for creating tables for structures. Continue working in 1309_PipeTable.dwg (1309_PipeTable_METRIC.dwg):

1. Press Esc to clear the current selection and click any sanitary sewer pipe.
2. From the Pipe Networks contextual tab➢ Labels & Tables panel, select Add Tables ➢ Add Pipe.

    The Pipe Table Creation dialog opens (Figure 13.74).

FIGURE 13.74
The Pipe Table Creation dialog

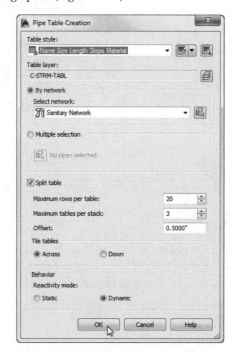

3. Verify that the By Network option is set to Sanitary Network. Click OK to accept the default settings.
4. Place the table to the right of the structure table.

    The pipe table should look similar to Figure 13.75.

The Table Panel Tools

When you click on a table, the Table contextual tab opens and has several tools available (Figure 13.76). You'll look at each in this section.

THE GENERAL TOOLS PANEL

The tools here are the same as mentioned earlier when the pipe network tools were discussed.

FIGURE 13.75
The finished pipe table

Pipe Table				
NAME	SIZE	LENGTH	SLOPE	MATERIAL
Pipe 1	8"	139.54'	4.22%	PVC
Pipe 2	8"	177.33'	4.91%	PVC
Pipe 3	8"	248.11'	7.00%	PVC
Pipe 4	8"	122.89'	6.00%	PVC
Pipe 5	8"	105.12'	3.00%	PVC
Pipe 6	8"	80.50'	1.00%	PVC

FIGURE 13.76
The Table contextual tab

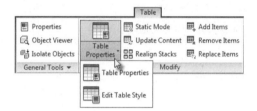

The Modify Panel

The Modify panel contains the following tools:

Table Properties This tool opens the Table Properties dialog. With it, you can set the table style and choose whether to split the table with all the options mentioned earlier. In addition, you can force realignment of stacks, and if the reactivity mode is set to Static, you can force content updating.

Edit Table Style This tool is located under the Table Properties drop-down and opens the Table Style dialog. For more on editing table styles, see Chapter 18.

Static Mode This tool turns a dynamic table into a static table.

Update Content This tool forces an update on a table.

Realign Stacks This tool readjusts the table columns back to the default setting.

Add Items This tool adds pipe data to the table that was added after the table was created.

Remove Items This tool removes pipe or structure objects from a table.

Replace Items This tool allows you to swap pipe or structures in the table.

Under Pressure

Pressure pipes work differently than gravity-flow pipe systems within Civil 3D. Much of the need for custom parts such as valves or hydrants is eliminated with these systems. In this section, you'll learn how easy it is to model water, gas, or other pressurized systems in 3D.

Pressure Network Parts List

Like gravity-based networks, a pressure network starts with a parts list. All of the parts available in Civil 3D are based on standards established by the American Water Works Association (AWWA) and are listed in both inches and millimeters.

Pressure parts lists contain pipes, fittings, and appurtenances. You'll find the style for each object in the parts list, but instead of rules, pressure pipe design checks are tucked into the command settings.

Examples of Pressure Pipe Network parts lists include:

- Water main and service connections
  - Ductile iron pipe
  - Tees, elbows, and crosses
  - Valves
- Gas main
  - PVC pipe
  - Valves

Under the Hood of the Pressure Network

Before you can create a Pressure parts list, you must determine the catalog from which you will be working. Set the pressure network catalog by going to the Home tab ➢ Create Design panel (expand the panel to view the additional tools) and selecting Set Pressure Network Catalog, as shown in Figure 13.77.

FIGURE 13.77
Setting the pressure network catalog

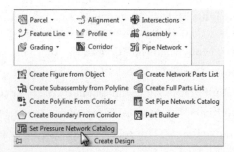

Set the path to the Pressure Pipes Catalog folder to C:\ProgramData\Autodesk\C3D 2014\enu\Pressure Pipes Catalog. Click the folder icon to choose either the Metric or the Imperial database, depending on your needs.

The catalog database file determines the join type between pressure network parts. In modern water main construction, the most commonly used join type is the push-on type, which is the default pressure database. As shown in Figure 13.78, with Imperial units you have three options:

- Imperial_AWWA_Flanged
- Imperial_AWWA_Mechanical
- Imperial_AWWA_PushOn

FIGURE 13.78
Setting your catalog database file

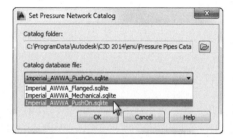

Metric_AWWA_PushOn is currently the only available option for metric users.

These differ slightly in their options for pipes, fittings, and appurtenances. Only one type of pressure network catalog can be active at a time. A parts list can have parts from only one catalog in it; for example, you cannot mix and match push-on with mechanical parts. You can, however, have multiple Pressure parts lists in your template; each can pull parts from the various catalog database files. You can place parts from different parts lists into the same pressure network, as long as the parts are meant for the same diameter and pressure.

Creating a Pressure Parts List

In the Settings tab of Toolspace, you will find the listing for pressure networks. The Pressure Networks ➢ Parts List branch is where you will create a parts list. In the case of pressure networks, a parts list contains three components:

Pressure Pipes   Ductile iron of various sizes can be added. Like the gravity networks, each part can have a style. Furthermore, different sizes within the part families can have styles, which can help you identify them in the graphic.

Fittings   Fittings such as tees, crosses, and elbows are specified in this tab.

Appurtenances   Valves are specified in the Appurtenances tab.

Bursting with Parts

To create a complete list of all pressure network parts available in your active part catalog, type **CreatePressurePartListFull** at the command line and press ↵.

Keep in mind that the three different catalogs available to US Imperial units vary with regard to which parts are available. For example, the following graphic shows the variation of the fittings available to Push-on, Flanged, and Mechanical catalogs.

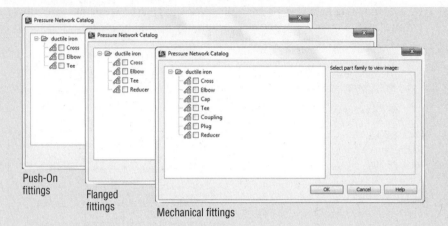

Push-On fittings
Flanged fittings
Mechanical fittings

Your next question is probably, "Okay, so can I make my own parts?" The answer is, "Well, sort of, but it is currently unsupported."

Pressure pipe network parts are 3D solids with a few extra bells and whistles that tell the program how they connect to other parts. If you are comfortable drafting 3D solids using the tools from the 3D modeling workspace, the procedure is fairly painless.

The Part Publishing Wizard tools built into AutoCAD will allow you to create a `*.CONTENT` package—for example, `Hydrant.CONTENT`. Once the content package is created, you can import it into the pressure pipe catalog of your choice by launching the Content Catalog Editor.

The Content Catalog Editor is automatically installed with Civil 3D 2014 and can be found from your Windows Start menu along with the other AutoCAD Civil 3D 2014 utilities.

For more information on creating parts through the Part Publishing Wizard, see `PartPublishingWizardUsersGuide.docx` located in `C:\Program Files\Autodesk\Autodesk AutoCAD Civil 3D 2014\Sample\Civil 3D API\Part Publishing Wizard`.

In the following exercise, you will create a Pressure Network parts list:

1. Open the `1310_Pressure.dwg` (`1310_Pressure_METRIC.dwg`) drawing, which you can download from this book's web page.

    This file is set up with a layer state that makes other objects gray. This will help you focus on the placement of pressure pipe network objects.

2. On the Home tab ➢ Create Design panel, click the arrow to view additional tools and select Set Pressure Network Catalog.

3. Verify that the Catalog Database File is set to `Imperial_AWWA_PushOn.sqlite` (`Metric_AWWA_PushOn.sqlite`). Click OK.

4. On the Settings tab of Toolspace ➢ Pressure Network, expand the Pressure Network branch, right-click Parts Lists, and select New.

5. In the Pressure Network Parts List dialog, switch to the Information tab. Rename the Pressure Network parts list **Watermain**.

6. Switch to the Pressure Pipes tab. Right-click New Parts List and select Add Material.

7. In the Pressure Network Catalog dialog, place a check mark next to Ductile Iron and click OK.

8. The name of the Pipe Parts list will update to Watermain. Expand the Watermain branch.

9. Right-click on Ductile Iron and select Add Size.

10. Set the Nominal Diameter value to **10″** (**250** mm).

11. Set the Cut Length value to **20′** (**6** m). Leave all other defaults values, as shown in Figure 13.79, and click OK.

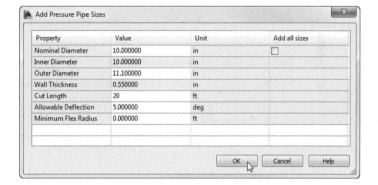

FIGURE 13.79
Adding ductile iron pipe to the Pressure Network parts list

12. Switch to the Fittings tab, right-click New Parts List, and select Add Type.

13. In the Add Fitting Sizes dialog, place a check mark next to all three fitting types—Cross, Elbow, and Tee—as shown in Figure 13.80, and then click OK.

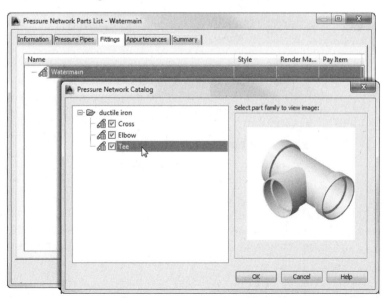

FIGURE 13.80
Adding fittings to the Pressure Network parts list

Metric users will have only the options for Elbow and Tee, and can skip to step 16.

14. Expand the new Watermain branch if it is not already expanded.
15. Right-click Ductile Iron Cross and select Add Size.
16. Change Nominal Diameter to **10 × 10 × 14 × 14**. Leave the allowable deflection as 5 degrees, and click OK.
17. Right-click Ductile Iron Elbow, and click Add Size.
18. Set the bend angle to **11.25** and set the nominal diameter to **10 × 10 (250 × 250** mm). Leave the Allowable Deflection value at 5 degrees, which is the default for the size you picked. Click OK.
19. Repeat steps 17 and 18 for 22.5-degree and 45-degree elbows, with a nominal diameter of **10 × 10 (250 × 250** mm).
20. Right-click Ductile Iron Tee and select Add Size.
21. Set the nominal diameter to **10″ (250** mm). Leave the allowable deflection as 5 degrees. Click OK.
22. Switch to the Appurtenances tab.
23. Right-click New Parts List and select Add Type.
24. Place a check mark next to Gate Valve – Push-On – Ductile Iron – 200 psi (Gate Valve – Push-On – Ductile Iron – 16 Bar) and click OK.
25. Expand the Appurtenance listing.
26. Right-click the new gate valve and select Add Size.
27. Change the Nominal Diameter value to **10 × 10 (250 × 250** mm) and click OK.
28. Click OK again to finish creating the Watermain Pressure Network parts list, and then save the drawing for use in the next exercise.

Creating a Pressure Network

After you have set your pressure network catalog, created your Pressure Network parts list, and set your design parameters, it is time to draw your first network.

Pressure Networks in Plan View

As you work with pressure pipes, you will see some useful glyphs appear as you draw.

As shown in Figure 13.81, selecting a pressure pipe will give you tools to modify and continue your design.

Location The Location glyph moves the pipe both horizontally and vertically and will disconnect it from the adjoining fittings or appurtenances. Changing the location of a fitting or appurtenance will move the pipe and maintain the connection.

Deflection The Deflection glyph will change the angle at which the pipe sits in the adjoining fitting. When this glyph is active, you will see a fan-shaped guide indicating the allowable deflection from the fitting properties in the Pressure Network Parts list. You are able to move the pipe beyond the guide, but will receive design check errors when analyzing the network. You will take a closer look at design checks later in this chapter.

Continue Layout Continue Layout will help you pick up where you left off when working with pressure pipes. When you use this glyph from the end of a pipe, it will create a bend using the elbows from your Pressure Parts list. When you use this grip from a fitting or appurtenance, you are restricted to creating your pipe within the object's deflection tolerance.

Lengthen Lengthen will allow you to stretch or shorten a pipe. When used with dynamic input, you can set pipes to a specified 3D length. See the section on working with pipes and dynamic input earlier in the "Editing with the Network Layout Tools Toolbar" section.

FIGURE 13.81
Glyphs on a pressure pipe end

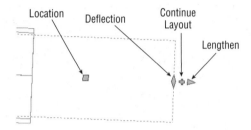

As shown in Figure 13.82, you will encounter more glyphs when working with fittings and appurtenances.

FIGURE 13.82
Glyphs on a cross fitting

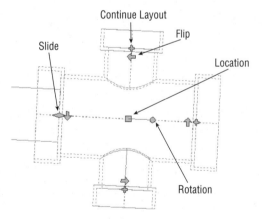

Slide The Slide glyph is similar to lengthen, but it can modify a pipe's length when a fitting or appurtenance is already attached.

Flip The Flip glyph will mirror the part at its center. Flip glyphs are especially handy when working with tees, as it is often necessary to change the outlet direction of a tee after it is inserted. Be cautious when using the Flip glyph with elbows, because it may disconnect the adjoining pipe.

In this exercise, you will create a pressure network. Use the Xs as guides for placement, but don't worry if your pipe network is slightly off from the guides. Due to the 3D nature of the pipes, the restrictions on placement angles within the pressure network parts, and object snap behavior, duplicating an example network exactly would be quite tedious. Get a feel for the pressure network creation tools and have fun! You need to have completed the previous exercise before continuing.

1. Continue working in the drawing 1310_Pressure.dwg (1310_Pressure_METRIC.dwg).

   As you work through this exercise, you will get the best results if you turn off object snaps, object snap tracking, polar tracking, and/or ortho. Because the pressure pipe tools already have restrictions on how they can be drafted, sometimes these tools conflict with where you want to place the pipe.

2. From the Home tab ➢ Create Design panel, click Pipe Network ➢ Pressure Network Creation Tools, as shown in Figure 13.83.

FIGURE 13.83
Selecting Pressure Network Creation Tools

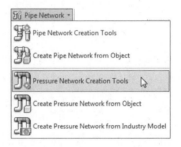

3. In the Create Pressure Pipe Network dialog, name the network **Watermain North**, and then do the following:

   a. Set the Parts list to Watermain.

   b. Set the surface name to Composite.

   c. Set Alignment Name to Syrah Way.

   d. Set the pipe, fitting, and appurtenance labels as shown in Figure 13.84.

4. Click OK.

   The ribbon will change to show you the Pressure Network Plan Layout contextual tab, as shown in Figure 13.85.

5. On the Pressure Network Plan Layout contextual tab ➢ Network Settings panel, set the default cover to **4.5′ (1.5 m)**.

6. On the Layout panel, set your Size and Material drop-down to **Pipe – 10 In – Push-On – Ductile Iron – 350 psi – AWWA C251 (Pipe – 250 mm – Push-On – Ductile Iron – 25 Bar – AWWA C251)**.

7. Start to place a waterline by clicking the X labeled 1 toward the east end of Syrah Way.

8. Place the first bend by clicking near the X labeled 2, to the left.

At the bend, you are restricted to the bend angles listed in your part network. The Compass glyph (Figure 13.86) that appears represents the elbow angles to the left and right of your pipe. If you had not included multiple elbow angles in your Pressure Pipe Network parts list, only the default elbow angle of 11.25 degrees would be available.

FIGURE 13.84
Creating your new Watermain system using pressure pipe network tools

FIGURE 13.85
The Pressure Network Plan Layout toolbar

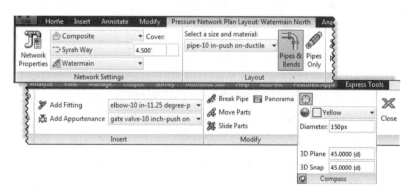

FIGURE 13.86
The Pressure Pipe Fitting glyph reflects your elbow angles.

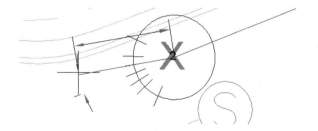

9. Click the X labeled 3 to place the next pipe end.
10. Continue working east to west until you click near the X labeled 8. Press the Esc key.

    To keep the pressure pipe on track, you will use the allowable deflection of the elbow to move the pipe closer to the edge of the road.

    Make sure your object snaps off for the next steps, as they will interfere with the pipe modification glyphs.

11. Select the pipe section between the X labeled 7 and the X labeled 8, and then use the Deflection glyph to move the end of the pipe up to the approximate center of the green circle, as shown in Figure 13.87.

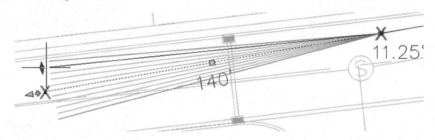

FIGURE 13.87
Use the Deflection glyph to move the pipe.

12. Click the Continue layout glyph to continue the layout.
13. Place the next pipe end at the X labeled 9.
14. Continue working west until you reach near the X labeled 10.

    Don't worry if you are off the desired location; you can always use the glyphs to edit the pipe location after the fact.

15. Press Esc after placing this pipe.
16. On the Pressure Network Plan Layout contextual tab ➤ Insert panel, change the fitting to **Tee–10 in × 10"** (**Tee–250 mm × 250 mm**).
17. Click Add Fitting.

    As you hover your cursor near the end of the pipe, you will see the Add Fitting glyph, as shown in Figure 13.88.

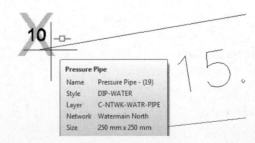

FIGURE 13.88
Add the fitting to the end of the pipe.

18. Click to add the tee.
19. Press Esc to complete the task.

   At this point, the tee is in the graphic but it is not positioned in such a way that would make it useful in continuing the design north and south along the intersecting road (Frontenac Drive). To fix this situation, you will disconnect it from the pipe, rotate it, and then reconnect the pipe.

20. Click the tee in the graphic to select it, and then right-click the part and select Disconnect From Pressure Part, as shown in Figure 13.89. (Note that this command is not available in the contextual tab.)

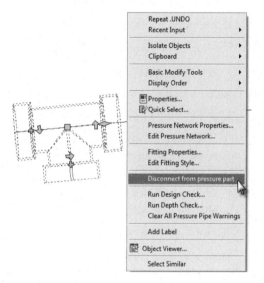

FIGURE 13.89
Disconnecting the part in preparation for rotating it

21. At the Select connected pressure part: prompt, select the pipe connected to the tee.

   Now that the part is disconnected, you are free to rotate it into place. Select the part to reveal its glyphs. A Rotation glyph is visible on the object, which you will use to rotate the part 90 degrees counterclockwise.

22. Press F12 to turn on dynamic input if it is not already on.
23. Select the tee and then, as shown in Figure 13.90, click the Rotation glyph and enter **-90**.

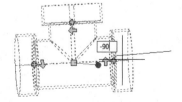

FIGURE 13.90
Rotate the tee to correct its position before reconnecting.

The tee is now in the correct position, but it must be reconnected to the pipe.

24. Select the Location glyph, and move the tee to the east until you see the Connection glyph similar to the one you saw in Figure 13.88.

25. When you see the correct glyph, click to set the tee.

   You will know the tee is connected properly when the Rotation glyph no longer appears on the selected fitting. Another graphic indication that the tee is connected will be the reappearance of the Slide glyph on the connected end.

26. Select the tee to reveal the grips. Working north from the tee, click the + glyph to continue the layout.

27. Click to place an elbow at the X labeled 11 and finally the X labeled 12. Press Esc.

28. Working south from the tee, click the + glyph to continue the layout.

29. Click to place a pipe ending at the X labeled 13, and press Esc.

30. From the Home tab ➢ Layers panel, click Layer Freeze.

31. Click one of the Xs to freeze the _PLACEMENT SYMBOLS layer, and press ↵ to complete.

   If you no longer see the Pressure Network Plan Layout contextual tab, you can get back to it by selecting any pressure network part, and from the Pressure Networks contextual tab ➢ Modify panel, clicking Edit Network ➢ Plan Layout Tools.

32. On the Pressure Network Plan Layout contextual tab ➢ Insert panel, verify that the Appurtenance is set to the 10″ gate valve (250 mm gate valve).

33. Click Add Appurtenance.

34. Place the appurtenance in the drawing by clicking near the end of the north pipe. Be sure to look for the single square glyph that indicates you are connecting to the end of the pipe. If you see the double "boxing gloves" glyph, it means you are about to break the pipe.

   Be sure to look for the Attachment glyph before clicking, as shown previously in Figure 13.88. Don't be shy about zooming in close to get a good look at the object with which you are working.

35. Place another valve at the south end. Press the Esc key when you're done.

36. Save and close the drawing.

Completed versions of this drawing can be found with the dataset for comparison: `1310_Pressure_FINISHED.dwg` (`1310_Pressure_METRIC_FINISHED.dwg`).

Pressure Pipe Networks in Profile View

Pressure pipe networks can do things in profile view that gravity pipes cannot. With pressure pipes, the profile view can be used to change straight pipes to curves and delete parts from the project altogether. It is not a good idea to attempt to add parts to your pressure network in profile view, because the resulting location in plan cannot be controlled.

Importing an Industry Model

One of the options you have with Civil 3D 2014 is to create a pressure network from an industry model.

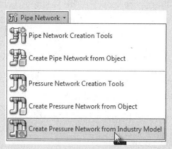

Industry models are a type of AutoCAD drawing file that contain data from one of the handful of Autodesk-created templates, most commonly from Autodesk® Map 3D. The Map 3D standalone product comes with premade infrastructure data classes such as wastewater, water, gas, and electric, just to name a few. Not just any Map 3D file will do. A drawing containing pipe and valve information will be recognized as an industry model only if it was created based on one of the industry-specific templates. The ability to create an industry model is not available on the Map 3D workspace that is part of Civil 3D.

To access these tools, select any pressure part; and from the Pressure Networks contextual tab ➢ Modify panel, choose Edit Network ➢ Profile Layout Tools, as shown in Figure 13.91.

FIGURE 13.91
Locating the Profile Layout tools

In the following exercise, you will draw the pressure pipe network in profile view and modify the layout using the Follow Surface command.

1. Open the `1311_PressureProfile.dwg` (`1311_PressureProfile_METRIC.dwg`) drawing, which you can download from this book's web page.

2. Select any pressure network part in the drawing.

3. From the Pressure Networks contextual tab ➢ Launch Pad panel, select Alignment From Network.

4. At the `Select first Pressure Network Part (Pipe or Fitting or Appurtenance):` prompt, click the Gate Valve at the far right of the drawing.

5. At the Select next Pressure Network Part or [Undo]: prompt, select the Gate Valve at the northwest part of the project.

6. Press ↵ to continue.

   The next few steps are exactly the same as when you created an alignment and profile from a gravity pipe network. You will be prompted to create an alignment, sample the surface, and create a profile view.

7. In the Create Alignment – From Pressure Network dialog, change the name to **Syrah Water**.

8. Verify that Create Profile And Profile View is checked. Leave all other styles and settings at their defaults and click OK.

9. In the Create Profile From Surface dialog, highlight the Composite surface and click Add.

10. Click Draw In Profile View.

11. In the Create Profile View Wizard, leave all settings at their defaults and click Create Profile View.

12. Place the view by clicking a location in the drawing off to the side of the project.

    You should see the profile view with your pressure pipe network present in all its glory. As you can see in Figure 13.92, the pipe looks good, except it appears that the pipe cover is inadequate toward the end of the alignment. You can fix this in the steps that follow.

FIGURE 13.92
Pressure network in profile

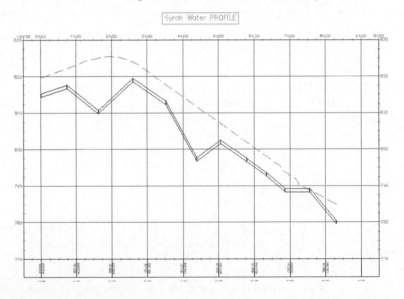

13. Select a part from the network if the Pressure Networks contextual tab is not already visible.

14. From the Pressure Networks contextual tab ➢ Modify panel, select Edit Network ➢ Profile Layout Tools.

The Pressure Network Profile Layout contextual tab will appear, as shown in Figure 13.93.

FIGURE 13.93
The Pressure Network Profile Layout contextual tab

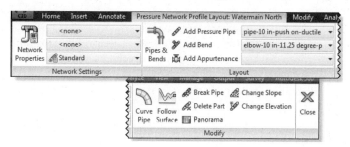

Follow Surface

15. From the Pressure Network Profile Layout contextual tab ➢ Modify panel, select Follow Surface.
16. At the Select first pressure part in profile: prompt, select the leftmost valve in the profile view.
17. At the Select next pressure part in profile [Enter to finish]: prompt, click the rightmost valve in the profile view and then press ↵ to finish selecting parts. All connected parts in between will become selected.
18. At the Enter depth below surface <0.0000>: prompt, enter **4.5′** (**1.5 m**).

Your profile view will change to resemble Figure 13.94.

FIGURE 13.94
The pressure pipe follows the surface.

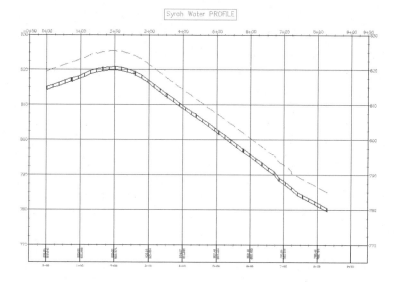

19. Save and close the drawing.

The 1311_PressureProfile.dwg and 1311_PressureProfile_METRIC.dwg files show the completed exercise.

Design Checks

Pressure networks differ from the networks you created earlier in this chapter. Because the fluid in a pressure network can go uphill, the rules you saw in gravity systems no longer apply. The main concerns for a pressure network are pressure loss and depth of cover.

You can locate the Depth Check values on the Settings tab of Toolspace. Locate and expand the Pressure Network branch and expand the Commands branch. Double-click RunDepthCheck to edit the command settings. A dialog like the one in Figure 13.95 will open.

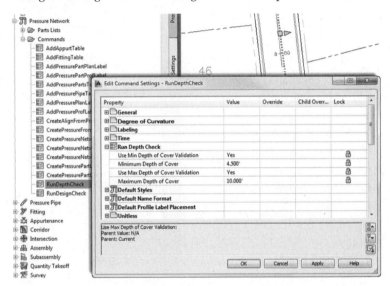

FIGURE 13.95
Edit the RunDepthCheck command settings to validate your design.

Also in the command settings you will find a separate listing for RunDesignCheck. Double-click (or right-click and click Edit) to enter these settings. You can set an acceptable range of values for pipe bends and radius of curvature for curved pipes. The Deflection Validation settings are found under RunDesignCheck, as shown in Figure 13.96.

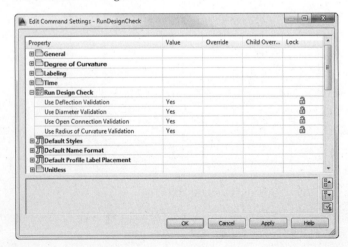

FIGURE 13.96
Turning on the Deflection Validation settings

Once you have created your pressure network, you should check your initial design for flaws. From the Pressure Networks contextual tab, you can check your design to see if it meets the requirements you set up in the command settings.

Depth Check

Depth Check verifies that all pipes and fittings are within the acceptable range of values for depth.

Design Check

Design Check will check for improperly terminating pipes, mismatched pipe and fitting diameters, any curved pipe whose radius has exceeded acceptable values, and pipes that have exceeded the maximum deflection you set up in the parts list.

In the following exercise, you will modify the command settings and run a depth check on the pipe network:

1. Open the 1312_DesignCheck.dwg (1312_DesignCheck_METRIC.dwg) drawing, which you can download from this book's web page.

2. In the Settings tab of Toolspace, expand Pressure Network ➢ Commands, right-click RunDepthCheck, and select Edit Command Settings.

3. Expand Run Depth Check, and verify that Minimum Depth Of Cover is set to 4.5′ (1.5 m).

4. Click in the Value column and change Use Max Depth Of Cover Validation to Yes.

5. Click OK.

6. Select a pressure network part if you do not already see the Pressure Networks contextual tab.

7. From the Pressure Networks contextual tab ➢ Analyze panel, select Depth Check.

   The Depth Check command will allow you to perform the analysis in either plan or profile view. Either way you choose to select your pressure pipe network, the result will be the same.

8. At the `Select a path along a Pressure Network in plan or profile view:` prompt, click the first pressure network object to the left in the profile view.

9. At the `Select next point on path [Enter to finish]:` prompt, select the Gate Valve to the far right in the profile view and press ↵.

   The Depth Check dialog will appear. The settings should be the same values as those in steps 3 and 4.

10. Click OK.

    In both plan view (Figure 13.97, left) and profile view (Figure 13.97, right), warnings will appear if any Depth Check violations are found.

11. Save and close the drawing.

The completed files are available for your reference: 1312_DesignCheck_FINISHED.dwg (1312_DesignCheck_METRIC_FINISHED.dwg).

FIGURE 13.97
Depth Check result in plan (left) and profile (right) views

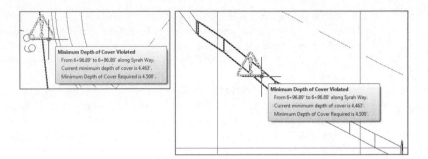

Part Builder

Part Builder is an interface that allows you to build and modify pipe network parts for gravity systems. You access Part Builder by selecting the Create Design drop-down from the Home tab. At first, you may use Part Builder to add a few missing pipes or structure sizes. As you become more familiar with the environment, you can build your own custom parts from scratch.

Parts created by Part Builder will not be available for use in pressure networks, so this section applies only to parts needed for gravity systems.

> **Real World Scenario**
>
> ### A Word of Caution about Part Builder
>
> Every time one of our clients asks about Part Builder, we cringe. Our advice is, "If all you need to do is add a size to an existing part, no problem. If you want to create new geometry from scratch? Run. Run, screaming."
>
> This is not a tool you can master by dabbling in it for a few hours. Even after you understand Part Builder, it can take days to build a complex part that functions correctly. Think of Part Builder as a half-step away from programming, and you will have more realistic expectations.
>
> Before you buckle down to learn about Part Builder, ask yourself a few questions:
>
> ◆ Do I really understand pipes and structures well enough to tackle this?
>
> ◆ Is it worth taking extra time from a billable project to learn this feature?
>
> ◆ Is the part I need unique and not available from another source (such as colleagues or purchased from an Autodesk reseller)?
>
> If you can answer yes to all these questions, you have our blessing. In many cases, the new pressure network tools eliminate the need for custom parts.

This section is intended to be an introduction to Part Builder and a primer in some basic skills required to navigate the interface. It isn't intended to be a robust "how-to" for creating custom parts. Civil 3D includes three detailed tutorials for creating three types of custom structures. The tutorials lead you through creating a cylindrical manhole structure, a drop inlet manhole structure, and a vault structure. You can find these tutorials by going to Help ➢ Tutorials and then navigating to Part Builder Tutorials.

> **Back Up the Part Catalogs**
>
> Here's a warning: Before you explore Part Builder in any way, it's critical that you make a backup copy of the part catalogs. Doing so will protect you from accidentally removing or corrupting default parts as you're learning, and will provide a means of restoring the original catalog.
>
> The catalog (as discussed in the previous section) can be found by default at:
>
> `C:\ProgramData\Autodesk\C3D 2014\enu\Pipes Catalog\`
>
> To make a backup, copy this entire directory and then save that copy to a safe location, such as another folder on your hard drive or network, or to a CD.
>
> We recommend that you do this and use the backup file for the exercises here. To do this, you will have to point to the new location by clicking the drop-down in the Home tab ➢ Create Design panel and selecting Set Pipe Catalog Location. Select the icon next to the `Catalog` folder and point it to your saved location. When you locate the backup folder, pick either Imperial or Metric and then click OK.

The parts in the Civil 3D pipe network catalogs are *parametric*. Parametric parts are dynamically sized according to a set of variables, or parameters. In practice, this means you can create one part and use it in multiple situations.

You can create one parametric model that understands how the different dimensions of the pipe are related to each other and what sizes are allowable. When a pipe is placed in a drawing, you can change its size. The pipe will understand how that change in size affects all the other pipe dimensions such as wall thickness, outer diameter, and more; you don't have to sort through a long list of individual pipe definitions.

Part Builder Orientation

Each drawing "remembers" which part catalog it is associated with. If you're in a metric drawing, you need to make sure the catalog is mapped to metric pipes and structures, whereas if you're in an imperial drawing, you'll want the Imperial catalog. By default, the Civil 3D templates should be appropriately mapped, but it's worth the time to check. Set the catalog by changing to the Home tab and selecting the drop-down on the Create Design panel. Verify the appropriate folder and catalog for your drawing units in the Pipe Network Catalog Settings dialog (see Figure 13.98), and you're ready to go.

Understanding the Organization of Part Builder

The vocabulary used in the Part Builder interface is different from the rest of Civil 3D, so we will first examine the basics.

FIGURE 13.98
Choose the appropriate folder and catalog for your drawing units.

Open Part Builder by going to the Home tab ➢ Create Design panel and selecting the Part Builder icon from the drop-down.

The first screen that appears when you start the Part Builder is Getting Started – Catalog Screen (see Figure 13.99).

FIGURE 13.99
The Getting Started – Catalog Screen dialog

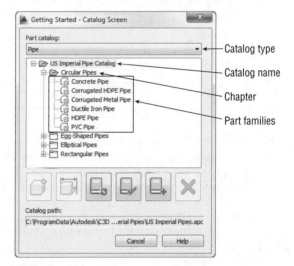

At the top of this window is a drop-down for selecting the part catalog. Depending on what you need to modify, you can set this to either Pipe or Structure. Once you pick the catalog type, you will see the main catalog name. In Figure 13.99, you see US Imperial Pipe Catalog.

Below the part catalog is a listing of chapters. In Part Builder vocabulary, a *chapter* is a grouping based on the shape. Inside the chapters, you will see the part families listed. The US Imperial Pipe Catalog has four default chapters:

- Circular Pipes
- Egg-Shaped Pipes
- Elliptical Pipes
- Rectangular Pipes

You can create new chapters for different-shaped pipes, such as Arch Pipe.

The US Imperial Structure Catalog also has four default chapters: Inlets-Outlets, Junction Structures With Frames, Junction Structures Without Frames, and Simple Shapes. You can create new chapters for custom structures. You can expand each chapter folder to reveal one or more part families. For example, the US Imperial Circular Pipe Chapter has six default families:

- Concrete Pipe
- Corrugated HDPE Pipe
- Corrugated Metal Pipe
- Ductile Iron Pipe
- HDPE Pipe
- PVC Pipe

Pipes that reside in the same family typically have the same parametric behavior, with differences only in size.

As Table 14.1 shows, a series of buttons on the Getting Started – Catalog Screen dialog lets you perform various edits to chapters, families, and the catalog as a whole.

TABLE 14.1: The Part Builder catalog tools

ICON	FUNCTION
	The New Parametric Part button creates a new part family.
	The Modify Part Sizes button allows you to edit the parameters for a specific part family.
	The Catalog Regen button refreshes all the supporting files in the catalog when you've finished making edits to the catalog.
	The Catalog Test button validates the parts in the catalog when you've finished making edits to the catalog.
	The New Chapter button creates a new chapter.
	The Delete button deletes a part family. Use this button with caution, and remember that if you accidentally delete a part family, you can restore your backup catalog as mentioned in the beginning of this section.

Exploring Part Families

The best way to become oriented to the Part Builder interface is to explore one of the standard part families:

1. Open Part Builder; and then in the Getting Started – Catalog Screen dialog, click the Part Catalog drop-down and select Pipe.

2. Expand the US Imperial Pipe Catalog ➢ Circular Pipe ➢ Concrete Pipe family, select Concrete Pipe, and click the Modify Part Sizes button.

3. It is not unusual to receive a message stating that the file contains previous version AEC objects. If you get such a message, click Close to continue.

   A Part Builder pane will appear with `AeccCircularConcretePipe_Imperial.dwg` on the screen, along with the Content Builder Toolspace, as shown in Figure 13.100.

FIGURE 13.100
DWG file as shown in Content Builder

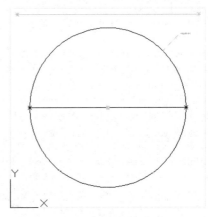

4. Close Part Builder by clicking on the X on the Part Builder pane.

5. Click No when asked to save the changes to the concrete pipe.

6. Click No when asked to save the drawing.

The Part Builder pane, or Content Builder (Figure 13.101), is well documented in the *Civil 3D User's Guide*. Please refer to the *User's Guide* for detailed information about each entry in Content Builder.

Adding a Part Size Using Part Builder

The hypothetical municipality requires a 12" (300 mm) sanitary sewer cleanout. After studying the catalog, you decide that Concentric Cylindrical Structure NF is the appropriate shape for your model, but the smallest inner diameter size in the catalog is 48" (1,200 mm). The following exercise gives you some practice in adding a structure size to the catalog—in this case, adding a 12" (300 mm) structure to the US Imperial Structures catalog.

FIGURE 13.101
Content Builder

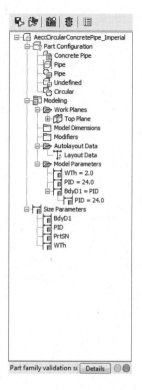

You can make changes to the US Imperial Structures catalog from any drawing that is mapped to that catalog, which is probably any imperial drawing you have open, as follows:

1. For this exercise, start a new drawing from _AutoCAD Civil 3D (Imperial) NCS.dwt (_AutoCAD Civil 3D (Metric) NCS.dwt).

2. On the Home tab, select the drop-down on the Create Design panel and select Part Builder.

3. In the Getting Started – Catalog Screen dialog, choose Structure from the drop-down in the Part Catalog selection box.

4. Expand the Junction Structures Without Frames chapter.

5. Highlight the Concentric Cylindrical Structure NF (no frames) part family.

6. Click the Modify Part Sizes button.

    The Part Builder interface opens AeccStructConcentricCylinderNF_Imperial.dwg along with the Content Builder pane.

7. Zoom extents if necessary.

8. In the Context Builder pane, expand the Size Parameters branch.

9. Right-click the SID (Structure Inner Diameter) parameter, and choose Edit.

    The Edit Part Sizes dialog appears.

10. Locate the SID column (see Figure 13.102), and double-click inside the box.

FIGURE 13.102
Examining the available part sizes

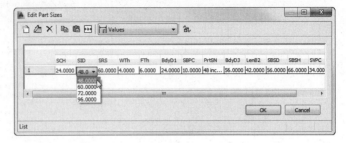

Note that a drop-down shows the available inner diameter sizes 48″, 60″, 72″, and 96″ (1,200 mm, 1,400 mm, 1,600 mm, and 1,800 mm).

11. Locate the Edit button. Make sure you're still active in the SID column cell, and then click Edit.

    The Edit Values dialog appears.

12. In the Edit Values dialog, click Add and type **12** (metric users type **300**), as shown in Figure 13.103.

FIGURE 13.103
Add the 12″ (300 mm) value to the Edit Values dialog

13. Click OK to close the Edit Values dialog, and click OK again to close the Edit Part Sizes dialog.

14. Click the small X in the upper-right corner of the Content Builder pane to exit Part Builder.

15. The message "Save Changes To Concentric Cylindrical Structure NF?" will appear. Click Yes.

    You could also click Save in Content Builder to save the part and remain active in the Part Builder interface.

You're back in your original drawing. If you created a new parts list in any drawing that references the US Imperial Structures catalog, the 12″ (300 mm) structure will be available for selection.

Sharing a Custom Part

You may need to go beyond adding pipe and structure sizes to your catalog and build custom part families or even whole custom chapters. Perhaps instead of building them yourself, you'll be able to acquire them from an outside source.

The following section can be used as a reference for adding a custom part to your catalog from an outside source, as well as sharing custom parts that you've created. The key to sharing a part is to locate the three files mentioned earlier.

Adding a custom part size to your catalog requires these steps:

1. Locate the `partname.dwg`, `partname.xml`, and (optionally) `partname.bmp` files of the part you'd like to obtain.

2. Make a copy of the `partname.dwg`, `partname.xml`, and (optionally) `partname.bmp` files.

3. Insert the `partname.dwg`, `partname.xml`, and (optionally) `partname.bmp` files in the correct folder of your catalog.

4. Run the **PARTCATALOGREGEN** command in Civil 3D.

Adding an Arch Pipe to Your Part Catalog

This exercise will teach you how to add a premade custom part to your catalog. You can make changes to the US Imperial Pipes catalog from any drawing that is mapped to that catalog, which is probably any imperial drawing you have open.

1. For this exercise, start a new drawing from the _AutoCAD Civil 3D (Imperial) NCS.dwt file (_AutoCAD Civil 3D (Metric) NCS.dwt).

2. In Windows Explorer, create a new folder called **Arch Pipes** in your backup directory.

   For the purposes of the exercise, you should add the new pipe files into a backup directory. Remember the real path is located here:

   `C:\ProgramData\Autodesk\C3D 2014\enu\Pipes Catalog\US Imperial Pipes\`

   `(C:\ProgramData\Autodesk\C3D 2014\enu\Pipes Catalog\Metric Pipes\)`

   This directory should now include five folders: `Arch Pipes, Circular Pipes, Egg-Shaped Pipes, Elliptical Pipes,` and `Rectangular Pipes`.

3. Copy the `Concrete Arch Pipe.dwg (Concrete Arch Pipe_METRIC.dwg), Concrete Arch Pipe.bmp (Concrete Arch Pipe_METRIC.bmp),` and `Concrete Arch Pipe.xml(Concrete Arch Pipe_METRIC.xml)` files into the `Arch Pipes` folder.

4. Return to your drawing, and enter **PARTCATALOGREGEN** at the command line.

5. Type P and press ↵ to regenerate the Pipe catalog, and then press ↵ again to exit the command.

   If you created a new parts list at this point in any drawing that references the US Imperial Pipes catalog, the arch pipe would be available for selection.

6. To confirm the addition of the new pipe shape to the catalog, locate the catalog HTML file at:

   `C:\ProgramData\Autodesk\C3D 2014\enu\Pipes Catalog\US Imperial Pipes\Imperial Pipes.htm`

The Bottom Line

Create a pipe network by layout. After you've created a parts list for your pipe network, the first step toward finalizing the design is to use Pipe Network By Layout.

Master It Open the MasterIt1301.dwg or MasterIt1301_METRIC.dwg file. From the Home tab ➢ Create Design panel ➢ Pipe Network drop-down, select Pipe Network Creation Tools to create a sanitary sewer pipe network named Mastering. Use the Composite surface, and name only structure and pipe label styles. Don't choose an alignment at this time. Create 8" (200 mm) PVC pipes and a manhole called SMH. There are blocks in the drawing to assist you in placing manholes. Begin at the START HERE marker, and place a manhole at each marker location. You can erase the markers when you've finished.

Create an alignment from network parts and draw parts in profile view.

Once your pipe network has been created in plan view, you'll typically add the parts to a profile view based on either the road centerline or the pipe centerline.

Master It Continue working in the MasterIt1301.dwg or MasterIt1301_METRIC.dwg file. Create an alignment from your pipes so that station zero is located at the START HERE structure. Create a profile view from this alignment, and show the pipes on the profile view.

Label a pipe network in plan and profile. Designing your pipe network is only half of the process. Engineering plans must be properly annotated.

Master It Continue working in the MasterIt1301.dwg or MasterIt1301_METRIC.dwg file. Add the Length Description And Slope style label to profile pipes and the Data With Connected Pipes (Sanitary) style to profile structures. Add the alignment created in the previous Master It to all pipes and structures.

Create a dynamic pipe table. It's common for municipalities and contractors to request a pipe or structure table for cost estimates or to make it easier to understand a busy plan.

Master It Continue working in the MasterIt1301.dwg or MasterIt1301_METRIC.dwg file. Create a pipe table for all pipes in your network. Use the default table style.

Chapter 14

Grading

Beyond creating streets, sewers, cul-de-sacs, and inlets, much of what happens to the ground as a site still needs to be determined. Planning for the earthwork of a site is a crucial part of bringing a project together. This chapter examines feature lines and grading groups, which are the two primary tools of site design. Feature lines and grading groups work in tandem, providing the site designer with tools to completely model the land.

In this chapter, you will learn to:

- Convert existing linework into feature lines
- Model a simple breakline with a feature line
- Model planar site features with grading groups

Working with Grading Feature Lines

There are two types of feature lines: corridor feature lines and grading feature lines. Corridor feature lines are discussed in Chapter 9, "Basic Corridors," and grading feature lines are the focus of this chapter. It's important to note that grading feature lines can be extracted from corridor feature lines, and you can choose whether or not to dynamically link them to the corridor.

Terrain modeling can be defined as the manipulation of triangles created by connecting points and vertices to achieve Delaunay triangulation, as discussed in Chapter 4, "Surfaces." In AutoCAD® Civil 3D®, the creation of the feature line object adds a level of control and complexity not available to 3D polylines. In this section, you'll look at the feature line, various methods of creating feature lines, some simple elevation edits, planar grading and editing functionality, and labeling of the newly created feature lines.

Accessing Grading Feature Line Tools

The Feature Line creation tools can be accessed from the Home tab's Create Design panel, as shown in Figure 14.1.

FIGURE 14.1
The Feature Line drop-down on the Create Design panel

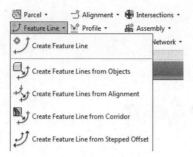

The Feature Line editing tools can be accessed via the Feature Line contextual tab (see Figure 14.2). To activate the Feature Line contextual tab, just select a feature line, or from the Modify tab ➢ Design panel, choose Feature Line.

FIGURE 14.2
The Feature Line contextual tab can be accessed by selecting an existing feature line.

When you're working with feature lines, one thing to remember is that they belong in a site, as shown in Figure 14.3. Feature lines within the same site snap to each other vertically, which can cause some confusion when you're trying to build surfaces.

FIGURE 14.3
Feature lines are located in the Sites branch in Prospector.

> **UNDERSTANDING PARENT SITES**
>
> If you notice some weird elevation data along your feature line, be sure to check out the parent site. If the concept of a site or sites in general doesn't make much sense to you, be sure to review Chapter 5, "Parcels," before going too much further. Sites are a major part of the way feature lines interact with each other, and many users who have problems with grading are the same users who ignore sites.

The next few sections break down the various tools in detail. You'll use almost all of them in this chapter, so in each section you'll spend some time getting familiar with these tools and the basic concepts behind them.

Creating Grading Feature Lines

There are five primary methods for creating feature lines, as shown previously in Figure 14.1. They generate similar results but have some key differences:

Create Feature Line The Create Feature Line tool allows you to create a feature line from scratch, assigning elevations as you go. These elevations can be based on direct data input at the command line, slope information, or surface elevations.

Create Feature Lines From Objects The Create Feature Lines From Objects tool converts lines, arcs, polylines, and 3D polylines into feature lines. This process also allows you to assign elevations. You can assign a single elevation to all vertices on the feature line. You can also acquire elevations from grading groups and surfaces if the feature line has been drawn over either. You also have the option to delete the original object or weed the vertices of the new feature line.

Create Feature Lines From Alignment The Create Feature Lines From Alignment tool allows you to build a new feature line from an alignment, using a profile to assign elevations. This feature line can be dynamically tied to the alignment and the profile, which limits your ability to edit it directly, but makes it easy to generate feature lines based on horizontal and vertical controls of alignments and profiles. If the feature line created from an alignment is not dynamically linked, any of the feature-line editing commands can be used.

Create Feature Line From Corridor The Create Feature Line From Corridor tool is used to export a grading feature line from a corridor feature line. The feature line created from a corridor feature line can be dynamically linked to the corridor.

Create Feature Line From Stepped Offset The Create Feature Line From Stepped Offset tool is used to create an offset feature line, allowing you to specify a vertical offset value in terms of elevation difference, grade, slope, actual elevation for the entire feature line or each vertex. This tool can also be used on survey figures, polylines, or 3D polylines. The feature line created from a stepped offset does not keep a dynamic link to the original feature line.

You'll explore each of these methods over the next few exercises. In this first exercise, you will create a swale from feature lines.

1. Open the `CreatingFeatureLines.dwg` or `CreatingFeatureLines_METRIC.dwg` file. (Remember, all data can be downloaded from www.sybex.com/go/masteringcivil3d2014.)

   This drawing has a polyline inset from the outer perimeter of the subdivision and a polyline at the proposed centerline of an overland swale.

2. From the Home ➢ Create Design panel, choose Feature Line ➢ Create Feature Line to display the Create Feature Lines dialog.

   You need to create a new site in which to put the swale. Just as parcels do, grading objects will react with like objects in a site if they touch. So by isolating this swale, you can ensure that everything will be drawn properly before committing it to a site.

3. Click the Create New button to the right of the site name. The Site Properties dialog appears.

4. Enter **Swale** for the name of this site, and click OK to dismiss the Site Properties dialog.

The Create Feature Lines dialog should now look like Figure 14.4.

FIGURE 14.4
The Create Feature Lines dialog

5. Click OK to accept the Create Feature Lines dialog.

6. At the `Specify start point:` prompt, use an Endpoint Osnap to pick the right end of the line (closest to the red inset line).

7. At the `Specify elevation or [Surface]:` prompt, enter **S** ↵ to use a surface to set the elevation.

   If there were multiple surfaces in the drawing, the Select Surface dialog would appear and you would use the drop-down to choose EG, and click OK. However, because you currently have only one surface in this drawing, there is no need to state what surface to use.

8. At the `Surface elevation:` prompt, press the Enter key to accept the default surface elevation offered on the command line.

9. At the `Specify the next point or [Arc]:` prompt, use an Endpoint Osnap to pick the left end of the line.

10. At the `Specify grade or [SLope Elevation Difference SUrface Transition]:` prompt, enter **-5.2** ↵ to set this grade between these two points.

11. Press the Enter key to end the command.

    Your screen should look like Figure 14.5 with a green feature line drawn over the top of the red polyline.

    This method of creating a feature line by picking each vertex and specifying its elevation before picking the next may seem tedious to some users. In the next portion of the exercise, you will convert an existing polyline to a feature line and acquire elevations for the vertices from a surface.

FIGURE 14.5
Setting the grade between points

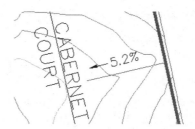

> **ASSIGNING NAMES**
>
> Looking back at Figure 14.4, note that there is an unused option for assigning a Name value to each feature line. Using names won't make it easier to pick feature line objects if you decide to use them for building a corridor object. The Name option is available for each method of creating feature line objects, but will generally be ignored for this chapter, except for one exercise covering the renaming tools available. The name option is a way to keep track of feature lines listed in the site branch on Prospector and is required when feature lines are used for targeting when building corridors.

12. From the Home tab ➢ Create Design panel, choose Feature Line ➢ Create Feature Lines From Objects.

13. At the Select lines, arcs, polylines or 3d polylines to convert to feature lines or [Xref]: prompt, notice the options for selection and note that by typing **X** and then pressing the Enter key you may pick these types of entities from an external reference. This will save you the trouble of opening the external reference, copy clipping, closing the reference, and pasting to original coordinates. Select the red closed inset polyline which represents the limits of grading.

14. Right-click and select the Enter option, or press the Enter key.

    The Create Feature Lines dialog will appear.

15. Click the Create New button to the right of the site name to display the Site Properties dialog.

16. Enter **Rough Grading** for the name of this site, and click OK to dismiss the Site Properties dialog.

    This Create Feature Lines dialog has some differences from that shown in Figure 14.4. Notably, the Conversion Options near the bottom of the dialog are now active, so take a look at the options presented:

    Erase Existing Entities This option deletes the object and replaces it with a feature line object. This avoids the creation of duplicate linework, but could be harmful if you wanted your linework for planimetric purposes.

    Assign Elevations This option lets you set the feature line elevations from a surface or grading group, essentially draping the feature line on the selected object.

    Weed Points Weed Points decreases the number of nodes along the object. This option is handy when you're converting digitized information into feature lines.

17. Check the Assign Elevations box.

    The Erase Existing Entities option is checked by default, and you will not select the Weed Points option.

18. Click OK to dismiss the Create Feature Lines dialog.

    The Assign Elevations dialog appears. Here you can assign a single elevation for the feature line, assign the elevations from a grading if one is present in the drawing, or select a surface to pull elevation data from.

19. Verify that the EG surface is selected.

20. Verify that the Insert Intermediate Grade Break Points option is checked.

    This inserts a vertical point of intersection (VPI) at every point along the feature line where it crosses an underlying TIN line.

21. Click OK to dismiss the Assign Elevations dialog.

22. Select the feature line that was just created, and the grips will look like Figure 14.6.

FIGURE 14.6
Conversion to a feature line object

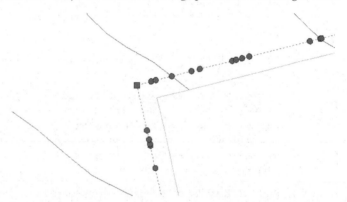

When this exercise is complete, you may close the drawing. A saved copy of this drawing (CreatingFeatureLines_FINISHED.dwg or CreatingFeatureLines_METRIC_FINISHED.dwg) is available from the book's web page.

> ### Square Grips vs. Circular Grips
>
> Note that Figure 14.6 shows two types of grips: square and circular. Feature lines offer feedback via the grip shape:
>
> **Square Grips** Square grips are PI points. They can be moved in the x, y, and z directions, manipulating both the horizontal and vertical design.
>
> **Circular Grips** Circular grips are elevation points. Elevation points can be slid along a given feature line, adjusting the vertical design, but cannot be moved in the x or y direction. In this part of the exercise, the elevation points are located where the original polyline intersected the TIN lines of the EG surface representing the intermediate grade break points you created.
>
> This combination of PIs and elevation points makes it easy to modify a feature line with a number of changes in design grade, enabling it to maintain its linear design intent if the endpoints are moved.

Both of the methods used so far assume static elevation assignments for the feature line. They're editable but are not physically related to other objects in the drawing. This is generally acceptable, but sometimes it's necessary to have a feature line that is dynamically related to another object. Sometimes it's necessary to create a profile and profile view representing the centerline of a minor road or ditch. Instead of building a corridor model as discussed in Chapter 9, "Basic Corridors," and Chapter 10, "Advanced Corridors, Intersections, and Roundabouts," a dynamic feature line can be extracted from a profile along an alignment, offset both horizontally and vertically to the right and left, and used for grading in a finished grade surface. In the following example, you will create roadway grading using this method.

1. Open the SteppedOffset.dwg or SteppedOffset_METRIC.dwg file.

2. From the Home tab ➢ Create Design panel, choose Feature Line ➢ Create Feature Lines From Alignment.

3. At the Select an alignment <or press enter key to select from list>: prompt, select the Cabernet Court alignment at the lower portion of the site (the north-south alignment) to display the Create Feature Line From Alignment dialog.

4. From the Create Feature Line From Alignment dialog, create a new site named **Grading Road**.

5. In the Create Feature Line From Alignment dialog, deselect Weed Points.

6. Leave the other default settings as shown in Figure 14.7.

    Note that the Create Dynamic Link To The Alignment check box is selected near the bottom.

FIGURE 14.7
The Create Feature Line From Alignment dialog

7. Click OK to dismiss the Create Feature Line From Alignment dialog. Note that if you select the feature line you just created, it will be listed as an Auto Feature Line (dynamic to the alignment and profile) in the Properties, whereas the feature line created in the first exercise is just a Feature Line.

8. From the Home tab ➢ Create Design panel, choose Feature Line ➢ Create Feature Line From Stepped Offset.

9. At the Specify offset distance or [Through Layer]: prompt, enter 25 ↵ (or 7.5 ↵ for metric users).

10. At the Select an object to offset: prompt, select the auto feature line along the alignment.

11. At the Specify side to offset or [Multiple]: prompt, select a point to the right of the alignment.

12. At the Specify elevation difference or [Grade Slope Elevation Variable]: prompt, type G ↵ for Grade and enter -2 ↵ at the command line.

This offsets the line from the road centerline to the location of the edge of the pavement at –2 percent.

13. Repeat steps 10 through 12, but this time, pick a point to the left of the alignment.

14. Once complete, press the Enter key to end the command.

Your results should appear as shown in Figure 14.8.

FIGURE 14.8
The completed alignment with offsets in place

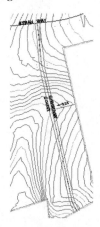

Keep this drawing open to use in the next portion of the exercise.

If you click on the alignment now, you will select the feature line that is on top of the alignment. The feature line created by the alignment won't have any grips available. This is because the feature line is dynamically linked to the design profile along the alignment and can't be modified. If either the alignment or the profile changes, the dynamic feature line will automatically update. Simply repeat the preceding procedure to create new offsets if needed. These three feature lines can be included in a new surface definition as breaklines, as discussed in Chapter 4.

Because dynamically linked auto feature line objects are slightly different, you'll look at them in this next portion of the exercise.

15. Pan or zoom to view the feature line along the Cabernet Court alignment.

16. Select the auto feature line created from the alignment to activate the Feature Line contextual tab.

17. From the Feature Line contextual tab ➢ Modify panel, choose Feature Line Properties to display the Feature Line Properties dialog, as shown in Figure 14.9.

FIGURE 14.9
The Information tab (left) and Statistics tab (right) of the Feature Line Properties dialog for an alignment-based feature line

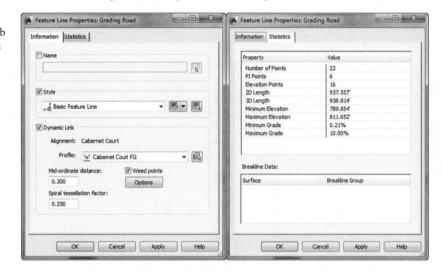

The information displayed on the Information tab is unique to the dynamic feature line, and you still have some level of control over the linking options. Notice that you can set the feature line to dynamically reference another profile based on the same alignment. You can also set tessellation factors for curves by mid-ordinate distance and spirals.

18. On the Information tab, uncheck the Dynamic Link option and click OK to dismiss the Feature Line Properties dialog.

Notice the grips appear; they weren't previously visible when the feature line was dynamically linked. The dynamic relationship between the feature line and the alignment has been broken.

19. With the feature line still selected, from the Feature Line contextual tab ➢ Modify panel, choose Feature Line Properties and notice the Dynamic Link options have disappeared.

Once the dynamic link is broken, it cannot be reconnected.

20. Click OK to dismiss the Feature Line Properties dialog and press Esc to deselect the feature line.

21. Select the two feature lines offset from the alignment feature line.

Notice that with two feature lines selected, the Feature Line contextual tab appears with only two panels.

22. With the feature lines still selected, from the Feature Line contextual tab ➢ Modify panel, choose Apply Feature Line Styles to display the Apply Feature Line Style dialog.

23. Select the Grading EP style from the Style drop-down and click OK to dismiss the dialog.

24. Press Esc to deselect the feature lines.

> **USING STYLES WITH FEATURE LINE OBJECTS**
>
> Although using styles with feature lines is optional, doing so makes it easy to recognize the type of breaklines being represented (flowline, crown, etc.). Using styles with feature lines provides another advantage: Split Point Resolution.
>
> When two feature lines in the same site cross, the point of intersection on both becomes a single split point because there cannot be more than one elevation at the same point when grading. The last feature line modified will have the winning elevation. The other feature line will spike up or down to match or join to the other's elevation at that point. One way to control this is by managing the Split Point Resolution hierarchy, which is simply a list of your feature line styles in the order of priority. The style at the top of the list will always have the winning elevation. The style at the bottom of the list will always lose.
>
> Split Point Resolution can be managed on the Prospector by expanding Sites, expanding the Feature Line site, right-clicking on the Feature Line branch, and clicking Properties. Split Point Resolution is managed on the Options tab.
>
>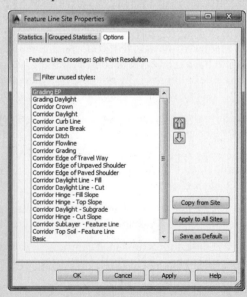

When this exercise is complete, you may close the drawing. A saved copy of this drawing (SteppedOffset_FINISHED.dwg or SteppedOffset_METRIC_FINISHED.dwg) is available from the book's web page.

Now that you've created a few feature lines, you can edit and manipulate them some more.

Editing Feature Line Information

From the Feature Line contextual tab, you can examine several more commands on the Modify panel, as shown in Figure 14.10, before you get into editing objects.

FIGURE 14.10
The Modify panel on the Feature Line contextual tab

The Modify panel of the Feature Line contextual tab provides commands for editing various properties of the feature line, the feature line style, and the feature line geometry as follows:

Feature Line Properties The top half of the Feature Line Properties tool is a button that will access the Feature Line Properties dialog. The Feature Line Properties dialog has two tabs: Information and Statistics, as shown previously in Figure 14.9. The Information tab of the Feature Line Properties dialog allows you to edit the name or feature line style and, if present, remove the dynamic link. The Statistics tab of the Feature Line Properties dialog allows you to access various physical properties such as minimum and maximum grade.

The bottom half of the Feature Line Properties tool is a drop-down containing two commands:

> **Feature Line Properties** This command accesses the same Feature Line Properties dialog as discussed earlier when you click the button.
>
> **Edit Feature Line Style** This command is used to access various display characteristics of the feature line such as color and linetype. Feature line styles will be discussed further in Chapter 19, "Object Styles."

Edit Geometry The Edit Geometry toggle opens and closes the Edit Geometry panel on the Feature Line contextual tab (see Figure 14.11). This panel will remain open until the Edit Geometry button is toggled off (it's highlighted when toggled on). The Edit Geometry panel will be discussed later in this section.

FIGURE 14.11
The Edit Geometry panel on the Feature Line contextual tab

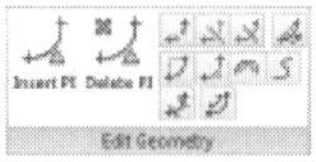

Edit Elevations The Edit Elevations toggle opens the Edit Elevations panel on the Feature Line contextual tab (see Figure 14.12). This panel will remain open until the Edit Elevations button is toggled off (it's highlighted when toggled on). The Edit Elevations panel will be discussed later in this chapter in the "Editing Feature Line Elevations" section.

FIGURE 14.12
The Edit Elevations panel on the Feature Line contextual tab

Add To Surface As Breakline   The Add To Surface As Breakline tool allows you to select a feature line or feature lines to add to a surface as breaklines. Once feature lines are added to the surface, they will be listed as a Breakline Set in Prospector ➢ Surface ➢ Definition ➢ Breakline.

Apply Feature Line Names   The Apply Feature Line Names tool allows you to change a series of feature lines en masse based on a new naming template. This tool can be helpful when you want to rename a group or just assign names to feature line objects. This tool cannot be used on an auto feature line that is dynamically linked to an alignment and profile.

Apply Feature Line Styles   The Apply Feature Line Styles tool allows you to change feature line objects and their respective styles en masse.

Move To Site   If you expand the Modify panel, you will notice two additional commands. The first command is Move To Site. This command allows you to associate the selected feature line with a new site.

Copy To Site   The other command in the extended Modify panel is Copy To Site. This command allows you to duplicate the selected feature with a new site while leaving the original feature line in its current site. The two feature lines do not remain dynamically linked.

Once the Feature Line contextual tab has been activated, the Quick Profile tool is available on the Launch Pad panel. The Quick Profile tool generates a temporary profile of the feature line based on user parameters found in the Create Quick Profiles dialog (Figure 14.13).

FIGURE 14.13
The Create Quick Profiles dialog

A few notes on this operation:

♦ Civil 3D creates a temporary phantom alignment that will not display in Prospector as the basis for a quick profile. A unique alignment number is assigned to this alignment.

◆ Panorama will display a message to tell you that a quick profile has been generated and to remind you that "this is a temporary object and will be deleted on save command or on exit from drawing." You can close Panorama or move the Panorama palette out of the way if necessary.

Feature lines aren't the only things from which you can create a quick profile. You can also create a quick profile for 2D or 3D lines or polylines, lot lines, survey figures, or even a series of points. When a quick profile is created from 3D objects, there is an additional option to draw the 3D entity profile.

The Edit Geometry and Edit Elevations toggles provide even more commands, which make them considerably more powerful than standard 3D polylines. The Edit Geometry functions and the Edit Elevations functions are described in the next sections. While these sections reference the Edit Geometry and Edit Elevations panels of the Feature Line contextual tab, these panels are also available on the Modify tab. When used from the Modify tab, many of these commands can also be used to edit parcel lines, survey figures, 3D polylines, and 2D polylines, in addition to feature lines as discussed in the sections that follow.

Editing Feature Line Geometry

Editing feature-line geometry grading revisions often requires adding PIs, breaking apart feature lines, trimming, and performing other horizontal operations without destroying the vertical information. To access the commands for editing feature-line horizontal information, select the feature line to access the Feature Line contextual tab and toggle on the Edit Geometry panel, as shown in Figure 14.14.

FIGURE 14.14
The Edit Geometry panel on the Feature Line contextual tab

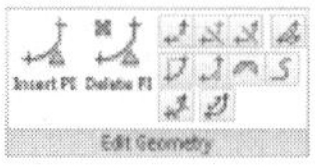

The first two tools are designed to manipulate the PI points that make up a feature line:

Insert PI   The Insert PI tool allows you to insert a new PI, controlling both the horizontal and vertical design.

Delete PI   The Delete PI tool removes a PI. If the PI is located at the beginning or end of a feature line, the beginning or end segment is removed. If the PI is removed at a PC, the arc is removed and the segment leading into the arc gets stretched to the next PI. If the PI is removed at a PT the arc segment gets stretched to the next PI.

The next few tools act like their AutoCAD® counterparts, but they are used specifically with feature lines because elevations are involved and they will add PIs accordingly:

Break   allowing two feature lines to be created from one. Additionally, if a feature line is part of a surface definition, both new feature lines are added to the surface definition to maintain integrity. Elevations at the new PIs are assigned on the basis of an interpolated elevation.

Trim   The Trim tool operates much like the AutoCAD Trim command, trimming a feature line and adding a new end PI on the basis of an interpolated elevation.

Join The Join tool creates one feature line from two, making editing and control easier. You can set the tolerance distance from the settings associated with the Join tool on the Settings tab of Toolspace by doing the following:

1. Expand the Grading ➢ Commands branch.
2. Right-click JoinFeatures and select Edit Command Settings.
3. In the Edit Command Settings – JoinFeatures dialog, expand the Feature Line Join property to change the tolerance.

Reverse The Reverse tool changes the direction of a feature line. This will change the labeling.

Edit Curve The Edit Curve tool allows you to modify the radius that has been applied to a feature line object. Once the feature line is selected, the Edit Feature Line Curve dialog will display. This dialog will allow you to step through each of the curves along the feature line. For each curve, you can modify the radius while viewing information on the curve length, chord length, and tangent length. There is also an option to maintain tangency if the curve is not already tangent to attached linework.

Fillet The Fillet tool inserts a curve at PIs along a feature line and will join feature lines sharing a common PI that are not actually connected. You can apply a radius to a single PI or to all PIs on a feature line.

The last few tools refine feature lines, making them easier to manipulate and use in surface building:

Fit Curve The Fit Curve tool analyzes a number of elevation points and attempts to define a working arc through them all. This tool is often used when the corridor utilities are used to generate feature lines. These derived feature lines can have a large number of unnecessary PIs in curved areas. You can modify the tolerance and the minimum number of segments by entering **O** to select Options on the command line during the prompt, and display the Fit Curve Options dialog.

Alternatively, you can set the default values for these options from the settings associated with the Fit Curve tool on the Settings tab of Toolspace by doing the following:

1. Expand the Grading ➢ Commands branch.
2. Right-click FitCurveFeature and select Edit Command Settings.
3. In the Edit Command Settings – FitCurveFeature dialog, expand the Feature Line Fit Curve property to change the tolerance and specify the minimum number of segments.

Smooth The Smooth tool takes a series of disjointed feature line segments and creates a best-fit curve. This tool is great for creating streamlines or other natural terrain features that are known to curve, but there's often not enough data to fully draw them that way. You can also straighten previously smoothed feature lines on the Modify tab ➢ Edit Geometry panel by choosing the Smooth command and typing **S** for Straighten. Notice that the Smooth command accessed through the Feature Line contextual tab does not give you the Straighten option.

Weed The Weed tool allows the user to remove elevation points and PIs on the basis of various criteria. Once you select the feature line or multiple feature lines, or a partial feature line, the Weed Vertices dialog will display. This dialog will allow you to weed based on any combination of angle, grade, or length; in addition, you can remove points based on their 3D distance between one another.

- Weeding by angle considers the horizontal angles at each bend of the feature line and will delete a vertex smaller than the value entered.
- Weeding by grade looks at three points, considering the middle point as a PVI. If the difference in grade is less than the entered amount, the PVI will be deleted.
- Weeding by length looks at three points and will delete the middle point if the distance between the outside points is less than the value entered.
- Weeding by 3D distance looks at the slope distance between two adjacent points and will delete the following vertex if the slope distance is less than the value entered.

At the bottom of the Weed Vertices dialog, it states how many of the total number of vertices will be weeded. This is great for cleaning up corridor-generated feature lines as well.

Similar to the Join and Fit Curve tools, you can set the default settings associated with the Weed command on the Settings tab of Toolspace by doing the following:

1. Expand the Grading ➢ Commands branch.
2. Right-click WeedFeatures and select Edit Command Settings.
3. In the Edit Command Settings – WeedFeatures dialog, expand the Feature Line Weed property to change the various values.

Stepped Offset   As discussed in detail earlier, the Stepped Offset tool allows offsetting in a horizontal and vertical direction, making it easy to create stepped features such as stairs or curbs.

In this exercise, you'll manipulate a number of feature lines that were created by corridor operations:

1. Open the FeatureLineGeometry.dwg or FeatureLineGeometry_METRIC.dwg file.

   This drawing is a continuation of the SteppedOffset.dwg file but has been populated with some more feature lines.

2. Select the southern east-west feature line (which is the offset from the Syrah Way alignment feature line), as shown in Figure 14.15.

FIGURE 14.15
Picking the southern feature line on Syrah Way

3. If necessary, from the Feature Line contextual tab ➢ Modify panel, choose Edit Geometry to toggle on the Edit Geometry panel.

4. Click the Break tool.

5. At the `Select an object to break:` prompt, select the southern east-west feature line again.

6. At the `Specify second break point or [First point]:` prompt, enter **F** ↵ to pick the first point of the break.

7. At the `Specify first break point:` prompt, using an Intersection Osnap, select the intersection of the east-west feature line to the south of Syrah Way, and the north-south feature line to the west of Cabernet Court, as shown in Figure 14.16.

FIGURE 14.16
Using the Intersection object snap to select a point

8. At the `Specify second break point:` prompt, using a Nearest Osnap, select a point on the east-west feature line to the east of the eastern north-south feature line, leaving a gap, as shown in Figure 14.17.

FIGURE 14.17
The feature line after executing the Break command

9. Select the left east-west feature line that was previously broken and notice the large number of grips.

10. From the Feature Line contextual tab ➤ Edit Geometry panel, click the Weed tool.

11. At the `Select a feature line, 3d polyline or polyline or [Multiple Partial]:` prompt, select the feature line again.

    The Weed Vertices dialog will be displayed.

12. Change the Angle to **0.2** and the Grade to **0.25**. Click in the Length field to set the Grade value. Leave other values as default.

    Watch the glyphs on the feature line and notice the number of vertices that will be removed from the feature line. Additionally, the glyphs on the feature line itself will change from green to red to reflect nodes that will be removed under the current setting, as shown in Figure 14.18.

FIGURE 14.18
Feature lines to be weeded

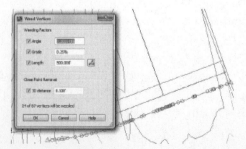

13. Click OK to complete the command and dismiss the Weed Vertices dialog. Press Esc to deselect the feature line.

14. Using a standard AutoCAD Extend command, extend the right east-west feature line to the eastern Cabernet Court offset feature line.

    There is no unique feature-line-extending tool.

15. Select the left east-west feature line that was previously broken to activate the Feature Line contextual tab.

16. From the Feature Line contextual tab ➢ Edit Geometry panel, click the Trim tool from the Edit Geometry panel.

    A standard AutoCAD trim will not work in this case.

17. At the `Select cutting edges:` prompt, select the left east-west feature line as the cutting edge and press the Enter key.

18. At the `Select objects to trim:` prompt, select the western Cabernet Court offset feature line above the cutting edge and press the Enter key to end the command and review the results (see Figure 14.19).

FIGURE 14.19
The left east-west feature line trimmed

19. From the Feature Line contextual tab ➢ Edit Geometry panel, click the Join tool.

20. At the `Select the connecting feature line, polyline or 3d polyline or [Multiple]:` prompt, select the left north-south feature line and press the Enter key.

    Notice the grips and that the two feature lines have been joined.

21. From the Feature Line contextual tab ➢ Edit Geometry panel, click the Fillet tool. Notice that the feature line is giving you a preview of the fillet with the current default radius value.

22. At the `Specify corner or [All Join Radius]:` prompt, enter **R** ↵ to adjust the radius value.

23. At the `Specify radius:` prompt, enter **15** ↵ (or **4.5** ↵ for metric users).

24. Move your cursor toward the corner created by joining the feature line until an arc-shaped glyph appears. Click to fillet the corner.

25. Press the Enter key to end the command.

    Notice the grips again. PIs (square grips) have been placed at the PC and the PT of the new curve.

26. From the Feature Line contextual tab ➢ Edit Geometry panel, click the Edit Curve tool.

27. At the Select feature line curve to edit or [Delete]: prompt, select the curve you just created.

The Edit Feature Line Curve dialog opens, as shown in Figure 14.20.

FIGURE 14.20
The Edit Feature Line Curve dialog

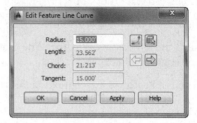

28. Enter a Radius value of **34** ↵ (or **10** ↵ for metric users), and click OK to close the dialog.
29. Press the Enter key to end the command. The drawing should now look similar to Figure 14.21.

FIGURE 14.21
Filleted feature lines

When this exercise is complete, you may close the drawing. A saved copy of this drawing (FeatureLineGeometry_FINISHED.dwg or FeatureLineGeometry_METRIC_FINISHED.dwg) is available from the book's web page.

> **MODIFYING FEATURE LINES**
>
> When you're modifying the radius of a feature line curve, it's important to remember that you must have enough tangent length on either side of the curve segment to make a curve fit; if you don't, the program will not make the change or it will create a nontangent curve. If this is the case, adjust the feature line's geometry on either side of the arc until there is a mathematical solution.
>
> Sometimes you may need to use the Weed Vertices or Delete PI tool to remove vertices and create enough room to fillet feature lines. When you're creating feature lines, you may need to plan ahead to ensure that vertices will not be placed too closely together.

EDITING FEATURE LINE ELEVATIONS

To access the commands to edit feature line elevation, from the Feature Line contextual tab ➢ Modify panel, choose the Edit Elevations toggle to open the Edit Elevations panel.

The first tool in this panel is the Elevation Editor, which will activate a palette in Panorama where you can edit station, elevation, length, and grade information about the feature line selected in a tabular grid format. When you are working in a row in the Elevation Editor, the corresponding point will be shown with a temporary triangular glyph on the plan.

A number of tools are available for modifying and manipulating feature line elevations. You will be using a few of them in the upcoming exercises. In this first exercise, you'll take a brief look at the Grading Elevation Editor tools:

1. Open the `EditingFeatureLineElevations.dwg` or `EditingFeatureLineElevations_METRIC.dwg` file.

   This drawing contains a sample layout with some curb and gutter work.

2. Select the feature line representing the left flowline of the curb and gutter area to activate the Feature Line contextual tab.

3. On the Feature Line contextual tab ➢ Edit Elevations panel, choose the Elevation Editor tool. Then press Esc to deselect the feature.

   The Grading Elevation Editor in Panorama will open, as shown in Figure 14.22.

FIGURE 14.22
The Grading Elevation Editor

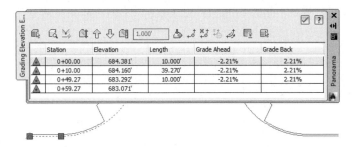

A series of symbols appears in the far left column of the Grading Elevation Editor; these are the same glyph symbols used on the feature line grips. A triangular symbol denotes a PI, and a circular symbol denotes an elevation point. In this example, you don't have any elevation points.

4. Click in the Grade Ahead column for the first PI at Station 0+00.00 (or 0+000.00).

   It's hard to see in the images, but as a row is selected in the Grading Elevation Editor, the PI or the elevation point that was selected will be highlighted on the screen with a small glyph.

   You can use the Grading Elevation Editor to make changes to the Station, Elevation, Length, Grade Ahead, and Grade Back settings. The exception is that you cannot edit stationing for primary geometry points, as indicated by the triangle glyph in the Grading Elevation Editor.

   Across the top of the Grading Elevation Editor are multiple tools that will be used as you edit the data in the table:

   Select Clicking the Select tool allows you to select the feature line on the screen for editing in the Grading Elevation Editor.

Zoom To The Zoom To tool will do exactly as it says. If a station is highlighted in the Panorama and you select the Zoom To tool, your plan view will be zoomed to that station on the feature line on the screen. When multiple stations are selected, the Zoom To tool will fit those multiple stations into your drawing window.

Quick Profile The Quick Profile tool will generate a temporary profile based on the feature line selected. This is the same tool discussed earlier that was available on the Feature Line contextual tab ➢ Launch Pad panel. The Create Quick Profiles dialog will open, allowing you to select which surface(s) you want to display, as well as what profile view style and 3D entity profile you want.

 Raise/Lower Clicking the Raise/Lower tool will activate the Set Increment text box. This will allow you to raise or lower selected station points, or if no station points are selected, it will raise or lower the entire feature line to the elevation displayed in the text box. This is an actual elevation, not an elevation difference.

 Raise Incrementally and Lower Incrementally The Raise Incrementally/Lower Incrementally tools will raise or lower the station point or points by the amount listed in the Set Increment text box. This is an elevation difference. A positive value will raise selected station points, and a negative value will lower them.

Set Increment The Set Increment tool activates the text box used with the Raise Incrementally and Lower Incrementally tools. This text box must be activated and supplied a value before proceeding with these tools.

Flatten Grade Or Elevations The Flatten Grade Or Elevations tool enables you to assign either one elevation to selected station points or a constant slope between selected station points. If you want to flatten the elevations, the elevation of the first selected station point will be applied to the other selected points. If nothing is selected, the elevation of the first point on the list will be used. If you want to apply a constant grade across a selected group of station points, the elevation of the first and last point in the selection will be held and all points in between will be flattened to the grade calculated between the first and last points. When this tool is selected, the Flatten dialog will open asking if you want to flatten by constant elevation or by constant grade.

 Insert Elevation Point The Insert Elevation Point will let you select a spot on the feature line, and will create an elevation point. The Insert PVI dialog will open, allowing you to fine-tune the station and elevation values.

 Delete Elevation Point The Delete Elevation Point will delete a point or points that are highlighted in the Grading Elevations Editor. Note that this tool will allow you to delete only elevation points.

 Elevations From Surface Clicking the Elevations From Surface tool will open the Select Surface dialog if there are multiple surfaces from which to choose. If you have selected a station point or points, it will assign elevations to only those points. If no station points are selected, the entire feature line will be draped over the surface assigning elevations to all station points.

 Reverse The Direction The Reverse The Direction tool will do exactly as it says; it will reverse the direction of the feature line, thereby changing the direction of the stationing and the grade ahead/grade back directions.

Show Grade Breaks Only The Show Grade Breaks Only tool is a toggle (click, it's on and click, it's off) that will display only the rows where the grade breaks on the feature line. Therefore, if two adjacent segments share the same grade value, the station point that they share will be hidden because the grade doesn't break on that point.

Unselect All Rows The Unselect All Rows tool does exactly as it says; it will deselect any rows that have been highlighted for editing.

5. Click the green check mark in the upper-right corner to dismiss the Panorama.

Using the Grading Elevation Editor is the most basic way to manipulate elevation information. Keep this drawing open for use in the next exercise.

Using Other Feature Line Elevation Editing Tools

To access the commands for editing feature line elevation information, select a feature line to access the Feature Line contextual tab and toggle on the Edit Elevations panel, as shown in Figure 14.23. Many of the tools in the Elevation Editor may seem redundant from the Grading Elevation Editor tools discussed earlier, but they are placed here for ease of use.

FIGURE 14.23
The Edit Elevations panel on the Feature Line contextual tab

Moving across the panel beyond the Elevation Editor tool, which was just discussed, you find the following tools for modifying or assigning elevations to feature lines:

Insert Elevation Point The Insert Elevation Point tool inserts an elevation point at the point selected or multiple elevation points at a specified increment. Note that elevation points can control only elevation information; an elevation point cannot be moved to alter the original orientation of the feature line.

Delete Elevation Point The Delete Elevation Point tool deletes the selected elevation point; the points on either side then become connected linearly on the basis of their current elevations. You can also delete all elevation points on a feature line with this command.

Quick Elevation Edit The Quick Elevation Edit tool allows you to use onscreen cues to set elevations and slopes quickly between PIs on any feature lines.

- Hover over a PI or an elevation point and a triangular glyph appears showing the elevation of that station point at the cursor. Click to trigger the prompt at the command line and type in a new elevation.

- Hover over a segment on the feature line and an arrowhead glyph appears showing the direction and grade of the segment. Click to trigger the prompt at the command line and type in a new grade.

Edit Elevations The Edit Elevations tool steps through the selected feature line, much like working through a polyline edit at the command line, allowing you to change elevations and grades or insert, move, and delete elevation points. A triangular glyph shows up at the current vertex. To skip editing the vertex and move on to the next, press the Enter key to accept the current value.

Set Grade/Slope Between Points   The Set Grade/Slope Between Points tool sets a continuous slope along the feature line between selected points. As you select the points, you will be asked to specify or verify the elevation at each point.

Insert High/Low Elevation Point   The Insert High/Low Elevation Point tool places a new elevation point on the basis of two picked points, the grade ahead of the first point, and the grade behind the second point. This calculated point is placed at the intersection of two vertical slopes.

Raise/Lower By Reference   The Raise/Lower By Reference tool allows you to adjust a feature-line station point vertically based on a grade, slope, or elevation difference in relation to a point on another 3D object. This relationship isn't dynamic!

Set Elevation By Reference   The Set Elevation By Reference tool at first looks like the same command as the Raise/Lower By Reference tool. The differences are that you can execute this tool on multiple points on the feature line without leaving the command and you can insert new elevation points. This relationship isn't dynamic!

Adjacent Elevations By Reference   The Adjacent Elevations By Reference tool allows you to adjust the elevation of multiple vertices on a feature line by specifying grade, slope, or elevation difference from another 3D object. If the 3D object has length, elevations are projected to the feature-line station points perpendicularly from the 3D object. This relationship isn't dynamic!

Grade Extension By Reference   The Grade Extension By Reference tool allows you to set the grade, slope, or elevation difference between feature line segments across a gap. For example, you might use this tool along the back of curbs at locations such as driveways or intersections. This relationship isn't dynamic!

Elevations From Surface   The Elevations From Surface tool sets the elevation at each PI and elevation point on the basis of the selected surface. It will optionally add elevation points at any point where the feature line crosses a surface TIN line.

Raise/Lower   The Raise/Lower tool simply moves the entire feature line in the *z* direction by an amount entered at the command line. A positive number raises and a negative number lowers.

Some of the relative elevation tools are a bit harder to understand, so you'll look at them in the next exercise and see how they function in some basic scenarios:

1. If not opened for the previous exercise, open the EditingFeatureLineElevations.dwg or EditingFeatureLineElevations_METRIC.dwg file.

   This drawing contains a sample layout with curb and handicap ramps.

2. Zoom to the ramp shown to the west of the intersection.

3. Select the feature line representing the west ramp to activate the Feature Line contextual tab.

4. If the Edit Elevations panel isn't already displayed, from the Feature Line contextual tab ➢ Modify panel, choose Edit Elevations.

5. From the Feature Line contextual tab ➤ Edit Elevations panel, choose the Elevation Editor tool to display the Grading Elevation Editor tab in Panorama.

   Notice that the entire feature line is at elevation 0.000′ (0.000 m).

6. Click the green check mark in the upper right to close the Panorama.

7. Press Esc to deselect the left ramp feature line.

8. To activate the Feature Line contextual tab, select the feature line representing the flow-line of the curb to the west.

9. From the Feature Line contextual tab ➤ Edit Elevations panel, choose the Adjacent Elevations By Reference tool.

10. At the Select object to edit or [Name]: prompt, select the left ramp feature line.

    Civil 3D will display a number of glyphs and lines to represent the points along the flow-line from which it is establishing elevations.

11. At the Specify elevation difference or [Grade Slope]: prompt, enter **0.5** ↵ (or **0.15** ↵ for metric users) at the command line to update the ramp elevations.

12. Press the Enter key again to end the command.

13. Press Esc to deselect the flowline of the curb feature line, and then select the ramp feature line.

14. From the Feature Line contextual tab ➤ Edit Elevations panel, choose the Elevation Editor tool to display the Grading Elevation Editor tab in Panorama.

    Notice that each of the PIs now has an elevation, as shown in Figure 14.24.

FIGURE 14.24
Completed editing of the curb ramp feature line

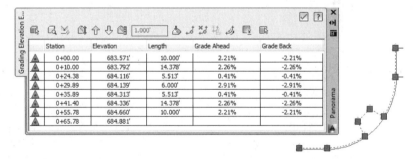

15. Click the green check mark in the upper right of the Panorama to dismiss it.

16. Next, you'll need to extend the grade on the flowline of the west curb toward the flowline of the east curb. You'll use the Grade Extension By Reference tool in the following portion of the exercise to accomplish this.

17. Select the feature line representing the west flowline.
18. From the Feature Line contextual tab ➤ Edit Elevations panel, choose the Grade Extension By Reference tool.
19. At the `Select reference segment:` prompt, select the west flowline feature line again (see Figure 14.22).

    This tool will evaluate the feature line as if it were three separate components (two lines and an arc). Because you are extending the grade of the flowline to the east and across the intersection, it is important to select the tangent segment, as shown in Figure 14.25. If you select the wrong segment, press Esc and then select the correct segment.

FIGURE 14.25
Selecting the flowline of the east curb

20. At the `Select object to edit or [Name]:` prompt, select the line representing the west flowline of the curb east, as shown in Figure 14.26.

FIGURE 14.26
Selecting the flowline of the west curb segment

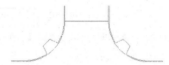

21. At the `Specify point:` prompt, pick the PI at the left side of the tangent, as shown in Figure 14.27.

FIGURE 14.27
Selecting the PI at the left side of the tangent segment representing the flowline of the east curb segment

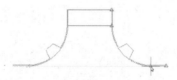

22. At the `Specify grade or [Slope Elevation Difference] <-2.21>:` prompt, press the Enter key to accept the default value of -2.21 (this is the grade of the flowline of the west curb segment).
23. Press the Enter key to end the command, and press Esc twice to cancel the command and deselect the feature line.
24. Select the feature line representing the flowline of the curb on the east side of the intersection to enable grips.
25. Move your cursor over the top of the PI, as shown in Figure 14.28, but do not click.

FIGURE 14.28
The x-, y-, and z-coordinates of the PI displayed on the status bar

Your cursor will automatically snap to the grip, and the grip will change color. Notice that the elevation of the PI is displayed on the status bar, as shown in Figure 14.28. This is a quick way to check elevations of vertices when modeling terrain.

If your coordinates are not displayed,

   a. Enter the AutoCAD command **COORDS**.

   b. Set the value to **1** to display them.

With the elevation of a single point determined, the grade of the feature line representing the flowline of the curb can be modified to ensure positive water flow. Next, you'll use the Set Grade/Slope Between Points tool to modify the grade of this feature line.

26. Select the feature line representing the flowline of the curb to the east of the intersection.

27. From the Feature Line contextual tab ➢ Edit Elevations panel, choose Set Grade/Slope Between Points.

28. At the `Specify the start point:` prompt, select the PI as shown previously in Figure 14.27.

The elevation of this point has been established and will be used as the basis for grading the entire feature line.

29. At the `Specify elevation:` prompt, press the Enter key to accept the default value.

This is the current elevation of the PI as established earlier.

30. At the `Specify the end point:` prompt, select the PI as shown in Figure 14.29.

FIGURE 14.29
Specifying the PI to establish the elevation at the flowline of the curb and gutter section

31. At the `Specify grade or [Slope Elevation Difference]:` prompt, enter 2 ↵ to set the grade between the points at 2 percent. Do not end the command.

32. At the `Select object:` prompt, select the feature line representing the flowline of the curb to the right again.

33. At the Specify the start point: prompt, pick the PI (shown earlier in Figure 14.27) again.

34. At the Specify elevation: prompt, press the Enter key to accept the default value.

    This is the current elevation of the PI as established earlier.

35. When you see the Specify the end point: prompt, pick the PI at the far bottom right along the feature line currently being edited.

36. At the Specify grade or [Slope Elevation Difference]: prompt, enter 2 to set the grade between the points at negative 2 percent.

37. Press the Enter key to end the command but do not cancel grips.

38. Select the Elevation Editor tool from the Edit Elevations panel to display Panorama.

    Notice the values in both the Grade Ahead and Grade Back columns, as shown in Figure 14.30.

FIGURE 14.30
The grade of the feature line set to 2 percent

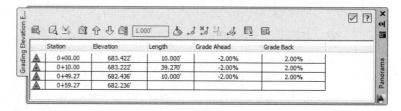

To finish the intersection, you could use the Adjacent Elevations By Reference tool again to define the right ramp.

When this exercise is complete, you may close the drawing. A saved copy of this drawing (EditingFeatureLineElevations_FINISHED.dwg or EditingFeatureLineElevations_METRIC_FINISHED.dwg) is available from the book's web page.

Using these feature line elevation editing tools, the possibilities are endless. When using feature lines to model proposed features, you are limited only by your creative approach. You've seen many of the tools in action, so you can now put a few more of them together and grade a pond.

Draining the Pond

A combination of feature line tools can be used to arrive quickly at a grading solution. These tools provide a lot of flexibility when grading site features such a ponds, berms, or parking lots. In this exercise, you will use feature lines to design your pond:

1. Open the PondDrainageDesign.dwg or PondDrainageDesign_METRIC.dwg file.

   The engineer gave us a bit more information about the pond design, as shown in Figure 14.31.

FIGURE 14.31
A feature line at the pond basin and a feature line for the outlet channel

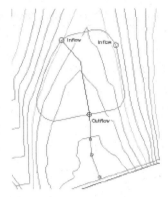

There is a feature line outlining the pond basin, which has been assigned a constant elevation of 0, and a feature line representing the outlet channel, which has been assigned elevations from the existing surface. In addition, the engineer has provided us with the Inflow and Outflow locations.

2. Select the pond-basin feature line to activate the Feature Line contextual tab.
3. From the Feature Line contextual tab ➢ Edit Elevation panel (toggle it on if needed), choose the Insert Elevation Point tool.
4. Use the Center Osnap to insert an elevation point in the center of the circle at the east inflow point.
5. Enter **779.5′ (237.6 m)** as the elevation.
6. Press the Enter key to end the command, and press Esc to cancel the grips.
7. From the Home tab ➢ Create Design panel, choose Feature Line ➢ Create Feature Lines From Objects.
8. Pick the polyline that represents the pilot channel from the northwestern inflow to the outflow channel centerline and press the Enter key.

   The Create Feature Lines dialog appears.

9. Verify that Site is set to Pond and that the Assign Elevations check box is unchecked, and click OK.

   Because this feature line was created in the same site as the bottom of the pond, the elevation of the PI at the southernmost PI will reset to match the elevation of the shared point of the outflow channel.

10. Select the feature line you just created representing the pilot channel to activate the Feature Line contextual tab.

11. From the Feature Line contextual tab ➤ Edit Geometry panel (toggle it on if needed), choose the Fillet tool.
12. At the Specify corner or [All Join Radius]: prompt, enter **R** ↵ and then enter **25** ↵ (**7.5** ↵ for metric users) for the radius.
13. Select the corner of the feature line in the center of the pond to fillet the PI.
14. Press the Enter key to exit the Fillet command.
15. From the Feature Line contextual tab ➤ Edit Elevations panel, choose the Set Grade/Slope Between Points tool.
16. At the Specify the start point: prompt, select the PI at the inflow.
17. At the Specify elevation: prompt, enter **779.5** ↵ (**237.6** ↵ for metric users) to set this elevation.
18. At the Specify the end point: prompt, select the other end of the feature line, as shown in Figure 14.32.

FIGURE 14.32
Setting grade between points

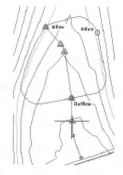

All the PIs will highlight, and Civil 3D will display the total length, elevation difference, and average slope at the command line.

19. At the Specify grade or [Slope Elevation Difference]: prompt, press the Enter key again to accept the grade as shown.

This completes a linear slope from one end of the feature line to the other, ensuring drainage through the pond and outfall structure.

20. Press the Enter key to end the command.
21. Press Esc to cancel the grips.
22. Select the feature line representing the pond basin to activate the Feature Line contextual tab.
23. From the Feature Line contextual tab ➤ Edit Elevations panel, choose the Elevation Editor tool to display the Grading Elevation Editor in Panorama.

24. Find the elevation point representing the outfall in the Elevation Editor. Click on it in the Station column to display the glyph on the feature line to ensure you are at the correct location and observe the elevation, as shown in Figure 14.33.

FIGURE 14.33
The Grading Elevation Editor

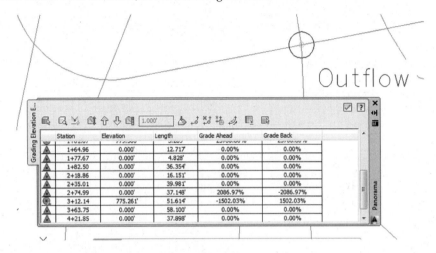

In this case, the elevation is 775.261′ (236.302 m). When done observing the elevations, you may close the Panorama.

25. From the Feature Line contextual tab for the feature line representing the pond basin, choose the Set Grade/Slope Between Points tool from the Edit Elevations panel.

26. At the `Specify the start point:` prompt, move your cursor over the elevation point at the northwest inflow and select it.

27. Press the Enter key to accept the default elevation.

28. At the `Specify the end point:` prompt, move your cursor counterclockwise (the glyphs will appear at the vertices while tracing your path) and select the elevation point at the outflow.

29. Press the Enter key to accept the grade.

This sets the elevations between the one inflow and the outflow elevation points to follow a constant grade.

30. At the `Select object:` prompt, select the feature line representing the pond bottom again.

31. Repeat steps 26 to 29 for the elevation point on the other inflow (located on the northeast side of the pond) and the outflow located clockwise from the start point.

This sets the elevations between the other inflow and the outflow elevation points to a constant slope.

32. When done, press the Enter key to end the command.

    The entire outline of the pond bottom is graded except the area between the two inflows, as shown in Figure 14.34.

    Because you want to avoid a low spot between the two inflows, you'll now force a high point:

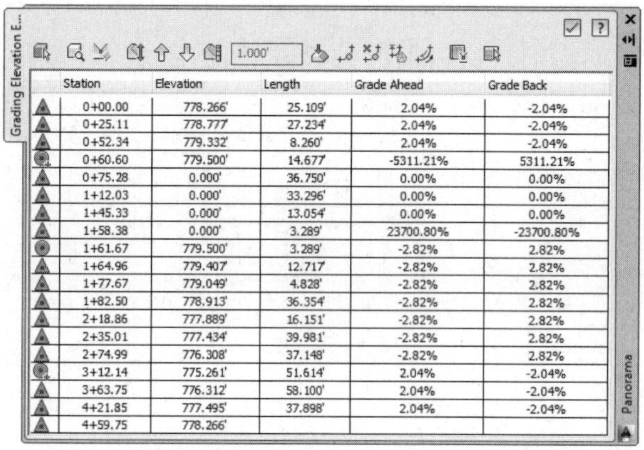

FIGURE 14.34
Zero elevation remains between the two inflows

Station	Elevation	Length	Grade Ahead	Grade Back
0+00.00	778.266'	25.109'	2.04%	-2.04%
0+25.11	778.777'	27.234'	2.04%	-2.04%
0+52.34	779.332'	8.260'	2.04%	-2.04%
0+60.60	779.500'	14.677'	-5311.21%	5311.21%
0+75.28	0.000'	36.750'	0.00%	0.00%
1+12.03	0.000'	33.296'	0.00%	0.00%
1+45.33	0.000'	13.054'	0.00%	0.00%
1+58.38	0.000'	3.289'	23700.80%	-23700.80%
1+61.67	779.500'	3.289'	-2.82%	2.82%
1+64.96	779.407'	12.717'	-2.82%	2.82%
1+77.67	779.049'	4.828'	-2.82%	2.82%
1+82.50	778.913'	36.354'	-2.82%	2.82%
2+18.86	777.889'	16.151'	-2.82%	2.82%
2+35.01	777.434'	39.981'	-2.82%	2.82%
2+74.99	776.308'	37.148'	-2.82%	2.82%
3+12.14	775.261'	51.614'	2.04%	-2.04%
3+63.75	776.312'	58.100'	2.04%	-2.04%
4+21.85	777.495'	37.898'	2.04%	-2.04%
4+59.75	778.266'			

33. From the Feature Line contextual tab for the pond feature line, choose the Insert High/Low Elevation Point tool from the Edit Elevations panel.

34. At the `Specify the start point:` prompt, select the elevation point at the northwest inflow.

35. At the `Specify the end point:` prompt, select the elevation point at the northeast inflow as the endpoint.

36. At the `Specify grade ahead or [Slope]:` prompt, enter **1.0** ↵.

37. At the `Specify grade back or [Slope]:` prompt, enter **1.0** ↵ as the grade behind. A new elevation point will be created, as shown in Figure 14.35.

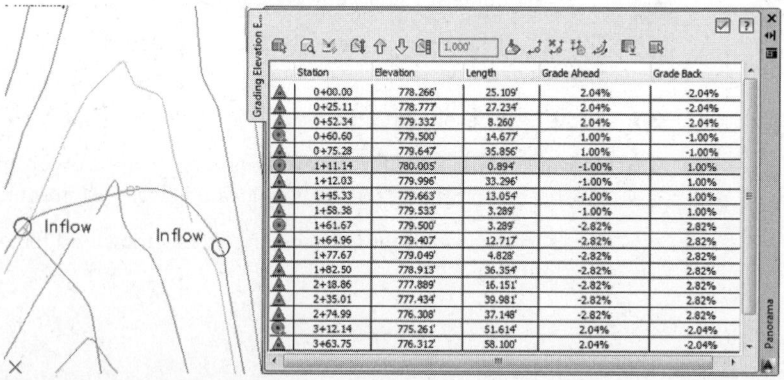

FIGURE 14.35
Using the Grade Back column in the Grading Elevation Editor

Station	Elevation	Length	Grade Ahead	Grade Back
0+00.00	778.266'	25.109'	2.04%	-2.04%
0+25.11	778.777'	27.234'	2.04%	-2.04%
0+52.34	779.332'	8.260'	2.04%	-2.04%
0+60.60	779.500'	14.677'	1.00%	-1.00%
0+75.28	779.647'	35.856'	1.00%	-1.00%
1+11.14	780.005'	0.894'	-1.00%	1.00%
1+12.03	779.996'	33.296'	-1.00%	1.00%
1+45.33	779.663'	13.054'	-1.00%	1.00%
1+58.38	779.533'	3.289'	-1.00%	1.00%
1+61.67	779.500'	3.289'	-2.82%	2.82%
1+64.96	779.407'	12.717'	-2.82%	2.82%
1+77.67	779.049'	4.828'	-2.82%	2.82%
1+82.50	778.913'	36.354'	-2.82%	2.82%
2+18.86	777.889'	16.151'	-2.82%	2.82%
2+35.01	777.434'	39.981'	-2.82%	2.82%
2+74.99	776.308'	37.148'	-2.82%	2.82%
3+12.14	775.261'	51.614'	2.04%	-2.04%
3+63.75	776.312'	58.100'	2.04%	-2.04%

38. When done, press the Enter key to end the command, and press Esc to deselect the feature line.

When this exercise is complete, you may close the drawing. A saved copy of this drawing (`PondDrainageDesign_FINISHED.dwg` or `PondDrainageDesign_METRIC_FINISHED.dwg`) is available from the book's web page.

By using all the tools in the Feature Lines toolbar, you can quickly grade elements of your design and pull them together in a surface. If you have difficulty getting all the elevations in this exercise to set as they should, slow down, and make sure you are moving your mouse in the right direction when setting the grades by slope. It's easy to get the calculation performed around the other direction—that is, clockwise versus counterclockwise. It helps to trace your mouse along the feature line in the direction you want the slope to be maintained. This procedure seems to involve a lot of steps, but it takes less than a minute in practice.

There are roughly 25 ways to modify feature lines using both the Edit Geometry and Edit Elevations panels of the Feature Line contextual tab. Take a few minutes and experiment with them to understand the options and tools available for these essential grading elements. By manipulating the various pieces of your Feature Line collection, it's easier than ever to create dynamic modeling tools that match the designer's intent.

Labeling Feature Lines

Feature lines can be labeled with their overall length, segment length, vertex elevation, and segment grade or slope. This is helpful to the designer because as you change your design, you can monitor geometry values, grades, and elevations since the labels update as you go. In the next couple of exercises, you'll label a few critical points on your pond design to help you better understand the drainage patterns.

Feature lines do not have their own unique label styles, but do use Line and Curve labels under General styles. You can learn more about label styles in Chapter 18, "Label Styles." The templates that ship with Civil 3D contain styles for labeling segment slopes, so you'll label the grades of feature line segments in the following exercise:

1. Open the `LabelingFeatureLines.dwg` or `LabelingFeatureLines_METRIC.dwg` file.

2. From the Annotate tab ➢ Labels & Tables panel, click the Add Labels button to display the Add Labels dialog.

3. In the Add Labels dialog, do the following:

    a. Set the Feature drop-down to Line And Curve.

    b. Set Label Type to Single Segment.

    c. Set Line Label Style to Grade Only.

    d. Set Curve Label Style to Grade Only.

    When complete, the dialog should look like Figure 14.36.

FIGURE 14.36
Adding feature line grade labels

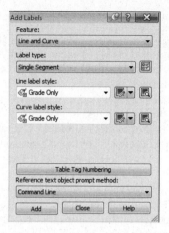

4. Click the Add button.
5. At the `Select point on entity:` prompt, pick a few points along the pilot channel feature line (the feature line that goes across the bottom of the pond) tangents to create labels, as shown in Figure 14.37.

FIGURE 14.37
Feature line grade labels in the Imperial drawing

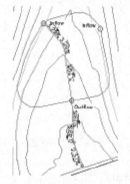

You can also try to change the slopes of some of the segments in Elevation Editor to see what happens to the labels.

When this exercise is complete, you may close the drawing. A saved copy of this drawing (`LabelingFeatureLines_FINISHED.dwg` or `LabelingFeatureLines_METRIC_FINISHED.dwg`) is available from the book's web page.

Feature line vertex elevations can be labeled with a general line or curve label style. However, applying these labels can be tedious if you want them anchored at the vertex. If you are simply using these labels as a design check, Figure 14.38 shows intelligent line and curve labels that are easily applied and communicate pertinent design properties.

FIGURE 14.38
Example of design check labels for feature lines

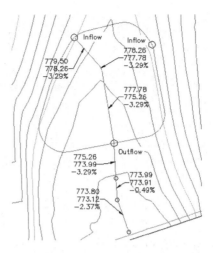

Grading Objects

Once a feature line is created for site grading, there are two main uses. One is to incorporate the feature line itself directly into a surface object as a breakline; the other is to create a grading object (referred to hereafter as simply a *grading* or *gradings*) using the feature line as a baseline. A grading consists of a baseline with elevation information and grading criteria, which describes how a slope projects from the baseline feature line. This projection can be defined by distance, slope or grade, relative elevation, or actual elevation. Surfaces can also be part of the criteria. Grading criteria can be defined and stored in grading criteria sets for ease of management. Gradings can be stylized to conform to the aesthetics of your CAD standards or convey information such as cut or fill.

Gradings are stored on the Prospector within Sites in a branch called *grading groups*. Gradings produce feature lines along the limits of each projected slope. These feature lines, like corridor feature lines, cannot be edited. They can be added to surfaces and also used to create other gradings. When these grading-dependent feature lines are used to create other gradings, the additional gradings become part of the grading group. These gradings are joined and will change when the baseline feature line or a parent grading is modified.

In this section, you'll use a number of methods to create gradings, edit those gradings, and finally convert the grading group into a surface.

Creating Gradings

Let's look at grading the pond as designed. In this section, you'll look at grading groups and then create the individual gradings within the group. Grading groups act as a collection mechanism for dependent gradings, and they let Civil 3D keep track of which gradings are to act in sync with each other.

One thing to be careful of when you're working with gradings is that they are part of a site. Any feature line within that same site will react with the feature lines created by the grading. For that reason, the exercise drawing has another site called Pond Grading to be used for just pond grading.

1. Open the `GradingThePond.dwg` or `GradingThePond_METRIC.dwg` file.

2. Select the feature line representing the pond basin to activate the Feature Line contextual tab.

3. From the Feature line contextual tab ➢ expand Modify panel, choose Move To Site to display the Move To Site dialog.

4. Set the Destination Site drop-down to the site named Pond Grading and click OK.

   This will avoid interaction between the pond gradings and the pilot-channel feature line you laid out earlier.

5. With the feature line still selected, on the Feature Line contextual tab ➢ Launch Pad panel, choose Grading Creation Tools.

   The Grading Creation Tools toolbar, shown in Figure 14.39, appears. The left section is focused on settings, the middle on creation, and the right on editing.

FIGURE 14.39
The Grading Creation Tools toolbar

6. On the Grading Creation Tools toolbar, click the Set The Grading Group tool to the far left of the toolbar to display the Site dialog.

7. Choose the Pond Grading site and click OK.

   The Create Grading Group dialog is displayed.

8. Enter **Pond Grading** in the Name text box, as shown in Figure 14.40, and click OK.

   You'll revisit the surface creation options in a bit.

9. On the Grading Creation Tools toolbar, click the Select Target Surface tool located next to the Select Grading Group tool to display the Select Surface dialog.

10. Select the EG surface and click OK.

FIGURE 14.40
Assign the name Pond Grading in the Create Grading Group dialog.

11. On the Grading Creation Tools toolbar, verify that the Grading Criteria is set to Grade To Elevation and click the Create Grading tool, or click the down arrow next to the Create Grading tool and select Create Grading, as shown in Figure 14.41.

FIGURE 14.41
Creating a grading using the Grade To Elevation criteria

12. At the Select the feature: prompt, select the pond outline.

    If you get a dialog asking you to weed the feature line,

    a. Select the Weed The Feature Line option.

    b. Click OK in the Weed Vertices dialog.

13. At the Select the grading side: prompt, pick a point on the outside of the pond basin to indicate the direction of the grading projections.

14. At the Apply to entire length? [Yes No]: prompt, enter Y ↵ to apply the grading to the entire length of the pond outline.

15. Enter 784 ↵ (or 239 ↵ for metric users) at the command line as the target elevation.

16. At the `Cut Format [Grade Slope]:` prompt, enter **S** ↵.

17. At the `Cut Slope:` prompt, enter **3** ↵ for a 3 horizontal to 1 vertical slope.

18. At the `Fill Format [Grade Slope]:` prompt, enter **S** ↵.

19. At the `Fill Slope:` prompt, enter **3** ↵ for a 3 horizontal to 1 vertical slope.

    The first grading is complete. The lines onscreen are part of the Grading style. In grading situations where you are projecting a slope and a surface is not involved, a cut slope implies grading upward and a fill slope implies grading downward. Because we are grading upward to an elevation from the pond basin, the value entered for Fill Slope is irrelevant.

20. Press Esc to end the command.

21. On the Grading Creation Tools toolbar, verify that the Grading Criteria is set to Grade To Distance, and click the Create Grading tool.

22. At the `Select the feature:` prompt, select the upper boundary of the grading made in step 19.

23. At the `Apply to entire length? [Yes/No]:` prompt, enter **Y** ↵ to apply to the whole length. Notice that you did not have to select a side, because there is already a grading object on one side of the selected feature line.

24. At the `Specify distance:` prompt, enter **10** ↵ (**3** ↵ for metric users) for the target distance to build the safety ledge.

25. At the `Format [Grade Slope]:` prompt, enter **G** ↵.

26. At the `Grade:` prompt, enter **0** ↵.

27. Press Esc to end the command. Your pond should look like Figure 14.42.

FIGURE 14.42
Creating a grading group

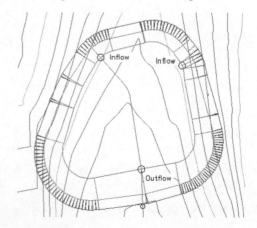

28. On the Grading Creation Tools toolbar, verify that Grading Criteria is set to Grade To Surface, and click the Create Grading tool.

29. At the Select the feature: prompt, select the outer edge of the safety ledge just created.
30. At the Apply to entire length? [Yes/No]: prompt, enter Y ↵ to apply to the whole length.
31. At the Cut Format [Grade Slope]: prompt, enter S ↵.
32. At the Cut Slope: prompt, enter 3 ↵ for a 3 horizontal to 1 vertical slope.
33. At the Fill Format [Grade Slope]: prompt, enter S ↵.
34. At the Fill Slope: prompt, enter 3 ↵ for a 3 horizontal to 1 vertical slope.
35. Press Esc to end the command.

Your drawing should look similar to Figure 14.43.

FIGURE 14.43
Completed feature line grading for the pond

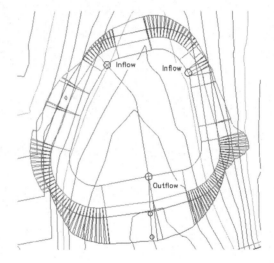

When this exercise is complete, you may close the drawing. A saved copy of this drawing (GradingThePond_FINISHED.dwg or GradingThePond_METRIC_FINISHED.dwg) is available from the book's web page.

Each piece of this pond is tied to the next, creating a dynamic model of your pond design on the basis of the designer's intent. What if that intent changes? The next section describes editing the various gradings.

Editing Gradings

Once you've created a grading, you often need to make changes. A change can be as simple as changing the slope or changing the geometric layout. In this exercise, you'll make a simple change, but the concept applies to all the gradings you've created in your pond.

1. Open the EditingGrading.dwg or EditingGrading_METRIC.dwg file.
2. Pick one of the interior projection lines or the red diamond on the interior slope of the pond in order to select the grading object.

3. From the Grading contextual tab ➤ Modify panel, choose the Grading Editor tool to display the Grading Editor in Panorama. If the correct grading object was selected in step 2, then at the top of Panorama it should state Criteria: Grade To Elevation.

4. Change the Fill Slope Projection and Cut Slope Projection both to **4**. Note that after you click out of the cell, the software will change this to read as 4.00:1 and recalculate this grading object and any other gradings based on the resulting feature line from this grading object, as shown in Figure 14.44.

FIGURE 14.44
Editing the Cut and Fill Slope values

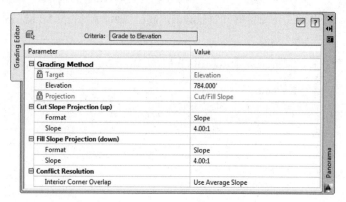

The grading may take a few seconds to update between edits.

5. Close Panorama.

Your display should look like Figure 14.45. (Compare this to Figure 14.43 if you'd like to see the difference.)

FIGURE 14.45
Completed grading edit

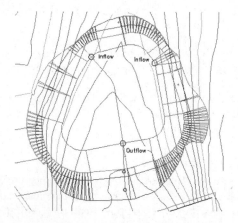

When this exercise is complete, you may close the drawing. A saved copy of this drawing (`EditingGrading_FINISHED.dwg` or `EditingGrading_METRIC_FINISHED.dwg`) is available from the book's web page.

Editing any aspect of the grading will reflect instantly, and if other gradings within the group are dependent on the results of the modified grading, they will recalculate as well.

Creating Surfaces from Grading Groups

Grading groups work well for creating the model, but you will eventually need a TIN surface to work these features into your overall grading scheme. In this section, you'll look at the conversion process and then use the built-in tools to understand the impact of your grading group on site volumes.

1. Open the CreatingGradingSurfaces.dwg or CreatingGradingSurfaces_METRIC.dwg file.
2. Pick one of the diamonds in the grading group.
3. From the Grading contextual tab ➢ Modify panel, choose the Grading Group Properties tool to display the Grading Group Properties – Pond Grading dialog.
4. On the Information tab, check the box for Automatic Surface Creation. The Create Surface dialog appears.
5. Click in the Style field, and then click the ellipsis button.
6. Select Contours 1' And 5' (Design) [or Contours 1 m And 5 m (Design) for metric users] from the drop-down in the selection box, and click OK to return to the Create Surface dialog.
7. Click OK to accept the settings in the Create Surface dialog.
8. In the Grading Group Properties – Pond Grading dialog, check the Volume Base Surface option and select the EG surface to perform a volume calculation, as shown in Figure 14.46.

FIGURE 14.46
Automatic surface creation through the Grading Group Properties

Note that this does not generate a volume surface; you will see what this check box does in a few steps.

9. Click OK to dismiss the Grading Group Properties – Pond Grading dialog. If Panorama appears, you may close it.

   You're going through this process now because you didn't turn on the Automatic Surface Creation option when you created the grading group. If you're performing straightforward gradings, that option can be a bit faster and simpler. Two options are available when you're creating a surface from a grading group. They both control the creation of projection lines in a curved area:

   - The Tessellation Spacing value controls how frequently along an arced feature line TIN points are created and projection lines are calculated. A TIN surface cannot contain any true curves the way a feature line can, because it is built from triangles. The default values typically work for site mass grading, but might not be low enough to work with things such as parking lot islands where the 10′ (3 m) value would result in too little detail.

   - The Tessellation Angle value is the degree measured between outside corners in a feature line. Corners with no curve segment must have a number of projections swung in a radial pattern to calculate the TIN lines in the surface. The tessellation angle is the angular distance between these radial projections. The typical values work most of the time, but in large grading surfaces a larger value might be acceptable, lowering the amount of data to calculate without significantly altering the final surface created.

   There is one small problem with this surface. If you examine the bottom of the pond, you'll notice there are no contours running through this area. If you move your mouse to the middle, you also won't see any Tooltip elevation because there is no surface data in the bottom portion of the pond. To fix that (and make the volumes accurate), you need a grading infill.

10. Pick one of the projection lines, or the red diamond on the inside of the pond basin grading in order to select the grading object.

11. From the Grading contextual tab ➢ Modify panel, choose the Create Grading Infill tool.

    The Select Grading Group dialog appears.

12. Verify that Site Name is set to Pond Grading, and Group Name is also set to Pond Grading, and click OK.

    The Grading Style dialog appears.

13. Verify that Grading Style is set to Cut Slope Display, and click OK.

14. At the `Select an area to infill:` prompt, hover your cursor over the middle of the pond and the pond basin feature line will be highlighted, indicating a valid area for infill.

15. Click once to create the infill, and press the Enter key to apply.

    If Panorama appears, you may close it. You should have some contours running through the pond base area, as shown in Figure 14.47.

FIGURE 14.47
The pond after applying an infill grading

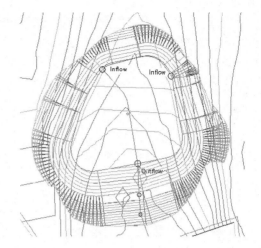

16. Zoom in if needed, and pick one of the grading diamonds again to select one of the gradings.

    Make sure you grab one of the gradings, and not the surface contours that are being drawn on top of them.

17. From the Grading contextual tab ➢ Modify panel, choose Grading Group Properties to display the Grading Group Properties – Pond Grading dialog.

18. Switch to the Properties tab to display the volume information for the pond, as shown in Figure 14.48.

FIGURE 14.48
Reviewing the grading group volumes

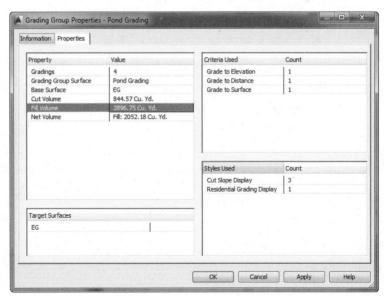

This tab also allows you to review the criteria and styles being used in the grading group.

19. Click OK to dismiss the Grading Group Properties - Pond Grading dialog.

This new surface is listed in Prospector and is based on the gradings created. A change to the gradings would affect the grading group, which would, in turn, affect the surface and these volumes.

When this exercise is complete, you may close the drawing. A saved copy of this drawing (`CreatingGradingSurfaces_FINISHED.dwg` or `CreatingGradingSurfaces_METRIC_FINISHED.dwg`) is available from the book's web page.

In the last exercise, you'll wrap things up and generate a composite surface from your grading surface and existing surface:

1. Open the `CreatingCompositeSurfaces.dwg` or `CreatingCompositeSurfaces_METRIC.dwg` file.
2. Right-click Surfaces in Prospector and select Create Surface.
3. In the Create Surface dialog, enter **Composite** in the Name text box.
4. Click in the Style field, and then click the ellipsis to display the Select Surface Style dialog.
5. Select the Elevation Banding (2D) style from the drop-down, and click OK to dismiss the Select Surface Style dialog.
6. Click OK to dismiss the Create Surface dialog and create the surface in Prospector.
7. In Prospector, expand the Surfaces ➢ Composite ➢ Definition branch.
8. Right-click Edits and select Paste Surface.

   The Select Surface To Paste dialog appears.

9. Select EG from the list and click OK. Dismiss Panorama if it appears.
10. Right-click Edits again and select Paste Surface one more time.
11. Select Pond Grading and click OK.
12. Select the new Composite surface in the drawing area, right-click, and select Display Order ➢ Send To Back.

    The drawing should look like Figure 14.49.

When this exercise is complete, you may close the drawing. A saved copy of this drawing (`CreatingGradingSurfaces_FINISHED.dwg` or `CreatingGradingSurfaces_METRIC_FINISHED.dwg`) is available from the book's web page.

FIGURE 14.49
Completed composite surface

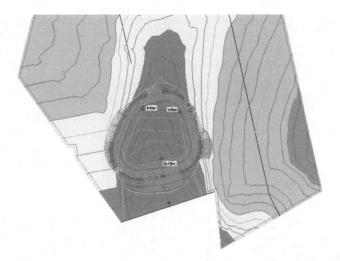

By creating a composite surface consisting of pasted-together surfaces, the TIN triangulation cleans up any gaps in the data, making contours that are continuous from the original grade, through the pond, and out the other side. With the grading group still being dynamic and editable, this composite surface reflects a dynamic grading solution that will update with any changes. Using a composite surface such as this as a reference for pipe networks is very useful, because the network may occur under both existing and proposed surfaces.

The Bottom Line

Convert existing linework into feature lines.   Many site features are drawn initially as simple linework for the 2D plan. By converting this linework to feature line information, you avoid a large amount of rework. Additionally, the conversion process offers the ability to drape feature lines along a surface, making further grading use easier.

> **Master It**   Open the MasteringGrading.dwg or MasteringGrading_METRIC.dwg file from the book's web page. Convert the magenta polyline, describing a proposed temporary swale, into a feature line and drape it across the EG surface to set elevations, and set intermediate grade break points.

Model a simple linear grading with a feature line.   Feature lines define linear slope connections or, in other words, breaklines. This can be the flow of a drainage channel, the outline of a building pad, or the back of a street curb. These linear relationships can help define grading in a model or simply enhance understanding of design intent.

> **Master It**   Edit the radius of the curve on the feature line you just created to be 100′ (30 m). Set the grade from the west end of the feature line to the next PI to 4 percent, and the remainder to a constant slope to be determined in the drawing. Draw a temporary profile view to verify the channel is below grade for most of its length.

Model planar site features with grading groups. Once a feature line defines a linear feature, gradings collected in grading groups model the slope projections from that line to other points in space. These projections can be combined to model a site much like a TIN surface, producing a dynamic design tool.

> **Master It** Use grading criteria to define the pilot channel, with grading on both sides of the sketched centerline. Define the channel using a Grading to Distance of 5′ (1.5 m) with a slope of 3:1 and connect the channel to the EG surface using a grading with slopes that are 4:1. Generate a surface from the grading group. If prompted, do not weed the feature line.

Chapter 15

Plan Production

So you've toiled for days, weeks, or maybe months creating your design in the AutoCAD® Civil 3D® program, and now it's time to share it with the world—or at least your corner of it. Even in this digital age, paper plan sets still play an important role. You generate these sheets in Civil 3D using the Plan Production feature. This chapter takes you through the steps necessary to create a sheet set, including viewport planning, generating sheets, data management, and publishing.

In this chapter, you will learn to:

- Create view frames
- Edit view frames
- Generate sheets and review Sheet Set Manager
- Create section views

Preparing for Plan Sets

Plan Production tools enable you to quickly create sheet files by automating a process you have been doing manually for years. Creating layouts, creating and orienting your viewports, inserting and filling in title blocks, establishing external and data references, creating match lines, inserting north arrows, and setting up Sheet Set Manager are all menial but necessary tasks you must undergo in order to publish your design to paper. Before you can put Plan Production tools into action, you need to address some prerequisites. Let's examine those components first.

Prerequisite Components

The Plan Production feature uses several components to create a sheet set. Here is a list of these components and a brief explanation of each. Later, this chapter will explore these elements in greater detail:

Drawing Template Plan Production creates new layouts for each sheet in a plan set. To do this, the feature uses drawing templates with predefined layout tabs. These layout tabs contain a suitable border, and up to two scaled viewports with configured types of plan, profile, or section.

For the exercises in this chapter, the default location for the final sheets will be `C:\Mastering\CH 15\Final Sheets`. It is recommended that you download all of the files (including the files in the `Final Sheets` folder) for this chapter from the book's web page (www.sybex.com/go/masteringcivil3d2014) and place them in the `C:\Mastering\CH 15` folder.

Object and Display Styles   Plan Production generates the following objects: view frames, view frame groups, match lines, and section sheets. Before creating plan sheets, you'll want to make sure you have styles set up for each of these objects. Section view group plot styles are associated with section sheet objects.

Alignments and Profiles   The Plan Production feature is used primarily for creating plan and profile views. For this to happen, your drawing must contain (or data reference) at least one alignment and profile. You must also have a sheet template ready with associated plan and profile viewports. You can produce plan-only or profile-only sheets if desired as long as you prepare a sheet template layout that supports single-plan or -profile viewports. Plan Production tools cannot produce sheet files with two or more viewports of the same type.

Sample Lines and Sections   Creating section sheets requires an alignment, a sample line group, cross sections, a group plot style, and a sheet template with associated section viewports.

With these elements in place, you're ready to dive in and create some sheets. The general steps in creating a plan set are as follows:

1. Meet the prerequisites listed previously.
2. Create view frames.
3. Create plan or plan-profile sheets.
4. (Optional) Create section view groups.
5. (Optional) Create section sheets.
6. Manage the sheets using the Sheet Set Manager.
7. Plot or publish (hard copy or digitally).

The next sections describe this process in detail and the tools used in Plan Production. The Sheet Set Manager, which is found in basic AutoCAD, is an integral part of this process.

Using View Frames and Match Lines

When you create sheets using the Plan Production tools, Civil 3D first automatically helps you divide your alignment into areas that will fit on your plotted sheet and display at the desired scale. To do this, Civil 3D creates a series of rectangular frames placed end to end (or slightly overlapping) along the length of alignment, like those in Figure 15.1. These rectangles are referred to as *view frames* and are automatically sized and positioned to meet your plan sheet requirements. This collection of view frames is referred to as a *view frame group*. Where the view frames overlap one another, Civil 3D creates *match lines* that establish continuity from frame to frame by referring to the previous or next sheet in the completed plan set. View frames and match lines are created in modelspace, using the prerequisite elements described in the previous section.

FIGURE 15.1
View frames and match lines

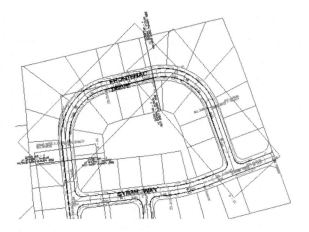

The Create View Frames Wizard

Certification Objective

The first step in the process of creating plan sets is to generate view frames. Civil 3D provides an intuitive wizard that walks you through each step of the view frame creation process. Let's look at the Create View Frames wizard and the various page options. After you've seen each page, you'll have a chance to put what you've learned into practice in an example.

From the Output tab ➢ Plan Production panel, choose Create View Frames to launch the Create View Frames Wizard (Figure 15.2). The wizard consists of several pages. A list of these pages is shown along the left sidebar of the wizard, and an arrow indicates which page you're currently viewing. You move among the pages using the Next and Back navigation buttons along the bottom of each page. Alternatively, as with all wizards, you can jump directly to any page by clicking its name in the list on the left. The following sections walk you through the pages of the wizard and explain their features.

FIGURE 15.2
The Create View Frames – Alignment wizard page

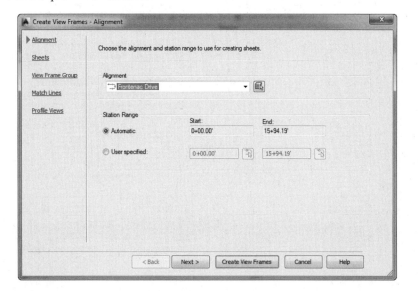

Create View Frames – Alignment Page

You use the first page of the Create View Frames Wizard (shown previously in Figure 15.2) to select the alignment and station range along which the view frames will be created.

Alignment In the top area of this page, you select the alignment along which you want to create view frames. You can either select it from the drop-down list or click the Select From The Drawing button to select the alignment on screen.

Station Range In the Station Range area of the page, you define the station range over which the frames will be created. Selecting Automatic creates frames from the alignment start to the alignment end. Selecting User Specified lets you define a custom range, by either keying start and end station values in the appropriate box or by clicking the button to the right of the station value fields and graphically selecting the station from the drawing.

An example of when you would want to select specific stations is if you have a subdivision that will be constructed in phases. You have designed an entire roadway but only need to create specific sheets for a specific phase.

Create View Frames – Sheets Page

You use the second page of the Create View Frames Wizard (Figure 15.3) to establish the sheet type and the orientation of the view frames along the alignment. A plan production *sheet* is a layout tab in a drawing file. To create the sheets, Civil 3D references a predefined drawing template (with the filename extension .dwt). As mentioned earlier, the template must contain layout tabs, and in each layout tab the viewport's Viewport Properties options must be set to either Plan or Profile. Each viewport must have an appropriate scale assigned. Later in this chapter, you'll learn about editing and modifying templates for use in Plan Production.

FIGURE 15.3
Create View Frames
– Sheets wizard page

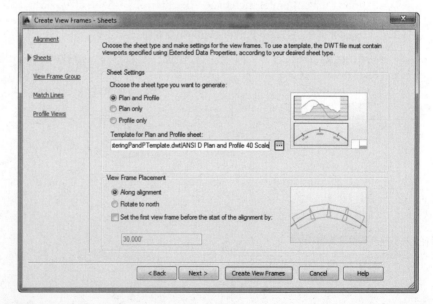

Sheet Settings

The Plan Production feature provides options for creating three types of sheets:

Plan And Profile   This option generates a sheet with two viewports; one viewport shows a plan view and the other shows a profile view of the section of the selected alignment segment.

Plan Only   As the name implies, this option creates a sheet with a single viewport showing only the plan view of the selected alignment segment.

Profile Only   Similar to Plan Only, this option creates a sheet with a single viewport, showing only the profile view of the selected alignment segment.

> **INFORMATIONAL GRAPHICS**
>
> Did you notice the nifty graphic to the right of the sheet-type options in Figure 15.3? This image changes depending on the type of sheet you've selected. It provides a schematic representation of the sheet layout to further assist you in selecting the appropriate sheet type. You'll see this type of graphic image throughout the Create View Frame Wizard and in other wizards used in Civil 3D.

After choosing the sheet type, you must define the template file and the layout tab within the selected template that Civil 3D will use to generate your sheets. Several predefined templates ship with Civil 3D and are part of the default installation. Be sure to choose the sheet type before selecting the template so that layouts associated with that sheet type will be displayed and selectable.

1. Click the ellipsis button on this page to display the Select Layout As Sheet Template dialog, shown in Figure 15.4.

FIGURE 15.4
Use the Select Layout As Sheet Template dialog to choose which layout you would like to apply to your newly created sheets.

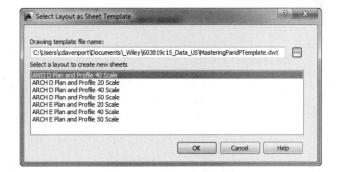

This dialog provides the option to select the DWT file and the layout tab within the template.

2. Click the ellipsis button in the Select Layout As Sheet Template dialog to browse to the desired template location.

   Typically the default template location is

   ```
   C:\Users\<username>\AppData\Local\Autodesk\C3D2014\
   enu\Template\Plan Production\
   ```

Alternatively, if you are working in a network environment, your templates can be kept in a common folder on the network.

After you select the template you want to use, a list of the layouts contained in the DWT file appears in the Select Layout As Sheet Template dialog.

3. Choose the appropriate layout.

Notice that in the template selected (see Figure 15.4) there are layouts for various sheet sizes as well as various scales that are included in the Plan Production templates that ship with Civil 3D.

View Frame Placement

Your view frames can be placed in one of two ways: either along the alignment or rotated to north. Use the bottom area of the Sheets page of the wizard to establish the placement.

Along Alignment   This option aligns the long axes of the view frames parallel to the alignment. Refer to the graphic to the right of the radio buttons in the dialog for a visual representation. This graphic is shown at the left in Figure 15.5.

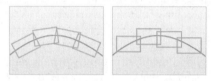

FIGURE 15.5
View Frame Placement shown using the Along Alignment option (left) and the Rotate To North option (right)

Rotate To North   As the name implies, this option aligns the view frames so they're all rotated to the north direction, regardless of the changing rotation of the alignment centerline. *North* is defined by the orientation of the drawing. This graphic is shown at the right in Figure 15.5.

> **TWISTED NORTH**
>
> If you want the north arrow to rotate according to the view twist of the viewport, the block that is being used for the north arrow must be included in the template and must be located in the layout.

Set The First View Frame Before The Start Of The Alignment By

Regardless of the view frame placement you choose, you have the option to place the first view frame some distance before the start of the alignment. This option is useful if you want to show a portion of the site, such as an existing offsite road, in the plan view. When this option is selected, the text box becomes active, letting you enter the desired distance.

CREATE VIEW FRAMES – VIEW FRAME GROUP PAGE

You use the third page of the Create View Frames Wizard (Figure 15.6) to define creation parameters for your view frames and the view frame group to which they'll belong. The page is divided into two areas: the top for the view frame group and the bottom for the view frames themselves.

FIGURE 15.6
Create View Frames – View Frame Group wizard page

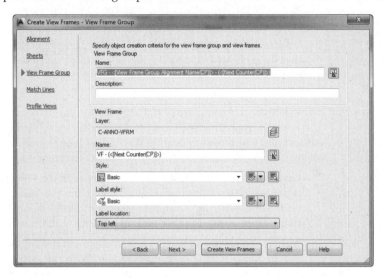

Use this area of the View Frame Group page to set the name and an optional description for the view frame group. The name can consist of manually entered text, text automatically generated based on the Name template settings, or a combination of both.

NAME TEMPLATE

To adjust the Name template settings, in the Create View Frames - View Frame Group dialog, click the Edit View Frame Group Name button to open the Name Template dialog, shown here.

> With the settings shown previously in Figure 15.6, the name will include manually defined prefix text (VFG -) followed by automatically generated text, which inserts the view frame group alignment name and a sequential counter number. This will result in a view frame group name of, for example, VFG - Frontenac Drive - 1.
>
> The Name Template dialog box isn't unique to the Plan Production feature. However, the property fields available vary depending on the features to be named. If you need to reset the incremental number counter, use the text box in the lower area of the Name Template dialog box. You can change the increment value in the Name Template dialog box as well.

View Frame Use this area of the View Frame Group page to set various parameters for the view frames, including the layer for the frames, view frame names, view frame object and label styles, and the label location. Each view frame can have a unique name (using an incremental counter), but the other parameters are the same for all view frames.

Layer This option defines the layer on which the view frames are created. This layer is defined in the Drawing Settings dialog, but you can override it by clicking the Layer button and selecting a different layer. Setting the view frame layer to No-Plot will ensure that your drawing does not end up plotting with unwanted rectangles.

Name The Name setting is nearly identical in function to that of the View Frame Group Name setting discussed earlier. With the settings shown previously in Figure 15.6, the default naming results would be VF - 1, VF - 2, and so on.

Style Like nearly all objects in Civil 3D, view frames have styles associated with them. The view frame style is simple, with only one component: the view frame border. You use the drop-down list to select a predefined style.

Label Style Also like most other Civil 3D objects, view frames have label styles associated with them. And like other label styles, the view frame labels are created using the Label Style Composer and can contain a variety of components. The label style used in Figure 15.6 includes the view frame name placed at the top of the frame.

Label Location The last option on this page lets you set the label location. The default feature setting places the label at the top left of the view frame.

For view frame labels placed at the top of the frame, the term *top* is relative to the frame's orientation. For alignments that run left to right across the page, the top of the frame points toward the top of the screen. For alignments that run right to left, the top of the frame points toward the bottom of the screen.

CREATE VIEW FRAMES – MATCH LINES PAGE

You use the fourth page of the Create View Frames Wizard (Figure 15.7) to establish settings for match lines. Match lines are used to maintain continuity from one sheet to the next. They're typically placed at or near the edge of a sheet, with instructions like "See Sheet XX" for continuation.

FIGURE 15.7
Create View Frames – Match Lines wizard page

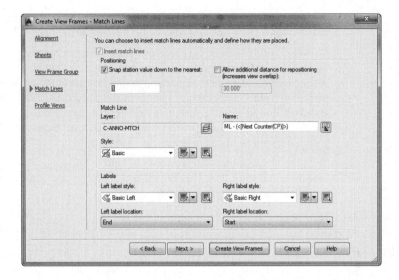

Insert Match Lines   You have the option to automatically insert match lines. Match lines are used only for plan views, so if you're creating Plan And Profile or Plan Only sheets, the option is automatically selected and can't be deselected.

Positioning   Use this area of the Match Lines page to define the initial location of the match lines and provide the ability to later move or reposition the match lines.

Snap Station Value Down To The Nearest   By selecting this option, you override the drawing station settings and define a rounding value specific to match line placement. With the settings shown previously in Figure 15.7, a value of 1 is entered, resulting in the match lines being placed at the nearest whole station. For example, if the station was 8+14.83 and the value was 1, it would round down to 8+14, but for the same station, if the value was set to either 50 or 100, it would round down to 8+00.

This feature always rounds down (snap station down as opposed to snap station up). The exception to this is that if the rounding would put the match line at an undesirable location (such as before the previous match line or before the beginning of the alignment), then no rounding would be performed and the calculated station would be used.

Allow Additional Distance For Repositioning (Increases View Overlap)   Selecting this option activates the text box, allowing you to enter a distance by which the views on adjacent sheets will overlap and the maximum distance that you can move a match line from its original position within the overlap area. While the match line locations are originally created automatically, there are going to be instances when you want to move the match line if it bisects a critical location. Any value entered will decrease the station range between match lines.

Match Line   Use this area of the Match Lines page to provide the settings for the match line. This area is similar to those for view frames on the previous page of the wizard. You can define the layer, the name, and the style.

With the settings shown previously in Figure 15.7, the match lines will be named using a predefined text (ML -) and a next counter: ML - 1, ML - 2, ML - 3, and so on.

Labels   The options in this area of the Match Line page are also similar to those for view frames. Different label styles are used to annotate match lines located at the left and right side of a frame. This lets you define match-line label styles that reference either the previous or next station adjacent to the current frame. You can also set the location of each label independently using the Left and Right Label Location drop-down lists. You have options for placing the labels at the start, end, or middle of the match line or at the point where the match line intersects the alignment.

With the settings shown previously in Figure 15.7, the label style for use at the left match line is shown in Figure 15.8. This Basic Left label style uses the Match Line Number, Match Line Station value, and Previous Sheet Number.

FIGURE 15.8
An example match line label style

Create View Frames – Profile Views Page

The final page of the Create View Frames Wizard (Figure 15.9) is optional and will be disabled and skipped if you chose to create Plan Only sheets on the Sheets page of the wizard. Use the drop-down lists to select both the profile view style and the band set style. These styles will be discussed in Chapter 19, "Object Styles."

FIGURE 15.9
Create View Frames – Profile Views wizard page

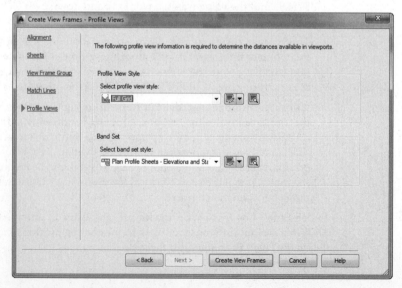

Civil 3D Plan Production tools use the settings in the profile view style and band styles to determine placement of the profile view in the viewport, sometimes disregarding station labels below the profiles. Sometimes it is necessary to force the program to boost the profile view higher in the viewport by adding a "white space" band. This would be an invisible band of a certain height. Even though you can't see the band, Civil 3D provides space for it in the viewport.

This last page of the wizard has no Next button. To complete the wizard, click the Create View Frames button.

Creating View Frames

Now that you understand the wizard pages and available options, you'll try them out in this exercise:

1. Open the `ViewFrameWizard.dwg` file or `ViewFrameWizard_METRIC.dwg` file. (Remember, all data can be downloaded from www.sybex.com/go/masteringcivil3d2014.)

    This drawing contains several alignments and profiles as well as styles for view frames, view frame groups, and match lines.

2. To launch the Create View Frames Wizard, from the Output tab ➢ Plan Production panel, choose Create View Frames.

3. On the Alignment page, do the following:

    a. Select Frontenac Drive from the Alignment drop-down list.

    b. For Station Range, verify that Automatic is selected.

    c. Click Next to advance to the next page.

4. On the Sheets page, do the following:

    a. In the Sheet Settings area, select the Plan And Profile option.

    b. Under Template For Plan And Profile Sheet, click the ellipsis button to display the Select Layout As Sheet Template dialog.

    c. In the Select Layout As Sheet Template dialog, click the ellipsis button and browse to the templates in the chapter folder: `C:\Mastering\Ch 15\`.

    d. Select the template named `MasteringPandPTemplate.dwt` (or `MasteringPandPTemplate_Metric.dwt`), and click Open.

    A list of the layouts in the DWT file appears in the Select Layout As Sheet Template dialog.

    e. Select the layout named ARCH D Plan And Profile 20 Scale (or ISO A1 Plan and Profile 1 to 500 for metric users), and click OK to dismiss the Select Layout As Sheet Template dialog.

    f. In the View Frame Placement area, select the Along Alignment option.

    g. Select the Set The First View Frame Before The Start Of The Alignment By option.

    Note that the default value for this particular drawing is 30′ (or 10m for metric users).

    h. Click Next to advance to the next page.

5. On the View Frame Group page, confirm that all settings are as follows (these are the same settings shown previously in Figure 15.6), and then click Next to advance to the next page:

Setting	Value
View Frame Group Name	VFG - <[View Frame Group Alignment Name(CP)]> - (<[Next Counter(CP)]>)
View Frame Name	VF - (<[Next Counter(CP)]>)
Style	Basic
Label Style	Basic
Label Location	Top Left

6. On the Match Lines page, confirm that all settings are as follows (these are the same settings shown previously in Figure 15.7), and then click Next to advance to the next page.

Setting	Value
Snap Station Value Down To The Nearest	1
Layer	C-ANNO-MTCH
Name	ML - (<[Next Counter(CP)]>)
Style	Basic
Left Label Style	Basic Left
Left Label Location	End
Right Label Style	Basic Right
Right Label Location	Start

7. On the Profile Views page, confirm that the settings are as follows (these are the same settings shown previously in Figure 15.9), and then click Create View Frames:

Setting	Value
Select Profile View Style	Full Grid
Select Band Set Style	Plan Profile Sheets - Elevations And Stations

The view frames and match lines are created as shown in Figure 15.10.

Due to the sheet sizes and scales, the Imperial drawing in this example generates four view frames while the metric drawing generates only two view frames.

FIGURE 15.10
Finished view frames and match lines in the drawing

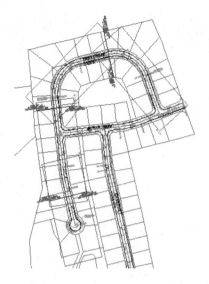

> **THE EFFECTS OF INCREMENTAL COUNTING**
>
> The numbering for your view frames, view frame groups, and match lines may not identically match that shown in the images. This is due to the incremental counting Civil 3D performs in the background. As previously mentioned, each time you create one of these objects, the counter increments. You can reset the counter by modifying the Name template.

When this exercise is complete, you may close the drawing. A saved copy of this drawing is available from the book's web page with the filename `ViewFrameWizard_FINISHED.dwg` or `ViewFrameWizard_METRIC_FINISHED.dwg`.

Editing View Frames and Match Lines

After you've created view frames and match lines, you may need to edit them. Edits to some view frame and match line properties can be made via the Prospector tab in the Toolspace palette by expanding the View Frame Groups branch, as shown in Figure 15.11.

You can change some information in the Preview area of Prospector when you highlight either the View Frames branch or the Match Lines branch. Alternatively, you can make further edits from the View Frame Properties dialog or the Match Line Properties dialog. One way of accessing these dialogs is through the View Frame contextual tab or the Match Line contextual tab. Another method is by right-clicking the desired object in Prospector and selecting Properties. For both view frames and match lines, you can only change the object's name and/or style via the Information tab in their Properties dialog. All other information displayed on the other tabs is read-only.

FIGURE 15.11
View Frame Groups in Prospector

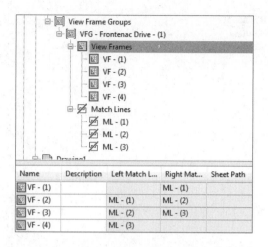

You make changes to geometry and location graphically using special grip edits (Figure 15.12). Like many other Civil 3D objects with special editing grips (such as profiles and pipe network objects), view frames and match lines have editing grips you use to modify the objects' location, rotation, and geometry. Let's look at each separately.

FIGURE 15.12
View frame and match line grips

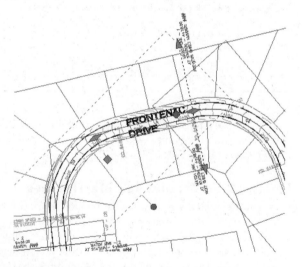

View frames can be graphically edited in three ways. Once you select a view frame object (the rectangular object selected at the left in Figure 15.12), you can move them, slide them along the alignment, and rotate them as follows:

To Move a View Frame The standard square grip is used for most typical edits, including moving the object.

To Slide a View Frame The diamond-shaped grip at the center of the frame lets you move the view frame in either direction along the alignment while maintaining the orientation (Along Alignment or Rotated North) you originally established for the view frame when it was created.

To Rotate a View Frame The circular handle grip works like the one on pipe-network structures. Using this grip, you can rotate the frame about its center.

DON'T FORGET YOUR AUTOCAD® FUNCTIONS!

While you're getting wrapped up in learning all about Civil 3D and its great design tools, it can be easy to forget you're sitting on an incredibly powerful AutoCAD application.

AutoCAD features add functionality beyond what you can do with Civil 3D commands alone. First, make sure the Dynamic Input option is enabled by either clicking the button at the bottom left of your screen as shown here or by pressing F12.

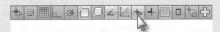

This gives you additional functionality when you're moving a view frame. With Dynamic Input enabled, you can enter an exact station value to precisely locate the frame where you want it. Similar to moving a view frame, with Dynamic Input active you can enter an exact rotation angle. Note that this rotation angle is relative to your drawing settings (for example, 0 degrees is to the left, 90 degrees is straight up, and so on).

Also, selecting multiple objects and then selecting their grips while holding Shift makes each grip "hot" (usually a red color). This allows you to grip-edit one object and all of the "hot" objects will also experience the same grip edit, like sliding a group of view frames along the alignment. You can edit a match line's location and length using special grips. As with view frames, you can slide them along the alignment and rotate them. They can also be lengthened or shortened. Unlike view frames, they can't be moved to an arbitrary location. Once you select a match line object (the object selected at the right in Figure 15.12), you can edit it as follows:

To Slide a Match Line The diamond-shaped grip at the center of the match line lets you move the match line in either direction along the alignment while maintaining the orientation (Along Alignment or Rotated North) that you originally established for the view frame.

Note that the match line can only be moved in either direction a distance equal to or less than that entered on the Match Line page of the wizard at the time the view frames were created. For example, if you entered a value of 50′ (15 m) for the Allow Additional Distance For Repositioning option, your view frames are overlapped 50′ (15 m) to each side of the match line, and you can slide the match line only 50′ (15 m) in either direction from its original location.

To Rotate a Match Line The circular handle grip enables you to rotate the match line just as you can with the circular handle grip on a view frame.

To Change a Match Line's Length When you select a match line, a triangular grip is displayed at each end. You can use these grips to increase or decrease the length of each half of the match line. For example, moving the grip on the top end of the match line changes the length of only the top half of the match line; the other half of the match line remains unchanged. See the sidebar "Don't Forget Your AutoCAD Functions!" for tips on using AutoCAD features. If you select a match line, click one of the triangular grips, and then hold Shift and select the other triangular grip, the match line will shorten on one end as you lengthen the other end, and vice versa.

The following exercise lets you put what you've learned into practice as you change the location and rotation of a view frame and change the location and length of a match line:

1. Open the EditViewFramesAndMatchLines.dwg or EditViewFramesAndMatchLines_METRIC.dwg file from this book's web page. This drawing contains view frames and match lines.
2. Confirm that Dynamic Input is enabled; if it is not, press F12.
3. Select the view frame, which consists of stations 9+56 to 14+34 (or 0+267 to the end for metric users), and select its diamond-shaped sliding grip.
4. Slide this diamond grip so the overlap with the lower view frame isn't so large. Either graphically slide it to station 11+00 (or 0+360 for metric users) or enter **1100** ↵ (**150** ↵ for metric users) in the Dynamic Input text box.
5. Press Esc to clear your selection.
6. Select the view frame, which consists of stations 14+34 to the end (or 0+267 to the end for metric users), and select the circular rotation grip.
7. Rotate the view frame slightly to better encompass the road. Enter **8** ↵ (**268** ↵ for metric users).
8. Press Esc to clear your selection in the Dynamic Input text box.

Next you will adjust the match line's location.

9. Select the match line, which is presently at station 14+34 (or 0+267 for metric users), and then select its diamond sliding grip.
10. Either graphically slide it to station 13+25 (or 0+250 for metric users) or enter **1325** ↵ (**250** ↵ for metric users) in the Dynamic Input text box.

Notice that the match line label is updated with the revised station.

Next, you'll adjust the length of the match line to fit within the view frame extents.

11. Select the triangular lengthen grip at the west end of the match line, and either graphically shorten it to 75′ (or 50 m for metric users) or enter **75** ↵ (or **50** ↵ for metric users) in the Dynamic Input text box.
12. Select the triangular lengthen grip at the east end of the same match line, and either graphically shorten it to 75′ (or 50 m for metric users) or enter **75** ↵ (or **50** ↵ for metric users) in the Dynamic Input text box.
13. Press Esc to clear your selection.

When this exercise is complete, you may close the drawing. A saved copy of this drawing is available from the book's web page with the filename EditViewFramesAndMatchLines_FINISHED.dwg or EditViewFramesAndMatchLines_METRIC_FINISHED.dwg.

Now that you have generated the view frames and match lines in the drawing and you have placed them where you want them, let's look at using these objects to generate sheets.

Creating Plan and Profile Sheets

The Plan Production feature uses the concept of *sheets* to generate the pages that make up a set of plans. Simply put, *sheets* are layout tabs with viewports showing a given portion of your design model, based on the view frames previously created. The viewports have special viewport properties set that define them as either plan or profile viewports. These viewports must be predefined in a drawing template (DWT) file to be used with the Plan Production feature. You manage the sheets using the standard AutoCAD Sheet Set Manager feature.

The Create Sheets Wizard

After you've created view frames and match lines, you can proceed to the next step of creating sheets. Like view frames, sheets are created using a wizard. Let's look at the Create Sheets Wizard and the various page options. After you've seen each page, you'll have a chance to put what you've learned into practice in an example.

From the Output tab ➢ Plan Production panel, you launch the Create Sheets wizard by choosing Create Sheets. As with the Create View Frames Wizard, a list of the Create Sheets Wizard's pages is shown along the left sidebar, and an arrow indicates which page you're currently viewing. You move among the pages using the Next and Back navigation buttons along the bottom of each page. Alternatively, you can jump directly to any page by clicking its name in the list on the left. The following sections walk you through the pages of the wizard and explain their features.

CREATE SHEETS – VIEW FRAME GROUP AND LAYOUTS PAGE

You will use the first page of the Create Sheets Wizard (Figure 15.13) to select the view frame group for which the sheets will be created. It's also used to define how the layouts for these sheets will be generated and which sheets will be created (all or a range).

FIGURE 15.13
Create Sheets – View Frame Group And Layouts wizard page

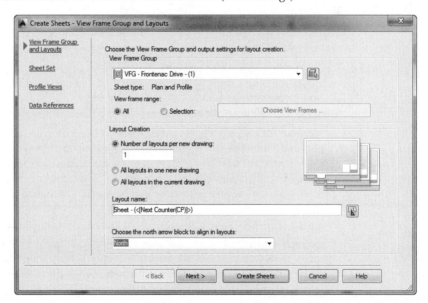

View Frame Group In the top area of this page, you select the view frame group. You can either select it from the drop-down list or click the Select From The Drawing button to select the view frame group onscreen. After you've selected the group, you use the View Frame Range option to create sheets for all frames in the group or only for specific frames of your choosing.

All Select this option when you want sheets to be created for all view frames in the view frame group.

Selection Select this option to activate the Choose View Frames button. Click this button to display the Select View Frames dialog, where you can select specific view frames from a list. You can select a range of view frames by using the standard Windows selection technique of clicking the first view frame in the range and then holding Shift while you select the last view frame in the range. You can also select individual view frames in nonsequential order by holding Ctrl while you make your view frame selections. Figure 15.14 shows two of the four view frames selected in the Select View Frames dialog.

FIGURE 15.14
Select view frames by using standard Windows selection techniques.

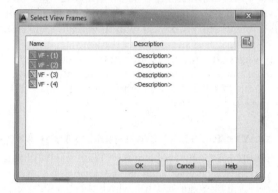

Layout Creation In this area, you define where and how the new layouts for each sheet are created as well as the name format for these sheets and information about the alignment of the north arrow block.

There are three options for creating layout sheets: the layouts are created in multiple new drawing files (with a limit to the maximum number of layout sheets created in each file), all the new layouts are created in a new drawing file, or all the layout tabs are created in the current drawing (the drawing you're in while executing the Create Sheets Wizard).

Number Of Layouts Per New Drawing This option creates layouts in new drawing files and limits the maximum number of layouts per drawing file to the value you enter in the text box. For best performance, Autodesk recommends that a drawing file contain no more than 10 layouts. On the last page of this wizard you're given the option to select the objects for which data references will be made. These data references are then created in the new drawings.

All Layouts In One New Drawing As the name implies, this option creates all layouts for each view frame in a single new drawing. Use this option if you have fewer than 10 view frames, to ensure best performance. If you have more than 10 view frames, the previous option is recommended. On the last page of this wizard you're given the option to select the objects for which data references will be made. These data references are then created in the new drawing.

All Layouts In The Current Drawing When you choose this option, all layouts are created in the current drawing. You need to be aware of two scenarios when working with this option. (As explained later in Chapter 16, "Advanced Workflows," you can share a view frame group via data shortcuts and link it to other drawings as a data reference.)

- When creating sheets, it's possible that your drawing references the view frame group from another drawing (rather than having the original view frame group in your current drawing). In this case, you're given the option to select the additional objects for which data references will be made (such as alignments, profiles, pipes, and so on). These data references are then created in the current drawing. You select these objects on the last page of the wizard.

- If you're working in a drawing where the view frames were created (therefore, you're in the drawing in which the view frame group exists), the last page of this wizard is disabled. This is because in order for you to create view frames (and view frame groups), the alignment (and possibly the profile) must either exist in the current drawing or be referenced as a data reference (recall the prerequisites for creating view frames, mentioned earlier).

Layout Name Use this text box to enter a name for each new layout. As with other named objects in Civil 3D, you can use the Name template to create a name format that includes information about the object being named. With the settings shown previously in Figure 15.13, the layouts will be named using a predefined text (Sheet -) and a next counter: Sheet - 1, Sheet - 2, Sheet - 3, and so on. Using the Name template, you could alternatively use the Parent Drawing Name, View Frame Start/End Raw Station, View Frame Start/End Station Value, View Frame Group Alignment Name, and View Frame Group Name options.

 **Real World Scenario**

WHERE AM I?

We strongly recommend that you set up the Name template for the layouts so that it includes the View Frame Group Alignment Name option and the station range (View Frame Start Station Value and View Frame End Station Value). This conforms to the way many organizations create sheets, helps automate the creation of a sheet index, and generally makes it easier to navigate a DWG file with several layout tabs.

This would result in a name that is listed as follows:

```
<[View Frame Group Alignment Name]> <[View Frame Start Station Value]>
to <[View Frame End Station Value]>
```

This will generate a layout named similar to Frontenac Drive 0 + 00.00 to 4 + 78.00.

Choose The North Arrow Block To Align In Layouts If the template file you've selected contains a north arrow block, it can be aligned so that it points north on each layout sheet. The block must exist in the template and be located in the layout. If there are multiple blocks, select the one you want to use from the drop-down list.

CREATE SHEETS – SHEET SET PAGE

You use the second page of the Create Sheets Wizard (Figure 15.15) to determine whether a new or existing sheet set (with the filename extension .dst) is used and the location of the DST file. The sheet file storage location and sheet file name are also defined here. Additionally, on this page you decide whether to add the sheet set file (with the filename extension .dst) and the sheet files (with the filename extension .dwg) to the project vault.

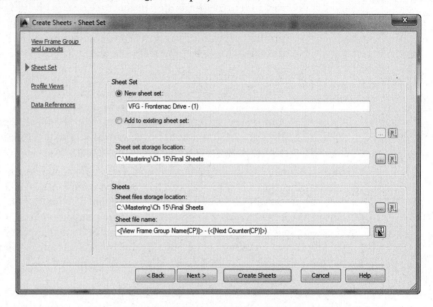

FIGURE 15.15
Create Sheets – Sheet Set wizard page

Sheet Set   In the Sheet Set area of the page, you select whether to create a new DST file or add the sheets created by this wizard to an existing DST file.

New Sheet Set   Selecting this option allows you to generate a new sheet set. In the text box, you can enter a name for the sheet set or accept the name that is generated for you.

Add To Existing Sheet Set   Selecting this option enables you to add to a sheet set that has already been generated. Click the ellipsis button to browse to the location of the DST file. If you are working in a Vault project, you can click the Vault button to browse to the file in the vault project. If Autodesk® Vault is not installed, the Vault button will be inactive.

Sheet Set Storage Location   This text box is active only if you have selected New Sheet Set. When it's active, you can type the path to the folder where the DST file will be created or you can use the ellipsis to browse to the storage location. If you are working in a vault project, you can click the Vault button to browse to the storage location in the vault project. If Autodesk Vault is not installed, the Vault button will be inactive.

Sheets   This area is active only if you are creating layouts in new drawings. If you choose to create layouts in the current drawing, the area will be inactive.

Sheet Files Storage Location   Type the path to the folder where the sheet files will be created, or you can use the ellipsis to browse to the storage location. If you are working in a vault project, you can click the Vault button to browse to the storage location in the vault project. If Autodesk Vault is not installed, the Vault button will be inactive.

Sheet File Name   As with many of the previous names, you can use a Name template to specify how the sheet-file name is generated. With the settings shown previously in Figure 15.15, the sheet files will be named using the View Frame Group Name entry and a next counter. Using the Name template, you could alternatively use the Parent Drawing Name, View Frame Start/End Raw Station, View Frame Start/End Station Value, and View Frame Group Alignment Name options.

> ### What Is a Sheet?
>
> The Create Sheets – Sheet Set page can be a little confusing due to the way the word *sheets* is used. In some places, *sheets* refers to layout tabs in a given drawing (DWG) file. On this page, the word *sheet* is being paired with the word *file(s)* and refers to DWG files that contain a collection of *sheet* layouts.

Create Sheets – Profile Views Page

The third page of the Create Sheets Wizard (Figure 15.16) lists the profile view style and the band set selected in the Create View Frames Wizard. You can't change these selections. You can, however, make adjustments to other profile settings.

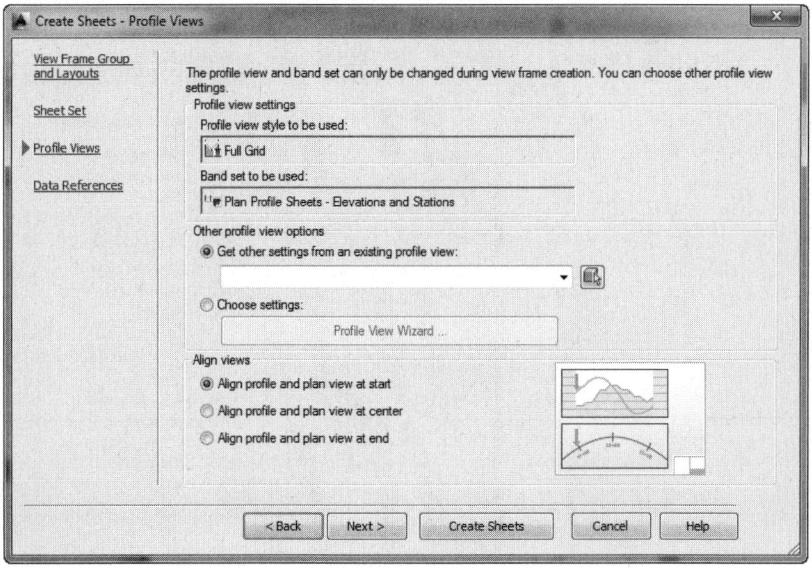

FIGURE 15.16
Create Sheets – Profile Views wizard page

Other Profile View Options The Other Profile View Options area lets you modify certain profile view options either by using an existing profile view in your drawing as an example or by running the Profile View Wizard. Regardless of what option you choose, the "other options" you can change are limited to the following in the Profile View wizard:

- Profile View Datum By Minimum Elevation or Mean Elevation on the Profile View Height page
- Split Profile View options from the Profile View Height page
- All options on the Profile Display Options page
- If available, all options of the Pipe Network Display page
- Most of the settings on the Data Bands page
- All options on the Profile Hatch Options page
- All settings on the Multiple Plot Options page

See Chapter 7, "Profiles and Profile Views," for details on each of these settings. The inactive settings are ones that were previously set when you were going through the settings of the Create View Frames Wizard. If you need to change these settings, you will have to delete your current view frame group and regenerate the view frames with the desired settings.

Align Views In the Align Views area of the page, you can choose Align Profile And Plan View At Start, Align Profile And Plan View At Center, or Align Profile And Plan View At End.

Create Sheets – Data References Page

The final page of the Create Sheets Wizard (Figure 15.17) is used to create data references in the drawing files that contain your layout sheets.

FIGURE 15.17
Create Sheets – Data References wizard page

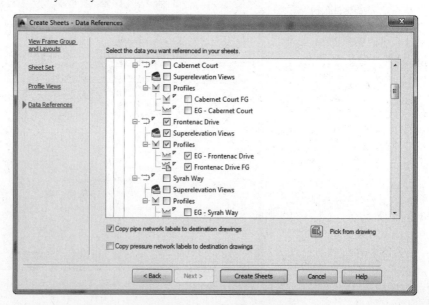

Select The Data You Want Referenced In Your Sheets   This area contains a list of objects in the current drawing that can be data referenced into the sheet files. Depending on the way sheet settings were configured in the Create View Frames Wizard (Plan And Profile, Plan Only, or Profile Only), certain objects will be selected for you. If there are other data references you want referenced into the drawing, you may use the check boxes here to select them.

Pick From Drawing   Use the Pick From Drawing button to select those objects you want to data reference from the drawing area.

Copy Pipe Network Labels To Destination Drawings   It's common to create references to pipe networks that are to be shown in plan and/or profile views. If you choose to create references for pipe network objects, you can also copy the labels for those network objects into the sheet file. This saves you from having to relabel your pipe networks.

Copy Pressure Network Labels To Destination Drawings   This option is a new feature in AutoCAD Civil 3D 2014. When this check box is selected, pressure network labels will be copied to the sheet file.

Managing Sheets

After you've completed all pages of the Create Sheets Wizard, you create the sheets by clicking the Create Sheets button. Doing so completes the wizard and starts the creation process. If you're creating sheets with profile views, you're prompted to select a profile view origin. Civil 3D then displays several dialogs, indicating the process status for the various tasks, such as creating the new sheet drawings and creating the DST file. Once that's complete, the Panorama Event Viewer vista will list two new events, in this example one stating "Sheets created were added to the sheet set file C:\Mastering\CH 15\Final Sheets\VFG - Frontenac Drive - (1).dst" and one stating "4 layout(s) created in path C:\Mastering\CH 15\Final Sheets."

If the Sheet Set Manager isn't currently open, it opens with the newly created DST file loaded. The sheets are listed, and the details of the drawing files for each sheet appear (Figure 15.18).

FIGURE 15.18
New sheets in the Sheet Set Manager

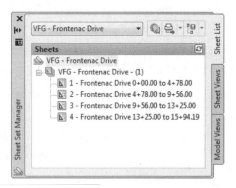

If you double-click to open the new drawing file that contains the newly created sheets, you'll see layout sheet tabs created for each of the view frames as selected in the Create Sheets Wizard. The sheets are named using the Name template as defined in the Create Sheets Wizard. Figure 15.19 shows the names that result from the following template:

```
<[View Frame Group Alignment Name]> <[View Frame Start Station Value]>
to <[View Frame End Station Value]>
```

FIGURE 15.19
The template produces the Frontenac Drive tab names shown here.

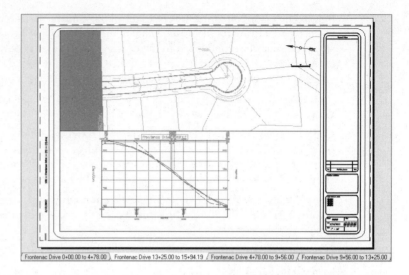

To create the final sheets in this new drawing, Civil 3D creates external references (XRefs) the drawing containing the view frames; creates data references (DRefs) for the alignments, profiles, and any additional objects you selected in the Create Sheets Wizard; and, if profile sheet types were selected in the Create View Frames Wizard, creates profile views in the final sheet drawing.

The following exercise pulls all these concepts together:

1. Open the SheetsWizard.dwg or SheetsWizard_METRIC.dwg file.

   This drawing contains the view frame group, alignment, and profile for Frontenac Drive. Note that the drawing doesn't have profile views.

2. To launch the Create Sheets Wizard, from the Output tab ➢ Plan Production panel, choose Create Sheets.

3. On the View Frame Group And Layouts page, do the following:

   a. Verify that View Frame Range is set to All.

   b. Verify that Number Of Layouts Per New Drawing is set to **10**.

   Since this view frame group has fewer than 10 view frames, only one drawing will be generated, but it is good practice to set this value to 10 nonetheless.

   c. Set Layout Name to <[View Frame Group Alignment Name]> <[View Frame Start Station Value]> to <[View Frame End Station Value]>.

   d. From the Choose The North Arrow Block To Align In Layouts drop-down list, select North.

   e. Click Next to advance to the next page.

> **USE A NAME TEMPLATE**
>
> To set the layout name in step 3 or any of the other steps that require the use of the Name template, don't type all that text into the text box. Instead, follow these simple steps:
>
> 1. Click the Edit Layout Name button to the right of the Layout Name text box to display the Name Template dialog.
> 2. Delete all of the text in the Name text box that you do not want.
> 3. Using the Property Fields drop-down list, select the first piece of text that you want. In this example it will be View Frame Group Alignment Name.
> 4. Click Insert.
> 5. Repeat steps 3 and 4 for each of the property fields that you want to use. If using start or end stations, use Start Station Value or End Station Value properties instead of their raw station counterparts.
> 6. Once complete, click OK to dismiss the Name Template dialog.
>
> The Layout Name option is now filled in based on the Name template.

4. On the Sheet Set page, do the following:
   a. Select the New Sheet Set option and set Name to **VFG - Frontenac Drive** (or **VFG - Frontenac Drive_METRIC** for metric users).
   b. For the Sheet Set File Storage Location option, use the ellipsis to browse to `C:\Mastering\CH 15\Final Sheets`, and click Open to dismiss the Browse For Sheet Set Folder dialog.

   Notice that by changing this location, the Sheet Files Storage Location entry automatically changes to match.

   c. Verify that Sheet File Name is set to

   `<[View Frame Group Name(CP)]> - (<[Next Counter(CP)]>)`

   for Imperial users, or

   `<[View Frame Group Name(CP)]> - (<[Next Counter(CP)]>)_METRIC`

   for metric users.

   d. Click Next to advance to the next page.

5. On the Profile Views page, do the following:
   a. For Other Profile View Options, select Choose Settings and then click the Profile View Wizard button.

   The Create Multiple Profile Views dialog opens.

b. On the left side of the Create Multiple Profile Views dialog, click Profile Display Options to jump to that page.

c. On the Create Multiple Profile Views – Profile Display Options page, verify that the Draw option is selected for only the EG - Frontenac Drive profile and the Frontenac Drive FG profile.

d. Scroll to the right and verify that the Labels setting for the EG – Frontenac Drive profile is set to No Labels and that the Frontenac Drive FG Profile is set to Complete Label Set.

e. On the left side of the Create Multiple Profile Views dialog, click Data Bands to jump to that page.

f. On the Create Multiple Profile Views – Data Bands page, change Profile2 to Frontenac Drive FG for both bands.

g. Click Finish to dismiss the Create Multiple Profile Views Wizard and return to the Create Sheets Wizard.

h. Verify that Align Views is set to Align Profile And Plan View At Start.

i. Click Next to advance to the next page.

6. On the Data References page, confirm that at minimum, the Frontenac Drive alignment and both the EG - Frontenac Drive and Frontenac Drive FG profiles are selected, and then click Create Sheets to complete the wizard.

Before creating the sheets, Civil 3D must save your current drawing.

7. Click OK when prompted to save.

The drawing is saved, and you're prompted for an insertion point for the profile view. The location you pick represents the lower-left corner of the profile view grid.

8. Select an open area in the drawing, above the right side of the site plan.

Civil 3D displays a progress dialog box, and then the Panorama palette is displayed with information about the results of the sheet creation process.

9. Close the Panorama palette.

> **INVISIBLE PROFILE VIEWS**
>
> Note that the profile views are created in the current drawing only if you selected the option to create all layouts in the current drawing. Because you didn't do that in this exercise, the profile views aren't created in the current drawing. Rather, they're created in the sheet files in modelspace in a location relative to the point you selected in this step.

After the sheet creation process is complete, the Sheet Set Manager window opens.

10. Click the first sheet, as shown in Figure 15.20, named Frontenac Drive 0+00.00 to 4+78.00 (or Frontenac Drive 0+000.00 to 0+250.00_METRIC for metric users).

FIGURE 15.20
The Sheet Set Manager once the sheet creation process is complete

Notice that the name conforms to the Name template and includes the alignment name and the station range for the sheet.

11. Review the details listed for the sheet by hovering over the sheet on the Sheet List tab of Sheet Set Manager or by right-clicking in the white space of the dialog and selecting Preview/Details Pane. In particular, note the filename and storage location.

12. Double-click this sheet to open the new sheets drawing and display the layout tab for Frontenac Drive 0+00.00 to 4+78.00 (or Frontenac Drive 0+000.00 to 0+250.00_METRIC for metric users).

13. Review the multiple tabs created in this drawing file as previously shown in Figure 15.19. If layout tabs are not displayed, then you may turn them on in the Options dialog, by going to the Display tab ➢ Layout Elements area, and filling the Display Layout And Model Tabs check box.

The template used also takes advantage of AutoCAD fields, some of which don't currently have values assigned.

When this exercise is complete, you may close the drawing. A saved copy of this drawing is available from the book's web page with the filename `SheetsWizard_FINISHED.dwg` or `SheetsWizard_METRIC_FINISHED.dwg`. In addition, the downloadable dataset includes the `Final Sheets` folder, which includes additional files created in this exercise.

Now that you have the plan and profile sheets generated, let's look at generating some section sheets using the Plan Production feature.

Creating Section Sheets

Creating section sheets is a two-step process; it's just like creating plan and profile sheets but with one difference: there are no view frames to tweak. Instead you will have section sheets, which are objects. Section sheets have group plot styles assigned to them that control the arrangement of section arrays per sheet. These section sheets are created when you run the Create Multiple Section Views command using the Production option. If you have already created multiple section views using the Draft option, you will have to re-create your section views using the Production option. This is a prerequisite for running the Create Section Sheets command. So here is the two-step process:

1. Generate multiple section views with the Placement option set to Production. You will have the opportunity to select styles to use for section view, section, and group plot styles as you work through the wizard. All section views created with this command become part of a section view group that can be managed from Prospector.

2. Generate section sheets, which will create your layouts and your sheet set. Section sheet layouts are created in the same drawing as your production section views; there is no option to create them in new drawings.

Creating Multiple Section Views

Before creating multiple section views for the purpose of sheet creation, be sure you have section, section view, and group plot styles in place. If you have any questions about these styles, refer to Chapter 19, "Object Styles."

In this exercise, you'll walk through creating multiple section views for the main road of our sample set:

1. Open MultipleSectionViews.dwg or MultipleSectionViews_METRIC.dwg from the provided dataset.

   In this drawing, sample lines have been added along the Frontenac Drive alignment. These lines are sampling the existing and proposed surfaces.

2. From the Home tab ➢ Profile & Section Views panel, choose Section Views ➢ Create Multiple Views to display the Create Multiple Section Views Wizard, shown in Figure 15.21.

3. On the General page, do the following:

   a. Verify that Section View Style is set to Road Section.

   b. Click Next to advance to the next page.

   The Section Placement page (Figure 15.22) is where you configure your placement options. If you select Production, you must use the ellipsis to browse to the DWT file containing your section layouts and choose the layout desired based on sheet border and viewport scale. If you select Draft, section views will be laid out in a single grid format.

FIGURE 15.21
Create Multiple Section Views – General wizard page

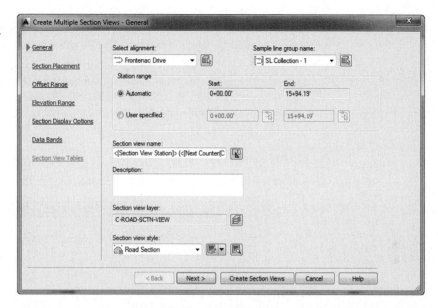

FIGURE 15.22
Create Multiple Section Views – Section Placement wizard page

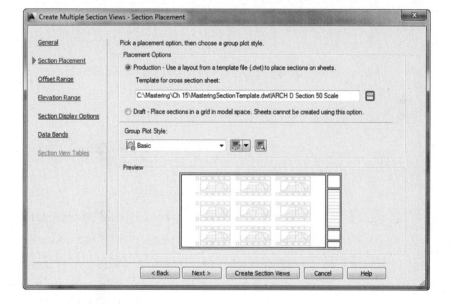

4. On the Section Placement page, do the following:

   a. In the Placement Options area, verify that Production is selected.

   b. For Template For Cross Section Sheet, use the ellipsis to browse to C:\Mastering\CH 15 and select MasteringSectionTemplate.dwt (or MasteringSectionTemplate_METRIC.dwt for metric users). Under Select A Layout To Create New Sheets, choose ARCH D Section 50 Scale (or ISO A1 Section 1 to 500 for metric users).

   c. Verify that Group Plot Style is set to Basic.

   d. Click Next to advance to the next page.

   The Offset Range page (Figure 15.23) determines the width of the section views.

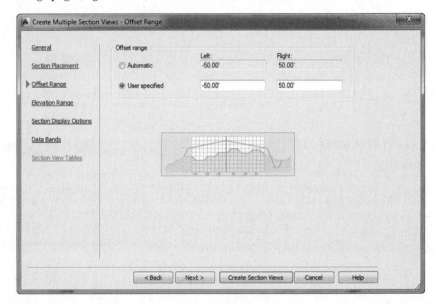

FIGURE 15.23
Create Multiple Section Views – Offset Range wizard page

5. On the Offset Range page, do the following:

   a. Using the User Specified option, set the left offset range to **-50′** (or **-15** m for metric users).

   b. Set the right offset range to **50′** (or **15** m for metric users).

   c. Click Next to advance to the next page.

   The Elevation Range page (Figure 15.24) determines the height of the section views. The options on the Elevation Range page help if you have extra tall sections, allowing you to set some limits manually.

6. On the Elevation Range page, click Next to advance to the next page.

FIGURE 15.24
Create Multiple Section Views – Elevation Range wizard page

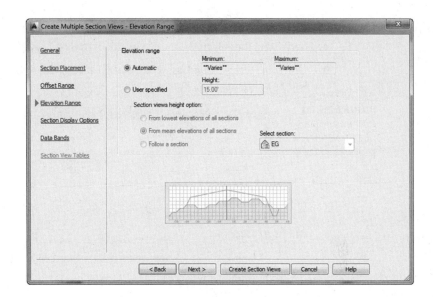

7. On the Section Display Options page, do the following:

   a. Verify that in the Change Labels column the EG labels are set to _No Labels and the FG surface labels are set to _No Labels.

   b. Verify that in the Style column the style for EG is set to Existing Ground, the style for FG is set to Finished Ground, and each of the corridor surface styles are set to Basic, as shown in Figure 15.25.

FIGURE 15.25
Changing styles in the Create Multiple Section Views – Section Display Options wizard page

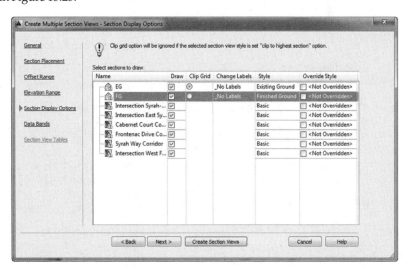

   c. Click Next to advance to the next page.

8. On the Data Bands page, do the following:

   a. Verify that the Select Band Set drop-down list is set to Offsets Only.

   b. Under Set Band Properties, verify that Surface1 is set to EG and Surface2 is set to FG, as shown in Figure 15.26.

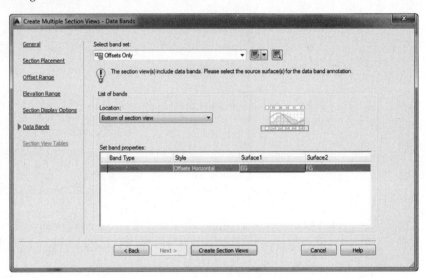

FIGURE 15.26
Create Multiple Section Views – Data Bands wizard page

   c. Click Create Section Views to dismiss the wizard and place your section views in the drawing.

9. Click a point to the east of the plan view to draw the section views and sheet outlines.

   Your drawing should look something like Figure 15.27.

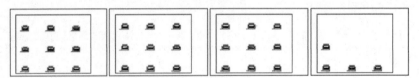

FIGURE 15.27
The finished multiple section views operation

When this exercise is complete, you may close the drawing. A saved copy of this drawing is available from the book's web page with the filename MultipleSectionViews_FINISHED.dwg or MultipleSectionViews_METRIC_FINISHED.dwg.

Now that you have your production section views, you can begin the process of creating section sheets for plotting.

Creating Section Sheets

Many long transportation projects such as highways, light-rail, or canals require the production of many section sheets. While Civil 3D could produce the views prior to Autodesk Civil 3D 2012, the sheet creation process improved greatly in that release. In this exercise, you'll convert a section view group into a collection of sheets and place them in a new sheet set.

1. Open the CreatingSectionSheets.dwg or CreatingSectionSheets_METRIC.dwg file.

    This file is the result of the previous exercise and contains the section view group for the Frontenac Drive alignment.

2. From the Output tab ➢ Plan Production panel, choose Create Section Sheets to display the Create Section Sheets dialog.

3. In the Create Section Sheets dialog, do the following:

    a. Verify that New Sheet Set name is set to **CreatingSectionSheets** (or **CreatingSectionSheets_METRIC**).

    b. Click the ellipsis to set Sheet Set Storage Location to C:\Mastering\CH 15\Final Sheets.

    The Create Section Sheets dialog should now look similar to Figure 15.28.

FIGURE 15.28
The Create Section Sheets dialog

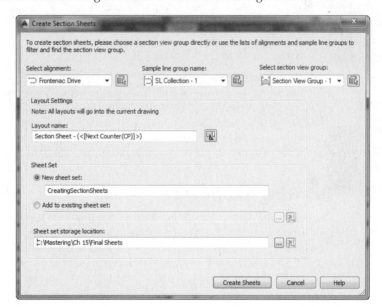

Note that when you are creating section sheets in a drawing where there is more than one section view group, you would have to repeat this command for each section view group.

4. Click Create Sheets to dismiss the dialog and generate sheets.

    Before creating the sheets, Civil 3D must save your current drawing.

5. Click OK when prompted to save.

    The drawing is saved and Civil 3D will generate new layouts in the drawing and sheets in a sheet set. The Sheet Set Manager will appear.

6. Switch to the Section Sheet - (2) layout tab.

   Your layout should look something like Figure 15.29.

FIGURE 15.29
A completed section sheet

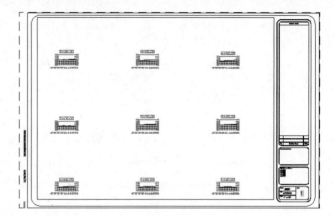

7. Close the Sheet Set Manager palette.

When this exercise is complete, you may close the drawing. A saved copy of this drawing is available from the book's web page with the filename CreatingSectionSheets_FINISHED.dwg or CreatingSectionSheets_METRIC_FINISHED.dwg.

While there are still some tweaks to be made to any sheet, large portions of the mundane details are handled by the wizards and tools. There are some elements that you can modify to customize these details for your organization, and you'll look at those in the next section.

Drawing Templates

The beginning of this chapter mentioned that there are several prerequisites to using the Plan Production tools in Civil 3D. The list includes drawing templates (DWT) set up to work with the Plan Production feature, and styles for the objects generated by this feature. In this section of the chapter, you'll learn how to prepare drawing templates for use in creating your finished sheets.

Civil 3D ships with several predefined template files for various types of sheets that Plan Production can create. By default, these templates are installed in a subfolder called Plan Production, which is located in the standard Template folder. You can see the Template folder location by opening the Files tab of the Options dialog, as shown in Figure 15.30.

Figure 15.31 shows the default contents of the Plan Production subfolder. Notice the templates for Plan, Profile, and Plan And Profile sheet types. There are Imperial and metric versions of each.

As previously discussed, each template contains layout tabs with pages set to various sheet sizes and plan scales. For example, the Civil 3D (Imperial) Plan and Profile.dwt template has layouts created at various ANSI and ARCH sheets sizes and scales, as shown in Figure 15.32.

FIGURE 15.30
Template files location

FIGURE 15.31
Plan Production DWT files

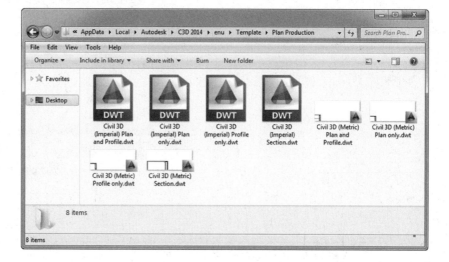

FIGURE 15.32
Various predefined layouts in standard DWT

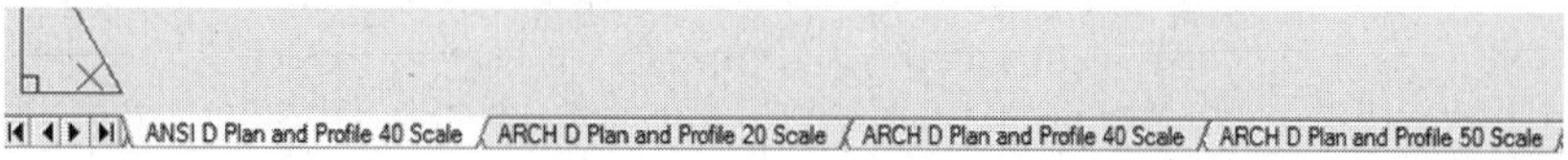

If you decide to make your own Plan Production templates, it is good practice to provide multiple drawing sizes and scales so that you have them available when you go to make your sheets. But beyond just having them available, make sure that the layout names that you provide in your Plan Production template are descriptive enough that you know which one to select. You may also insert a border on each layout tab or opt to externally reference the border in later, after sheets are created.

The viewports in these templates must be rectangular in shape and must have Viewport Type set to Plan, Profile, or Section, depending on the intended use. You set Viewport Type on the Design tab of the Properties dialog, as shown in Figure 15.33.

FIGURE 15.33
Viewport Properties – Viewport Type

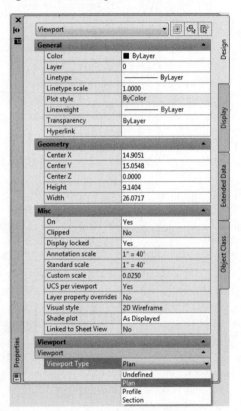

Irregular Viewport Shapes

Just because the viewports must start out rectangular doesn't mean they have to stay that way. Experiment with creating viewports from rectangular polylines that have vertices at the midpoint of each side of the viewport (not just at the corners). After you've created your sheets using the Plan Production tool, you can stretch your viewport into irregular shapes. You can also convert a rectangular viewport into an irregular viewport by selecting the viewport, right-clicking, selecting Viewport Clip, and following the prompts.

The Bottom Line

Create view frames.   When you create view frames, you must select the template file that contains the layout tabs that will be used as the basis for your sheets. This template must contain predefined viewports. You can define these viewports with extra vertices so you can change their shape after the sheets have been created.

Master It   Open the `MasteringPlanProduction.dwg` or `MasteringPlanProduction_METRIC.dwg` file. Run the Create View Frames Wizard to create view frames for Alignment A in the current drawing. (Accept the defaults for all other values.) These view frames will be used to generate Plan and Profile sheets on ARCH D (ISO A1 for metric users) sheets at 20 scale (1:200 scale for metric users) using the plan and profile template `MasteringPandPTemplate.dwt` or `MasteringPandPTemplate_METRIC.dwt`. All files should be saved in `C:\Mastering\CH 15\`.

Edit view frames.   The grips available to edit view frames allow the user some freedom on how the frames will appear.

Master It   Open the `MasteringEditViewFrames.dwg` or `MasteringEditViewFrames_METRIC.dwg` file, and move the VF- (1) view frame to Sta. 2+20 (or Sta. 0+050 for metric users) to lessen the overlap. Then adjust Match Line 1 (or Match Line 2 for metric users) so that it is now at Sta. 4+25 (or Sta. 0+200 for metric users) and shorten it so that the labels are completely within the view frames.

Generate sheets and review Sheet Set Manager.   You can create sheets in new drawing files or in the current drawing. The resulting sheets are based on the template you chose when you created the view frames. If the template contains customized viewports, you can modify the shape of the viewport to better fit your sheet needs.

Master It   Open the `MasteringCreateSheets.dwg` or `MasteringCreateSheets_METRIC.dwg` file. Run the Create Sheets Wizard to create plan and profile sheets in the current drawing for Alignment A using the plan and profile template `MasteringPandPTemplate.dwt` or `MasteringPandPTemplate_METRIC.dwt`. Make sure to choose a north arrow. (Accept the defaults for all other values.) All files should be saved in `C:\Mastering\CH 15\`.

Create section views.   More and more municipalities are requiring section views. Whether this is a mile-long road or a meandering stream, Civil 3D can handle it nicely via Plan Production.

Master It   Open the `MasteringSectionSheets.dwg` or `MasteringSectionSheets_METRIC.dwg` file. Create section views and Plan Production section sheets in a new sheet set for Alignment A using the Road Section section view style and the section sheet template `MasteringSectionTemplate.dwt` or `MasteringSectionTemplate_METRIC.dwt`. Make sure the sections are set to be generated on ARCH D (ISO A1 for metric users) sheets at 20 scale (1:200 scale for metric users). (Accept the defaults for all other values.) All files should be saved in C:\Mastering\CH 15\.

Chapter 16

Advanced Workflows

The AutoCAD® Civil 3D® program is unique in that with a few exceptions, the data you create is stored in the DWG file. Using data shortcuts will allow you to collaborate with coworkers and prevent the DWG size from becoming unwieldy. Even if you work on small projects, data shortcuts can make creating cross-section views a more manageable process.

You will also look at LandXML as a tool to share data. Using this file format, you can transfer "intelligent" data without a DWG file.

In this chapter, you will learn to:

- Create a data shortcut folder
- Create data shortcuts
- Export to earlier releases of AutoCAD
- Export to LandXML

Data Shortcuts

Certification Objective

A *data shortcut* is a link between drawings that allows specific types of Civil 3D data to be shared. The shortcut itself does not contain data, but it is a pointer, directing Civil 3D to read information from a common pool of data. A *data shortcut* is created in the source drawing and a *data reference* is the manifestation of the data in a recipient drawing.

There are many situations in which you need data or information to link between drawings. Connections between drawings can be in the form of external references (XRefs), data shortcuts, or a combination of the two. These two options are similar but not the same. Let's compare (Table 16.1).

TABLE 16.1: XRef vs. data shortcut

XREF	DATA SHORTCUT
For most objects, XRef is a graphic-only representation of objects created in another drawing.	Information-only link to Civil 3D data created in another DWG file.
Any objects (base AutoCAD or Civil 3D) are displayed in XRefs.	Only specific types of Civil 3D data can be used with data shortcuts.

TABLE 16.1: XRef vs. data shortcut *(continued)*

XREF	DATA SHORTCUT
Visibility of objects controlled by original drawing layers.	Visibility of data controlled by host drawing styles and layers.
With the exception of Catchment and Intersection objects, you can use the Civil 3D Add Labels commands on items in an XRef. All other objects (e.g., Surfaces, Alignments, pipes, profile views, etc.) can be labeled through an XRef.	To use Civil 3D object data in design (i.e., using a surface to create an existing ground profile, or using an alignment as a corridor baseline) the object must be data referenced.
A drawing containing a Civil 3D corridor can be XRef'ed into a host drawing. Sample lines can pull corridor data from the XRef.	A data shortcut to an alignment in conjunction with an XRef to the corridor enables sample lines to be created in a separate drawing.

A new feature in Civil 3D 2014 brings another important exception to the general rule that XRefs are "graphics only." Users now have the ability to use figures, lines, polylines, and feature lines as corridor targets through an XRef.

As noted, only Civil 3D objects can be used with data shortcuts. Not all object types are available through shortcuts. The following objects are available for use through data shortcuts:

- Alignments
- Surfaces
- Profile data
- Pipe networks
- Pressure networks (new to Civil 3D 2014)
- View frame groups

> **A Note about the Exercises in This Chapter**
>
> This chapter is about workflow—therefore, it is difficult to jump in partway through the chapter. In order to get the most out of the exercises, you should work from the beginning to end, as all of the exercises build on each previous ones. If you skip any exercises (including the Real World Scenario, "Sending Cross-Section Sheets to Their Own Drawing"), the subsequent steps will not work.
>
> Use the recommended names for objects and drawings you create on your own. Doing so will make things go much smoother for you.
>
> Remember that you can always get the original files from the chapter folder that you can download from www.sybex.com/go/masteringcivil3d2014.

Getting Started

Before making your first project, you should make a *project template*. A project template is simply a set of folders, subfolders, and files. When you start a new project using data shortcuts, we recommend that you use a project template to keep files organized.

In the following exercise, you will create folders for use when creating a data shortcut project:

1. Open Windows Explorer and navigate to C:\Civil 3D Project Templates.

2. Create a new folder titled **MasteringC3D2014**.

3. Inside MasteringC32014, create subfolders called **Design**, **Documents**, **Sheets**, and **Survey**, as shown in Figure 16.1.

FIGURE 16.1
Folder structure and file to be included in the project template

4. Locate the file 1601_Project Checklist.doc from this chapter's files that you downloaded from www.sybex.com/go/masteringcivil3d2014.

5. Place the file in the Documents folder you created in step 3.

6. Locate the file 1601_SheetSetTemplate.dst from this chapter's files.

7. Place the file in the Sheets folder you created in step 3.

This structure will appear inside Civil 3D and in the working folder when a project is created. A Documents folder is included as an example of other, non–Civil 3D–related folders you might have in your project. The 1601_Project Checklist.doc and 1601_SheetSetTemplate .dst files will also be copied automatically to each project you create from this template.

A completed version of this set of folders is in the dataset for your reference and is called 1601_MasteringC3D2014_COMPLETE.

Setting a Working Folder and Data Shortcut Folder

You can think of the working folder as a project directory. The working folder can contain a number of projects, each with a data shortcut folder where the shortcut files reside.

In this exercise, you'll set the working folder and create a new project:

1. Create a new blank drawing using the template of your choice. You won't save this file, but you do need to have a file open to see the Prospector tab of Toolspace.

2. From the Manage tab ➢ Data Shortcuts panel, click Set Working Folder, as shown in Figure 16.2.

FIGURE 16.2
Creating a new working folder

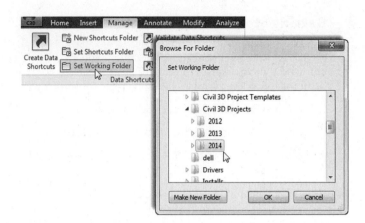

3. Browse to your local C drive's `Civil 3D Projects` folder.

   This folder is created automatically on your local drive when Civil 3D is initially installed.

4. With the `Civil 3D Projects` folder highlighted, click Make New Folder, and make a new folder called **2014** (also shown in Figure 16.2).

5. Highlight the new folder, and click OK.

6. From the Manage tab ➢ Data Shortcuts panel, click New Shortcuts Folder to display the New Data Shortcut Folder dialog shown in Figure 16.3.

FIGURE 16.3
Creating a new shortcut folder

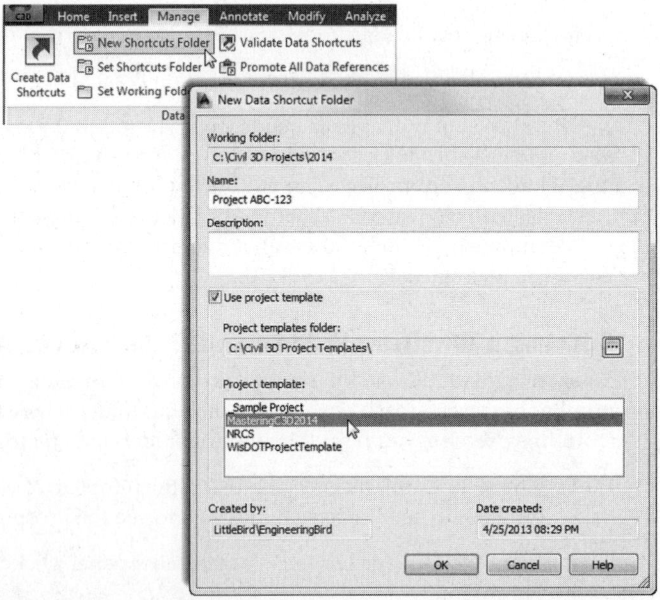

7. Type **Project ABC-123** for the Name, and toggle on the Use Project Template option, as shown in Figure 16.3.

8. Select the MasteringC3D2014 folder from the list, and click OK to dismiss the dialog.

   Notice that the Data Shortcuts branch in Prospector now reflects the path of the Project ABC-123 project, as shown in Figure 16.4.

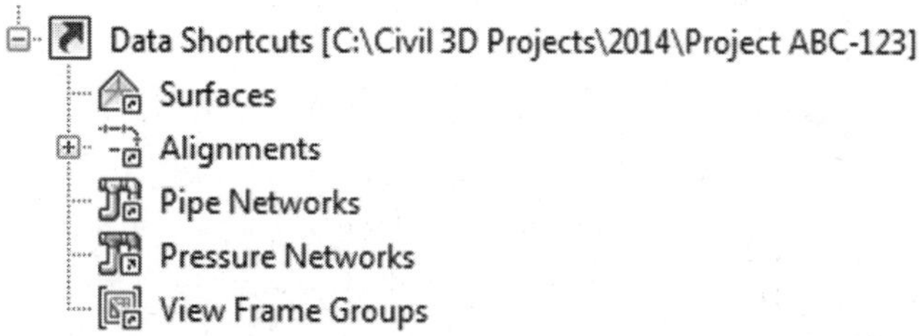

FIGURE 16.4
The Data Shortcut area listed in Prospector

If you open Windows Explorer and navigate to C:\Civil 3D Projects\2014\Project ABC-123, you'll see the folder from the Mastering project template plus a special folder named _Shortcuts, as shown in Figure 16.5. Civil 3D creates this folder, and it is where the data shortcuts will reside.

FIGURE 16.5
Your new project shown in Windows Explorer; "2014" is the working folder. Project ABC-123 is the data shortcut folder.

In real life, your working folder and data shortcut folder should be on a local network server. If you have an established workflow for creating project folders and don't want to use the Civil 3D project template, you can manually create a folder named _Shortcuts. This _Shortcuts folder needs to be inside the folder set as your data shortcut folder in Civil 3D (i.e., one folder level down, as shown in Figure 16.5). To force Civil 3D to recognize the manually created folder, you will need to reset the working folder.

Creating Data Shortcuts

With a shortcut folder in place, it's time to use it. It's a best practice to keep the drawing files in the same location as your shortcut files, just to make things easier to manage. In this exercise, you'll publish data shortcuts for the alignments and layout profiles in your project:

1. Open the 1602_Points-Surface.dwg (1602_Points-Surface_METRIC.dwg) file, which you can download from this book's web page.

   This drawing contains points and an existing ground surface.

2. From the Application menu, click Save As, and save a copy of this drawing to C:\Civil 3D Projects\2014\Project ABC-123\Survey.

LOCATION, LOCATION, LOCATION

When you're working with data shortcuts, location matters! You save the project in its permanent home. Making a data shortcut to the surface and moving the file afterward would cause a broken reference. Later in this chapter, the "Fixing Broken References" section will explain what to do if relocating or renaming a project is unavoidable.

Create Data Shortcuts

3. From the Manage tab ➤ Data Shortcuts panel, click Create Data Shortcuts.

4. In the Create Data Shortcuts dialog, place a check mark next to Surfaces and EXISTING SURFACE, as shown in Figure 16.6, and click OK.

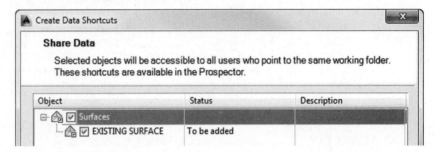

FIGURE 16.6
Adding surface data to data shortcuts

5. Save and close the current drawing file.

6. Open the 1602_Alignments-Profiles.dwg (1602_Alignments-Profiles_METRIC.dwg) file from your class data.

   This file contains three alignments that you will add to the list of data shortcuts.

7. From the Application menu, click Save As, and save a copy of this file to C:\Civil 3D Projects\2014\Project ABC-123\Design.

8. From the Manage tab ➤ Data Shortcuts panel, click Create Data Shortcuts.

9. Place a check mark next to all the alignments (as shown in Figure 16.7), and click OK.

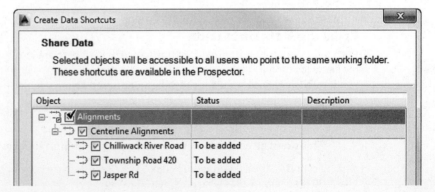

FIGURE 16.7
Adding alignment data to the pool of data shortcuts

DATA SHORTCUTS | **757**

You should now have surfaces and alignments available for use in the data shortcuts listing in Prospector, as shown in Figure 16.8. Notice in Figure 16.8 how highlighting the data shortcut in the listing reveals the Source Location at the bottom of Toolspace.

FIGURE 16.8
List of Civil 3D data shortcuts available to the current project

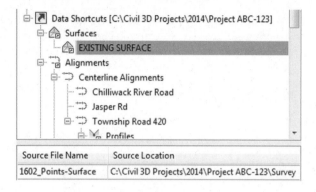

Keep 1602_Alignments-Profiles.dwg (1602_Alignments-Profiles_METRIC.dwg) open for use in the next exercise.

Creating a Data Reference

Now that you've created the shortcut files to act as pointers back to the original drawing, you'll use them in other drawings. In this section, you will use and create data references. You will also use data references in conjunction with a traditional XRef to create cross-section sheets.

Shortcut references are made using the Data Shortcuts branch within Prospector. In this exercise, you'll create references to the surface you previously shared to the project.

You need to have completed the previous exercises to continue.

1. Keep working in 1602_Alignments-Profiles.dwg (1602_Alignments-Profiles_METRIC.dwg).

2. Expand Prospector ➢ Data Shortcuts ➢ Surfaces collection, right-click EXISTING SURFACE and select Create Reference, as shown in Figure 16.9.

FIGURE 16.9
The Create Surface Reference dialog

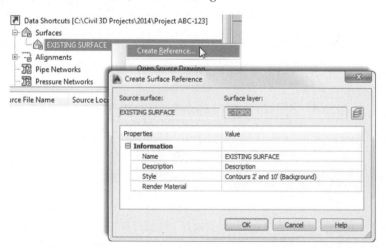

3. In the Create Surface Reference dialog, leave all surface options at their defaults, and click OK. Zoom extents to view the surface.

4. Save the current drawing and keep it open for the next exercise.

You should now see the surface in the drawing. At first glance, it does not look different from other surfaces you've worked with within Civil 3D. If you examine the Surfaces branch of Prospector, however, you'll see that the Definition category under EXISTING SURFACE is missing (as shown in Figure 16.10). This is because you cannot edit the surface data in the recipient drawing. If a change needs to be made to the surface, you need to open the file where EXISTING SURFACE was originally created.

FIGURE 16.10
You can see and use the surface, but you cannot edit the surface definition.

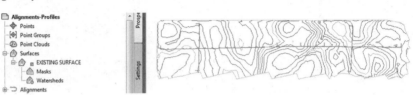

Any changes to the surface will be communicated through the data shortcut to the drawings where it is referenced. If the drawing is open when a data reference has been changed, a bubble message will pop up to inform the user that the item must be synchronized to view the most up-to-date information. If a recipient drawing is closed at the time the change takes place, Civil 3D will automatically synchronize all data references as the file opens.

> **RENAMING OBJECTS ON REFERENCE CREATION**
>
> Avoid it. Not only is it confusing to people who open the file after you, it is simply bad practice. You may experience broken references if the name of an object changes during the Create Reference part of the process.

In the exercise that follows, you will use the skills you learned in Chapter 7, "Profiles and Profile Views," to create an existing ground profile and a design profile. After you have all the information, you will use skills from Chapter 9, "Basic Corridors," to put the information together. You need to have completed the previous exercises to continue.

1. Keep working in `1602_Alignments-Profiles.dwg` (`1602_Alignments-Profiles_METRIC.dwg`).

2. Select the longest alignment in the drawing, Township Road 420.

3. In the Alignment contextual tab ➢ Launch Pad panel, click Surface Profile.

4. In the Create Profile From Surface dialog, do the following:

    a. Click Add.

    b. Click Draw In Profile View.

DATA SHORTCUTS | 759

5. In the Create Profile View dialog, click Create Profile View, and click to place the profile view anywhere in the drawing.

6. Select the profile view. From the Profile View contextual tab ➢ Launch Pad panel, click Profile Creation Tools.

7. In the Create Profile dialog, name the new profile **TR 420 Design**.

8. Leave all other options at their defaults, and click OK.

9. From the Profile Layout Tools, click Insert PVIs Tabular.

10. Enter the data for your unit system, as shown in Figure 16.11, then click OK.

FIGURE 16.11
Profile data
Imperial (left) and
metric (right)

You can close the Profile Layout Tools toolbar when complete.

11. Save the drawing.

12. From the Manage tab ➢ Data Shortcuts panel, click Create Data Shortcuts.

13. Place a check mark next to Profiles, as shown in Figure 16.12, and click OK.

FIGURE 16.12
Adding additional
drawing data to the
data shortcuts

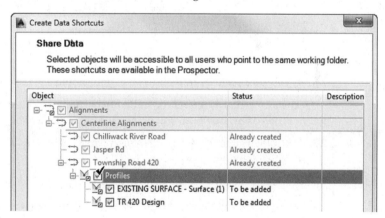

The alignments that you added earlier are listed but are grayed out. Once you add an item to the Data Shortcut list, you do not need to add it again. Only new data will be added to the list of data shortcuts. Shown in gray will be the list of items that are already published as data shortcuts. If you find the list of already published items distracting, you can hide all of the published items by clicking the check box at the bottom of the dialog.

14. Save and close 1602_Alignments-Profiles.dwg (1602_Alignments-Profiles_METRIC.dwg).

15. From your class files, open 1603_Corridor.dwg (1603_Corridor_Metric.dwg).

16. From the Application menu, click Save As, and save the drawing to C:\Civil 3D Projects\2014\Project ABC-123\Design.

17. From the Prospector tab of Toolspace, expand the Data Shortcuts branch, and then expand Surfaces.

18. Right-click EXISTING SURFACE, and click Create Reference.

19. In the Create Surface Reference dialog, leave all default settings, and click OK.

20. From the Prospector tab of the toolbar, Data Shortcuts branch, expand Alignments ➤ Centerline Alignments ➤ Township Road 420 ➤ Profiles.

21. Right-click TR 420 Design and select Create Reference, as shown in Figure 16.13.

FIGURE 16.13
Creating a data reference to a profile will automatically bring along its alignment.

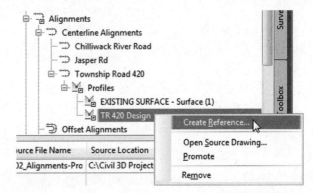

22. In the Create Profile Reference dialog, leave all settings at their default, and click OK.

    If you zoom extents in the drawing, you should see the alignment and existing surface.

    The profile data is referenced but is not currently visible. If you wanted to, you could create a profile view. However, you do not need the profile to be visible to create a corridor.

23. From the Home tab ➤ Create Design panel, click Corridor. In the Create Corridor dialog, do the following:

    a. Name the corridor **Township Road 420**.
    b. Verify the alignment is set to Township Road 420.
    c. Verify that the profile is set to TR 420 Design.
    d. Set the assembly to Typical Section.
    e. Set Target Surface to EXISTING SURFACE.
    f. Clear the check box for Set Baseline And Region Parameters.
    g. Click OK.

24. After the corridor builds successfully, save and close the drawing; you will come back to it later.

> ### Real World Scenario
>
> **SENDING CROSS-SECTION SHEETS TO THEIR OWN DRAWING**
>
> Unlike plan and profile sheets, cross-section sheets do not give you any options for automatically creating layouts in new sheets.
>
> It is a great idea to create cross sections to a drawing separate from the corridor. If you don't use data shortcuts for anything else, we implore you to use them for cross sections.
>
> There are two main reasons for separating sections from the herd:
>
> - You don't need to worry that the annotation scale that looks good for cross sections looks strange for other modelspace items.
> - Corridor drawings already contain lots of data; create section sheets in their own drawing to keep file size down.
>
> The following exercise builds from what you have already created in this chapter. You will step through the procedure of creating cross-section sheets using a combination of XRef and data shortcut functionality. You may want to review Chapter 12, "Cross Sections and Mass Haul," if you have trouble completing the steps in this exercise.
>
> 1. Start a new drawing with the Civil 3D template of your choice.
> 2. Save the drawing to the folder `C:\Civil 3D Projects\2014\Project ABC-123\Sheets` as **Sections.dwg**.
> 3. From the Prospector tab of Toolspace ➢ Expand Data Shortcuts ➢ Alignments ➢ Centerline Alignments, right-click Township Road 420 and click Create Reference.
> 4. In the Create Alignment Reference dialog, leave all the defaults and click OK.
> 5. From the Insert tab ➢ Reference panel, click Attach.
> 6. At the bottom of the Select Reference File dialog, set the Files Of Type to Drawing (*.dwg).
> 7. Browse to `C:\Civil 3D Projects\2014\Project ABC-123\Design`, select `1603_Corridor.dwg` (`1603_Corridor_METRIC.dwg`), and click Open. In the Attach Reference dialog:
>   a. Set the Reference Type to Overlay.
>   b. Set the Path Type to Relative Path.
>   c. Clear the check box for Insertion Point Specify On Screen.
>   d. Click OK.
>
>   You will receive an Unreconciled New Layers message, which you can dismiss by clicking the X.
>
>   Zoom extents to get a look at your drawing so far. You now have everything you need to create sample lines in this new drawing.
>
> 8. From the Home tab ➢ Profile & Section Views panel, click Sample Lines, press Enter to pick Township Road 420 from a list in the Select Alignment dialog, and click OK.

9. In the Create Sample Line Group dialog, keep all default options and styles.

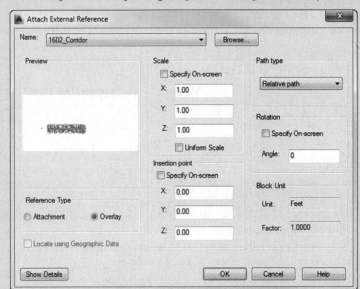

Pause here and take note of a few peculiar things. The EXISTING SURFACE is appearing even though there is no data reference to it. Additionally, you should see the corridor listed. Data references to corridors do not exist in Civil 3D, but you can still link to the data without them. Sample lines are "strong enough" to pull this data out of an external drawing reference. If you had created corridor surfaces, those too would appear in this dialog.

10. Click OK when you are done having your mind blown.
11. From the Sample Line Tools ➤ Sample Line Creation Methods drop-down, click the By Range Of Stations tool.
12. Set the left and right swath widths to **100′** (**30** m).
13. Keep the default sampling increments, and click OK.
14. Press Esc to complete the command.
15. Set your annotation scale to 1″ = 20′ (1:500).
16. From the Home tab ➤ Profile & Section Views panel ➤ Section Views, click Create Multiple Views.
17. In the Create Multiple Section Views dialog, leave all the defaults, and click Create Section Views.
18. Click to place the sheets in the drawing.
19. Save the drawing and keep it open for the next exercise.

After you have successfully created sample lines, the procedure for creating section sheets is the same as you learned in previous chapters.

Updating References

When you need to make a change, you can use the tools in the Data Shortcut menu to jump back to that file, make the changes, and refresh the reference:

1. If it is not open from the exercise in the Real World Scenario "Sending Cross-Section Sheets to Their Own Drawing," open the file Sections.dwg.

   Hopefully, you saved it to C:\Civil 3D Projects\2014\Project ABC-123\Sheets.

2. Keep this drawing open and also open the file 1603_Corrior.dwg.

3. In Prospector, expand Data Shortcuts ➢ Surfaces.

4. Right-click EXISTING SURFACE and select Open Source Drawing, as shown in Figure 16.14.

FIGURE 16.14
Open Source Drawing is a fast way to jump to the drawing you want.

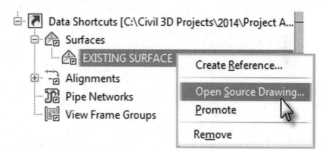

At this point, you should have three drawings open (Corridors, Points-Surface, and Sections). You can use the tabs across the top of the screen to switch drawings (which is a new feature in AutoCAD). You can also use the Quick View Drawings button at the bottom of the AutoCAD window to switch between open drawings.

The 1602_Points-Surface.dwg (1602_Points-Surface_METRIC.dwg) file should be active and ready to make updates. Remember that the surface is read-only in all other project files; this is the only drawing in which changes can be made to the surface.

In this drawing, you will add points to the surface definition. The change to the surface will affect all drawings where the surface is used as a data reference. The dynamic communication between drawings shows the power of the Data Shortcut tool.

5. On the Insert tab ➢ Import panel, click Points From File.

6. In the Import Points dialog, click the plus sign (+) and browse for 1604_NewSurfacePoints.txt (1604_NewSurfacePoints_METRIC.txt). Select this file and click Open.

   This file is part of the chapter data set and can be downloaded from this book's web page.

7. Set Point File Format to PNEZD (Comma Delimited), leave all other settings at their defaults, and click OK.

8. In Prospector, right-click Point Groups and select Update.

9. Expand Prospector ➢ Surfaces, right-click EXISTING SURFACE and select Rebuild.

10. Save and close 1602_Points-Surface.dwg (or 1602_Points-Surface_METRIC.dwg).
11. Switch drawings so that 1603_Corridor.dwg (1603_Corridor_METRIC.dwg) is the active drawing.

   Shortly after bringing this drawing to the forefront, you should see a bubble message appear indicating that Data Shortcut Definitions may have changed, as shown in Figure 16.15.

FIGURE 16.15
Thanks for the head's up! Civil 3D will send the user a message when a data referenced object has changed.

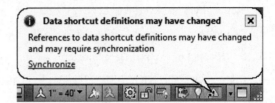

12. Click the Synchronize link in the message.

   In some situations, you may not see the large bubble message. You may just see a warning symbol in the reference indicator at the bottom of the screen. In that case, you can always synchronize directly from the right-click menu of the referenced object, as shown in Figure 16.16.

FIGURE 16.16
Synchronizing from the object's right-click menu

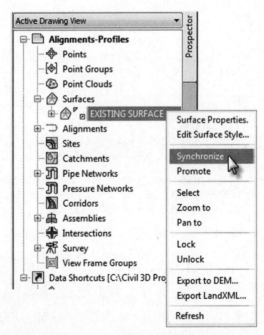

   You may receive a message in Panorama indicating that the item was synchronized.

13. Dismiss this message by clicking the green check box.

The surface that changed is used as a target in the corridor; therefore, the corridor needs to be rebuilt.

14. Expand Prospector ➢ Corridors, right-click the Township Road 420 corridor, and select Rebuild.

15. Save and close 1603_Corridor.dwg (1603_Corridor_METRIC.dwg).

16. Switch to Sections.dwg.

    This time, you will receive a message that your XRef is out of date, as shown in Figure 16.17. If you accidentally dismiss this message without updating, a warning symbol will remain in the tray area at the bottom of your screen. You can always right-click to reload the reference from this icon.

FIGURE 16.17
Reload the XRef to ensure your cross sections reflect the design update.

17. Click Reload 1603_Corridor.

    Your sections should now reflect all the design changes. If you receive a message regarding unreconciled layers, click the X to dismiss it.

18. Save and close all drawings.

    A completed version of the project folder with updates can be found with this chapter's dataset. The folder is called 1603_Project ABC-123_COMPLETE (1603_Project ABC-123_METRIC_COMPLETE).

 Real World Scenario

BEST PRACTICES FOR EMPLOYING DATA SHORTCUTS

When people first add data shortcuts into their workflow, they often have the following questions:

I only work on small sites by myself. Do I need to use data shortcuts?   Even if you work on very small projects, you should use data shortcuts for cross-section sheets.

How should I break up my project?   Every project is different, of course, but use these guidelines:

Existing Survey and Surface   The first drawing you create is usually the existing conditions. Keep survey points, figures, and the existing surface together in the first drawing. Create a data shortcut to the existing surface. Points can't be shared via data shortcut, but you can always use a point query to pull specific points between drawings. For this reason, we highly recommended that you use the Survey Database functionality for all existing survey data.

Alignments and Profiles While the existing surface is being fine-tuned, you can get to work on developing the alignments and profiles right away. Alignments and profiles are fairly lightweight in terms of how much data they consume. You can usually store all of your project's alignment and profile data in the same DWG file.

Corridors When in doubt, use multiple corridor drawings. This way several designers can work on different segments of the project at the same time. Be mindful not to put too much data in one file—you may want to keep interchanges and roundabouts in their own drawings. At the end of the process, data shortcuts will allow you to bring proposed surface data back together.

Pipe Networks and Pressure Networks Pipe networks and pressure pipe networks can be used with data shortcuts. It is OK to put a large network in one file, but you may want to split the type of pipe network (storm versus sanitary, for example) if different designers are working on portions concurrently. Interference checks work between gravity pipe networks even if they are data references.

Site Grading Parking lots, building pads, and other site items are good candidates for data shortcuts.

Where is the data stored? The data for a data shortcut is stored in the source DWG file. The data shortcut file itself is simply a pointer that keeps track of filenames and paths. No object data is stored with the data shortcut.

Can I store my working folder and data shortcut folder on a network? Yes! This is an excellent way to work. You will want to keep your working folder, data shortcut folder, and project drawings on a local network drive, however (i.e., within your building).

What if I need to share the project data with another office? This question is where data shortcuts get a little tricky. How to handle multiple offices depends on the scenario:

Example 1 Your Houston office is working on the road design while the Chicago office is working on grading. You want a live, remote workflow with frequent syncing between offices. For this type of scenario, you would be better served by Autodesk® Vault Collaboration: http://www.autodesk.com/products/autodesk-vault-family/overview.

Example 2 Your Houston office has completed this stage of the road design, and the next phase will be done in the Chicago office. In other words, no live data exchange needs to take place. You can use eTransmit (discussed later in this chapter), or you can move the entirety of the project's data shortcut folder to the new location.

Fixing Broken References

A broken reference is fairly easy to fix, but you will not be able to continue working in your design until broken references are rectified.

Actions that will cause broken references include:

- Renaming a source file
- Moving a source file
- Missing a source file
- Renaming project folders

In the following exercise, you will connect to an existing project and fix the broken data references that you find. You *do not* need to have completed the previous exercises to continue.

1. Using Windows Explorer, locate the folder called 1604_Project XYZ-789.

   This is part of the download for this chapter at the book's web page.

2. Copy this folder and its contents to C:\Civil 3D Projects\2014.

3. Open the XYZ-789 Corridor.dwg file, located in C:\Civil 3D Projects\2014\1604_Project XYZ-789\Design.

   This file contains a number of references pointing to a file that was renamed. Panorama will appear with the messages regarding the problems that it found.

4. Close the Panorama window by clicking the green check mark.

   Expand Prospector ➤ Alignments ➤ Centerline Alignments. You will see that all three alignments have a warning chevron next to them.

5. Right-click Donner Pass and select Repair Broken References, as shown in Figure 16.18.

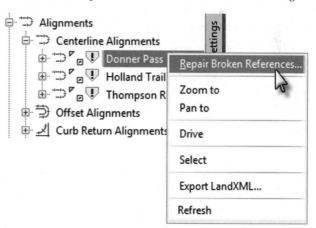

FIGURE 16.18
Choosing Repair Broken References

6. Navigate to the XYZ-789 Alignment.dwg file in the C:\Civil 3D Projects\2014\1604_Project XYZ-789\Design folder, and click Open.

After fixing the first broken reference, Civil 3D determines whether this drawing can be used to fix other reference issues. If it finds more fixes, you will see the message shown in Figure 16.19.

FIGURE 16.19
The Additional Broken References message

7. Click Repair All Broken References.

   Civil 3D will match the Civil 3D objects with broken references to objects in the selected drawing.

The ability to repair broken links helps make file management a bit easier, but there will be times when you need to completely change the path of a shortcut to point to a new file. To do so, you must use the Data Shortcuts Editor.

THE DATA SHORTCUTS EDITOR

Civil 3D tenaciously tries to hang on to links to the information it uses. The Data Shortcuts Editor is used to update or change the file to which a shortcut points. You may want to do this when an alternative design file is approved, or when you move from preliminary to final design.

In the following exercise, you'll move the example project and ensure that the data shortcuts are updated (the procedure is the same for both metric and Imperial units):

1. Save and close any drawings you have open, and close Civil 3D.

2. Using Windows Explorer, browse to `C:\Civil 3D Projects\2014`.

3. Inside the 2014 folder, create a new folder called **Submitted**.

4. In the download for this chapter is a folder called `1604_Project XYZ-789`. Copy `Project XYZ-789` to the Submitted folder. Rename the folder `1604_Project XYZ-789` to **1604_Project XYZ-789 SUB**.

   Your resulting folder structure should look like Figure 16.20.

   You are keeping the original intact for archiving purposes. If you attempted to open drawings that referenced data in the new location, you would see that the data references are looking at the original path.

FIGURE 16.20
Simulating a new project phase by copying the example project to a new folder

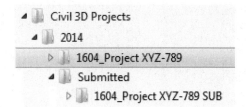

Next, you will ensure that Civil 3D will use the new path in the Submitted version.

5. In Windows, choose Start ➢ All Programs ➢ Autodesk ➢ Autodesk Civil 3D 2014, and click the Data Shortcuts Editor to load it.

6. Select File ➢ Open Data Shortcuts Folder to display the Browse For Folder dialog.

7. Navigate to C:\Civil 3D Projects\2014\Submitted\1604_Project XYZ-789 SUB, and with the 1604_Project XYZ-789 SUB folder highlighted, click OK.

Your Data Shortcuts Editor should resemble Figure 16.21.

FIGURE 16.21
Inside the Data Shortcuts Editor

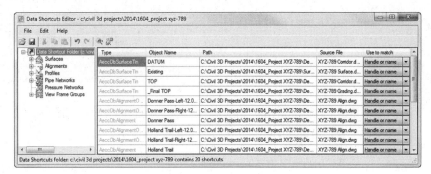

Several things need to be changed before the new phase of the project can begin. Some of the data references are looking for the incorrect file.

Notice that the Path column of the table still refers to the old path. In order to make a clean break from the old project to move forward into the Submitted version, you need to change all these.

8. Choose Edit ➢ Find And Replace.

Unfortunately, there is no browse option here, but you can use the basic Windows Copy and Paste tools to make this a little easier. You can also click your cursor in one of the fields you want to change for a starting point.

9. Type `C:\Civil 3D Project\2014\1604_Project XYZ-789\` in the Find field, and `C:\Civil 3D Projects\2014\Submitted\1604_Project XYZ-789 SUB\` in the Replace With field, as shown in Figure 16.22.

FIGURE 16.22
Updating the paths to the new project

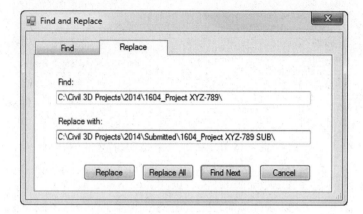

10. Click Replace All.

    Your data shortcut paths and source files are now correct (as shown in Figure 16.23) and ready for more action. Click Cancel to close the Find And Replace tool.

FIGURE 16.23
Updated paths

Object Name	Path	Source File	Us
DATUM	C:\Civil 3D Projects\2014\Submitted\1604_Project XYZ-789 SUB\Design	XYZ-789 Corridor.dwg	Han
Existing	C:\Civil 3D Projects\2014\Submitted\1604_Project XYZ-789 SUB\Survey	XYZ-789 Surface.dwg	Han
TOP	C:\Civil 3D Projects\2014\Submitted\1604_Project XYZ-789 SUB\Design	XYZ-789 Corridor.dwg	Han
_Final TOP	C:\Civil 3D Projects\2014\Submitted\1604_Project XYZ-789 SUB\Design	XYZ-789 Grading.dwg	Han

11. Highlight the Alignments branch on the left side of the Data Shortcuts Editor.

    This will narrow the focus to alignment information.

12. Choose Edit ➢ Find And Replace.

13. In the Find And Replace dialog, in the Find field, enter `XYZ-789 Align.dwg`.

14. In the Replace field, enter `XYZ-789 Alignment.dwg`, as shown in Figure 16.24.

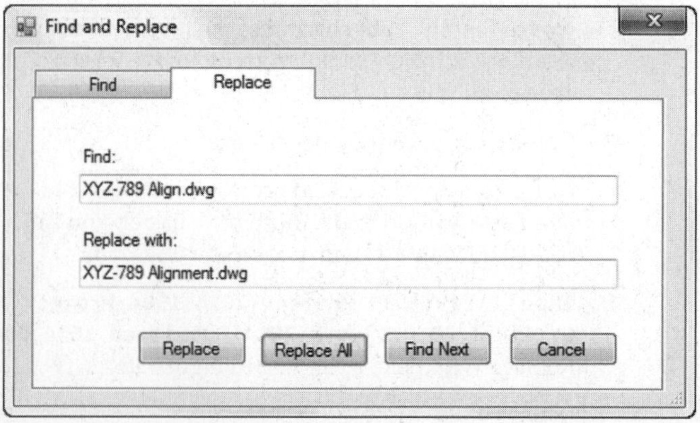

FIGURE 16.24
Using Find And Replace to ensure the new project does not contain references to an old filename

15. Click Replace All.

    All of the source file paths for alignments should be updated.

16. Click the Save icon to commit your changes.

17. You can now close the Data Shortcuts Editor and return to Civil 3D.

---

eTransmit + Data Shortcuts = Awesomesauce!

If you need to pass a project on to another designer with all the information they need, use the eTransmit command.

To find the eTransmit command, select the Application menu ➢ Publish ➢ eTransmit. If a file contains a data shortcut, the eTransmit command will recognize the link and include the necessary drawings. You can even include non-AutoCAD files such as Word or Excel documents by clicking the Add File button.

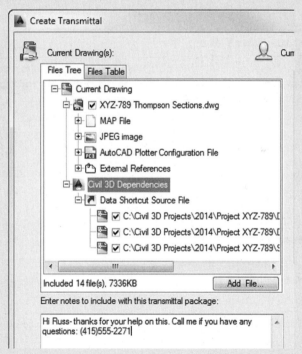

AutoCAD will generate a zip file that contains any XRef files (including DWF, DGN, PDF, or TIF references), plot configurations, images, and templates used in sheet creation. When your recipient unzips the file, the connections between drawings are maintained.

eTransmit is the best way to send Civil 3D files with the assurance that you are not sending a pile of broken references.

Using LandXML

As crazy as it sounds, not everybody is using Civil 3D 2014, so you need to know how to work with the "world outside."

Files created in Civil 3D 2014 are backward compatible to Civil 3D 2013. Currently, these are the only two versions that will work together without losing significant amounts of data.

If you do go back to Civil 3D 2013 from Civil 3D 2014, there are some things that will not work perfectly. Objects created with new functionality will behave as proxy objects when opened in the older software. For example, pressure pipe networks that are present as a result of data shortcuts will be visible, but not editable in the older file format. In the case of corridors that use XREF data as targets, the targets will be removed in the previous version.

The Save As command can be very misleading for Civil 3D users. Keep in mind that the Save As command works only on AutoCAD objects such as lines, arcs, and circles—*not* Civil 3D objects. To save to a version of Civil 3D 2012 or prior, you will need to use the Application Menu ➢ Export ➢ DWG ➢ 2010 command. The result will be exploded Civil 3D objects and lost "intelligence."

> **What Is LandXML?**
>
> LandXML is not specific to Civil 3D. It is a file format that allows users to share data in a nonproprietary format. You can use LandXML to archive projects and to send information to non-Civil 3D users or users of Civil 3D using 2012 or earlier.
>
> A drop-down in the Import LandXML dialog lets you select the version of LandXML. This is a result of that consortium. The latest, as of this writing, is the 1.2 schema. Most CAD programs have methods of importing and exporting LandXML files, and this is one way to tackle that barrier.

Alas, the need to send data to older versions of Civil 3D is not always avoidable. In the following exercise, you will step through an example of what needs to happen to force Civil 3D to previous versions:

1. Open 1605_LandXML-OUT.dwg (1605_LandXML-OUT_METRIC.dwg), which can be found at this book's web page.

2. From the Output tab ➢ Export panel, select Export To LandXML.

   The Export To LandXML dialog shown in Figure 16.25 opens.

3. Click OK.

4. In the Export To LandXML dialog, browse to the same folder as the source drawing.

5. Keep the default name (the dialog will pick up the name of the drawing you are exporting from), and click Save.

FIGURE 16.25
The Export To LandXML dialog

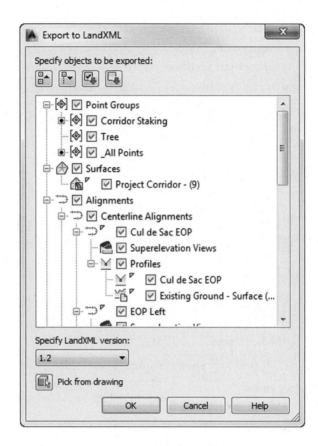

You now have as much Civil 3D as possible packed up into the LandXML file. This file contains:

- Alignments
- Point groups and points
- Surfaces
- Profiles
- Pipe data

The next few steps will break the Civil 3D objects down into base AutoCAD components and save them to a previous version format.

6. From the Application menu ➢ Export ➢ DWG, select 2010.
7. In the Export Drawing Name dialog, browse to the same folder as the source drawing.
8. Keep the default name (the dialog will pick up the name of the drawing you are exporting and add an ACAD- prefix), and click Save.

The resulting file is a "dumbed-down" version of your original Civil 3D file.

Next, you will go through the process of reassembling the data as if you were working on a previous version of Civil 3D.

9. Start a new drawing based on the template of your choice.
10. Save the file as **LandXML-IN.dwg** in the location of your choice.

    You need to start with a Civil 3D template, because exporting to earlier versions will destroy Civil 3D–specific styles and settings.

Insert

11. On the Insert tab ➢ Block panel, click Insert.
12. In the Insert dialog, clear the Insertion Point, Scale, and Rotation check boxes.
13. Place a check mark next to Explode.
14. Click Browse and locate the file LandXML-OUT.dwg or LandXML-OUT_METRIC.dwg that you created in steps 7 and 8 and click Save.
15. Click OK.
16. On the Insert tab ➢ Import panel, click LandXML.
17. In the Import LandXML dialog, browse for the XML file you created in steps 4 and 5, and click Open.

    Figure 16.26 shows the Import LandXML dialog.

FIGURE 16.26
Importing LandXML data

18. Click OK.

19. Dismiss Panorama, and save and close the drawing.

Completed versions of the resulting file after LandXML import can be found with this book's dataset. Compare your work with `1605_LandXML-IN_FINISHED.dwg`. or `1605_LandXML-IN_METRIC_FINISHED.dwg`.

The Bottom Line

Create a data shortcut folder. The ability to load design information into a project environment is an important part of creating an efficient team. The main design elements of the project are available to the data shortcut mechanism via the working folder and data shortcut folder.

Master It Using the `MasterIt1601.dwg` (`MasterIt1601_METRIC.dwg`) drawing, create a new data shortcut folder called Master Data Shortcuts. Use the _Sample Project project template.

Create data shortcuts. To allow sharing of the data, shortcuts must be made before the information can be used in other drawings.

Master It Save the drawing to the Source Drawings folder in the Master Data Shortcuts project you created in the previous exercise. Create data shortcuts to all the available data in the `MasterIt1601.dwg` (`MasterIt1601_METRIC.dwg`) file.

Export to earlier releases of AutoCAD. Being able to export to earlier base AutoCAD versions is sometimes necessary.

Master It Using `MasterIt1601.dwg` (`MasterIt1601_METRIC.dwg`), export the Civil 3D file so it can be used by a user working in base AutoCAD 2010.

Export to LandXML. Being able to work with outside clients or even other departments within your firm who do not have Civil 3D is an important part of collaboration.

Master It Using `MasterIt1601.dwg` (`MasterIt1601_METRIC.dwg`), create a LandXML file with all of the exportable information.

Chapter 17

Quantity Takeoff

The goal of every project is eventually construction. Before the first bulldozer can be fired up and the first pile of dirt moved, the owner, city, or developer has to know how much all of this paving, pipe, and dirt are going to cost. Although contractors and construction managers are typically responsible for creating their own estimates for contracts, the engineers often perform an estimate of cost to help judge and award the eventual contract. To that end, many firms have entire departments that spend their days counting manholes, running planimeters around paving areas, and measuring street lengths to figure out how much striping will be required.

The AutoCAD® Civil 3D® software includes a Quantity Takeoff (QTO) feature to help relieve that tedious burden. You can use the model you've built as part of your design to measure and quantify the pieces needed to turn your project from paper to reality. You can export this data to a number of formats and even to other applications for further analysis.

In this chapter, you will learn to:

- Open and review a list of pay items along with their categorization
- Assign pay items to AutoCAD objects, pipe networks, and corridors
- Use QTO tools to review what items have been tagged for analysis
- Generate QTO output to a variety of formats for review or analysis

Employing Pay Item Files

Before you can begin running any sort of analysis or quantity, you have to know what items you are trying to quantify. Various municipalities, states, and review agencies have their own lists of items and methods of breaking down the quantities involved in a typical development project.

There are three main files associated with quantity takeoff in Civil 3D; the *pay item list*, the *pay item categorization file*, and the *formula file*. Each file serves a different purpose and is stored externally to the project. Once a file has been associated to a DWG, the path to these files is stored with the DWG.

Pay Item List   When preparing quantities, different types of measurements are used based on the items being counted. Some are simple individual objects such as light posts, fire hydrants, and manholes. Only slightly more complicated are linear objects such as road striping and area items such as grass cover. These measurements are also part of the pay item list. At minimum, the pay item list will provide you with three main pieces of information for each item:

- The pay item number
- The pay item description
- The pay item unit of measure

There are three file types allowable for the pay item list file:

- CSV (Comma Delimited)
- AASHTO TransXML
- Florida DOT

Pay Item Categorization File   In any project, there can be thousands of items to tabulate. To make this process easier, most organizations have built pay categories, and Civil 3D makes use of this system in a categorization file. Civil 3D includes an option to create favorite items that are used most frequently. This file, which is always formatted as an XML file, is optional.

Formula File   The formula file is used when the unit of measure for the pay item needs to be converted or calculated based on another value. For example, bituminous concrete for a parking lot is usually paid for by the ton, but you will generally be obtaining an area from CAD. You can set up a formula that converts square feet into tons, taking into account an assumed depth and density. Later in this chapter, you will create a formula that relates length to light poles along the road.

A command is available if you need to disconnect your pay item and formula files from the drawing. At the command line you can type **DETACHQTOFILES**. The Undo command will not reconnect them if you type this in accidentally.

In the next section, you will learn how to connect the needed pay item files and create favorites.

Pay Item Favorites

Your Civil 3D file will "remember" which files it needs to correctly assess quantities. The Favorites list is a separate category that can be populated for quick access to commonly used pay items. This list is saved inside the DWG.

In this exercise, you'll look at how to open a pay item and categorization files and add a few items to the Favorites list for later use:

1. Create a new drawing using the _AutoCAD Civil 3D (Imperial) NCS template. Metric users should use the _AutoCAD Civil 3D (Metric) NCS template.

2. From the Analyze tab ➢ QTO panel, choose QTO Manager (see Figure 17.1) to display the QTO Manager.

FIGURE 17.1
The QTO tools on the Analyze tab

3. Click the Open button at the top left of the QTO Manager to display the Open Pay Item File dialog.

4. In the Open Pay Item File dialog, verify that the Pay Item File Format drop-down list is set to CSV (Comma Delimited).

5. Click the Open button next to the Pay Item File text box to browse for the file.

In Windows Vista, Windows 7, and Windows 8, this file is found in `C:\ProgramData\Autodesk\C3D 2014\enu\Data\Pay Item Data\Getting Started\`.

> **C:\ProgramData**
>
> If you do not see the `ProgramData` folder listed in `C:\`, follow these simple steps:
>
> 1. Open Windows Explorer.
> 2. Click the Organize button at the upper left.
> 3. Select Folder And Search Options to display the Folder Options dialog.
> 4. On the View tab, in Advanced Settings select the radio button Show Hidden Files, Folders, And Drives.
> 5. Click OK to accept the settings in the Folder Options dialog.

Metric users should use the `Getting Started_METRIC.csv`, downloadable from the book's web page (www.sybex.com/go/masteringcivil3d2014).

6. Select the `Getting Started.CSV` file.
7. Click Open to select this CSV pay item file.
8. Click the Open button next to the Pay Item Categorization File text box to browse for the file.
9. Navigate back to the `Getting Started` folder and select the `Getting Started Categories.xml` file.

   This file can be used by both Imperial and metric users.

10. Click Open to select the file.

    Your display should look similar to Figure 17.2.

FIGURE 17.2
Open Pay Item File dialog

11. Click OK to accept the settings in the Open Pay Item File dialog.

    The QTO Manager will now be populated with a collection of divisions, as shown in Figure 17.3. These divisions came from the `Getting Started Categories.xml` file.

FIGURE 17.3
The QTO Manager in Panorama populated within categories (Divisions and Groups)

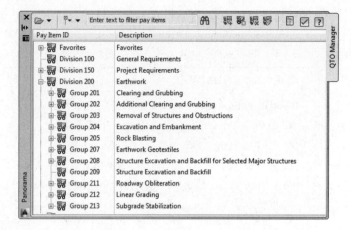

12. Expand the Division 200 ➤ Group 201 ➤ Section 20101 branches, and select the item 20101-0000 CLEARING AND GRUBBING.

13. Right-click and select Add To Favorites List, as shown in Figure 17.4.

FIGURE 17.4
Selecting a pay item to add as a favorite

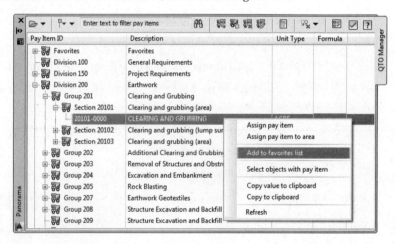

14. Expand a few other branches to familiarize yourself with these pay items.

15. Add the following to your Favorites list in preparation for the next exercise:

    ◆ 60404-2000 CATCH BASIN, TYPE 2
    ◆ 61106-0000 FIRE HYDRANT
    ◆ 61203-0000 MANHOLE, SANITARY SEWER
    ◆ 63401-0600 PAVEMENT MARKDINGS, TYPE C, BROKEN

16. Save the drawing as **QTOPractice.dwg** or **QTOPractice_METRIC.dwg** and keep it open for the next exercise.

When complete, scroll to the top of the Pay Item ID list and expand the Favorites category. Your QTO Manager should look similar to Figure 17.5.

FIGURE 17.5
A list of favorites within the QTO Manager

Civil 3D ships with a number of pay item list categorization files but only one actual pay item list. This avoids any issue with out-of-date data. One commonly used categorization type found in the `C:\ProgramData\Autodesk\C3D 2014\enu\Data\Pay Item Data\CSI\` folder is MasterFormat. In the `United States` folder, you can also find categorization files for AASHTO and Federal Highway Administration. The pay item files that install with the software are intended to be examples. Contact your reviewing agency for access to its pay item list and categorization files if they're not already part of the Civil 3D product.

Once you have pay items to choose from, it's time to assign them to your model for analysis.

Searching for Pay Items

In the top of the QTO Manager is an extremely handy filter tool. If you don't know what category an item is listed under, you can type it in the field to the left of the binoculars icon.

When you first click the binoculars icon to execute the filter, it will appear as if nothing happened. That's because the category headings are getting in the way of the listing. To see an uncategorized version of the list your filter produced, select Turn Off Categorization from the drop-down to the left of the filter field, as shown in Figure 17.6.

FIGURE 17.6
Choose Turn Off Categorization to see the results of your filter.

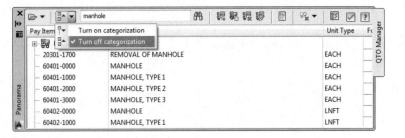

In the following exercise, you will use the filter functionality to add items to your Favorites list:

1. If it's not still open from the previous exercise, open the QTOPractice.dwg or the QTOPractice_METRIC.dwg file.

2. If the QTO Manager is not already open, from the Analyze tab ➢ QTO panel, choose QTO Manager to open it.

3. To the left of the filter field, click the drop-down to select Turn Off Categorization.

4. In the filter field of the QTO Manager, type **grandi** and then click the binoculars icon.

   Your filter should result in three types of trees.

5. Right-click the first item, Fagus Grandifloria, and select Add To Favorites List.

6. Using the same filter technique, add the following item to the Favorites list: 63620-0500 Pole, Type Washington Globe No. 16 Light Standard.

You will find that by adding your most frequently used items to the Favorites list, they will be at your fingertips instead of you having to dig through branch after branch of pay items.

When this exercise is complete, you may close the drawing. A saved copy of this drawing is available from the book's web page with the filename QTOPractice_FINISHED.dwg or QTOPractice_METRIC_FINISHED.dwg.

> **Real World Scenario**
>
> ### Creating Your Own Categorization File
>
> One of the challenges in automating any quantity takeoff analysis is getting the pay item list and categories to match up with your local requirements. Getting a pay item list is pretty straightforward—many reviewing agencies provide their own list for public use to keep all bidding on an equal footing.
>
> Creating the category file can be slightly more difficult, and it will probably require experimentation to get it just right. In this example, you'll walk through creating a couple of categories to be used with a provided pay item list.
>
> Note that modifying the QTO Manager affects all open drawings. You might want to finish the other exercises and come back to this when you need to make your own file in the real world.
>
> 1. Create a new drawing using the _AutoCAD Civil 3D (Imperial) NCS or _AutoCAD Civil 3D (Metric) NCS template.
>
> 2. From the Analyze tab ➢ QTO panel, choose QTO Manager to display the QTO Manager.
>
> 3. Click the Open button at the top left of the QTO Manager to display the Open Pay Item File dialog.
>
> 4. Verify that Pay Item File Format is set to CSV (Comma Delimited).

5. Click the Open button next to the Pay Item File text box and navigate to open the Mastering.csv or Mastering_METRIC.csv file.

   Remember, all files can be downloaded from www.sybex.com/go/masteringcivil3d2014.

6. Click the Open button next to the Pay Item Categorization File text box to display the Open Pay Item Categorization File dialog.

7. Browse to open the MasteringCategories.xml file (the same file is acceptable for both Imperial and metric users), and then click OK to accept the settings in the dialog.

   Your QTO Manager should look like this:

   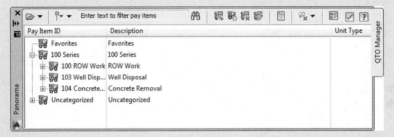

8. Launch XML Notepad. (This is a simple XML editor that you can download for free from Microsoft.)

9. Browse to and open the MasteringCategories.xml file from within XML Notepad.

10. Expand the PayItemCategorizationRules ➤ Categories branch, right-click the Category branch, and select Duplicate, as shown here.

    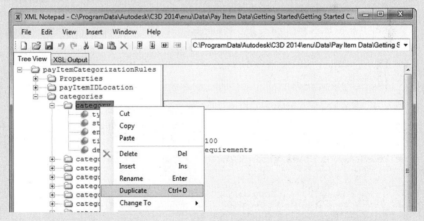

    Note that the duplicated category branch is made current.

11. Modify the values on the right side of the dialog for the duplicate category to match the following.

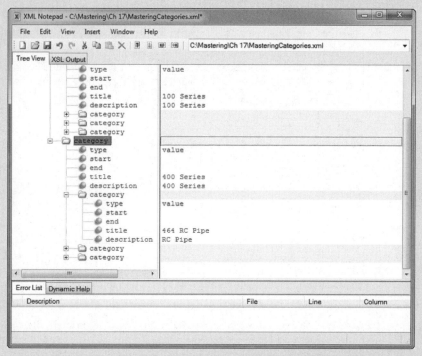

12. Delete the extra category branches under the new branch by right-clicking the two extra category listings and choosing Delete.
13. Save and close the modified `MasteringCategories.xml` file.
14. Switch back to the Civil 3D QTO Manager.

    Next, you will need to reload the categorization file to view the changes.

15. Click the Open drop-down menu in the top left of the QTO Manager, and select Open ➢ Categorization File.
16. Browse to the `MasteringCategories.xml` file again and open it.

    Once Civil 3D processes your XML file, your QTO Manager should look similar to this:

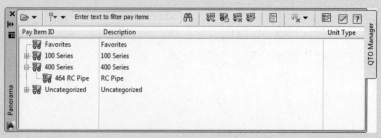

> When this exercise is complete, you may close the drawing. A saved copy of this drawing is available from the book's web page with the filename `QTOCategories_FINISHED.dwg` or `QTOCategories_METRIC_FINISHED.dwg`. In addition, a saved copy of the modified XML file is available with the filename `MasteringCategories_modified.xml`.
>
> Creating a fully developed category list is a bit time consuming, but once it's done, the list can be shared with your entire office so everyone has the same data to use.

Keeping Tabs on the Model

Once you have a list of pay items that must be quantified in your project, you have to assign these pay items to items in your drawing file. You can do so in any of the following ways:

- Assign pay items to simple AutoCAD® objects such as blocks and lines.
- Assign pay items to corridor components.
- Assign pay items to pipes and structures in pipe networks.
- Assign pay items to pressure network pipes, fittings, and appurtenances.

In the next few sections, you'll explore each of these methods, along with some formula tools that can be used to convert things such as linear items to individual quantity counts.

AutoCAD Objects as Pay Items

The most basic use of the QTO tools is to assign pay items to things like blocks and linework within your drawing file. The QTO tools can be used to quantify any number of things, including tree plantings, signposts, and area items such as clearing and grubbing. In the following exercise, you'll assign pay items to blocks as well as to some closed polylines. Be sure the `Getting Started.csv` (or `Getting Started_METRIC.csv`) and `Getting Started Categories.xml` files are loaded as described in the first exercise.

1. Open the `AcadObjectsInQTO.dwg` or `AcadObjectsInQTO_METRIC.dwg` file.
2. From the Analyze tab ➢ QTO panel, choose QTO Manager to display the QTO Manager.
3. Expand the Favorites branch, right-click Clearing And Grubbing, and select Assign Pay Item To Area, as shown in Figure 17.7.

FIGURE 17.7
Assigning an area-based pay item

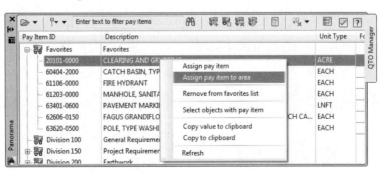

4. At the `select Point or [select Object]:` prompt, type **O** ↵ to activate the Object option for assignment.

5. Click on the closed polyline around the outer edge of the site, as shown in Figure 17.8.

FIGURE 17.8
Selecting a closed polyline for an area-based quantity

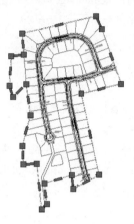

Notice the closed polyline fills with a solid hatch pattern indicating what area is being used. The command line should also echo `Pay item 20101-0000 assigned to area` when you pick the polyline.

6. Press ↵ again to end the command.

7. Select the hatch to activate the Hatch Editor contextual tab.

8. From the Hatch Editor contextual tab ➤ Properties panel, change Hatch Transparency to **80**.

9. Right-click the hatch and select Display Order ➤ Send To Back.

10. Move the QTO Manager to the side and zoom in to the south end of Cabernet Court where you can see one or more of the blocks representing trees.

11. Select one of the tree blocks and then right-click and choose Select Similar. All of the tree blocks are selected.

12. Back in the QTO Manager, under the Favorites branch select 62606-0150 Fagus Grandifloria, American Beech.

13. Near the top of the panel, click the Assign The Selected Pay Items To Object(s) In The Drawing button.

   Notice the great tooltips on these buttons.

14. Press ↵ to end the assigning command.

15. Repeat steps 11 through 14 to assign the pay item 61106-0000 Fire Hydrant to all of the fire hydrant blocks on the site.

When this exercise is complete, you may close the drawing. A saved copy of this drawing is available from the book's web page with the filename `AcadObjectsInQTO_FINISHED.dwg` or `AcadObjectsInQTO_METRIC_FINISHED.dwg`.

Assigning Pay Items

The two assignment methods described in this section, by object or by area, are essentially interchangeable. Pay items can be assigned to any number of AutoCAD objects, meaning you don't have to redraw the planners' or landscape architects' work in Civil 3D to use the QTO tools.

Keep in mind that the pay item tag is saved with the block. This is a good thing if you assign a pay item to a block and copy the block because the copies will be automatically tagged. If you assign a pay item to an object and use the **WBLOCK** command to copy that object out of your current drawing and into another, the pay item assignment goes along as well.

You'll find out how to unassign pay items after we discuss all the ways to assign them.

Pricing Your Corridor

The corridor functionality of Civil 3D is invaluable. You can use it to model everything from roads to streams to parking lots. With the QTO tools, you can also use the corridor object to quantify much of the project construction costs.

Assemblies and QTO Are Related

Be mindful of what parts of an assembly are available when creating quantity takeoff assignments in the code set style.

If you plan to price an item based on the centerline of your road, make sure you have set a crown point code in your assembly properties. If you used the LaneSuperelevationAOR subassembly, this is defined by the Inside Point Code value if your slope direction is set to Away From Crown.

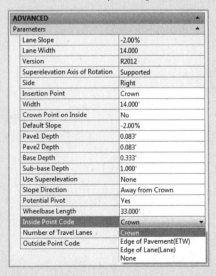

Changes to QTO information will not cause your corridor to flag itself as out of date. Before you run a takeoff report, it is a good idea to rebuild your corridors to ensure that the most up-to-date information is accounted for.

In this example, you'll use the pay item list along with a formula to convert the linear curb measurement to an incremental count of light poles required for the project:

1. Open the QTOCorridors.dwg or QTOCorridors_METRIC.dwg file.
2. From the Analyze tab ➢ QTO panel, choose QTO Manager.
3. Expand the Favorites branch to display your 63620-0500 light standard pay item.
4. Scroll to the right to the Formula column and click in the empty cell on the light standard row.

   When you do, Civil 3D will display the alert box shown in Figure 17.9, which warns you that formulas must be written to an external file.

FIGURE 17.9
Click in the Formula cell to display this warning dialog.

5. Click OK to dismiss the warning, and Civil 3D will present the Select A Quantity Takeoff Formula File dialog.
6. Navigate to C:\Mastering\, change the filename to **Mastering** (or **Mastering_METRIC**), and click Save.

   Civil 3D will display the Pay Item Formula: 63620-0500 dialog, as shown in Figure 17.10. (The Expression box will be empty when you first open it, but you'll take care of that in the next steps.)

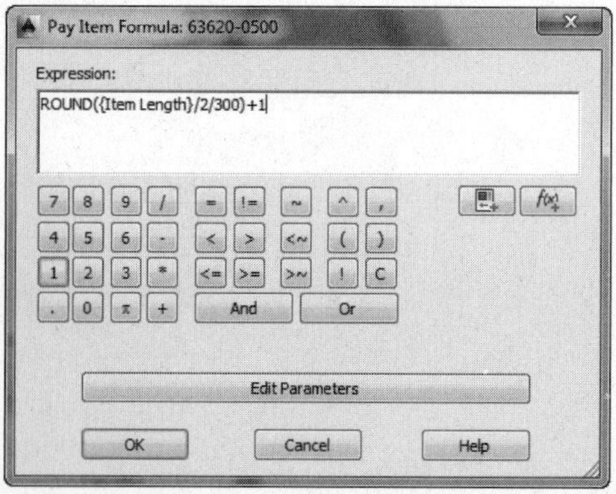

FIGURE 17.10
The completed pay item expression

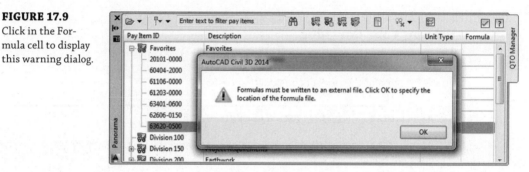

Assume that you need a street light every 300 feet (or 100 m for metric), but only on one side of the street. To do this, you add up all the lengths of curb and then divide by 2 because you want only one half of the street to have lights. You then divide by 300 (or 100 for metric) because you are running lights in an interval. Finally, you round to the nearest integer and add 1 to make the number conservative.

7. Enter the formula using the following steps:

    a. Click the Property button and select Item Length from the drop-down list.

    b. Use the buttons in the dialog or your keyboard to divide this value by 2 to get half of the curb length.

    c. Use the buttons in the dialog or your keyboard to divide this value by 300 (or 100), which is the spacing of the light standards.

    d. To round this value, move your cursor to the beginning of the expression, click the Function button, and select ROUND.

    Note that there are also ROUNDUP and ROUNDDOWN expressions available if these were better suited for your calculations.

    e. Move your cursor to the end of the expression and add the closing parenthesis for the ROUND command.

    f. Use the buttons in the dialog or your keyboard to add 1 to this count to be conservative.

    The formula should now match the one shown previously in Figure 17.10.

8. Click OK.

    Note that the pay item list now shows a small calculator icon on that row to indicate a formula is in use.

    Now that you've modified the way the light poles will be quantified from your model, you can assign the pay items for light poles and road striping to your corridor object. This is done by modifying the code set, as you'll see in the next steps.

9. In the Toolspace Settings tab, expand General ➢ Multipurpose Styles ➢ Code Set Styles, right-click All Codes, and select Edit to display the Code Set Style – All Codes dialog.

10. On the Codes tab, scroll to the Point branch and find the row for Crown, as shown in Figure 17.11.

    As always, you may need to widen some of the columns to view the necessary text.

11. Click the truck icon in the Pay Item column, shown in Figure 17.11, to open the Pay Item List dialog.

FIGURE 17.11
Select the Crown row in the Point section in the Code Set Style – All Codes dialog.

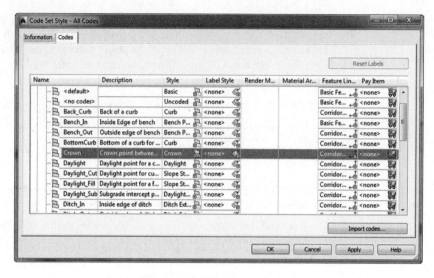

12. Expand the Favorites branch, and select Pavement Markings, Type C, Broken.

13. Click OK to return to the Code Set Style – All Codes dialog.

    You should see the pay item number of 63401-0600 in the Pay Item column of your dialog.

14. Again in the Point branch, find the row for Back_Curb, and repeat steps 10 through 12 to assign 63620-0500 to the Back_Curb code.

    Remember, this is your light-standard pay item with the formula from the previous steps. Your Code Set Style – All Codes dialog should look like Figure 17.12, which lists a pay item for both of these point codes.

FIGURE 17.12
Completed code set editing for pay items

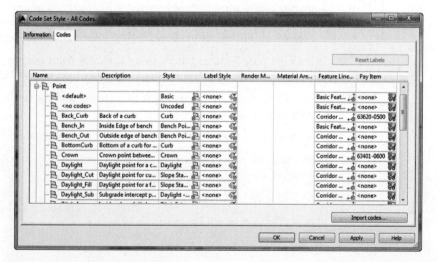

15. Click OK to accept the settings in this dialog.

16. Switch to Prospector, and expand the Corridor branch.

17. Right-click Frontenac Drive Corridor, and select Properties to open the Corridor Properties dialog.

18. On the Codes tab, scroll down to the Point branch.

    Notice that Back_Curb and Crown codes reflect pay items in the far-right column.

19. Click OK to accept the settings in the dialog.

When this exercise is complete, you may close the drawing. A saved copy of this drawing is available from the book's web page with the filename QTOCorridors_FINISHED.dwg or QTOCorridors_METRIC_FINISHED.dwg.

> **BEST PRACTICE: STORING A FORMULA FILE WITH THE PROJECT FILES**
>
> Every drawing in which you utilize QTO tools is intended to have its own, unique formula file. We recommend that when you create a new formula file, you store it in the same folder as the rest of the project.
>
> If you need to send a drawing to another firm and you want them to have your quantity takeoff formulas, use eTransmit to export everything they need to work with the drawing. You can access the eTransmit command by clicking the Application menu in the upper-left corner and selecting Publish ➤ eTransmit. The eTransmit command will attach the pay item file, the categorization file, and the formula file as part of the transmission.

Corridors can be used to measure a large number of items. You've always been able to manage pure quantities of material, but now you can add to that the ability to measure linear and incremental items as well. Although we didn't explore every option, you can also use link codes to assign pay items to your corridor models. Point codes measure the length of the associated feature line. Link codes measure cumulative area between assemblies.

Now that you've looked at AutoCAD objects and corridors, it's time to examine the pipe network objects in Civil 3D as they relate to pay items.

Pipes and Structures as Pay Items

One of the easiest items to quantify in Civil 3D is the pipe network. There are numerous reports that will generate pipe and structure quantities. This part of the model has always been fairly easy to account for; however, with the ability to include it in the overall QTO reports, it's important to understand how parts get pay items assigned. There are two methods: via the parts lists and via the part properties. These methods can also be applied to pressure network pipes, fittings, and appurtenances.

Assigning Pay Items in the Parts List

Ideally, you'll build your model using standard Civil 3D parts lists that you've set up as part of your template. These parts lists contain information about pipe sizes, structure thicknesses, and so on. They can also contain pay item assignments. This means that the pay item property will be assigned as each part is created in the model, skipping the assignment step later.

In this exercise, you'll see how easy it is to modify parts lists to include pay items:

1. Open the `QTOPipeNetworks.dwg` or `QTOPipeNetworks_METRIC.dwg` file.

2. On the Settings tab of Toolspace, expand the Pipe Network ➢ Parts Lists branch.

3. Highlight and right-click Sanitary Sewer and select Edit to display the Network Parts List – Sanitary Sewer dialog.

4. On the Pipes tab, expand the Sanitary Sewer ➢ PVC Pipe (PVC Pipe SI for metric) part family.

   Notice that the far-right column is the Pay Item assignment column.

5. Click the truck icon in the 8-inch PVC (or 200 mm PVC Pipe) row to display the pay item list.

6. Turn off categorization and enter **PVC** in the text box, as shown in Figure 17.13, and press ↵ to filter the dialog.

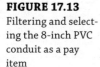

FIGURE 17.13
Filtering and selecting the 8-inch PVC conduit as a pay item

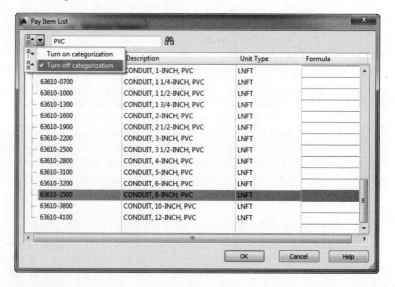

7. Select the CONDUIT, 8-INCH, PVC item (CONDUIT, 200 MM PVC) as shown, and click OK to assign this pay item to the applicable pipe part.

8. Repeat steps 4 through 7 to assign pay items to 10- and 12-inch (250 mm and 300 mm) PVC conduits.

   Your dialog should look like Figure 17.14.

FIGURE 17.14
Completed pipe parts pay item assignment

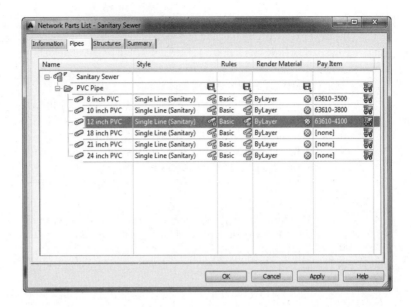

9. On the Structures tab of the Networks Parts List dialog, expand the Sanitary Sewer ➢ Concentric Cylindrical Structure branch.

10. Click the truck icon on the Concentric Structure row to display the Pay Item List dialog.

11. In the Pay Item List dialog under Favorites, select MANHOLE, SANITARY SEWER, and click OK.

The Network Parts List should now look similar to Figure 17.15.

12. Click OK to accept the settings in the Network Parts List.

FIGURE 17.15
Completed structure parts pay item assignment

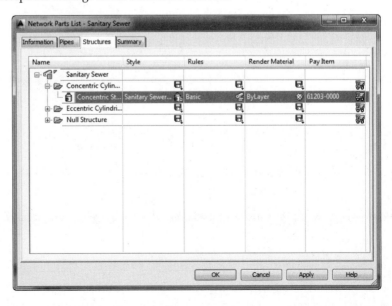

When this exercise is complete, you may save the drawing and keep it open to continue on to the next exercise. Or you may use the saved copy of this drawing available from the book's web page (QTOPipeNetworksPart1_FINISHED.dwg or QTOPipeNetworksPart1_METRIC_FINISHED.dwg).

PAY ITEMS AS PART PROPERTIES

If you have existing Civil 3D pipe networks that were built before your parts list had pay items assigned, or if you change out a part during your design, you need a way to review and modify the pay items associated with your network. Unfortunately, doing so isn't as simple as just telling Civil 3D to reprocess some data, but it's not too complicated either. You simply remove the pay item association and then add new ones.

In this exercise, you'll add pay item assignments to a number of parts already in place in the drawing:

1. Continue working in your file, or open the PipeNetworksPart1_FINISHED.dwg or QTOPipeNetworksPart1_METRIC_FINISHED.dwg file.

2. If the QTO Manager is not already open, from the Analyze tab ➢ QTO panel choose QTO Manager to open it.

3. Slide the QTO Manager to one side, and then select one of the manholes in the Sanitary Network (they have an S symbol on them).

4. Right-click and choose Select Similar. Twenty manholes should highlight, as shown in Figure 17.16.

FIGURE 17.16
Use Select Similar to find all sanitary manhole structures.

5. In the QTO Manager, expand Favorites and select MANHOLE, SANITARY SEWER.
6. Click the Assign The Selected Pay Items To Object(s) In The Drawing button.
7. At the command line, press ↵ to complete the assignment.

You can pause your cursor over one of the manholes and the tooltip will reflect a pay item now, in addition to the typical information found on a manhole.

When this exercise is complete, you may save the drawing and keep it open to continue on to the next exercise. Or you may use the saved copy of this drawing available from the book's web page (QTOPipeNetworksPart2_FINISHED.dwg or QTOPipeNetworksPart2_METRIC_FINISHED.dwg).

HEADS UP ON PIPE NETWORK AND PRESSURE NETWORK QTO ASSIGNMENTS

Pay items change when network parts change if the new part has a pay item assigned in the parts list. However, if you swap to a part that does not have a pay item assigned, Civil 3D drops the QTO tag.

The best way to keep a proper count of your pipe network or pressure network parts is to make sure every part in your network parts list has an appropriate pay item.

If you graphically change a pipe property (such as its diameter), this will not cause a change in the pay item assignment. You should always use Swap Part to change sizes, as you learned in Chapter 13, "Pipe Networks."

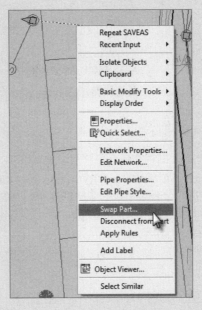

This is something you definitely want to keep an eye on. So, if you do need to change a pay item assignment to a part that's already in the network, how do you do it? You'll find that out in the next section.

Assigning pay items to existing structures and pipes is similar to adding data to standard AutoCAD objects. As mentioned before, the pay item assignments sometimes get confused in the process of changing parts and pipe properties, and they should be manually updated. To do so, you'll need to remove pay item data and then add it back in, as demonstrated in this exercise:

1. Continue working with your file or open the QTOPipeNetworksPart2_FINISHED.dwg or QTOPipeNetworksPart2_METRIC_FINISHED.dwg file.
2. Pan to the northwest area of the site where the sanitary sewer network terminates.
3. Pause your cursor over the pipe connected to the structure at the termination of the sewer network.

   The tooltip will appear indicating the pipe information but no pay item, as shown in Figure 17.17.

FIGURE 17.17
Tooltip for a pipe without an assigned pay item

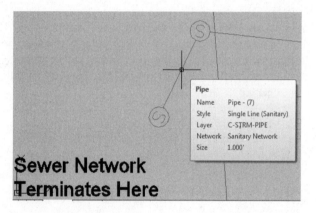

4. Select the pipe, right-click, and select Swap Part.

   This will display the Swap Part Size dialog.

5. Select 12 Inch PVC (or 300 mm PVC Pipe) from the list of sizes (the same size of pipe that it was originally—you are not changing the diameter), and then click OK.

6. Pause near the newly sized pipe and notice that the tooltip now shows the pay item, as shown in Figure 17.18.

FIGURE 17.18
Tooltip after the pipe part has been swapped to a part with an assigned pay item

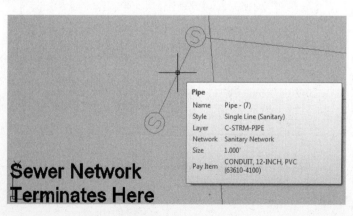

When this exercise is complete, you may close the drawing. A saved copy of this drawing is available from the book's web page with the filename QTOPipeNetworksPart3_FINISHED.dwg or QTOPipeNetworksPart3_METRIC_FINISHED.dwg.

You might wonder why Civil 3D allows you to have multiple pay items on a single object. For example, linear feet of striping and tree counts can both be derived from street lengths; bedding and pipe material can both be calculated from pipe objects. You can also add related tasks to an item. For instance, a tree is usually a pay item by itself, but the labor to install the tree may be treated as a separate pay item.

You've now built up a list of pay items, tagged your drawing a number of ways, updated and modified pay item data, and looked at formulas in pay items. In the next section, you'll make a final check of your assignments before running reports.

Highlighting Pay Items

Before you run any reports, it's a good idea to make a cursory pass through your drawing and look at what items have had pay items assigned and what items have not. This review will allow you to hopefully catch missing items (such as hydrants added after the pay item assignment was done) as well as see any items that perhaps were blocked in with unnecessary pay items already assigned. In this exercise, you'll look at tools for highlighting objects with and without pay item assignments:

1. Open the QTOHighlighting.dwg or the QTOHighlighting_METRIC.dwg file.

2. From the Analyze tab ➢ QTO panel, choose QTO Manager to display the QTO Manager.

3. In the QTO Manager, select Highlight Objects With Pay Items, as shown in Figure 17.19.

FIGURE 17.19
Turning on highlighting for items with pay items assigned

Notice that from this same drop-down menu you can also choose Highlight Objects Without Pay Items, Highlight Objects With Selected Pay Items, or Clear Highlight.

4. Pan around the drawing and zoom in on a tree.

5. In the QTO Manager, select Highlight Objects Without Pay Items.

Notice that the trees turn from green to muted. This means they have a pay item assigned.

6. In the QTO Manager, switch back to Highlight Objects With Pay Items.

Next, you will change the pay item assignment for the trees at the cul-de-sac since the planner has decided to use a different tree type.

7. In the QTO Manager, click the Remove Pay Item(s) From Specified Objects button.

8. At the Select object(s): prompt, select the tree to the southwest of the cul-de-sac at the end of Frontenac Drive and press ↵ to end the command.

Notice that the tree goes from green to a muted gray, indicating that it no longer has a pay item associated with it since the highlighting is enabled.

9. In the QTO Manager, enter **Maple** in the text box to filter.
10. Turn the categorization option off to more easily see the filtered results.
11. Select item 62601-0100.

12. Click the Assign The Selected Pay Items To Objects In The Drawing button.
13. At the `Select pay item(s) from master pay item list or [Enter]:` prompt, press ↵.
14. At the `Select object(s):` prompt, select the same tree to the southwest of the cul-de-sac at the end of Frontenac Drive that you just unassigned a pay item from, and press ↵ to end the command.

    Notice that the tree goes from a muted gray to green, indicating that it once again has a pay item associated with it since the highlighting is enabled.

    Unassigning and then reassigning a pay item may seem cumbersome. Instead, you may find it simpler to edit the pay item.

15. In the QTO Manager, click the Edit Pay Items On Specified Object button.
16. At the `Select object(s):` prompt, select the other tree to the southeast of the cul-de-sac at the end of Frontenac Drive to display the Edit Pay Items dialog.

    While we are selecting only one tree for this exercise, you could select multiple objects.

17. Select the Fagus Grandifloria row, and then click the red X in the upper right to remove this pay item from the tree, as shown in Figure 17.20.

FIGURE 17.20
Editing pay item assignments: deleting the tree pay item

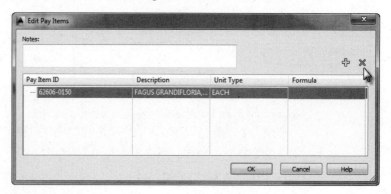

18. Click the plus in the upper right to display the Pay Item List dialog in order to add a new pay item to the tree.
19. Enter **Maple** in the text box to filter.
20. Select item 62601-0100 and click OK.
21. Click OK to accept the new pay item designation shown in the Edit Pay Items dialog.

22. In the QTO Manager, select Highlight Objects Without Pay Items from the drop-down menu shown previously in Figure 17.19.

    This switches the highlighting from objects that do have pay items to objects that do not have pay items.

23. In the QTO Manager, select Clear Highlight to return the drawing view to normal.

When this exercise is complete, you may close the drawing. A saved copy of this drawing is available from the book's web page with the filename QTOHighlighting_FINISHED.dwg or QTOHighlighting_METRIC_FINISHED.dwg.

While highlighting objects with QTO manager, you can add, remove, or edit pay items using the tools at the top of the QTO Manager. You can leave the objects highlighted while performing any other AutoCAD. This makes it easier to correct any mistakes made during the assignment phase of the process. Finally, always be sure to clear highlighting before exiting the drawing or your peers might wind up awfully confused when they open the file!

Inventorying Your Pay Items

At the end of the process, you need to generate some sort of report that shows the pay items in the model, the quantities of each item, and the units of measurement. This data can be used as part of the plan set in some cases, but it's often requested in other formats to make further analysis possible. In this exercise, you'll look at the Quantity Takeoff tool that works in conjunction with the QTO Manager to create reports:

1. Open the QTOReporting.dwg or QTOReporting_METRIC.dwg file.

2. From the Analyze tab ➢ QTO panel, choose Takeoff to display the Compute Quantity Takeoff dialog, shown in Figure 17.21.

FIGURE 17.21
The Compute Quantity Takeoff dialog with default settings

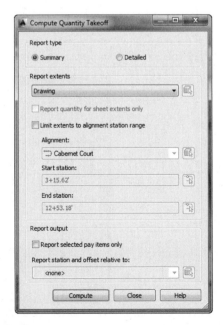

> **LIMITING THE REPORT EXTENTS**
>
> Note that you can limit the report extents by drawing; by sheets, if done from paperspace; by selection set; or by alignment station ranges. Most of the time, you'll want to run the full drawing. You can set the report output for only selected pay items if, for instance, you just want a table of pipe and structures.

3. Click Compute to open the Quantity Takeoff Report dialog, as shown in Figure 17.22.

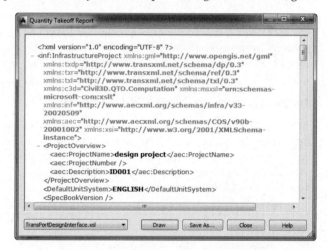

FIGURE 17.22
Quantity Takeoff Report in the default XSL format

The report is shown in the default Extensible Stylesheet Language (XSL) format.

4. From the drop-down menu on the lower left of the dialog, select Summary (TXT).xsl to change the format to something more understandable, as shown in Figure 17.23.

FIGURE 17.23
Quantity Takeoff Report in the Summary TXT format

At this point, you can export this data out as a text file, but for the purpose of this exercise, you'll simply insert it into the drawing.

5. Click the Draw button at bottom of the dialog, and Civil 3D will prompt you to pick a point in the drawing for the report table.

6. Click near some clear space, and you'll be returned to the Quantity Takeoff Report dialog.

7. Click Close to dismiss this dialog, and then click Close again to dismiss the Compute Quantity Takeoff dialog.

8. The program zooms into where you clicked in step 6, and you should see something like Figure 17.24.

FIGURE 17.24
Summary takeoff data inserted into the drawing

When this exercise is complete, you may close the drawing. A saved copy of this drawing is available from the book's web page with the filename QTOReporting_FINISHED.dwg or QTOReporting_METRIC_FINISHED.dwg.

That's it! The hard work in preparing QTO data is in assigning the pay items. The reports can be saved to HTML, TXT, or XLS format for use in almost any analysis program.

The Bottom Line

Open and review a list of pay items along with their categorization. The pay item list is the cornerstone of quantity takeoffs. You should download and review your pay item list and compare it against the current reviewing agency list regularly to avoid any missed items.

> **Master It** Using the template of your choice, open the Getting Started.csv (or Getting Started_Metric.csv) pay item file that you have been working with in the previous exercises and add the 12-, 18-, and 24-Inch Pipe Culvert (or 300 mm, 450 mm, and 600 mm Pipe Culvert) pay items to your Favorites list in the QTO Manager.

Assign pay items to AutoCAD objects, pipe networks, and corridors. The majority of the work in preparing quantity takeoffs is in assigning pay items accurately. By using the linework, blocks, and Civil 3D objects in your drawing as part of the process, you reduce the effort involved in generating accurate quantities.

> **Master It** Open the MasteringQTO.dwg or MasteringQTO_Metric.dwg file and assign the CLEARING AND GRUBBING pay item to the polyline that was originally extracted from the border of the corridor. Change the hatch to have a transparency of 80.

Use QTO tools to review what items have been tagged for analysis. By using the built-in highlighting tools to verify pay item assignments, you can avoid costly errors when running your QTO reports.

> **Master It** Verify that the area in the previous exercise has been assigned a pay item.

Generate QTO output to a variety of formats for review or analysis. The quantity takeoff reports give you a quick understanding of what items have been tagged in the drawing, and they can generate text in the drawing or external reports for uses in other applications.

Master It Display the length of Type C Broken markings in a Quantity Takeoff Report with the Summary (TXT) report style using the `MasteringQTOReporting.dwg` or `MasteringQTOReporting_Metric.dwg` file.

Chapter 18

Label Styles

The creation of proper styles and settings can make or break your experience with AutoCAD® Civil 3D® software. Styles control the display properties of Civil 3D objects and labels. Understanding and applying styles correctly can mean the difference between getting a job out in several hours and fighting with your project drawing for days.

This chapter is organized by style type. First, read the chapter to be introduced to the style concepts in a logical manner. Later, when you use this chapter as a reference to build styles, you will likely jump around to the examples that meet your needs.

In this chapter, you will learn to:

- Override individual labels with other styles
- Create a new label set for alignments
- Create and use expressions
- Apply a standard label set to profiles

Label Styles

The best design in the universe is not worth anything unless it is labeled properly. Civil 3D labels are smart objects that are dynamically linked to the object they are labeling. Civil 3D labeling is customizable to fit your design needs and local requirements.

When talking about labels in Civil 3D, keep in mind that you are not just talking about text. Labels can contain lines and blocks if desired. Label styles control the plotted height, contents, and precision of text. They also control how leaders are applied when the label is dragged away from its initial position. You will even work with some labels that contain no text at all!

General Labels

Certification Objective

On the Settings tab, you'll see a complete list of objects that Civil 3D uses to build its design model. Each of them has special features unique to the object being described, but there are some common features as well. The General collection contains settings and styles that are applied to various objects across the entire product.

The General collection serves as the catchall for styles that apply to multiple objects and for settings that apply to *no* objects. For instance, the Civil 3D General Note object doesn't really belong with the Surface or Pipe collections. It can be used to relate information about those objects, but because it can also relate to something like "Don't Dig Here!" or a northing-easting of an arbitrary location, it falls into the General category.

The Label Styles collection allows Civil 3D users to place general text notes or label single entities while still taking advantage of the flexibility and scaling properties. The various label styles shown in Figure 18.1 can give you some idea of their uses.

FIGURE 18.1
Line label styles

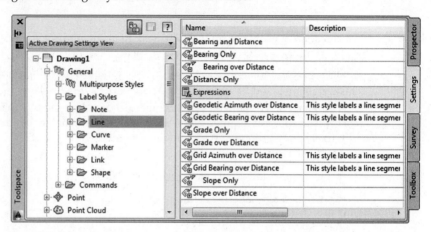

Label styles are a critical part of producing plans with Civil 3D. In this chapter, you'll learn how to build a new basic label and explore some of the common components that appear in every label style throughout the product.

Frequently Seen Tabs

To get into a label style, find the appropriate label type in the Settings tab of Toolspace. The Settings tab is organized into object collections. The Label Styles branch inside the object collections may be organized by type, which makes it very easy to find the label style you need to edit. When you locate the desired label style, right-click the style to display Label Style menu options:

Edit   Opens the label style for modification.

New   Copies the label style as a child style, a style that inherits its settings from a parent.

Copy   Creates a duplicate label style.

Delete   Erases the label style.

Find References   Visible only if label style is in use. Searches settings and objects and lists where label styles are in use.

Replace With   Visible only if label style is in use. Finds a label style in use in the drawing and replaces it with another.

All styles, regardless of type, have a few things in common:

Information Tab   The Information tab, shown in Figure 18.2, controls the name of the style.

FIGURE 18.2
The Information tab exists for all object and label styles.

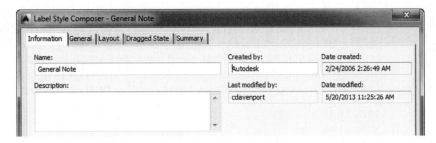

If desired, you can create a description. This information will appear as a tooltip as you browse through the Settings tab.

On the right side of the dialog, you will see the name of the person who originally created the style, the date when it was created, and the last person who modified the style. These names are initially pulled from the Windows login information, and only the Created By field can be edited.

General Tab The General tab (Figure 18.3) contains basic settings for the label style and consists of the following:

FIGURE 18.3
The General tab

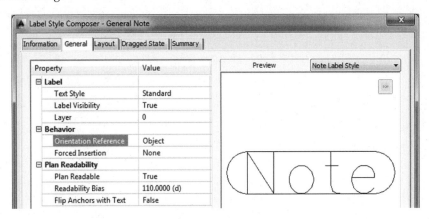

Text Style This option refers to the AutoCAD text style used in the label. Text styles must already be created in the drawing.

Layer In Chapter 1, "The Basics," you learned that the Object Layer tab in the Drawing Settings dialog is where you configure insertion layers for object types, labels, and tables. Freezing an object layer will, of course, make any object contained on that layer disappear. The Layer setting here on the General tab allows you to specify a layer that will override the layer properties of the object layer and also add another layer for controlling visibility of specific label types. For example, the object layer for alignment labels could be set to C-ALIGN-LABL. However, Major Station label styles could be configured to place the label on C-ROAD-STAN-MAJR. When C-ALIGN-LABL is frozen, all alignment labels disappear. When C-ROAD-STAN-MAJR is frozen, only Major Station labels disappear. If this setting is left set to layer 0, then the label using this style will default to the object layer display properties.

Orientation Reference This option controls how text rotation is controlled. Figure 18.4a shows the label aligned with the object. Figure 18.4b shows the text rotated to the view. Even though the view has been rotated, the text still appears parallel with the bottom of the screen. The last, and least used, option for orientation reference is the World Coordinate System option, which is shown in Figure 18.4c. The view is rotated, and the direction of the text rotates with the World Coordinate System.

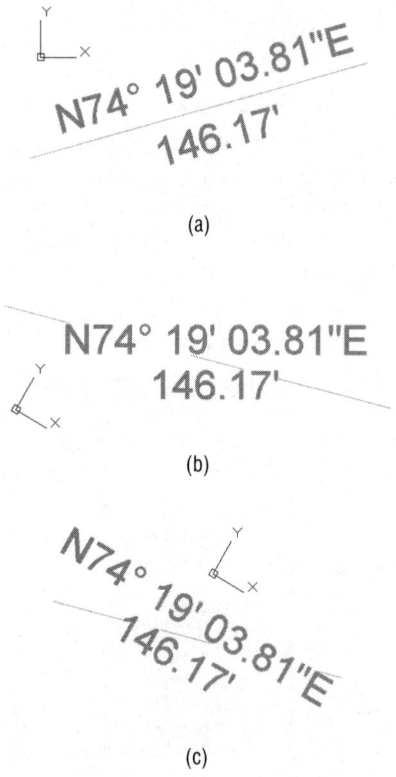

FIGURE 18.4
Orientation reference options set to Object (a), View (b), and World Coordinate System (c)

N74° 19' 03.81"E
146.17'

(a)

N74° 19' 03.81"E
146.17'

(b)

N74° 19' 03.81"E
146.17'

(c)

Forced Insertion This option makes more sense in other objects and will be explored further. The Forced Insertion feature allows you to dictate the insertion point of a label on the basis of the object being labeled. Figure 18.5 shows the effects of the various options on a bearing and distance label.

Plan Readable When this option is enabled, text maintains the up direction in spite of view rotation. Rotating 100 labels is a tedious, thankless task, and this option handles it automatically.

Readability Bias This option specifies the angle at which readability kicks in. This angle is measured from the 0 degree of the x-axis that is common to AutoCAD angle measurements. When a piece of text goes past the readable bias angle, the text flips 180 degrees to maintain vertical orientation, as shown in Figure 18.6. The default Readability angle is 110. A common desired default is 90.1, which would flip any text just after passing a vertical direction.

FIGURE 18.5
Forced Insertion options for parcel segments

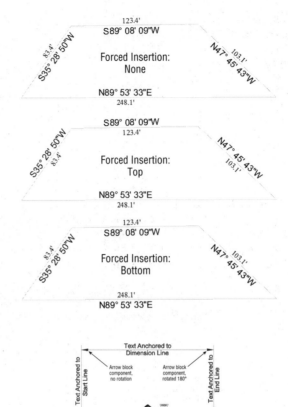

FIGURE 18.6
Plan-readable text shown on contours; note the difference in direction between the top grouping and the bottom grouping.

Flip Anchors With Text Most users leave this setting at its default, False, and never give it another thought. Figure 18.7a shows what happens to the text insertion points when this option is set to False and readability kicks in. The SW bearing is followed by its distance as originally configured. Readability has been applied to the NW bearing because it was configured to be oriented to the object it is labeling. Without readability, the label would be upside down. Distance and bearing have been flipped individually so it appears the two values are out of order. When this setting is set to True, the distance text flips to the other side of the bearing, as shown in Figure 18.7b. You'll have the opportunity to work with this setting in a later exercise.

FIGURE 18.7
Flip anchors with text when readability kicks in: set to False (a); set to True (b)

Layout Tab Each label can be made up of several components. A Label component can be text, a block, or a line. Depending on the type of label style, other options can include Reference Text, Ticks (station labels), Direction Arrow (line labels), or Text For Each (structure styles only). The top row of buttons controls the selection, creation, and deletion of these components, as shown in Figure 18.8.

FIGURE 18.8
The Layout tab

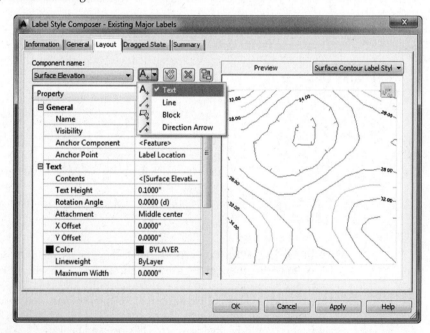

Component Name From the Component Name drop-down, choose the component you want to modify. The components are listed in the order in which they were created. When you make changes to properties on the Layout tab, pay attention to which component is active. Any changes you make to the properties will apply only to the active component.

Create Component The Create Component button lets you add new components to enhance your labels. A component can be Text, Line, Block, Tick, Direction Arrow, Reference Text, or Text For Each. The options will vary depending on the object you are labeling. Not every option is available in every label. For example, Text For Each is available only in structure labels.

The ability to label one object while referencing another (reference text) is one of the most powerful labeling features of Civil 3D. This is what allows you to label a spot elevation for both an existing and a proposed surface at the same time, using the same label. Alignments, COGO points, parcels, profiles, surfaces, and survey figures can all be used as reference text. Text For Each is a type of reference text that can label properties of connected pipes in a structure label.

Copy Component The Copy Component button copies the component currently selected in the Component Name drop-down. This will be helpful when you're creating label styles that contain multiple pieces of similar information.

Delete Component The Delete Component button deletes components. Components that are configured as anchors for other components can be deleted, but you will receive a warning. Table tag components, found in curve, line, alignment segment, and parcel segment labels, cannot be deleted.

Component Draw Order The Component Draw Order button lets you shuffle components up and down within the label. This feature is especially important when you're using masks or borders as part of the label.

Once a component is added, the component properties can be addressed one by one by starting at the top and working your way down to the bottom of the dialog. There are three groups of property types: General, Text, and Border.

The General properties consist of the following:

Name This option defines the name used in the Component Name drop-down and when selecting other components for anchor components. Once a component is configured as an anchor component, the name cannot be changed until the reference is removed. When you're building complicated labels, a descriptive name goes a long way.

Visibility When this option is set to True, the component can be seen on screen. When a component can't be seen, you can do some cool tricks with styles, as you'll see in the "Pipe Labels" section later in this chapter.

Anchor Component This setting allows you to position the component relative to the feature being labeled or to another component.

Anchor Point The options here may vary depending on the Anchor Component setting. When the feature is being used as an Anchor component, the anchor point can be a location relative to the feature offering combinations of middle, top, and bottom vertical positions with left, right, and center horizontal positions. Line components offer start, middle, and end anchor points.

The Layout tab of the Label Style Composer dialog is displayed in Figure 18.9.

FIGURE 18.9
The circle and square indicate the anchor point and attachment point, respectively.

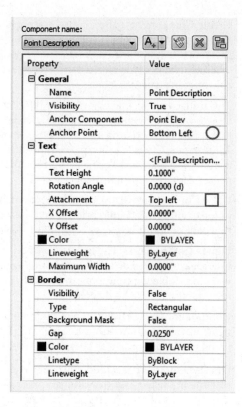

The Text properties consist of the following:

Contents   This is where the information contained in the label is composed. Click the Value field to reveal an ellipsis. If the ellipsis is clicked, the Label Style Composer dialog opens to where property fields and text can be configured.

Text Height   This option determines the plotted height of the label. This setting overrides the height configured in the text style. Regardless of how the text style is configured in the label, text placed by Civil 3D is always annotative. The two viewports in Figure 18.10 show some COGO points along a road. Even when the viewport scales differ, the text is the same size.

Rotation Angle, X Offset, and Y Offset   These options give you the ability to adjust the placement of the component by rotating or displacing the text in an x or y direction. Set your text as close as possible using the anchors and attachments, and use the offsets as additional spacing.

Attachment   This option determines which of the nine points on the Label Components bounding box are attached to the anchor point. Figure 18.11 illustrates the relationship between the anchor points and attachments.

FIGURE 18.10
Annotative text shown at multiple scales

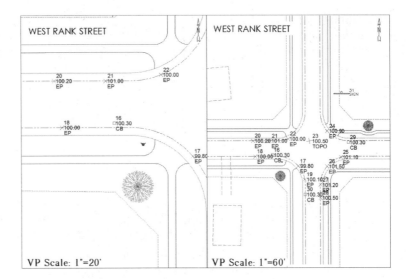

FIGURE 18.11
Schematic showing the relationship between anchor points (circles) and attachments (squares)

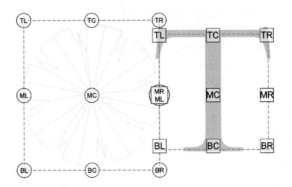

Color and Lineweight   These options allow you to override the color and lineweight assigned by the layer configured on the General tab.

Maximum Width   Some labels can be rather lengthy. Instead of letting the Label component continue indefinitely across the page, you can use Maximum Width to force word wrap after a specified plotted length. The default setting of 0.00 will not force word wrap.

The Border properties consist of the following:

Visibility   Use this option to turn the border on and off for the component. Remember that component borders shrink to the individual component; if you're using multiple components in a label, each will have its own Border component.

Type   This option allows you to select a rectangle, a rounded rectangle (slot), or a circle border. Figure 18.12 shows examples of the three types of borders.

FIGURE 18.12
Border types shown on various surface label styles

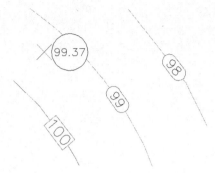

Background Mask This option lets you determine whether linework and text behind this component are masked. This option can be handy for construction notes in place of the usual wipeout tools. The surface labels in Figure 18.12 show the background mask in action.

Gap This option determines the offset from the Component bounding box to the outer points on the border. Setting this to half of the text size usually creates a visually pleasing border.

Color, Linetype, and Lineweight These options allow you to override the display properties assigned by the layer configured on the General tab.

Dragged State Tab Typically, dragging a label in Civil 3D creates a leader and rearranges the text. The settings that control these two actions appear on this tab (see Figure 18.13).

FIGURE 18.13
The Dragged State tab

Property	Value
Leader	
Arrow Head Style	Closed filled
Arrow Head Size	0.1000"
Visibility	True
Type	Straight Leader
Color	BYLAYER
Linetype	ByBlock
Lineweight	ByLayer
Dragged State Componen...	
Display	As Composed
Border Visibility	False
Border Type	Rectangular
Background Mask	False
Border and Leader Gap	0.0250"
Text Height	0.1000"
Leader Attachment	Middle
Leader Justification	True
Color	BYLAYER
Linetype	ByBlock
Lineweight	ByLayer
Maximum Text Width	0.0000"

Many of the property types found on this tab appeared on the Layout tab, so they don't need to be defined again. There are two groups of property types: Leader and Dragged State Components.

The Leader properties include some of the following settings:

Arrow Head Style This allows you to set the appearance of the leader. The same Arrow Head styles that appear in Dimension styles are available in Dragged State.

Arrow Head Size This setting controls the length of the arrow head as well as the length of the landing leading to the text object.

Type This option specifies the leader type. The options are Straight Leader and Spline Leader. Civil 3D labels can have only one leader.

The Dragged State Component properties include some of the following settings:

Display In this option, you can choose Stacked Text or As Composed. The Stacked Text option will remove any blocks, ticks, lines, or borders that are components of the label, and it will stack the Text components as a list in the order they were created, realign the text to horizontal, and justify based on the direction you drag. The As Composed option leaves blocks, ticks, lines, text, or borders intact, and it adds a leader next to the attached component that is anchored to the feature.

Text Height Text Height was mentioned for the previous setting. However, there is a separate setting for Text Height when in dragged state and the label style Display property is configured to Stacked Text.

Leader Attachment When the label style Display property is configured to Stacked Text, this property sets the location of the landing relative to the top line, bottom line, or middle of the stacked text.

Leader Justification When the label style Display property is configured to Stacked Text, this property determines if and how text is justified. When set to True and the leader is to the left of the text, the text will be left justified; when set to True and the leader is to the right of the text, the text will be right justified. When set to False, the leader is always left justified.

Maximum Text Width This property is the same as Maximum Width on the Layout tab.

Figure 18.14a shows an alignment label as it was originally placed. Figure 18.14b shows the same label in a dragged state with the Stacked Text option set. Figure 18.14c shows the label in a dragged state with the As Composed option set.

Summary Tab The Summary tab is exactly what it sounds like—that is, a summary of all other settings that exist in the style. The information from other tabs in list form, as well as their override status, is shown in Figure 18.15.

ALIGNMENT=BOURBON STREET
STATION=2+70.18
OFFSET=90.68L
NORTHING=2006.37
EASTING=889.94
(a)
ALIGNMENT=BOURBON STREET
STATION=2+70.18
OFFSET=90.68L
NORTHING=2006.37
EASTING=889.94
(b)
ALIGNMENT=BOURBON STREET
STATION=2+70.18
OFFSET=90.68L
NORTHING=2006.37
EASTING=889.94
(c)

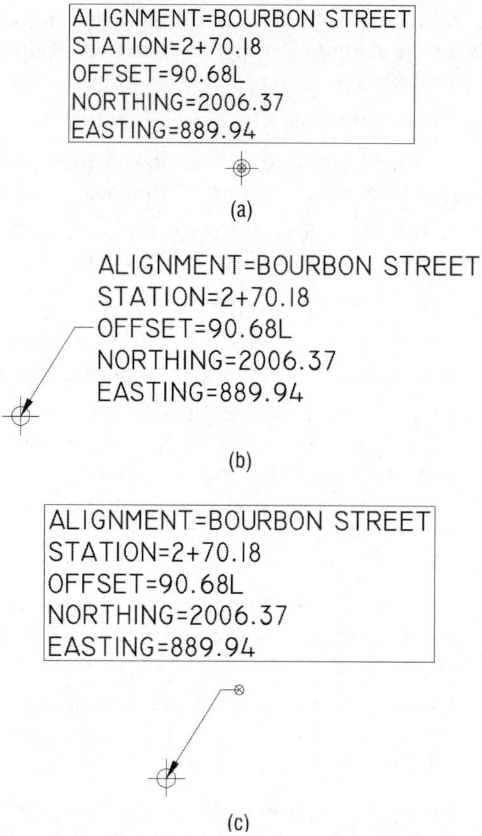

FIGURE 18.14
An alignment label as originally placed (a); dragged state, Stacked Text (b); and dragged state, As Composed (c)

Label Style Composer - Point#-Elevation-Description
Information
General
Layout
Dragged State
Summary
Property
Value
Override
Child Override
Lock
Component 1
Component 2
Component 3
Information
Label
Behavior
Orientation Reference
Object
Forced Insertion
None
Plan Readability
Leader
Dragged State Compone...

FIGURE 18.15
The Summary tab of a label style

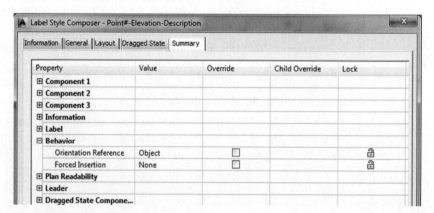

Certification
Objective

As with other settings in Civil 3D, a hierarchy helps determine which styles take precedence over which other styles. There are also defaults that can be set or changed at a drawing-wide level and overridden at an object-level. Make sure to put a lot of thought into using these hierarchical settings and relationships. By using them efficiently, you can save a lot of time because you won't need to tweak every single setting in each and every label you create—and you will be creating lots of labels.

In this first exercise, you will set all the labels to use the same initial text style:

1. Open the drawing `Label Basics.dwg` (`Label Basics_METRIC.dwg`), which you can download from this book's web page at www.sybex.com/go/masteringcivil3d2014.

2. From the Settings tab of Toolspace, right-click on the name of the drawing and select Edit Label Style Defaults, as shown in Figure 18.16.

FIGURE 18.16
Accessing the global label settings

3. Expand the Label category and change Text Style to Arial.

4. Click the arrow in the Child Override column to force all the label styles in this drawing to use the same text style, as shown in Figure 18.17.

FIGURE 18.17
The label placement options at the drawing level

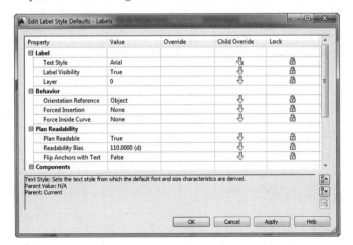

5. Do not make any more changes to this dialog. Examine the various options and settings, and click OK.

6. Save the drawing to use for the next exercise.

> ### Getting to Know the Text Component Editor
>
> The Text Component Editor dialog is where you can combine property fields from objects with static text to create dynamic object labels. The interface is very logical and will do exactly what you tell it to do—but not necessarily what you *want* it to do. However, after you see the reasoning behind its behavior you will soon master it!
>
> To enter the Text Component Editor dialog, from the Layout tab of any text style, click the ellipsis that pops up when you click the Value column of the Text Contents area.
>
>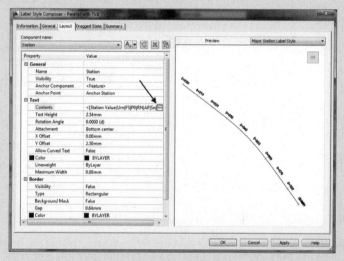
>
> Within the Text Component Editor dialog, the Properties tab has two main areas:
>
> ♦ The left side is where you select the Property field of the Civil 3D object you want to pull into your label.
>
> ♦ The right side indicates the fields and static text already in the label.
>
> To modify an existing Property field, highlight the field on the right side of the dialog. Property values will always highlight as a unit. Each code in the label represents a setting for units, precision, or other format the text can display. What appears on the left is a decoded list of what is currently highlighted on the right. After you make any changes, don't forget to click the arrow to update the information on the right.
>
> Before adding new text to a label, make sure that no existing text is highlighted on the right, so you won't overwrite it.

The Properties list shows everything that can be included in the label; therefore, depending on the object and label type you are working with, the contents of the list will vary. Here you see the available properties for a contour (left) and a pipe-network structure (right):

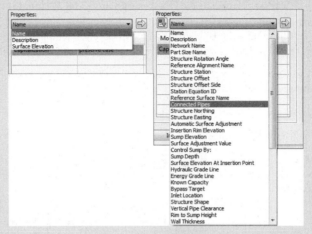

Once you select a Property field, you can configure the units, precision, and any other special rounding or formatting you would like to see.

When everything is set, click the arrow next to the Properties list.

The Format tab is used to add special symbols or override the color, justification, or font.

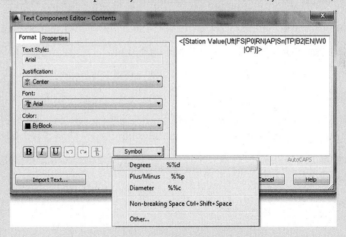

Inevitably, you will forget to click the arrow that adds or updates the text. You may even click it twice and end up with duplicates! Now that you've been given the heads-up, you'll be able to laugh it off, knowing that it happens to even the most seasoned users.

General Note Labels

General Note labels are versatile, non-object-specific labels that can be placed anywhere in the drawing. There are several advantages to using these instead of base-AutoCAD Mtext. Notes will leader and scale the same as the rest of your Civil 3D labels—and best of all, they can contain reference text.

In the following example, you'll create an alternative parcel label that contains reference text. You don't need to have completed the previous exercise to proceed.

1. Continue working in Label Basics.dwg (Label Basics_METRIC.dwg), which you can download from this book's web page.
2. From the Settings tab of Toolspace, choose General ➢ Label Styles ➢ Note.
3. Right-click Note and select New (as shown in Figure 18.18).

FIGURE 18.18
Creating your first new style from the Settings tab

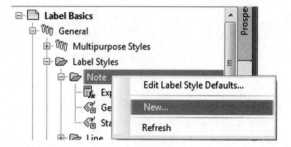

4. On the Information tab, name the style **Easement Parcel Text**.
5. On the General tab, set Layer to C-PROP-TEXT and set Orientation Reference to View.
6. On the Layout tab, click the Value field next to Contents under the Text category.

   Clicking this will cause an ellipsis button to appear.

7. Click the ellipsis button to enter the Text Component Editor dialog.
8. On the right side of the Text Component Editor dialog, delete the existing text and replace it with **Drainage Easement**, as shown in Figure 18.19. Click OK.

FIGURE 18.19
Entering the Text Component Editor dialog for basic text

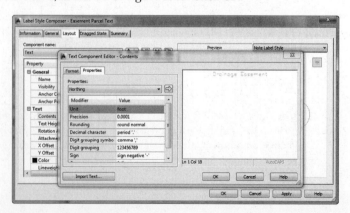

9. Back in the Layout tab, set Border Visibility to False.
10. Click the flyout next to Create Text Component and select Reference Text.

    You will be prompted to select the type of reference text, as shown in Figure 18.20.

FIGURE 18.20
Picking the reference type

11. In the Select Type dialog, select Parcel and click OK.
12. Click in the Value field next to Name under the General category, rename Reference Text.1 to **Parcel Area** and make the following changes:
    a. Set Anchor Component to Text.
    b. Set Anchor Point to Bottom Center.
    c. Set Attachment to Top Center.
    d. To enter the Text Component Editor dialog, click the Value field next to Contents and click the ellipsis.
13. In the Text Component Editor dialog, delete the default label text.
14. On the left side of the Text Component Editor dialog, do the following:
    a. Set the Properties drop-down to Parcel Area.
    b. Set Unit to Acre (Hectares for metric users).
    c. Set Precision to 0.01.

15. When you have set the properties, click the arrow to place the text in the right side of the editor.
16. Click to place your cursor after the previously inserted text. Type **ACRES** (or **HECTARES** for metric users) after the coding.

    The Text Component Editor dialog will resemble Figure 18.21.

FIGURE 18.21
Adding "smart" text to the Text Component Editor dialog

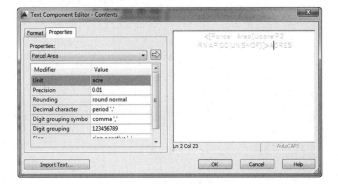

17. Click OK.

    You will still have question marks in the preview, but this is completely expected.

18. Click OK to complete the command.

19. On the Annotate tab ➢ Labels & Tables panel, click the top half of the Add Labels button.

20. With Feature and Label Type set to Note, change Note Label Style to Easement Parcel Text, and click Add.

21. At the `Pick Label Location:` prompt, click anywhere in the example parcel.

22. At the `Select parcel for label style component Parcel Area:` prompt, click the label on Property : 1.

23. Press Esc to finish placing the labels.

    Your completed and placed label should resemble Figure 18.22.

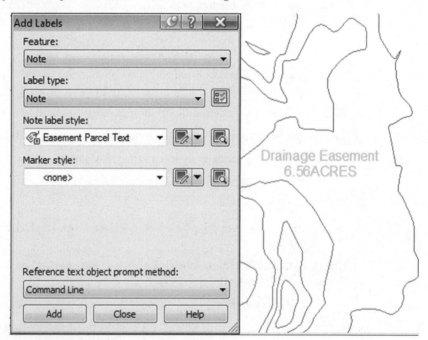

FIGURE 18.22
Your first label! Referencing a parcel area

Save and close the drawing. A saved copy of this drawing (with the filename `Label Basics_FINISHED.dwg` or `Label Basics_METRIC_FINISHED.dwg`) is available from the book's web page.

Point Label Styles

Certification Objective

If you have used other software packages for surveying work, odds are that you controlled the display of point label text with layers, but Civil 3D is different. In Civil 3D, you can control what information is showing next to a point by swapping the label style applied to a group of points.

In the following exercise, you will create a new point label style. Your first point label style will show only Point Number and Description, so you will need to delete the default Elevation component.

1. Open the drawing `Point Labels.dwg` (`Point Labels_METRIC.dwg`), which you can download from this book's web page.

2. From the Settings tab of Toolspace, expand Point ➤ Label Styles.

3. Right-click Label Styles and select New.

4. On the Information tab, name the style **Point Number & Description**.

5. On the General tab, set the layer to V-NODE-TEXT, and leave all other General tab options at their defaults.

6. On the Layout tab, do the following:

    a. Set the Active Component to Point Number.

    b. Change the Anchor Component to <Feature>.

    c. Set Anchor Point to Middle Right.

    d. Set Attachment to Middle Left.

    e. Change the Active Component to Point Elevation.

    f. Click the red X to delete this component.

       You will receive a warning that reads "This label component is used an as anchor in this style or in a child style. Do you want to delete it?"

    g. Click Yes.

    h. Change the Active Component to Point Description.

    i. Change the Anchor Component to Point Number.

    j. Change the Anchor Point to Middle Right.

    k. Change the Attachment to Middle Left.

    l. Change the X Offset to **0.05"** (**1 mm** for metric users).

7. Click OK to complete the label style.

8. On the Prospector tab of Toolspace, locate the point group named TOPO; right-click TOPO and select Properties.

9. On the Information tab, set the Point Label Style to **Point Number & Description**.
10. Click OK to close the Point Group.

All the points in the group will change to resemble Figure 18.23.

FIGURE 18.23
Completed point label style

In the previous exercise, you created a simple new label style and made some modifications to the default components. In the following exercise, you will remove all the default components and add Northing and Easting values to the label using the Text Component Editor dialog.

1. Continue working in the drawing Point Labels.dwg (Point Labels_METRIC.dwg).
2. From the Settings tab of Toolspace, choose Point ➢ Label Styles.
3. Right-click Label Styles, and select New.
4. On the Information tab, name the style **Northing & Easting**.
5. On the General tab, set the layer to V-NODE-TEXT, and leave all other General tab options at their defaults.
6. On the Layout tab, do the following:
    a. Click the red X until all three default components are gone.

    For the second component being deleted, you will receive a warning that reads "This label component is used an as anchor in this style or in a child style. Do you want to delete it?"

    b. Click Yes and delete the last component.
    c. Click the Create Text Component button to create a new Text component, and rename the component to **N-E**.
    d. Set the Anchor Point to Bottom Center.
    e. Set the Attachment to Top Center.
    f. Click the Value field next to Contents and click the ellipsis to enter the Text Component Editor dialog.
7. In the Text Component Editor dialog, do the following:
    a. Highlight the default label text and delete it by pressing the Delete key on your keyboard.
    b. From the Properties list, select Northing.
    c. Set the Precision to **0.01** (two decimal places).
    d. Click the arrow to place the text to the right.

e. After the label text, place an **N** as a static text after the Northing value. Press the Enter key to move to the next line.

f. From the Properties list, select Easting, and set the Precision to **0.01** (two decimal places).

g. Click the arrow to place the text to the right.

h. After the label text, place an **E** as a static text after the Easting value.

The Text Component Editor dialog will now resemble Figure 18.24.

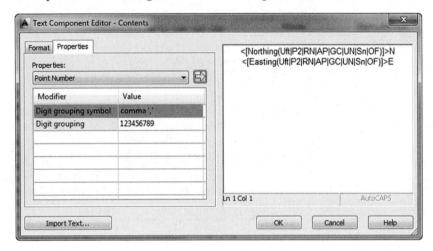

FIGURE 18.24
The Northing & Easting label in progress

i. Click OK to dismiss the Text Component Editor dialog, and click OK to complete the style.

8. On the Prospector tab of Toolspace, locate the point group named Group2; right-click it and select Properties.

9. Set the Point Label Style to **Northing & Easting**, and then click OK.

The labels will resemble Figure 18.25.

FIGURE 18.25
Northing & Easting in the completed exercise

33952.26N
16959.68E

Save and close the drawing. A saved copy of this drawing (Point Labels_FINISHED.dwg or Point Labels_METRIC_FINISHED.dwg) is available from the book's web page.

> ### Sanity-Saving Settings
>
> The Civil 3D template has AutoCAD styles, linetypes, layers, Civil 3D styles, and a plethora of helpful goodies that make doing your job easier. There are a few drawing-specific AutoCAD variables you may not have thought of that will improve your relationship with Civil 3D:
>
> **MSLTSCALE**   This variable stands for modelspace linetype scale. We strongly recommend that you have this set to 1 in your template. This setting makes linetypes react to your annotation scale. All your other Civil 3D objects are doing it, so having your linetypes follow suit will help! A general rule of thumb is that MSLTSCALE, PSLTSCALE, and LTSCALE should be set to 1. For more information on what these control, check out the Help system.
>
> **LAYEREVALCTL**   Set this to 0 to avoid the annoying pop-up that flags users when there are new layers in a drawing. Civil drafters are constantly using XRefs and inserting blocks, both of which cause the pop-up to occur.
>
> **GEOMARKERVISIBILITY**   This is the control for that weird thing in your drawing that appears any time a coordinate system is set up on your drawing. The geomarker looks like a block, but you can't select it and it keeps rescaling as you zoom. Set the GEOMARKERVISIBILITY variable to 0 in your template and the geomarker won't appear. Be aware that if you set it to 0, the option for Locate Using Geographic Data will not be available in XRef and block insert commands.
>
> **AUNITS**   This variable defines the angular units for the base AutoCAD part of the world. Keep this set at 0 (decimal degrees) to help differentiate base AutoCAD angular entry from Civil 3D angular entry.

Line and Curve Labels

You can add bearing and distance labels to many Civil 3D objects. Anything from plain lines and polylines, to parcels and alignment tangent segments, can use nearly identical label types.

The examples in the following exercises will use parcels for labeling, but the tools can be applied to all other types of line labels.

Single Segment Labels

In the following exercise, you will create a new line label style that uses default components. You will remove the direction arrow and change the display precision of the direction component.

1. Open the `Line and Curve Labels.dwg` (`Line and Curve Labels_METRIC.dwg`) drawing file, which you can download from this book's web page.
2. From the Settings tab of Toolspace, expand Parcel ➢ Label Styles ➢ Line; right-click Line and select New.
3. On the Information tab, name the style **Parcel Segment**.
4. On the General tab, set the layer to C-PROP-LINE-TEXT.

   Leave all other General tab options at their defaults.

5. On the Layout tab, do the following:
   a. Change the Active Component to Direction Arrow, and click the red X to delete this component.
   b. Change the Active Component to Distance.
   c. Click the Value field next to Contents, and click the ellipsis to enter the Text Component Editor dialog.
6. In the Text Component Editor dialog, do the following:
   a. Click on the text to highlight the Segment Length contents on the right.

   All of the text should highlight as a unit.
   b. On the left side of the Text Component Editor dialog, change the Precision value to **0.01**.
   c. Click the arrow to update the text.
   d. Click OK to dismiss the Text Component Editor dialog, and click OK to complete the style.
7. Select the Annotate tab ➢ Labels & Tables panel and click the top of the Add Labels button to add labels to the parcels. In the Add Labels dialog, do the following:
   a. Set Feature to Parcel.
   b. Set Label Type to Single Segment.
   c. Set Line Label Style to Parcel Segment.
   d. Click Add and select several line segments for labeling.
   e. Close the Add Labels dialog.

The completed labels will resemble Figure 18.26.

FIGURE 18.26
Your new bearing and distance line label style in action on parcel segments

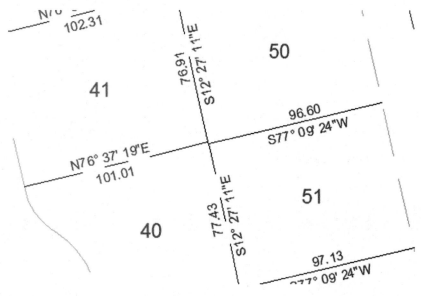

Spanning Segment Labels

When you look at the labels you created in the previous exercise, you may notice that they stop at each parcel vertex. If there is a series of back lot lines that share the same bearing, most plats show this using a single label outside the overall property line displaying the bearing and the combined distance of these lot lines. To label this in Civil 3D, a separate label style is needed.

Spanning labels can be used in both line and curve parcel labels. In the following exercise, you will create line labels that span across multiple parcel segments:

1. Continue working in the drawing Line and Curve Labels.dwg (Line and Curve Labels_METRIC.dwg). You need to have completed the previous exercise to continue.

2. From the Settings tab of Toolspace, expand Parcel ➢ Label Styles ➢ Line, right-click Parcel Segment, and select Copy.

3. On the Information tab, name the style **Spanning Segment**.

4. On the General tab, under Plan Readability, change Readability Bias to **90.1** and change Flip Anchors With Text to True.

5. On the Layout tab:

    a. Change the Active Component name to Table Tag.

    b. Change the Span Outside Segments setting to True.

    c. Change the Active Component name to Bearing.

    d. Change the Span Outside Segments setting to True.

    e. Change the Active Component name to Distance.

    f. Change the Anchor Component to Bearing.

    g. Change the Anchor Point to Bottom Right.

    h. Change the Span Outside Segments setting to True.

    i. Change Attachment Point to Bottom Left.

    j. Change X Offset to **0.2″** (3 mm for metric users).

6. Click OK to complete the style.

7. Using a technique similar to the one you used in the previous exercise, add labels to the parcels along both sides of the Right-of-Way. This time use Spanning Segment as the line label style. Your completed labels will resemble Figure 18.27.

FIGURE 18.27
A spanning label shown on the outside of parcel segments

Flip Label

When you place spanning labels on segments, the Distance component must be forced to the outside of the parcel line. You may need to use the Flip Label command on occasion to move the Distance component to the right location. To access the Flip Label command, select the label and then click the contextual tab ➤ Modify panel.

Curve Labels

In base AutoCAD, creating curve labels is a chore. If you want text to align to curved objects, you won't be able to use it as traditional Mtext. Luckily, Civil 3D gives you the ability to add curved text without compromising its usability.

In the following exercise, you will create a curve label style with a delta symbol (Δ) and text that curves with the parcel segment:

1. Continue working in the drawing file Line and Curve Labels.dwg (Line and Curve Labels_METRIC.dwg). You need to have completed the previous exercise to continue.

2. From the Settings tab of Toolspace, expand Label Styles ➤ Curve, right-click Curve, and select New.

3. On the Information tab, rename the style to **Delta Length & Radius**.

4. On the General tab, set Layer to C-PROP-LINE-TEXT.

5. On the Layout tab, do the following:

    a. Change the Active Component to Distance And Radius.

    b. Click the Value field next to Contents and click the ellipsis to enter the Text Component Editor dialog.

    c. Highlight the segment length property on the right and change the precision to **0.01**.

    d. Click the arrow to update the style.

    e. Highlight the Segment Radius property and change the precision to **0.01**.

    f. Click the arrow to update the text.

    g. Delete the comma that appears as static text in the Text Component Editor dialog, and click OK.

6. Click the Create Text Component button and do the following:

    a. Rename the new text component to **Delta**.

    b. Change Attachment to Bottom Center.

    c. Change the Y Offset to **0.025"** (**1 mm** for metric users).

    d. Set Allow Curved Text to True.

7. Enter the Text Component Editor dialog, and do the following:

   a. Remove the default label text.

   b. Switch to the Format tab, click the Symbol button, and select Other.

      You should now see the Character Map dialog (Figure 18.28).

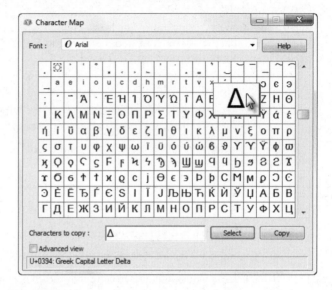

FIGURE 18.28
Browse for special symbols using the Windows character map.

   c. Browse through the symbols of the dialog to find the delta symbol; when you locate it, click on the symbol, click the Select button, and then click Copy. Click the red X to close the Character Map dialog.

   d. Back in the Text Component Editor dialog, click the right side of the dialog, right-click, and select Paste.

   e. Press the Backspace key if the cursor jumps to the next line of text.

      You should see the delta symbol appear in the Text Component Editor dialog.

   f. Enter the equal sign (=) as static text after the delta.

   g. Switch back to the Properties tab.

   h. From the Properties drop-down, select Segment Delta Angle, and do the following:

      ♦ Set the Format to DD°MM′SS.SS″.

      ♦ Set the Precision value to **1 Second**.

      ♦ Click the arrow to place the text in the right side.

      ♦ Click OK to exit the Text Component Editor dialog.

8. Click OK to complete the style.

9. Add labels to the parcels using the same technique you used in the previous exercise.
10. Set the active curve label style to Delta Length & Radius and add labels to some curves.

When it is applied to the design, your completed label should resemble Figure 18.29.

FIGURE 18.29
Completed curve labels with delta symbol and curved text

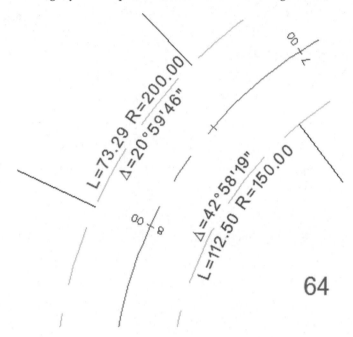

Save and close the drawing. A saved copy of this drawing (`Line and Curve Labels_FINISHED .dwg` or `Line and Curve Labels_METRIC_FINISHED.dwg`) is available from the book's web page.

Pipe and Structure Labels

No two municipalities seem to label their pipes and sewer structures exactly the same way. Fortunately, Civil 3D offers a lot of flexibility in how you label these items.

Pipe Labels

Pipe labels have two separate label types: Plan Profile and Crossing Section. Both label types have many of the same options, but those options are used in different view directions.

In the following exercise, you will use a common trick. When a nonvisible component acts as an anchor to visible objects, you can use the flow direction arrow to force text to be placed at the ends and middle of the pipe regardless of the pipe length.

1. Open the `Pipe and Structure Labels.dwg` (`Pipe and Structure Labels_METRIC.dwg`) drawing file, which you can download from this book's web page.
2. From the Settings tab of Toolspace, choose Pipe ➢ Label Styles ➢ Plan Profile, and then right-click Plan Profile and select New.

3. On the Information tab, name the style **Length Diameter Slope**.
4. On the General tab, set the layer to C-STRM-TEXT.
5. On the Layout tab, delete the existing Pipe Text component by clicking the red X.
6. Expand the Create New Component drop-down and select Flow Direction Arrow; then adjust these settings:
    a. Set Visibility to False.
    b. Set Anchor Point to Top Outer Diameter.
    c. Set a Y offset of **0.1″** (0.3 mm).
7. Expand the Create New Component drop-down and select Text. Change these settings:
    a. Rename the Text Component to **Length**.
    b. Set Anchor Component to Flow Direction Arrow.1.
    c. Set Anchor Point to Start.
    d. Set Attachment to Bottom Left.
8. Enter the Text Component Editor dialog and do the following:
    a. Delete the default label text.
    b. From the Properties list, set 2D Length – Center to Center current.
    c. Set Precision to **1** and Rounding to round up. This causes the pipe length value to round to the next highest whole unit.
    d. Click the arrow to place the text in the editor.
    e. Add a foot symbol (or **m** for meters) after the text component.
    f. Click OK.

9. With Length as the current component, click Copy Component. Then change these settings:
    a. Rename the component to **Diameter**.
    b. Change Anchor Point to Middle.
    c. Change Attachment to Bottom Center.
10. Enter the Text Component Editor dialog and do the following:
    a. Delete all of the text.
    b. From the Properties list, select Inner Pipe Diameter.
    c. Set the Precision to **1** and click the arrow to add the text.

d. Add the inch symbol (or **mm** for millimeters) after the Text component.

e. Click OK to exit the Text Component Editor dialog.

11. With Diameter as the current component, click Copy Component. Then change these settings:

    a. Rename the component to **Slope**.

    b. Set Anchor Point to End.

    c. Set Attachment to Bottom Right.

12. Enter the Text Component Editor dialog and do the following:

    a. Delete all of the text.

    b. Set Pipe Slope as the current property.

    c. Click the arrow to add the text.

    d. Click OK to exit the Text Component Editor dialog.

13. Click OK to complete the style.

14. Add the label to the pipe by doing the following:

    a. On the Annotate tab ➢ Labels & Tables panel, click the top half of the Add Labels button.

    b. Set the Feature option to Pipe Network and Label Type to Single Part Plan.

    c. Set the pipe label style to Length Diameter Slope

    d. Click Add. Select any pipe in plan view.

Your labeled pipe will resemble Figure 18.30.

FIGURE 18.30
Use the invisible arrow trick to label pipe.

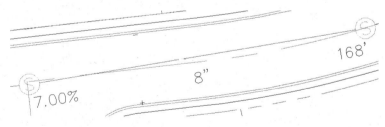

Structure Labels

Structure labels are unique because they possess a special component that references information from each connected pipe, allowing you to place pipe properties in structure labels. Examples would be pipe invert, pipe direction (NE, SW, etc.), and flow direction (in or out).

Adding Existing Ground Elevation to Structure Labels

In design situations, it's often desirable to track not only the structure rim elevation at finished grade, but also the elevation at existing ground. This gives the designer an additional tool for optimizing the earthwork balance.

This exercise will lead you through creating a structure label that includes surface-reference text. It assumes you're familiar with Civil 3D label composition in general:

1. Continue working in the file `Pipe and Structure Labels.dwg` (`Pipe and Structure Labels_METRIC.dwg`).
2. From the Settings tab of Toolspace, choose Structure ➢ Label Styles, and then right-click Label Styles and select New.
3. In the Label Style Composer dialog, do the following:
   a. On the Information tab, name the label **Structure w Surface**.
   b. On the General tab, set the layer to C-STRM-TEXT.
   c. On the Layout tab is a default Text component called Structure Text. Set the Y offset to **-0.25″** (**-1** mm).
4. Click in the Contents box, click the ellipsis to bring up the Text Component Editor dialog, and do the following:
   a. Delete the `<[Description(CP)]>` text string.
   b. Set the Properties drop-down to Name and click the arrow.
   c. Place the cursor after the new Property field. Press the Enter key to move to the next line and then type **RIM** with a space after it.
   d. Set the Properties drop-down to Insertion Rim Elevation, set the Precision to two decimal places, and click the arrow.
   e. Click OK to dismiss the Text Component Editor dialog.
5. In the Label Style Composer, choose Reference Text from the Add Component drop-down.
6. In the Select Type dialog, choose Surface, and then click OK.
7. Rename the component from Reference Text.1 to **Existing Ground**.
8. Enter the Text Component Editor dialog and do the following:
   a. Delete the Label Text string.
   b. Type **EG** and a space in the text window.
   c. Use the Properties drop-down to select Surface Elevation.
   d. Set Precision to **0.01**.
   e. Click the arrow.
   f. Click OK to dismiss the Text Component Editor dialog.

9. In the Label Style Composer, do the following:
   a. Change Anchor Component For Existing Ground to Structure Text.
   b. Change Anchor Point to Bottom Center.
   c. Change Attachment to Top Center.
10. Choose the Text For Each option from the Create drop-down.
11. In the Select Type dialog, choose Structure All Pipes, and then click OK.
12. Enter the Text Component Editor dialog and do the following:

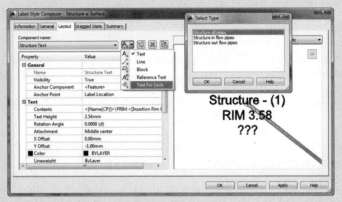

   a. Delete the default Label Text string.
   b. Type **INV** and a space in the text window.
   c. Use the Properties drop-down to select Connected Pipe Invert Elevation.
   d. Set Precision to **0.01**.
   e. Click the arrow.
   f. Click OK to dismiss the Text Component Editor dialog.
13. In the Label Style Composer, do the following:
   a. Change Anchor Component For Text for Each.1 to Existing Ground.
   b. Change Anchor Point to Bottom Center.
   c. Change Attachment to Top Center.
14. Click OK to dismiss the Label Style Composer.
15. Select the Annotate tab ➢ Labels & Tables panel and click the top half of the Add Labels button.
   a. In the Add Labels dialog, set the Feature option to Pipe Network.
   b. Set the Label Type to Single Part Plan.
   c. Set the Structure Label Style to Structure w Surface.
   d. Click Add.

> 16. At the `Select structure or pipe:` prompt, click the structure you want to label.
>
>   You will immediately see a prompt at the command line that reads `Select surface for label style component Existing Ground:`.
>
> 17. Press the Enter key to select EG from the surface listing and then click OK. Press the Enter key to close the Add Labels dialog.
>
> Your label is now complete.
>
> ```
> 1
> RIM 250.60
> EG 250.91
> INV 249.47
> ```
>
> Save and close the drawing. A saved copy of this drawing (`Pipe and Structure Labels_FINISHED.dwg` or `Pipe and Structure Labels_METRIC_FINISHED.dwg`) is available from the book's web page.

Profile and Alignment Labels

Profile and alignment labels can take on many forms. On an alignment you may want to show station labels every 100′ (25 m) in addition to PC, PT, and PI information. On a profile, you will want tangent grades, curve information, and grade breaks.

Many types of label styles can be applied to an alignment and profile. Each type has a unique set of properties. Fortunately, these styles can be applied in sets to expedite labeling.

Label Sets

A *label set* is a grouping of labels that apply to the same object. In lieu of having one big style that accounts for multiple aspects of an object, the labels are broken out into specific types to allow you more control.

Label sets come into play with alignments and design profiles. When you look at an alignment or profile and see labels, you are usually seeing multiple label styles in action.

Consider the alignment shown in Figure 18.31. How many labels are on this alignment? The geometry points, the major stations, the minor ticks, superelevation critical points, and design speed are all different label types.

FIGURE 18.31
One alignment, five label styles in play

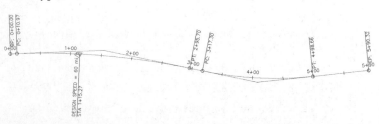

How did those labels get there? When you first created the alignment or profile, one of the options was to specify a label set (as shown in Figure 18.32a). All label types included in the label set by default are placed on the alignment at all locations where they can be applied with their associated style.

FIGURE 18.32
Specifying an alignment label set upon creation (a) and accessing the Label list after creation (b)

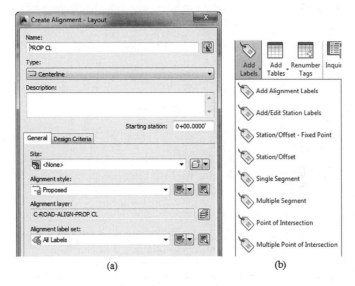

To edit which labels are applied to an alignment or profile, click on one of the alignment labels; and from the Labels - Alignment Geometry Point Label Group contextual tab ➤ Modify panel, select Edit Label Group (as shown in Figure 18.32b).

Label sets also control some aspects of the location of the annotation. An alignment label set controls the major and minor station labeling increment and the type of geometry points that are labeled. A profile label set can control whether labels are positioned with respect to the graph top or bottom edge or to the profile. Figure 18.33 shows the two columns, Dim Anchor Opt and Dim Anchor Val, where this positioning is configured.

FIGURE 18.33
Profile labels and placement

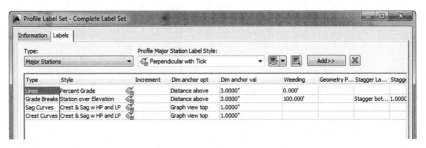

Alignment Labels

You'll create individual label styles over the next couple of exercises and then pull them together with a label set. At the end of this section, you'll apply your new label set to the alignments.

Major Station

Major station labels typically include a tick mark and a station callout. In this exercise, you'll build a style to show only the station increment and run it parallel to the alignment:

1. Open the `Alignment&ProfileLabels.dwg` (`Alignment&ProfileLabels_METRIC.dwg`) file, which you can download from this book's web page.

2. Switch to the Settings tab, and expand the Alignment ➢ Label Styles ➢ Station ➢ Major Station branch.

3. Right-click the Parallel With Tick and select Copy.

   The Label Style Composer dialog appears.

4. On the Information tab, type **Index Station** in the Name field.

5. Switch to the Layout tab.

6. Click in the Value field next to Contents and click the ellipsis button to enter the Text Component Editor dialog.

7. Click in the preview area, and delete the text that's already there.

8. With Station Value as the active property, click the Output Value field, and click the down arrow to open the drop-down.

   You will need to scroll down to see the Output option.

9. Select the Left Of Station Character option, as shown in Figure 18.34, and click the arrow. (Metric users, the resulting label will be more interesting if you use the Right Of Station Character option instead. Because this alignment is less than 300 m, the left option will give you all 0s!)

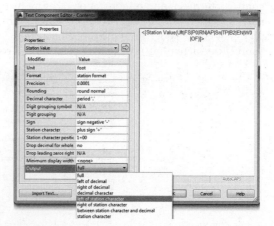

FIGURE 18.34
Modifying the Station Value Output value in the Text Component Editor dialog

10. Click OK to close the Text Component Editor dialog.

11. Click OK to close the Label Style Composer dialog.

    The label style now shows in your label styles, but it hasn't been applied to any alignments yet.

Geometry Points

Geometry points reflect the PC, PT, and other points along the alignment that define the geometric properties. The existing label style was not configured to be plan-readable, so you'll copy it and make a minor change in this exercise:

1. Continue working in the `Alignment&ProfileLabels.dwg` (`Alignment&ProfileLabels_METRIC.dwg`) file.
2. Expand the Alignment ➢ Label Styles ➢ Station ➢ Geometry Point branch.
3. Right-click the Perpendicular With Tick, and select Copy to open the Label Style Composer.
4. On the Information tab, change the name to **Perpendicular with Line**.
5. Switch to the General tab.
6. Change the Readability Bias setting to **90**.

    This value will force the labels to flip at a much earlier point.

7. Switch to the Layout tab and make these changes:
    a. Set the Component Name field to Tick.
    b. Click the Delete Component button (the red X).
    c. Choose Line from the Create Component drop-down.
    d. Change Angle to **90**.
    e. Change back to the Geometry Point & Station component.
    f. Change the Anchor Component to Line.1.
    g. Change the Anchor Point to End.
    h. Change Rotation Angle to **0**.
8. Click OK to close the Label Style Composer dialog.

This new style flips the plan-readable labels sooner and includes a line with the label. Next, you will put the styles together in a set.

Alignment Label Set

Once you have several labels you want to use on an alignment, it is time to save them as an alignment label set.

1. Continue working in `Alignment&ProfileLabels.dwg` (`Alignment&ProfileLabels_METRIC.dwg`). You must have completed the previous exercise to continue. On the Settings tab of Toolspace, expand the Alignment ➢ Label Styles ➢ Label Sets branch.
2. Right-click Label Sets, and select New to open the Alignment Label Set dialog.
3. On the Information tab, change the name to **Paving**.

4. Switch to the Labels tab.

5. Set the Type drop-down to the Major Stations option and the Major Station Label Style drop-down to the Index Station style you just created; then click the Add button.

6. Set the Type drop-down to the Minor Stations option and the Minor Station Label Style drop-down to the Tick option; then click the Add button.

7. Set the Type drop-down to Geometry Points and the Geometry Point Label Style drop-down to Perpendicular With Line. Then click the Add button to open the Geometry Points dialog, as shown Figure 18.35.

FIGURE 18.35
Deselecting the Alignment Beginning and Alignment End geometry point check boxes

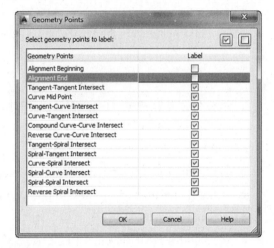

8. Deselect the Alignment Beginning and Alignment End check boxes as shown, and then click OK to dismiss the dialog.

   Three label types will appear in the Alignment Label Set dialog.

9. Click OK to dismiss this dialog.

In the next exercise, you'll apply your label set to the example alignment and then see how an individual label can be changed from the set.

1. Continue working in the Alignment&ProfileLabels.dwg (Alignment&ProfileLabels_METRIC.dwg) file. You need to have completed the previous exercise. Select one of the labels on the Example alignment on screen.

2. On the Labels - Alignment Geometry Point Label Group contextual tab ➢ Modify tab, click Edit Label Group to display the Alignment Labels dialog.

   This dialog shows which labels are currently applied to the alignment. Initially, it will be empty.

3. Click the Import Label Set button near the bottom of this dialog.

Any labels appearing in this listing will be replaced by the labels in the set that is imported.

4. In the Select Label Set dialog, use the drop-down to select the Paving label set and click OK.

   The Alignment Labels list populates with the option you selected.

5. Click OK to dismiss the dialog.

6. When you finish, press Esc twice to be sure you have deselected your alignment labels, and zoom in on any of the major station labels.

7. Hold down the Ctrl key, and select one of the major station labels.

   Notice that a single label is selected, not the label set group.

8. Now that the single label is selected, drag the grip to place it into dragged state. Then click the circular grip to reset the label to its original position.

9. While the label is still selected, on the Labels - Alignment Geometry Point Label Group contextual tab ➢ Modify tab, click Edit Label Group to display the Alignment Labels dialog.

   The Properties palette appears, allowing you to pick another label style from the Major Station Label Style drop-down.

10. Change the Label Style value to Parallel With Tick, and change the Flip Label value to True, as shown in Figure 18.36.

FIGURE 18.36
Modifying a single label's properties through base AutoCAD properties

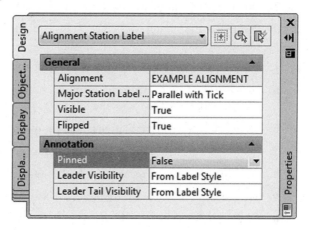

11. Press Esc to deselect the label item.

If you add labels to an alignment and like the look of the set, use the Save Label Set option. By using alignment label sets, you'll find it easy to standardize the appearance of labeling and stationing across alignments. Building label sets can take some time, but it's an easy, effective way to enforce standards.

Station Offset Labeling

Beyond labeling an alignment's basic stationing and geometry points, you may want to label points of interest in reference to the alignment. Station offset labeling is designed to do just that. In addition to labeling the alignment's properties, you can include references to other object types in your station-offset labels. The objects available for referencing are as follows:

- Other alignments
- COGO points
- Parcels
- Profiles
- Surfaces
- Survey figures

In Chapter 10, "Advanced Corridors, Intersections, and Roundabouts," you used special alignment labels that referenced other alignments to make adjusting your design easier. In this exercise, you will make a similar type of label. The label you create in the following exercise finds the intersection of two alignments:

1. Continue working in the `Alignment&ProfileLabels.dwg` (`Alignment&ProfileLabels_METRIC.dwg`) drawing file.
2. On the Settings tab, expand Alignment ➢ Label Styles ➢ Station Offset.
3. Right-click the Station And Offset style, and select Copy to open the Label Style Composer dialog.
4. On the Information tab, change the name of your new style to **Alignment Intersection**.
5. Switch to the Layout tab. In the Component Name drop-down, select Marker, and click the red X to delete the component.
6. Change the name of the Station Offset component to **Main Alignment**.
7. In the Contents field, click the ellipsis button to bring up the Text Component Editor dialog.
8. Select the text in the preview area and delete it all.
9. Type **Sta.** in the preview area; be sure to leave a space after the period.
10. In the Properties drop-down field, select Station Value, and set the Precision to **0.01**.
11. Click the arrow in the Text Component Editor dialog; add a space after the inserted text.
12. In the Properties drop-down field, select Alignment Name.
13. Click the arrow to add this bit of code to the preview.
14. Click your mouse in the preview area, add a space after the inserted text and an equal sign (=).

    Your Text Component Editor dialog should look like Figure 18.37.

FIGURE 18.37
The start of the alignment label style

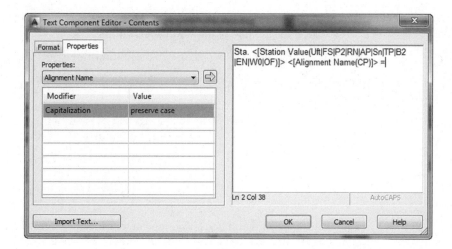

15. Click OK to return to the Label Style Composer dialog.
16. Under the Border Property, set the Visibility field to False.
17. Click the Create Component drop-down and select Reference Text.
18. In the Select Type dialog that appears, select Alignment and click OK.
19. Change the name to **Intersecting Alignment**.
20. In the Anchor Component field, select Main Alignment.
21. In the Anchor Point field, select Bottom Left.
22. In the Attachment field, select Top Left.

    When you choose the anchor point and attachment point in this fashion, the bottom left of the Main Alignment text is linked to the top left of the Intersection Alignment text.

23. Click in the Contents field, and click the ellipsis button to open the Text Component Editor dialog.
24. Delete the default label text that currently appears in the preview area.
25. Type **Sta.** in the preview area; be sure to leave a space after the period.
26. In the Properties drop-down, select Station Value, and set the Precision to **0.01**.
27. Click the insert arrow in the Text Component Editor dialog and add a space after the inserted text.
28. In the Properties drop-down, select Alignment Name.
29. Click the insert arrow.
30. Click OK to exit the Text Component Editor dialog, and click OK again to exit the Label Style Composer dialog.

31. Add the label to the drawing by selecting the Annotate tab ➤ Labels & Tables panel and clicking the top half of the Add Labels button.

32. Change the label settings to match those shown in Figure 18.38, and click Add.

FIGURE 18.38
Adding the new alignment label

33. Watch the command line for placement instructions. You will be prompted to select the main alignment, the station along the alignment, which is the intersection point of both alignments, and the offset. You will then be prompted to select the intersecting alignment.

34. Press Esc to complete the labeling command.

35. Click the label to select it and reveal the grips. Select the square grip and drag it away from the current location to form a leader.

Your completed label should look like Figure 18.39.

FIGURE 18.39
The completed alignment label with reference text

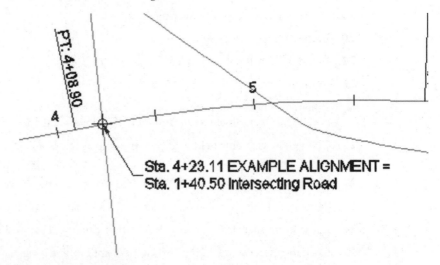

Profile Labels

It's important to remember that the profile and the profile view aren't the same thing. The labels discussed in this section are those that relate directly to the profile. This usually means station-based labels, individual tangent and curve labels, or grade breaks. You'll look at individual label styles for these components and then at the concept of the label set.

Profile Label Sets

As with alignments, you apply labels to profiles in the form of profile label sets. In this exercise, you'll learn how to add labels along a profile object:

1. Continue working in the Alignment&ProfileLabels.dwg (Alignment&ProfileLabels_METRIC.dwg) file.

2. Pick the blue layout profile (the profile with two vertical curves) to activate the profile object.

Edit Profile Labels

3. From the Profile contextual tab ➤ Labels panel, select Edit Profile Labels to display the Profile Labels dialog (see Figure 18.40).

FIGURE 18.40
An empty Profile Labels dialog

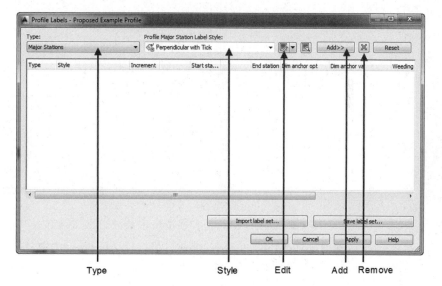

Selecting the type of label from the Type drop-down changes the Style drop-down to include styles that are available for that label type. Next to the Style drop-down are the usual Style Edit/Copy button and a preview button. Once you've selected a style from the Style drop-down, clicking the Add button places it on the profile. The middle portion of this dialog displays information about the labels that are being applied to the profile selected; you'll look at that in a moment.

4. Choose the Major Stations option from the Type drop-down.

   The name of the second drop-down changes to Profile Major Station Label Style to reflect this option.

5. Set the style to Perpendicular With Tick in this drop-down.

6. Click Add to apply this label to the profile.

7. Choose Horizontal Geometry Points from the Type drop-down.

   The name of the Style drop-down changes to Profile Horizontal Geometry Point.

8. Select the Horizontal Geometry Station style, and click Add again to display the Geometry Points dialog shown in Figure 18.41.

FIGURE 18.41
The Geometry Points dialog appears when you add labels to horizontal geometry points.

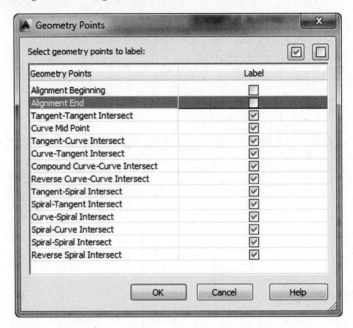

This dialog lets you apply different label styles to different geometry points if necessary.

9. Deselect the Alignment Beginning and Alignment End rows, as shown in Figure 18.41, and click OK to close the dialog.

10. Click the Apply button, and then drag the dialog out of the way to view the changes to the profile (see Figure 18.42).

FIGURE 18.42
Labels applied to major stations and alignment geometry points

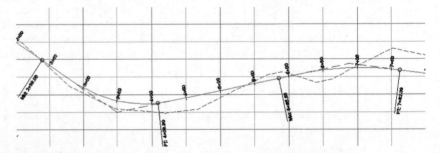

11. In the middle of the Profile Labels dialog, change the Increment value in the Major Stations row to **50′ (20 m)**, as shown in Figure 18.43.

FIGURE 18.43
Modifying the Major Stations labeling Increment

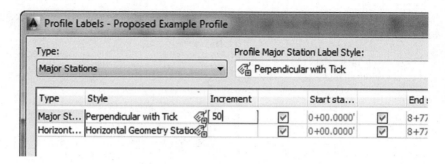

This modifies the labeling increment only, not the grid or other values.

12. Click OK to close the Profile Labels dialog.
13. Press Esc to deselect the layout profile.

As you can see, applying labels one at a time could turn into a tedious task. After you learn about the types of labels available, you'll revisit this dialog and look at the two buttons at the bottom for dealing with label sets.

Line Labels

Line labels in profiles are typically used to convey the slope or length of a tangent segment. In this exercise, you'll add a length and slope to the layout profile:

1. Continue working in the `Alignment&ProfileLabels.dwg` (`Alignment&ProfileLabels_METRIC.dwg`) file. You must have completed the previous exercise to continue.
2. Switch to the Settings tab of Toolspace.
3. Expand the Profiles ➢ Label Styles ➢ Line branch.
4. Right-click Percent Grade, and select the New option to open the Label Style Composer dialog and create a child style.
5. On the Information tab, change the name to **Length and Percent Grade**.
6. Switch to the Layout tab and make these changes:
    a. Change Attachment to Top Center.
    b. Set the Y-offset to **-0.025″** (**-1 mm**).
    c. Set Background Mask to True.
7. Click the Value field next to Contents and click the ellipsis to enter the Text Component Editor dialog.
8. Change the Properties drop-down to the Tangent Slope Length option and the Precision value to **0.01**, as shown in Figure 18.44.

FIGURE 18.44
The Text Component Editor dialog with the values for the Tangent Slope Length entered

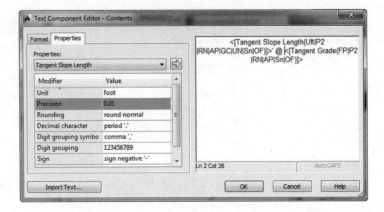

9. Put your cursor at the beginning of the existing text in the preview windows, click the arrow, and then add a foot (or **m** for meter) symbol, a space, an @ symbol, and another space in the editor's preview pane so that it looks like Figure 18.44.

10. Click OK to close the Text Component Editor dialog, and click OK again to close the Label Style Composer.

11. Pick the layout profile. From the Profile contextual tab ➢ Labels panel, select Edit Profile to display the Profile Labels dialog.

12. Change the Type drop-down to the Lines option. The name of the Style drop-down changes to Profile Tangent Label Style. Select the Length And Percent Grade option.

13. Click the Add button, and then click OK to exit the dialog.

14. Press Esc to deselect the layout profile.

The profile view should look like Figure 18.45.

FIGURE 18.45
A new line label applied to the layout profile

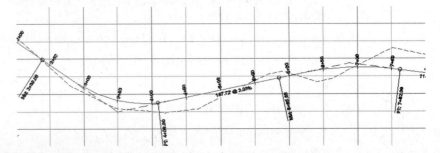

WHERE IS THAT DISTANCE BEING MEASURED?

The *tangent slope length* is the distance along the vertical geometry between vertical curves. This value doesn't include the tangent extensions. There are a number of ways to label this length; be sure to look in the Text Component Editor if you want a different measurement.

Curve Labels

Vertical curve labels are one of the most confusing aspects of profile labeling. Many people become overwhelmed rapidly, because there's so much that can be labeled and there are so many ways to get all the right information in the right place. In this quick exercise, you'll look at some of the special label anchor points that are unique to curve labels and how they can be helpful:

1. Continue working in the Alignment&ProfileLabels.dwg (Alignment&ProfileLabels_METRIC.dwg) file. You must have completed the previous exercise to continue.

2. Pick the layout profile. From the Profile contextual tab ➢ Labels panel, select Edit Profile Labels to display the Profile Labels dialog.

3. Choose the Crest Curves option from the Type drop-down.

   The name of the Style drop-down changes to Profile Crest Curve Label Style.

4. Select the Crest Only option.

5. Click the Add button to apply the label to the list.

6. Choose the Sag Curves option from the Type drop-down.

   The name of the Style drop-down changes to Profile Sag Curve Label Style.

7. Select and add the Sag Only label style. Click the Add button to apply the label.

8. Click OK to close the dialog.

9. Press Esc to deselect the layout profile.

Your profile should look like Figure 18.46.

FIGURE 18.46
Curve labels applied with default Dim Anchor values

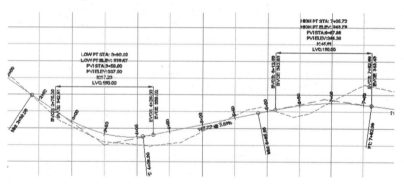

Most labels are applied directly on top of the object being referenced. Because typical curve labels contain a large amount of information, putting the label right on the object can yield undesired results. In the following exercise, you'll modify the label settings to review the options available for curve labels:

1. Continue working in the Alignment&ProfileLabels.dwg (Alignment&ProfileLabels_METRIC.dwg) file. You must have completed the previous exercise to continue.

2. Pick the layout profile. From the Profile contextual tab ➢ Labels panel, select Edit Profile Labels to display the Profile Labels dialog.

3. Scroll to the right, and change both Dim Anchor Opt values for the Crest and Sag Curves to Graph View Top.

4. Change the Dim Anchor Val for both curves to **-2.25"** (**-40 mm**), and click OK to close the dialog.

5. Press Esc to deselect the layout profile.

Your drawing should look like Figure 18.47.

FIGURE 18.47
Curve labels anchored to the top of the graph

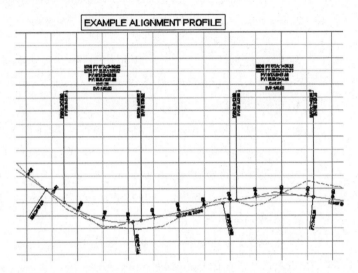

The labels can also be grip-modified to move higher or lower as needed. By using the top or bottom of the graph as the anchor point, you can apply consistent and easy labeling to the curve, regardless of the curve location or size.

> **THOSE CRAZY CURVE LABELS!**
>
> A profile curve label can be as intricate or simple as you desire. Civil 3D gives you many options for where along the curve feature you want your Label component to appear. The following illustration shows where some of the commonly used curve locations are in a label:
>
>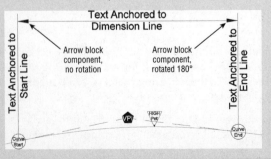

GRADE BREAKS

The last label style typically involved in a profile is a grade-break label at PVI points that don't fall inside a vertical curve, such as the beginning or end of the layout profile. Additional uses include things like water-level profiling, where vertical curves aren't part of the profile information or existing surface labeling. In this exercise, you'll add a grade-break label and look at another option for controlling how often labels are applied to profile data:

1. Continue working in the `Alignment&ProfileLabels.dwg` (`Alignment&ProfileLabels_METRIC.dwg`) file. You must have completed the previous exercise to continue.

2. Pick the green surface profile (the irregular profile). From the Profile contextual tab ➢ Labels panel, select Edit Profile Labels to display the Profile Labels dialog.

3. Choose Grade Breaks from the Type drop-down.

   The name of the Style drop-down changes.

4. Select the Station Over Elevation style and click the Add button.

5. Click Apply, and drag the dialog out of the way to review the change.

   A sampled surface profile has grade breaks every time the alignment crosses a surface TIN line. Why wasn't your view coated with labels?

6. Scroll to the right, and change the Weeding value to **150′** (**45 m**). Click OK.

7. Select one of the new grade-break labels. Use the square grip at the location where the label touches the profile to form a leader and clean up any labels that overlap.

   The profile labels should appear as shown in Figure 18.48.

FIGURE 18.48
The grade-break labels on a sampled surface are starting to get crowded.

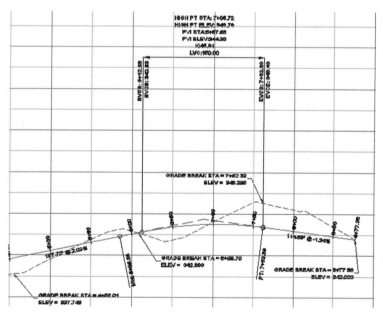

8. Click OK to dismiss the dialog.

9. Press Esc to deselect the layout profile.

Weeding lets you control how frequently grade-break labels are applied. This makes it possible to label profiles with frequent grade breaks, such as a surface profile, at even increments instead of at every PVI.

As you've seen, there are many ways to apply labeling to profiles, and applying these labels to each profile individually could be tedious. In the next section, you'll build a label set to make this process more efficient.

Profile Label Sets

Applying labels to both crest and sag curves, tangents, grade breaks, and geometry with the label style selection and various options can be monotonous. Thankfully, Civil 3D gives you the ability to use label sets, as in alignments, to make the process quick and easy. In this exercise, you'll apply a label set, make a few changes, and export a new label set that can be shared with team members or imported to the Civil 3D template. Follow these steps:

1. Continue working in the `Alignment&ProfileLabels.dwg` (`Alignment&ProfileLabels_METRIC.dwg`) file. You must have completed the previous exercise to continue.

2. To tidy things up, select one of the grade-break labels from the previous exercise. Press Delete on your keyboard to remove the labels.

3. Pick the layout profile. From the Profile contextual tab ➢ Labels panel, select Edit Profile Labels to display the Profile Labels dialog.

4. Click the Import Label Set button near the bottom of the dialog to display the Select Label Style Set dialog.

5. Select the Complete Label Set option from the drop-down, and click OK.

6. Click OK again to close the Profile Labels dialog and see the profile view.

    The label set you chose contains curve labels, grade-break labels, and line labels.

7. Pick the layout profile. From the Profile contextual tab ➢ Labels panel, select Edit Profile Labels to display the Profile Labels dialog.

8. Click Import Label Set to display the Select Label Style Set dialog.

9. Select the No Labels option from the drop-down, and click OK.

    All the labels from the listing will be removed.

    In the next steps, you will add labels to the listing and save the listing as its own label set for future use.

10. Set the Active Type to Lines. Set Profile Tangent Label Style to Length And Percent Grade, and click Add.

11. Set the Active Type to Grade Breaks. Set Profile Grade Break Label Style to Station Over Elevation, and click Add.
12. Set the Type to Crest Curves. Set Profile Crest Curve Label Style to Crest Only, and click Add.
13. Set the Type to Sag Curves. Set Profile Sag Curve Label Style to Sag Only, and click Add.
14. Set the Crest Curve and Sag Curve label types to use the Graph View Top as the Dim Anchor Opt.
15. Set both Dim Anchor Val fields to **-1.5"** (**-40** mm), as shown in Figure 18.49.

FIGURE 18.49
Four label types and dimension anchor settings in the label set to be saved

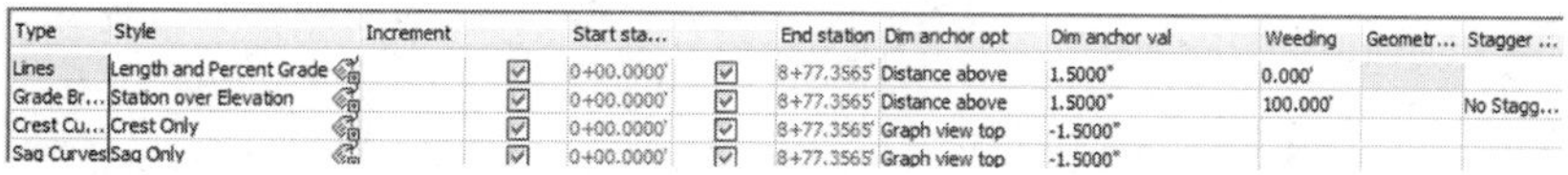

Type	Style	Increment	Start sta...		End station	Dim anchor opt	Dim anchor val	Weeding	Geometr...	Stagger...
Lines	Length and Percent Grade		0+00.0000'	☑	8+77.3565'	Distance above	1.5000"	0.000'		
Grade Br...	Station over Elevation		0+00.0000'	☑	8+77.3565'	Distance above	1.5000"	100.000'		No Stagg...
Crest Cu...	Crest Only		0+00.0000'	☑	8+77.3565'	Graph view top	-1.5000"			
Sag Curves	Sag Only		0+00.0000'	☑	8+77.3565'	Graph view top	-1.5000"			

16. Click the Save Label Set button to open the Profile Label Set dialog and create a new profile label set.
17. On the Information tab, change the name to **Road Profile Labels**.
18. Click OK to close the Profile Label Set dialog.
19. Click OK to close the Profile Labels dialog.
20. Press Esc to deselect the layout profile.
21. On the Settings tab of Toolspace, select Profile ➢ Label Styles ➢ Label Sets.

    Note that the Road Profile Labels set is now available for sharing or importing to other profile label dialogs.

Save and close the drawing. A saved copy of this drawing (Alignment&ProfileLabels_FINISHED.dwg or Alignment&ProfileLabels_METRIC_FINISHED.dwg) is available from the book's web page.

Label sets are the only way to apply profile labeling uniformly. When you're working with a well-developed set of styles and label sets, going from sketched profile layout to plan-ready output is quick and easy.

Advanced Style Types

Now that you are familiar with the basics of label styles, you are ready to take your skills to the next level. The styles in the following section combine aspects of label styles and object styles. You will cover object styles more in depth in Chapter 19, "Object Styles."

You have a great deal of control over every detail, even ones that may seem trivial. Instead of being bogged down trying to understand every option, don't be afraid to try a "trial and error" approach. If you make a change you don't like, you can always edit the style until you get it right.

Table Styles

Civil 3D does a beautiful job of placing dynamically linked data tables that relate to your objects. The tables use the Text Component Editor to grab dynamic information from your objects. You also have control over fill colors, table headings, and how the data is sorted.

For the table style, the Data Properties tab contains all the column information. You can add columns by clicking the plus sign. You can remove columns by highlighting the column you want to remove and clicking the Delete button. Change column order by dragging them around and dropping them where you want them to go, as shown in Figure 18.50.

FIGURE 18.50
Modifying table columns

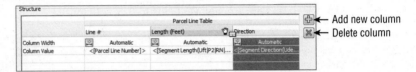

In the following exercise, you will see the basic steps of modifying a table style. Now that you understand the ins and outs of the Label Style Composer, this procedure should be a breeze:

1. Open the Tables.dwg (Tables_METRIC.dwg) file, which you can download from this book's web page.

   This file contains a parcel line table whose style you will modify.

2. Zoom into the parcel line table.

   Notice there are several things you will want to change:

   ♦ The line numbers are out of order.

   ♦ The directions have far too much precision.

   ♦ The length does not display units.

   All that is about to change.

Edit Table Style

3. Click the table to select it. From the Table contextual tab ➢ Modify panel, select the Table Properties drop-down and choose Edit Table Style.

4. Switch to the Data Properties tab.

   The Data Properties tab (Figure 18.51) is the main control area for all table styles. This is where you set behavior, text styles, and sizes for fields.

5. Place a check mark next to Sort Data, and set the Sorting column to **1**.

   Column 1 corresponds to the Column Value containing the Parcel Line Number. Set the order option to Ascending. The Ascending option will ensure that the parcel numbers are listed in the table from the lowest to highest value.

6. Double-click the column heading field for Length.

   Doing so opens a stripped-down version of the Text Component Editor dialog. Headings and table titles are static text only; therefore, only the text formatting tools are shown.

FIGURE 18.51
The Data Properties tab for table styles

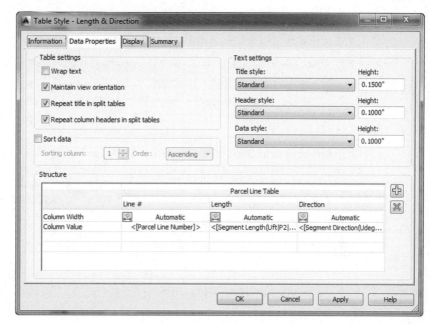

7. Add **(Feet) ((meters))** after the column heading to add a proper units heading to the column, as shown in Figure 18.52, and click OK.

FIGURE 18.52
Adding static text to a table column heading

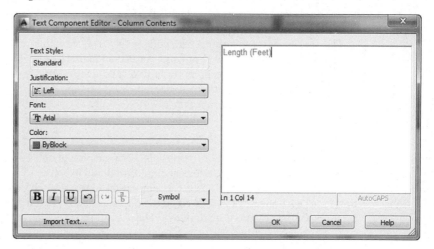

8. Double-click the Direction Column Value field below the column heading.

   Doing so opens the Text Component Editor dialog, similar to what you've used in earlier exercises.

9. Click the text in the preview area so that it becomes highlighted.

10. Change Precision to **1 Second**.

11. Click the arrow to update the text. Click OK.

12. Click OK to complete the table style modifications.

Save and close the drawing. A saved copy of this drawing (`Tables_FINISHED.dwg` or `Tables_METRIC_FINISHED.dwg`) is available from the book's web page.

Code Set Styles

Code set styles determine how your assembly design will appear, and they are used in many places. A code set style is in play when you first create your assembly. One is used in corridor creation and in the Section Editor. The most apparent use of a code set style is in section views.

A code set style is a collection of many other styles. In a code set style you will find:

- Link styles and link label styles
- Points styles, point label styles, and feature line styles
- Shapes styles and shape label styles
- Quantity takeoff pay items
- Render materials for visualization tasks

When naming your code set style, you'll find it helpful to have the name of the set reflect its use. Multiple code set styles are needed because of different applications of their use. When you are designing an assembly, you may want to see more labels than when you are getting ready to plot the assembly in a cross-section sheet. Labeling that is useful in a cross-section sheet may obstruct your view of the design when working with it in the corridor cross-section editor.

The hardest part of working with code set styles is figuring out the name of the link or point you want to label. Luckily, it is unusual for users to label shapes, so you won't need to worry about those. The names of each point or link can be found in the subassembly properties. Most of the links and points are logically named, but there's no harm in a little trial and error if you are not sure.

SHAPES

Shapes are the areas that define materials. Because people don't usually label these materials in a section view, you won't be experiencing these in an exercise.

One heads-up, however: Resist the temptation to use a hatch pattern on shapes where multiple cross-section views will be created. Solid fills and no patterns are your best bet to avoid performance issues and the annoying "Hatch pattern is too dense" warning.

LINKS AND LINK LABELS

You learned in Chapter 8, "Assemblies and Subassemblies," that a link is the linear part of a subassembly. The object style for the link itself is very simple—just a single linear component. The label for a link is usually expressed as a percent grade or as a slope ratio.

In the following exercise, you will modify a code set style to apply link labels to an assembly.

1. Open the Code Set Styles.dwg (Code Set Styles_METRIC.dwg) file, which you can download from this book's web page.

   This file contains corridor and cross-section views. Zoom into one of the cross-section views so you can observe the changes as you apply them to the code set style.

2. From the Settings tab of Toolspace, expand General ➢ Multipurpose Styles ➢ Code Set Styles; right-click on All Codes, and select Edit.

3. On the Codes tab under the Link category, select Pave, and click the Label Tag icon in the Label Style column, as shown in Figure 18.53.

FIGURE 18.53
Adding labels to the link codes in the code set style

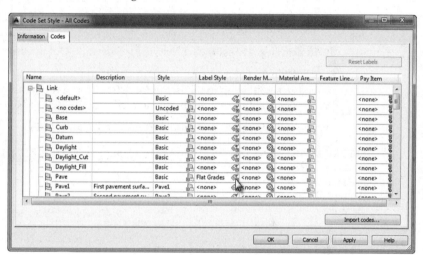

4. In the Pick Style dialog, select the Flat Grades label style and click OK; then click Apply to examine the change on the cross sections.

   You should see that the lanes now have slope information labeled.

5. Under the Link category, locate Daylight; click the Label Tag icon in the Label Style column, and select Steep Grades as the label style.

6. Click OK, and then click OK to dismiss the All Codes label style and see what is happening with the cross section.

   The cross sections should resemble Figure 18.54.

FIGURE 18.54
Cross section with link labels applied to pave and daylight links

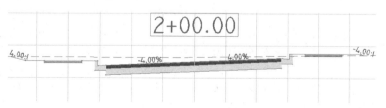

POINTS AND POINT LABELS

A common frustration for new users of Civil 3D are the marker styles and their labels. For cross-section views, you may not want points to display at all. In the following exercise, you will create a new code set style, modify point codes, and add more labels to the sections:

1. Continue working in Code Set Styles.dwg (Code Set Styles_METRIC.dwg).

2. From the Settings tab of Toolspace, expand General ➢ Multipurpose Styles ➢ Code Set Styles. Right-click on All Codes and select Copy.

3. On the Information tab, rename the style to **All Codes-Plotting**.

4. On the Codes tab, locate the Point category. Locate the Back_Curb point and click the Label Tag icon in the Label Style column. In the Pick Style dialog, set the style to Offset Elevation. Click OK.

5. Repeat step 4 to set the label style for Sidewalk_Out to Offset Elevation.

6. Click the first point name, <default>. While holding down the Shift key on your keyboard, scroll down to the last point listing, Top_Curb. With all of the points selected, click the Tag icon in the style column and change the style to No Markers. Click OK.

The Code Set Style dialog should resemble Figure 18.55.

FIGURE 18.55
Points set to No Markers and labels set to Offset Elevation

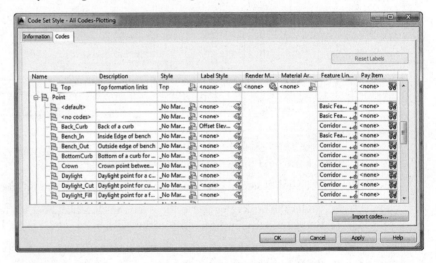

7. Click OK.

You can't see any changes to your cross sections yet because the style is not active.

8. Select a section view. From the Section Views contextual tab ➢ Modify View panel, select View Group Properties.

9. On the Sections tab, change the style of the Example Corridor to the All Codes-Plotting style you created, as shown in Figure 18.56, and click OK.

FIGURE 18.56
Setting the code set style current on the section views

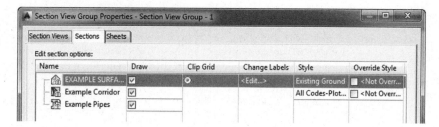

10. Click OK again to complete the changes to the Section View Group Properties dialog.

Your section view should resemble Figure 18.57.

FIGURE 18.57
New code set style applied to the section view

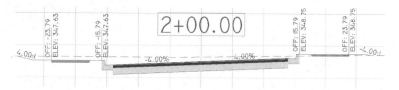

Save and close the drawing. A saved copy of this drawing (Code Set Styles_FINISHED.dwg or Code Set Styles_METRIC_FINISHED.dwg) is available from the book's web page.

> **It's Just an Expression**
>
> Expressions allow you to use properties available in a label and perform calculations. The resulting expression can be used as part of a displayed label, or as any of the numerical text settings such as text height, rotation angle, or width.
>
> A frequently seen use of expressions is with surface spot elevations. In the following image, an expression called Slant Curb contains {Surface Elevation}-0.125. The label style then uses the Slant Curb as one of the properties displayed in the spot elevation label as shown here:
>
>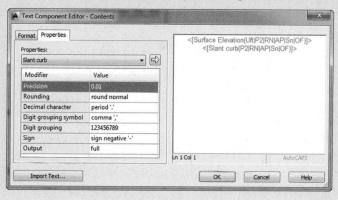

You may encounter uses for expressions on nontraditional Civil 3D projects. For instance, let's look at projects dealing with natural gas pipelines. These pipelines do not fall under the category of Civil 3D pressure pipes. Pipeline vertical bends are not standard; they are manufactured in the field. When designing the vertical conditions for a pipeline, Civil 3D pipeline users turn to profiles. How are these nonstandard bends identified?

1. Open the Expression.dwg (Expression_METRIC.dwg) drawing.
2. Zoom into the profile view.

   This blue profile represents the top of a natural gas pipeline.

3. In the Settings tab of Toolspace ➢ Profile ➢ Label Styles ➢ Curve, right-click Expressions and select New.

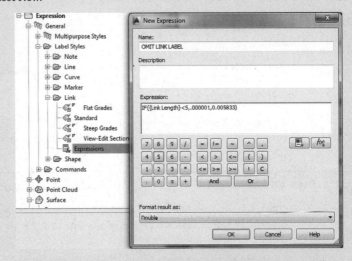

Vertical bends in profiles can be measured in profile views if a small vertical curve is applied. In a profile curve expression, the mathematical relationship between the two tangents coming together can be examined. This is a great example of how the tools in Civil 3D can be molded to take on project types other than land development.

4. Name the expression **Overbend**.
5. Click the Function button to the right and select ATAN.

   You are going to take a look at the mathematical relationship between the tangent coming into the curve and the tangent leaving the curve and determine the field value of the vertical bend.

6. Click the Property button and select Grade In.
7. Enter a close parenthesis and a + sign, or click the) and then the + operator button in the New Expression dialog.
8. Click the Function button and select ABS, which turns the next value into its absolute value.
9. Click the Function button and select ATAN.
10. Click the Property button and select Grade Out.

11. Add a close parenthesis. Add a close parenthesis again and change the Format Result As to Angle.

    Changing the format results allows you to take advantage of unit-specific formatting in the Text Component Editor.

12. When your new expression resembles the following image, click OK.

13. In the Settings tab of Toolspace ➤ Profile ➤ Label Styles ➤ Curve, right-click Curve and click New.
14. On the Information tab, type **OverBend** for the new style name.
15. On the Layout tab of the Label Style Composer, delete the following components: Dimension Line, Start Line, End Line, and Length. If a dialog opens warning about a Label component being used as an anchor, click Yes to dismiss.
16. Create a new text component.
17. Click into the Contents box and click the ellipsis to open the Text Component Editor dialog.
18. Delete the Label Text text.
19. Under Properties, at the top of the list should be the expression you created called Overbend. Select it.
20. Set the precision to **0.1**.
21. Click the arrow to set the text in the label.
22. Type the following behind the code just entered: **%%d OB.**
23. Click OK to close the Text Component Editor dialog.
24. Click OK to close the Label Style Composer.

25. Click on the blue layout profile in the profile view.
26. Pick the layout profile. From the Profile contextual tab ➢ Labels panel, select Edit Profile Labels to display the Profile Labels dialog.
27. Set the Label Type to Crest Curves and under Profile Crest Curve Label Style, use the drop-down to select OverBend. Click the Add button.
28. Click OK to close the Profile Labels dialog.

Drag the label to a more aesthetically pleasing location.

The Bottom Line

Override individual labels with other styles. In spite of the desire to have uniform labeling styles and appearances between alignments within a single drawing, project, or firm, there are always exceptions. Using the Ctrl+click method for element selection, you can access commands that let you modify labels and even change their styles.

Master It Open the drawing MasteringLabelStyles.dwg (MasteringLabelStyles_METRIC.dwg). Create a copy of the Perpendicular With Tick Major Station style called **Major With Marker**. Change Tick Block Name to **Marker Pnt**. Replace some (but not all) of your major station labels with this new style.

Create a new label set for alignments. Label sets let you determine the appearance of an alignment's labels and quickly standardize that appearance across all objects of the same nature. By creating sets that reflect their intended use, you can make it easy for a designer to quickly label alignments according to specifications with little understanding of the requirement.

Master It Within the MasteringLabelStyles.dwg (MasteringLabelStyles_METRIC.dwg) file, create a new label set containing only major station labels and apply it to all the alignments in that drawing.

Create and use expressions. Expressions give you the ability to add calculated information to labels or add logic to label creation.

Master It In the MasteringLabelStyles.dwg (MasteringLabelStyles_METRIC.dwg) file, create an expression called Top of Curb that adds 0.5′ (0.15 m) to a surface elevation. Use the expression in a spot elevation label that shows both the surface elevation and the expression-based elevation.

Apply a standard label set to profiles. Standardization of appearance is one of the major benefits of using Civil 3D styles in labeling. By applying label sets, you can quickly create plot-ready profile views that have the required information for review.

Master It In the MasteringLabelStyles.dwg (MasteringLabelStyles_METRIC.dwg) file, apply the Road Profiles label set to all layout profiles.

Chapter 19

Object Styles

As you learned in the previous chapter, styles control the display properties of labels, but they also control the display properties of AutoCAD® Civil 3D® objects such as points, surfaces, alignments, profile views, pipes, sections, and so on.

Alignments are composed of components such as lines, curves, and spirals. Surfaces contain contours, triangles, and points. A point has a very important marker component. Object styles enable you to control which components are displayed and how they are displayed. Traditionally, you control the display of such items with layers and AutoCAD properties. *Styles* offer a quick way to change an object's display state for the purpose of plotting, editing, or analysis.

In this chapter, you will learn to:

- Override object styles with other styles
- Create a new surface style
- Create a new profile view style

Getting Started with Object Styles

Before you get your hands on specific object styles and begin working with them, you should understand some general things all styles have in common.

There are several ways to enter the various dialogs used for editing styles. The easiest, most direct way to access any Style dialog is from the Settings tab. Right-click any style you see listed and select Edit, as shown in Figure 19.1.

FIGURE 19.1
Every style can be edited by right-clicking the style name from the Settings tab of Toolspace.

You can also enter a Style dialog from the properties of any object by clicking the Edit button. Figure 19.2 shows the active style on a surface, with the Edit button to the right. Editing at this level affects all objects that use the style, the same as it would if you had entered the style from the

Settings tab. The downside to editing a style in this manner is that you will not immediately be able to click Apply to see your change because the Object Properties dialog is open beneath it. You'll need to exit the Style dialog and click Apply at the object level before you'll see your style update.

FIGURE 19.2
An object's properties reveal the current style, which can be edited, as in this Surface Properties dialog.

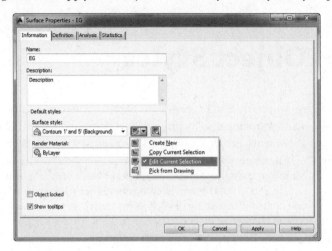

Real World Scenario

MAKING SENSE OF CHILD STYLES AND OVERRIDES

Civil 3D styles are configurable at several levels. In Chapter 1, "The Basics," you learned that the Drawing Settings dialog was where you make adjustments to the overall settings that affect all styles, such as setting precision values and object layers. On the object level in the Settings tab, you will find Feature Settings, which you can use to override any settings being applied by Drawing Settings. The Feature Settings dialog is also used to assign default styles. Finally, inside the object level you will find Command Settings, where you can override settings applied by Drawing or Feature Settings at the command level. For example, when using the CreateProfileFromSurface command, you are usually creating an existing profile so you could configure the settings of that command to use an existing surface style. These layers of parent-child settings are often referred to as hierarchical settings.

Styles themselves can be used to override other styles. These other styles are referred to as child styles. For example, a precision of two decimal places (0.00) may be adequate for station labels. However, when creating corridors, you may find it advantageous to see more decimal places of precision. Set the station precision to 0.00 in Drawing Settings. Then further down on the Settings tab, right-click Corridors and select Edit Feature Settings.

Inside the Edit Feature Settings – Corridor dialog, you might feel déjà vu from when you edited the Ambient Settings tab in the Drawing Settings dialog. All the same settings (and a few more object-specific ones) are here. The difference is that changing the setting here will affect only corridors.

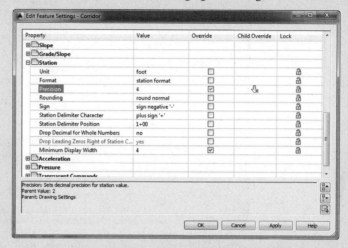

The check mark in the Override column indicates that this setting differs from settings higher up in Drawing Settings. At the bottom of the dialog further information is provided regarding the parent setting and its value.

An arrow in the Child Override column indicates that further down the chain of command a style or setting differs. To force these subordinate styles or Command Settings to match the style of the parent, click the arrow so that a red X appears. The red X indicates that the change you make, in the dialog you are looking at, will be pushed to its child styles or Command Settings.

The Lock column allows you to lock a setting, which prevents it from being changed from Command Settings.

As you are designing the styles for use in your office, be sure to keep an eye on the overrides that are set.

Frequently Seen Tabs

When creating or editing an object style, you will be opening a Style dialog containing several tabs. Some of the tabs contain settings that are unique to the object. Some tabs are common to all or several types of objects. In this section, you will learn about those tabs.

Information Tab

The Information tab (Figure 19.3) contains the field where the name of the style is entered or changed.

FIGURE 19.3
The Information tab exists for all object styles.

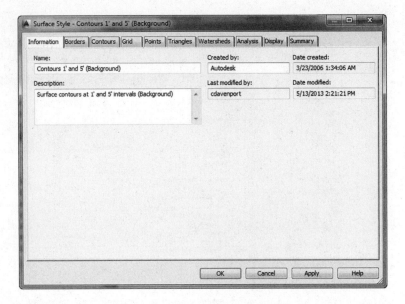

The description is optional. However, a description may be helpful to others who will be using your template. If you do use descriptions, be sure to revise them when you copy your styles for other purposes to avoid confusing your users. The description can be seen in tooltip form as you search through the Settings tab, as shown in Figure 19.4.

FIGURE 19.4
Tooltip showing style information, including the description

On the right side of the dialog, you will see the name of the user who created the style, the date when it was created, the last person who modified the style, and the date it was modified. These names are initially pulled from the Windows login information and only the Created By field can be edited.

Display Tab

On the Display tab, you will find a list of components available for display within the object. You will see this tab in every Object Style dialog. The Display tab controls the look of the object in Plan, Model, Profile, and Section view directions. Not every style type will have all of these view directions available. In Figure 19.5, you can see that a general-purpose marker contains Plan, Model, Profile, and Section view directions; and only one component (the marker itself) and a surface model have Plan, Model, and Section view directions and components for contours, triangles, points, and more.

FIGURE 19.5
The display of a simple marker style (above) and a surface style (below)

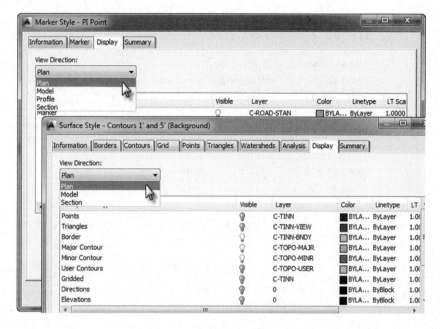

This is 3D software, so objects exist in three dimensions. View Direction controls the display of an object depending on how you are looking at it. Certain items, such as a profile view, are intended to be seen only in plan, so they do not have multiple view directions listed. Alignments can be seen in plan, model, and section views. Profiles can be seen in profile, model, and section views, as shown in Figure 19.6

FIGURE 19.6
The View Direction options for a profile style

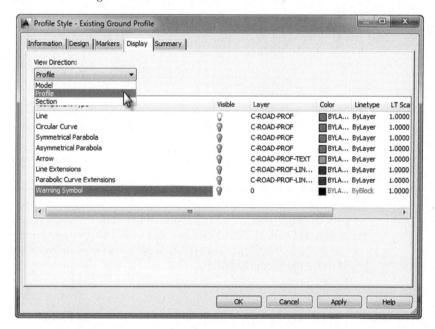

While other tabs in an object style control the specifics of *how* certain components look or behave, the Display tab controls *if* the component displays at all. The Visibility lightbulb indicates whether the component will be displayed when the style is applied to the object. In the Surface Style dialog shown in Figure 19.5, you can see that border, major contours, and minor contours will all display for the surface object style, but the other components will not be displayed in plan view. In the Profile Style dialog shown in Figure 19.6, only line components will display.

Each component can have a layer designation. The component layer will override the display properties of the object layer, which was set initially in the Drawing Settings ➢ Object Layer tab. When the component is set to layer 0, it will display using the display properties of the object layer as long as all other properties on the style's Display tab are set to ByLayer. This is another example of the parent-child settings.

Summary Tab

The Summary tab contains a list of all the settings configured on the other tabs of the style's dialog. Settings are editable on the Summary tab. Figure 19.7 shows the Summary tab of a profile style.

FIGURE 19.7
The Summary tab of a profile style

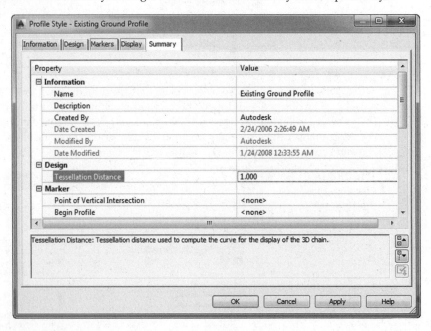

You can click the plus (+) or minus (–) button next to each category branch to expand to see further settings. At the bottom-right corner of the dialog are three buttons. The top button collapses all of the category branches, and the middle button expands all of the category branches. The bottom Override All Dependencies button can also be found on a Label Style Summary tab, but it is inactive for object styles. At the bottom of this dialog, additional information will be shown for the property selected.

General Settings

On the Settings tab of the Toolspace, the General collection (or branch) contains settings and styles that can be applied to multiple object types in various scenarios. The General collection has three collections:

- Multipurpose Styles
- Label Styles
- Commands

You learned about the Label Styles collection in Chapter 18, "Label Styles," and now you will look at the Multipurpose Styles and Commands collections.

Multipurpose Styles

If you expand the Multipurpose Styles collection, you will see seven folders, as shown in Figure 19.8.

FIGURE 19.8
General ➢ Multipurpose Styles

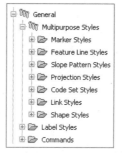

These style types are used to control the display of components in various objects. For example, Marker Styles, Link Styles, and Shape Styles are typically used in section views containing corridors and assemblies; whereas Feature Line Styles, shown in Figure 19.9, are used when grading linear features and displaying corridors in plan.

FIGURE 19.9
The Feature Line Styles collection

868 | CHAPTER 19 OBJECT STYLES

> **TOGGLE TOOLSPACE ORIENTATION**
>
> Toolspace is shown horizontally in Figure 19.9 and in other figures throughout this chapter for illustration purposes. If you like the way this looks, you can set your Toolspace like this by clicking the Orientation toggle at the top of Toolspace when it is floating (not docked). This will bring the item view to the right side of Toolspace instead of the default location at the bottom.

Commands

A Commands folder will reside in almost every object branch of the Settings tree. Each item listed in this folder represents a command in that object category and is named after the keystrokes necessary to execute that command at the command line, as shown in Figure 19.10.

FIGURE 19.10
The Commands folder

Point and Marker Object Styles

Markers are used in many places throughout Civil 3D. They are called from other styles to show vertices on Civil 3D objects such as feature lines, alignments, profiles, and figures. They can also be attached to the origination point on labels such as alignment station offset labels or surface spot elevations. Markers can even be used to indicate the start of a flow path.

Marker Tab

Marker styles and point styles both contain a Marker tab (Figure 19.11). The Marker tab controls what symbol or block is used, its rotation, and how it should be sized when it is placed in the drawing.

Three symbol types can be used. Use Custom Marker and Use AutoCAD BLOCK Symbol For Marker are the most popular options. Use AutoCAD POINT For Marker doesn't actually produce an AutoCAD point but uses the AutoCAD point style and size specified by the DDPTYPE dialog. When you choose the Use Custom Marker option, you can produce a marker similar in appearance to an AutoCAD point but will use the size options on the right side of this tab instead of the DDPTYPE dialog.

GETTING STARTED WITH OBJECT STYLES | 869

FIGURE 19.11
The Marker tab for the PI Point Marker style

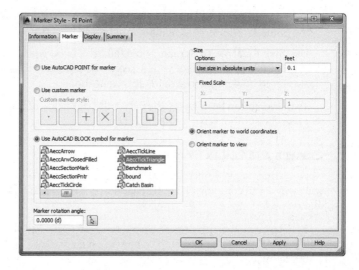

When you choose the Use AutoCAD BLOCK Symbol For Marker option, you will be able to access a listing of blocks in your drawing. If the block you want to use does not yet exist in the drawing, you can right-click in the block listing and choose Browse, as shown in Figure 19.12.

FIGURE 19.12
Right-click to browse for a block if it is not already defined in your drawing.

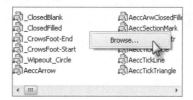

The Size options control how the marker is scaled when inserted in the drawing (Figure 19.13):

FIGURE 19.13
The Size options for marker display

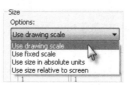

Use Drawing Scale Use Drawing Scale allows you to specify the plotted size of the symbol. The modelspace size of the symbol will be the size specified in the style multiplied by your annotation scale.

Use Fixed Scale Use Fixed Scale will scale the symbol based on the X, Y, and Z scale entered in the Fixed Scale area, just like a block. This option will also apply the Fixed Scale factor in the description key set when configured to a point.

Use Size In Absolute Units Use Size In Absolute Units is the option you will use to specify a real-world size for the symbol regardless of scale. For example, if you want a manhole to always show as 5 feet in diameter, you could use this option. In the case of survey points used with description keys, if this option is set to .0833 (1 inch) and a size parameter is included in the point description, the symbol will be scaled to the value measured in the field.

Use Size Relative To Screen   Size Relative To Screen allows you to specify the size of the symbol as a percentage of your screen. The marker will change size as you zoom in or out similar to how points resize in GIS applications.

Two orientation buttons (as shown on the right side of the dialog in Figure 19.11) controls whether the symbol stays rotated with the world coordinate system or the view.

CREATE A MARKER STYLE

Now it's time to get your hands on some object styles. You will start with simple styles and work your way up in complexity as this chapter progresses. In this first exercise, you will create a marker style.

1. Start a new blank drawing from the _AutoCAD Civil 3D (Imperial) NCS template that ships with Civil 3D. For metric users, use the _AutoCAD Civil 3D (Metric) NCS template.

2. From the Settings tab of Toolspace, expand General ➢ Multipurpose Styles ➢ Marker Styles.

3. Right-click Marker Styles and select New.

4. On the Information tab, do the following:

    a. Set Name to **PVI Marker**.

    b. Add the description **Use to indicate PVI in profiles**.

    Your login name will be listed in the Created By and Last Modified By fields.

5. On the Marker tab, do the following:

    a. Click the Use AutoCAD BLOCK Symbol For Marker radio button.

    b. In the block listing, highlight STA by clicking on it.

    c. Verify that Size Options is set to Use Drawing Scale.

    d. Set the size to **0.2"** (or **5** mm).

    e. Leave all other Marker settings at their defaults.

6. On the Display tab with the View Direction set to Plan, do the following:

    a. Click in the Layer column for the Marker component to display the Layer Selection dialog.

    b. In the Layer Selection dialog, set the layer to C-ROAD-PROF-STAN-GEOM and click OK.

    Note that you may need to widen the column to view the full layer name.

- c. Click in the Color column for the Marker component to display the Select Color dialog.
- d. In the Select Color dialog, click the ByLayer button and click OK.
7. Repeat step 6 for View Direction set to Model, Profile, and Section.
8. Click OK to complete the creation of a new marker style.
9. From the Application menu, select Save As ➢ Drawing Template.
10. Set File Name to `PointObjectTemplate.dwt` (or `PointObjectTemplate_METRIC.dwt`) and click Save.
11. Set the description in the Template Options dialog to **Mastering Template with Point Object Styles**, verify that Measurement is set to either English or metric as applicable, and click OK.

Your PVI marker will now be listed in the General ➢ Multipurpose Styles ➢ Marker Styles branch. By creating this marker style in a drawing template, you have made it accessible for any future projects that use this template to begin. Keep this drawing template open for the next portion of the exercise.

Create a Survey Point Style

Survey point styles contain many of the same options as marker styles. As you work through the following example, you will perform many of the same steps you did in the previous exercise:

1. Continue working in the drawing template from the previous exercise.
2. From the Settings tab of Toolspace, expand Point ➢ Point Styles.
3. Right-click Point Styles and select New.
4. On the Information tab, do the following:
    - a. Set Name to **TRUNK**.
    - b. Add the description **Simple circle representing trunk diameter in inches** (or **Simple circle representing trunk diameter in mm**).
5. On the Marker tab, do the following:
    - a. Click the Use Custom Marker radio button.
    - b. Click the Blank Marker option from the group of symbols on the left.
    - c. Click to add the Circle option on the right.
    - d. Verify that Size Options is set to Use Size In Absolute Units and set the size to **0.0833** (or **0.001** for metric).

    This value will scale down the symbol so the trunk diameter represents inches (or millimeters). Leave all other Marker settings at their defaults. Instead of entering the value as a decimal, you could alternatively enter it as a fraction such as 1/12 or 1/1000.

6. On the Display tab with View Direction set to Plan, do the following:

   a. Click in the Layer column for the Marker component to display the Layer Selection dialog.

   b. In the Layer Selection dialog, set the Marker layer to V-NODE-TREE and click OK.

   c. Use steps a and b to set the Layer for the Label component to V-NODE-TREE. Leave the other settings at their defaults.

   Hint! To save time in this step, you could alternatively use the Shift key to multiselect the Marker and Label components, as shown in Figure 19.14. By selecting both first, when you select the layer for one component, it will apply to both components.

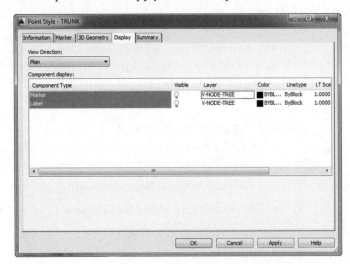

FIGURE 19.14
Use the Shift key on your keyboard as you click the components to multiselect.

7. Repeat step 6 for View Direction set to Model, Profile, and Section turning on the Visibility lightbulb for the Marker component in Model.

8. Click OK to complete the creation of a new point style.

You can save and keep this drawing template open to continue to the next exercise, or use the saved copy of this drawing template (PointObjectTemplate_FINISHED.dwt or PointObjectTemplate_METRIC_FINISHED.dwt) available from the book's web page at www.sybex.com/go/masteringcivil3d2014.

Linear Object Styles

In this section, you will see some linear styles such as alignments, profiles, and parcels. Hopefully, you are already seeing that concepts from one type of style often apply to other types of styles. For alignment styles and profile styles, this is especially true.

Both alignment styles and profile styles have a Design tab, as shown in Figure 19.15. In the case of alignment styles, the Enable Radius Snap option restricts the grip-edit behavior of alignment curves. If you enable this option and set a value of 0.5', the resulting radius value of curves will be rounded to the nearest 0.5'. In the case of profiles, the curve tessellation distance is a little more abstract. *Curve tessellation* refers to the smoothing factor applied to the profile when viewing it in 3D. Most users leave these settings at their default values.

FIGURE 19.15
Design tabs exist in both alignment and profile object styles.

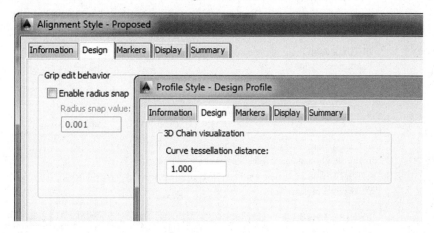

Alignment styles and profile styles have very similar Markers tabs. This tab is where you can place markers (like the one you created in the first exercise of the chapter) at specific geometry points. Figure 19.16 shows the markers assigned to various locations along an alignment.

FIGURE 19.16
Markers for an alignment style

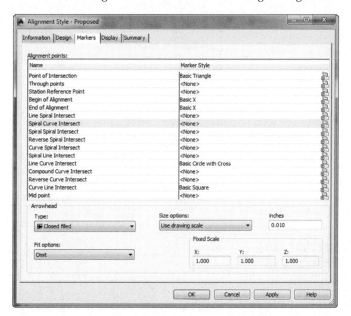

At the bottom of Figure 19.16, you see Arrowhead information. Both alignment styles and profile styles have the option of showing a direction arrow on each segment. You may omit this by turning the Arrow component off in the Display tab or by setting the component to a layer that is set to No Plot. When designing roundabouts, consider the latter because knowing the direction of an alignment comes in handy, as you saw in Chapter 10, "Advanced Corridors, Intersections, and Roundabouts."

Figure 19.17 shows some commonly highlighted geometry points and components in an alignment. The markers in Figure 19.17 correspond to the markers defined in Figure 19.16.

FIGURE 19.17
Example alignment with alignment points labeled

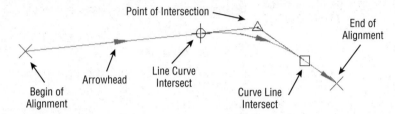

Figure 19.18 shows the Markers tab for the profile style. It looks pretty similar to the alignment Markers tab.

FIGURE 19.18
Markers for a profile style

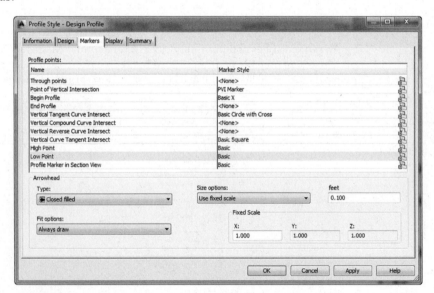

The markers in Figure 19.19 correspond to the markers defined in Figure 19.18.

FIGURE 19.19
Example profile with profile points labeled

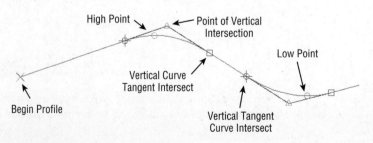

Alignment Style

Alignment styles can be helpful in identifying key design components, as well as showing the stationing direction. Use multiple alignment styles to visually differentiate centerline alignments from supplemental alignments such as offset alignments and curb return alignments.

In the following exercise, you will create a style that restricts radius grip edits to five-foot increments and displays basic alignment components:

1. If not still open from the previous exercise, open PointObjectTemplate_FINISHED.dwt or PointObjectTemplate_METRIC_FINISHED.dwt.

    You can download either file from this book's web page.

2. From the Settings tab of Toolspace, expand Alignment ➢ Alignment Styles.

3. Right-click Alignment Styles and select New.

4. On the Information tab, set Name to **Centerline**.

5. On the Design tab, do the following:

    a. Verify that Enable Radius Snap is selected.

    b. Verify that Radius Snap is set to **5** (or **1** for metric).

    To see the Help document about any dialog, press F1 on your keyboard to be taken directly to the help section for that topic or click on the Help button on the tab when you have a question regarding a specific tab on a dialog.

6. On the Markers tab, do the following:

    a. Set the Point Of Intersection marker by double-clicking the current value of <None>.

    b. In the Pick Marker Style dialog, select PI Marker from the marker listing, and click OK.

    c. Using the same procedure, set the Through Points marker and the Station Reference Point marker both to <None>.

    Note that <None> is located near the top of the list.

7. On the Display tab with the View Direction set to Plan, do the following:

    a. Using the Shift key to select Arrow, Line Extensions, and Curve Extensions together, use one of their lightbulb icons to turn off the display of these three components.

    b. Using the Shift key to select Line, Curve, Spiral, and Warning Symbol together, click in the Layer column to display the Layer Selection dialog.

    c. In the Layer Selection dialog, set the layer to C-ROAD-CNTR and click OK. This will set the layer for all four of the selected components.

    The Warning Symbol displays when design criteria or design checks are violated only if its display is turned on in this dialog. Design criteria and design checks were covered in Chapter 6, "Alignments."

    Leave the Model and Section view directions at their defaults.

8. Click OK to finish creating a new alignment style.

This alignment will display the line, curve, and spiral on layer C-ROAD-CNTR, and a marker will be shown at the Point Of Intersection, as shown in Figure 19.20.

FIGURE 19.20
Centerline alignment style

When this exercise is complete, you can close the drawing template. A saved copy of this drawing template (with the filename `AlignmentObjectTemplate_FINISHED.dwt` or `AlignmentObjectTemplate_METRIC_FINISHED.dwt`) is available from the book's web page.

Parcel Styles

The parcel styles have several unique features that make them different from other styles.

In the Design tab of a parcel (shown at the top in Figure 19.21), you see parcel-specific options. A fill distance can be specified to place a hatch pattern along the perimeter of the parcel. This setting is used to help differentiate "special" parcels such as parks, limits of disturbance, or environmentally sensitive areas.

FIGURE 19.21
Parcel style options (above) and the resulting parcel graphic (below)

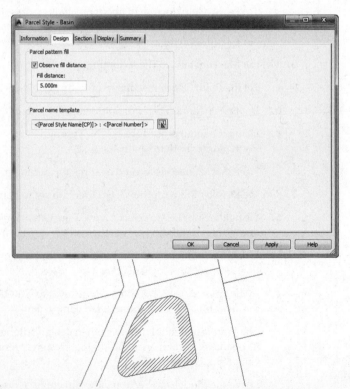

The *fill distance* indicates the width of the hatch pattern. The Component Hatch Display characteristics, including the pattern, angle, and scale, are specified at the bottom of the Display tab. Be sure to turn on Parcel Area Fill and configure a layer for it so it can be turned off independently from the parcel itself, as hatches tend to slow down the graphics. The bottom of Figure 19.21 shows the parcel graphic resulting from the design settings shown.

Feature Line Styles

Feature lines are found in quite a few different places. They are created automatically as part of corridors, created as a result of generating a grading group, or can be created independently by the user. Because their scope crosses functionality, you will find feature line styles in the General ➢ Multipurpose Styles collection in Settings.

By definition, a feature line is a 3D object; therefore, its style can be controlled in plan, model, profile, and section. As shown in Figure 19.22, a feature line can use markers at geometry points.

FIGURE 19.22
Feature-line profile marker options (left) and section marker option (right)

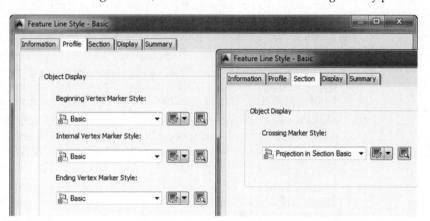

Surface Styles

Certification Objective

Surface styles are the most widely used styles in any Civil 3D project. Depending on what objects are visible in the style that is active, certain editing options may be restricted. For example, points need to be visible before the Civil 3D software will let you use the Delete Point command on a surface.

There are seven tabs in the Surface Style dialog. Each tab corresponds to components listed in the Display tab. However, the Analysis tab covers the following components: Directions, Elevations, Slope, and Slope Arrows. While the Display tab is where the basic AutoCAD properties of each component are set, these other tabs allow you to configure in detail how each component is displayed. Some of these settings include applying marker styles to watersheds, setting 3D parameter borders, configuring Contour intervals and depression ticks, and configuring color ranges for the different analysis tools.

Once a surface is created, you can display information in many ways. The most common so far have been contours and triangles, but these are the basics. By using varying styles, you can show a large amount of data with one single surface. Not only can you do simple things such as adjust the contour interval, but the Civil 3D program can apply a number of analysis tools to any surface:

Contours Allows the user to apply to multiple ranges a color scheme or linetype as opposed to the typical minor-major scheme. Commonly used in cut-fill maps to color negative contours one way, positive contours another, and the balance or zero contours yet another color.

Directions Draws arrows showing the normal direction of the surface face. This tool is typically used for aspect analysis, helping site planners review the way a site slopes with regard to cardinal directions and the sun.

Elevations Creates bands of color to differentiate various ranges of elevations. You can use this tool to create a simple weighted distribution to help create marketing materials, hard-coded elevations to differentiate floodplain, and other elevation-driven site concerns, or ranges to help a designer understand the earthwork involved in creating a finished surface.

Slopes Colors the face of each triangle on the basis of the assigned slope values. While a distributed method is the normal setup, a common use is to check site slopes for compliance with the Americans with Disabilities Act of 1990 (ADA) requirements or other site slope limitations, including vertical faces (where slopes are abnormally high).

Slope Arrows Displays the same information as a slope analysis, but instead of coloring the entire face of the TIN, this option places an arrow pointing in the downhill direction and colors that arrow on the basis of the specified slope ranges. This is useful in confirming surface flow direction for site drainage.

User-Defined Contours Refers to contours that typically fall outside the normal intervals. These user-defined contours are useful for drawing lines on a surface that are especially relevant but don't fall on one of the standard levels. A typical use is to show the normal water surface elevation on a site containing a pond or lake.

Watersheds Used for watershed analysis, this style allows you to examine how water flows along and off of a surface. Using the surface TIN, the drain targets and watersheds are defined. An example of creating a watersheds surface style is provided later in this section.

In the following exercises, you will walk through the steps of creating and modifying surface styles.

Contour Style

Contouring is the standard surface representation on which land development plans are built. In this example, you'll create a new surface contouring style and modify the interval to a setting more suitable for commercial site design review:

1. Open the SurfaceStylesContours.dwg or SurfaceStylesContours_METRIC.dwg file, which you can download from this book's web page.

2. From the Settings tab of Toolspace, expand Surface ➢ Surface Styles.

3. Right-click Surface Styles and select New.

4. On the Information tab, do the following:

    a. Set Name to **Exaggerated Existing Contours**.

    b. Add the description **2′ minor contours with a 5× exaggeration when viewed in 3D** (or **1m minor contours with a 5× exaggeration when viewed in 3D** for metric).

5. On the Contours tab, do the following:

    a. Expand the Contour Intervals category.

    b. Set the Minor Interval to **2′** (or **1** m).

    Notice that the Major Interval automatically adjusts to **10′** (or **5** m).

    c. Expand the Contour Smoothing category (you may have to scroll down).

    d. Set Smooth Contours to True, which activates the Contour Smoothing slider bar near the bottom.

    Don't change this Smoothing value, but keep in mind that this gives you a level of control over how much Civil 3D modifies the contours it draws.

> **SURFACE VS. CONTOUR SMOOTHING**
>
> Remember, contour smoothing is not surface smoothing. *Contour smoothing* applies smoothing at the individual contour level but not at the surface level. If you want to make your surface contouring look fluid, you should be smoothing the surface.

The Contours tab will now look similar to Figure 19.23.

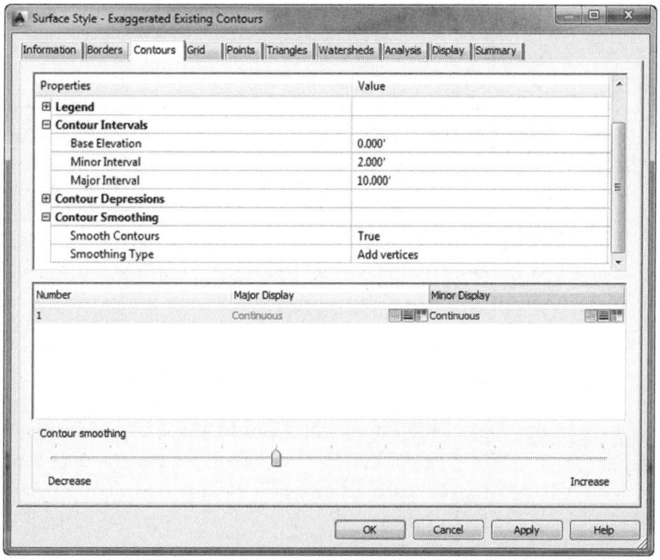

FIGURE 19.23
The Contours tab in the Surface Style dialog

6. On the Triangles tab, do the following:

   a. Change Triangle Display Mode to Exaggerate Elevation.

   b. Set Exaggerate Triangles By Scale Factor to **5** (see Figure 19.24).

FIGURE 19.24
Exaggerate the elevations shown in the Object Viewer.

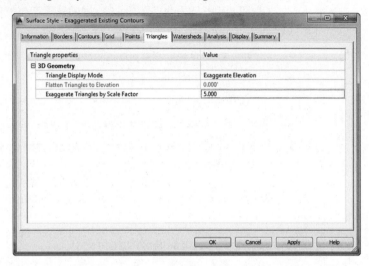

7. On the Display tab with the View Direction set to Plan, do the following:

   a. Shift+click to highlight all the components.

   b. Click in the Color column of one of the components to display the Select Color.

   c. In the Select Color dialog, click the ByLayer button and then click OK.

   d. While all the components are still highlighted, set Linetype to ByLayer using a similar procedure to steps b and c.

> **BYLAYER OR BYSTYLE**
>
> If you are a true CAD stickler, you will try to make as many items as possible set to ByLayer. This approach greatly simplifies things if you need to change color or linetype using the Layer Manager. Alternatively, you can define the color and linetype independently through the style.

You will be using only the Minor Contour in this style, but later on you will copy this style and your efforts will be carried forward.

   e. Select the Points component at the top of the list, then scroll down, and while pressing Shift, select Watersheds. This selects all components. Click one of the lightbulbs that are on to turn off all components. Then select the Minor Contour component and turn on this lightbulb.

f. Set the Minor Contour layer to C-TOPO-MINR.

   Your Display tab should now resemble Figure 19.25.

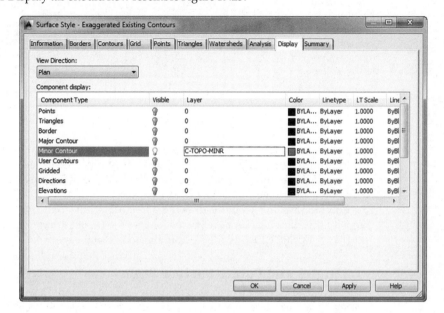

FIGURE 19.25
Minor Contour flying solo in plan

8. On the Display tab, change View Direction to Model and do the following:

   a. Shift+click to highlight all the components.

   b. Using the same procedure previously used, set all the Colors and Linetypes to ByLayer.

   c. Verify that Triangles is the only component turned on.

   d. Set the Triangles layer to C-TINN-VIEW.

      To see the surface in the Object Viewer or in any other 3D view, you must have triangles set to display in the Model View Direction.

9. Click OK to complete the style. Save the drawing.

10. From the Prospector tab of Toolspace, expand Surfaces, right-click the EG surface, and select Surface Properties.

11. On the Information tab, set Surface Style to Exaggerated Existing Contours, and click OK.

The surface should be rendered faster than you can read this sentence, even with the contour interval you've selected with only the minor contours displayed. After the style is applied to the surface model, you should see simple contours in plan view (the left image in Figure 19.26) and an exaggerated surface model in the Object Viewer (the right image in Figure 19.26).

882 | **CHAPTER 19** OBJECT STYLES

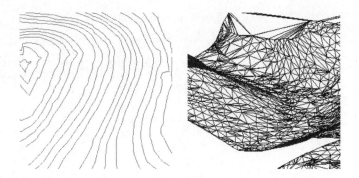

FIGURE 19.26
A portion of the surface showing your new style shown in plan (left) and in model (right), as shown in the Object Viewer

You skipped over one portion of the surface contours that many people consider a great benefit of using Civil 3D: depression contours. If this option is turned on via the Contours tab, ticks will be added to the downhill side of any closed contours leading to a low point. This is a stylistic option, and usage varies widely.

You can save and keep this drawing open to continue to the next exercise or use the saved copy of this drawing (SurfaceStylesContours_FINISHED.dwg or SurfaceStylesContours_METRIC_FINISHED.dwg) available from the book's web page.

Triangles and Points Surface Style

The next style you create will help facilitate surface editing. To work with the Swap Edge or Delete Line Surface edits, you must be able to see triangles. To work with the Delete Point, Modify Point, and Move Point commands, you must be able to see surface points.

It is important to note that the points you see in the surface style do not refer to survey points. The points you are working with in this exercise are *triangle vertices*. The triangle vertices and survey points will initially be in the same locations for a surface built from points. However, as breaklines are added or edits are made to the triangle vertices, the surface model will differ from the original survey.

1. If it isn't still open from the previous exercise, open SurfaceStylesContours_FINISHED.dwg or SurfaceStylesContours_METRIC_FINISHED.dwg.

   You can download either file from this book's web page.

2. From the Settings tab of Toolspace, expand Surface ➢ Surface Styles.

3. Right-click the Exaggerated Existing Contours style you created in the previous exercise, and select Copy, as shown in Figure 19.27.

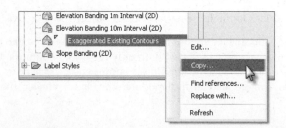

FIGURE 19.27
Access Copy by right-clicking directly on the style name.

4. On the Information tab, do the following:
   a. Rename the surface style to **Surface Editing**.
   b. Remove the text in the description file and type the description **Points and triangles**.
5. On the Points tab, do the following:
   a. Expand Point Size and set the Point Units to **3′** (or **1** m).
   b. Expand Point Display and click the ellipsis for Data Point Symbol.
   c. Set the point type to the X symbol, as shown in Figure 19.28.

FIGURE 19.28
Change the Point Display so triangle vertices stand out.

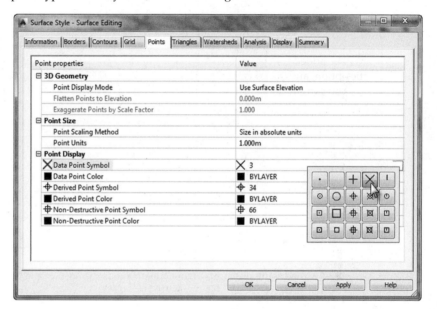

6. On the Triangles tab, remove the exaggeration by setting the Triangle display mode back to Use Surface Elevation.
7. On the Display tab with the View Direction set to Plan, do the following:
   a. Use the lightbulb icons to turn off visibility for Minor Contour, and turn it on for Points and Triangles.
   b. Click in the Layer column for the Points components to display the Layer Selection dialog.
   c. In the Layer Selection dialog, click the New button to create a new layer for the Points.
   d. In the Create Layer dialog, set Layer Name to **C-TINN-PNTS**, set Color to Red, and click OK.
   e. In the Layer Selection dialog, set the layer to C-TINN-PNTS and click OK.

f. Click in the Layer column for the Triangles components to display the Layer Selection dialog.

g. In the Layer Selection dialog, set the Triangles layer to C-TINN-VIEW and click OK.

Your Display tab will resemble Figure 19.29.

FIGURE 19.29
Change the Points and Triangles layers in the Display tab.

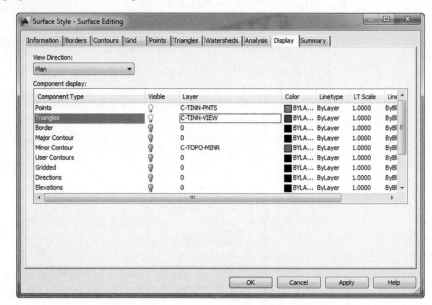

8. Click OK to complete the style.

9. Use the same procedure from the previous exercise to apply the Surface Editing style to the surface.

Your surface will resemble Figure 19.30.

FIGURE 19.30
Triangles and points shown using the new surface style

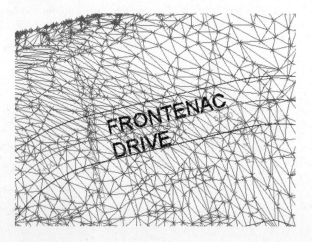

With the points and triangles displayed, you can edit the surface, such as deleting points or swapping edges.

You can save and keep this drawing open to continue to the next exercise or use the saved copy of this drawing (`SurfaceStyleTriangles_FINISHED.dwg` or `SurfaceStyleTriangles_METRIC_FINISHED.dwg`) available from the book's web page.

Analysis Styles

Analysis styles are unique in several ways. To see the style applied to your design, you must run the analysis in the surface properties in addition to applying the style to the surface. Although layers can now be configured to these styles, colors and behavior are configured on the Analysis tab. Visibility can be controlled by turning off the layer assigned on the Display tab of the style, turning off the Analysis component in the style, or by assigning another style that doesn't display that component.

You can choose distribution methods to apply to your analysis on the Analysis tab. The distribution method is selected by configuring the Group By option under each analysis type. Here's what they mean:

Equal Interval This method uses a stepped scale, created by taking the minimum and maximum values and then dividing the delta into the number of selected ranges. For example, if the surface has elevations from 0 to 10, with four ranges they will be 0 to 2.5, 2.5 to 5.0, 5.0 to 7.5, and 7.5 to 10. This method can create real anomalies when extremely large or small values skew the total range so that much of the data falls into one or two intervals, with almost no sampled data in the other ranges.

Quantile This method is often referred to as an *equal count distribution* and will create ranges that are equal in sample size. These ranges will not be equal in linear size but in distribution across a surface. For example, if the surface has elevations from 0 to 10 with most of the surface at the higher elevations, with four ranges they may be 0 to 4.78 (25 percent of the surface), 4.78 to 6.25 (25 percent of the surface), 6.25 to 7.95 (25 percent of the surface), and 7.95 to 10 (25 percent of the surface). This method is best used when the values are relatively equally spaced throughout the total range, with no extremes to throw off the group sizing.

Standard Deviation Standard Deviation is the bell curve that most engineers are familiar with, suited for when the data follows the bell distribution pattern. It generally works well for slope analysis, where very flat and very steep slopes are common, and would make another distribution setting unwieldy.

You looked at an elevations, slopes, and slope arrows analysis earlier in Chapter 4, "Surfaces." In the following exercise, you will create a surface style for watershed analysis. To apply the new style to the surface, you must also run the analysis.

1. If not still open from the previous exercise, open `SurfaceStylesTriangles_FINISHED.dwg` or `SurfaceStylesTriangles_METRIC_FINISHED.dwg`.

   You can download either file from this book's web page.

2. From the Settings tab of Toolspace, expand Surface ➢ Surface Styles.

3. Right-click the Exaggerated Existing Contours style you created in the earlier exercise and select Copy.

4. On the Information tab, do the following:

   a. Rename the surface style to **Watershed Analysis**.

   b. Remove the current description and type **Display watersheds and slope arrows**.

5. On the Triangles tab, change Triangle Display Mode to Use Surface Elevation.

6. On the Watersheds tab, do the following:

   a. Expand the Boundary Segment Watershed category.

   b. Set Use Hatching to False.

   c. Expand the Multi-Drain Watershed category.

   d. Set Use Hatching to False.

   Your Watersheds tab will look like Figure 19.31.

FIGURE 19.31
Changing the hatch options for watershed areas

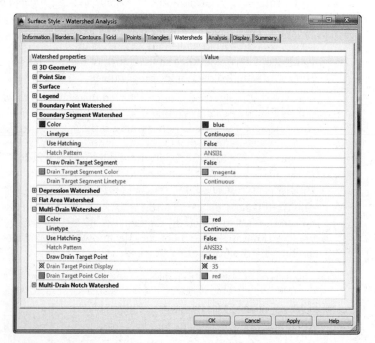

7. On the Analysis tab, do the following:

   a. Expand the Slope Arrows category.

b. Change Scheme to Hydro.

  c. Change Arrow Length to **2′** (or **1 m**).

  The Analysis tab will resemble Figure 19.32.

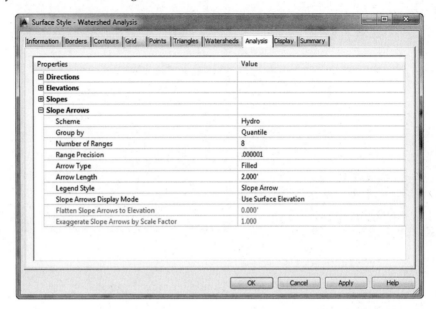

FIGURE 19.32
Set the color scheme and arrow length on the Analysis tab.

8. On the Display tab with the View Direction set to Plan, do the following:

    a. Turn off visibility for the Minor Contour component, and turn it on for Slope Arrows and Watersheds—you may have to scroll.

    b. Set the Watershed layer to C-TINN-VIEW.

9. Click OK to complete the surface style.

10. Select the surface and open Surface Properties.

11. On the Information tab, set Surface Style to Watershed Analysis and click Apply.

12. On the Analysis tab, do the following:

    a. Set Analysis type to Watersheds.

    b. Set Merge Depressions to **0.4′** (or **0.1 m**).

    c. Place a check mark next to Merge Adjacent Boundary Watersheds.

    d. Click the Run Analysis arrow in the middle of the dialog to populate the Range Details area.

    Your Surface Properties Analysis tab should resemble Figure 19.33.

FIGURE 19.33
You must run the analysis in Surface Properties before the Watershed style kicks in.

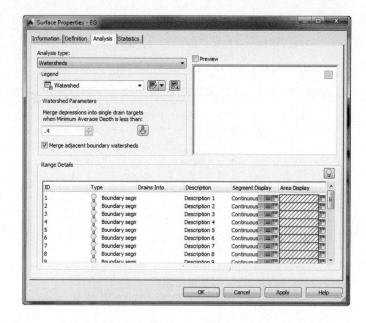

13. Click OK to close the Surface Properties dialog.

    Your surface model should resemble Figure 19.34.

FIGURE 19.34
The isolated surface using the Watershed Analysis style

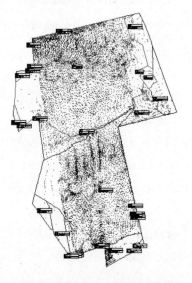

Now that the watershed analysis has been run, you could provide further information by generating a dynamic Watershed table similar to how you generated a table showing other surface information in Chapter 4.

When this exercise is complete, you can close the drawing. A saved copy of this drawing (SurfaceStylesWatershed_FINISHED.dwg or SurfaceStylesWatershed_METRIC_FINISHED.dwg) is available from the book's web page.

Pipe and Structure Styles

In Chapter 13, "Pipe Networks," you learned that the first step to managing pipes and structures was to build a Parts list. One of the functions of a Parts list is to associate styles to pipes and structures. In this section, you will learn how to create the pipe and structure styles that are used by a Parts list.

In your template, you will have many styles assigned to the various Parts lists. You will want to have separate styles for water systems, storm sewers, and sanitary sewers. Additionally, you may want to have separate styles for existing and proposed systems. The main difference between the styles for the different systems will be the layers you set in the Display tab.

Pipe Styles

It seems like no two municipalities want pipes displayed the same way on construction documents. Fortunately, Civil 3D offers many variations for pipe display that will satisfy miscellaneous submittal requirements. With one pipe style, you can control how a pipe is displayed in plan, profile, and section views. You can use multiple pipe styles to graphically differentiate larger pipes from smaller ones. This section explores all the options.

Plan Tab The tab you see in Figure 19.35 controls how your pipe is represented when you're working in plan view.

FIGURE 19.35
The Plan tab in the Pipe Style dialog

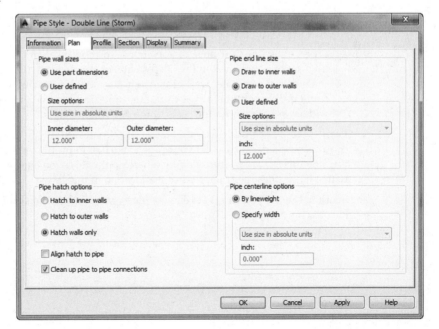

Options on the Plan tab include the following:

Pipe Wall Sizes   You can choose between having the program apply the part size directly from the part catalog (that is, the literal pipe dimensions as defined in the catalog) and specifying your own constant or scaled dimensions.

Pipe Hatch Options   If you choose to show pipe hatching, this part of the tab gives you options to control that hatch. You can hatch the entire pipe to the inner or outer walls, or you can hatch the space between the inner and outer walls only, as shown in Figure 19.36. There are also options to align the hatch to the pipe and to clean up pipe-to-pipe connections.

FIGURE 19.36
Pipe hatch to inner walls (a), outer walls (b), and hatch walls only (c)

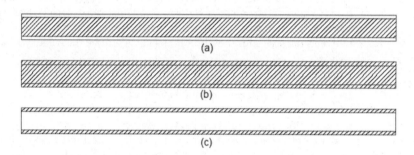

Pipe End Line Size   If you choose to show an end line, you can control its length with these options. An end line can be drawn connecting the outer walls or the inner walls, or you can specify your own constant or scaled dimensions. The pipes from Figure 19.36 are all shown with pipe end lines drawn to outer walls.

Pipe Centerline Options   If you choose to show a centerline, you can display it by the lineweight established in the Display tab, or you can specify centerline widths based on the pipe inner or outer wall dimension, a scaled dimension, or a constant dimension. You could use this option for your sanitary pipes in places where the width of the centerline widens or narrows on the basis of the pipe diameter.

Profile Tab   The Profile tab (see Figure 19.37) is almost identical to the Plan tab, except the controls here determine what your pipe looks like in profile view. The only additional settings on this tab are the crossing-pipe hatch options. If you choose to display crossing pipe with a hatch, these settings control the location of that hatch.

Section Tab   If you choose to show a hatch on your pipes in section, you control the hatch location on this tab (see Figure 19.38).

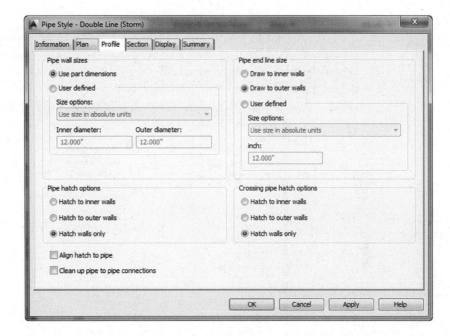

FIGURE 19.37
The Profile tab in the Pipe Style dialog

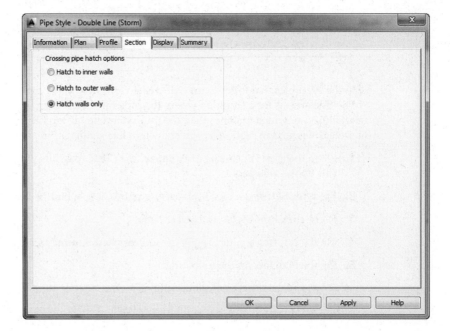

FIGURE 19.38
The Section tab in the Pipe Style dialog

> **PRESSURE PIPE STYLES**
>
> Pressure pipes are relatively new in Civil 3D, and with new objects come new styles. The Pressure Pipe Style dialog looks very similar to the Pipe Style dialog, with a Plan and Profile tab that have similar behavior to that discussed for regular pipes.
>
>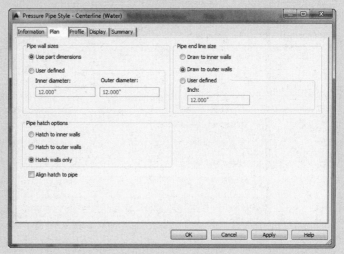
>
> The two biggest differences are that there are no Section tab or Crossing Pipe Hatch options on the Plan tab. There are a few other minor changes to how the user can define sizes.

In the examples that follow, you will create various types of pipe styles.

The first style is for a situation where the pipe must be shown in plan view with a single line, the thickness of which matches the pipe inner diameter. In profile, the pipe will show the inner diameter lines; and in section, it will show as a hatch-filled ellipse.

1. Open the `PipeStyle.dwg` or `PipeStyle_METRIC.dwg` file, which you can download from this book's web page.

2. From the Settings tab of Toolspace, expand Pipe ➢ Pipe Styles.

3. Right-click Pipe Styles and select New.

4. On the Information tab, set Name to **Proposed Sanitary CL**.

5. On the Plan tab, do the following:

    a. Verify that Pipe Centerline Options is set to Specify Width.

    b. Set Specify Width to Draw To Inner Walls.

6. On the Profile tab, no changes are needed.
7. On the Section tab, verify that Crossing Pipe Hatch Options is set to Hatch To Inner Walls.
8. On the Display tab with the View Direction set to Plan, do the following:
   a. Turn off the display for all components except Pipe Centerline.
   b. Set the Pipe Centerline layer to C-SSWR-PIPE.
9. On the Display tab, change View Direction to Profile and do the following:
   a. Turn off the display for all components except Inside Pipe Walls.
   b. Set the layer to C-SSWR-PROF.
10. On the Display tab, change View Direction to Section and do the following:
    a. Set Crossing Pipe Inside Wall to the C-SSWR-PIPE layer and make it visible.
    b. Set Crossing Pipe Hatch to the C-SSWR-PIPE-PATT layer and make it visible.
    c. Turn off Crossing Pipe Outside Wall.
    d. At the bottom of the dialog, click in the Pattern column for the Crossing Pipe Hatch component type to display the Hatch Pattern dialog.
    e. Set Type to Solid Fill, as shown in Figure 19.39, and click OK.

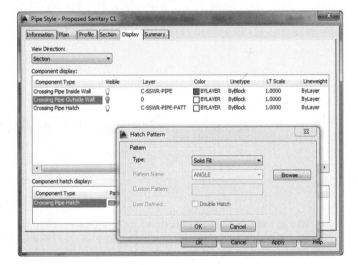

FIGURE 19.39
Setting the Hatch Pattern display for the section View Direction

11. Click OK to finish creating a new pipe style.
12. From the Prospector tab of Toolspace, expand Pipe Networks ➤ Networks ➤ Sanitary Network ➤ Pipes.

13. Using the Shift key, select all of the pipes in the item list.
14. Right-click the heading of the Style column and select Edit.
15. In the Select Pipe Style dialog, select Proposed Sanitary CL and click OK.
16. Examine the pipe in plan, profile, and cross section, as shown in Figure 19.40.

FIGURE 19.40
Proposed Sanitary CL pipe style shown in plan (a), profile (b), and section (c)

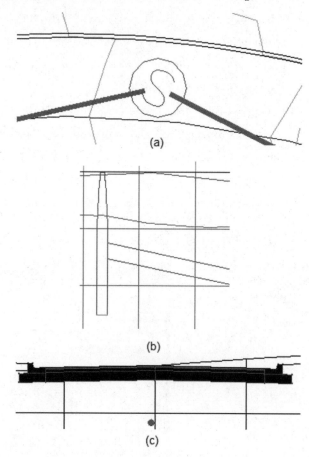

You can save and keep this drawing open to continue to the next exercise or use the saved copy of this drawing (`PipeStyle_FINISHED.dwg` or `PipeStyle_METRIC_FINISHED.dwg`) available from the book's web page.

> **HEY! WHY DOES MY PIPE LOOK LIKE AN OCTAGON?**
>
> The Civil 3D program helps itself perform better on large drawings by knocking down the resolution of 3D curved objects, such as the pipe in the cross-section view in the previous steps.
>
>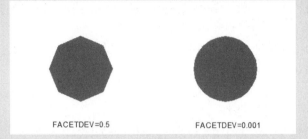
>
> FACETDEV=0.5   FACETDEV=0.001
>
> The system variable you can use to make these pipes look nicer is Facet Deviation, or `FACETDEV`. The default `FACETDEV` value for any Imperial unit drawing is 0.5 inches. In metric drawings, the default is 10 millimeters. The lower the `FACETDEV` value, the smoother the 3D curve.
>
> Note that another variable, `FACETMAX`, controls the maximum number of facets on any curved object. The Civil 3D default of 500 facets is usually more than enough to display a Civil 3D pipe smoothly.

In the next pipe style example, you will create a style that uses several options for hatching pipe walls for a pipe:

1. If not still open from the previous exercise, open `PipeStyle_FINISHED.dwg` or `PipeStyle_METRIC_FINISHED.dwg`.
2. From the Settings tab of Toolspace, expand Pipe ➢ Pipe Styles.
3. Right-click Pipe Styles and select New.
4. On the Information tab, set Name to **Proposed Hatch Wall**.
5. On the Plan tab, do the following:
    a. Verify that Pipe Hatch Options is set to Hatch Walls Only.
    b. Verify that Align Hatch To Pipe is selected.
6. On the Profile tab, do the following:
    a. Verify that Pipe Hatch Options is set to Hatch Walls Only.
    b. Verify that Align Hatch To Pipe is selected.
7. On the Display tab with View Direction set to Plan, do the following:
    a. Verify that the only components turned on are Inside Pipe Walls, Outside Pipe Walls, Pipe End Line, and Pipe Hatch components.

b. Set all four layers to C-SSWR-PIPE.

  c. Click on the Component Hatch Display Scale field and type in a new value of **0.1**.

8. On the Display tab, change View Direction to Profile and do the following:

  a. Verify that the only components turned on are Inside Pipe Walls, Outside Pipe Walls, Pipe End Line, and Pipe Hatch.

  b. Set all four of these layers to C-SSWR- PROF.

9. Click OK to complete creation of a new pipe style.

10. From the Prospector tab of Toolspace, expand Pipe Networks ➢ Networks ➢ Sanitary Network ➢ Pipes.

11. Select all of the pipes in the Item list using the Shift key.

12. Right-click the heading of the Style column and select Edit.

13. In the Select Pipe Style dialog, select Proposed Hatch Wall, and click OK.

14. Examine the pipe in plan and profile, as shown in Figure 19.41. They look very similar in both views.

FIGURE 19.41
Proposed Hatch Wall pipe style shown in plan

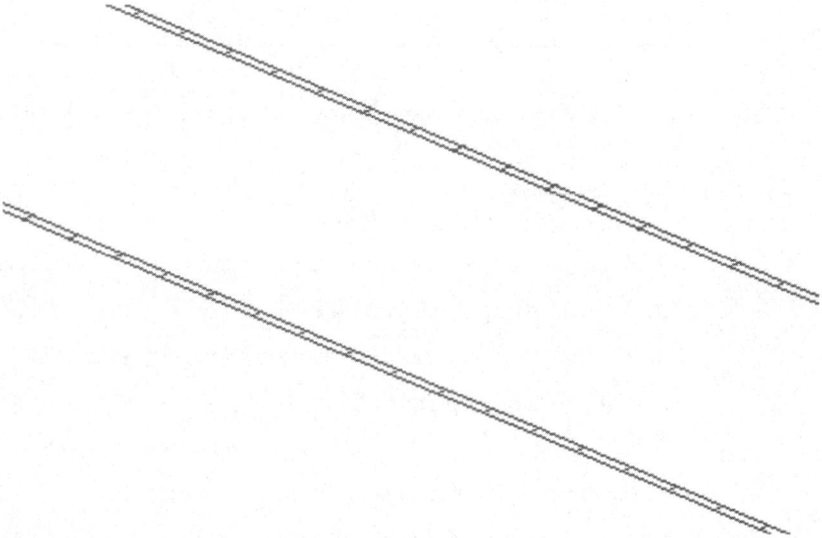

As you can see depending on your drawing's scale, it may or may not be worth it to you to add a pipe wall hatch because with thin walls it may be hard to see, as shown in Figure 19.41.

When this exercise is complete, you can close the drawing. A saved copy of this drawing (`PipeStyleHatch_FINISHED.dwg` or `PipeStyleHatch_METRIC_FINISHED.dwg`) is available from the book's web page.

You could do this same exercise for a Pressure Pipe style. The only difference would be setting the layers to C-WATR-PIPE instead of C-SSWR-PIPE, or C-WATR-PROF instead of C-SSWR-PROF.

Structure Styles

The following tour through the structure-style interface can be used for reference as you create company-standard styles:

Model Tab   The Model tab (Figure 19.42) controls what represents your structure when you're working in 3D. Typically, you want to leave the Use Catalog Defined 3D Part radio button selected so that when you look at your structure, it looks like your flared end section or whatever you've chosen in the Parts list.

FIGURE 19.42
The Model tab in the Structure Style dialog

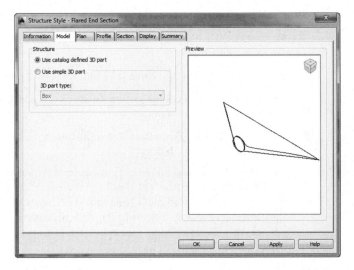

Plan Tab   The Plan tab (Figure 19.43) enables you to compose your object style to match any particular standard.

FIGURE 19.43
The Plan tab in the Structure Style dialog

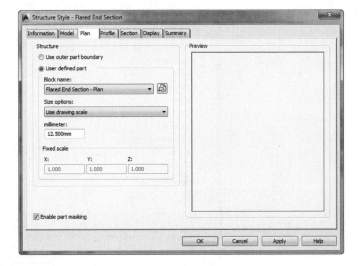

Options on the Plan tab include the following:

Use Outer Part Boundary This option uses the limits of your structure from the Parts list and shows you an outline of the structure as it would appear in the plan.

User Defined Part This option uses any block you specify. In the case of your flared end section, you chose a symbol to match the CAD standard. When using a User Defined Part, you also must provide Size Options. The options in this drop-down are similar to what you see in other styles, such as point or marker styles in Civil 3D.

- Use Drawing Scale will treat the object like an annotative block.
- Use Fixed Scale allows you to enter X, Y, and Z scale factors as you would do when inserting a block.
- Use Size In Absolute Units is a common way to represent a manhole at actual size.
- Use Size As Percentage Of Screen keeps the part the same size whether you zoom close or are far away.
- Use Fixed Scale From Part Size will stretch the block around the part even if the dimensions of both don't match.

Enable Part Masking This option creates a wipeout or mask inside the limits of the structure. Any pipes that connect to the center of the pipe appear trimmed at the limits of the structure. This will mask the pipes entering and exiting the structure based on where they intersect the actual walls of the structure as modeled. Keep in mind that when using a block, these locations may not match the extents of the block.

Profile Tab The Profile tab (Figure 19.44) is where you configure what your structure will look like in profile view.

FIGURE 19.44
The Profile tab in the Structure Style dialog

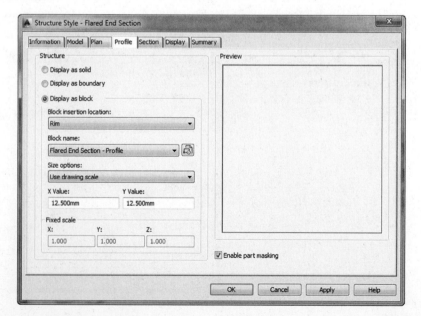

Options on the Profile tab include:

Display As Solid   This option uses the limits of your structure from the Parts list and shows you the mesh of the structure as it would appear in profile view.

Display As Boundary   This option uses the limits of your structure from the Parts list and shows you an outline of the structure as it would appear in profile view. You'll use this option for the sanitary manhole.

Display As Block   This option uses any block you specify. When using a Structure displayed as a block, you also must provide Size Options. The size options are the same as the ones described for the Plan tab.

Enable Part Masking   This option creates a wipeout or mask inside the limits of the structure. Any pipes that connect to the center of the pipe appear trimmed at the limits of the structure.

Section Tab   The Section tab (Figure 19.45) is where you configure what your structure will look like in section view.

FIGURE 19.45
The Section tab in the Structure Style dialog

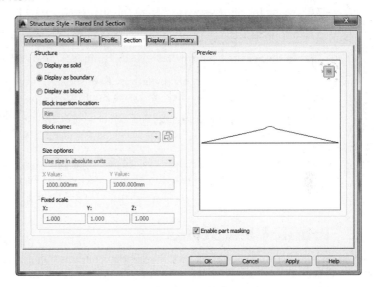

These options look (and behave) very much like the Profile tab options listed earlier.

In the following exercise, you'll create a new structure style that uses a block in plan view to represent a sanitary manhole. Because the block is drawn at actual size, you will use the size option Use Fixed Scale.

1. Open the `StructureStyle.dwg` or `StructureStyle_METRIC.dwg` file, which you can download from this book's web page.

2. From the Settings tab of Toolspace, expand Structure ➢ Structure Styles.

3. Right-click Structure Styles and select New.

4. On the Information tab, rename the style to **Simple Sanitary Manhole**.

5. On the Plan tab, do the following:
   a. Verify that User Defined Part is selected.
   b. Set the Block Name to **_Wipeout_Circle** using the list box.
   c. Set Size to Use Fixed Scale.
   d. Set the X and Y scale factors set to **3** (or **1** for metric).
6. On the Display tab with View Direction to Plan, set the Structure layer to C-SSWR-STRC.
7. Repeat step 6 with View Direction set to Profile and then repeat with View Direction set to Section.
8. Click OK to complete the style.
9. From the Prospector tab of Toolspace, expand Pipe Networks ➣ Networks ➣ Sanitary Network ➣ Structures.
10. Using the Shift key, select all of the structures in the Item list.
11. Right-click the heading of the Style column and select Edit.
12. In the Select Structure dialog, select Simple Sanitary Manhole and click OK.
13. Press Esc to deselect all structures. To observe the style change in plan, select any one of the structures, right-click, and click Zoom To.

When this exercise is complete, you can close the drawing. A saved copy of this drawing (StructureStyle_FINISHED.dwg or StructureStyle_METRIC_FINISHED.dwg) is available from the book's web page.

While this example discussed creating object styles for a structure, you will find that the same procedure is applicable to the Appurtenances and Fittings styles used in pressure pipe networks.

Profile View Styles

When you are looking at a profile view that contains data, you are seeing many styles displayed. The profiles themselves (existing and proposed) have a profile object style applied to them. The labels consist of many types of styles, as you learned in Chapter 18. Additionally, there are profile view styles and band styles to consider.

This section focuses on the profile view. A profile view controls many aspects of the display. The profile view style consists of some of the following properties:

- Vertical exaggeration
- Grid spacing
- Elevation and station annotation
- Title annotation

Figure 19.46 shows a profile view with some of its basic components labeled. There are many more components in a profile view style.

FIGURE 19.46
Profile view style with some of its basic components

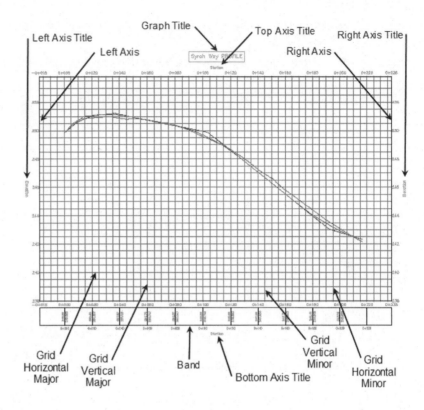

In the example that follows, you will be making major modifications to a profile view style. The profile view you will be practicing with does not contain any bands. Later on in this section, you will learn the ins and outs of band creation and modification.

1. Open the ProfileViewStyles.dwg or ProfileViewStyles_METRIC.dwg file.

   This file contains a profile view of Cabernet Court with a very ugly style applied to it. You will perform a complete makeover on this style.

2. Zoom in as close as possible so you can see the entire profile view and select the profile view by clicking anywhere on a grid line or axis.

3. From the Profile View contextual tab ➢ select Profile View Properties ➢ Edit Profile View Style, as shown in Figure 19.47.

FIGURE 19.47
Accessing the profile view style

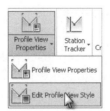

4. Position the dialog on your screen so you can make changes to the style and observe the changes in the profile view behind it.

5. On the Graph tab, change the Vertical Exaggeration value to **10**, as shown in Figure 19.48, and click Apply.

FIGURE 19.48
Change Vertical Exaggeration on the Graph tab of the Profile View Style dialog.

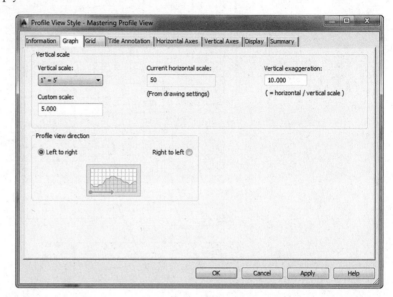

When you do, you will notice that the Vertical Scale listed in the dialog automatically changes from 1" = 50' to 1" = 5' (or from 1:500 to 1:50 for metric users). In addition, if you can see your profile view in the background, you should notice that it has expanded vertically by a factor of 10.

6. On the Grid tab, do the following:

   a. Verify that both Clip Vertical Grid and Clip Horizontal Grid options are unchecked.

   If you need additional information on any of the controls on this tab, click the Help button at the bottom of the dialog.

   b. Set Grid Padding (Major Grids) to **0.5** for Above Maximum Elevation and **0.5** for Below Datum.

   This setting will create additional space above and below the design data at 0.5 times the vertical major tick interval (you will set the major tick interval later).

   c. Set all values for Axis Offset (Plotted Units) to **0**.

   d. Click Apply to review changes on the profile view.

   This will ensure that the axes and the grid lines coincide around the edges of the view. The settings on the Grid tab should match what is shown in Figure 19.49.

FIGURE 19.49
The Grid tab of the Profile View Style dialog

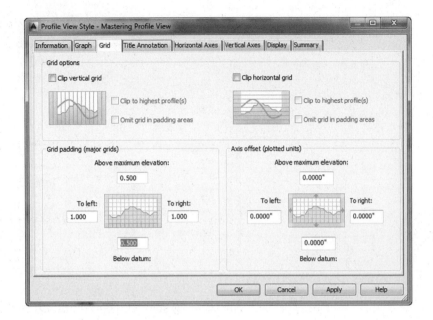

7. On the Title Annotation tab, do the following:

   a. In the Graph View Title area, change Text Height to **0.4"** (or **10** mm).

   b. In the Graph View Title area, click the Edit Mtext button.

   c. In the Text Component Editor dialog, remove all the text in the Text Component Editor window.

   You will be starting over with a blank Text Component Editor dialog.

   d. From the Properties drop-down, select Parent Alignment and make these changes:

   ♦ Set Capitalization to Upper Case.

   ♦ Click the arrow button to add the Property field to the Text Component Editor window.

   ♦ Click in the Text Component Editor window behind the Property field, add a space, and type **PROFILE VIEW**.

   e. Click OK to accept the entry in the Text Component Editor dialog.

   f. Change the Y Offset for the Title Position to **2"** (or **10** mm).

8. Click Apply and examine your changes in the background.

   The Title Annotation tab should match the settings shown in Figure 19.50.

FIGURE 19.50
Working with the Graph View Title size and placement

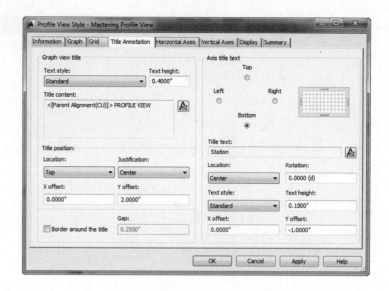

Do not bother to adjust any settings for the Axis title text on the right side of Figure 19.50 because the display will be turned off for all four of these possible elements.

9. On the Horizontal Axes tab, do the following:

   a. Verify that the Axis To Control radio button is set to Bottom.

   b. In the Major Tick Details area, set Interval to **100′** (or **20** m).

   c. In the Major Tick Details area, click the Edit Mtext button.

   d. In the Text Component Editor dialog, do the following:

   ♦ Remove all the text in the Text Component Editor window.

   ♦ From the Properties drop-down, select Station Value.

   ♦ Change the precision to **1**.

   ♦ Click the arrow button to add the Property field to the Text Component Editor window.

   e. Click OK to accept the entry in the Text Component Editor dialog.

   f. In the Major Tick Details area, change Rotation to **90**.

   g. In the Major Tick Details area, change the X offset to **0″** (**0** mm) and the Y offset to **−0.25″** (**−10** mm).

   h. In the Minor Tick Details area, set Interval to **50′** (**10** m).

   No other changes are needed in the Minor Tick Details area.

   The Horizontal Axes tab should match the settings shown in Figure 19.51.

FIGURE 19.51
The bottom axis controls grid spacing.

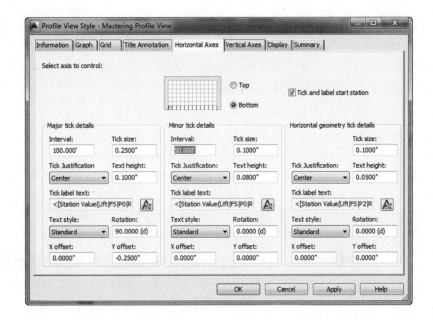

10. On the Vertical Axes tab, do the following:

    a. Verify that the Select Axis To Control radio button is set to Left.

    b. In the Major Tick Details area, click the Edit Mtext button.

    c. In the Text Component Editor dialog, do the following:

    ♦ Remove all the text in the Text Component Editor window.

    ♦ From the Properties drop-down, select Profile View Point Elevation.

    ♦ Change Precision to **1**.

    ♦ Click the arrow button to add the Property field to the Text Component Editor window.

    d. Click OK to exit the Text Component Editor dialog.

    e. In the Major Tick Details area, change the X offset to **–0.25"** (**–5 mm**) and the Y offset to **0"** (**0 mm**).

    All the changes made up to this point on the Vertical Axes tab apply to the Left axis. You will now do similar modifications to the Right axis.

    f. Change the Select Axis To Control radio button to Right.

    g. In the Major Tick Details area, click the Edit Mtext button.

h. In the Text Component Editor dialog, do the following:
- Remove all the text in the Text Component Editor window.
- From the Properties drop-down, select Profile View Point Elevation.
- Change Precision to **1**.
- Click the arrow button to add the property value to the Text Component Editor window.

i. Click OK to exit the Text Component Editor.

j. In the Major Tick Details area, change the X offset to **0.25″** (**5 mm**) and the Y offset to **0″** (**0 mm**).

11. Click Apply and examine your changes.

Figure 19.52 shows the Vertical Axes tab as yours should look at this point in the exercise.

FIGURE 19.52
Don't forget to change the settings for both the Left and Right axes in this tab.

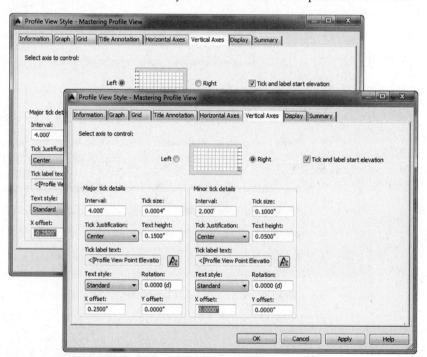

12. On the Display tab, do the following:

a. Select all of the components and use the lightbulb icon to turn off visibility for all of the components.

b. Turn back on the visibility for the following components:
- Graph Title
- Left Axis
- Left Axis Annotation Major
- Right Axis
- Right Axis Annotation Major
- Top Axis
- Bottom Axis
- Bottom Axis Annotation Major
- Grid Horizontal Major
- Grid Horizontal Minor
- Grid Vertical Major
- Grid Vertical Minor

13. Click OK to complete the profile view style.

Your profile view should resemble Figure 19.53.

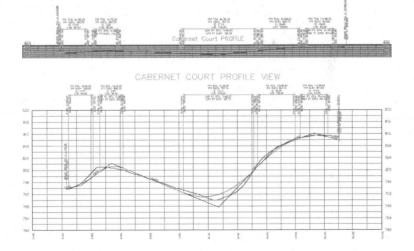

FIGURE 19.53
The profile view you started with (above) and after the style is completed (below)

When this exercise is complete, you can close the drawing. A saved copy of this drawing (`ProfileViewStyles_FINISHED.dwg` or `ProfileViewStyles_METRIC_FINISHED.dwg`) is available from the book's web page.

> ### What Drives Profile View Grid Spacing?
>
> On the Horizontal Axes tab and Vertical Axes tab of the Profile View Style dialog, you may notice that you can control opposing axes separately. Each tab has a toggle for Select Axis To Control: Top or Bottom and Left or Right.
>
> It is the Bottom and Left options in the respective tabs that control grid spacing. For both axes, you will find options for Major Tick Details. The Interval values for the major ticks are the key to getting the grid spacing to look the way you want. Changes to either horizontal or vertical major tick intervals will affect the height and length of the profile view, as well as grid spacing. Changing the Interval on Minor Tick Details will affect the grid spacing, but will not affect the aspect ratio of the profile view.
>
> Here is a common practice for setting major tick interval values for your horizontal and vertical axes: Vertical Exaggeration value (Grid tab) × Major Tick Interval value (Vertical Axes tab – right or left) = Major Tick Interval value (Horizontal Axes tab – top or bottom). Therefore, a profile view style with Vertical Exaggeration set to **10** and Vertical Major Tick Intervals set to **5'** would yield horizontal major tick intervals set to **50'**.
>
> Even if you don't turn on the ticks or grid lines on these axes, the spacing increment will be reflected in the profile view.

Profile View Bands

Data bands are strips of labels and/or schematics that display additional information about the profile or alignment that is referenced in a profile view. The most common band type is the profile data band.

Bands can be applied to both the top and bottom of a profile view, and there are six band types: Profile Data Bands, Vertical Geometry Bands, Horizontal Geometry Bands, Superelevation Data Bands, Sectional Data Bands, and Pipe Network Bands. These band types were discussed in Chapter 7, "Profiles and Profile Views," but graphic reminders of the various band types are shown in Figure 19.54 through Figure 19.59.

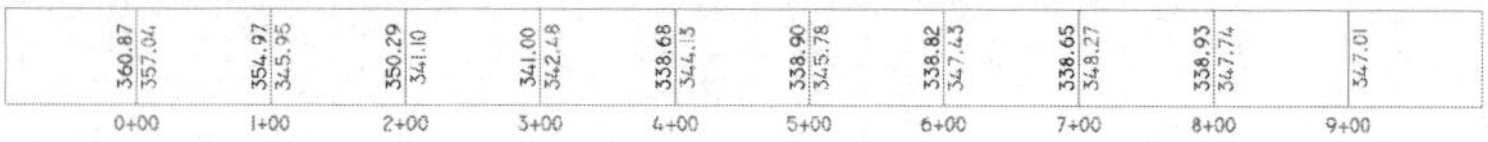

FIGURE 19.54
Profile Data Band showing existing and proposed elevation in addition to major stations

FIGURE 19.55
Vertical Geometry Band

FIGURE 19.56
Horizontal Geometry Band

FIGURE 19.57
Superelevation Data Band

FIGURE 19.58
Section Data Band

FIGURE 19.59
Pipe Data Band showing invert elevations and slope schematic

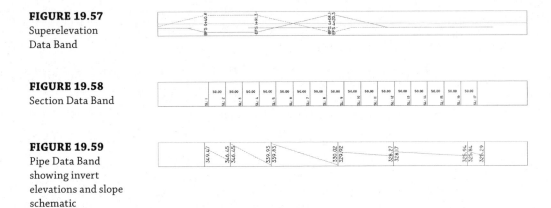

Bands can be assigned to band sets. Like alignment label sets and profile label sets, a *band set* determines which bands are applied to a profile view and how they are positioned. The most common use for a band set is to create a single, viewable band and an additional nonvisible band for spacing purposes in plan and profile sheet generation. However, depending on your jurisdictional requirements, you may have to put together a more extensive band set.

In the following exercise, you will create a band that contains existing and proposed profile elevations:

1. Open the ProfileBands.dwg or ProfileBands_METRIC.dwg file.

   This file contains the profile view from the previous exercise and an empty band on the bottom of the view. You will be adding information to the profile data band.

2. From the Settings tab of Toolspace, expand Profile View ➢ Band Styles ➢ Profile Data.

3. Right-click the style called Mastering Band and select Edit.

4. On the Band Details tab, do the following:

   a. Change Band Height to **1.0″** (or **25 mm**).

   b. On the right side of the Band Details tab, highlight Major Station, and click the Compose Label button, as shown in Figure 19.60.

FIGURE 19.60
The Band Details tab

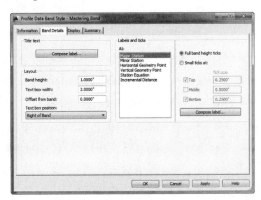

c. On the Layout tab of the Label Style Composer – Major Station dialog, click the Create Text Component button and make these changes:

- Set Name to **Station**.
- Use the drop-down to change Anchor Point to Band Bottom.
- Use the drop-down to change Attachment to Top Center.
- Change the Y offset to **–0.02"** (or **–0.5 mm**).
- Click the Contents value, which currently reads Label Text, and click the ellipsis to display the Text Component Editor dialog.
- In the Text Component Editor, remove all text in the Text Component Editor window and select Station Value from the Properties list.
- Set Precision to **1**.
- Click the arrow button to add the Property field to the Text Component Editor window.

d. Click OK to exit the Text Component Editor.

e. Click the Create Text Component button again, and make these changes:

- Set Name to **Existing El**.
- Use the drop-down to change Anchor Point to Band Middle.
- Set Rotation Angle to **90**.
- Use the drop-down to change Attachment to Bottom Center.
- Change the X offset to **–0.02"** (or **–0.5 mm**).
- Click the Contents value, which currently reads Label Text, and click the ellipsis to display the Text Component Editor dialog.
- In the Text Component Editor window, remove the Property field and select Profile1 Elevation from the Properties list.
- Set Precision to **0.01** (or **0.001** for metric).
- Click the arrow button to add the Property field to the Text Component Editor window.

f. Click OK to exit the Text Component Editor.

g. Click the Copy Component button and make these changes:

- Change Name to **Proposed El**.
- Verify that Anchor Point is set to Band Middle.
- Verify that Rotation Angle is set to **90**.
- Use the drop-down to change Attachment to Top Center.

- Change the X offset to **0.02"** (or **0.5** mm).
- Click the Contents value and click the ellipsis to display the Text Component Editor dialog.
- In the Text Component Editor window, double-click on the existing Property field and select Profile2 Elevation from the Properties list.
- Verify the Precision is **0.01** (or **0.001** for metric).
- Click the arrow button to add the Property field to the Text Component Editor window.

  h. Click OK to exit the Text Component Editor.

  i. Click OK to finish working with the Major Stations Label Composer and return to the Band Details tab.

5. On the Display tab, use the lightbulb icon to turn off visibility for Minor Tick, and click OK.

The completed band should resemble Figure 19.61.

FIGURE 19.61
Text along the bottom of your profile view in the form of a band

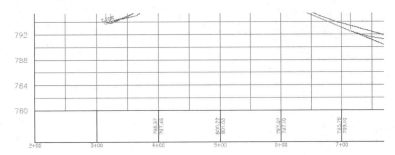

As you can see, profile bands can provide a lot of information in a compact manner. In this example, you provided information only at the major stations, but you could also provide information at minor stations, horizontal geometry points, vertical geometry points, station equations, and incremental distances.

When this exercise is complete, you can close the drawing. A saved copy of this drawing (`ProfileBands_FINISHED.dwg` or `ProfileBands_METRIC_FINISHED.dwg`) is available from the book's web page.

Section View Styles

Section view styles share many of the same concepts as creating profile view styles. In fact, the Section View Style dialog has all the same tabs and looks nearly identical to the profile view style.

In this section, you will walk through the creation of a section view style suitable for creating a section sheet:

1. Open the `SectionStyles.dwg` or `SectionStyles_METRIC.dwg` file.

This file contains section views created with the default settings for section views.

2. From the Settings tab of Toolspace, expand Section View ➤ Section View Styles ➤ Road Section.

3. Right-click Road Section and select Edit.

4. On the Grid tab, set Grid Padding (Major Grids) to **0** for the Above Maximum Elevation and Below Datum options.

5. On the Display tab, turn off the visibility for all components except:

    ◆ Graph Title

    ◆ Left Axis Annotation Major

    ◆ Right Axis Annotation Major

    ◆ Bottom Axis Annotation Major

6. Click OK.

Your section views should resemble Figure 19.62.

FIGURE 19.62
Yes, this is correct! It is a very stripped-down section view.

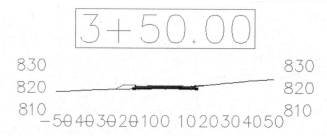

7. Select one of the views by clicking on the station label.

8. From the Section View contextual tab ➤ Modify View panel, select Update Group Layout.

    The section views will rearrange to fit more sections per page.

The section view is so bare bones because the section view grid will come from the group plot style, so the only information you really need is in this simple style.

To continue to the next exercise, you can save and keep this drawing open or use the saved copy of this drawing (SectionStyles_FINISHED.dwg or SectionStyles_METRIC_FINISHED.dwg) available from the book's web page.

Group Plot Styles

Group plot styles determine how sections are arranged on a sheet. When multiple section views are created, the group plot style uses the Section template file discussed in Chapter 15, "Plan Production," and places sections inside the paperspace viewport:

1. If not still open from the previous exercise, open SectionStyles_FINISHED.dwg or SectionStyles_METRIC_FINISHED.dwg.

    You can download either file from this book's web page. This file contains section views created with the default settings.

2. From the Settings tab of Toolspace, expand Section View ➢ Group Plot Styles.

3. Right-click Basic and select Edit.

4. On the Array tab, change the column spacing to **4″** (or **100** mm). Change the row spacing to **2″** (or **50** mm), as shown in Figure 19.63.

FIGURE 19.63
The Array tab controls section view spacing.

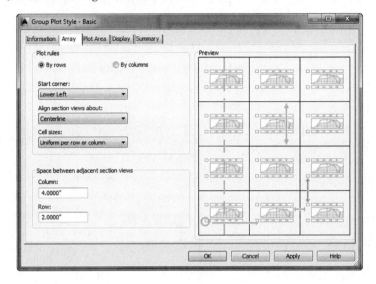

These spacing changes should allow a more aesthetic arrangement of cross sections per page and ample room for moving the views up and right to fit on the page better in the upcoming steps.

5. On the Plot Area tab, leave all the default settings as shown in Figure 19.64.

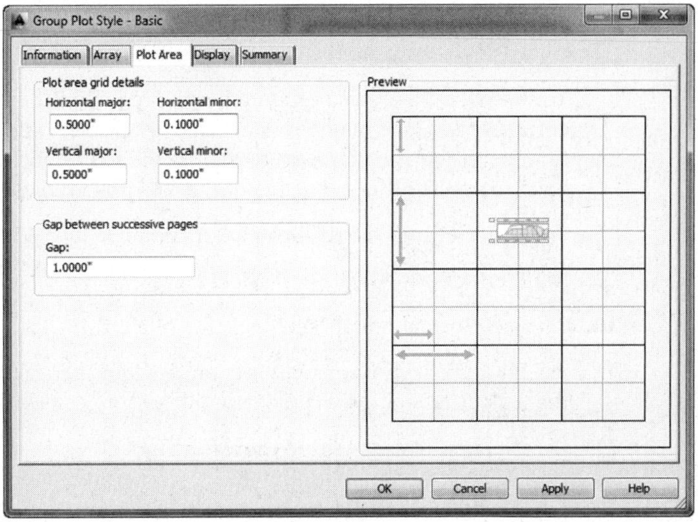

FIGURE 19.64
Grid spacing on sheets is specified on the Plot Area tab.

This is where you configure grid spacing per sheet.

6. On the Display tab, do the following:

   a. Verify that the visibility is turned on for Major Horizontal Grid and Major Vertical Grid.

   b. Verify that the visibility is turned off for Minor Horizontal Grid and Minor Vertical Grid.

   c. Verify that the layers for all of the grid components are set to C-ROAD-SCTN-GRID.

   d. Verify that the layers for Print Area and Sheet Border are set to C-ROAD-SCTN-TTLB.

   e. Set Color for Major Horizontal Grid and Major Vertical Grid to **9**.

   f. Set Color for Print Area to Cyan and Sheet Border to Green. Your Display tab will look like Figure 19.65.

FIGURE 19.65
The Display components for the group plot style

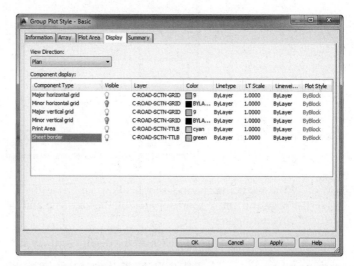

7. Click OK to complete the group plot style edits.

   Your section view sheets should be shaping up to the point where you could almost generate sheets. There may be instances when some text is placed outside of the cyan line that represents the viewport border. In the next steps, you will use a nonvisible section band to prevent this from happening and push the views onto the page.

8. From the Settings tab of Toolspace, expand Section View ➢ Band Styles ➢ Section Data.

9. Right-click Section Data and select New.

10. On the Information tab, set Name to **_NO DISPLAY**.

    Prefixing the style name with the underscore ensures it will be alphabetized to the top of the Style list.

11. On the Band Details tab, do the following:

a. Set Band Height to **0.2"** (or **5** mm).
b. Set Text Box Width to **0.2"** (or **10** mm).
c. Set Offset From Band to **0"** (or **0** mm).

Even though the band will not be not visible, the Civil 3D program still accounts for this spacing when placing the views on the sheet. In this step, you are using this to your advantage.

12. On the Display tab, verify that the visibility is turned off for all components.
13. Click OK to finish creating a new section data band style.
14. Select any section view by clicking on the station label or the elevation labels.
15. From the Section View contextual tab ➢ Modify View panel, choose View Group Properties to display the Section View Group Properties dialog.
16. On the Section Views tab, do the following:
    a. Click the ellipsis in the Change Band Set column, as shown in Figure 19.66. You may need to widen the columns to view the full titles.

    The Section View Group Bands dialog will appear.

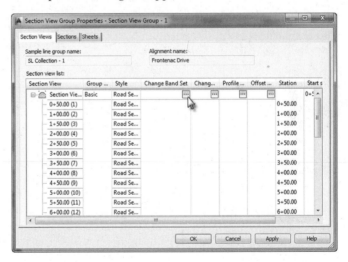

FIGURE 19.66
Changing the band set in use for all section views

b. Verify Band Type is set to Section Data.
c. Set the band style as _NO DISPLAY and click Add.
d. Set the Gap distance to **0**, as shown in Figure 19.67, and click OK to dismiss the Section View Group Bands dialog.

FIGURE 19.67
Add the data band and set the gap to 0.

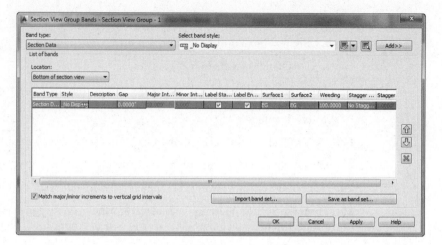

  e. Click OK to exit the View Group Properties dialog.

17. From the Section View contextual tab ➢ Modify View panel, select Update Group Layout.

The cross-section sheets should look like Figure 19.68.

FIGURE 19.68
The completed exercise

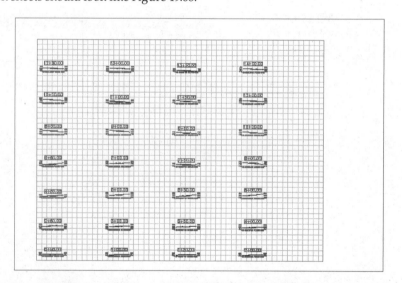

When this exercise is complete, you can close the drawing. A saved copy of this drawing is available from the book's web page with the filename `GroupPlotStyles_FINISHED.dwg` or `GroupPlotStyles_METRIC_FINISHED.dwg`.

With all of the object styles, you have a great deal of control over every detail, even ones that may seem trivial. Instead of being bogged down trying to understand every option, don't be afraid to use a "trial and error" approach. If you make a change you don't like, you can always edit the style until you get it right.

The Bottom Line

Override object styles with other styles. In spite of the desire to have uniform styles and appearances between objects within a single drawing, project, or firm, there are always going to be changes that need to be made.

Master It Open the `MasteringStyles.dwg` or `MasteringStyles_METRIC.dwg` file and change the alignment style associated with Alignment B to Layout. In addition, change the surface style used for the EG surface to Contours And Triangles, but change the contour interval to be **1'** and **5'** (or 0.5 m and 2.5 m) and the color of the triangles to be yellow.

Create a new surface style. Almost every set of plans that you send out of the office is going to include a surface, so it is important to be able to generate multiple surface styles that match your company standards. In addition to surface styles for production, you may find it helpful to have styles to use when you are designing that show a tighter contour spacing as well as the points and triangles needed to make some edits.

Master It Open the `MasteringSurfaceStyle.dwg` or `MasteringSurfaceStyle_METRIC.dwg` file and create a new surface style named **Micro Editing**. Set this style to display contours at **0.5'** and **1.0'** (or 0.1 m and 0.2 m), as well as triangles and points. Set the EG surface to use this new surface style.

Create a new profile view style. Everyone has their preferred look for a profile view. These styles can provide a lot of information in a small space, so it is important to be able to create a profile view that will meet your needs.

Master It Open the `MasteringProfileViewStyle.dwg` or `MasteringProfileViewStyle_METRIC.dwg` file and create a new profile view style named **Mastering Profile View**. Set this style to not clip the vertical or horizontal grid. Set the bottom horizontal ticks at 50' and 10' intervals (25 m and 5 m). Set the left and right vertical ticks at 10' and 2' intervals (5 m and 1 m). In addition, turn off the visibility of the Graph Title, Bottom Axis Annotation Major, and Bottom Axis Annotation Horizontal Geometry Point. Set the profile view in the drawing to use this new profile view style.

Appendix A

The Bottom Line

Each of The Bottom Line sections in the chapters suggests exercises to deepen skills and understanding. Sometimes there is only one possible solution, but often you are encouraged to use your skills and creativity to create something that builds on what you know and lets you explore one of many possibilities.

Chapter 1: The Basics

Find any Civil 3D object with just a few clicks. By using Prospector to view object data collections, you can minimize the panning and zooming that are part of working in a CAD program. When common subdivisions can have hundreds of parcels or a complex corridor can have dozens of alignments, jumping to the desired one nearly instantly shaves time off everyday tasks.

Master It Open 0103_Example.dwg (0103_Example_METRIC.dwg) from www.sybex.com/go/masteringcivil3d2014, and find parcel number 6 without using any AutoCAD commands or scrolling around on the drawing screen. (Hint: Take a look at Figure 1.5.)

Solution

1. In Prospector, expand Sites ➢ Proposed ➢ Parcels.
2. Right-click on Special ROW: 6 and select Zoom To.

Modify the drawing scale and default object layers. Civil 3D understands that the end goal of most drawings is to create hard-copy construction documents. When you set a drawing scale, Civil 3D removes a lot of the mental gymnastics that other programs require when you're sizing text and symbols. When you set object layers for the entire drawing, Civil 3D makes uniformity of drawing files easier than ever to accomplish.

Master It Change the Annotation scale in the model tab of 0103_Example.dwg from the 50-scale drawing to a 100-scale drawing. (For metric users: Use 0103_Example_METRIC.dwg and change the scale from 1:250 to 1:1000.)

Solution

1. In the lower-right corner of the application window, select 1" = 100' (1:1000) from the Annotation Scale list.
2. Type **REA** and press ↵ to regenerate the screen and show the labels at the new scale.

Navigate the ribbon's contextual tabs. As with AutoCAD, the ribbon is the primary interface for accessing Civil 3D commands and features. When you select an AutoCAD Civil 3D object, the ribbon displays commands and features related to that object. If several object types are selected, the Multiple contextual tab is displayed.

Master It Continue working in the file `0103_Example.dwg` (`0103_Example_METRIC.dwg`). It is not necessary to have completed the previous exercise to continue. Using the ribbon interface, access the Alignment properties for QuickStart Alignment and rename it **Existing CL**.

Solution

1. Select QuickStart Alignment to display the contextual Alignment tab on the ribbon.
2. From the Alignment contextual tab ➢ Modify panel, click Alignment Properties.
3. In the Alignment Properties menu, select the Information tab.
4. Rename the alignment **QuickStart CL**.

Create a curve tangent to the end of a line. It's rare that a property stands alone. Often, you must create adjacent properties, easements, or alignments from their legal descriptions.

Master It Open the drawing `MasterIt0101.dwg` (`MasterIt0101_METRIC.dwg`). Create a curve tangent to the east end of the line labeled in the drawing. The curve should meet the following specifications:

- Radius: 200.00′ (60 m)
- Arc Length: 66.580′ (20 m)

Solution

1. Select Home ➢ Draw ➢ Curves ➢ Create Curve From End Of Object.
2. Select the east side of the line that is labeled "Create a curve tangent to this line."
3. On the command line, press ↵ to confirm that you will enter a radius value.
4. On the command line, type **200.00** (**60**), and then press ↵.
5. Type **L** to specify the length, and then press ↵.
6. Type **66.580** (**20**), and then press ↵.

Label lines and curves. Although converting linework to parcels or alignments offers you the most robust labeling and analysis options, basic line- and curve-labeling tools are available when conversion isn't appropriate.

Master It Add line and curve labels to each entity created in `MasterIt0101.dwg` or `MasterIt0101_METRIC.dwg`. It is recommended that you complete the previous exercise so you will have a curve to work with. Choose a label that specifies the bearing and distance for your lines and length, radius, and delta of your curve.

Solution

1. Go to the Annotate tab in the ribbon ➢ Labels & Tables panel and click the Add Labels button. Then do the following:
   - Set Feature to Line And Curve.
   - Set Label Type to Single Segment.
   - Set Line Label Style to Bearing Over Distance.
   - Set Curve Label Style to Distance – Radius And Delta.
2. Click Add and then select each line and curve by clicking it in the drawing.

The default label should be acceptable. If not, perform the following steps:

1. Select one of the newly placed labels.
2. On the Labels: General Segment Label contextual tab ➢ Modify panel, click Label Properties.
3. In the resulting AutoCAD Properties palette, select an alternative label in the General section.

Chapter 2: Survey

Properly collect field data and import it into Civil 3D. Once survey data has been collected, you will want to pull it into Civil 3D via the survey database. This will enable you to create lines and points that correctly reflect your field measurements.

Master It Create a new drawing based on the template of your choice and a new survey database and import the `MasterIt_0201.txt` (or `MasterIt_0201_METRIC.txt`) file into the drawing. The format of this specific file is PNEZD (comma-delimited).

Solution

1. Create a new drawing using a template of your choice.
2. On the Survey tab, create a new local survey database.
3. Create a new network in the newly created survey database.
4. Import the `MasterIt_0201.txt` (or `MasterIt_0201_METRIC.txt`) file and edit the options to insert both the figures and the points.

Set up description key and figure databases. Proper setup is key to working successfully with the Civil 3D survey functionality.

Master It Create a new description key set and the following description keys using the default styles. Make sure all description keys are going to layer V-Node:

- CL*
- EOP*
- TREE*
- BM*

Change the description key search order so that the new description key set takes precedence over the default.

Create a figure prefix database called **MasterIt** containing the following codes:

- CL
- EOP
- BC

Test the new description key set and figure prefix database by importing the file `MasterIt_CodeTest_0202.txt` (use the same file for both US and metric units). Note that this file is a comma-delimited PNEZD file.

Solution

1. Open a new drawing based on a Civil 3D template or continue working in the drawing from the previous exercise.
2. On the Settings tab, locate the description key sets area. Right-click Description Key Sets and select New. Give the description key set the name of your choice and click OK.
3. Right-click the new description key set and select Edit Keys. Add the description keys, including the asterisk, as shown in the list.
4. Set the layer for each item in the table to V-NODE.
5. Close the Description Key Editor and save the drawing.
6. Right-click Description Key Sets and select Properties.
7. Move your new description key set to the top of the listing using the arrows. Click OK.
8. In the Toolspace ➢ Survey tab, right-click Figure Prefix Databases and select New.
9. Create a new figure prefix database called **MasterIt**.
10. Add the required codes to the list. Leave all options as default and click OK.
11. Create a new local survey database and import the file MasterIt_CodeTest_0202.txt. When importing, verify that the current figure prefix database is set to **MasterIt**. Be sure that Process Linework During Import, Insert Figure Objects, and Insert Survey Points are each set to Yes.

    Your file should now contain linework that reflects your efforts.

Translate surveys from assumed coordinates to known coordinates. Understanding how to manipulate data once it is brought into Civil 3D is important to making your field measurements match your project's coordinate system.

Master It Create a new drawing based on the template of your choice and start a new survey database. Import 0203_traverse.fbk (or 0203_traverse_METRIC.fbk). When you import the file, turn on the Insert Network Object option. Translate the database based on the following settings:

- Base Point 1
- Rotation Angle of 10.3053°

Solution

1. Create a new drawing and survey database and import the 0203_traverse.fbk (or 0203_traverse_METRIC.fbk) file into a network.
2. Using the Translate Survey Network command, rotate the network based on the point number and rotation you were given.
3. Save the drawing and leave it open.

Your drawing should look like MasterIt_0203.dwg (MasterIt_0203_METRIC.dwg), which you can download from this book's web page, www.sybex.com/go/masteringcivil3d2014.

Perform traverse analysis. Traverse analysis is needed for boundary surveys to check for angular accuracy and closure. Civil 3D will generate the reports that you need to capture these results.

Master It Use the survey database and network from the previous Master It exercise. Analyze and adjust the traverse using the following criteria:

- Use an Initial Station value of 2 and an Initial Backsight value of 1.
- Use the Compass Rule option for Horizontal Adjustment.
- Use Length Weighted Distribution Method for Vertical Adjustment.
- Use a Horizontal Closure Limit value of 1:25,000.
- Use a Vertical Closure Limit value of 1:25,000.

Solution

1. Continue working in the drawing from the previous Master It exercise.
2. Create a new traverse from the four points using the initial station and backsight point numbers given.
3. Perform a traverse analysis on the newly created traverse using the method and closure limits given and apply the changes to the survey database.

Chapter 3: Points

Import points from a text file using description key matching. Most engineering offices receive text files containing point data at some time during a project. Description keys provide a way to automatically assign the appropriate styles, layers, and labels to newly imported points.

Master It Create a new drawing from _AutoCAD Civil 3D (Imperial) NCS.dwt or _AutoCAD Civil 3D (Metric) NCS.dwt. Revise the Civil 3D description key set to contain only the parameters listed here:

Code	Point style	Point label style	Format	Layer
GS*	Basic	Elevation Only	Ground Shot	V-NODE
GUY*	Guy Pole	Elevation and Description	Guy Pole	V-NODE
HYD*	Hydrant (existing)	Elevation and Description	Existing Hydrant	V-NODE-WATR
TOP*	Basic	Point#-Elevation-Description	Top of Curb	V-NODE
TREE*	Tree	Elevation and Description	Existing Tree	V-NODE-TREE

Import the PNEZD (space delimited) file `MasterIt0301.txt` (`MasterIt0301_METRIC.txt`). Confirm that the description keys made the appropriate matches by looking at a handful of points of each type. Do the trees look like trees? Do the hydrants look like hydrants?

Save the resulting file for use in the remaining exercises.

Solution

1. Choose the Application menu ➢ New, and create a drawing from `_AutoCAD Civil 3D (Imperial)NCS.dwt` or `_AutoCAD Civil 3D (Metric) NCS.dwt`.
2. Switch to the Settings tab of Toolspace, and locate the description key set called Civil 3D.
3. Right-click this set and choose Edit Keys.
4. Delete the first two keys in this set by right-clicking each one and choosing Delete (you may need to close Panorama to see the result).
5. Revise the remaining key to match the GS specifications listed under the Master It instructions.
6. Right-click the GS key, and choose Copy.
7. Create the four additional keys listed in the instructions, and exit Panorama.
8. On the Home tab of the ribbon ➢ Create Ground Data panel, select Points ➢ Point Creation Tools and then click the Import Points button on the toolbar.
9. Navigate out to the `MasterIt_0301.txt` (`MasterIt_0301_METRIC.txt`) file and click Open.
10. Select PNEZD (Space Delimited) from the listing, check Add Points To Point Group, and create a point group with the name of your choosing and click OK. Click OK again to complete the command.
11. Zoom extents to see the points.
12. Save the file as `MasterIt_0301.dwg` for use in the next Master It exercise.

Note that each description key parameter (style, label, format, and layer) has been respected. Your hydrants should appear as hydrants on the correct layer, your trees should appear as trees on the correct layer, and so on. Compare your work to `MasterIt0301_FINISHED.dwg` (`MasterIt0301_METRIC_FINISHED.dwg`).

Create a point group. Building a surface using a point group is a common task. Among other criteria, you may want to filter out any points with zero or negative elevations from your Topo point group.

Master It Create a new point group called **Topo** that includes all points *except* those with elevations of zero or less. Use the DWG created in the previous Master It exercise or start with `MasterIt0301_FINISHED.dwg` (`MasterIt0301_METRIC_FINISHED.dwg`).

Solution

1. In Prospector, right-click Point Groups and choose New.
2. On the Information tab, enter **Topo** as the name of the new point group.

3. Switch to the Exclude tab.
4. Click the With Elevations Matching check box to turn it on, and enter **<=0** in the field.
5. Click OK to close the box.

Export points to LandXML and ASCII format. It's often necessary to export a LandXML or ASCII file of points for stakeout or data-sharing purposes. Unless you want to export every point from your drawing, it's best to create a point group that isolates the desired point collection.

> **Master It** Create a new point group that includes all the points with a raw description of TOP. Export this point group via LandXML to a PNEZD comma-delimited text file.
>
> Use the DWG created in the previous Master It exercise or start with `MasterIt0302_FINISHED.dwg` (`MasterIt0302_METRIC_FINISHED.dwg`).

Solution

1. In Prospector, right-click Point Groups and choose New.
2. On the Information tab, enter **Top of Curb** as the name of the new point group.
3. Switch to the Include tab.
4. Select the With Raw Descriptions Matching check box, and type **TOP** in the field.
5. Click OK, confirm in Prospector that all the points have the description TOP, and click OK.
6. Right-click the Top Of Curb point group, and choose Export LandXML.
7. Click OK in the Export To LandXML dialog.
8. Choose a location to save your LandXML file, and then click Save.
9. Navigate out to the LandXML file to confirm it was created.
10. Right-click the Top Of Curb point group, and choose Export Points.
11. Choose the PNEZD (comma-delimited) format and a destination file, and confirm that the Limit To Points In Point Group check box is selected for the Top of Curb point group. Click OK.
12. Navigate out to the ASCII file to confirm it was created.
13. Save the file for use in the next Master It exercise.

Create a point table. Point tables provide an opportunity to list and study point properties. In addition to basic point tables that list number, elevation, description, and similar options, you can customize point table formats to include user-defined property fields.

> **Master It** Use the DWG created in the previous Master It exercise or start with `MasterIt0303_FINISHED.dwg` (`MasterIt0303_METRIC_FINISHED.dwg`). Create a point table for the Topo point group using the PNEZD format table style.

Solution

1. Change to the Annotate tab of the ribbon, and select Add Tables ➢ Add Point Table.
2. Choose the PNEZD format for the table style.

3. Click the Point Groups button, choose the Topo point group, and click OK.

   The command line prompts you to choose a location for the upper-left corner of the point table.

4. Choose a location on your screen somewhere to the right of the project.

5. Zoom in, and confirm your point table.

6. Compare your work to the file `MasterIt0304_Finished.dwg` (`MasterIt0304_METRIC_FINISHED.dwg`).

Chapter 4: Surfaces

Create an existing ground surface using points.   The most common way to create a surface model is by adding point data to the definition of a surface.

Master It   Open the `MasterIt0401.dwg` or the `MasterIt0401_METRIC.dwg` file. Create a new surface called **Existing**. Add the point group Topo to its definition. Leave the default styles.

Solution

1. On the Home tab ➢ Create Ground data panel of the ribbon, click Surfaces ➢ Create Surface.

2. Name the surface **Existing** and click OK.

3. In Prospector, expand the Surfaces ➢ Existing ➢ Definition branch.

4. Right-click Point Groups and select Add.

5. Select the point group Topo and click OK.

Modify and update a TIN surface.   TIN surface creation is mathematically precise, but sometimes the assumptions behind the equations leave something to be desired. By using the editing tools built into Civil 3D, you can create a more realistic surface model.

Master It   Continue working in the file from the previous exercise or open the `MasterIt0402.dwg` or the `MasterIt0402_METRIC.dwg` file. Use the irregular-shaped polyline and apply it to the surface as an outer boundary of the surface. Make the boundary a destructive breakline.

Solution

1. Expand the Surfaces ➢ Existing ➢ Definition branch.

2. Right-click Boundaries and select the Add option.

3. Verify that the check box by Non-Destructive Breakline is unchecked and click OK. Select the magenta colored polyline to complete the boundary addition.

Prepare a slope analysis.   Surface analysis tools allow users to view more than contours and triangles in Civil 3D. Engineers working with nontechnical team members can create strong meaningful analysis displays to convey important site information using the built-in analysis methods in Civil 3D.

Master It Open the `MasterIt0403.dwg` or the `MasterIt0403_METRIC.dwg` file. Create a slope banding analysis showing slopes under and over 10 percent and insert a dynamic slope legend to help clarify the result of the analysis.

Solution

1. Select the surface. From the TIN Surface contextual tab ➢ Modify panel, click Surface Properties.
2. Set the Surface Style field to Slope Banding (2D).
3. Switch to the Analysis tab for the Slopes analysis type.
4. Set Ranges Number to **2**, and then click the Run Analysis arrow.
5. Change both the maximum slope for ID 1 and the minimum slope for ID 2 to **10%**.
6. Click OK to close the Surface Properties dialog.
7. Select the surface to display the TIN Surface contextual tab.
8. From the TIN Surface contextual tab ➢ Labels & Tables panel, choose Add Legend Table.
9. Enter **S ↵** and then **D ↵** at the command line and pick a placement point on the screen to create a dynamic elevations legend.

Label surface contours and spot elevations. Showing a stack of contours is useless without context. Using the automated labeling tools in Civil 3D, you can create dynamic labels that update and reflect changes to your surface as your design evolves.

Master It Open the `MasterIt0404.dwg` or the `MasterIt0404_METRIC.dwg` file. Label the major contours on the surface at 2′ and 10′ (Background) or 1 m and 5 m (Background).

Solution

1. Change the Surface Style to Contours 2′ And 10′ (Background) or Contours 1 m And 5 m (Background).
2. From the Annotate tab ➢ Labels & Tables panel, click the Add Labels button.
3. Set Feature to Surface and Label Type to Contour – Multiple.
4. Set Major Contour Label Style to Existing Major Labels and Minor Contour Label Style to <none>.
5. Click Add.
6. Pick a point on one side of the site, and draw a contour label line across the entire site.

Import a point cloud into a drawing and create a surface model. As laser scan data collection becomes more common and replaces other large-scale data-collection methods, the ability to use point clouds in Civil 3D is critical. Intensity helps postprocessing software determine the ground cover type. While Civil 3D can't do postprocessing, you can see the intensity as part of the point cloud style.

Master It Import an LAS format point cloud file (MasterIt0405_Denver.las) into the Civil 3D template (with a coordinate system) of your choice. As you create the point cloud file, set the style to Elevation Ranges. Use a portion of the file to create a Civil 3D surface model. No coordinate system needs to be set for this example.

Solution

1. Start a new file by using the default Civil 3D template of your choice. Save the file before proceeding as `MasterIt0405_DenverUSA.dwg`.
2. In Prospector, right-click Point Clouds and select the Create Point Cloud option to display the Create Point Cloud Wizard.
3. Set the name of the point cloud to **Denver**.
4. Set the point cloud style to **Elevation Ranges**, and click the Next button.
5. Use the white plus sign to browse to the LAS file.
6. Select `MasterIt0405_Denver.las` and select Open. Click Finish.

   This file contains 4.7 million data points, so be patient while the file imports.

7. When the point cloud has completed processing, zoom extents. Select the bounding box representing the point cloud to display the Point Cloud contextual tab.
8. Select the Add Points To Surface command.
9. Name the surface, set a surface style, and click the Next button.
10. Choose the Window radio button, and click Define Region In Drawing.
11. Define the region by creating a window around the western half of the point cloud.
12. Click Next to see the Summary page and click the Finish button. Close Panorama. Due to the large size of point cloud files, no finished drawing is available on the web page.

Chapter 5: Parcels

Create a boundary parcel from objects. The first step in any parceling project is to create an outer boundary for the site.

Master It Open the `MasteringParcels.dwg` (`MasteringParcels_METRIC.dwg`) file, which you can download from www.sybex.com/go/masteringcivil3d2014. Convert the line segments in the drawing to a parcel.

Solution

1. From the Home tab ➢ Create Design panel, select Parcel ➢ Create Parcel From Objects.
2. At the `Select lines, arcs, or polylines to convert into parcels or [Xref]:` prompt, pick the lines that represent the site boundary, and press ↵.

   The Create Parcels – From Objects dialog appears.

3. From the drop-down menus, select Subdivision, Property, and Parcel Name in the Site, Parcel Style, and Area Label Style selection boxes, respectively.

   Keep the default values for the remaining options.

4. Click OK to dismiss the dialog.

   The boundary polyline forms parcel segments that react with the alignment. The label is placed at the newly created parcel centroid.

5. Save the drawing but keep it open for the next exercise. If you would like to see what the drawing should look like at this point, you can open MasteringParcels1.dwg (MasteringParcels1_METRIC.dwg), available from the book's website.

Create a right-of-way parcel using the right-of-way tool. For many projects, the ROW parcel serves as frontage for subdivision parcels. For straightforward sites, the automatic Create ROW tool provides a quick way to create this parcel. A cul-de-sac serves as a terminal point for a cluster of parcels.

Master It Continue working in the Mastering Parcels.dwg (MasteringParcels_METRIC.dwg) file or you can open MasteringParcels1.dwg (MasteringParcels1_METRIC.dwg), available from the book's website. Create a ROW parcel that is offset by 25′ (10 m) on either side of the road centerline with 25′ (10 m) fillets at the parcel boundary and alignment ends. Then add the circles representing the cul-de-sac as a parcel. Edit the cul-de-sac area to remove unwanted parcel lines.

Solution

1. From the Home tab ➢ Create Design panel, select Parcel ➢ Create Right Of Way.
2. At the Select parcels: prompt, pick the area label of your newly created parcel.
3. Press ↵ to stop picking parcels.

   The Create Right Of Way dialog appears.

4. Expand the Create Parcel Right Of Way branch, and enter **25′ (10** m) in the Offset From Alignment text Value field.
5. Expand the Cleanup At Parcel Boundaries branch. Enter **25′ (10** m) in the Fillet Radius At Parcel Boundary Intersections Value field.
6. Select Fillet from the drop-down menu in the Cleanup Method selection box.
7. Expand the Cleanup At Alignment Intersections branch. Set the Fillet Radius At Alignment Intersections to **25′ (10** m). Verify that Cleanup Method is set to Fillet.
8. Click OK to dismiss the dialog and create the ROW parcels.
9. Trim the two circles at the ROW line to create arcs.
10. From the Home tab ➢ Create Design panel, select Parcel ➢ Create Parcel From Objects.
11. Pick the two arcs and accept the default settings.

    Two new parcels are created.

12. Click any parcel line in the drawing. From the Parcel Segments contextual tab ➢ Launch Pad panel, select Parcel Layout Tools.

13. Select the Delete Sub-Entity tool and pick the two ROW lines and ROW arc, leaving the outer arc alone.

    The cul-de-sac is created and is part of the ROW parcel.

14. Repeat steps 11 and 12 for the other cul-de-sac arc. Press Esc twice when complete.

15. Save the drawing but keep it open for the next exercise. If you would like to see what the drawing should look like at this point, you can open `MasteringParcels2.dwg` (`MasteringParcels2_METRIC.dwg`), available from the book's website.

Create subdivision lots automatically by layout. The biggest challenge when creating a subdivision plan is optimizing the number of lots. The precise sizing parcel tools provide a means to automate this process.

Master It Continue working in the previous drawing or open `MasteringParcels2.dwg` (`MasteringParcels2_METRIC.dwg`), available from the book's website. Create a series of lots with a minimum of 8,000 sq. ft. (700 m^2) and 75′ (20 m) frontage. Set the Use Minimum Offset option to No. Leave all other options at their defaults.

Solution

1. From the Home tab's Create Design panel, select Parcel ➢ Parcel Creation Tools.

2. Expand the Parcel Layout Tools toolbar.

3. Change the value of the following parameters by clicking in the Value column and typing in the new values:

    - Minimum Area: **8,000** sq. ft. (**700** square meters)
    - Minimum Frontage: **75′** (**20 m**)

4. Change the following parameters by clicking in the Value column and selecting the appropriate option from the drop-down menu:

    - Automatic Mode: On
    - Remainder Distribution: Redistribute Remainder

5. Click the Slide Line – Create tool.

    The Create Parcels – Layout dialog appears.

6. Select Subdivision, Single Family, and Parcel Name from the drop-down menus in the Site, Parcel Style, and Area Label Style selection boxes, respectively.

    Keep the default values for the rest of the options.

7. Click OK to dismiss the dialog.

8. At the `Select Parcel to be subdivided or [Pick]:` prompt, pick the Property: 1 parcel area label.

9. At the Select start point on frontage: prompt, use your Endpoint Osnap to pick the point of curvature on the north end of the project along the ROW parcel segment. The side of the road you start with is up to you.

    The parcel jig appears.

10. Move your cursor slowly along the ROW parcel segment, and notice that the parcel jig follows the segment.

11. At the Select end point on frontage: prompt, loop back to the opposite side of the street from where you started. Use your Endpoint Osnap to pick the point of curvature along the ROW parcel segment.

12. At the Specify angle or [Bearing aZimuth]: prompt, type **90** and press ↵. If you receive the message No Solution Found on your command line, try again. This may mean you snapped to the wrong spot.

13. At the Accept Result? prompt, press ↵ to accept the lot layout. Press Esc twice to exit the command.

    Note that some of the parcels are not properly shaped. For extra credit, fix them to your liking.

14. Save the drawing but keep it open for the next exercise. If you would like to see what the drawing should look like at this point, you can open MasteringParcels3.dwg (MasteringParcels3_METRIC.dwg), available from the book's website.

Add multiple-parcel segment labels. Every subdivision plat must be appropriately labeled. You can quickly label parcels with their bearings, distances, direction, and more using the segment labeling tools.

Master It Continue working in the previous drawing, or you can open MasteringParcels3.dwg (MasteringParcels3_METRIC.dwg), available from the book's website. Place Bearing Over Distance labels on every parcel line segment and Delta Over Length And Radius labels on every parcel curve segment using the Multiple Segment Labeling tool.

Solution

1. From the Annotate tab ➢ Labels & Tables panel, select Add Labels ➢ Parcel ➢ Add Parcel Labels.

2. From the drop-down menus, in the Add Labels dialog, select Multiple Segment, Bearing Over Distance, and Delta Over Length And Radius in the Label Type, Line Label Style, and Curve Label Style selection boxes, respectively.

3. Click Add.

4. At the Select parcel to be labeled by clicking on area label: prompt, pick the area label for each of your single-family parcels. After each pick, press ↵ to accept Clockwise as the default.

5. Save the drawing. If you would like to see what the drawing should look like at this point, you can open MasteringParcels_FINISHED.dwg (MasteringParcels_METRIC_FINISHED.dwg), available from the book's website.

Chapter 6: Alignments

Create an alignment from an object. Creating alignments based on polylines is a traditional method of building engineering models. With built-in tools for conversion, correction, and alignment reversal, it's easy to use the linework prepared by others to start your design model. These alignments lack the intelligence of crafted alignments, however, and you should use them sparingly.

Master It Open the `MasteringAlignments-Objects.dwg` or `MasteringAlignments-Objects_METRIC.dwg` file, and create alignments from the linework found there with the All Labels label set.

Solution

1. From the Home tab ➤ Create Design panel, choose Alignment ➤ Create Alignment From Objects.

2. When prompted to select objects, pick the lines and arc, then press Enter twice to complete the selection process and accept the alignment direction.

3. In the Create Alignment From Objects dialog, verify that Alignment Label Set is set to All Labels.

4. Uncheck the Add Curves Between Tangents option and click OK.

Create a reverse curve that never loses tangency. Using the alignment layout tools, you can build intelligence into the objects you design. One of the most common errors introduced to engineering designs is curves and lines that aren't tangent, requiring expensive revisions and resubmittals. The free, floating, and fixed components can make smart alignments in a large number of combinations available to solve almost any design problem.

Master It Open the `MasteringAlignments-Reverse.dwg` or the `MasteringAlignments-Reverse_METRIC.dwg` file, and create an alignment using the linework on the right as a guide. Create a reverse curve with both radii equal to 200 (or 60 for metric users) and with a pass-through point at the intersection of the two arcs.

Solution

1. From the Home tab ➤ Create Design panel, choose Alignment ➤ Alignment Creation Tools.

2. In the Create Alignment – Layout dialog, accept the defaults and click OK to display the Alignment Layout Tools toolbar.

3. Use the Fixed Line (Two Points) tool to trace both lines and press ↵ when complete to end the command.

4. Use the Floating Curve (From Entity, Radius, Through Point) tool to draw an arc from the endpoint of the west line with a radius of 200 (or 60 for metric users) to a pass-through point at the intersection of the two sketched arcs.

5. Press ↵ when complete to end the command.

6. Use the Free Curve Fillet (Between Two Entities, Radius) tool to fillet the floating curve created in the previous step and the last fixed segment with a reverse curve with a radius of 200 (or 60 for metric users).

7. Close the Alignment Layout Tools toolbar.

Replace a component of an alignment with another component type. One of the goals in using a dynamic modeling solution is to find better solutions, not just the first solution. In the layout of alignments, this can mean changing components out along the design path or changing the way they're defined. The ability of Civil 3D to modify alignments' geometric construction without destroying the object or forcing a new definition lets you experiment without destroying the data already based on an alignment.

Master It Convert the reverse curve indicated in the `MasteringAlignments-Rcurve.dwg` or the `MasteringAlignments-Rcurve_METRIC.dwg` file to a floating arc that is constrained by the next segment. Then change the radius of the curves to **150** (or **45** for metric users).

Solution

1. Select the alignment to activate the Alignment contextual tab.

2. From the Alignment contextual tab ➢ Modify panel, choose Geometry Editor to display the Alignment Layout Tools toolbar.

3. Select the Alignment Grid View tool.

4. Starting with the first segment, click in the Tangency Constraint field and change it to **Constrained By Next (Floating)**.

5. Repeat for the other segments except the last one, which cannot be modified because it is dependent on the previous constraint.

6. Change the radii of the two curves to **150′** (or **45** m for metric users).

Create alignment tables. Sometimes there is just too much information displayed on a drawing, and to make it clearer, tables are used to show bearings and distances for lines, curves, and segments. With their dynamic nature, these tables are kept up to date with any changes.

Master It Open the `MasteringAlignments-Table.dwg` or `MasteringAlignments-Table_METRIC.dwg` file, and generate a line table, a curve table, and a segment table. Use whichever style you want to accomplish this.

Solution

For lines:

1. Select the alignment to activate the Alignment contextual tab.

2. From the Alignment contextual tab ➢ Labels & Tables panel, choose Add Labels ➢ Multiple Segments and select the alignment. Press Enter to end the selection process.

3. On the Alignment contextual tab ➢ Labels & Tables panel, choose Add Tables ➢ Add Line.

4. Using the Pick On-Screen button at the bottom of the dialog, select the line segment labels of the alignment. Press Enter to end the selection process.

If a warning comes up regarding child styles, select the Convert All Selected Label Styles To Tag mode.

5. Click OK to accept the settings in the dialog.
6. Place the table anywhere on your drawing. The bearings and distances are now replaced by tag labels.

For curves:

1. If not done during the lines portion of the exercise, select the alignment, and on the Alignment contextual tab ➢ Labels & Tables panel, choose Add Labels Multiple Segments and select the alignment. Press Enter to end the selection process.
2. From the Alignment contextual tab ➢ Labels & Tables panel, select Add Tables ➢ Add Curve.
3. Using the Pick On-Screen button at the bottom of the dialog, select the curve segment labels of the alignment. Press Enter to end the selection process.

   If a warning comes up regarding child styles, select the Convert All Selected Label Styles To Tag mode.

4. Click OK to accept the settings in the dialog.
5. Place the table anywhere on your drawing. The bearings and distances are now replaced by tag labels.

For segments:

1. If not done during the lines portion of the exercise, select the alignment, and on the Alignment contextual tab ➢ Labels & Tables panel, choose Add Labels Multiple Segments and select the alignment.
2. From the Alignment contextual tab ➢ Labels & Tables panel, choose Add Tables ➢ Add Segment.
3. In the By Alignment section, select the alignment you want to label and click OK.
4. Place the table anywhere on your drawing. The bearings and distances are now replaced by tag labels.

   If a warning comes up regarding child styles, select the Convert All Selected Label Styles To Tag mode.

Chapter 7: Profiles and Profile Views

Sample a surface profile with offset samples.   Using surface data to create dynamic sampled profiles is an important advantage of working with a three-dimensional model. Quick viewing of various surface centerlines and grip-editing alignments makes for an effective preliminary planning tool. Combined with offset data to meet review agency requirements, profiles are robust design tools in Civil 3D.

Master It Open the `MasteringProfiles.dwg` file (or `MasteringProfiles_METRIC.dwg` file) and sample the ground surface along Alignment A, along with offset values at 15' left and 15' right (or 4.5 m left and 4.5 m right) of the alignment. Generate a profile view showing this information using the Major Grids profile view style with no data band sets.

Solution

1. From the Home tab ➢ Create Design panel, choose Profile ➢ Create Surface Profile.
2. Verify that Alignment A and the EG surface are selected and then click the Add button to add the EG surface.
3. Check the Sample Offsets check box and enter **15, -15** (or **4.5, -4.5** for metric users) in the box below the sample offsets and then click the Add button.
4. Click the Draw In Profile View button to open the Create Profile View Wizard.
5. On the General page of the wizard, verify that the profile view style is set to Major Grids.
6. On the Data Bands wizard page, verify that Select Band Set is set to _No Bands.
7. Click the Create Profile View button.
8. Place the profile anywhere on the drawing.
9. Save the drawing but keep it open for the next exercise. If you would like to see what the drawing should look like at this point, you can open `MasteringProfiles_SolutionA.dwg` (`MasteringProfiles_SolutionA_METRIC.dwg`), available from the book's website.

Lay out a design profile on the basis of a table of data. Many programs and designers work by creating pairs of station and elevation data. The tools built into Civil 3D let you input this data precisely and quickly.

Master It Continue in the `MasteringProfiles.dwg` file (or the `MasteringProfiles_METRIC.dwg` file) or open `MasteringProfiles_SolutionA.dwg` (`MasteringProfiles_SolutionA_METRIC.dwg`). Create a layout profile on Alignment A using the Layout profile style and a complete label set with the following information for Imperial users:

Station	PVI Elevation	Curve Length
0+00	822.00	
1+80	825.60	300'
6+50	800.80	

Or the following information for metric users:

Station	PVI Elevation	Curve Length
0+000	250.400	
0+062	251.640	100 m
0+250	244.840	

Solution

1. Create a surface profile for Alignment A and generate a profile view (if not done in the previous example) or use the `MasteringProfiles_SolutionA.dwg` or `MasteringProfiles_SolutionA_METRIC.dwg` file.

2. From the Home tab ➢ Create Design panel, choose Profile ➢ Profile Creation Tools.

3. Select a grid line on the profile view that shows the surface profile. The Create Profile – Draw New dialog will open.

4. Verify that Profile Style is set to Layout and Profile Label Set is set to Complete Label Set.

5. Click OK to dismiss the Create Profile – Draw New dialog.

6. In the Profile Layout Tools toolbar, set the Length value of the Curve settings to the specified curve length.

7. Use the Draw Tangents With Curves button and the Transparent Commands toolbar to enter station elevation data.

8. If needed, you may move the labels to be legible.

9. Save the drawing but keep it open for the next exercise. If you would like to see what the drawing should look like at this point, you can open `MasteringProfiles_SolutionB.dwg` (`MasteringProfiles_SolutionB_METRIC.dwg`), available from the book's website.

Add and modify individual entities in a design profile. The ability to delete, modify, and edit the individual components of a design profile while maintaining the relationships is an important concept in the 3D modeling world. Tweaking the design allows you to pursue a better solution, not just a working solution.

Master It Continue in the `MasteringProfiles.dwg` file (or the `MasteringProfiles_METRIC.dwg` file) or you can open `MasteringProfiles_SolutionB.dwg` (`MasteringProfiles_SolutionB_METRIC.dwg`), available from the book's website. For the layout profile created in the previous exercise, modify the curve so that it is **200′** (or **60** m for metric users). Then insert a PVI at Station 4+90, Elevation 794.60 (or at Station 0+150, Elevation 242.840 for metric users) and add a **300′** (or **96** m for metric users) parabolic vertical curve at the newly created PVI.

Solution

1. Continue in the drawing you have open from the previous exercise or open the `MasteringProfiles_SolutionB.dwg` (or `MasteringProfiles_SolutionB_METRIC.dwg`) file. Pick the Design profile, and from the Profile contextual tab ➢ Modify Profile panel, select the Geometry Editor button.

2. In the Profile Layout Tools toolbar, select the Profile Grid View button.

3. In the Profile Entities tab of Panorama, change the Profile Curve Length field to **200** (or **60** for metric users).

4. In the Profile Layout Tools toolbar, select the Insert PVI button.

5. Using the Profile Station Elevation transparent command, select the profile grid, and enter **490** for the station and **794.60** for the elevation (or **150** for the station and **242.840** for the elevation for metric users). Press Esc twice.

6. Back in the Profile Layout Tools toolbar, click the drop-down arrow next to the Vertical Curve Creation button and select More Free Vertical Curves ➢ Free Vertical Parabola (PVI Based).

7. Pick the newly created PVI and enter **300** (or **96** for metric users) for Curve Length. Press ↵ twice.

8. Save the drawing but keep it open for the next exercise. If you would like to see what the drawing should look like at this point, you can open MasteringProfiles_SolutionC.dwg (MasteringProfiles_SolutionC_METRIC.dwg), available from the book's website.

Apply a standard band set. Standardization of appearance is one of the major benefits of using styles in labeling. By applying band sets, you can quickly create plot-ready profile views that have the required information for review.

Master It Continue in the drawing you have open from the previous exercise or open MasteringProfiles_SolutionC.dwg (MasteringProfiles_SolutionC_METRIC.dwg). Apply the Cut And Fill band set to the layout profile created in the previous exercise with the appropriate profiles referenced in each of the bands.

Solution

1. Using the file from the previous example or the MasteringProfiles_SolutionC.dwg or MasteringProfiles_SolutionC_METRIC.dwg file, pick a grid line in the profile view, and from the Profile View contextual tab ➢ Modify View panel, choose Profile View Properties to display the Profile View Properties dialog.

2. On the Bands tab, click Import Band Set, and select the Cut And Fill band set.

3. Select Bottom Of Profile View from the Location drop-down list.

4. Scroll over and change Profile2 to **Layout (1)** for both rows.

5. Select Top Of Profile View from the Location drop-down list.

6. Scroll over and change Profile2 to **Layout (1)**.

Chapter 8: Assemblies and Subassemblies

Create a typical road assembly with lanes, curbs, gutters, and sidewalks. Most corridors are built to model roads. The most common assembly used in these road corridors is some variation of a typical road section consisting of lanes, curbs, gutters, and sidewalks.

Master It Create a new drawing from either the Civil 3D metric or Imperial template. Build a symmetric assembly using LaneSuperelevationAOR, UrbanCurbGutterValley2, and LinkWidthAndSlope for terrace and buffer strips adjacent to the UrbanSidewalk. Use widths and slopes of your choosing.

Solution

1. Create a new drawing from the DWT of your choice.
2. From the Home tab ➢ Create Design panel, choose Assembly ➢ Create Assembly.
3. Name your assembly and set styles as appropriate.
4. Pick a location in your drawing for the assembly.
5. Locate the Lanes tab on the Tool Palettes window.
6. Click the LaneSuperelevationAOR button on the Lanes tab.
7. Use the AutoCAD Properties palette to edit the subassembly parameters, and follow the command-line prompts to set the LaneSuperelevationAOR on the left and right sides of your assembly.
8. Repeat the process with UrbanCurbGutterValley2, LinkWidthAndSlope, and UrbanSidewalk.
9. Complete this portion of the exercise by placing a final LinkWidthAndSlope on the outside of the UrbanSidewalk.
10. Save the drawing for use in the next Master It exercise.

Edit an assembly. Once an assembly has been created, it can be easily edited to reflect a design change. Often, at the beginning of a project you won't know the final lane width. You can build your assembly and corridor model with one lane width and then change the width and rebuild the model immediately.

Master It Working in the drawing from the preceding exercise, edit the width of each LaneSuperelevationAOR to **14′ (4.3 m)**, and change the cross slope of each LaneSuperelevationAOR to **-3.00%**.

Solution

1. Select the right lane subassembly. Be sure this is the only element selected.
2. From the Subassemblies contextual tab ➢ Modify Subassembly panel, choose Subassembly Properties.
3. On the Parameters tab, change the width to **14′ (4.3 m)**. Note that width will be listed twice. The topmost width reports the default value. You will change the second occurrence.
4. Change the Default Slope to **-3.00%**.
5. Repeat for the left lane subassembly. Save the drawing for use in the next Master It exercise.

Add daylighting to a typical road assembly. Often, the most difficult part of a designer's job is figuring out how to grade the area between the last engineered structure point in the cross section (such as the back of a sidewalk) and the existing ground. An extensive catalog of daylighting subassemblies can assist you with this task.

Master It Working in the drawing from the preceding exercise, add the DaylightMinWidth subassembly to both sides of your typical road assembly. Establish a minimum width between the outermost subassembly and the daylight offset of **10'** (**3 m**).

Solution

1. Locate the Daylight tab on the tool palette.
2. Click the DaylightMinWidth button on the tool palette.
3. Use the AutoCAD Properties palette to verify that Min Width is set to **10'** (**3 m**).
4. Follow the command-line prompts to set the DaylightMinWidth on the right and left side of your assembly.
5. Press Esc on your keyboard to complete the command.

You should now have daylighting subassemblies visible on both sides of your assembly. To see what the drawing should look like at this point, you can open `MasteringAssemblies_FINISHED.dwg` or `MasteringAssemblies_METRIC_FINISHED.dwg`.

Chapter 9: Basic Corridors

Build a single baseline corridor from an alignment, profile, and assembly. Corridors are created from the combination of alignments, profiles, and assemblies. Although corridors can be used to model many things, most corridors are used for road design.

Master It Open the `MasteringCorridors.dwg` or `MasteringCorridors_METRIC.dwg` file. Build a corridor named Corridor A on the basis of the Alignment A alignment, the FG profile, and the Basic Assembly. Set all frequencies to 10' (or 3 m for metric users).

Solution

1. From the Home tab ➢ Create Design panel, choose Corridor.
2. In the Name text box, name your corridor **Corridor A**.

   Keep the default values for Corridor Style and Corridor Layer.
3. Verify that Alignment is set to Alignment A and Profile is set to FG.
4. Verify that Assembly is set to Basic Assembly.
5. Verify that Target Surface is set to EG surface.
6. Verify that Set Baseline And Region Parameters is checked.
7. Click OK to accept the settings in the Create Corridor dialog and to display the Baseline And Region Parameters dialog.

8. Click the Set All Frequencies button to display the Frequency To Apply Assemblies dialog.
9. Change the value for all of the frequencies to **10′** (or **3 m** for metric users).
10. Click OK to accept the settings in the Frequency To Apply Assemblies dialog.
11. Click OK to accept the settings in the Baseline And Region Parameters dialog.
12. You may receive a dialog warning that the corridor definition has been modified. If you do, select the Rebuild The Corridor option.

Use targets to add lane widening. Targets are an essential design tool used to manipulate the geometry of the road.

Master It Open the `MasteringCorridorTargets.dwg` or `MasteringCorridorTargets_METRIC.dwg` file. Set Right Lane to target Alignment A-Right.

Solution

1. From the Corridor contextual tab ➢ Modify Corridor panel, choose Corridor Properties.
2. On the Parameters tab, in the Targets column in the baseline row, click the ellipsis button to display the Target Mapping dialog.
3. In the Target Mapping dialog, click <None> in the Width Alignment row for Right Assembly Group to display the Set Width Or Offset Target dialog.
4. Select Alignment A-Right and click Add.
5. Click OK to dismiss the Set Width Or Offset Target dialog.
6. Click OK to accept the settings in the Target Mapping dialog.
7. Click OK to accept the settings in the Corridor Properties dialog and allow the corridor to rebuild.

Create a corridor surface. The corridor model can be used to build a surface. This corridor surface can then be analyzed and annotated to produce finished road plans.

Master It Open the `MasteringCorridorSurface.dwg` or `MasteringCorridorSurface_METRIC.dwg` file. Create a corridor surface for the Alignment A corridor from Top links. Name the surface **Corridor A-Top**.

Solution

1. From the Corridor contextual tab ➢ Modify Corridor panel, choose Corridor Properties.
2. On the Surfaces tab, click the Create A Corridor Surface button in the upper-left corner of the dialog.
3. Click the surface item under the Name column and change the default name of your surface to **Corridor A-Top**.
4. Verify that Links has been selected from the drop-down list in the Data Type selection box.
5. Verify that Top has been selected from the drop-down list in the Specify Code selection box.

6. Click the Add Surface Item button to add Top Links to the surface definition.

7. Click OK to accept the settings in the dialog; then choose Rebuild The Corridor when prompted.

    The corridor and surface will build.

Add an automatic boundary to a corridor surface. A surface can be improved with the addition of a boundary. Single-baseline corridors can take advantage of automatic boundary creation.

Master It Open the `MasteringCorridorBoundary.dwg` or `MasteringCorridorBoundary_METRIC.dwg` file. Use the Automatic Boundary Creation tool to add a boundary using the Daylight code.

Solution

1. From the Corridor contextual tab ➢ Modify Corridor panel, choose Corridor Properties and switch to the Boundaries tab.

2. Right-click the surface entry and hover over Add Automatically ➢ Daylight.

3. Click OK to accept the settings in the dialog; then choose Rebuild The Corridor when prompted.

    The corridor and surface will build.

Chapter 10: Advanced Corridors, Intersections, and Roundabouts

Create corridors with non-centerline baselines. Although for simple corridors you may think of a baseline as a road centerline, other elements of a road design can be used as a baseline. In the case of a cul-de-sac, the EOP, the top of curb, or any other appropriate feature can be converted to an alignment and profile and used as a baseline.

Master It Open the `MasterIt1001.dwg` (`MasterIt1001_METRIC.dwg`) file, which you can download from www.sybex.com/go/masteringcivil3d2014. Add the cul-de-sac alignment and profile to the corridor as a baseline. Create a region under this baseline that applies the Intersection Typical assembly.

Solution

1. Select the corridor. From the Corridor contextual tab ➢ Modify Corridor panel, click Corridor Properties.

2. Switch to the Parameters tab.

3. Click Add Baseline, choose Cul de Sac EOP in the Create Corridor Baseline dialog, and click OK.

4. In the Profile column, click inside the <Click here…> box, choose Cul de Sac EOP FG in the Select A Profile dialog, and click OK.

5. Right-click the new baseline, and choose Add Region.
6. Select Intersection Typical in the Create Corridor Region dialog and click OK.
7. Click OK to leave the Corridor Properties dialog and rebuild the corridor.

   An example of the finished exercise can be found in `MasterIt1001_SolutionA.dwg` (`MasterIt1001_SolutionA_METRIC.dwg`) on the book's web page.

Add alignment and profile targets to a region for a cul-de-sac. Adding a baseline isn't always enough. Some corridor models require the use of targets. In the case of a cul-de-sac, the lane elevations are often driven by the cul-de-sac centerline alignment and profile.

Master It Continue working in the `MasterIt1001.dwg` (`MasterIt1001_METRIC.dwg`) file. You need to have completed the previous exercise before continuing. Add the Second Road alignment and Second Road FG profile as targets to the cul-de-sac region. Adjust Assembly Application Frequency to 5′ (1 m) for tangents and curves.

Solution

1. Select the corridor. From the Corridor contextual tab ➢ Modify Corridor panel, click Corridor Properties.
2. In the Corridor Properties dialog, switch to the Parameters tab.
3. Click the Target Mapping button in the RG – Intersection Typical region.
4. In the Target Mapping dialog:
   - Assign Second Road as Width Alignment for Lane – L.
   - Assign Second Road FG profile as Outside Elevation Profile.
   - Click OK.
5. Click OK to leave the Target Mapping dialog.
6. Click the Frequency button in the appropriate region.
7. Change the Along Curves value to 5′ (1 m).
8. Click OK to exit the Frequency To Apply To Assemblies dialog.
9. Click OK to leave the Corridor Properties dialog and rebuild the corridor.

   An example of the finished exercise can be found in `MasterIt1001_SolutionB.dwg` (`MasterIt1001_SolutionB_METRIC.dwg`) on the book's web page.

Create a surface from a corridor and add a boundary. Every good surface needs a boundary to prevent bad triangulation. Bad triangulation creates inaccurate contours and can throw off volume calculations later in the process. Civil 3D provides several tools for creating corridor surface boundaries, including an Interactive Boundary tool.

Master It You need to have completed the previous exercise before continuing. Keep working in the `MasterIt1001.dwg` (`MasterIt1001_METRIC.dwg`) file. Create an interactive corridor surface boundary for the entire corridor model.

Solution

1. Select the corridor. From the Corridor contextual tab ➢ Modify Corridor panel, click Corridor Properties.
2. In the Corridor Properties dialog, switch to the Boundaries tab.
3. Select the corridor surface, right-click, and choose Add Interactively.
4. Follow the command-line prompts to add a boundary based on feature lines all the way around the entire corridor.
5. Type **C** to close the boundary and then press ↵ to end the command.
6. Click OK to leave the Corridor Properties dialog and rebuild the corridor.

   An example of the finished exercise can be found in `MasterIt1001_SolutionC.dwg` (`MasterIt1001_SolutionC_METRIC.dwg`) on the book's web page.

Chapter 11: Superelevation

Add superelevation to an alignment.   Civil 3D has convenient and flexible tools that will apply safe, correct superelevation to an alignment curve.

Master It   Open the `MasterIt1101.dwg` (`MasterIt1101_METRIC.dwg`) file, which you can download from www.sybex.com/go/masteringcivil3d2014. Verify that the design speed of the road is 20 miles per hour (35 km per hour) and apply superelevation to the entire length of the alignment. Use AASHTO 2004 design criteria with an eMax of 6% 2-Lane. Use the option to automatically resolve overlap. For the remainder of the options, leave default settings unless otherwise directed.

Solution

1. Select the alignment. From the Alignment contextual tab ➢ Modify panel, choose Alignment Properties.
2. On the Design Criteria tab, verify that there is a check mark next to Use Criteria-Based Design.
3. Set the design criteria file and superelevation eMax from the right side of the dialog.
4. Verify that the design speed is set to **20** mph (**35** km/h) on the left side, and click OK.
5. From the Alignment contextual tab, click Superelevation ➢ Calculate/Edit Superelevation.
6. Click Calculate Superelevation Now.
7. Step through the superelevation wizard, taking all the defaults for pivot and shoulder control.
8. On the Attainment page, place a check mark next to Automatically Resolve Overlap. Click Finish.

   You should now have superelevation applied to the design with no overlap.

Create a superelevation assembly. For superelevation to happen, you need to have an assembly that is capable of superelevation.

Master It Continue working in MasterIt1101.dwg (MasterIt1101_METRIC.dwg). Create an assembly similar to the one in the top image shown earlier in the chapter in Figure 11.11. Set each lane to be **14′ (4.5 m)** wide and each shoulder to be **6′ (2 m)** wide. Leave all other options at their defaults. If time permits, build a corridor based on the alignment and assembly.

Solution

1. From the Home tab ➢ Create Design panel ➢ Assembly, choose Create Assembly.
2. Name the assembly **AOR** and set the assembly type to Undivided Crowned Road. Click OK.
3. Click to place the assembly in the graphic.
4. From the Lanes subassembly palette, click the LaneSuperelevationAOR subassembly and set the lane width in the Properties palette.
5. Click the assembly to place one lane on the right; click again to place the assembly to the left. Press Esc when complete.
6. Select the right subassembly and set its Use Superelevation parameter to Right Lane Outside. Press Esc to clear the selection.
7. Select the left subassembly and set its Use Superelevation parameter to Left Lane Outside. Press Esc to clear the selection.
8. Place the shoulders on each side using the default parameters for ShoulderExtendAll.
9. If time permits, create a corridor based on the alignment and assembly you just created.
10. Save the drawing.

Create a rail corridor with cant. Cant tools allow users to create corridors that meet design criteria specific to rail needs.

Master It In the drawing MasterIt1102.dwg (MasterIt1102_METRIC.dwg), create a Railway assembly with the RailSingle subassembly using the default parameters for width and depth. Add a LinkSlopetoSurface generic link with 50 percent slope to each side. Add cant to the alignment in the drawing using the default settings for attainment. Create a corridor from these pieces.

Solution

1. From the Home tab ➢ Create Design panel ➢ Assembly, choose Create Assembly.
2. Name the assembly **Rail** and set the type to Railway. Click OK and place the assembly in the graphic.
3. From the Bridge And Rail subassembly palette, click the Rail Single subassembly.
4. Click the assembly in the drawing to place the rail design.
5. From the Generic palette, click LinkSlopetoSurface.

6. Set the slope to **-50%** and click once on each side of the assembly to place the link.

7. Press Esc to complete the process.

8. Select the alignment. From the Alignment contextual tab ➢ Modify panel, click Cant ➢ Calculate/Edit Cant.

9. Click Calculate Cant Now, and click Finish.

10. Build a corridor from the alignment, assembly, and the proposed profile (which has been designed for you ahead of time).

11. Set the target surface to Existing Intersection and clear the check box for Set Baseline And Region Parameters.

   If you need assistance building your corridor, review Chapter 9, "Basic Corridors," and Chapter 10, "Advanced Corridors, Intersections, and Roundabouts." But hopefully you've figured it out by this point!

Create a superelevation view. Superelevation views are a great place to get a handle on what is going on in your roadway design. You can visually check the geometry as well as make changes to the design.

Master It Open the drawing `MasterIt1103.dwg` (`MasterIt1103_METRIC.dwg`). Create a superelevation view for the alignment. Show only the left and right outside lanes as blue and red, respectively.

Solution

1. Select the alignment and choose Superelevation ➢ Create Superelevation View.

2. In the Create Superelevation View dialog, toggle off Left Outside Shoulder and Right Outside Shoulder. The remaining check boxes will be for the lane views.

3. Set Left Outside Lane Color to Blue, set Right Outside Shoulder Color to Red, and click OK.

4. Place the view in the graphic.

Chapter 12: Cross Sections and Mass Haul

Create sample lines. Before any section views can be displayed, sections must be created from sample lines.

Master It Open `MasterIt1201.dwg` (`MasterIt1201_METRIC.dwg`) and create sample lines along the USH 10 alignment every 50′ (20 m). Sample all data, and set the left and right swath widths to **50′ (20 m)**.

Solution

1. From the Home tab ➢ Profile & Section Views panel, click Sample Lines.

2. Select the USH_10 alignment and sample all data; then click OK.

3. In the Sample Line tools dialog, click the By Range Of Station option.

4. Create sample lines by station range and set your sampling increments to **50' (20 m)**; then click OK and ↵ to complete the command.

5. Save the drawing and keep it open for use in the next exercise. (See `MasterIt1201_A.dwg` or `MasterIt1201_A_METRIC.dwg` for finished versions of this exercise.)

Create section views. Just as profiles can be shown only in profile views, sections require section views to be displayed. Section views can be plotted individually or all at once. You can break them up into groups for plotting into sheets.

Master It In the previous exercise, you created sample lines. In that same drawing, create section views for all the sample lines. For US units, use a cross section scale of 1" = 20' on an Arch D size layout sheet. For metric units, use a cross section scale of 1:500 on an ISO A0 size sheet. For all other options, use the default settings and styles. You need to have completed the previous exercise before continuing.

Solution

1. Continue working in `MasterIt1201.dwg` (`MasterIt1201_METRIC.dwg`). You need to have completed the previous exercise before continuing.

2. For US users, change the annotation scale to 1" = 20'. For metric users, make sure the annotation scale is 1:500.

3. Select one of the sample lines.

4. From the Sample Line contextual tab ➤ Launch Pad panel, click Create Section View ➤ Create Multiple Section Views.

5. On the section placement page, click the ellipsis to set the Arch D Section 20 Scale option (ISO A0 Section 1 to 500 for metric users). Click OK.

6. Leave all options at their defaults, and click Create Section Views.

7. Click in the graphic to place the views. Save the drawing and keep it open for use in the next exercise.

Define and compute materials. Materials are required to be defined before any quantities can be displayed. You learned that materials can be defined from surfaces or from corridor shapes. Corridors must exist for shape selection, and surfaces must already be created for comparison in materials lists.

Master It Using `MasterIt1201.dwg` (`Master It1201_METRIC.dwg`), create a materials list that compares Existing Intersection with HWY 10 DATUM Surface. Use the Earthworks Quantity takeoff criteria. You need to have completed the previous exercise before continuing.

Solution

1. Continue working in `MasterIt1201.dwg` (`MasterIt1201_METRIC.dwg`).

2. Select one of the sample lines.

3. From the Sample Line contextual tab ➤ Launch Pad panel, click Compute Materials. Select the alignment and sample line group, and then click OK.

4. In the upper-right corner of the Compute Materials dialog, set the quantity takeoff criteria to Earthworks.

5. Set the Existing Ground surface to Existing Intersection by clicking the <Click Here to Set All> field.

6. Set Datum to HWY 10 DATUM, and click OK. Save the drawing and keep it open for use in the next exercise.

    Graphically, nothing will appear. Continue to the next exercise to see the results of your work.

Generate volume reports. Volume reports give you numbers that can be used for cost estimating on any given project. Typically, construction companies calculate their own quantities, but developers often want to know approximate volumes for budgeting purposes.

Master It Continue using MasterIt1201.dwg (MasterIt1201_METRIC.dwg). Be sure you have completed all the previous Master It exercises before continuing. Use the materials list created earlier to generate a volume report. Create a web browser–based report and a total volume table that can be displayed on the drawing.

Solution

1. Continue working in MasterIt1201.dwg (MasterIt1201_METRIC.dwg).

2. Without any object selected, go to the Analyze tab ➢ Volumes And Materials panel, and click Volume Report.

3. Leave all options at their defaults, and click OK.

4. If asked "Do you want to allow scripts to run?" click Yes.

    Your report will be displayed.

5. Close the browser window.

6. In Civil 3D, from the Analyze tab ➢ Volumes And Materials panel, click Total Volume Table.

7. Leave all options at their defaults and click OK.

Chapter 13: Pipe Networks

Create a pipe network by layout. After you've created a parts list for your pipe network, the first step toward finalizing the design is to use Pipe Network By Layout.

Master It Open the MasterIt1301.dwg or MasterIt1301_METRIC.dwg file. From the Home tab ➢ Create Design panel ➢ Pipe Network drop-down, select Pipe Network Creation Tools to create a sanitary sewer pipe network named Mastering. Use the Composite surface, and name only structure and pipe label styles. Don't choose an alignment at this time. Create 8" (200 mm) PVC pipes and a manhole called SMH. There are blocks in the drawing to assist you in placing manholes. Begin at the START HERE marker, and place a manhole at each marker location. You can erase the markers when you've finished.

Solution

1. From the Home tab ➢ Create Design panel, select Pipe Network ➢ Pipe Network Creation Tools.
2. In the Create Pipe Network dialog, set the following parameters:
    - Network Name: **Mastering**
    - Network Parts List: Sanitary Sewer
    - Surface Name: **Composite**
    - Alignment Name: <none>
    - Structure Label Style: Name Only (Sanitary)
    - Pipe Label Style: Name Only
3. Click OK. The Network Layout Tools toolbar will appear.
4. Set the structure to SMH and the pipe to 8" PVC (200 mm PVC).
5. Click Draw Pipes And Structures, and use your Insertion Osnap to place a structure at each marker location.
6. Press ↵ to exit the command.
7. Move the structure labels as desired.
8. Select a marker, right-click, choose Select Similar, and click Delete.

    See `MasterIt1301_SolutionA.dwg` or `MasterIt1301_SolutionA_METRIC.dwg` for the result.

Create an alignment from network parts and draw parts in profile view.

Once your pipe network has been created in plan view, you'll typically add the parts to a profile view based on either the road centerline or the pipe centerline.

Master It Continue working in the `MasterIt1301.dwg` or `MasterIt1301_METRIC.dwg` file. Create an alignment from your pipes so that station zero is located at the START HERE structure. Create a profile view from this alignment, and show the pipes on the profile view.

Solution

1. Select the structure labeled START HERE to display the Pipe Networks contextual tab and select Alignment From Network on the Launch Pad panel.
2. Select the last structure in the pipe run, and press ↵ to accept the selection.
3. In the Create Alignment dialog, name the alignment **Mastering** and make sure the Create Profile And Profile View check box is selected.
4. Accept the other defaults, and click OK.
5. In the Create Profile From Surface dialog, sample both the EG and Corridor FG surfaces for the profile. Select the surfaces and click Add. Change the FG style to **FG Profile**.

6. Click Draw In Profile View.

7. In the Create Profile View dialog, click Create Profile View and choose a location in the drawing for the profile view.

   A profile view showing your pipes appears.

   See `MasterIt1301_SolutionB.dwg` or `MasterIt1301_SolutionB_METRIC.dwg` for the result.

Label a pipe network in plan and profile. Designing your pipe network is only half of the process. Engineering plans must be properly annotated.

Master It Continue working in the `MasterIt1301.dwg` or `MasterIt1301_METRIC.dwg` file. Add the Length Description And Slope style label to profile pipes and the Data With Connected Pipes (Sanitary) style to profile structures. Add the alignment created in the previous Master It to all pipes and structures.

Solution

1. Select one of the pipe or structure objects. From the Pipe Networks contextual tab ➢ Labels & Tables panel, select Add Labels ➢ Add Pipe Network Labels.

2. In the Add Labels dialog, change Feature to **Pipe Network**, and then change Label Type to **Entire Network Profile**.

   - For pipe labels, choose Length Description And Slope.
   - For structure labels, choose Data With Connected Pipes (Sanitary).

3. Click Add, and choose any pipe or structure in your profile view.

4. Drag or adjust any profile labels as desired.

5. In the Prospector tab of Toolspace, expand Pipe Networks ➢ Networks ➢ Mastering and select Pipes.

6. Select all pipes in Prospector, right-click on the Reference Alignment column header, and select Edit.

7. Choose the Mastering alignment and then click OK.

8. Repeat steps 5 through 7 but choose Structures.

   See `MasterIt1301_SolutionC.dwg` or `MasterIt1301_SolutionC_METRIC.dwg` for the result.

Create a dynamic pipe table. It's common for municipalities and contractors to request a pipe or structure table for cost estimates or to make it easier to understand a busy plan.

Master It Continue working in the `MasterIt1301.dwg` or `MasterIt1301_METRIC.dwg` file. Create a pipe table for all pipes in your network. Use the default table style.

Solution

1. Select one of the pipe or structure objects. From the Pipe Networks contextual tab, select Add Tables ➢ Add Pipe.

2. In the Pipe Table Creation dialog, make sure your pipe network is selected.
3. Accept the other defaults, and click OK.
4. Place the table in your drawing.

   See `MasterIt1301_SolutionD.dwg` or `MasterIt1301_SolutionD_METRIC.dwg` for the result.

Chapter 14: Grading

Convert existing linework into feature lines. Many site features are drawn initially as simple linework for the 2D plan. By converting this linework to feature line information, you avoid a large amount of rework. Additionally, the conversion process offers the ability to drape feature lines along a surface, making further grading use easier.

Master It Open the `MasteringGrading.dwg` or `MasteringGrading_METRIC.dwg` file from the book's web page. Convert the magenta polyline, describing a proposed temporary swale, into a feature line and drape it across the EG surface to set elevations, and set intermediate grade break points.

Solution

1. From the Home tab ➢ Create Design panel, choose Feature Lines ➢ Create Feature Lines From Objects.
2. Select the polyline. Press the Enter key.
3. Toggle on the Assign Elevations check box and then click OK.
4. Select the EG surface in the Assign Elevations dialog.
5. Verify that Insert Intermediate Grade Break Points is on.
6. Click OK to close the dialog and return to your model.

Model a simple linear grading with a feature line. Feature lines define linear slope connections or, in other words, breaklines. This can be the flow of a drainage channel, the outline of a building pad, or the back of a street curb. These linear relationships can help define grading in a model or simply enhance understanding of design intent.

Master It Edit the radius of the curve on the feature line you just created to be 100' (30 m). Set the grade from the west end of the feature line to the next PI to 4 percent, and the remainder to a constant slope to be determined in the drawing. Draw a temporary profile view to verify the channel is below grade for most of its length.

Solution

1. Select the feature line to activate the Feature Line contextual tab.
2. From the Feature Line contextual tab ➢ Modify panel, toggle on the Edit Geometry panel if not already visible.
3. From the Feature Line contextual tab ➢ Edit Geometry panel, choose the Edit Curve tool.

4. Select the feature line curve.
5. In the Edit Feature Line Curve dialog, change the radius to **100′ (30 m)** and click OK.
6. Press the Enter key to end the command.
7. From the Feature Line contextual tab ➢ Modify panel, toggle on the Edit Elevations panel if not already visible.
8. From the Feature Line contextual tab ➢ Edit Elevations panel, choose the Set Grade/Slope Between Points tool.
9. Click the triangular glyph on the western end of the feature, and press the Enter key to accept the elevation.
10. Using your cursor, trace your way up the feature line until you reach the PI at the start of the curve, which will be indicated by the appearance of another green triangular glyph. Click this glyph.
11. At the `Specify grade or [SLope Elevation Difference SUrface Transition]:` prompt, enter **4** ↵ to set the grade.
12. Select the feature line again.
13. Pick the PI at the start of the curve, and press the Enter key to accept the elevation.
14. Pick the PI at the upstream (eastern) end of the channel, and press the Enter key to accept the grade.
15. Press the Enter key to end the command.
16. Select the feature line, and from the Feature Line contextual tab ➢ Launch Pad panel, choose Quick Profile.
17. Click OK to accept the defaults and pick a point on the screen to draw the quick profile view.
18. Dismiss Panorama to view the Quick Profile.

Model planar site features with grading groups. Once a feature line defines a linear feature, gradings collected in grading groups model the slope projections from that line to other points in space. These projections can be combined to model a site much like a TIN surface, producing a dynamic design tool.

Master It Use grading criteria to define the pilot channel, with grading on both sides of the sketched centerline. Define the channel using a Grading to Distance of 5′ (1.5 m) with a slope of 3:1 and connect the channel to the EG surface using a grading with slopes that are 4:1. Generate a surface from the grading group. If prompted, do not weed the feature line.

Solution

1. From the Home tab ➢ Create Design panel, choose Grading ➢ Grading Creation Tools to activate the Grading Creation Tools toolbar.
2. Click the Create A Grading Group tool to create a grading group and name the group.
3. Verify that the Automatic Surface Creation option is checked, and click OK.

4. Click OK to accept the surface creation options.
5. Click the Set The Target Surface tool to set the target surface to EG, and click OK.
6. Change Grading Criteria to Grade To Distance.
7. Click the Create Grading tool, and pick the feature line.
8. If the Weed Feature Line dialog appears, select Continue Grading Without Feature Line Weeding.
9. Pick the left or right side, and press the Enter key to model the full length.
10. Enter 5' (1.5 m) for the distance.
11. Press the Enter key to accept entering the slope.
12. Enter 3 for the slope value.
13. Pick the main feature line again, and grade the other side using the same steps.
14. Change Grading Criteria to Grade To Surface, and then create a grading object on both the left and right outer feature lines with slopes of 4:1 for both cut and fill along the full length of the feature lines.
15. Press Esc to complete the gradings.

Chapter 15: Plan Production

Create view frames. When you create view frames, you must select the template file that contains the layout tabs that will be used as the basis for your sheets. This template must contain predefined viewports. You can define these viewports with extra vertices so you can change their shape after the sheets have been created.

Master It Open the MasteringPlanProduction.dwg or MasteringPlanProduction_METRIC.dwg file. Run the Create View Frames Wizard to create view frames for Alignment A in the current drawing. (Accept the defaults for all other values.) These view frames will be used to generate Plan and Profile sheets on ARCH D (ISO A1 for metric users) sheets at 20 scale (1:200 scale for metric users) using the plan and profile template MasteringPandPTemplate.dwt or MasteringPandPTemplate_METRIC.dwt. All files should be saved in C:\Mastering\CH 15\.

Solution

1. From the Output tab ➢ Plan Production panel, choose Create View Frames.
2. On the Alignment page, select Alignment A from the Alignment drop-down list and click Next.
3. On the Sheets page, select the Plan And Profile option.
4. Click the ellipsis button to display the Select Layout As Sheet Template dialog.

5. In the Select Layout As Sheet Template dialog, click the ellipsis button, browse to C:\Mastering\CH 15\, select the template named MasteringPandPTemplate.dwt (or MasteringPandPTemplate_METRIC.dwt for metric users), and click Open.

6. Select the layout named ARCH D Plan And Profile 20 Scale (or ISO A1 Plan And Profile 1 to 200 for metric users), and click OK.

7. Click Create View Frames.

Edit view frames. The grips available to edit view frames allow the user some freedom on how the frames will appear.

Master It Open the MasteringEditViewFrames.dwg or MasteringEditViewFrames_METRIC.dwg file, and move the VF- (1) view frame to Sta. 2+20 (or Sta. 0+050 for metric users) to lessen the overlap. Then adjust Match Line 1 (or Match Line 2 for metric users) so that it is now at Sta. 4+25 (or Sta. 0+200 for metric users) and shorten it so that the labels are completely within the view frames.

Solution

1. Click the VF- (1) view frame.
2. Make sure you have Dynamic Input on.
3. Click the diamond grip, and type **220** ↵ (or **50** ↵ for metric users).
4. Press Esc to clear the selection.
5. Click Match Line 1 (or Match Line 2 for metric users) to show its grips.
6. Click the diamond grip, and type **425** ↵ (or **200** ↵ for metric users).
7. Click the triangular grip on one end of the match line and shorten it so that the labels are completely within the view frames.
8. Repeat step 7 for the triangular grip on the opposite end of the match line.

The match line is now centered better between the two view frames.

Generate sheets and review Sheet Set Manager. You can create sheets in new drawing files or in the current drawing. The resulting sheets are based on the template you chose when you created the view frames. If the template contains customized viewports, you can modify the shape of the viewport to better fit your sheet needs.

Master It Open the MasteringCreateSheets.dwg or MasteringCreateSheets_METRIC.dwg file. Run the Create Sheets Wizard to create plan and profile sheets in the current drawing for Alignment A using the plan and profile template MasteringPandPTemplate.dwt or MasteringPandPTemplate_METRIC.dwt. Make sure to choose a north arrow. (Accept the defaults for all other values.) All files should be saved in C:\Mastering\CH 15\.

Solution

1. From the Output tab ➢ Plan Production panel, choose Create Sheets.
2. On the View Frame Group And Layouts page, in the Layout Creation area, select All Layouts In The Current Drawing

3. Verify that the north arrow is selected from the drop-down list.
4. Click Create Sheets.
5. Click OK to save the drawing.
6. Click a location as the profile origin.
7. Dismiss the Panorama Event Viewer.

Create section views. More and more municipalities are requiring section views. Whether this is a mile-long road or a meandering stream, Civil 3D can handle it nicely via Plan Production.

Master It Open the `MasteringSectionSheets.dwg` or `MasteringSectionSheets_METRIC.dwg` file. Create section views and Plan Production section sheets in a new sheet set for Alignment A using the using the Road Section section view style and the section sheet template `MasteringSectionTemplate.dwt` or `MasteringSectionTemplate_METRIC.dwt`. Make sure the sections are set to be generated on ARCH D (ISO A1 for metric users) sheets at 20-scale (1:200 scale for metric users). (Accept the defaults for all other values.) All files should be saved in `C:\Mastering\CH 15\`.

Solution

1. From the Home tab ➢ Profile & Section Views panel, choose Section Views ➢ Create Multiple Views.
2. On the General page, verify that Section View Style is set to Road Section and click Next.
3. On the Section Placement page, select the Production option.
4. Click the ellipsis button to display the Select Layout As Sheet Template dialog.
5. In this dialog, click the ellipsis button, browse to `C:\Mastering\CH 15\`, select the template named `MasteringSectionTemplate.dwt` (or `MasteringSectionTemplate_METRIC.dwt` for metric users), and click Open.
6. Select the layout named ARCH D Section 20 Scale (or ISO A1 Section 1 to 200 for metric users), and click OK.
7. Click Create Section Views.
8. Click a location as the section origin. The multiple section views are created.
9. From the Output tab ➢ Plan Production panel, choose Create Section Sheets.
10. In the Create Section Sheets dialog, verify that New Sheet Set is selected and set Sheet Set Storage Location to `C:\Mastering\CH 15\Final Sheets`.
11. Click Create Sheets.
12. Click OK to save the drawing.

Chapter 16: Advanced Workflows

Create a data shortcut folder. The ability to load design information into a project environment is an important part of creating an efficient team. The main design elements of the project are available to the data shortcut mechanism via the working folder and data shortcut folder.

Master It Using the `MasterIt1601.dwg` (`MasterIt1601_METRIC.dwg`) drawing, create a new data shortcut folder called `Master Data Shortcuts`. Use the `_Sample Project` project template.

Solution

1. Open the `MasterIt1601.dwg` (`MasterIt1601_METRIC.dwg`) drawing.
2. On the Manage tab, click New Shortcuts Folder.
3. Give the project a name of your choosing, such as **MasterIt**, and place a check mark next to Use Project Template.
4. With `_Sample Project` highlighted, click OK.

   The data shortcut folder is now complete.

Create data shortcuts. To allow sharing of the data, shortcuts must be made before the information can be used in other drawings.

Master It Save the drawing to the `Source Drawings` folder in the Master Data Shortcuts project you created in the previous exercise. Create data shortcuts to all the available data in the `MasterIt1601.dwg` (`MasterIt1601_METRIC.dwg`) file.

Solution

1. Continue working in the drawing from the previous Master It.
2. From the Application menu, use Save As to save the drawing to `C:\Civil 3d projects\2014\Master Data Shortcuts\Source Drawings`.
3. On the Manage tab, click Create Data Shortcuts.
4. Place a check mark next to all items that appear in the data listing, and click OK.

Export to earlier releases of AutoCAD. Being able to export to earlier base AutoCAD versions is sometimes necessary.

Master It Using `MasterIt1601.dwg` (`MasterIt1601_METRIC.dwg`), export the Civil 3D file so it can be used by a user working in base AutoCAD 2010.

Solution

1. Continue working in the drawing from the previous Master It.
2. From the Application menu, select Export ➤ DWG ➤ 2010.
3. Save the file with the default name in the same directory as `MasterIt1601.dwg` (`MasterIt1601_METRIC.dwg`).

Export to LandXML. Being able to work with outside clients or even other departments within your firm who do not have Civil 3D is an important part of collaboration.

Master It Using `MasterIt1601.dwg` (`MasterIt1601_METRIC.dwg`), create a LandXML file with all of the exportable information.

Solution

1. Continue working in the drawing from the previous Master It.
2. From the Output tab ➢ Export panel, click Export To LandXML. Use all the default settings and click OK.
3. Save the file with the default name in the same directory as `MasterIt1601.dwg` (`MasterIt1601_METRIC.dwg`).

Chapter 17: Quantity Takeoff

Open and review a list of pay items along with their categorization. The pay item list is the cornerstone of quantity takeoffs. You should download and review your pay item list and compare it against the current reviewing agency list regularly to avoid any missed items.

Master It Using the template of your choice, open the `Getting Started.csv` (or `Getting Started_Metric.csv`) pay item file that you have been working with in the previous exercises and add the 12-, 18-, and 24-Inch Pipe Culvert (or 300 mm, 450 mm, and 600 mm Pipe Culvert) pay items to your Favorites list in the QTO Manager.

Solution

1. Start a new file by using the default Civil 3D template of your choice.
2. Open the QTO Manager.
3. Click the Open button at the top left of the QTO Manager.
4. Verify that the Pay Item File Format drop-down list is set to CSV (Comma Delimited).
5. Click the Open button next to the Pay Item File text box.
6. Navigate to the `Getting Started` folder and select the `Getting Started.csv` file. Metric users should use the `Getting Started_METRIC.csv` file, which is downloadable from the book's web page.
7. Click Open to select this CSV pay item file.
8. Click OK.
9. Enter **12-Inch Pipe** (or **300 mm Pipe**) in the text box to filter.
10. Right-click the 12-Inch Pipe Culvert item (or the 300 mm Pipe Culvert item), and select Add To Favorites.
11. Repeat for the other sizes.

A saved copy of this drawing is available from the book's web page with the filename `MasteringPayItemList_FINISHED.dwg` or `MasteringPayItemList_METRIC_FINISHED.dwg`.

Assign pay items to AutoCAD objects, pipe networks, and corridors. The majority of the work in preparing quantity takeoffs is in assigning pay items accurately. By using the linework, blocks, and Civil 3D objects in your drawing as part of the process, you reduce the effort involved in generating accurate quantities.

Master It Open the `MasteringQTO.dwg` or `MasteringQTO_Metric.dwg` file and assign the CLEARING AND GRUBBING pay item to the polyline that was originally extracted from the border of the corridor. Change the hatch to have a transparency of **80**.

Solution

1. Open the QTO Manager.
2. Expand the Favorites branch and select the CLEARING AND GRUBBING item.
3. Right-click and select Assign Pay Item To Area.
4. Switch to the Object option by entering **O** ↵ at the command line.
5. Select the polyline representing the limits of corridor surface.
6. Press ↵ again to end the command.
7. Select the hatch to activate the Hatch Editor contextual tab.
8. From the Hatch Editor contextual tab ➢ Properties panel, change Hatch Transparency to **80**. Using Display Order, send the hatch to the back.

Use QTO tools to review what items have been tagged for analysis. By using the built-in highlighting tools to verify pay item assignments, you can avoid costly errors when running your QTO reports.

Master It Verify that the area in the previous exercise has been assigned a pay item.

Solution

1. Turn on Highlight Objects With Pay Items in the QTO Manager.
2. Pan and hover over the hatch to confirm that the tooltip indicates a pay item assignment.

A saved copy of this drawing is available from the book's web page with the filename `MasteringQTO_FINISHED.dwg` or `MasteringQTO_METRIC_FINISHED.dwg`.

Generate QTO output to a variety of formats for review or analysis. The quantity takeoff reports give you a quick understanding of what items have been tagged in the drawing, and they can generate text in the drawing or external reports for uses in other applications.

Master It Display the length of Type C Broken markings in a Quantity Takeoff Report with the Summary (TXT) report style using the `MasteringQTOReporting.dwg` or `MasteringQTOReporting_Metric.dwg` file.

Solution

1. From the Analyze tab ➢ QTO panel, choose Takeoff Command, and click Compute to run the report with default settings.
2. In the lower-left corner of the Quantity Takeoff Report dialog, change the report style to `Summary(TXT).xsl`.

3. Click the Draw button at the bottom of the dialog.

4. Click near some clean space and you'll be returned to the Quantity Takeoff Report dialog.

5. Click Close to dismiss this dialog, and then click Close again to dismiss the Compute Quantity Takeoff dialog.

6. The calculated amount for Type C Broken Pavement Markings should be 3,163.30′ (or 1,000.528 m).

A saved copy of this drawing is available from the book's web page with the filename MasteringQTOReporting _FINISHED.dwg or MasteringQTOReporting_METRIC_FINISHED.dwg.

Chapter 18: Label Styles

Override individual labels with other styles. In spite of the desire to have uniform labeling styles and appearances between alignments within a single drawing, project, or firm, there are always exceptions. Using the Ctrl+click method for element selection, you can access commands that let you modify labels and even change their styles.

Master It Open the drawing MasteringLabelStyles.dwg (MasteringLabelStyles_METRIC.dwg). Create a copy of the Perpendicular With Tick Major Station style called **Major With Marker**. Change Tick Block Name to **Marker Pnt**. Replace some (but not all) of your major station labels with this new style.

Solution

1. On the Settings tab, expand the Alignment ➢ Label Styles ➢ Station ➢ Major Station branch.

2. Right-click Perpendicular With Tick, and select Copy.

3. On the Information tab, change the name to **Major with Marker**.

4. Change to the Layout tab.

5. Select the Tick from the Component drop-down.

6. Click in the Value field next to Block name and click the ellipsis.

7. In the Select a Block dialog, select Marker Pnt and click OK.

8. Click OK to close the Label Style Composer dialog.

9. Open the AutoCAD Properties palette.

10. Ctrl+click one or more major station labels.

11. Change the major station label style to **Major with Marker**.

A saved copy of this drawing (MasteringLabelStyles_Solution.dwg or MasteringLabelStyles_METRIC_Solution.dwg) is available from the book's web page.

Create a new label set for alignments. Label sets let you determine the appearance of an alignment's labels and quickly standardize that appearance across all objects of the same nature. By creating sets that reflect their intended use, you can make it easy for a designer to quickly label alignments according to specifications with little understanding of the requirement.

Master It Within the MasteringLabelStyles.dwg (MasteringLabelStyles_METRIC.dwg) file, create a new label set containing only major station labels and apply it to all the alignments in that drawing.

Solution

1. On the Settings tab, expand the Alignments ➢ Label Styles ➢ Label Sets branch.
2. Right-click Major And Minor Only, and select Copy.
3. In the Alignment Label Set dialog, change the name to **Major Only** on the Information tab.
4. Delete the Minor Stations label on the Labels tab.
5. Click OK to close the Alignment Label Set dialog.
6. Select an alignment label and on the contextual tab ➢ Modify panel, click Edit Label Group.
7. In the Alignment Labels dialog, click Import Label Set.
8. In the Select Label Set dialog, use the drop-down to select the Major Only label set. Click OK to close the dialog. Click OK again to close the Alignment Labels dialog.
8. Repeat for each alignment.

A saved copy of this drawing (MasteringLabelStyles_Solution.dwg or MasteringLabelStyles_METRIC_Solution.dwg) is available from the book's web page.

Create and use expressions. Expressions give you the ability to add calculated information to labels or add logic to label creation.

Master It In the MasteringLabelStyles.dwg (MasteringLabelStyles_METRIC.dwg) file, create an expression called Top of Curb that adds 0.5′ (0.15 m) to a surface elevation. Use the expression in a spot elevation label that shows both the surface elevation and the expression-based elevation.

Solution

1. In the Settings tab of Toolspace, expand Surface Label Styles ➢ Spot Elevation, right-click Expressions, and click New.
2. In the New Expression dialog, name the expression **Top of Curb**.

    The Expression will read {Surface Elevation}+0.5 (for metric {Surface Elevation}+0.15).

3. Use the Properties button to select the Surface Elevation Property field and type **+0.5** (**+0.15** for metric) behind it.
4. Format as Double.
5. Click OK.

6. In the same branch of Settings, right-click Spot Elevation, and select New.
7. In the Label Style Composer dialog, name the label **Top of Curb** on the Information tab.
8. On the Layout tab of the Label Style Composer, click the Value field next to Contents and open the Text Component Editor dialog.
9. Without removing the existing text, add the new expression under the surface elevation text. Press the Enter key to ensure the text appears on two lines with the expression-based label on the bottom.
10. Click OK twice to exit both dialogs.
11. Place the new label in the drawing to check your work.

A saved copy of this drawing (MasteringLabelStyles_Solution.dwg or MasteringLabelStyles_METRIC_Solution.dwg) is available from the book's web page.

Apply a standard label set to profiles. Standardization of appearance is one of the major benefits of using Civil 3D styles in labeling. By applying label sets, you can quickly create plot-ready profile views that have the required information for review.

Master It In the MasteringLabelStyles.dwg (MasteringLabelStyles_METRIC.dwg) file, apply the Road Profiles label set to all layout profiles.

Solution

1. Pick one of the layout profiles, and on the contextual tab ➢ Label panel, select Edit Profile Labels.
2. In the Profile Labels dialog, click Import Label Set.
3. In the Select Label Set dialog, use the drop-down to select the Road Profile Label Set.
4. Click OK twice to exit both dialogs.
5. Repeat this procedure for all layout profiles.

A saved copy of this drawing (MasteringLabelStyles_Solution.dwg or MasteringLabelStyles_METRIC_Solution.dwg) is available from the book's web page.

Chapter 19: Object Styles

Override object styles with other styles. In spite of the desire to have uniform styles and appearances between objects within a single drawing, project, or firm, there are always going to be changes that need to be made.

Master It Open the MasteringStyles.dwg or MasteringStyles_METRIC.dwg file and change the alignment style associated with Alignment B to Layout. In addition, change the surface style used for the EG surface to Contours And Triangles, but change the contour interval to be 1' and 5' (or 0.5 m and 2.5 m) and the color of the triangles to be yellow.

Solution

1. Select Alignment B and from the Alignment contextual tab, and go to Modify tab ➢ Alignment Properties.

2. Set Alignment Object Style to Layout and click OK. Click Esc to deselect.
3. Select the EG surface and on the Surface contextual tab, go to Modify tab ➢ Surface Properties.
4. Set Surface Style to Contours And Triangles and click OK.
5. From the Settings tab of Toolspace, expand Surface ➢ Surface Styles.
6. Right-click Contours And Triangles and select Edit.
7. On the Contours tab, do the following:
   a. Expand the Contour Intervals category.
   b. Set the Minor Interval to **1′** (or **0.5 m**).

   Notice that the Major Interval automatically adjusts to 5′ (or 2.5 m).
8. On the Display tab with the View Direction set to Plan, set the color of the Triangles component to yellow.
9. Click OK to complete the revisions to the style.

Create a new surface style. Almost every set of plans that you send out of the office is going to include a surface, so it is important to be able to generate multiple surface styles that match your company standards. In addition to surface styles for production, you may find it helpful to have styles to use when you are designing that show a tighter contour spacing as well as the points and triangles needed to make some edits.

Master It Open the `MasteringSurfaceStyle.dwg` or `MasteringSurfaceStyle_METRIC.dwg` file and create a new surface style named **Micro Editing**. Set this style to display contours at 0.5′ and 1.0′ (or 0.1 m and 0.2 m), as well as triangles and points. Set the EG surface to use this new surface style.

Solution

1. From the Settings tab of Toolspace, expand Surface ➢ Surface Styles.
2. Right-click Surface Styles and select New.
3. On the Information tab, set Name to **Micro Editing**.
4. On the Contours tab, do the following:
   a. Expand the Contour Intervals category.
   b. Set the Minor Interval to **0.5′** (or **0.1 m**). Notice that the Major Interval automatically adjusts to **2.5′** (or **0.5 m**).
   c. Override the Major Interval to **1.0′** (or **0.5 m**).
5. On the Display tab with the View Direction set to Plan, verify that the only components turned on are Points, Triangles, Minor Contours, and Major Contours.
6. Click OK to complete creation of a new surface style.
7. Select and then right-click the EG surface and select Surface Properties.
8. Set Surface Style to Micro Editing, and click OK.

Create a new profile view style. Everyone has their preferred look for a profile view. These styles can provide a lot of information in a small space, so it is important to be able to create a profile view that will meet your needs.

Master It Open the `MasteringProfileViewStyle.dwg` or `MasteringProfileViewStyle_METRIC.dwg` file and create a new profile view style named **Mastering Profile View**. Set this style to not clip the vertical or horizontal grid. Set the bottom horizontal ticks at 50′ and 10′ intervals (25 m and 5 m). Set the left and right vertical ticks at 10′ and 2′ intervals (5 m and 1 m). In addition, turn off the visibility of the Graph Title, Bottom Axis Annotation Major, and Bottom Axis Annotation Horizontal Geometry Point. Set the profile view in the drawing to use this new profile view style.

Solution

1. From the Settings tab of Toolspace, expand Profile View ➢ Profile View Styles.
2. Right-click Profile View Styles and select New.
3. On the Information tab, set Name to **Mastering Profile View**.
4. On the Grid tab, verify that Clip Vertical Grid and Clip Horizontal Grid are not selected.
5. On the Horizontal Axes tab, do the following:
    a. Verify that the Axis To Control radio button is set to Bottom.
    b. In the Major Tick Details area, set Interval to **50′** (or **25** m).
    c. In the Minor Tick Details area, set Interval to **10′** (or **5** m).
6. On the Vertical Axes tab, do the following:
    a. Verify that the Axis To Control radio button is set to Left.
    b. In the Major Tick Details area, set Interval to **10′** (or **5** m).
    c. In the Minor Tick Details area, set Interval to **2′** (or **1** m).
    d. Change the Axis To Control radio button to Right.
    e. In the Major Tick Details area, set Interval to **10′** (or **5** m).
    f. In the Minor Tick Details area, set Interval to **2′** (or **1** m).
7. On the Display tab with View Direction set to Plan, turn off the visibility of Graph Title, Bottom Axis Annotation Major, and Bottom Axis Annotation Horizontal Geometry Point.
8. Click OK to complete creation of a new Profile View style.
9. Select the profile view in the drawing, and on the Profile View contextual tab, go to Modify View tab ➢ Profile View Properties.
10. Set Profile View Style to Mastering Profile View, and click OK.

Appendix B

Autodesk Civil 3D 2014 Certification

Autodesk® certifications are industry-recognized credentials that can help you succeed in your design career, providing benefits to both you and your employer. Getting certified is a reliable validation of skills and knowledge, and it can lead to accelerated professional development, improved productivity, and enhanced credibility.

This Autodesk Official Training Guide can be an effective component of your exam preparation. Autodesk highly recommends (and we agree!) that you schedule regular time to prepare, review the most current exam preparation roadmap and objectives available at http://www.autodesk.com/certification, use Autodesk Official Training Guides, take a class at an Authorized Training Center (find ATCs near you here: http://www.autodesk.com/atc), and use a variety of resources to prepare for your certification—including plenty of actual hands-on experience.

To help you focus your studies on the skills you'll need for these exams, Table B.1 shows the objectives that could potentially appear on an exam and in what chapter you can find information on that topic—and when you go to that chapter, you'll find certification icons like the one in the margin here. The sections and exam objectives listed in the table are from the Autodesk Certification Exam Guide.

Good luck preparing for your certification!

TABLE B.1: Certified Professional Exam sections and objectives

TOPIC	LEARNING OBJECTIVE	CHAPTER
User Interface	Navigate the user interface	1
	Use the functions on the Prospector tab	1
	Use the functions on the Settings tab	1
Styles	Create and use object styles	19
	Create and use label styles	18
Lines and Curves	Use the Line and Curve commands	1
	Use the Transparent command	1
Points	Create points using the Point Creation command	3
	Create points by importing point data	3
	Use point groups to control the display of points	3

TABLE B.1: Certified Professional Exam sections and objectives *(continued)*

TOPIC	LEARNING OBJECTIVE	CHAPTER
Surfaces	Create and edit surfaces	4
	Use styles and settings to display surface information	4 and 19
	Create a surface by assembling fundamental data	4
	Use styles to analyze surface display results	4
Parcels	Create parcels using the parcel layout tools	5
	Design a parcel layout	5
	Select parcel styles to change the display of parcels	5
	Select styles to annotate parcels	5
	Create alignments	6
Alignments	Design a geometric layout	6
Profiles and Profile Views	Create a surface profile	7
	Design a profile	7
	Create a layout profile	7
	Create a profile view style	19
	Create a profile view	7
Corridors	Design and create a corridor	9
	Derive information and data from a corridor	9
	Design and create an intersection	10
Sections and Section Views	Create and analyze sections and section views	12
Pipe Networks	Design and create a pipe network	13
Grading	Design and create a grading model	14
	Create a grading model feature line	14
Managing and Sharing Data	Use data shortcuts to share/manage data	16
	Create a data sharing setup	16

TABLE B.1: Certified Professional Exam sections and objectives *(continued)*

TOPIC	LEARNING OBJECTIVE	CHAPTER
Plan Production	Generate a sheet set using plan production	15
	Create a sheet set	15
Survey	Use description keys to control the display of points created from survey data	2
	Use figure prefixes to control the display of linework generated from survey data	2
	Create a boundary drawing from field data	2

Index

Note to the Reader: Throughout this index **boldfaced** page numbers indicate primary discussions of a topic. *Italicized* page numbers indicate illustrations.

A

abbreviations, **19–20**, *20*
activating description key sets, **68–70**, *69–70*
Active Drawing view, 4
Add An Offset option, 111
Add Appurtenance option, 654
Add Assembly Offset option, 402
Add Automatically boundary tool, **444**
Add Boundaries dialog, 159–161, *160–161*
Add Breaklines dialog, 9, 152–153, *153*
Add Contour Data dialog, 139, *139*
Add DEM File dialog, 132–134, *133*
Add Distances tool, 94
Add Dump Site option, 584
Add Existing option, 293
Add Existing And New option, 245, 293
Add Fitting Sizes dialog, 647, *647*
Add Fixed Curve (Three Point) tool, 258
Add From Polygon tool, **444**
Add Interactively boundary tool, **444**
Add Items tool, 643
Add Labels dialog
 alignments, 842
 contours, 185
 corridors, 488–489, *488*
 feature lines, 699–700, *700*
 General Note labels, 820, *820*
 line tables, 294
 lines and curves, 50–51, *50*
 parcels, 239, *239*, 242
 pipes, 833–834
 profile views, 354–355
 single segments, 825
 station offsets, 291–292, 294
Add Labels option for pipe networks, 617
Add Line option, 162
Add Part Family option, 597, *597–598*
Add Pipe Network Labels option, 617
Add Point File dialog, 143–144, *144*
Add Point option, 163
Add Points To Point Group option, 104–105
Add Points To Surface dialog
 Region Options page, 193–194, *194*
 Summary page, 194, *194*
 Surface Options page, 193, *193*
Add Rule dialog, 591, *591*, 595–596
Add Tables option, 617
Add To Existing Sheet Set option, **732**
Add To Favorites List option, 780, *780*
Add To Surface As Breakline option, 152, 680
Add Turn Slip Lane option, 515
Additional Broken References message, 768, *768*
Additional Sample Controls setting, 559
Adjacent Elevations By Reference tool, 690–691, 694
Advanced Parameters setting, 373, *373*, 376
Align Views settings, 734
Alignment Design Check Set dialog, 268, *268*
Alignment From Corridor tool, **501**
Alignment From Network tool, 620, 624
Alignment Label Set dialog, 837–838
Alignment Labels dialog, 289–290, *289–290*, 838–839
Alignment Layout Parameters dialog, 274–275, *274*
Alignment Layout Tools toolbar, 256–260, *256*
Alignment Properties dialog, 282
 check sets, 269, *269*
 constraints, 269, *269*, 278, *278*
 design speed, 287–288, *287–288*
 intersections, 278
 masking, 254, *254*
 offsets, 255
 stations, 284–2865, *285–286*

Alignment Style dialog
    Design page, 873, *873*
    Markers page, 873–875, *873–874*
Alignment Table Creation dialog, 293–296, *293, 296*
alignments, 247
    best fit, **261–265**, *261–264*
    cants, **548**
    components, **279–280**
    corridor targets, **430–433**, *430–433*
    creating, **249**, *249*
    data shortcuts, 766
    design constraints and check sets, **267–270**, *267–270*
    design criteria files, 528–529
    design speeds, **286–288**, *287–288*
    entities, **248–249**, *248–249*
    geometry, **271**
        component-level editing, **274–275**, *274*
        constraints, **275–279**, *275–279*
        grip editing, **271–273**, *271–272*
        tabular design, **273–274**, *273*
    labels
        geometry points, **837**
        label sets, **288–291**, *289 290*, **834–835**, *834–835*, **837–839**, *838–839*
        major station, **836**, *836*
        station offset, **291–292**, *292*, **840–842**, *841–842*
    from layout, **255–259**, *256–259*
    from lines, arcs, and polylines, **250–255**, *251, 253–254*
    from network parts, **624–625**, *625–626*
    as objects, **280**
    offset, 253–255, *253–254*, 494
    parcels, 198–199, *198*
    Plan Production, 714
    point settings, 107, *107*
    profiles, 301
    properties, **281–283**, *281–283*
    rail, 252, **260–261**, 545, **548**
    reverse curve creation, **265–267**, *266*
    roundabouts, **511–518**, *512–518*
    sanitary sewer networks, 602

    siteless, **201**, *201*
    and sites, 247
    stationing, **284–286**, *285–286*
    styles, 252, **872–876**, *873–874*
    superelevation, **531**
    tables
        design, **273–274**, *273*
        line, **294–295**, *295*
        overview, **292–294**, *293*
        segment, **295–296**, *296*
        settings, **295–296**, *296*
    view frames, 715–716, *715*
All Layouts In One New Drawing option, **730**
All Layouts In The Current Drawing option, **731**
_All Points group, 114
Allow Additional Distance For Repositioning (Increases View Overlap) option, **721**
Allow Crossing Breaklines option, 146
Along Alignment option, 718, *718*
alternative daylight subassemblies, **400–401**, *400–402*
Always Perform Implied Tangency Constraint Swapping option, 278
Ambient Settings tab, **20–24**, *21*
American Water Works Association (AWWA), 644
analysis
    pipe networks, **619–620**, *619*
    styles, **885–889**, *886–888*
Analyze panel, **619–620**, *619*
Anchor Component setting, 809
Anchor Point setting, 809
Angle Distance command, 53
Angle Information tool, 94
angled triangular grips, 312, *312*
angles
    AutoCAD, **93–94**, *93*
    convergence, 16
    deflection, 36
    entering, **34**
    surfaces, 146

tessellation, 708
text rotation, 810
annotation labels, **353–355**, *354–355*
annotation scale
    changing, 15
    section views, **570–572**
Apply 3D Proximity Check option, 638
Apply Curve Smoothing option, 540
Apply Feature Line Names tool, 680
Apply Feature Line Style dialog, 678
Apply Feature Line Styles tool, 680
    pipe networks, 593, 618
    sanitary sewer networks, 611
Apply Sea Level Scale Factor option, 15–16
Apply To X-Y field, 67
Apply To Z field, 67
approximations, surface, **138–143**, *139–142*
appurtenances in pressure networks, 645, 654
arcs
    alignments from, **250–255**, *251, 253–254*
    best fit, 48, *48*
area labels, **234–238**, *234, 236–267*
Arrow Head Size setting, 813
Arrow Head Style setting, 813
arrows
    alignment styles, 873–874
    north, 731
    surfaces, **176–178**, *177*, 878
assemblies, **365**
    building, **368–369**, *369*
    corridors, **411**, *411*
    editing, **384–388**, *385–387*
    intersections, **478–479**, *479*
    labels, **376–377**, 394
    marked points and friends, **402–404**, *403–404*
    multiple baselines, 456–457, *457*
    non-road uses, **388–390**, *389–390*
    offset, **402**, *402*, **493–497**, *494–497*
    organizing, **404–407**, *406–407*
    pipe trench, **390–393**, *391–392*
    Prospector for, **407**, *407*
    and QTO, **787–791**, *787–788, 790*

rail, **546–547**, *546–547*
renaming, **386**
road, **369–377**, *370–376*
roundabouts, **521**, *521*
storing, **405–407**, *406–407*
subassemblies. *See* subassemblies
Tool Palettes window, 372, *372*
Assembly Properties dialog
    conditional subassemblies, 508
    daylight subassemblies, 398
    entire assembly editing, 386–387, *386–387*
    generic links, 395
    non-road uses, 388–390
    pipe trenches, 391–392
    subassembly parameters, 384–385, *386*
assembly sets, **469**, *469*
Assign Elevations dialog, 674
Assign Elevations option, 673–674, 695
Assign Pay Item To Area option, 785, *785*
Assign The Selected Pay Items To Object(s) In The Drawing option, 786, 794, 798
asterisks (*)
    description key sets, 68
    layers, 17–19
    points, 115
Astronomic Direction Calculator, 86, *86*
At A Station option, 558
at symbols (@) in description key sets, 68
Attach External Reference dialog, 762, *762*
Attach Multiple Entities command, **49**, *49*
attached parcel segments, **214–215**, *214–215*
Attachment setting for label style text, 810, *811*
AUNITS variable, 824
AutoCAD
    angles, **93–94**, *93*
    for pay items, **785–787**, *785–786*
    view frame functions, **727**
    visual styles, **174**, *174*
AutoCAD Properties palette, 373, *373*, 376
AutoCAD Select Color dialog, 183
Automatic Begin On Figure Prefix Match option, 64

Automatic Layout for parcel sizing, **216**
Automatic - Object option, 108
Automatic Surface Adjustment setting, 628
automatic surface boundaries, **491–493**, *492–493*
Automatic Surface Creation option, 708
Automatically Resolve Overlap option
  rail alignments, 548
  superelevations, 537
AWWA (American Water Works Association), 644
axis of rotation (AOR) subassembly, **532–533**, *533*
Azimuth Distance command, 53

B

Background Mask setting, 812
backups for Part Builder, **661**
backward cul-de-sacs, 463
balanced state in mass haul diagrams, 582
Band Set - New Profile View Band Set dialog, 350–352, *351*
band sets, **350–352**, *351*, 909
bands
  elevations, **171–175**, *172–175*
  profile views, **346–352**, *347–351*, **908–911**, *908–909*, *911*
  section views, 565, *566*
base points in surveys, 85, *85*
Baseline And Region Parameters dialog
  corridor feature lines, 414–416, *414*
  cul-de-sacs, 459
  non-road corridors, 449
  offset assemblies, 497
  pipe trenches, 451
baselines
  corridors, **411**
  cul-de-sacs, **456–457**, *456–457*
  intersections, 467, **480**
  multiregion, **453–454**, *454–455*
  regions, **480**
BasicLane subassembly, 430
BasicLaneTransition subassembly, 478

BasicShoulder subassembly, **383**, *384*
BasicSideSlopeCutDitch subassembly, **400–401**, *401*
batch changes for pipe networks, 615
Bearing Distance command, 53
Begin Full Super (BFS) assemblies, 527, *527*
Begin Normal Crown (BNC) assemblies, 527
Begin Normal Shoulder (BNS) assemblies, 527
Begin Shoulder rollover (BSR) assemblies, 527, *527*
best fit alignments, **261–265**, *261–264*
best fit entities, **47–49**, *48–49*
best fit profiles, **320–321**, *320*
Best Fit Report dialog, 264, *264*
Bi-Directional pipe flow direction, 613
bike paths, **402**, *402*, 493–494, *493–494*
borders
  vs. boundaries, 128
  label styles, **811–812**
borrow value in mass haul diagrams, 582
boundaries
  vs. borders, 128
  Coordinate Geometry Editor, 89
  soil, 199, *199–200*
  surfaces, 127
    corridors, **442–446**, *443*
    destructive vs. non-destructive, **156–157**, *156–157*, 161, *161*
    interactive process, **491–493**, *492–493*
    overview, **155–162**, *156–161*
boundary parcels, **204–205**, *205*
  cul-de-sacs, **211–213**, *212–214*
  Parcel Layout tools, **206–208**, *206–208*
  right-of-way, **208–210**, *209–211*
bowties, **498–500**, *498–500*
Branching menu, 424, *424*
Break tool, 681
breaklines
  crossing, **154**, *154*
  Figure Prefix Database Manager, 62
  surfaces, **128**, **152–154**, *153–154*
breakover removal, 526
broken references, **766–768**, *767–768*

Browse For Folder dialog, 769
Browse For Sheet Set Folder dialog, 737
By Network option, 642
By Range Of Stations option, 558
By Slope pipe flow direction, 613
ByLayer contour style, 880
ByStyle contour style, 880

C

Calculate Superelevation dialog
    Attainment page, 536, 540, *540*
    Lanes page, 536, 538, *539*
    Roadway Type page, 538, *538*
    Shoulder Control page, 536, 538–539, *539*
calculations of expressions, **857–860**, *857–859*
cants
    alignments, **548**
    rails, 545, *545*
    superelevation, **548–550**, *549*
catalogs
    corridor modeling, **367**, *367*
    Part Catalog dialog, 597–600, *598*, *600*
    pipe networks, 662–665, *662*
    pressure networks, **644–646**, *644*
categorization files, **782–785**, *783–784*
center design for roundabouts, **518**, *519*
Center Pivots Not Applied When Only One Group warning, 533
centerlines
    alignments, 251
    pipes, 890
Change Flow Direction option, 612, 618
Change Reporting option, 60
Channel subassembly, 388–389, *389*
chapters in part catalogs, **662–663**
Character Map dialog, 828, *828*
Check Properties dialog
    Criteria page, 619
    Information page, 619
    Statistics page, 619
check sets for alignments, **267–270**, *267–270*
Child Override column, 23, *23–24*, 862

circular grips
    alignments, 272, *272*, 839
    feature lines, **674**
    PVI-based layouts, 312, *312*
circulatory roads for roundabouts, 510
circumcircles, 126–127
cleanouts, 589, *589*
Code field, **65**
code set styles, **854**
    assembly labels, 376–377
    corridors, 789–790, *790*
    links and link labels, **854–855**
    points and point labels, **856–857**, *856–857*
    shapes, **854**
coding diagrams, **380**, *381*
COGO line commands, **32–34**, *33*
COGO points, **98**, *98*, 503, *503*
collections, 3, *3*, **803–804**, *804*, 867, *867*
Color setting
    label borders, 812
    label text, 811
command settings, **98–99**
commands, transparent, **52**, *52*
    matching, **53–54**, *54*
    settings, **23**
    standard, **52–53**
Commands folder, **868**, *868*
comparing surfaces, **179–184**, *180–183*
Component Draw Order option, 809
Component Hatch Display characteristics, 877
component-level editing
    alignments, **274–275**, *274*
    profiles, **326–327**, *327*
Component Name setting, 808
Compute Materials dialog, 575, *575*
Compute Quantity Takeoff dialog, 799–801, *799*
conditional subassemblies, **506–510**, *507–509*
ConditionalCutOrFill subassembly, 507
ConditionalHorizontalTarget subassembly, 507–508
Connect and Disconnect Part tool, 618

Connect Extra Points option, 424
connection glyph for pressure networks, 654
Constraint Editing tab, 278
constraints with alignments, 250
    creating alignments with, **267–270**, *267–270*
    working with, **275–279**, *275–279*
Content Builder Toolspace, 664, *664*
Content Catalog editor, 646
Contents setting for label text, 810
contextual tabs, 1, 25, *25*
Continue Layout glyph, 649, *649*
Continuous Distance tool, 94
contours
    ByLayer or ByStyle, 880
    creating, **878–882**, *879–882*
    description, **128**
    labeling, **184–186**, *185–186*
    polylines, **138–143**, *139–142*
    surface vs. contour smoothing, 879
Contours tool, 878
control points in survey networks, 77
convergence angle, 16
Conversion Options for alignments, 252
Convert AutoCAD Line And Spline tool, 319
Convert Autodesk Land Desktop Points dialog, 105, *105*
Convert Proximity Breaklines To Standard option, 146
converting points, **104–106**, *105*
Coordinate Geometry Editor, 86, **89–91**, *89–90*
coordinates
    corridors, **438**, *438*
    elevations, 693, *693*
    surveys, 86, **89–91**, *89–90*
Copy option for labels, 804
Copy Component option, 809
Copy Deleted Dependent Objects option, **146**
Copy Pipe Network Labels To Destination Drawings option, 735
Copy Pressure Network Labels To Destination Drawings option, 735

Copy Profile Data dialog, 328, *328*
Copy Profile tool, 319
Copy To Site tool, 680
Corridor Extents As Outer Boundary option, **444**, 458
Corridor Modeling Catalogs, **367**, *367*
Corridor Properties dialog
    Boundaries page
        boundary methods, 443–445, *443*
        surface boundaries, 492–493
    Code page, 791
    Feature Lines page, 424–425, *424*
    Parameters page
        conditional subassemblies, 509–510
        cul-de-sacs, 459–460
        feature lines, 416, 418–419, *419*, 422–423, 504–505
        intersections, 474, *475*, 477–478, 480
        multiregion baselines, 453–454, *454*
        targets, 432–433, 481
    Surfaces page, 441, *441*, 446, *446*
Corridor Section Editor 435–436, *435–436*
corridors, **409**
    assemblies, **411**, *411*
    baseline, **411**
    bowties, **498–500**, *498–500*
    checking and fine-tuning, **486–491**, *487–488, 490–491*
    conditional subassemblies, 509–510
    coordinate systems, **438**, *438*
    cul-de-sacs
        EOP design profiles, **457–458**
        multiple baselines, **456–457**, *456–457*
        putting pieces together, **458–462**, *460–462*
        troubleshooting, **462–463**, *463–466*
    data shortcuts, 766
    daylighting, **423**, *423*
    feature lines, **412–419**, *412–419*, **424–428**, *424–429*
    frequency, **411–412**
    frequency stations, **421–422**, *422*
    intersections, 466, *466*
    locking regions, **414**, 417
    non-road, **448–450**, *448*

in Object Viewer, **418–419**, *418*
overview, **409–410**, *409–410*
pipe trenches, 390, **450–451**, *450–451*
properties, 434
QTO, **787–791**, *787–788, 790*
rebuilding, **420**, *420*, 445
regions, **411**
roundabouts. *See* roundabouts
sections, **435–437**, *435–437*
surfaces
    boundaries, **442–446**, *443*
    creating, **438–442**, *439–442*, **446–447**, *446*
    name templates, **442**
targets, **412**, **429–433**, *430–433*
tweaking, **421–423**, *421–423*
utilities, **500–503**, *500, 502–503*
volume calculations, **447–448**, *447*
waterfall effect, **421**
Counter option for points, 112
Cover And Slope rule
  pipe networks, **593**, *593*
  sanitary sewer networks, **607**, *607*
Cover Only rule, **593**, *594*
Create A Corridor Surface For Each Link option, 442
Create Alignment - From Pipe Network dialog, 624
Create Alignment - Layout dialog, 255–259, *256, 261, 265, 835, 835*
Create Alignment From Network Parts dialog, 634
Create Alignment From Objects dialog
    Design Criteria page, 252
    General page, 250–252, 426, *427*
Create Assembly dialog, 368–369, *369*
  non-road uses, 388
  pipe trenches, 391
  roads, 372–373, *373*
Create Best Fit Alignment command, 261
Create Best Fit Alignment dialog, 10, *10*, 263, *263*
Create Best Fit Arc command, 48, *48*
Create Best Fit Line command, **47–48**, *48*

Create Best Fit Parabola command, **48–49**, *49*
Create Best Fit Profile dialog, 11–12, *12*, 320–321, *320*
Create Breaklines dialog, 9
Create Breaklines option, 152
Create COGO Points dialog, 503, *503*
Create Component option, **809**
Create Corridor dialog, 12, *13*
    feature lines, 413–414, *413*
    non-roads, 449
    pipe trenches, 451
    references, 760
Create Corridor Baseline dialog
    cul-de-sacs, 460
    intersections, 476–477, 480–481
Create Corridor Region dialog
    cul-de-sacs, 460–461, *461*
    intersections, 477, 482
Create Corridor Surfaces dialog, 439
Create Cropped Surface command, 162
Create Curve Between Two Lines command, **42**, *42*
Create Curve From End Of Object command, 44, *44*
Create Curve On Two Lines command, **42–43**, *43*
Create Curve Through Point command, **43**, *43*
Create Curves options, 40, *40*
Create Data Shortcuts dialog, 756, *756*, 759, *759*
Create Dynamic Link To The Alignment option, 675
Create Dynamic Link To The Corridor option, 428, 502
Create Feature Line From Alignment dialog, 675–676, *675*
Create Feature Line From Corridor dialog, 428, *429*, 502–503, *502*
Create Feature Line From Corridor tool, 671
Create Feature Line From Stepped Offset tool, 671, 676
Create Feature Line tool, 671
Create Feature Lines dialog, 671–674, *672*

Create Feature Lines From Alignment tool, 671, 675
Create Feature Lines From Objects tool, 671, 673, 695
Create Full Parts List tool, 618
Create Grading Group dialog, 702, *703*
Create Grading tool, 704, *704*
Create Grading Infill tool, 708
Create Ground Data panel, 6, 8
Create Interference Check dialog, 637, *637*
Create Interference Check tool, 619
Create Intersection dialog
    Corridor Regions page, 469, *469*, 472, 474
    General page, 468–469, *470*
    Geometry Details page, 470–472, *471*
Create Layer dialog, 883
Create Line command, 32
Create Line By Angle command, 36, *36*
Create Line By Azimuth command, 35, *36*
Create Line By Bearing command, **35**, *35*
Create Line By Deflection command, 36, *36*
Create Line By Extension command, **38**, *38*
Create Line By Grid Northing/Grid Easting command, 34
Create Line By Latitude/Longitude command, 34
Create Line By Northing/Easting command, 34
Create Line By Point # Range command, **33**, *33*
Create Line By Point Name command, 34
Create Line By Point Object command, 33
Create Line By Side Shot command, 37–38, *38*
Create Line By Station/Offset command, **36–37**, *37*
Create Line From End Of Object command, 39, *39*
Create Line Perpendicular From Point command, 39, *39*
Create Line Tangent From Point command, 39
Create Mass Haul Diagram dialog
    Balancing Options page, 583

    General page, 582, *582*
    Mass Haul Display Options page, 583, *583*
Create Multiple Curves command, **43–44**, *44*
Create Multiple Profile Views dialog
    Data Bands page, 336, 738
    Data References page, 738
    General page, 335, *335*
    Multiple Plot Options page, 336, *336*
    Profile Display Options page, 336, *336*, 738
    Profile View Height page, 335
Create Multiple Section Views dialog
    Data Bands page, 744, *744*
    Elevation Range page, *743*
    General page, 567, 740, 741, 762
    Offset Range page, 742
    Section Display Options page, 743, *743*
    Section Placement page, 567, *568*, 740–742, *741*
Create New Definitions Automatically option, 60
Create New Offsets From Start To End Of Centerlines option, 471
Create New Survey Database option, 6, 73
Create Offset Alignments dialog, 253–255, *253*
Create Parcel From Objects command, 227
Create Parcel From Remainder option, 216
Create Parcels - From Objects dialog, 204, *205*, 212
Create Parcels - Layout dialog, 206, 217, 221
Create Parts List tool, 618
Create Pipe Network dialog, 601–602, *601*, 608
Create Pipe Network From Object dialog, 610–611, *611*
Create Point Cloud dialog
    Information page, 190, *190*
    Source Data page, 190–191, *190*
    Summary page, 191, *191*
Create Points dialog, 100, *100*, 109–110, *109*
Create Points toolbar, 99, **106–108**, *106–108*
Create Pressure Pipe Network dialog, 650–651, *651*
Create Profile - Draw New dialog
    entity layout, 315

feature lines, 426, *427*
General page, 309–310, *309*
profiles from files, 321
PVI layout, 309–313, *309*
Create Profile And Profile View option, 624
Create Profile from Surface dialog, 11, *11*
alignments, 624, *625*
data references, 758–759
EOP design profiles, 458
pipe networks, 634, 656
roundabouts, 518, *519*, 522, *523*
surface sampling, **301–302**, *301–302*
Create Profile Reference dialog, 760
Create Profile View dialog, 330, 458, 656, 758–759
Data Bands page, 304, *305*, 339
General page, 11, 302–303, *303*, 337, 625, *625*
Pipe/Pressure Network Display page, 625, 634
Profile Display Options page, 304, *304*, 338, *338*
Profile View Height page, 303, *304*, 333–334, *333*, 337
Stacked Profile page, 337, *338*
Station Range page, 303, *303*, 331–332, *332*, 335, 337
Create Quick Profiles dialog, 363, *363*, 680, *680*, 688
Create Reverse Or Compound Curve command, **45**, *45*
Create Right Of Way dialog, 210, *210*
Create Roundabout dialog
Approach Roads page, 513–514, *513*
Circulatory Road page, 512, *512*
Islands page, 514, *514*
Markings And Signs page, 515, *515*
Create Roundabout option, 511, *512*
Create ROW tool, 208–209
Create Sample Line Group dialog, 558, *558*, 560, 762
Create Sample Lines - By Station Range page, 559–560, *559*
Create Sample Lines utility, 501

Create Section Sheets dialog, 745, *745*
Create Section Sheets tool, 568
Create Section View dialog
Data Bands page, 565, *566*
Elevation page, 564, *565*
General page, 564, *564*
Offset Range page, 564, *565*
Section Display Options page, 565, *565*
Section View Tables page, 566, *566*
Create Sheets dialog
Data references page, **734–735**, *734*
Profile Views page, **733–734**, *733*, 737–738
Sheet Set page, **732–733**, *732*, 737
View Frame group and Layouts page, **729–731**, *729*, 736
Create Snapshot option, 145
Create Superelevation View dialog, 549, *549*
Create Superelevation View option, 340, 548, *549*
Create Superimposed Profile option, 359
Create Surface dialog, 8
contours, 139
DEM files, 132
external text files, 143
grading groups, 707, 710
TIN surfaces, 129, *129*, 180–181, *180*
volume calculations, 447
Create Surface From GIS Data dialog
Connect to Data page, 136, *136*
Data Mapping page, 137–138, *137*
Geospatial Query page, 137, *137*
Object Options page, 135, *135*
Schema And Coordinates page, 136, *136*
Create Surface Profile command, 301
Create Surface Reference dialog, 757–758, *757*, 760
Create Table - Convert Child Styles dialog, 295
Create Text Component option, 819
Create Total Volume Table dialog, 576, *576*
Create View Frames dialog
Alignment page, 715–716, *715*, 723
Match Lines page, **720–722**, *721*, 724
Profile Views page, **722–724**, *722*

Sheets page, **716–718**, *716–718*, 723
View Frame Group page, **719–720**, *719*, 724
Create Widening command, 255
CreatePressurePartListFull command, 646
criteria-based profile design, 310
Criteria dialog for interference checks, 638, *638*
criteria files for superelevation, **527–531**, *528–531*
critical stations, 529
cross sections. *See* section views
crossings
    breaklines, 146, **154**, *154*
    pipe networks, **631–635**, *631–632*
CSV files, **542**, *542*
Ctrl+click keys for labels, 290
cul-de-sac corridors
    combining components, **458–462**, *460–462*
    EOP design profiles, **457–458**
    multiple baselines, **456–457**, *456–457*
    troubleshooting, **462–463**, *463–466*
cul-de-sac parcels, **211–213**, *212–214*
curb island corridors, 510
curb return alignments, **251**
curbs, 375, *375*
    feature lines, 691, *691*
    roundabouts, 522
    subassemblies, **383–384**, *383–384*
Curve Calculator, **45–47**, *45*
curve tessellation, 873
curves, **40**, *40*
    Curve Calculator, **45–47**, *45*
    feature lines, **682**
    frontage offset, 220, *220*
    labels
        adding, **50–51**, *50–51*
        creating, **827–829**, *828–829*
        profile, **847–848**, *847–848*
        tags from, **242–244**, *243–244*
    standard, **42–45**, *42–45*
custom panorama views, **273–274**

custom parts, sharing, **666–667**
Customize Columns dialog, 273
Cut Area setting, 352
cut factor for soil, 577

D

data bands
    profile views, **346–352**, *347–351*, **908–911**, *908–909*, *911*
    section views, 565, *566*
data clip surface boundaries, 155, 162
data entry, **34**
data references, 751
    broken, **766–768**, *767–768*
    creating, **757–760**, *757–760*
    renaming objects, 758
    updating, **763–765**, *763–765*
data shortcuts, **751–752**
    best practices, **765–766**
    creating, **755–757**, *756–757*
    cross-sections, **761–762**, *761*
    Data Shortcuts Editor, 768–771, *769–770*
    eTransmit command, **771**, *771*
    folders, **753–755**, *754–755*
    location, 756
    references. *See* data references
    surfaces, 162
databases for surveys, **57–58**, *58*
    data for, **71–76**, *71–72*, *74–76*
    defaults, **58–60**, *59*
    equipment, **60–61**, *61*
    figure prefix, **61–63**, *61*
    figures in, **76–77**, *76–77*
    linework code set, **63–64**, *64*
daylight subassemblies
    alternative, **400–401**, *400–402*
    input parameters, **399**, *399*
    working with, **397–398**, *398*
DaylightBasin subassembly, **401**, *402*
daylighting
    corridors, **423**, *423*
    subassemblies, **396–397**

DaylightInsideROW subassembly, 398–400, *399–400*
DaylightMaxOffset subassembly, **371**, *371*, 376, *376*
DaylightToROW subassembly, **400**, *400*
DDPTYPE dialog, 868
deeds, **39–40**
defaults
    point, **100**
    survey database, **58–60**, *59*
definitions, surface, **127–129**
deflection angles, 36
Deflection Distance command, 53
deflection glyph, 649, *649*, 652, *652*
Deflection Validation settings, 658, *658*
Delaunay, Boris, 126
Delaunay triangulation algorithm, **126–127**
Delete Component option, 809
Delete Elevation Point tool, 688–689
Delete Entity tool, 320
Delete Line option, 163
Delete option for labels, 804
Delete PI tool, 681
Delete Pipe Network Object option, 605
Delete Point option, 163
Delete PVI tool, 318
Delete Sub-Entity tool, 213, 223–227, *225*
deleting
    parcel segments, **223–227**, *224–227*
    profile views, **331**
    section views, **571**, *571*
    subassemblies, 375
DEM (Digital Elevation Model) files, **128**, **131–134**, *131–133, 135*
dependencies, objects, 3
Depth Check settings, 658–659, *660*
Description Key Editor, **65–68**
Description Key Set dialog, 67
Description Key Sets Search Order dialog, 70, *70*
description keys and keysets
    activating, **68–70**, *69–70*
    creating, **67–68**, *67–68*

overview, **64–67**, *65*
wildcards, **68–69**
descriptions
    alignments, 252, 282
    vs. names, 101
    points, 34, **100–101**, *100*
    sanitary sewer networks, 602
design checks
    vs. design criteria, 269
    pressure networks, **658–659**, *658, 660*
design constraints, **267–270**, *267–270*
design criteria
    alignments, 252
    superelevation, 525, **527–531**, *528–531*
Design Criteria Editor, 527–529, *528, 531*
design speeds, **286–288**, *287–288*
destructive surface boundaries, **156–157**, *156–157*, 161, *161*
DETACHQTOFILES command, 778
diamond-shaped grips
    attached segments, 214–215, *215*
    cross sections, 557
    match lines, 727–728
    parcel segments, 240, *240*
    site plans, 220, 223
    superelevation, 550
    view frames, 726
Digital Elevation Model (DEM) files, **128**, **131–134**, *131–133, 135*
direction-based line commands, **34–39**, *35–39*
directions
    driving side, **22**
    flow, **612–613**
    surface styles, 878
    survey networks, 78
Disconnect From Pressure Part option, 653
Display As Block option, 899
Display As Boundary option, 899
Display As Solid option, 899
Display Change Report option, 60
display options
    label style dragged state, 813

Plan Production, 714
profile views, **345–346**, *345*
Display Warnings For Missing Required Fields option, 60
distribution methods in analysis, **885**
Divided Crown With Median setting, 536
Draft option for section views, 568
Dragged State Component properties, 813
Dragged State settings, **812–813**, *812*
drainage
pipe networks, **609–612**, *610–611*
roundabouts, **510–511**, *511*
Draw Order Icons option, 617
Draw Parts In Profile option, 626, 631
Draw Pipes And Structures tool, 608
Draw Slip Lane dialog, 516, *517*
Draw Tangent tool, **318**
Draw Tangent-Tangent With No Curves tool, 206
Draw Tangents With Curves tool, 313–314
drawing objects for surfaces, **128**
drawing parts in profile view, **626–632**, *627–632*
Drawing Scale setting, 22
Drawing Settings dialog
Abbreviations page, **19–20**, *20*
Ambient Settings page, **20–24**, *21*
Object Layers page, **16–19**, *17*, 805
object styles, 862
Transformation page, **15–16**, *16*
Units And Zone page, **14–15**, *14*, 131, *131*
drawing templates (DWT), 713, **746–748**, *747–748*
Drawing Unit setting, 22
driveways, **506–510**, *507–509*
Driving Direction setting, 22
Duplicate Point number dialog, 111–112, *111–112*
duplicate points, **111–112**, *111–112*
dynamic area labels, 197
dynamic pipe tables, 640

E

easements, 200
edge of pavement (EOP)
cul-de-sacs, **457–458**
intersections, 467, 480
Edit Alignment Labels option, 289
Edit Best Fit Data For All Entities tool, 320
Edit Command Settings dialog
CreateCorridor page, 411–412, 414
CreatePoints pages, 70, *70*
FitCurveFeature page, 682
General page, 99, *99*
JoinFeatures page, 682
RunDepthCheck page, 658–659, *658*
WeedFeatures page, 683
Edit Curve tool, 682
Edit Drawing Settings option, 14
Edit Elevations panel, 679–681, *680*, 686, **689**, *689*
Edit Feature Line Curve dialog, 682, 686, *686*
Edit Feature Line Style option, 679
Edit Feature Settings dialog
Corridor page, 862, *863*
Profile View page, 23, *24*
Edit Geometry panel, 679, *679*, 681, *681*
Edit In Storm And Sanitary Analysis tool, 620
Edit Interference Style tool, 620
Edit Label Style Defaults dialog, 815, *815*
Edit Linework Code Set dialog, 63, *64*
Edit Material List dialog, 577, *578*
Edit option for label styles, 804
Edit Parcel Properties dialog, 235–236, *236*
Edit Part Sizes dialog, 623–624, 665–666, *666*
Edit Parts List tool, 619
Edit Pay Items dialog, 798, *798*
Edit Pay Items On Specified Object option, 798
Edit Pipe Network option, 615, 618, 622, 635
Edit Pipe Style tool, 618
Edit Profile Labels option, 355
Edit Profile View Style option, 901, *901*
Edit Setups That Observe option, 78

Edit Structure Style tool, 618
Edit Survey Database Settings option, 73
Edit Table Style tool, 643
Edit Values dialog
    Part Builder, 666, *666*
    sumps, 624
Elevation Editor, 686–687, 691, 694
elevation factor, 15–16
Elevation To Use option, 146
elevations
    banding, **171–175**, *172–175*
    color, 878
    feature lines
        editing, **686–694**, *687, 689, 691–694*
        as targets, **504–506**, *505–506*
    points, **100–101**, *100*, **118–119**, *118–119*
    profile views, **342–345**, *343–344*
    profiles, **299–308**, *300–307*
    structure labels, **832–834**, *833*
    sump control, **629**
    superelevation. *See* superelevation
Elevations tool, 878
Elevations From Surface tool
    description, 690
    grading elevation, 688
    points, 119, *119*
Enable Part Masking options, 898–899
Enable Radius Snap option, 872, 875
End Full Super (EFS) assemblies, 527
end line size of pipes, 890
End Normal Crown (ENC) assemblies, 526, *526*
End Offset option, 595
End Shoulder Rollover (ESM) assemblies, 527
End To Start pipe network flow direction, 613
entering data, **34**
entity-based layouts, 308, **315–317**, *316–317*
EOP (edge of pavement)
    cul-de-sacs, **457–458**
    intersections, 467, 480
Equal Count distribution method, **885**
Equal Interval distribution method, **885**

equipment database, **60–61**, *61*
Equipment Database Manager dialog, 60–61, *61*
Erase Existing Entities option, 673
Error Tolerance settings, 59
eTransmit command and data shortcuts, **771**, *771*
events in survey databases, **71**
exclamation points (!) point status icon, **130**, *130*
Exclude Elevations Greater Than setting, 146
Exclude Elevations Less Than setting, 146
Export Drawing Name dialog, 773
Export Superelevation Data tool, 542
Export To LandXML dialog, 772, 773
exporting superelevation data, **542**, *542*
expressions, **857–860**, *857–859*
Extended Properties settings, 60
Extensible Stylesheet Language (XSL) format, 800
external references (XRefs), **751–752**
external text files, surfaces from, **143–144**, *144*
Extract Objects From Surface dialog, 157–158, *158*
Extract Objects From Surface utility, 155

F

FACETDEV system variable, 895
FACETMAX system variable, 895
families in Part Builder, **664**
favorites, pay item, **778–781**, *778–781*
Feature Line contextual tab, 670, *670*
Feature Line Properties dialog
    Information page, 677, *677*, 679
    Statistics page, 677, *677*, 679
Feature Line Properties tool, 679
Feature Line Site Properties dialog, 678, *678*
Feature Line Styles collection, 867, *867*
feature lines, **669**
    corridor surfaces from, **441**
    corridors, **412–419**, *412–419*, **424–428**, *424–429*

grading, **669**
  creating, **671–678**, *672–677*
  elevations, **686–694**, *687, 689, 691–694*
  geometry editing, **681–686**, *681, 683–686*
  grips, **674**
  information editing, **679–681**, *679–680*
  labeling, **699–700**, *700–701*
  names, **673**
  object style, **678**, *678*
  ponds, **694–699**, *695–698*
  tools, **669–670**, *670*
marker points, 378
storm drainage pipe networks from, **609–612**, *610–611*
styles, **877**, *877*
as width and elevation targets, **504–506**, *505–506*
Feature Lines From Corridor utility, 501, 503
Feature Settings dialog, 862
figure groups in survey databases, 72
figure prefix database, **61–63**, *61*
Figure Prefix Database Manager, 61–63, *61*
Figure Properties dialog, 77, *77*
Figure Survey Queries, **128**
figures in survey databases, 72, **76–77**, *76–77*
file formats for points, **102–103**, *103*
files, profiles from, **321–322**, *322*
Fill Area setting, 352
fill distance in parcel styles, 877
fill factor of soil, 577
Fillet tool for feature lines, 682
Find And Replace dialog, 769–771, *770*
Find References option, 804
Fit Curve tool, **682**
fittings in pressure networks, 645, 652, *652*
Fixed Line (Two Points) tool, 258, *258*, 275
Fixed Property field, 46
Fixed Scale Factor field, **66**
fixed segments for alignments, **248**, *248*
Fixed Tangent (Through Point) command, 317
flag symbols, **532–533**
flat areas in contours, 140–142, *140–141*
flat cul-de-sacs, 463, *466*
Flatten Grade Or Elevations tool, 688

Flip Anchors With Text option, **807**, *808*
flip glyphs, 649, *649*
Floating Curve tool, 258, *258*, 265
Floating Line tool, 266
floating segments, **248**, *248*
flow direction in sanitary sewer networks, **612–613**
fluff factor in soil, 577
folders for data shortcuts, **753–755**, *754–755*
Forced Insertion option, 806, *807*
forcing superelevation, **543–544**, *543–544*
formats
  Description Key Editor, **65–66**
  profiles from files, 321
formula files, 777–778, **791**
Free Form Create tool, **220–223**, *221–223*
free haul areas, 581
Free Haul Distance setting, 583
free haul in mass haul diagrams, 582
free segments, **248**, *249*
Free Vertical Curve options, 316–317
Free Vertical Parabola (PVI Based) option, 316–317
frequency
  corridors, **411–412**, 434
  cul-de-sacs, 462
  feature lines, 415, *415*
frequency lines, **555–556**
frequency stations for corridors, **421–422**, *422*
Frequency To Apply Assemblies dialog, 415, *415*, 434, 462
friends with marked points, **402–403**
From Criteria option, 352
From File option, 152
frontage offset of curves, 220, *220*
Function tool, 858

G

gapped profile views, **335–337**, *335–336*
gaps
  cul-de-sacs, 462–463
  label style borders, 812

General collection, **803–804**, *804*, 867
General Note labels, **818–820**, *818–820*
General Tools panel, **e617**, *617*
generic links, **394–397**, *396–397*
GenericPavementStructure subassembly, 478, 546–547
Geodetic Calculator, **86–87**, *87*
Geographic Information Systems (GIS), **135–138**, *135–138*
GEOMARKERVISIBILITY variable, 824
geometric markers, 15
geometry
    alignment, **271**
        component-level editing, **274–275**, *274*
        constraints, **275–279**, *275–279*
        grip editing, **271–273**, *271–272*
        tabular design, **273–274**, *273*
    feature line, **681–686**, *681*, *683–686*
Geometry Details page, 471–472, *471*
Geometry Editor, 273
Geometry Locking regions, 414
geometry points for labels
    alignments, **289–291**, *289–290*, 837–838, *838*
    applying, 356–357, *357*
    profile, 844, *844*
Geometry Points To Label In Band dialog, 347–348, *347*
Getting Started - Catalog Screen dialog, 662–665, *662*
GIS (Geographic Information Systems), **135–138**, *135–138*
grade breaks, **849–850**, *849*
Grade Extension By Reference tool, 690–692
Grading Creation Tools toolbar, 702–704, *702–703*
Grading Editor tool, 706
Grading Elevation Editor, **687–689**, *687*, 696–698, *697–698*
Grading Group Properties dialog, 707–710, *707*, *709*
grading groups, 701, **707–711**, *707*, *709*, *711*
gradings, **669**, **701**
    creating, **701–705**, *702–705*

editing, **705–707**, *706*
feature lines. *See* feature lines
roundabouts, **510–511**, *511*
sites, 200
Grid Easting command, 53
Grid Northing command, 53
Grid Scale Factor setting, 16
grids
    labels, **188**, *189*
    profile views, **908**
grips
    alignments, **271–273**, *271–272*
    attached segments, 214–215, *215*
    contours, 186, *186*
    corridors, 490
    cross sections, 556–557, *557*
    curves, 266, *266*
    feature lines, **674**
    grade breaks, 849
    match lines, 727–728
    offset alignments, 255
    parcel segments, 240, *240*
    pipe networks, **613–614**, *613–614*
    points, **112**, *113*
    profiles, **322–323**, *323*
    PVI-based layouts, 312, *312*
    site plans, 220, 223
    station offset labeling, 292, *292*, 842
    sump depth, 629, *629*
    superelevation, 550–551, *551–552*
    vertical movement edits, **628–629**, *628–629*
    view frames, 726, *726*
Group Plot Style dialog
    Array page, 913, *913*
    Display page, **914–915**, *914*
    Information page, 914
    Plot Area page, 913, *913*
groups
    grading, 701, **707–711**, *707*, *709*, *711*
    plot styles, 568, **912–916**, *913–916*
    points, **114–118**, *115–118*
    renaming, **387–388**
guardrails, 454, *455*

H

hashes (#) in description key sets, 68
Hatch Editor, 786
Hatch Pattern dialog, 893, *893*
hatching
    pay items, 786
    pipes, 890, 893, *893*
    profile views, 333–334, *333*, **352**, *352–353*
height
    profile views, 343, *343*
    text, 810, 813
help
    accessing, **4**
    subassemblies, **368**, **379–380**, *379–380*
Help icon, 4
Hide surface boundaries, 155
Highlight Objects With Pay Items option, 797
Highlight Objects Without Pay Items option, 797, 799
Home tab, 1–2, *1–2*
horizontal geometry bands, 346
Hydraflow program, 620
Hydraflow Express program, 620
Hydrographs tool, 620

I

Immediate And Independent Layer On/Off Control Of Display Components option, 18
Imperial To Metric Conversion option, 14, 22
impervious site areas, 200
Import Civil 3D Styles dialog, 30–31, *31*
import events in survey databases, **71**
Import Field Book dialog, 79, *80*
Import Label Set option, 838
Import LandXML dialog, 772, 774, *774*
Import Point File dialog, 84
Import Points dialog, 101–102, *101*, 763
Import Profile From File dialog, 321
Import Settings option, 30–31
Import Subassemblies dialog, 406
Import Survey Data dialog
    Import Options page, 7–8, *8*
    Import Survey Data page, 73–74, *74*
    Specify Data Source page, 6, *7*
    Specify Network page, 7, *7*
Import Survey Data option, 6, *6*, 71
importing
    industry models, **655**, *655*
    point clouds, **189–191**, *190–191*
    points, **101–104**, *101*
    styles, **30–32**, *31*
    superelevation data, **542**, *542*
incremental counting for view frames, **725**
Independent Layer On option, 22
industry models, importing, **655**, *655*
informational graphics, 717
input parameters
    daylight subassemblies, **399**, *399*
    help, **379–380**, *380*
inquiry commands for surveys, **91–94**, *91–93*
Inquiry Tool, 91, *92*
Insert dialog
    blocks, 54, *54*
    cul-de-sac parcels, 212, *212*
    non-Civil 3D drawings, 29, *29*
Insert Elevation Point tool, 688–689, 695
Insert High/Low Elevation Point tool, 690, 698
Insert Intermediate Grade Break Points option, 674
Insert Into Drawing option, 72
Insert Match Lines option, 721
Insert PI tool, 681
Insert PVI dialog, 688
Insert PVI tool, 318
Insert PVIs dialog, 491, *491*
Insert PVIs - Tabular tool, 319, 491
inserting
    non-Civil 3D drawings, **28–30**, *29–30*
    subassemblies, **374–375**
interface, **1–2**, *1–2*
    Panorama window, **25**
    ribbon, **25–26**, *25–26*
    Toolspace. *See* Toolspace
Interference Check Properties dialog, 619
Interference Check Properties tool, 619

interference checks, 619, **636–639**, *637–639*
Interference Properties dialog
    Information page, 620
    Statistics page, 620
Interference Style dialog, 620
Interpolation settings for points, 108, *108*
Intersection Curb Return Parameters dialog, 471, *472*
Intersection Curb Return Profile Parameters dialog, 472, *473*
Intersection Lane Slope Parameters dialog, 472, *473*
Intersection Offset Parameters dialog, 471
Intersection settings for points, **107**, *107*
Intersection Wizard, **468–476**, *468*, *470–476*
intersections
    assemblies for, **478–479**, *479*
    assembly sets, **469**, *469*
    baselines, 467, **480**
    Intersection Wizard, **468–476**, *468*, *470–476*
    modeling, **476–478**, *477–478*
    overview, **466–467**, *466–467*
    targets, **480–484**, *481–484*
    troubleshooting, **485–486**, *485–486*
inventory of pay items, **799–801**, *799–801*
invisible profile views, 738
irregular viewport shapes, 748
islands, roundabouts, 514, *514*, 522, *523*
Isolate Objects option, 617
Item Preview Toggle icon, 4

J

Join tool, 682

K

K value for layout profiles, 310–312
kriging surfaces, 163

L

Label Fixed Rotation field, 67
Label Location setting, 720
Label Rotate Parameter field, 67
Label Style Composer, 720
    Dragged State page, **812–813**, *812*
    General page, **805–807**, *805–808*, 832
    Information page, 804–805, *805*
        geometry points, 837
        major station labels, 836
        station offset labels, 840–841
        structure labels, 832
    Layout page
        expressions, 859
        settings, 808–812, *808*, *810*
        structure labels, 832–833, *833*
    Major Station page, 910
    Summary tab, **813–815**, *814*
Labels - Alignment Geometry Point Label Group tab, 835, 838–839
Labels - Parcel Segment Label tab, 240, *240*
labels and label styles, 13, **803**
    alignments
        geometry points, **837**
        label sets, **288–291**, *289–290*, **834–835**, *834–835*, **837–839**, *838–839*
        major station, **836**, *836*
        station offset, **291–292**, *292*, **840–842**, *841–842*
    assemblies, **376–377**, 394
    contours, **184–186**, *185–186*
    corridors, **488–490**, *488*
    creating, **815–816**, *816*
    cross sections, **580**, *580*
    curves
        adding, **50–51**, *50–51*
        creating, **827–829**, *828–829*
        profile, **847–848**, *847–848*
        tags from, **242–244**, *243–244*
    dynamic area, 197
    general, **803–804**
    General Note, **818–820**, *818–820*
    grids, **188**, *189*
    lines
        adding, **50–51**, *50–51*
        creating, **824–827**, *825–826*, **845–846**, *846*
        feature, **699–700**, *700–701*
        match, **722**, *722*
        tables, **242–244**, *243–244*

parcels
    areas, **234–238**, *234*, *236–267*
    segments, **238–245**, *239–241*, *243–245*
pipe networks
    adding, 617
    creating, **829–831**, *831*
    with crossings, **633–636**, *633*, *636*
points, **821–823**, *822–823*
profile views, **353–355**, *354–355*
profiles, **842**
    applying, **355–358**, *356–358*
    curve, **847–848**, *847–848*
    expressions, 860
    grade breaks, **849–850**, *849*
    label sets, **358**, **843–845**, *843–845*, **850–851**, *851*
    line, **845–846**, *846*
sanitary sewer networks, 602
Settings tab, 803–805
slopes, **186–188**, *187–188*
station offset, **291–292**, *292*
structure, **831–834**, *833*
surfaces, **184–188**, *185–189*
Text Component Editor, **816–817**, *816–817*
view frame groups, 720
Labels & Tables panel, **616–617**, *617*
Land Desktop points, **104–106**, *105*
LandXML, **772**
    overview, **772–775**, *773–774*
    parcels, 229
    surfaces, 130
LaneBrokenBack subassembly, **382**, *382*
LaneOutsideSuperWithWidening subassembly, 478
LaneParabolic subassembly, **382**, *382*
lanes
    intersections, **485**, *485–486*
    subassemblies, **382**, *382*
LaneSuperelevationAOR subassembly, 430
    intersections, 478–479, 483
    lanes, 382, *382*
    roads, **370**, *370*, 373, 377
    superelevation, 533–535, *533*
large surfaces, **162**

LAS (Log ASCII Standard) files, 189
Launch Pad panel, 11, *11*
    corridors, 425, *425*, 500, *500*, 503
    pipe networks, **620**
Layer Selection dialog, 18, *18*
    alignment styles, 875
    marker object styles, 870
    survey point styles, 872
    triangles surface style, 883–884
LAYEREVALCTL variable, 824
layers
    alignments, 252
    Description Key Editor, 66
    Figure Prefix Database Manager, 62
    label styles, **805**
    Object Layers tab, **16–19**, *17–19*
    pipe tables, 639
    points, **100**
    view frame groups, 720
Layout Creation settings, **730**
layouts
    alignments from, **255–259**, *256–259*
    label styles, **808–812**, *808*
    parcels, **206–208**, *206–208*
    profiles, **308–309**
        entity-based, **315–317**, *316–317*
        parameters, **323–325**, *324*
        PVI-based, **309–315**, *309–314*
        tools, **318–320**, *318–319*
    sheets, **730–731**
Leader Attachment setting, 813
Leader Justification setting, 813
Leader properties for label styles, 813
Least Squares Analysis Defaults settings, 59
Left Of Station Character option, 836
Length Check rule, 594, *594*
length of match lines, **727–728**
lengthen glyph, 649, *649*
Level Crown (LC) assemblies, 526, *526*
Level Of Detail setting, 134, *135*
Light Detection and Ranging (LiDAR), 189
Line And Arc Information tool, 93, *93*
Line By Bearing And Distance command, 89

linear object styles
    alignments, **875–876**, *876*
    feature lines, **877**, *877*
    overview, **872–874**, *873–874*
    parcel, **876–877**, *876*
lines, **32**, *32*
    alignments from, **250–255**, *251*, *253–254*
    best fit, **47–48**, *48*
    COGO commands, **32–34**, *33*
    for deeds, **39–40**
    direction-based commands, **34–39**, *35–39*
    feature. *See* feature lines
    frequency, **555–556**
    labels
        adding, **50–51**, *50–51*
        creating, **824–827**, *825–826*, **845–846**, *846*
        feature, **699–700**, *700–701*
        match, **722**, *722*
        tables, **242–244**, *243–244*
    match
        editing, **725–728**, *726*
        Plan Production, 714, *715*
        view frames, **720–722**, *721*
    tables, **294–295**, *295*
Linetype setting, 812
Lineweight setting
    label borders, 812
    label text, 811
linework code set database, **63–64**, *64*
links and link labels
    code set styles, **854–855**
    corridors, 411, **439–442**, *440*
    to marked points, **403–404**, *403*
    subassemblies
        description, **377–378**, *378*
        generic, **394–397**, *396–397*
LinkSlopesBetweenPoints subassembly, **496**, *496*
LinkSlopeToSurface subassembly, 496
LinkWidthAndSlope subassembly, 395
List Slope tool, 92
ListAvailablePointNumbers command, 111
location glyph, 648, *649*, 654
location of data shortcuts, 756

lockdown, PVIs in, 326
locking regions, **414**, 417
Log ASCII Standard (LAS) files, 189
lots, subdivision, 199, *199–200*, **214**
    attached segments, **214–215**, *214–215*
    sizing, **215–223**, *215*, *218–223*
Low Shoulder Match (LSM) assemblies, 527, *527*

M

major station labels, 289
    alignments, **836**, *836*
    profiles, 357
Major Tick Details settings, 908
Make Orthogonal option, 557, *557*
manholes, 589, *589*
manually limited profile views, **331–332**, *332*
Map 3D product, 655
Mapcheck reports, 86–87, *88*
Mark The Corridor As Out-Of-Date option, 416
marked points for assemblies, **402–404**, *403–404*
Marker Fixed Rotation field, 67
marker object styles
    creating, **870–871**
    settings, **868–870**, *869*
marker points
    corridors, 411
    subassemblies, **378**, *378*
Marker Rotate Parameter field, 67
Marker Style dialog
    Display page, **864–866**, *865*
    Marker page, **868–870**, *869*
MarkPoint subassembly, 495, *495*
masks
    alignment, 254, *254*
    part, 898–899
mass haul diagrams, **581**, *581*
    components, **581–582**, *581*
    creating, **582–583**, *582–583*
    editing, **583–584**, *584*

Mass Haul Line Properties dialog, 583–584, *584*
mass haul lines, 556
mass haul views, 556
Master view, 4
Match Length command, 53
Match Line Properties dialog, 725
match lines
    editing, **725–728**, *726*
    Plan Production, 714, *715*
    view frames, **720–722**, *721*
Match On Description Parameters setting, 70
Match Radius command, 53
matching transparent commands, **53–54**
materials lists
    creating, **575**, *575*
    soil factors, **577–578**, *578*
Maximum Change In Elevation setting, 169
Maximum Pipe Size Check rule, 591, *591*
Maximum Shoulder Rollover setting, 539
Maximum Text Width setting, 813
Maximum Width setting, 811
Measurement Corrections settings, 59
Measurement Type Defaults settings, 59
median islands for roundabouts, 522, *523*
Merge option for points, 111
Minimize Flat Areas option, 141–142, 163
Minimum Radius Tables setting, 529
Minimum Slope rule, 607
Minor Contour component, 880–881
Minor Tick Details settings, 908
Mirror Subassemblies command, 376
Miscellaneous alignments, 252
Miscellaneous point settings, 106, *107*
modeless toolbars, 106
Modify panel
    feature lines, 679, *679*
    pipe networks, **618**
    pipe tables, **643**
Modify Corridor Sections panel, 435
Modify Point option, 163
Move Point option, 163
Move PVI tool, **318**
Move To Site dialog, 702
Move To Site tool, 680
moving
    subassemblies, 374
    view frames, 726
MSLTSCALE variable, 824
multiple baselines, **456–457**, *456–457*
Multiple Boundaries setting, 352
multiple-parcel segments, **238–240**, *239–240*
multiple section views, **567–569**, *568–569*, **740–744**, *741–744*
Multiple Selection option, 640
Multiple tab, 25–26, *26*
multiple targets for intersections, **481**, *481*
Multipurpose Styles, **867**, *867*
multiregion baselines, **453–454**, *454–455*

N

name templates
    corridors, 442, *442*
    sheets, **737**
    view frames, **719–720**, *719*
names
    alignments, 251
    code set styles, 854
    with data references, 758
    vs. descriptions, 101
    feature lines, **673**
    Figure Prefix Database Manager, 62
    intersections, 480
    label styles, 809
    points, 34, 97, *97*, **100–101**, *100*
    sanitary sewer networks, 602
    sheet layouts, **731**
    subassemblies, 371
    view frame groups, 720
Natural Neighbor Interpolation (NNI), 163, **166–167**, *167*
net cut and net fill in mass haul diagrams, 581, *581*
network components in surveys, **77–85**, *78–83*, *85*

network groups in survey databases, 72
Network Layout Tools toolbar
    pipe networks, **621–623**
    sanitary sewer networks, 602, *602*, 608
network parts, alignments from, **624–625**, *625–626*
Network Parts List dialog
    Information page, 646
    Pipes page, 597–600, *597–600*, 792–793, *793*
    Sanitary Sewer page, 588, *588*
    Structures page, 624, 793, *793*
Network Properties tool, 618
Network Settings panel, 650
Network Tools panel, **618–619**, *619*
networks in survey databases, 72
New Data Shortcut Folder dialog, 754, *754*
New Design Check dialog, 267–268, *267–268*
New Entity Tooltip State setting, 22
New Expression dialog, 858–859, *858–859*
New Figure Prefix Database dialog, 62
New Local Survey Database dialog, 6, *6*, 79, 84
New Local Survey Database option, 60, 71
New Network dialog, 79, 84
New option for labels, 804
New Point Cloud Database dialog, 191
New Query dialog, 72, *72*
New Sheet Set option, 732
New Shortcuts Folder option, 754, *754*
New Traverse dialog, 80, *81*
New User-Defined Property dialog, 122, *122*
NNI (Natural Neighbor Interpolation), 163, **166–167**, *167*
No Center Pivots Found warning, 533
non-Civil 3D drawings
    point conversions from, **104–106**, *105*
    receiving, **28–30**, *29–30*
Non-Control Points Editor, 75
non-control points in survey networks, 77
non-destructive breaklines, 152
non-destructive surface boundaries, **156–157**, *156–157*, 160, *160*
non-road corridors, **448–450**, *448*

non-road uses of assemblies, **388–390**, *389–390*
north arrow in sheet layouts, 731
Northing Easting command, 53
null structures in pipe networks, **587**
Number Of Layouts Per New Drawing setting, **730**
numbers for points, 97, *97*

O

Object Layers tab, **16–19**, *17*
Object Properties dialog, 862
object snaps for feature lines, 684, *684*
Object Viewer
    contour styles, 881–882, *882*
    corridors, **418–419**, *418*
    pipe networks, 617
objects and object styles, 13, **861**
    analysis, **885–889**, *886–888*
    Commands folder, **868**, *868*
    dependencies, 3
    feature lines, **678**, *678*
    linear
        alignments, **875–876**, *876*
        feature lines, **877**, *877*
        overview, **872–874**, *873–874*
        parcel, **876–877**, *876*
    marker, **868–871**
    Multipurpose Styles, **867**, *867*
    overview, **861–862**, *861–862*
    pipe, **889–896**, *889–896*
    Plan Production, 714
    profile views
        bands, **908–911**, *908–909*, *911*
        grid spacing, **908**
        overview, **900–907**, *901–907*
    projecting
        profile views, **360–362**, *361–362*
        section views, **569–570**, *569–570*
    section views, **911–916**, *912–916*
    structure, **897–900**, *897–899*
    survey point, **871–872**, *872*
Offset Alignment Parameters palette, 431, *431*
offset alignments, **251**, 253–255, *253–254*

offset assemblies, **402**, *402*, **493–497**, *494–497*
one-point slope labels, 186–187, *187*
Open Pay Item Categorization File
 dialog, 783
Open Pay Item File dialog, 778–779, *779*
Open The Superelevation Curve Manager
 option, 543
Options dialog
 drawing templates, 746, *747*
 sheets, 739
 surfaces, 162
organizing assemblies, **404–407**, *406–407*
Orientation Reference setting, **806**, *806*
orientation toggle, 868
origin points in mass haul diagrams, 582
orthometric height scale, 15–16
Other Profile View Options settings, **734**
Outer surface boundaries, 155
output parameters for help, **380**
Outside Edge Shoulders options, 538
over haul in mass haul diagrams, 582
overhang correction for surfaces, **440**, *440*
Overkill command, **231–232**, *231–232*
overlap in transition station, **541–543**, *541*
Override All Dependencies option, 866
Override fields, 23–24, *23–24*, **862–863**, *862*
Overwrite option for points, 111

P

palettes, 1–2, *2*
panels, 1–2
Panorama Display Toggle icon, 4
Panorama window, 4
 custom views, **273–274**
 description, **25**
 feature lines, 681
 parabolas, 49, *49*
 point edits, **113–114**, *113–114*
parabolas, best fit, **48–49**, *49*
Parameter Editor, 437, *437*
parameters
 corridors, 437, *437*
 help, **379–380**, *380*
 subassemblies, 365, 375, **385**, *385*

parametric parts in Part Builder, 661
Parcel Layout tools, **206–208**, *206–208*
Parcel Properties dialog, 207–208
Parcel Style dialog
 Design page, 876, *876*
 Display page, 877
parcels, **197**
 area labels, **234–238**, *234, 236–267*
 boundary, **204–205**, *205*
  cul-de-sac, **211–213**, *212–214*
  Parcel Layout tools, **206–208**, *206–208*
  right-of-way, **208–210**, *209–211*
 feature lines, 670
 frontage offset, 220, *220*
 General Note labels, 820
 LandXML, 229
 Overkill command, **231–232**, *231–232*
 Parcel Layout tools, **206–208**, *206–208*
 segments
  deleting, **223–227**, *224–227*
  forming parcels from, **227–228**, *227–228*
  labeling, **238–245**, *239–241, 243–245*
  tables, **244–245**, *245*
  vertices, **233–234**, *233–234*
 sites, **197**
  alignments, **247**
  creating, **202–203**, *202–203*
  overview, **197–201**, *198–201*
  parcel reaction to objects, **228–231**, *228–231*
 styles, 198, **876–877**, *876*
 subdivision lots, 199, *199–200*, **214**
  attached segments, **214–215**, *214–215*
  sizing, **215–223**, *215, 218–223*
Parcels Layout Tools toolbar, 215, *215*
Part Builder
 arch pipes, **667**
 backups for, **661**
 limitations, **660**
 organization, **661–663**, *662*
 overview, **660–661**
 part families, **664**
 part sizes, **664–666**, *665–666*
 pipe networks, 619
 sharing custom parts, **666–667**
Part Catalog dialog, 597–600, *598, 600*

part properties, pay items as, **794–797**, *794, 796*
Part Publishing wizard, 646
part rules for pipe networks, **590**
    creating, **595–597**
    pipe rules, **593–595**, *593–595*
    structure rules, **590–592**, *591–592*
Part Size Creator dialog, 597, *598*
PARTCATALOGREGEN command, 667
Parts List dialog, 619
Parts List tool, 604
parts lists
    pay items in, **792–794**, *792–793*
    pipe networks, **597–600**, *597–600*, 604, 619
    pressure networks, **644–648**, *644–647*
    sanitary sewer networks, **588**, *588*, 602
parts placement in sanitary sewer networks, **605–606**, *606*
Paste Surface option, 163–164
Pay Item Formula dialog, 788, *788*
Pay Item List dialog, 789, 792, *792*, 798
pay items and pay item files, **777–778**
    assigning, 787, **792–794**, *792–793*
    AutoCAD objects as, **785–787**, *785–786*
    categorization files, 777–778, **782–785**, *783–784*
    favorites, **778–781**, *778–781*
    highlighting, **797–799**, *797–798*
    inventory, **799–801**, *799–801*
    as part properties, **794–797**, *794, 796*
    in parts lists, **792–794**, *792–793*
    pipe networks, 614
    pipes and structures as, **791–797**, *792–797*
    searching for, **781–782**, *781*
periods (.) in description key sets, 68
PI points in feature lines, 681
Pick From Drawing option, 735
Pick Marker Style dialog, 875
Pick Points On Screen option, 558
Pick Profile Style dialog, 302
Pick Style dialog
    assembly labels, 376–377
    link labels, 855
    point labels, 856

Pick Sub-Entity tool, 274
pipe-diameter grips, 614
Pipe Drop Across Structure rule, **592**, *592*
pipe-length grips, 614
pipe midpoint grips, 614, *614*
Pipe Network Catalog Settings dialog, 619, 661, *662*
Pipe Network Properties dialog, 618
    Information page, 603
    Layout Settings page, 603, *603*
    Profile page, 603, *603*
    Section page, 603, *604*
    Statistics page, 603, *604*
Pipe Network Vistas, 606, *606*, **615**, *615*
Pipe Properties dialog, 618
Pipe Rule Set dialog, 599
Pipe Style dialog
    Display page, 893, *893*, 895–896
    Information page, 892, 895
    Plan page, 889–890, *889*, 895
    Profile page, 890, *891*, 893, 895
    Section page, 890, *891*, 893
Pipe Table Creation dialog, 639–643, *642*
Pipe To Pipe Match rule, **594**, *594*
pipes and pipe networks, 587
    adding to part catalogs, 667
    alignments from parts, **624–625**, *625–626*
    Analyze panel, **619–620**, *619*
    contextual tabs, **616**, *616*
    crossing profile views, **631–632**, *631–632*
    data bands, 346
    data shortcuts, 766
    drawing parts in profile view, **626–632**, *627–632*
    editing, **612**
    General Tools panel, **617**, *617*
    interference checks, **636–639**, *637–639*
    labels
        adding, 617
        creating, **829–831**, *831*
        with crossings, **633–636**, *633, 636*
    Labels & Tables panel, **616–617**, *617*
    Launch Pad panel, **620**
    Modify panel, **618**

Network Layout Tools toolbar, **621–623**
Network Tools panel, **618–619**, *619*
Part Builder. *See* Part Builder
part rules, **590**
 creating, **595–597**
 pipe rules, **593–595**, *593–595*
 structure rules, **590–592**, *591–592*
parts lists, **588**, *588*, **597–600**, *597–600*
 as pay items, **791–797**, *792–797*
 planning, **588–590**, *589–590*
 pressure. *See* pressure networks
 QTO assignments, **795**, *795*
 sanitary sewer. *See* sanitary sewer networks
 setup, **587**
 styles, **889–896**, *889–896*
 sump, **623–624**
 tables, **639–643**, *640–643*
 trench assemblies, **390–393**, *391–392*
 trench corridors, **450–451**, *450–451*
Pipes Only option, 605
PIs in feature lines, 693–696, *693*
pivot points for superelevation, **532–536**, *533–537*
Place Remainder In Last Parcel option, 216
Plan And Profile option, 717
Plan Only option, 717
Plan Production feature, **713**
 plan sets, **713–714**
 sheets. *See* sheets
 templates, **746–748**, *747–748*
 view frames. *See* view frames
Plan Readable label option, 806
plan view
 pipe networks, **613–614**, *613–614*
 pressure networks, **648–654**, *649–653*
Plotted Unit Display Type setting, 20
point cloud surfaces
 creating, **193–195**, *193–194*
 description, **189**
 point clouds
 importing, **189–191**, *190–191*
 working with, **192–193**
Point Editor, **113–114**, *113–114*

Point File Format dialog, 103, *103*
Point File Formats dialog, 101, *101*
point files for surfaces, **128**
Point Group Properties dialog, 115–116, *115–116*, 123, *123*
Point Groups collection, 3, *3*
Point Inverse option, 92, *92*
Point Style dialog, 872, *872*
Point Survey Queries, **128**
Point Table Creation dialog, 120, *120*
POINTCLOUDDENSITY value, 192
points, **97**
 anatomy, **97–98**, *97–98*
 code set styles, **856–857**, *856–857*
 COGO vs. survey, **98**, *98*
 converting, **104–106**, *105*
 Create Points toolbar, **106–108**, *106–108*
 creating, **109–111**, *109*
 duplicate, **111–112**, *111–112*
 editing, **112–114**, *113–114*, **164–166**, *165*
 elevations, **118–119**, *118–119*
 file formats, **102–103**, *103*
 groups, **114–118**, *115–118*
 adding, 9
 surfaces, **128–129**, 158
 updating, 10
 importing, **101–104**, *101*
 labels, **65**, **821–823**, *822–823*
 settings, **98–101**, *99–101*
 surface styles, **882–885**, *882–884*
 in survey databases, **72**
 tables, **119–120**, *120*
 user-defined properties, **121–123**, *121–123*
Points From Corridor utility, **501**
Polyline From Corridor utility, 501
polylines
 alignments from, 250
 parcels from, 229
 surfaces from, **138–143**, *139–142*
ponds, **694–699**, *695–698*
Positioning setting for match lines, **721**
pound signs (#) in description key sets, 68
Precision settings in survey databases, 59
prefixes for layers, 17

Pressure Network Catalog dialog, 647, *647*
Pressure Network Parts List dialog, 647, *647*
Pressure Network Plan Layout toolbar, 650–651, *651*
Pressure Network Profile Layout toolbar, 657, *657*
pressure networks, **587**
    catalogs, **644–645**, *644*
    data shortcuts, 766
    design checks, **658–659**, *658*, *660*
    parts lists, **644–648**, *644–647*
    pipes, **644–645**, **892**, *892*
    plan view, **648–654**, *649–653*
    profile view, **654–657**, *655–657*
    QTO assignments, **795**, *795*
Pressure Pipe Style dialog, 892, *892*
Preview Area Display Toggle icon, 4
Process Linework During Import option, 74
Profile Data Band Style dialog, 909, *909*
Profile Display Options page, 358
Profile From Corridor utility, **501**
Profile Grade Length command, 314
Profile Grade Station command, 313–314
Profile Grid View tool, 320, **325–326**, *325–326*
Profile Label Set dialog, 851
Profile Layout Parameters dialog, 324–325, *324*
Profile Layout Parameters tool, 320
Profile Layout Tools toolbar
    component-level editing, 326–329
    corridors, 491, *491*
    layout by entity, 315–317
    layout by PVI, 310–311, *310*
    pressure networks, 655, *655*
    tools, **318–320**, *318*
Profile Only option, 717
Profile Properties dialog, 301, 421, *421*
Profile Station Elevation command, 313, *313*, 323
Profile Style dialog
    Display page, 865, *865*
    Summary page, 866, *866*
Profile View Properties dialog
    Bands page, **346–352**, *347*, *349*

Elevations page, 334, *334*, **342–345**, *343–344*
Hatch page, **352**, *352*
Information page, 308, 340, *340*
Pipe Networks page, 630–632, *630*, *632*
Profiles page, **345–346**, *345*
Projections page, 361–362, *362*
Stations page, **340–342**, *341*
Profile View Style dialog
    Display page, **906–907**
    Graph page, 308, 902, *902*
    Grid page, 902, *903*
    Horizontal Axes page, **904**, *905*, 908
    Title Annotation page, 903, *904*
    Vertical Axes page, **905–906**, *906*, 908
profile views, **299**
    bands, **346–352**, *347–351*, **908–911**, *908–909*, *911*
    creating, **329–330**, *330*
    deleting, **331**
    drawing parts in, **626–632**, *627–632*
    editing, **340**
    grid spacing, **908**
    intersections, 480
    labeling styles, **353–355**, *354–355*
    pipes crossing, **631–632**, *631–632*
    pressure networks, **654–657**, *655–657*
    projecting objects in, **360–362**, *361–362*
    properties, **340–350**
    removing parts, **630**, *630*
    snapping to, 354
    splitting, **331**
        gapped, **335–337**, *335–336*
        manually limited, **331–332**, *332*
        stacked, **337–339**, *338–339*
        staggered, **333–334**, *333–334*
    styles overview, **900–907**, *901–907*
    vertical movement edits, **628–629**, *628–629*
profiles, **299**
    best fit, **320–321**, *320*
    component-level editing, **326–327**, *327*
    corridor targets, **430–433**, *430–433*
    data bands, 346

data shortcuts, 766
design criteria files, 528
elevation, **299–308**, *300–307*
from files, **321–322**, *322*
grip-editing, **322–323**, *323*
labels, **842**
    applying, **355–358**, *356–358*
    curve, **847–848**, *847–848*
    expressions, 860
    grade breaks, **849–850**, *849*
    label sets, 358, **843–845**, *843–845*, **850–851**, *851*
    line, **845–846**, *846*
layouts, **308–309**
    entity-based, **315–317**, *316–317*
    parameters, **323–325**, *324*
    PVI-based, **309–315**, *309–314*
    tools, **318–320**, *318–319*
miscellaneous edits, **328–329**, *328–329*
offset assemblies, 494
Plan Production, 714
Profile Grid View command, **325–326**, *325–326*
quick, **363**, *363*
roundabouts, **520**, *520*
sheets. *See* sheets
superimposing, **359–360**, *359–360*
views. *See* profile views
ProgramData folder, 779
Project Objects To Multiple Section Views dialog, 569–570, *570*
Project Objects To Profile View dialog, 360–362, *361*
projecting objects
    profile views, **360–362**, *361–362*
    section views, 569–570, *570*
projects
    formula files in, **791**
    starting, **5–13**, *6–13*, **27–28**, *28*
    templates, **753**, *753*
Prompt For 3D Points option, 23
Prompt For Easting Then Northing option, 23
Prompt For Longitude Then Latitude option, 23
Prompt For Y Before X option, 23
prompts
    points, **100–101**, *100*
    transparent commands, 23
properties
    alignment, **281–283**, *281–283*
    corridors, 434
    label styles, **816–817**, *816–817*
    part, **747–749**, *794*, *796*
    pipe networks, 617
    points, **121–123**, *121–123*
    profile views, **340–350**
    surfaces, **145–149**, *147–149*
Prospector, **3–5**, *3*
    assemblies, **407**, *407*
    point edits, 113–114, *114*
    profiles, 305, *306*
    sample lines, 556, *556*
proximity breaklines, **152**
PVI-based layouts, **308–315**, *309–314*
PVI or Entity Based tool, 320
PVIs in lockdown, 326

Q

QSelect option, 617
QTO Manager, 778–782, 785
Quantile distribution method, analysis styles, **885**
Quantity Takeoff (QTO) feature, **777**
    corridors, **787–791**, *787–788*, *790*
    pay item files. *See* pay items and pay item files
    tools, 785
Quantity Takeoff Report dialog, 800–801, *800*
queries
    surfaces, **128–129**
    survey databases, **71–72**, *72*
question marks (?) in description key sets, 68
Quick Elevation Edit tool, 689
Quick Profile tool, 680, 688
quick profiles, **363**, *363*
Quick Select dialog, 617

R

rail
    alignments, 252, **260–261**, 545, **548**
    assemblies, **546–547**, *546–547*
    superelevation, **545–547**, *545–547*
Raise Incrementally/Lower Incrementally tools, 688
Raise/Lower By Reference tool, 690
Raise/Lower PVI Elevation dialog, 328–329, *329*
Raise/Lower PVIs tool, 319
Raise/Lower Surface option, 163
Raise/Lower tool, 688, 690
Reactivity Mode setting, 293
Readability Bias option, 806
Realign Stacks tool, 643
Rebuild All Corridors option, 420
Rebuild Corridor option, 419–420
Rebuild Snapshot option, 145
Rebuild The Corridor option, 416
Rebuild The Surface option, 148–151
rebuilding corridors, **420**, *420*, 445
ReCap product, **192**, *192*
rectangular grips, 550
Redistribute Remainder option, 216
Redo tool, 320
Reduction Options page, 168–169, *169*
Reference Point setting, 16
references, 751
    broken, **766–768**, *767–768*
    creating, **757–760**, *757–760*
    renamed objects, 758
    updating, **763–765**, *763–765*
refill factor for soil, 577
Region Options page, 168, *169*
regions
    baselines, **480**
    corridors, **411**
    intersections, 480
Regression Data tab, 262, *262*
Remainder Distribution parameter, 216
Remove Grade Break option, 551
Remove Items tool, 643
Remove Pay Item(s) From Specified Objects option, 797
Remove Snapshot option, 145
removing parts in profile view, **630**, *630*
Rename Pipe Network Parts dialog, 618
renaming
    assemblies, **386**
    groups and subassemblies, **387–388**
    objects with data references, 758
Renumber/Rename Parcels dialog, 235
reordering surface build operations, **150–151**, *151*
Repair All Broken References option, 768
Repair Broken References option, 767, *767*
Replace Items tool, 643
Replace With option, 804
replacing subassemblies, 375
report extents, limiting, 800
Report Quantities dialog, 578
Reset Labels option, 617
Resolve Crossing Breaklines tool, 154
Reverse Crown (RC) assemblies, 526, *526*
reverse curves
    alignments, **265–267**, *266*
    creating, **45**, *45*
Reverse tool, 682
Reverse The Direction tool, 688
Revision Direction tool, 522
ribbon, 1–2, *1–2*, **25–26**, *25–26*
Right Of Station Character option, 836
right-of-way parcels, **208–210**, *209–211*
Rim Insertion Point grips, 628
roads
    assemblies, **369–377**, *370–376*
    intersections. *See* intersections
    roundabouts, 510, 512, *512*
    sites, 200
rollover with superelevation, 526
Rotate To North option, 718, *718*
rotating
    match lines, 727
    text, 810
    view frames, 718, *718*, 727

Rotation Angle setting, 810
Rotation Direction field, 67
Rotation Point setting, 16
rotational grips, 613–614, 621, *621*
ROUND function, 789
roundabouts, **510**
    alignments, **511–518**, *512–518*
    center design, **518**, *519*
    combining components, **520–522**, *521*
    drainage, **510–511**, *511*
    finishing, **522**, *522–524*
    profiles, **520**, *520*
ROUNDDOWN function, 789
ROUNDUP function, 789
ROW lines in section views, **572–573**, *572–573*
rules for pipe networks, **590**
    creating, **595–597**
    pipe, **593–595**, *593–595*
    structure, **590–592**, *591–592*
RunDepthCheck command, 658–659
RunDesignCheck command, 658

S

Sag Curves option, 847–848
Sample Line Group Properties dialog, 561, *561*
Sample Line Tools toolbar, 558, *558*
Sample More Sources option, 579
samples and sample lines
    cross sections, **556–560**, *556–559*, 579
    vs. frequency lines, **555–556**
    Plan Production, 714
    profile view creation during, **329–330**, *330*
    surfaces, **300–308**, *300–307*
    swath width, **561–563**
Sampling Increments setting, 559
sanitary sewer networks, **600–601**, *601*
    Cover And Slope rule, **607**, *607*
    creating, **606–609**, *609*
    drainage networks from feature lines, **609–612**, *610–611*
    flow direction, **612–613**
    layout creation tools, **602–604**, *602–604*
    parameters, **601–602**, *601*
    parts lists, **588**, *588*
    parts placement, **605–606**, *606*
    Pipe Network Vistas, **615**, *615*
    plan view, **613–614**, *613–614*
    planning, **588–590**, *589–590*
Save Command Changes To Settings option, 21, 99
saving panorama views, 273
Scale Inserted Objects setting, 22
Scale Objects Inserted From Other Drawings option, 14–15
Scale Parameter field, **66**
SDTS (Spatial Data Transfer Standard) format, 131
sea level scale factor, 15–16
searching for pay items, **781–782**, *781*
Section Editor
    non-road corridors, 449
    offset assemblies, 497, *497*
    working with, **435–437**, *435*
section sheets, **740**
    creating, **744–746**, *745–746*
    multiple section views, **740–744**, *741–744*
Section Sources dialog, 562, *562*, 579, *579*
Section View Group Bands dialog, 915–916, *915*
Section View Group Properties dialog, 857, *857*
Section View Style dialog
    Display page, 912
    Grid page, 912
section views, **555–556**
    annotation scale, **570–572**
    corridors, **435–437**, *435–437*
    creating, **563–569**, *563–569*
    data shortcuts, **761–762**, *761*
    deleting, **571**, *571*
    labels, **580**, *580*
    materials lists, **575**, *575*, **577–578**, *578*
    multiple, **567–569**, *568–569*, **740–744**, *741–744*

object projection, **569–570**, *569–570*
pipe trench, 390
Plan Production, 714
ROW lines, **572–573**, *572–573*
sample lines, **579**
   creating, **556–560**, *556–559*
   vs. frequency lines, **555–556**
soil factors, **577–578**, *578*
styles
   group plot, **912–916**, *913–916*
   overview, **911–912**, *912*
volume reports, **574**, *574*, **578–579**
volume tables, **576–577**, *576*
sectional data bands, 346
segments, parcel
   deleting, **223–227**, *224–227*
   forming parcels from, **227–228**, *227–228*
   labeling, **238–245**, *239–241*, *243–245*
   tables, **244–245**, *245*
   vertices, **233–234**, *233–234*
Select A Feature Line dialog, 426, *426*, 428, 502
Select A Profile dialog
   cul-de-sacs, 460
   intersections, 477, 482
Select A Quantity Takeoff Formula File dialog, 788
Select A Sample Line Group dialog, 575, 577
Select Alignment dialog
   cross sections, 761
   pipe networks, 635
   sample lines, 560
Select Alignment tool for sanitary sewer networks, 604
Select Axis To Control setting, 908
Select Color dialog
   contour styles, 880
   linear objects, 871
   marker object styles, 871
Select Destination Pipe Network dialog, 612
Select Existing Polylines option, 558
Select First Pressure Network Part option, 655
Select Grading Group dialog, 708

Select Grading Group tool, 702
Select Label Set dialog, 289, 839
Select Label Style dialog, 237, 281, 283
Select Label Style Set dialog, 850
Select Layout As Sheet dialog, 568, *568*
Select Layout As Sheet Template dialog, 717–718, *717*, 723
Select Pipe Style dialog, 894, 896
Select PVI or Select Entity tool, 320
Select Reference File dialog, 761
Select Similar tool, 617
Select Structure dialog, 900
Select Surface dialog
   feature lines, 672
   grading elevation, 688
   gradings, 702
   grid labels, 188
Select Surface Style dialog, 710
Select Surface To Paste dialog, 710
Select Surface tool, 604
Select Target Surface tool, 702
Select Template dialog, 26, *27*
Select The Data You Want Referenced In Your Sheets settings, **735**
Select tool for grading elevation, 687
Select Type dialog
   alignment labels, 841
   pipe labels, *833*
   surfaces, 832–833
Select View Frames dialog, 730, *730*
Send To Back option, 488
Sequence From setting, 111
Set AutoCAD Units option, 21
Set AutoCAD Variables To Match option, 14–15
Set Baseline And Region Parameters option, 413, 457, 459, 496
Set Elevation By Reference tool, 690
Set Entities by Layer dialog, 509, *509*
Set Grade/Slope Between Points tool, 690, 693, 696–697
Set Increment tool, 688
Set Network Catalog tool, 619

Set Pipe End Location rule, **594–595**, *595*
Set Pressure Network Catalog dialog, 644–645, *645*
Set Slope Or Elevation Target dialog, 483, *483*, 505
Set Sump Depth rule, **592**, *592*
Set Width Or Offset Target dialog
　alignments, 432, *432*
　driveways, 509, *509*
　feature lines, 504, *505*
　intersections, 483, *483*
Set Working Folder setting, 71, *71*, 753
Settings tab, **13**
　contour style, 878
　description key sets, 67, *67*
　drawing settings, **14–24**, *14*, *16–17*, *20–21*
　expressions, 858–859
　General Note labels, 818
　group plot styles, 913–914
　labels
　　alignment, 836, 840
　　line, 845
　　pipe, 829
　　point, 821, 856
　　single segment, 824
　　spanning, 826
　　styles, 803–805
　links, 855
　marker object styles, 870
　object styles, 861, *861*, 864, **867**
　part rules, 595–596
　pay items, 792
　pipe styles, 892
　pressure networks, 646
　pressure pipe styles, 895
　profile view bands, 909
　section view styles, 912
　structure rules, 590, *591*
　structure styles, 899
　surfaces, **145–146**
　survey point styles, 871
　triangle surface style, 882
　watershed analysis, 885
　Weed tool, 683
Setups Editor, 78, *78*
sewer systems. *See* sanitary sewer networks
shapes
　code set styles, 854
　corridors, 411
　subassemblies, **378**, *379*
　viewport, 748
sharing custom parts, **666–667**
Sheet File Name setting, **733**
Sheet Files Storage Location setting, **733**
Sheet Set Manager, **735–739**, *735*, *739*, 745–746
Sheet Set Storage Location setting, **732**
sheets, **729**
　Create Sheets Wizard, **729–735**, *729*, *732–734*
　Create View Frames Wizard, **716–718**, *716*
　invisible profile views, 738
　managing, **735–739**, *735–736*, *739*
　name templates, **737**
　section, **740**
　　creating, **744–746**, *745–746*
　　multiple section views, **740–744**, *741–744*
shortcuts. *See* data shortcuts
ShoulderExtendAll subassembly, **384**, *384*, 536
ShoulderExtendSubbase subassembly, **384**, *384*
shoulders in subassemblies, **383–384**, *383–384*
Show Event Viewer option, 21
Show Grade Breaks Only tool, 689
Show Only Crossing Pipes option, 630
Show Only Parts Drawn In Profile View option, 630
Show surface boundaries, 155
Show Tooltips option, 21–22
Simplify Surface dialog, 168–170, *168–169*
Simplify Surface option, 164
simplifying surfaces, **168–170**, *168–170*
single-section views, **564–567**, *564–567*
single segment labels, **824–825**, *825*
Site dialog for gradings, 702
site geometry objects, 197

Site Properties dialog, 202–203, *203*, 671–673
Site setting, 252
siteless alignments, **201**, *201*
sites, **197**
   alignments, **247**, 252
   creating, **202–203**, *202–203*
   Figure Prefix Database Manager, 62
   grading, 702, 766
   overview, **197–201**, *198–201*
   parcel reaction to objects, **228–231**, *228–231*
size
   marker object styles, **869–870**, *869*
   parcels, **215–216**, *215*
      Automatic Layout, **216**
      tools, **217–223**, *218–223*
   parts, **664–666**, *665–666*
slide glyph, 649, *649*
Slide Line - Create tool, **217–220**, *218–220*
sliding
   match lines, 727
   view frames, 726–727
slip lanes, **515–518**
Slope Arrows tool, 878
slopes and slope arrows
   intersections, 479
   labeling, **186–188**, *187–188*
   points, 108, *108*
   sewer systems, 589
   surfaces, **176–178**, *177*, 878
Slopes tool, 878
Smooth Surface option, **163**
Smooth tool, **682**
smoothing
   feature lines, **682**
   surface vs. contour, 879
   surfaces, **166–167**, *167*
Snap Station Value Down To The Nearest option, **721**
snapping to profiles and profile views, 354
snapshots, surface, **145**
soils
   boring data for points, **121–123**, *121–123*
   boundaries, 199, *199–200*

cross sections, **577–578**, *578*
sites, 201
spanning pipe labels, **636**
spanning segment labels, **241–242**, *241*, **826–827**, *826*
Spatial Data Transfer Standard (SDTS) format, 131
specialized subassemblies, **393**, *393*
Specify Grid Rotation Angle setting, 16
Specify option for points, 112
Split and Merge Network tool, 618
split-created vertices, 233–234
Split Point Resolution setting, 678
Split Region option, 453
Split Table settings
   alignment tables, 293
   pipe tables, 640
splitting profile views, **331**
   gapped, **335–337**, *335–336*
   manually limited, **331–332**, *332*
   stacked, **337–339**, *338–339*
   staggered, **333–334**, *333–334*
square brackets ([]) in description key sets, 68
square grips
   corridors, 490
   cross sections, 557
   feature lines, **674**
   grade breaks, 849
   pipe networks, 613–614
   station offset labeling, 292, *292*, 842
   view frames, 726
stacked profile views, **337–339**, *338–339*
staggered profile views, **333–334**, *333–334*
standard breaklines, 152
standard curves, **42–45**, *42–45*
Standard Deviation distribution method, **885**
Start Offset option, 595
Start To End pipe network flow direction, 613
Starting Design Speed setting, 252
starting new projects, **27–28**, *28*

Starting Station setting, 252
Static Mode tool, 643
static pipe tables, 640
station limits in profile views, *341–342*
Station Locking regions, 414
station offset labeling, **291–292**, *292*, **840–842**, *841–842*
Station Range settings
    cross sections, 559
    view frames, 716
Station Selection panel, 436
stationing alignments, **284–286**, *285–286*
statistics
    feature lines, 677, *677*, 679
    interference locations, 620
    pipe networks, 603, *604*
    surfaces, 170, *170*
Stepped Offset tool, 683
storing assemblies, **405–407**, *406–407*
Storm and Sanitary Analysis (SSA) program, 620
storm drainage pipe networks, **609–612**, *610–611*
Storm Sewers tool, 620
Structure Properties dialog, 616, *616*, 618, 623
Structure Rule Set dialog, 591, *591*, 595
structure rules
    creating, **595–597**
    pipe networks, **590–592**, *591–592*
Structure Style dialog
    Display page, 631, 900
    Model page, 897, *897*
    Plan page, 897–898, *897*, 900
    Profile page, 898–899, *898*
    Section page, 899–900, *899*
Structure Table Creation dialog, 639–643, *641*
structures
    labels, **635–636**
    as pay items, **791**
    pipe networks, **587**
    styles, **831–834**, *833*, **897–900**, *897–899*
Structures Only option, 605

Style dialog, 861
Style Override option, 632
Style setting, 720
styles, 26
    alignments, 252
    AutoCAD, **174**, *174*
    code set, **854–860**
    Description Key Editor, **65**
    feature line, **678**, *678*
    Figure Prefix Database Manager, 62
    importing, **30–32**, *31*
    labels. *See* labels and label styles
    object. *See* objects and object styles
    parcels, 198
    Plan Production, 714
    table, **852–854**, *852–853*
Sub-Entity Editor tool, 274
subassemblies, **365**
    commonly used, **381**
    components, 377, *377*
    conditional, **506–510**, *507–509*
    Corridor Modeling Catalogs, **367**, *367*
    daylighting
        alternative, **400–401**, *400–402*
        input parameters, **399**, *399*
        working with, **397–398**, *398*
    deleting, 375
    help, **368**, **379–380**, *379–380*
    inserting, **374–375**
    lanes, **382**, *382*
    links
        description, **377–378**, *378*
        generic, **394–397**, *396–397*
    marker points, **378**, *378*
    moving, 374
    names, 371
    parameters, 375, **385**, *385*
    renaming, **387–388**
    replacing, 375
    shapes, **378**, *379*
    shoulder and curb, **383–384**, *383–384*
    specialized, **393**, *393*
    storing, **405**
    superelevation, **531–532**, *531–532*

Tool Palettes window, **366–368**, *366*, *368*
   zero values, **390**
Subassembly Composer, 365
Subassembly contextual tab, 376, *376*
Subassembly Properties dialog, 384–385, *385*
Subassembly Reference page, 389, 398
subdivision lots, 199, *199–200*, **214**
   attached parcel segments, **214–215**, *214–215*
   sizing, **215–216**, *215*
      Automatic Layout, **216**
      tools, **217–223**, *218–223*
suffixes for layers, 17
sump depth
   pipe networks, **592**, *592*, **623–624**
   profile view, **628–629**
super assemblies, **531–532**, *531–532*
superelevation, **525**
   alignments, **531**
   applying, **537–540**, *538–541*
   axis of rotation subassembly, **532–533**, *533*
   bands, 346
   cant views, **548–550**, *549*
   design criteria files, **527–531**, *528–531*
   editing, **550–552**, *551–552*
   forcing, **543–544**, *543–544*
   importing and exporting, **542**, *542*
   preparing, **525–527**, *526–527*
   rail, **545–547**, *545–547*
   super assemblies, **531–532**, *531–532*
   transition station overlap, **541–543**, *541*
   views, 340
Superelevation Curve Manager, 543–544, *544*
Superelevation Tabular Editor, 537, 543–544
Superelevation utility, 501
Superelevation Wizard, 544
Superimpose Profile Options dialog, 359, *359*
superimposing profiles, **359–360**, *359–360*
surface creation problems in corridors, **446–447**, *446*
surface name templates for corridors, **442**
Surface Properties dialog
   Analysis page, 181–183, *182*, 887, *888*

Definition page
   boundaries, 155
   contours, 147, *147*, 149
   rebuilding surfaces, 150–151, *151*
Information page
   boundaries, 159
   contours, 142, 148
   DEM files, 134
   elevation banding, 171, 175, *175*
   objects 862, *862*
   slopes and slope arrows, 176–177
   volume surfaces, 184
   watershed, 887
Statistics page, 170, *170*, 181
Surface Style dialog, 877
   Analysis page, 885–886, *887*
   Contours page, 879, *879*
   Display page, 880–881, *881*, 883–885, *884*, 887
   Information page, 863–864, *864*, 879, 883, 886
   Points page, 883, *883*
   Triangles page, 880, *880*, 883, 886
   Watersheds page, 886, *886*
Surface Style Editor, 184
surfaces, **125**
   analysis, **171**
   approximations, **138–143**, *139–142*
   basics, **125–127**, *126*
   boundaries, 127
      corridors, **442–446**, *443*
      destructive vs. non-destructive, **156–157**, *156–157*, 161, *161*
      interactive process, **491–493**, *492–493*
      overview, **155–162**, *156–161*
   breaklines, **152–154**, *153–154*
   build operation order, **150–151**, *151*
   build options, **145–149**, *147–149*
   comparing, **179–184**, *180–183*
   for corridors, **438–442**, *439–442*
   creating, **127–129**, *127*, *129*
   definitions, **127–129**
   elevation banding, **171–175**, *172–175*
   from external text files, **143–144**, *144*
   from GIS Data, **135–138**, *135–138*

from grading groups, **707–711**, *707, 709, 711*
labeling, **184–188**, *185–189*
LandXML files, 130
large, **162**
manual edits, **162–164**
point cloud, **189–195**, *190–194*
points
    editing, **164–166**, *165*
    settings, 108, *108*
sampling for elevations, **300–308**, *300–307*
sanitary sewer networks, 602
simplifying, **168–170**, *168–170*
slopes and slope arrows, **176–178**, *177, 878*
smoothing, **166–167**, *167*
snapshots, **145**
styles
    contour, **878–882**, *879–882*
    overview, **877–878**
    triangles and points, **882–885**, *882–884*
triangle editing, **164–166**, *165*
Visibility Checker, **178–179**
volume, **180–184**, *180–183*
Survey Command window, 59
Survey Database Settings dialog, 58–60, *59*
Survey Database Settings Path setting, 58
survey points, **98**, *98*
    style, **871–872**, *872*
    in survey databases, **72–73**
Survey User Settings dialog, 57–58, *58*
surveys, **57**
    Coordinate Geometry Editor, **89–91**, *89–90*
    databases. *See* databases for surveys
    description keys
        activating, **68–70**, *69–70*
        creating, **67–68**, *67–68*
        overview, **64–67**, *65*
        wildcards, **68–69**
    inquiry commands, **91–94**, *91–93*
    miscellaneous features, **86–87**, *86–87*
    network components, **77–85**, *78–83, 85*

queries, **71–72**, *72*
tools, 24
Swap Edge option, 163
Swap Part option, 616
Swap Part Size dialog, 618, 796
Swap Part tool, 618
swath widths, 556
    cross sections, 559
    sample line groups, **561–563**
Swing Line - Create tool, **220**
synchronizing data references, 764, *764*

T

Table Creation dialog, 244–245, *245*, 294–295, *295*
Table Layer option, 639
Table panel tools, **642–643**, *643*
Table Properties dialog, 643
Table Style dialog, 643, 852, *853*
Table Style option, 639
Table Tag Numbering dialog, 243–244, *244*
tables
    alignments
        design, **273–274**, *273*
        line, **294–295**, *295*
        overview, **292–294**, *293*
        segment, **295–296**, *296*
        settings, **295–296**, *296*
    labels, **242–244**, *243–244*
    parcel segments, **244–245**, *245*
    pipe, **639–643**, *640–643*
    points, **119–120**, *120*
    styles, **852–854**, *852–853*
    volume, **576–577**, *576*
tangency constraints, 275–276, *275*
Tangent By Best Fit dialog, 261, *261*
Tangent Creation tool, **318–319**
tangent slope length, 846
Tangent-Tangent (With Curves) tool, 256, *257*, 270
Target Mapping dialog
    conditional subassemblies, 510

corridors, 431–432, *433*
cul-de-sacs, 462–463
feature lines, 415–416, *416*, 504
intersections, 483
multiregion baselines, 454, *455*
target parameters, help, **380**
Target To Nearest Offset setting, 483
targets
    corridors, **412**, **429–433**, *430–433*
    cul-de-sacs, 462–463, *463–464*
    feature lines as, **504–506**, *505–506*
    intersections, **480–484**, *481–484*
    marker points, 378
    multiregion baselines, 454
tees in pressure networks, **653–654**, *653*
Template Options dialog
    linear objects, 871
    marker object styles, 871
templates, **26**, *27*
    name
        corridor surfaces, 442, *442*
        sheets, **737**
        view frames, **719–720**, *719*
    Plan Production, **746–748**, *747–748*
    projects, **27–28**, *28*, **753**, *753*
    styles, **30–32**, *31*
terrain modeling. *See* gradings
tessellation
    curve, 873
    settings, 708
Text Component Editor
    expressions, 857, *857*, 859
    labels
        alignment, 836, *836*, 840, *841*
        curve, 827–828
        General Note, 818–820, *818–819*
        line, 845–846, *846*
        pipe, 830–833
        point, 822–823, *823*
        single segment, 825
        styles, **816–817**, *816–817*
    profile view bands, 910–911
    profile view styles, 903–906
    table styles, 852–853, *853*

text files
    importing points from, **101–104**, *101*
    surfaces from, **143–144**, *144*
Text Height setting
    label style dragged state, 813
    label style text, 810
Text properties for label styles, **810–811**
Text Style setting, 805
tildes (~) in description key sets, 68
TINs (triangulated irregular networks), **125–127**, *126*
    surfaces from, 708
    volume surface, **180–184**, *180–183*
Toggle Upslope/Downslope tool, 605, *606*, 612
tool palettes, assemblies and subassemblies on
    adding, **368**, *368*, **405**
    changing sets, 372, *372*
    predefined, **366–367**, *366*
    storing, **405–407**, *406–407*
Tool Properties dialog
    assemblies, 405, *406*
    subassemblies, 405
Toolbox tab, **24–25**
Toolspace, **2–3**, *2*
    help, **4–5**
    orientation toggle, 868
    Prospector tab, **3–5**, *3*
    Settings tab. *See* Settings tab
    Survey tab, 24
    Toolbox tab, **24–25**
Transformation tab, **15–16**, *16*
transition station overlap, **541–543**, *541*
Translate Survey Database command, **83**, *83*
Translate Survey Database dialog, **85**, *85*
transparent commands, **52**, *52*
    matching, **53–54**, *54*
    settings, **23**
    standard, **52–53**
Transparent Commands toolbar, 312–314, *312*
Traverse Analysis Defaults settings, 59

Traverse Analysis dialog, 81, *81*
Traverse Editor, 78–79, *79*
traverse reports, 91, *91*
Traverses settings, **78–83**, *79–83*
trenches
    assemblies, **390–393**, *391–392*
    corridors, **450–451**, *450–451*
triangles
    surface editing, **164–166**, *165*
    surface style, **882–885**, *882–884*
    vertices, 882
triangular grips
    cross sections, 557
    PVI-based layouts, 312, *312*
    vertical movement edits, 628
triangulated irregular networks (TINs), **125–127**, *126*
    surfaces from, 708
    volume surface, **180–184**, *180–183*
Trim tool, 681
troubleshooting
    cul-de-sacs, **462–463**, *463–466*
    intersections, **485–486**, *485–486*
Turn Off Categorization option, 781–782, *781*
Twisted North option, 718
two-point slope labels, 186–187, *188*
Type setting
    alignments, **251–252**
    label style borders, 811
    label style leaders, 813

U

Uncheck Added option, 31
Uncheck Conflicting option, **31**
Uncheck Deleted option, 31
Undo tool for layouts, 320
United States Geological Survey (USGS), 131
units
    design criteria files, 528
    survey databases, **58**, *59*
Units And Zone tab, **14–15**, *14*
Unselect All Rows tool, 689

Unsupported in Assemblies Containing Offsets warning, 533
Unsupported Subassemblies warning, 533
Update Content tool, 643
updating data references, **763–765**, *763–765*
UrbanCurbGutterGeneral subassembly, 374–375, *374–375*, 395–396
    description, **370**, *370*
    help, 379–380, *379–381*
    intersections, 479
UrbanCurbGutterValley subassembly, **383**, *383*
UrbanSidewalk subassembly, 375, *375*
    description, **371**, *371*
    intersections, 479
    links, 377, *378*
    shapes, 378, *379*
US Imperial Circular Pipe Chapter, 662–663
US Imperial Structures catalog, 664–665
Use AutoCAD BLOCK Symbol For Marker symbol, 868
Use AutoCAD POINT For Marker symbol, 868
Use Catalog Defined 3D Part option, 897
Use Criteria-Based Design option, 252, 270
Use Custom Marker symbol, 868
Use Design Check Set option, 252, 270
Use Design Criteria File option, 252
Use Drawing Scale option
    Description Key Editor, **66**
    marker object styles, 869
    structure styles, 898
Use Fixed Scale option
    marker object styles, 869
    structure styles, 898–899
Use Fixed Scale From Part Size option, 898
Use Maximum Angle option, 146
Use Maximum Triangle Length option, 146, 148–149, *149*
Use Name Template option, 112
Use Next Point Number option, 111
Use Outer Part Boundary option, 898
Use Project Template option, 755

Use Size As Percentage Of Screen option, 898
Use Size In Absolute Units option
    marker object styles, 869–870
    structure styles, 898
Use Size Relative To Screen option, 870
Use Superelevation setting, 535, *535*
Use Vertex Elevations option, 611
User-Defined Contours tool, 878
user-defined criteria files for superelevation, **528–531**, *528–531*
User Defined Part option, **898**
user-defined point properties, **121–123**, *121–123*
USGS (United States Geological Survey), 131

V

Vertical Curve Creation tool, **319**
Vertical Curve Settings dialog, 310–311, *310*
Vertical Exaggeration setting
    profile view grid spacing, 908
    profile view styles, 902, *902*
vertical geometry bands, 346
vertical movement edits, **628–629**, *628–629*
vertical triangular grips
    profiles, 322–323, *323*
    PVI-based layouts, 312, *312*
vertices
    parcel segments, **233–234**, *233–234*
    triangle, 882
    weed, **683–685**, *684*
View Direction options, 865, *865*
View Frame Properties dialog, 725
view frames, **714**, *715*
    alignments, **716**
    creating, **723–725**, *725*
    editing, **725–728**, *726*
    groups
        description, 714
        settings, **719–721**, *719*, **730**
    match lines, **720–722**, *721–722*
    profile views, **722–723**, *722*
    sheets, **716–718**, *716–718*
View Group Properties dialog, 916

Viewport Configuration dialog, 435
Viewport Properties dialog, 748, *748*
viewport shapes, 748
views
    profile. *See* profile views
    section. *See* section views
violation only rules, 593
Visibility Checker, **178–179**
Visibility setting for labels, 809, 811
visual styles in AutoCAD, **174**, *174*
volume
    corridor calculations, **447–448**, *447*
    cross sections
        reports, **574**, *574*, **578–579**
        tables, **576–577**, *576*
    mass haul diagrams, 581, *581*
    TIN surfaces, **180–184**, *180–183*

W

wall breaklines, **152**
wall size for pipes, 890
waste value in mass haul diagrams, 582
waterfall effect in corridors, **421**
watershed analysis, 878, **885–889**, *886–889*
WBLOCK command, 787
Weed Points option, 673
Weed tool, **682–683**
Weed Vertices dialog, 683–685, *684*
Weeding setting for grade breaks, 849–850
Widen Turn Lane For Incoming Road option, 471
Widen Turn Lane For Outgoing Road option, 471
widening offset alignments, 255
width targets, feature lines as, **504–506**, *505–506*
wildcards
    in description key sets, 65, **68–69**
    layers, 17–19
    points, 115
workflows, **751**
    data shortcuts. *See* data shortcuts
    LandXML, **772–775**, *773–774*

working folders, **753–755**, *754–755*
World Coordinate System, 806, *806*

X

X Offset setting for label text, 810
XML Notepad, 783–784, *783–784*
XRefs (external references), **751–752**
XSL (Extensible Stylesheet Language) format, 800

Y

Y Offset setting for label text, 810
yellow exclamation points, **130**, *131*

Z

zero values for subassemblies, **390**
Zoom To tool, 688